THIRD EDITION

Systems Analysis and Design
in a Changing World

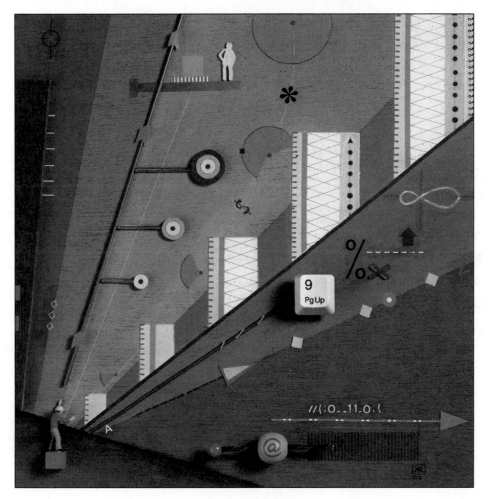

John W. Satzinger
Southwest Missouri State University

Robert B. Jackson
Brigham Young University

Stephen D. Burd
University of New Mexico

THOMSON

COURSE TECHNOLOGY

Systems Analysis and Design in a Changing World, Third Edition
by John W. Satzinger, Robert B. Jackson, Stephen D. Burd

Executive Editor:
Mac Mendelsohn

Development Editor:
Karen Hill, Elm Street Publishing
Services

Marketing Manager:
Amy Yarnevich

Senior Product Manager:
Tricia Boyle

Production Editor:
Christine Freitas

Associate Product Manager:
Mirella Misiaszek

Editorial Assistant:
Amanda Piantegosi

Manufacturing Coordinator:
Laura Burns

Cover Designer:
Betsy Young

Cover Image:
Rakefet Kenaan

Text Designer:
Lou Ann Thesing

Compositor:
Pre-Press Company, Inc.

Disclaimer

Course Technology reserves the right to revise this publication and make changes from time to time in its content without notice.

ISBN 0-619-16031-4
Instructor's Edition

ISBN 0-619-21325-6
Student Edition

ISBN 0-619-21371-X
International Student Edition

To JoAnn, Brian, and Kevin—JWS

To Anabel and my children for their continued support—RBJ

To Dee, Amelia, and Alex—SDB

BRIEF CONTENTS

CONTENTS

Features

Systems Analysis and Design in a Changing World, Third Edition was written and developed with both instructor and student needs in mind. Here is just a sample of the unique and exciting features that help bring the field of systems analysis and design to life.

By the early 2000s, RMO had grown to become a significant regional sports clothing distributor in the Rocky Mountain and Western states. The states of Arizona, New Mexico, Colorado, Utah, Wyoming, Idaho, Oregon, Washington, and Nevada, and the eastern edge of California had seen tremendous growth in recreation activities. Along with the increased interest in outdoor sports, the market for both winter and summer sports clothes had exploded. Skiing, snowboarding, mountain biking, water skiing, jet skiing, river running, jogging, hiking, ATV biking, camping, mountain climbing, and rappelling had all seen a tremendous increase in interest in these states. Of course, people needed appropriate sports clothes for their activities, so RMO expanded its line of sportswear to respond to this market. It also added a line of casual and active wear to round out its offerings to the expanding market of active people. The current RMO catalog offers an extensive selection (see Figure 1-9).

FIGURE *1-9*
Current RMO catalog cover (Fall 2005).

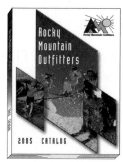

Rocky Mountain Outfitters now employs more than 600 people and produces almost $100 million annually in sales. The mail-order operation is still the major source of revenue, at $60 million. In-store retail sales have remained a modest part of the business, with sales of $2.5 million at the Park City retail store and $5 million at the recently opened Denver store. In the early 1990s, the Blankenses added a phone-order operation that now accounts for $30 million in sales. To the customer, this service was a natural extension, but to RMO, it meant considerable changes in transaction processing systems to handle the phone orders.

RMO Strategic Issues
Rocky Mountain Outfitters was also one of the first sports clothing distributors to provide a Web site featuring its products. The site originally gave RMO a simple Web presence to enhance its image and to allow potential customers to request a copy of the catalog. It also served as a portal for links to all sorts of outdoor sports Web sites. The first RMO Web site enhancement added more specific product information, including

The text uses an **integrated case study** of moderate complexity—Rocky Mountain Outfitters (RMO)—to illustrate key concepts and techniques.

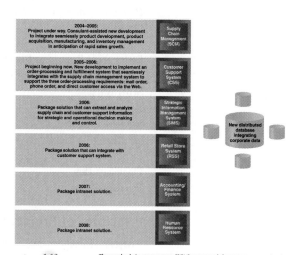

FIGURE *1-13*
The timetable for RMO's application architecture plan.

The supply chain management (SCM) system and the customer support system (CSS) require custom software development. Custom development is advisable when the system must match very specific company requirements. Consultants have been called in to help define requirements and develop the integration plan for supply chain management. Several leading consulting firms specialize in supply chain management.

The customer support system will probably be developed in-house by the information systems staff, although final decisions about the implementation approach will be deferred until a more complete analysis is completed.

The other systems in the plan will probably be software package solutions selected from among the best-rated software currently available. Package solutions are desirable for standard business systems such as accounting and human resources. The key requirement is that any package must integrate seamlessly with other RMO systems and use intranet technology.

The Customer Support System
The RMO system development project described in this text is the customer support system (CSS). Rocky Mountain Outfitters has always prided itself on its customer orientation. One of the core competencies of RMO has been its ability to develop and main-

An overview of the strategic systems plan for RMO is presented in Chapter 1 to place the project in context. The planned system architecture provides for rich examples—a client-server Windows-based component, as well as a Web-based, e-commerce component with direct customer interaction via the Internet.

The new customer support system (CSS) is the system development project used throughout the text for examples and explanations. It is strategically important to RMO, and the company must integrate the new system with legacy systems and other planned systems.

Details about the RMO case are integrated directly into each chapter to make a point or to illustrate a concept—just-in-time examples—rather than isolating the case study in separate sections of the chapters.

After the previous project planning activities are complete, it is time to launch the project. The scope of the new system is defined, the risks have been identified, the project has been found feasible both economically and otherwise, a detailed schedule has been developed, team members have been identified and are ready, and it is now time to start. Two final tasks usually occur at this point. First, the membership of the oversight committee is finalized, and it meets to give final go-ahead for the project, including releasing the necessary funds. Second, the organization makes a formal announcement through its standard communication channels that gives credence to the project and solicits cooperation from all involved parties in the organization. In other words, the project gets the blessing and visible support of the senior executives of the organization. No project should begin without these two events.

The key question to be answered when launching the project is, *are we ready to start?*

RECAP OF PROJECT PLANNING FOR RMO

Barbara and Steve spent the entire month of February putting together the schedule and plans for the CSS. Even though Barbara was the project manager, she and Steve worked together as peers. As a team, they could brainstorm and double-check each other's work. They had worked together before and had an excellent relationship—one based on mutual respect and trust. They could be candid and knew how to work through disagreements as well as how to come to consensus on important issues. Barbara also knew that the work Steve produced was always well thought out and very professionally done. He was a skilled systems analyst and would help make sure that the work done in the planning phase was solid.

The success of the overall project depended heavily on the planning Barbara and Steve did during this phase. The foundation for all other project activities is established during the project planning phase. As Barbara planned for the kickoff meeting to launch the project officially, she reviewed the areas of project management to make sure that she had addressed all of the critical issues.

For project scope management, she developed a list of business benefits, a list of system capabilities, and a context diagram. At this point in the project, the scope definition was still very general. She would make sure the project's scope was precisely defined during the information-gathering activities of the analysis phase.

She and Steve had developed a detailed work breakdown structure and entered the information into MS Project. The schedule was very detailed for the analysis phase, but less so for the design and implementation phases. She would add those details as decisions were made about the implementation approach. She thought that her approach to project time management had been established, and she would have the tools necessary to track the schedule as the project progressed.

The costs and potential benefits had been estimated and used to develop an NPV estimate. She would redo the NPV when she redid the schedule at the end of the analysis phase to ensure that the costs and schedule were within the allowed budget. The other part of cost management was to monitor the costs during the life of the project. MS Project would help her track the costs of each task.

Steve had done a lot of the work to identify and assess risks during the feasibility analysis. Barbara knew that they would both continue to look for risks and assess potential problems during the project. She asked Steve to take time each week to assess the risks and update the list of the highest risks for the project. She felt confident that she would not be blindsided by some unexpected problem.

For project communication and project quality, Barbara established procedures for the project. She set up a central database to post the project's status, decisions, and

February 17, 2005

To: CSS Project Oversight Committee
           William McDougal, Chair

MEMO

From: , Barbara Halifax, Project Manager

Following instructions from John MacMurty, I will send a status report memo to the oversight committee every two weeks during the first two months of the project. Thereafter, I will provide it monthly.

Completed during the last period (two weeks)

Steve Deerfield and I have worked on two major items. First, we have completed a high-level statement of the project scope. As you are aware, we interviewed each of you, as well as other major stakeholders, to determine the business needs and overall scope of the project.

Second, we have developed a project schedule, with a detailed WBS for the planning and analysis phases and a less detailed WBS for the design and implementation phases. I have attached a copy of the project schedule. We anticipate having the system operational before the end of the year.

Plans for the next period (two weeks)

During the next two weeks, we will finalize project feasibility. Primary focus will be on developing a preliminary cost/benefit analysis. As more information is gathered during the analysis phase, this financial analysis will become more precise.

We will also develop a feasibility analysis of the project. Our risk analysis will determine whether the project is feasible and identify areas of high risk. Finally, we will develop a staffing plan, identify team members—both technical staff and users—and begin staffing the project.

Problems, Issues, Open Items -- None

working documents to make sure that all the team members were kept well informed. She established a routine and format for weekly status reports from the team leaders and a status report to the oversight committee. An example of one of her status report memos to the oversight committee follows. These status reports all follow a standard format. In addition to the formal status memos, she would also write more informal memos to John MacMurty. For project quality, internal procedures required that team members and RMO users review all work products. Other quality procedures, such as the test plan, would be established as the project progressed.

She and Steve had identified the other team members they would like to have on the team. John had been especially helpful in finding solid analysts who were available or who would be available soon. In fact, Barbara had already interviewed all of the members who were coming on board. Recognizing the importance of having a team whose members could work together, she had scheduled several days for the team members to get to know each other, to refine their internal working procedures, and to teach them about the tools and techniques that would be used on the project.

All in all, it had been a very hectic but productive month. A lot of work had been done, and a solid foundation had been established for a successful project.

Project management aspects of the case are reinforced throughout by use of **RMO memos** describing the status of the project in every chapter. The same system project is used to illustrate traditional and object-oriented models and solutions, so both approaches can be understood and directly compared.

If the driver fails to pay the fine within the required period, the ticket processing system produces a warrant request notice and sends it to the court. This happens if the driver does not return the original envelope within two weeks or does not return the court-supplied envelope within two weeks of the trial date. What happens then is in the hands of the court. Sometimes the court requests that the driver's license be suspended, and the system that processes drivers' licenses handles the suspension.

1. To what events must the ticket processing system respond? Create a complete event table listing the event, trigger, source, activity, response, and destination for each event.
2. Draw an entity-relationship diagram to represent the data storage requirements for the ticket processing system, including the attributes mentioned. Explain why it is important to understand how the system is integrated with other State Patrol systems.
3. Draw a class diagram that corresponds to the ERD but assumes there are different types of drivers. Classifications of types of drivers vary by state. Some states have restricted licenses for minors, for example, and special licenses for commercial vehicle operators. Research your state's requirements, and create a generalization/specialization hierarchy for the class Driver, showing the different attributes each special type of driver might have. Consider the same issues for types of tickets. Include some special types of tickets in a generalization/specialization hierarchy in the class diagram.

Rethinking Rocky Mountain Outfitters

When listing nouns and making some decisions about the initial list of things (see Figure 5-17), the RMO team decided to research Customer Account as a possible data entity or class if the system included an RMO payment plan (similar to a company charge account plan). Many retail store chains have their own charge accounts for the convenience of the customer—to increase sales to the customer and to better track customer purchase behavior.

Consider the implications to the system if management decided to incorporate an RMO charge account and payment plan as part of the customer support system.

1. Discuss the implications that such a change would have on the scope of the project. How might this new capability change the list of stakeholders the team would involve when collecting information and defining the requirements? Would the change have any effect on other RMO systems or system projects planned or under way? Would the change have any effect on the project plan originally developed by Barbara Halifax? In other words, is this a minor change or a major change?
2. What events need to be added to the event table? Complete the event table entries for these additional events. What activities or use cases for existing events might be changed because of a charge account and payment plan? Explain.
3. What are some additional things and relationships among things that the system would be required to store because of the charge account and payment plan? Modify the entity-relationship diagram and the class diagram to reflect these charges.

ELECTRONICS UNLIMITED, INC.: INTEGRATING THE SUPPLY CHAIN

Electronics Unlimited is a warehousing distributor that buys electronic equipment from various suppliers and sells it to retailers throughout the United States and Canada. It has operations and warehouses in Los Angeles, Houston, Baltimore, Atlanta, New York, Denver, and Minneapolis. Its customers range from large nationwide retailers, such as Target, to medium-sized independent electronics stores.

Many of the larger retailers are moving toward integrated supply chains. Information systems used to be focused on processing internal data; however, today these retail chains want suppliers to become part of a totally integrated supply chain system. In other words, the systems need to communicate between companies to make the supply chain more efficient.

To maintain its position as a leading wholesale distributor, Electronics Unlimited has to convert its system to link with both its suppliers—the manufacturers of the electronic equipment—and its customers—the retailers. It is developing a completely new system utilizing object-oriented techniques to provide these links. Object-oriented techniques facilitate system-to-system interfaces by using predefined components and objects to accelerate the development process. Fortunately, many of the system development staff have recently begun learning about object-oriented development and are eager to apply the techniques and models to a system development project.

William Jones is explaining object-oriented development to the group of systems analysts who are being trained in this approach. "We're developing most of our new systems using object-oriented principles. The complexity of the new system, along with its interactivity, makes the object-oriented approach a natural way to

develop requirements. It takes a little different thought process than you may be used to, but the object-oriented models track very closely with the new object-oriented programming languages."

William continued, "This way of thinking about a system in terms of objects is very interesting. It also is consistent with the object-oriented programming techniques you learned in your programming classes. You probably first learned to think about objects when you developed screens for the user interface. All of the controls on the screen, such as buttons, text boxes, and drop-down boxes, are objects. Each has its own set of trigger events that activate its program functions.

"Now you just extend that same thought process so that you think of things like purchase orders and employees as objects, too. We can call them *problem domain* or sometimes *business objects* to differentiate them from screen objects such as windows and buttons. During analysis, we have to find out all of the trigger events and methods associated with each business object."

"How do we do that?" one of the analysts asked.

"You continue with your fact-finding activities and build a scenario for each business process. The way the business objects interact with each other in the scenario determines how you identify the initiating activity. We refer to those activities as the *messages* between objects. The tricky part is that you need to think in terms of objects instead of just processes. Sometimes it helps me to pretend I am an object. I will say, 'I am a purchase order object. What functions and services are other objects going to ask me to do?' Once you get the hang of it, it works very well, and it is enlightening to see how the system requirements unfold as you develop the diagrams."

OVERVIEW

The basic objective of requirements definition is understanding—understanding users' needs, understanding how the business processes are carried out, and understanding how the system will be used to support those business processes. As we indicated in Chapter 2, system developers use a set of tools and techniques to discover and understand the requirements for a new system. This activity is a key part of the systems analysis phase of the systems development life cycle. In object-oriented development, the set of analysis activities is more specifically referred to as *object-oriented analysis* (OOA). The first step in the process for developing this understanding requires the fact-finding skills you learned in Chapter 4. Fact-finding activities are also called *discovery activities*, and obviously discovery must precede understanding. In this chapter, you learn to take discovery to the next level—to build understanding.

Chapter 5 introduced the concepts of models and modeling activities as a way to define and document system requirements. The models introduced in Chapter 5 focus on two primary aspects of functional requirements: the events and the things involved in

---

Every chapter follows up on the RMO case details by adding an end of chapter case study named **Rethinking Rocky Mountain Outfitters**. Each case extends an example in the chapter or poses additional questions to consider about the RMO system project.

Short stand-alone **opening case studies** describe a real-world situation relevant to the material in each chapter. A variety of companies and situations is included to provide the reader with a broad view of the problems and opportunities found in the real world.

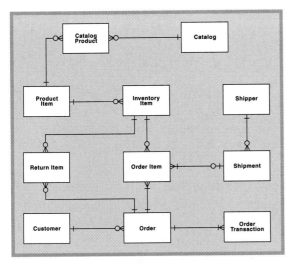

FIGURE *5-28*

Rocky Mountain Outfitters customer support system entity-relationship diagram (ERD) without attributes.

The ERD shown in Figure 5-28 contains a lot of very specific information about the requirements for the system. Be sure that you can trace through all of the relationships shown and try to describe a specific example of each data entity involved in one specific order. Try listing the key attributes for each data entity to check your understanding. Draw a sketch similar to that shown in Figure 5-25 to show some actual data this ERD describes. You can check your understanding by looking ahead to the class diagram in the next section and to the relational database design in Chapter 13.

Once it is developed, a model like this entity-relationship diagram needs to be walked through carefully, as you would walk through the logic of a program. Being able to walk through and "debug" any model is a very important skill in system develop

180

Margin definitions of key terms are placed in the text when the term is first used.

FIGURE *7-14*

Sample system sequence diagram (SSD).

SSD Notation

Figure 7-14 shows a generic SSD. As with a use case diagram, the stick figure represents an actor—a person (or role) that interacts with the system. In a use case diagram, the actor "uses" the system, but the emphasis in an SSD is on how the actor "interacts" with the system by entering input data and receiving output data. The idea is the same with both diagrams; the level of detail is different.

The box labeled :System is an object that represents the entire automated system. In SSDs, and all interaction diagrams, instead of using class notation, analysts use object notation. Object notation indicates that the box refers to an individual object and not the class of all similar objects. The notation is simply a rectangle with the name of the object underlined. The colon before the underlined class name is a frequently used, but optional, part of the object notation. In an interaction diagram, the messages are sent and received by individual objects, not by a class. In an SSD, the only object included is one representing the entire system.

Underneath the actor and the :System are vertical dashed lines called *lifelines*. A *lifeline*, or *object lifeline*, is simply the extension of that object, either actor or object, throughout the duration of the SSD. The arrows between the lifelines represent the messages that are sent or received by the actor or the system. Each arrow has an origin and a destination. The origin of the message is the actor or object that sends it, as indicated by the lifeline at the arrow's tail. Similarly, the destination actor or object of a message is indicated by the lifeline that is touched by the arrowhead. The purpose of lifelines is to indicate the sequence of the messages sent and received by the actor and object. The sequence of messages is read from top to bottom in the diagram.

A message is labeled to describe both the message's purpose and any input data being sent. The syntax of the message label has several options; the simplest forms are shown in Figure 7-14. Remember that the arrows are used to represent both a message and input data. But what is meant by the term *message* here? In a sequence diagram, a message is considered to be an action that is invoked on the destination object, much like a command. Notice in Figure 7-14 that the input message is called inquireOnItem. The clerk is sending a request, or a message to the system, to find an item. The input data that is sent with the message is contained within the parentheses, and in this case it is data to identify

lifeline, or object lifeline
the vertical line under an object on a sequence diagram to show the passage of time for the object

taken, they both result in this common activity. Finally, the custom
and decides whether it needs changes or is acceptable. In this simple
always does buy something, so this workflow is obviously not compl

Notice that an activity diagram focuses on the sequence of activiti
straightforward and quite easy to understand. In fact, one of the stren
ity diagrams to document workflows is that users also find them v
stand. You can use graphical representations such as this diagram to
standing of the particular workflow procedure with the user.

Figure 4-13 illustrates another workflow. This diagram demonstra
cepts. Let's assume that the customer from the previous example di
with an order. This next figure shows the workflow that is required to
uled for production. The salesperson sends to engineering the printe
now become an order. This example emphasizes the fact that a docu
mitted. To indicate that a document is being passed, you place the do
the end of the connecting arrow, and the arrow now becomes a cond
a document, not just a flow of activities. After engineering develops
two concurrent activities happen: purchasing orders the materials, and
the program for the automated milling machines. These two activities
dependent and can occur at the same time. Notice that one synchro
path into two concurrent paths, and another synchronization bar reconnects them.
Finally, scheduling puts the order on the production schedule.

FIGURE *4-13*

An activity diagram showing concurrent paths.

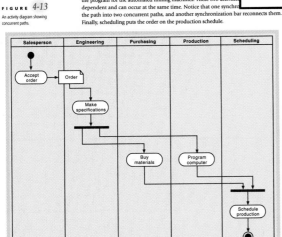

136   Part 2   Systems Analysis Tasks

Each chapter includes extensive **figures and illustrations** designed to clarify and summarize key points and to provide examples of models and other deliverables produced by an analyst. **Color coding** is used to differentiate traditional models (diagrams with blue backgrounds), object-oriented models (diagrams with light green backgrounds), and models used with both approaches (diagrams with yellow backgrounds).

SUMMARY

The object-oriented approach has a complete set of diagrams that together document the user's needs and define the system requirements. These requirements are specified using the following models:

- domain model class diagrams
- use case diagrams
- use case detailed models, either descriptive format or activity diagram
- system sequence diagrams (SSDs)

A use case diagram documents the various ways that the system can be used. It can be developed independently or in conjunction with the event table where one event triggers one use case. A use case consists of actors, use cases, and connecting lines. A use case identifies a single function that the system supports. An actor represents a role of someone or something that uses the system. The connecting lines indicate which actors invoke which use cases. Use cases can also invoke other use cases as a common subroutine. This type of connection between use cases is called the «includes» relationship.

The internal activities of a use case are first described by an internal flow of activities. It is possible to have several different internal flows, which represent different scenarios of the same use case. Thus, a use case may have several scenarios. These details are documented either in descriptive format or with activity diagrams to describe the workflows. Preconditions and post-conditions can be defined for each use case to assist in understanding other effects on the system of executing a use case.

Another diagram that provides more details of the processing requirements of a use case is a system sequence diagram, or SSD. An SSD documents the inputs and outputs of the system. The scope of each SSD is usually a use case or a scenario within a use case. The components of an SSD are the actor—the same actor identified in the use case—and the system. The system is treated as a black box, in that the internal processing is not addressed. Messages, which represent the inputs, are sent from the actor to the system. Output messages are returned from the system to the actor. The sequence of messages is indicated by the top-to-bottom sequence of messages.

The domain model class diagram continues to be refined when defining requirements. Often the class symbol includes just two sections to indicate the diagram represents a domain model, leaving out the methods. This is because the classes in the domain model represent the real-world objects involved in the users' work; they do not represent software classes. Additional class diagram notation can be used, such as labeling association relationships, indicating class attributes, and identifying properties.

KEY TERMS

domain model class diagram, or domain model, p. 267
interaction diagram, p. 254
lifeline, or object lifeline, p. 259
message, p. 244
postcondition, p. 256
precondition, p. 256

scenario, or use case instance, p. 251
statechart diagram, p. 244
system sequence diagram, p. 244
true/false condition, p. 260
use case diagram, p. 244

Chapter 7   The Object-Oriented Approach to Requirements

New to this edition is the **Focusing on Reliable Pharmaceutical Service** case study, which is included at the end of every chapter to provide additional experience with problem-solving techniques and issues addressed in the chapter. Reliable is a smaller company than RMO, and the strategic information system plan and specific system development project provide a different perspective to analysis and design.

End-of-chapter material includes a **detailed summary,** an **indexed list of key terms,** and a list of **additional resources** and **references.**

---

4. How might graphics be used? What about drill-down capabilities?
5. How would you prepare a mock-up of each report, assuming a printed output and also an on-line output?
6. What output controls should be associated with each report?

Reliable
PHARMACEUTICALS   Focusing on Reliable Pharmaceutical Service

One of the challenges of a pharmaceutical company is keeping current with new drugs and changes to existing drugs. New drugs are continually being developed and approved. In addition, generic drugs are often available to compete with brand-name drugs. One of the services that Reliable provides is to try to find the least expensive alternative to fulfill a prescription. This cost-saving service is one of the marketing advantages that the nursing homes can use to promote their services. Obviously, this service builds tremendous loyalty between Reliable and its customers.

To keep current with these changes, Reliable subscribes to an on-line drug-update service. The service provides updates in several formats, one of which is an XML file.

1. Based on the content of your design class diagrams that you developed in Chapter 11, illustrate a sample XML input file that could be used to update drug information in the Reliable database.
2. In earlier chapters, the case description indicated that a case manifest was produced for each patient whenever prescriptions needed to be filled and delivered. Based on the data found in your class diagrams, design a case manifest. Consider that a patient may have multiple prescriptions that are being filled on the same delivery.
3. Each month, Reliable produces a statement for each nursing home. The statement lists each patient who received prescriptions during the month. All the filled prescriptions are listed. For each prescription, the following information is listed: the price, the amount billed to the patient's insurance provider, the amount paid by the insurance provider, and the co-pay amounts due from the patients. Design this monthly statement. Also, identify and highlight output controls that you believe are appropriate for this type of report.
4. In the preceding chapter, you defined an input form to be used to collect orders from the nursing homes. Go back and analyze that input form and identify all of the input controls that you think are necessary to ensure that the prescriptions are correct. What other procedures or controls would you recommend to make sure that there are no mistakes on the prescriptions?

FURTHER RESOURCES

David Benyon, Diana Bental, and Thomas Green, *Conceptual Modeling for User Interface Development.* Springer-Verlag, 1999.

Elfriede Dustin, Jeff Rashka, Douglas McDiarmid, and Jakob Nielson, *Quality Web Systems: Performance, Security, and Usability.* Addison-Wesley, 2001.

Simson Garfinkel, Gene Spafford, and Debby Russell, *Web Security, Privacy, & Commerce.* O'Reilly Publishing, 2001.

Anup K. Ghosh, *E-Commerce Security: Weak Links, Best Defenses.* John Wiley & Sons, 1997.

IS Audit and Control Association, *IS Audit and Control Journal,* Volume I. 1995.

Brenda Laurel, *The Art of Human-Computer Interface Design.* Addison-Wesley, 1990.

Ben Shneiderman, *Designing the User Interface: Strategies for Effective Human-Computer Interaction.* Addison-Wesley-Longman, 1998.

Donald Warren Jr., and J. Donald Warren, *The Handbook of IT Auditing.* Warren Gorham & Lamont, 1998.

Donald A. Wayne and Peter B. B. Turney, *Auditing EDP Systems.* Prentice Hall, 1990.

Chapter 15   Designing System Interfaces, Controls, and Security   **621**

REVIEW QUESTIONS

1. Give an example of a business problem.
2. What are the main steps followed when solving a problem?
3. Define *system*.
4. Define *information system*.
5. What are the types of information systems found in most organizations?
6. List the six fundamental technologies an analyst needs to understand.
7. List four types of tools the analyst needs to use to develop systems.

8. List five types of techniques used during system development.
9. What are some of the things an analyst needs to understand about businesses and organizations in general?
10. What are some of the things an analyst needs to understand about people?
11. What are some of the types of technology an analyst might encounter?
12. List 10 job titles that involve analysis and design work.
13. How might an analyst become involved with executives and strategic planning relatively early in his or her career?

THINKING CRITICALLY

1. Describe a business problem your university has that you would like to see solved. How can information technology help solve it?
2. Describe how you would go about solving a problem you face. Is the approach taken by a systems analyst, as described in the text, any different?
3. Many different types of information systems were described in this chapter. Give an example of each type of system that might be used by a university.
4. What is the difference between technical skills and business skills? Explain how a computer science graduate might be strong in one area and weak in another. Discuss how the preparation for a CIS or MIS graduate is different from that for a computer science graduate.
5. Explain why an analyst needs to understand how people think, how they learn, how they react to change, how they communicate, and how they work.

6. Who needs greater integrity to be successful, a salesperson or a systems analyst? Or does every working professional need integrity and ethical behavior to be successful? Discuss.
7. Explain why developing an information system requires different skills if the system is a client-server rather than a large-scale centralized mainframe architecture.
8. How might working for a consulting firm for a variety of companies make it difficult for the consultant to understand the business problem a particular company faces? What might be easier for the consultant to understand about a business problem?
9. Explain why a strategic information systems planning project must involve people outside the information systems department. Why would a consulting firm be called in to help organize the project?
10. Explain why a commitment to enterprise resource planning (ERP) would be very difficult to undo once it has been made.

EXPERIENTIAL EXERCISES

1. It is important to understand the nature of the business you work for as an analyst. Contact some information systems developers and ask them about their employers. Do they seem to know a lot about the nature of the business? What types of classes did they take to prepare themselves (for example, banking classes, insurance classes, retail management classes, hospital administration classes, manufacturing technology classes)? Do they plan to take additional courses? If so, which?
2. Think about the type of position you want (working for a specific company, working for a consulting firm, or working for a software package vendor). Do some research on each

job by looking at companies' recruiting brochures or Web sites. What do they indicate are the key skills they look for in a new hire? Are there any noticeable differences between consulting firms and the other organizations?
3. You have read an overview of the Rocky Mountain Outfitters' strategic information systems plan, including the technology architecture plan and the application architecture plan. Research system planning at your university. Is there a plan for how information technology will be used over the next few years? If so, describe some of the key provisions of the technology architecture plan and the application architecture plan.

CASE STUDIES

Association for Information Technology Professionals Meeting

"I'll tell you exactly what I look for when I interview a new college grad," Alice Adams volunteered. Alice, a system development manager at a local bank, was talking with several professional acquaintances at a monthly dinner meeting of the Association for Information Technology Professionals (AITP). AITP provides opportunities for information systems professionals to get together occasionally and share experiences. Usually a few dozen professionals from information systems departments at a variety of companies attend the monthly meetings.

"When I interview students, I look for problem-solving skills," continued Alice. "Every student I interview claims to know all about Java and .NET and Dreamweaver and XYZ, or whatever the latest development package is. But I always ask interviewees one thing: 'How do you generally approach solving problems?' And then I want to know if they have even thought much about banks like mine and financial services generally, so I ask, 'What would you say are the greatest problems facing the banking industry these days?'"

Jim Parsons, a database administrator for the local hospital, laughed. "Yes, I know what you mean. It really impresses me if they seem to appreciate how a hospital functions, what the problems are for us—how information technology can help solve some of our problems. It is the ability to see the big picture that really gets my attention."

"Yeah, I'm with you," added Sam Young, the manager of marketing systems for a retail store chain. "I am not that impressed with the specific technical skills an applicant has. I assume they have the aptitude and some skills. I do want to know how well they can communicate. I do want to know how much they know about the nature of our business. I do want to know how interested they are in retail stores and the problems we face."

"Exactly," confirmed Alice.

1. Do you agree with Alice and the others about the importance of problem-solving skills? Industry-specific insight? Communication skills? Discuss.
2. Should you research how a hospital is managed before interviewing for a position with an information systems manager at a hospital? Discuss.
3. In terms of your career, do you think it really makes a difference whether you work for a bank, a hospital, or a retail chain? Or is an information systems job going to be the same no matter where you work? Discuss.

Rethinking Rocky Mountain Outfitters

RMO's strategic information systems plan calls for building a new supply chain management (SCM) system prior to building the customer support system (CSS). John Blankens has stated often that customer orientation is the key to success. If that is so, why not build the CSS first, so customers can immediately benefit from improved customer ordering and fulfillment? Wouldn't that increase sales and profits faster? RMO already has factories that produce many items RMO sells, and RMO has longstanding relationships with suppliers around the globe. The product catalog is well established, and the business has existing customers who appear eager and willing to shop on-line. Why wait? Perhaps John Blankens has made a mistake in planning.

1. What are some of the reasons that RMO decided to build the supply chain management system prior to the customer support system?
2. What are some of the consequences to RMO if it is wrong to wait to build the customer support system?
3. What are some of the consequences to RMO if the owners change their minds and start with the customer support system before building the supply chain management system?

Each chapter also includes ample **review questions**, problems and exercises to get the student **thinking critically**, a collection of **experiential exercises** involving additional research or problem solving, and end-of-chapter **case studies** that invite students to practice completing analysis and design tasks appropriate to the chapter.

We have been very gratified as authors to receive so many supportive and enthusiastic comments about *Systems Analysis and Design in a Changing World* since it was first published just a few years ago. In the last few years, the field of systems analysis and design has continued to evolve and mature. Our innovative and truly balanced coverage of traditional structured approaches and newer object-oriented approaches has kept pace with changes in the field. The recent IS 2002 model curriculum now suggests including a balanced coverage of both traditional and objected-oriented analysis and design, something this text has supported from the very beginning. In this third edition, we continue to lead the way by making it feasible to cover object-oriented analysis and design in much greater depth using the latest OO models and design patterns.

OBJECTIVES AND VISION

This text is designed for use in undergraduate and graduate courses that teach systems analysis and design. Systems analysis and design is a practical field that relies on a core set of concepts and principles, as well as what sometimes seems an eclectic collection of rapidly evolving tools and techniques. Learning analysis and design today therefore requires an appreciation of the tried-and-true techniques widely embraced by experienced analysts plus mastery of new and emerging tools and techniques that recent graduates are increasingly expected to apply on the job. It is not easy to develop information systems in today's rapidly changing environment, but the satisfaction and rewards for a job well done are substantial.

This text was developed by a team who was committed to producing an analysis and design text that was different—a text that is flexible and innovative, yet comprehensive and deep. We were guided by the belief that the text must be flexible enough to appeal to instructors emphasizing more traditional approaches to systems analysis and design and to those emphasizing the latest object-oriented techniques. At the same time, we did not want to oversimplify the problem of system development. There are many new developments affecting systems analysis and design, and we wanted to include key trends—packaged solutions, enterprise resource planning (ERP), components, rapid development, the Internet, and so on.

We also wanted the text to teach the key concepts and techniques, not just describe them. Therefore, we focus on fundamentals of lasting value and then show how these fundamentals apply to all development approaches. We explore both traditional structured analysis and design and object-oriented analysis and design in depth. Flexible and innovative? Comprehensive and deep? We think you will agree these objectives have been achieved with this text.

INNOVATIONS

This text is unique in its integration of key systems modeling concepts that apply to both the traditional structured approach and the newer object-oriented approach—events that trigger system activities and objects/entities that are part of the system's problem domain. We devote one chapter to event partitioning and modeling key objects/entities. After completing that chapter, instructors can emphasize structured analysis and design or object-oriented analysis and design, or both. The object-oriented approach is not added as an afterthought—it is assumed from the beginning that everyone should understand the key object-oriented concepts. The traditional approach is not discarded—it is assumed from the beginning that everyone should understand the key structured concepts.

Full Coverage of OO Approach

The object-oriented approach presented in this text is based on the Unified Modeling Language (UML) from the Object Management Group as originated by Grady Booch, James Rumbaugh, and Ivar Jacobson. A model-driven approach to analysis starts with use cases and scenarios and then defines problem domain classes involved in the users' work. We include requirements modeling with use case diagrams, use case descriptions, activity diagrams, and system sequence diagrams. Design models are also discussed in detail, with particular attention to detailed sequence diagrams, design class diagrams, and package diagrams. Design principles and design patterns are discussed throughout. Our database design chapter covers two approaches to object persistence—a hybrid approach using relational database management and a pure approach using object database management systems (ODBMS). Instructors who emphasize the object-oriented approach will not be disappointed by the presentation and depth of coverage in this text.

Full Coverage of Traditional Approach

The traditional approach presented in this text is based on modern structured analysis and design as refined by Stephen McMenamin and John Palmer, Ed Yourdon, and Meilir Page-Jones. Modern structured analysis is an integrated, model-driven approach that includes event partitioning, data modeling with entity-relationship diagrams (ERDs), and process modeling with data flow diagrams (DFDs). Modern structured design is also based on event partitioning and uses the structure chart for software design. Database design using relational database management techniques is featured. Instructors who emphasize the structured approach to development will be satisfied by the presentation and depth of coverage in this text.

Emerging Tools and Trends

Additional concepts and techniques are included in response to the realities of system development today. First, system development and the system development life cycle (SDLC) are explicitly defined as highly iterative. Although the text is organized as a sequential series of phases, the actual development project and the project plan are iterative. Second, emerging techniques and approaches that use an iterative approach are introduced, including the Unified Process (UP), eXtreme Programming (XP), and Agile Modeling. Third, rapid application development and component-based development are covered in depth. Finally, packaged solutions and enterprise resource planning (ERP) are described as alternatives to custom development throughout the book and in detail in a separate ERP chapter.

CHANGES FOR THE THIRD EDITION

As we began considering updates to include in the third edition, we focused on refining some of the presentation and pedagogy, tightening some of the examples, and updating the material to reflect ongoing changes in analysis and design theory and practice. We also made some major changes based on our current research and feedback from instructors using the book.

The balanced coverage of the structured approaches and newer object-oriented approaches remains intact. This text can be used to emphasize the traditional structured approach with data flow diagrams, entity-relationship diagrams, structure charts, and relational databases; to focus on the object-oriented approach with use case modeling, domain and design class diagrams, interaction diagrams, package diagrams, and statecharts; or to cover and compare both approaches in depth.

Increased Emphasis on Iteration and Architecture

We did not reduce the amount of attention paid to the traditional approach to development. Many instructors choose to emphasize the traditional approach, but they also now cover the object-oriented approach to varying degrees. For both the traditional and the OO approach, however, we increased the emphasis on iterative development and three-layer architecture throughout.

Improved Organization

We changed the organization and order of some material in Part 1. Approaches to system development, including the SDLC, tools, techniques, models, and methodologies, are now discussed in Chapter 2. Chapter 3 applies SDLC concepts when discussing project management and the project planning phase of the SDLC. We also divided some chapters to provide a clearer separation of traditional and object-oriented approaches to system design in Part 2.

Enhanced and Expanded OO Coverage

Probably the most noticeable change in the third edition is the extensive enhancement and expansion of the coverage of the object-oriented approach, specifically,

- Providing a more gradual introduction to the process of creating OO requirements models and design models
- Doubling the number of chapters covering OO concepts and modeling
- Increasing the emphasis on use case descriptions for defining system requirements
- Adding system sequence diagrams (SSDs) to model input and output for each use case or scenario when defining requirements
- Emphasizing three-layer architecture for design
- Including multiple design micro-iterations that develop sequence diagrams showing the user interface (view) layer, the problem domain (business logic) layer, and the data access layer for use case realization
- Introducing and applying key OO design principles and design patterns
- Addressing OO development for Web-based applications

The text's original four-part structure remains intact, although the expansion of OO coverage and the separation of traditional and OO design coverage increases the number of chapters to 18.

Additional End-of-Chapter Cases

We made many specific adopter-suggested improvements throughout the text. For example, the Rocky Mountain Outfitters case is now more tightly integrated with the material in all chapters, and other end-of-chapter cases have been added. Instructors also requested a second end-of-chapter running case to explore both traditional and OO

design are reviewed, including networks, client-server architecture, and three-layer design. Chapter 10 discusses the traditional approach to design, including the latest thinking on three-layer designs. Chapters 11 and 12 address object-oriented design. Chapter 11 teaches students how to design the interaction details for each use case—use case realization. Implementation issues for the three-layer design architecture are also discussed. Chapter 12 discusses more advanced design patterns and design principles, including OO design for enterprise-level and Web-based systems. Additionally, state transitions and the statechart diagram are discussed in detail. Instructors can simply choose to emphasize Chapter 10 or Chapters 11 and 12 to focus the course on either the traditional or the object-oriented approach, or both. Chapter 13 covers database design—relational, hybrid, and object-oriented databases. Chapter 14 covers user interfaces and human-computer interaction, and we include general principles and concepts of dialog design in addition to using UML diagrams to model the dialog. Chapter 15 discusses system interfaces, with particular attention to system controls and system security.

Part 4: Implementation and Support

Systems implementation is increasingly technology specific, and because of the diverse development environments in the real world, we decided to streamline the discussion of implementation. Chapter 16 provides an overview of implementation and support that addresses traditional technology and object technology. However, we did include two chapters on important alternative approaches to implementation. Although the text emphasizes iteration and prototyping throughout, we include a comprehensive discussion of rapid application development and some emerging approaches to system development in Chapter 17, including the Unified Process (UP), eXtreme Programming (XP), Agile Modeling, and others. Similarly, although packaged solutions are discussed as viable alternatives throughout, we include a detailed discussion of packages and enterprise resource planning (ERP) in Chapter 18, including specific examples from SAP.

DESIGNING YOUR ANALYSIS AND DESIGN COURSE

As discussed earlier, there are many approaches to teaching analysis and design courses, and the objectives of the course differ considerably from college to college. In some IS departments, the analysis and design course is a capstone course where students apply the material learned in prior database, telecommunications, and programming courses to a real analysis and design project. In other IS departments, analysis and design is used as an introduction to the field of system development, taken prior to more specialized courses. Some IS departments offer a two-course sequence emphasizing analysis in the first semester and design and implementation in the second semester. Some IS departments have only one course that covers both analysis and design.

The design of the analysis and design course, always difficult, is complicated even more by the choice of emphasizing either the traditional structured approach or the newer object-oriented approach, again depending on local curriculum priorities. Additionally, the more iterative approach to development, in general, has made choices about sequencing the analysis and design topics more difficult. For example, with iterative development, a two-course sequence cannot be divided into analysis and then design as easily.

Given these issues, it is not practical to offer sample syllabi that will work for all of these options. The objectives, course content, assignments, and projects have too many variations. What we can offer are some suggestions for using the text in various approaches to the course.

Traditional Analysis and Design Course

A traditional analysis and design course provides coverage of both systems analysis and systems design activities and tasks using structured analysis and structured design, with database design, input/output/controls design and dialog (interface) design. It is usually assumed the project will use custom development, including Web development. The course emphasizes the SDLC, project management, information gathering, and management reporting. One-semester courses are usually limited to completing some prototypes of the user interface to give students closure. Sometimes this course is spread over two semesters, with some implementation of an actual system going on in the second semester for a more complete development experience.

For this approach to the analysis and design course, a reasonable outline would omit chapters and sections detailing OO, current trends, and packages (these concepts are introduced throughout the text, however, so students will still be familiar with them). Additionally, because of the amount of material to cover, the appendices detailing project management, financial feasibility, scheduling, and presentations might be omitted.

A suggested outline for a course emphasizing the traditional approach follows:

Chapter 1: The World of the Modern Systems Analyst
Chapter 2: Approaches to System Development
Chapter 3: The Analyst as a Project Manager
Chapter 4: Beginning the Analysis: Investigating System Requirements
Chapter 5: Modeling System Requirements: Events and Things
Chapter 6: The Traditional Approach to Requirements
Chapter 8: Evaluating Alternatives for Requirements, Environment, and Implementation
Chapter 9: Moving to Design
Chapter 10: The Traditional Approach to Design
Chapter 13: Designing Databases (skip OO design sections)
Chapter 14: Designing the User Interface (skip UML examples)
Chapter 15: Designing System Interfaces, Controls, and Security (skip OO sections)
Chapter 16: Making the System Operational (skip OO sections)

Object-Oriented Analysis and Design Course

This course is similar to the coverage of both analysis and design in the traditional course, except that object-oriented models and techniques are emphasized exclusively. Object-oriented analysis and object-oriented design, with database design, input/output/controls design and dialog (interface) design, are covered. It is usually assumed the projects will use custom development, including Web development. The course emphasizes iterative development with three-layer architecture, project management, information gathering, and management reporting. One-semester courses are usually limited to completing some prototypes of the user interface to give students closure. Sometimes this course is spread over two semesters, with some implementation of an actual system going on in the second semester for a more complete development experience. Iterative development is usually emphasized.

For this approach to the analysis and design course, a reasonable outline would omit chapters and sections detailing structured analysis and structured design. The Current Trends chapter might be included to cover components and iteration, but packages probably would not be covered. Additionally, because of the amount of material to cover, the appendices detailing project management, financial feasibility, scheduling, and presentations might be omitted.

A suggested outline for a course emphasizing object-oriented development follows:

Chapter 1: The World of the Modern Systems Analyst
Chapter 2: Approaches to System Development
Chapter 3: The Analyst as a Project Manager
Chapter 4: Beginning the Analysis: Investigating System Requirements
Chapter 5: Modeling System Requirements: Events and Things
Chapter 7: The Object-Oriented Approach to Requirements
Chapter 8: Evaluating Alternatives for Requirements, Environment, and
    Implementation
Chapter 9: Moving to Design
Chapter 11: The Object-Oriented Approach to Design: Use Case Realization
Chapter 12: Advanced Topics in Object-Oriented Design
Chapter 13: Designing Databases
Chapter 14: Designing the User Interface
Chapter 15: Designing System Interfaces, Controls, and Security
Chapter 16: Making the System Operational
Chapter 17: Current Trends in System Development

Traditional Course with In-Depth Analysis and Project Management

Some courses delve more deeply into systems analysis methods and emphasize project management. Sometimes these courses are graduate courses, and sometimes they assume design and implementation are covered in more technical courses. In some cases, it might be assumed that packages are likely solutions rather than custom development, so defining requirements and managing the process are more important than design activities.

The appendices covering project management, financial feasibility, scheduling, and presentations should be included. Chapters on detailed design might be omitted. The packages/ERP chapter (Chapter 18) might be included, if appropriate.

A suggested outline for courses emphasizing the traditional approach, with in-depth coverage of analysis and project management follows:

Chapter 1: The World of the Modern Systems Analyst
Chapter 2: Approaches to System Development
Chapter 3: The Analyst as a Project Manager
Appendix A: Principles of Project Management
Appendix B: Project Schedules with PERT/CPM Charts
Appendix C: Calculating Net Present Value, Payback Period, and Return on Investment
Chapter 4: Beginning the Analysis: Investigating System Requirements
Appendix D: Presenting the Results to Management
Chapter 5: Modeling System Requirements: Events and Things
Chapter 6: The Traditional Approach to Requirements
Chapter 8: Evaluating Alternatives for Requirements, Environment, and
    Implementation
Chapter 9: Moving to Design
Chapter 18: Packaged Software and Enterprise Resource Planning

Object-Oriented Course with In-Depth Analysis and Project Management

Some courses cover object-oriented systems analysis methods in more depth—but not OO design—and emphasize project management. Sometimes these courses are graduate courses, and sometimes they assume design and implementation are covered in more technical courses. In some cases, it might be assumed that packages are likely solutions rather than custom development, so defining requirements and managing the process are more important than design activities.

The appendices covering project management, financial feasibility, scheduling, and presentations should be included. Chapters on detail design might be omitted. The packages/ERP chapter (Chapter 18) might be included, if appropriate.

A suggested outline for a course covering object-oriented analysis, with in-depth coverage of project management follows:

Chapter 1: The World of the Modern Systems Analyst
Chapter 2: Approaches to System Development
Chapter 3: The Analyst as a Project Manager
Appendix A: Principles of Project Management
Appendix B: Project Schedules with PERT/CPM Charts
Appendix C: Calculating Net Present Value, Payback Period, and Return on Investment
Chapter 4: Beginning the Analysis: Investigating System Requirements
Appendix D: Presenting the Results to Management
Chapter 5: Modeling System Requirements: Events and Things
Chapter 7: The Object-Oriented Approach to Requirements
Chapter 8: Evaluating Alternatives for Requirements, Environment, and Implementation
Chapter 9: Moving to Design
Chapter 18: Packaged Software and Enterprise Resource Planning

Comparative Analysis and Design Course

Some courses survey the field of analysis and design to provide a comprehensive exposure to major approaches. Sometimes these courses are graduate courses for experienced developers, and sometimes they emphasize concepts over detailed hands-on experience with techniques. A reading knowledge of the key models might be the objective. However, often the instructor will require hands-on projects using both traditional and object-oriented techniques for the same system in the same course.

The entire book can be covered for the most complete treatment. Alternatively, sections of material that cover details about some of the techniques can be omitted. A fast-paced survey course can cover the chapters quickly for recognition and reading knowledge of models. Chapters 17 and 18 might directly follow Chapter 8, as shown in the following outline, and then the course can continue surveying design. If the comparative course emphasizes systems analysis and project management, it might end after Chapter 18 without covering design. There are many possibilities to consider.

A suggested outline for a comparative course follows:

Chapter 1: The World of the Modern Systems Analyst
Chapter 2: Approaches to System Development

An Iterative Approach to the Analysis and Design Course

One of the biggest challenges facing analysis and design instructors is how to handle iterative development. This is an issue for both the traditional approach and the object-oriented approach. Textbooks can teach analysis techniques and then design techniques sequentially, but that is not the way the techniques are used in practice. Students do not always appreciate this point. One way to make the course resemble real-world practice is to teach iteratively. The idea that no one gets it right the first time certainly applies to learning analysis and design.

As with iterative development, the course could survey analysis and design techniques rapidly, perhaps obtaining a reading knowledge of the models, and then go back over the analysis and design material in more depth. Some sections of chapters might be skipped the first time through. But there is a great difference between understanding and interpreting analysis and design models and actually creating analysis and design models. Therefore, it might make sense to go through the techniques with a reading knowledge as a goal for the first iteration. Then students can be asked to reconsider the models as they create new ones based on a course project.

It would be difficult to ask the students to read everything once and then to reread it all again. Therefore, one approach might be to rapidly survey the field without digressing into specifics. Then the second iteration could add new material while going into depth in prior material. For example, the first time through might emphasize Chapter 5 but skim through either 6 or 7 (depending on whether traditional or OO is emphasized). The overview of design in Chapter 9 might be covered, but the rest of the design chapters might be limited to Chapter 10 or Chapter 11 (depending on whether traditional or OO is emphasized). The second iteration could go into requirements models in depth and design chapters in depth.

There are many other possibilities to consider. What is important is that you consider this issue in some way when designing your course. We would appreciate any feedback you can provide on ideas you have considered or tried with an iterative approach to teaching analysis and design.

AVAILABLE SUPPORT

Systems Analysis and Design in a Changing World, Third Edition, includes teaching tools to support instructors in the classroom. The ancillaries that accompany the textbook include an Instructor's Manual, solutions, Test Banks and Test Engine, Distance Learning content, PowerPoint presentations, and Figure Files. Please contact your Course Technology sales representative to request the Teaching Tools CD-ROM, if you have not already received it. Or, go to the Web page for this book at www.course.com to download many of these items.

The Instructor's Manual

The Instructor's Manual includes suggestions and strategies for using the text, including course outlines for instructors emphasizing the traditional structured approach or the object-oriented approach, as well as for those teaching graduate courses on analysis and design.

Solutions

We provide instructors with answers to review questions and suggested solutions to chapter exercises and cases.

ExamView®

This objective-based test generator lets the instructor create paper, LAN, or Web-based tests from test banks designed specifically for this Course Technology text. Instructors can use the QuickTest Wizard to create tests in fewer than five minutes by taking advantage of Course Technology's question banks or create customized exams.

Distance Learning Content

Course Technology, the premiere innovator in management information systems publishing, is proud to present on-line courses in WebCT and Blackboard, as well as at My-Course 2.0 to provide the most complete and dynamic learning experience possible.

- **MyCourse 2.0.** MyCourse 2.0 is a flexible, easy-to-use management tool that gives instructors true customization over the on-line components of their course. It allows them to personalize their course home page, schedule course activities and assignments, post messages, administer tests, and much more. MyCourse 2.0 is hosted by Thomson Learning, allowing for hassle-free maintenance and student access at all times.
- **Blackboard and WebCT Level 1 Online Content.** If you use Blackboard or WebCT, the test bank for this textbook is available at no cost in a simple, ready-to-use format. Go to www.course.com and search for this textbook to download the test bank.
- **Blackboard and WebCT Level 2 Online Content.** Blackboard Level 2 and WebCT Level 2 are also available for *Systems Analysis and Design in a Changing World, Third Edition*. Level 2 offers course management and access to a Web site that is fully populated with content for this book. Students purchase the Blackboard User Guide (ISBN 0-7895-6165-4) or the WebCT User Guide (0-7895-6163-8). The User Guides include a password that allows student access to Level 2.

For more information on how to bring distance learning to your course, instructors should contact their Course Technology sales representative.

PowerPoint Presentations

Microsoft PowerPoint slides are included for each chapter. Instructors might use the slides in a variety of ways, including their use as teaching aids during classroom presentations or as printed handouts for classroom distribution. Instructors can add their own slides for additional topics introduced to the class.

Figure Files

Figure Files allow instructors to create their own presentations using figures taken directly from the text.

Software Bundling Options

Many instructors like to include software for students to use for exercises and course projects, and this text offers many bundling possibilities. Some instructors like to emphasize CASE tools, and Course Technology can bundle several popular CASE tools with the text, including Oracle Designer, Describe Enterprise, and Visible Analyst. In addition, we offer Microsoft Visio, Microsoft Project 2003, Edge Diagrammer, and Popkin's System Architect. Instructors can contact their Course Technology representative for the latest information.

CREDITS AND ACKNOWLEDGMENTS

This book was originally launched following some extensive brainstorming by senior vice president and publisher Kristen Duerr of Course Technology and lead author John Satzinger. We agreed that an analysis and design text required a major commitment from the publisher to be competitive. We also agreed that no one person could complete a text that met the objectives—flexible and innovative, yet comprehensive and deep. Therefore, Course Technology took an active role in assembling a team of authors who shared the vision. The managing editor brought in to direct the project was Jennifer Locke, who had a major role in bringing the authors together and shaping the direction and final form of the text. Her contributions overseeing the second and third edition have also been substantial.

We were very fortunate to have senior product manager Barrie Tysko again placed in charge of managing the revision project, and she was instrumental in helping us formulate objectives for the revision and for making countless improvements and additions that make the third edition even better than we had hoped. We were also pleased to have senior product manager Tricia Boyle step in to assist in the final stages of production.

Another essential member of the team is developmental editor Karen Hill of Elm Street Publishing Services, who returned to guide us through the third edition. She collected and digested the comments and reactions of reviewers, provided guidance and design for the features and chapter pedagogy, suggested improvements and refinements to the organization and content, read each draft of each chapter from the perspective of the student to help us be consistent and clear, and edited the chapters to provide a consistent style.

Many other people were involved in the production of this text. Amanda Young Shelton of Course Technology provided substantial support for the first edition. Christine Freitas, associate production manager, worked carefully with Barrie Tysko to help sort out the subtleties of traditional and UML diagrams that probably appeared rather primitive to the artists. The production team went the extra mile to be true to the diagramming conventions and standards to the extent that we could define them. Mirella Misiaszek did a thorough job assembling the instructor's materials for the third edition.

We also want to thank some other key people for their specific contributions— Richard A. Johnson of Southwest Missouri State University for writing Chapter 18 on packages and ERP and William Baker for contributing material on presentation techniques. Many other colleagues and friends at SMSU, Brigham Young University, the University of New Mexico, and elsewhere contributed to and supported our work in one way or another. Special thanks also go to Lavette Teague, Lorne Olfman, and Paul Gray for guidance and inspiration.

Last, but certainly not least, we want to thank all of the reviewers who worked so hard for us, beginning with an initial proposal and continuing all the way through to the completion of the first, second, and third editions of this text. We were lucky enough to have reviewers with broad perspectives, in-depth knowledge, and diverse preferences. We listened very carefully, and the text is much better as a result of their input. The reviewers for the first, second, and third edition included:

Rob Anson, *Boise State University*

Marsha Baddeley, *Niagara College*

Teri Barnes, *DeVry Institute—Phoenix*

Robert Beatty, *University of Wisconsin—Milwaukee*

Anthony Cameron, *Fayetteville Technical Community College*

Genard Catalano, *Columbia College*

Paul H. Cheney, *University of Central Florida*

Jung Choi, *Wright State University*

Jon D. Clark, *Colorado State University*

Lawrence E. Domine, *Milwaukee Area Technical College*

Jeff Hedrington, *University of Phoenix*

Ellen D. Hoadley, *Loyola College* in Maryland

Norman Jobes, *Conestoga College,* Waterloo, Ontario

Gerald Karush, *Southern New Hampshire University*

Robert Keim, *Arizona State University*

Rajiv Kishore, *The State University of New York,* Buffalo

Rebecca Koop, *Wright State University*

Hsiang-Jui Kung, *Georgia Southern University*

James E. LaBarre, *University of Wisconsin—Eau Claire*

Tsun-Yin Law, *Seneca College*

David Little, *High Point University*

George M. Marakas, *Indiana University*

Roger McHaney, *Kansas State University*

Cindi A. Nadelman, *New England College*

Bruce Neubauer, *Pittsburgh State University*

Michael Nicholas, *Davenport University—Grand Rapids*

George Pennells

Julian-Mark Pettigrew

Mary Prescott, *University of South Florida*

Alex Ramirez, *Carleton University*

Eliot Rich, *The State University of New York,* Albany

Robert Saldarini, *Bergen Community College*

Laurie Schatzberg, *University of New Mexico*

Deborah Stockbridge, *Quincy College*

Jean Smith, *Technical College of the Lowcountry*

Peter Tarasewich, *Northeastern University*

Craig VanLengen, *Northern Arizona University*

Bruce Vanstone, *Bond University*

Terence M. Waterman, *Golden Gate University*

All of us involved in the development of this text wish you all the best as you take on the challenge of analysis and design in a changing world.

—John Satzinger
—Bob Jackson
—Steve Burd

The Modern Systems Analyst

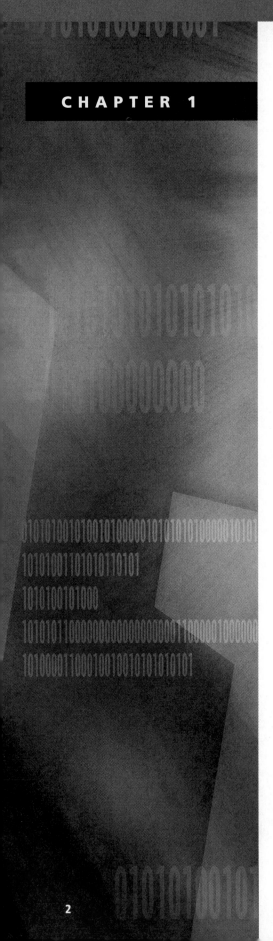

CHAPTER 1

The World of the Modern Systems Analyst

LEARNING OBJECTIVES

After reading this chapter, you should be able to:

- Explain the key role of a systems analyst in business

- Describe the various types of systems an analyst might work on

- Explain the importance of technical, people, and business skills for an analyst

- Explain why ethical behavior is crucial for a systems analyst's career

- Describe the many types of technology an analyst needs to understand

- Describe various job titles in the field and places of employment where analysis and design work is done

- Discuss the analyst's role in strategic planning for an organization

- Describe the analyst's role in a system development project

CHAPTER OUTLINE

The Analyst as a Business Problem Solver

Systems That Solve Business Problems

Required Skills of the Systems Analyst

The Environment Surrounding the Analyst

The Analyst's Role in Strategic Planning

Rocky Mountain Outfitters and Its Strategic Information Systems Plan

The Analyst as a System Developer (the Heart of the Course)

Mary Wright thought back about her two-year career as a programmer analyst. She had been asked to talk to visiting computer information system (CIS) students about life on the job. "It seems like yesterday that I finally graduated from college and loaded up a U-Haul to start my new job at Consolidated," she began.

Consolidated Refineries is an independent petroleum refining company in west Texas. Consolidated buys crude oil from freelance petroleum producers and refines it into gasoline and other petroleum products for sale to independent distributors. Demand for refined petroleum products had been increasing rapidly, and Consolidated was producing at maximum capacity. Capacity planning systems and refining operations systems were particularly important computer information systems for Consolidated, since careful planning and process monitoring resulted in increased production at reduced costs. This increasing demand, and other competitive changes in the energy industry, made information systems particularly important to Consolidated.

Mary continued her informal talk to visiting students. "At first I did programming, mainly fixing things that end users wanted done. I completed some training on Java and object-oriented analysis to round out my experience. The job was pretty much what I had expected at first until everything went crazy over the IPCS project."

The Integrated Process Control System (IPCS) project was part of the company's information systems plan drawn up the year before. Edward King, the CEO of Consolidated Refineries, had pushed for more strategic planning at the company from the beginning, including drawing up a five-year strategic plan for information systems. The IPCS development project was scheduled to begin in the third or fourth year of the plan, but suddenly priorities changed. Demand for petroleum products had never been higher, and supplies of crude oil were becoming scarce. At the same time, political pressure was making price increases an unpopular option.

Something had to be done to increase production and reduce costs. It would be years before an additional refinery could be built, and additional crude oil supplies from new oil fields were years away. The only option for Consolidated's growth and increased profits was to do a better job with the plants and supplies they had. So, top executives decided to make a major commitment to implementing the IPCS project, with the goal of radically improving capacity planning and process monitoring. Everyone at Consolidated also wanted access to this information anywhere and anytime.

"It seemed like the IPCS project was the only thing the company cared about," continued Mary. "I was assigned to the project as the junior analyst assisting the project manager, so I got in on everything. Suddenly I was in meeting after meeting, and I had to digest all kinds of information about refining and distribution, as if I were a petroleum engineer. I met with production supervisors, suppliers, and marketing managers to learn about the oil business, just as if I were taking business school courses. I traveled all over to visit oil fields and pipelines—including a four-day trip to Alaska on about two days' notice! I interviewed technology vendors' representatives and consultants who specialized in capacity planning and process control systems. I've been spending a lot of time at my computer, too, writing reports, letters, and memos—not programming!

"We've been working on the project for seven months now, and every time I turn around, Mr. King, our CEO, is saying something about how important the IPCS project is to the future of the company. He repeats the story to employees and to the stockholders. Mr. King attends many of our status meetings, and he even sat next to me the day I presented a list of key requirements for the system to the top management team.

"This is not at all the way I thought it would be."

OVERVIEW

Information systems are crucial to the success of modern business organizations, and new systems are constantly being developed to make businesses more competitive. People today are attracted to information systems careers because information technology can have a dramatic impact on productivity and profits. Most of you regularly use the latest technologies for on-line purchases and reservations, on-line auctions and customer support, and e-mail and wireless messaging. But it is not the technology itself that increases productivity and profits; it is the people who develop information system solutions that harness the power of the technology that makes these benefits possible. The challenges are great because more and more people expect to have information systems that provide access to information anywhere and anytime.

The key to successful system development is thorough systems analysis and design to understand what the business requires from the information system. *Systems analysis*

systems analysis

the process of understanding and specifying in detail what the information system should do

systems design

the process of specifying in detail how the many components of the information system should be physically implemented

systems analyst

a business professional who uses analysis and design techniques to solve business problems using information technology

means understanding and specifying in detail what the information system should do. *Systems design* means specifying in detail how the many components of the information system should be physically implemented. This text is about systems analysis and design techniques used by a *systems analyst*, a business professional who develops information systems.

This chapter describes the world of the systems analyst—the nature of the work, the knowledge and skills that are important, and the types of systems and special projects an analyst works on. First, we define the analyst's work as problem solving for an organization, so the problem-solving process the analyst follows is described. Next, since most problems an analyst works on are solved in part by an information system, the chapter reviews the types of information systems that businesses use. A systems analyst is a business professional who requires extensive technical, business, and people knowledge and skills, so these skills are reviewed next. Then we survey the types of technology used for information systems and the variety of workplaces and positions where analysis work is done. Sometimes an analyst works on special projects such as strategic planning, business process reengineering, and enterprise resource planning. An analyst's work is really not at all the way most CIS students think it will be.

Finally, the chapter introduces Rocky Mountain Outfitters (RMO), a regional sports clothing distributor headquartered in Park City, Utah. RMO is following a strategic information systems plan that calls for a series of information system development and integration projects over the next several years. The project that RMO is about to launch is a system development project for a new customer support system that will integrate phone orders, mail orders, and direct customer orders via the Internet. The Rocky Mountain Outfitters case is used throughout the text to illustrate analysis and design techniques.

THE ANALYST AS A BUSINESS PROBLEM SOLVER

Systems analysis and design is, first and foremost, a practical field grounded in time-tested and rapidly evolving knowledge and techniques. Analysts must certainly know about computers and computer programs. They possess special skills and develop expertise in programming. But they must also bring to the job a fundamental curiosity to explore how things are done and the determination to make them work better.

Developing information systems is not just about writing programs. Information systems are developed to solve problems for organizations, as the opening case study demonstrated, and a systems analyst is often thought of as a problem solver rather than a programmer. So, what kinds of problems does an analyst typically solve?

- Customers want to order products anytime of the day or night. So, the problem is how to process those orders round the clock without adding to the selling cost.
- Production needs to plan very carefully the amount of each type of product to produce each week. So, the problem is how to estimate the dozens of parameters that affect production and then allow planners to explore different scenarios before committing to a specific plan.
- Suppliers want to minimize their inventory holding costs by shipping parts used in the manufacturing process in smaller daily batches. So, the problem is how to order in smaller lots and accept daily shipments to take advantage of supplier discounts.
- Marketing wants to anticipate customer needs better by tracking purchasing patterns and buyer trends. So, the problem is how to collect and analyze information on customer behavior that marketing can put to use.
- Management continually wants to know the current financial picture of the company, including profit and loss, cash flow, and stock market forecasts. So, the prob-

lem is how to collect, analyze, and present all of the financial information management wants.

- Employees demand more flexibility in their benefits programs, and management wants to build loyalty and morale. So, the problem is how to process transactions for flexible health plans, wellness programs, employee investment options, retirement accounts, and other benefit programs offered to employees.

Information system developers work on problems such as these—and many more. Some of these problems are large and strategically important. Some are much smaller, affecting fewer people, but important in their own way. All programming for the information system that solves the business problem is important, but solving each of these problems involves more than programming. Programming comes much later in the development process.

How does an analyst solve problems? Systems analysis and design focuses on understanding the business problem and outlining the approach to be taken to solve it. Figure 1-1 shows a general approach to problem solving that can be adapted to solving business problems using information technology. Obviously, part of the solution is a new information system, but that is just part of the story.

The analyst must first understand the problem and learn everything possible about it—who is involved, what business processes come into play, what other systems would be affected by solving the problem. Then the analyst needs to confirm for management that the benefits of solving the problem outweigh the costs. Sometimes it would cost a fortune to solve the problem, so it might not be worth solving.

If solving the problem is feasible, the analyst defines in detail what is required to solve it—what specific objectives must be satisfied, what data need to be stored and used, what processing must be done to the data, what outputs must be produced. *What* needs to be done must be defined first. *How* it will be done is not important yet.

Once detailed requirements are defined, the analyst develops a set of possible solutions. Each possible solution (an alternative) needs to be thought through carefully. Usually, an information system alternative is defined as a set of choices about physical components that make up an information system—*how* it will be done. Many choices must be made, involving questions such as these:

- What are the needed components?
- What technology should be used to build the different components?
- Where are the components located?
- How will components communicate over networks?
- How are components configured into a system?
- How will people interact with the system?
- Which components are custom-made, and which are purchased from vendors?
- Who should build the custom-made components?
- Who should assemble and support the components?

Many different alternatives must be considered, and the challenge is to select the best—that is, the solution with the fewest risks and most benefits. Alternatives for solving the problem must be cost-effective, but they also must be consistent with the corporate strategic plan. Does the alternative contribute to the basic goals and objectives of the organization? Will it integrate seamlessly with other planned systems? Does it use technology that fits the strategic direction that management has defined? Will end users be receptive to it? Analysts must consider many factors and make tough decisions.

Once the systems analyst has determined, in consultation with management, which alternative to recommend and management has approved the recommendation, the details must be worked out. Here the analyst is concerned with creating a blueprint (design

FIGURE *1-1*

The analyst's approach to problem solving.

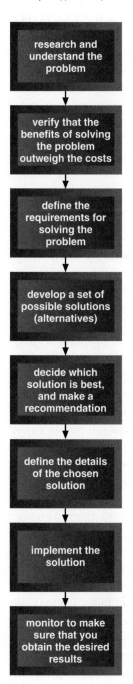

specifications) for how the new system will work. Systems design specifications cover databases, user interfaces, networks, operating procedures, conversion plans, and, of course, program modules. Once the design specifications are complete, the actual construction of the system can begin, including the programming.

An information system can cost a lot of money to build and install—perhaps millions of dollars—so detailed plans must be drawn up. It is not unusual for dozens of programmers to work on programs to get a system up and running, and those programmers need to know exactly what the system is to accomplish—thus, detailed specifications are required. We present in this text the tools and techniques that an analyst uses during system development to create the detailed specifications. Some of these specifications are the result of systems analysis, and some are the result of systems design.

Although this text is oriented toward potential systems analysts, it also provides a good foundation for others who will deal with business problems that could be solved with the help of an information system. Managers throughout business must become more and more knowledgeable about using information technology to solve business problems. Many general business students take a systems analysis and design course to round out their background in two-year and four-year degree programs. Many graduate programs, such as master of business administration (MBA) and master of accountancy (M.Acc) programs, have technology tracks with courses that use this book. Remember that systems analysis and design work is not just about developing systems; it is really about solving business problems using information technology. So even though they never build information systems, managers need to gain expertise in these concepts to be effective in their jobs.

SYSTEMS THAT SOLVE BUSINESS PROBLEMS

We described the systems analyst as a business problem solver. We said that the solution to the problem is usually an information system. Before we talk about how you learn to be a systems analyst, let's quickly review some information systems concepts.

Information Systems

system

a collection of interrelated components that function together to achieve some outcome

information system

a collection of interrelated components that collect, process, store, and provide as output the information needed to complete business tasks

subsystem

a system that is part of a larger system

supersystem

a larger system that contains other systems

A *system* is a collection of interrelated components that function together to achieve some outcome. An *information system* is a collection of interrelated components that collect, process, store, and provide as output the information needed to complete a business task. Completing a business task is usually the "problem" we talked about earlier.

A payroll system, for example, collects information on employees and their work, processes and stores that information, and then produces paychecks and payroll reports (among other things) for the organization. A sales management system collects information about customers, sales, products, and inventory levels, and then processes and stores the information. Then information is provided to manufacturing so the department can schedule production.

What are the interrelated components of an information system? There are several ways to think about components. Any system can have subsystems. A *subsystem* is a system that is part of another system, so subsystems might be one way to think about the components of a system. For example, a customer support system might have an order-entry subsystem that creates new orders for customers. Another subsystem might handle fulfilling the orders, including shipping and back orders. A third subsystem might maintain the product catalog database. The view of a system as a collection of subsystems is very useful to the analyst. These subsystems are interrelated components that function together.

Every system, in turn, is part of a larger system, called a *supersystem*. So, the customer support system is really just a subsystem of the production system. The production sys-

functional decomposition

dividing a system into components based on subsystems that in turn are further divided into subsystems

tem includes other subsystems, such as inventory management and manufacturing. Figure 1-2 shows how one system can be divided, or decomposed, into subsystems, which in turn can be further decomposed into subsystems. This approach to dividing a system into components is referred to as *functional decomposition.*

FIGURE *1-2*

Information systems and subsystems.

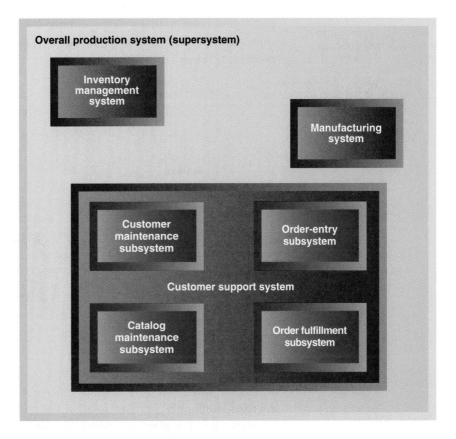

Another way to think about the components of a system is to list the parts that interact. For example, an information system includes hardware, software, inputs, outputs, data, people, and procedures. This view is also very useful to the analyst. These interrelated components function together in a system, as shown in Figure 1-3.

FIGURE *1-3*

Information systems and component parts.

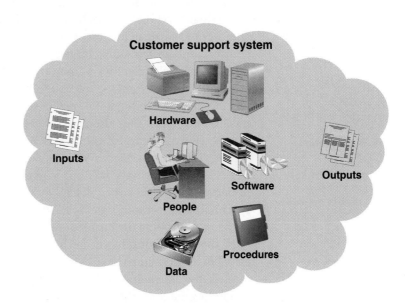

system boundary

the separation between a system and its environment that inputs and outputs must cross

automation boundary

the separation between the automated part of a system and the manual part of a system

FIGURE *1-4*

The system boundary versus the automation boundary.

Every system has a boundary between it and its environment. Any inputs or outputs must cross the *system boundary*. Defining what these inputs and outputs are is an important part of systems analysis and design. In an information system, people are also key components, and these people do some of the system's work. So there is another boundary that is important to a systems analyst—the *automation boundary*. On one side of the automation boundary is the automated part of the system, where work is done by computers. On the other side is the manual part of the system, where work is done by people (see Figure 1-4).

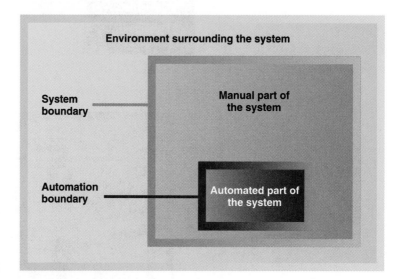

Types of Information Systems

Because organizations perform many different types of activities, there are many different types of information systems—all of which can be innovative and use the latest technologies. The types of systems found in most businesses include transaction processing systems, management information systems, executive information systems, decision support systems, communication support systems, and office support systems (see Figure 1-5). You learned about these types of systems in your introductory information systems course, so we will briefly review only the most common ones here.

Transaction processing systems (TPS) capture and record information about the transactions that affect the organization. A transaction occurs each time a sale is made, supplies are ordered, or an interest payment is made. Usually these transactions create credit or debit entries in accounting ledgers. They eventually end up on accounting statements used for financial accounting purposes, such as the income statement. Transaction processing systems were among the first to be automated by computers. Newer TPSs use state-of-the-art technology and present some of the greatest challenges to information systems developers. They also present some of the greatest competitive advantages and returns on investments for companies. Business-to-consumer (B2C) and business-to-business (B2B) e-commerce systems are the latest challenges in transaction processing. Newer TPSs are often called on-line transaction processing (OLTP) systems.

Management information systems (MIS) are systems that take information captured by transaction processing systems and produce reports that management needs for planning and controlling the business. Management information systems are possible because the information has been captured by the transaction processing systems and placed in organizational databases.

Executive information systems (EIS) provide information for executives to use for monitoring the competitive environment and for strategic planning. Some of the infor-

transaction processing systems (TPS)

information systems that capture and record information about the transactions that affect the organization

management information systems (MIS)

information systems that take information captured by transaction processing systems and produce reports that management needs for planning and control

executive information systems (EIS)

information systems for executives to use for monitoring the competitive environment and for strategic planning

FIGURE *1-5*

Types of information systems.

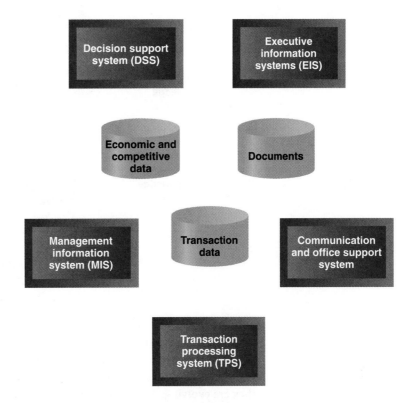

mation comes from the organizational databases, but much of the information comes from external sources—news about competitors, stock market reports, economic forecasts, and so on.

Decision support systems (DSS) allow a user to explore the impact of available options or decisions. Sometimes this process is referred to as "what if" analysis, because the user asks the system to answer questions such as, "What if sales dip below $100 million during the third quarter and interest rates rise to 7.5 percent?" Financial projections made by the DSS can then explore the results. Some decision support systems are used to make routine operational decisions, such as how many rental cars to move from one city to another for a holiday weekend based on estimated business travel patterns.

Communication support systems allow employees to communicate with each other and with customers and suppliers. Communication support now includes wireless personal digital assistants (PDAs), cell phones with messaging and PDA features, anywhere anytime e-mail, broadband Internet access, and desktop video conferencing.

Office support systems help employees create and share documents, including reports, proposals, and memos. Office support systems also help maintain information about work schedules, appointments, and meetings.

decision support systems (DSS)

support systems that allow a user to explore the impact of available options or decisions

communication support systems

support systems that allow employees to communicate with each other and with customers and suppliers

office support systems

support systems that help employees create and share documents, including reports, proposals, and memos

REQUIRED SKILLS OF THE SYSTEMS ANALYST

Systems analysts (or any professionals doing systems analysis and design work) need a great variety of special skills. First, they need to be able to understand how to build information systems, which requires quite a bit of technical knowledge. Then, as discussed previously, they have to understand the business they are working for and how the business uses each of the types of systems. Finally, the analysts need to understand quite a bit about people and the way they work, since people will use the information systems. These three types of knowledge and skills are summarized in Figure 1-6.

FIGURE *1-6*

Required skills of the systems analyst.

Knowledge and Skills Required of a Systems Analyst

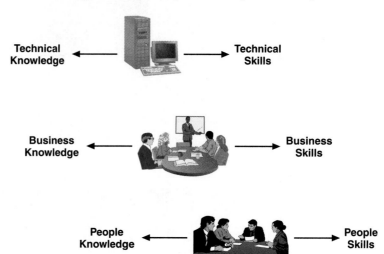

Technical Knowledge and Skills

It should not be surprising that a systems analyst needs technical expertise. Even if an analyst is not involved in programming duties, it is still crucial to have an understanding of different types of technology—what they are used for, how they work, and how they are evolving. No one person can be an expert at all types of technology; there are technical specialists to consult for the details. But a systems analyst should understand the fundamentals about:

- Computers and how they work
- Devices that interact with computers, including input devices, storage devices, and output devices
- Communications networks that connect computers
- Databases and database management systems
- Programming languages
- Operating systems and utilities

A systems analyst also needs to know a lot about tools and techniques for developing systems. *Tools* are software products that are used to develop analysis and design specifications and completed system components. Some tools used in system development include the following:

- Software packages such as Microsoft Access, Oracle Developer, and IBM WebSphere Studio that can be used to develop systems
- Integrated development environments (IDEs) for specific programming languages, such as Sun ONE Studio for Java or Microsoft Visual Studio .NET for VB .NET and C# .NET
- Computer-aided system engineering (CASE) tools, such as Rational XDE Modeler, Visible Analyst, or Embarcadero Describe, that store information about system specifications created by analysts and sometimes generate program code
- Program code generators, testing tools, configuration management tools, software library management tools, documentation support tools, project management tools, and so on

Techniques are used to complete specific system development activities. How do you plan and manage a system development project? How do you complete systems analy-

---

tools

software products used to help develop analysis and design specifications and completed system components

techniques

strategies for completing specific system development activities

---

sis? How do you complete a systems design? How do you complete implementation and testing? How do you install and support a new information system? Much of this text explains how to use specific techniques for planning, analysis, and design. But it also covers some aspects of implementation and support. Some examples of techniques include:

- Project planning techniques
- Systems analysis techniques
- Systems design techniques
- System construction and implementation techniques
- System support techniques

Business Knowledge and Skills

Other knowledge and skills that are crucial for an analyst include those that apply to understanding business organizations in general. After all, the problem to be solved is a business problem. What does the analyst need to know? The following are examples:

- What business functions do organizations perform?
- How are organizations structured?
- How are organizations managed?
- What type of work goes on in organizations (finance, manufacturing, marketing, customer service, and so on)?

Systems analysts benefit from a fairly broad understanding of businesses in general, so they typically study business administration in college. In fact, computer information systems (CIS) or management information systems (MIS) majors are often included in the college of business for that reason. The accounting, marketing, management, and operations courses taken in a CIS or MIS degree program serve a very important purpose of preparing the graduate for the workplace.

Systems analysts also need to understand the type of organization for which they work. Some analysts specialize in a specific industry for their entire career—perhaps in manufacturing, retailing, financial services, or aerospace. The reason for this business focus is simple: It takes a long time to understand the problems of a specific industry. An analyst with deep understanding of a specific industry can solve complex problems for companies in the industry.

Familiarity with a specific company also provides important guidance on system needs and changes. Often, just knowing the people who work for a company and understanding subtleties of the company culture can make a big difference in the effectiveness of an analyst. It takes years of experience working for a company to really understand what is going on. The more an analyst knows about how an organization works, the more effective he or she can be. Some specifics the analyst needs to know about the company include the following:

- What the specific organization does
- What makes it successful
- What its strategies and plans are
- What its traditions and values are

People Knowledge and Skills

Because systems analysts usually work on development teams with other employees, they need to understand a lot about people and possess many interpersonal skills. An analyst spends a great deal of time working with people, trying to understand their perspectives on the problems they are trying to solve. It is critical that the analyst understand how people:

- Think
- Learn
- React to change
- Communicate
- Work (in a variety of jobs and levels)

Analysts must understand how people think to anticipate the way they will want to interact with the computer system—for example, to have the computer appear to anticipate their actions. Understanding how people learn is important when designing training materials and help tools for new systems. When a new system is implemented, it can change the way people do their jobs, and employees must be prepared for the change and helped to see the benefits of the change. An analyst must use a variety of people skills—including interpersonal skills and communication skills—to get the required information and to influence and motivate people to cooperate. Finally, since the information systems are designed to support the work of people, it is necessary to understand the work that people perform in a variety of jobs and at a variety of levels, ranging from clerical and factory workers to managers and executives. Because analysts come into contact with so many people throughout an organization, they have a unique opportunity to influence the organization as a whole.

A Few Words about Integrity and Ethics

One aspect of a career in information systems that students often underestimate is the importance of personal integrity and ethics. A systems analyst is asked to look into problems that involve information from many different parts of an organization. Especially if it involves individuals, the information might be very private, such as salary, health, and job performance. The analyst must have the integrity to keep this information private.

But the problems the analyst works on can also involve confidential corporate information, including proprietary information about products or planned products, strategic plans or tactics, and even top-secret information involving government military contracts. Sometimes a company's security processes or specific security systems can be involved in the analyst's work.

Some system developers work for consulting firms that are called in to work on specific problems for clients. They are expected to uphold the highest ethical standards when it comes to private proprietary information they might encounter on the job. Any appearance of impropriety can ruin an analyst's career.

THE ENVIRONMENT SURROUNDING THE ANALYST

Types of Technology Encountered

Most students are familiar with personal computers. In many university MIS or CIS degree programs, students complete modest course projects that run on PCs. But not all businesses function with desktop systems, and sometimes students don't recognize how different the large-scale systems are that they will work on when they are out in the real world.

Many basic systems now run with PCs on desktops, but behind the scenes, these computers may be connected to data centers through very complex networks. A company's "simple" on-line order-processing application might involve a system with thousands of users spread over hundreds of locations. The database might contain hundreds of tables with millions of records in each table. The system might have taken years to complete, costing millions of dollars. If the system fails for even an hour, the company could lose millions of dollars in sales. Such a system is critical for the business, so the programmers and analysts who support and maintain it work in round-the-clock shifts, often wearing pagers in case of a problem. The importance of these systems to business cannot be overstated.

Different configurations of information systems that future analysts may encounter, which are discussed in later chapters, include the following:

- Desktop systems
- Networked desktop systems that share data
- Client-server systems
- Large-scale centralized mainframe systems
- Systems using Internet, intranet, and extranet technology

Just as an organization's business environment continually changes, so does the technology used for its information systems. Enterprise-level systems increasingly require flexible development environments provided by Web-based technology so that employees and customers can get access to systems and data anywhere and anytime. The rapid change in technology often drives needed changes. Thus, it is important for all people involved in information system development to upgrade their knowledge and skills continually. Those who don't will be left behind.

Typical Job Titles and Places of Employment

We have discussed the work of a systems analyst in terms of problem solving and system development. But it is important to recognize that many different people do systems analysis and design work, and not all of them have the job title of systems analyst. Sometimes analysis and design work is done by end users assigned to the project to provide expertise. Often analysis and design work is combined with other tasks (such as programming or end-user support). A recent graduate employed as a programmer analyst will typically work on system maintenance and support projects doing programming. Even so, maintenance programming requires analysis of the requirements for the change, design of the solution, and implementation of the change. In this regard, even beginning programmers are involved in analysis and design.

Here are some job titles you may encounter:

- Programmer analyst
- Business systems analyst
- System liaison
- End-user analyst
- Business consultant
- Systems consultant
- System support analyst
- Systems designer
- Software engineer
- System architect
- Webmaster
- Web developer

Sometimes systems analysts might also be called project leaders or project managers. Be prepared to hear all kinds of titles for people who are involved in analysis and design work.

People who do analysis and design work are also found at organizations of all sizes—small businesses, medium-sized regional businesses, national Fortune 500 businesses, and multinational corporations. The type of technology used and the nature of the projects differ quite a bit, depending on the size of the organization. In addition, some businesses have very centralized information systems departments or divisions, while others have smaller information systems units that serve specialized parts of the organization. And the size of the organization may not always correspond to the size of the systems the analyst works on. Some analysts work at smaller companies with a large-company feel, and others work at a large company with a small-company feel.

But not all analysts work directly for the company with the problem to solve. Analysis and design work for a company is accomplished in many different work arrangements, including:

- Programmer analysts working for the company (part of the time is spent in analysis and design work)
- Systems analysts (who specialize in analysis and design) working for the company
- Independent contractors (who work as analysts or programmer analysts) working as needed under the direction of company managers
- Outsource provider employees (who work on- or off-site) working on a project under contract for the company
- Consultants (who are employed by a consulting firm) working on a specific project for the company
- Software development firm employees (who work on developing and supporting software packages purchased by the company)
- Application service provider (ASP) employees (who work on developing and supporting ASP system solutions contracted for by the company)

Analysis and design work is also done in other situations. Naturally, companies such as Microsoft, Sun Microsystems, and IBM hire software developers to create or adapt software packages such as Office XP and operating systems such as Windows XP. These developers do analysis and design work, too, but the problem to be solved is generally different. Computer science graduates are more likely to work on new versions of Windows or Office, where their technical skills can be used to best advantage. In this text, though, we assume the analyst is working on information systems that solve business problems, not on operating systems or software packages.

THE ANALYST'S ROLE IN STRATEGIC PLANNING

We have described a systems analyst as someone who solves specific business problems by developing or maintaining information systems. The analyst might also be involved with senior managers on strategic management problems—that is, problems involving the future of the organization and plans and processes to ensure its survival and growth. Sometimes an analyst only a few years out of college can be summoned to meet with top-level executives and even be asked to present recommendations to achieve corporate goals. How might this happen?

Special Projects
First, the analyst might be working to solve a problem that affects executives, such as designing an executive information system. The analyst might interview the executives to find out what information they need to do their work. An analyst may be asked to spend a day with an executive or even travel with an executive to get a feel for the nature of the executive's work. Then the analyst might develop and demonstrate prototypes of the system to get more insight into the needs of the executives.

Another situation that could involve an analyst in strategic management problems is a business process reengineering study. *Business process reengineering* seeks to alter radically the nature of the work done in a business function. The objective is radical improvement in performance, not just incremental improvement. Therefore, the analyst might be asked to participate in a study that carefully examines existing business processes and procedures and then to propose information system solutions that can have a radical impact. Many tools and techniques of analysis and design are used to analyze business processes, redesign them, and then provide computer support to make them work.

Strategic Planning

Most business organizations invest considerable time and energy completing strategic plans that typically cover five or more years. During the *strategic planning* process, executives ask themselves such fundamental questions about the company as where the business is now, where they want the business to be, and what they have to do to get there. A typical strategic planning process can take months or even years, and often plans are continually updated. Many people from throughout the organization are involved, all completing forecasts and analyses, which are combined into the overall strategic plan. Once a strategic plan is set, it drives all of the organization's processes, so all areas of the organization must participate and coordinate their activities. Therefore, a marketing strategic plan and a production strategic plan must fit within the overall strategic plan.

Information Systems Strategic Planning

One major component of the strategic plan is the *information systems strategic plan*. Today, information systems are so tightly integrated into an organization that nearly any planned change calls for new or improved information systems. Beyond that, the information systems themselves often drive the strategic plan. For example, after some chaotic early years involving the Internet, many new Internet-based companies have survived (such as Amazon.com and eBay), and many other companies have altered their business processes and developed new markets in which to compete. In other cases, the opportunities presented by new information systems technology have led to new products and markets with more subtle impacts. Information systems and the possibilities that they present play a large role in the strategic plans of most organizations.

Information systems strategic planning sometimes involves the whole organization. Usually at the recommendation of the chief information systems executive, top management will authorize a major project to plan the information systems for the entire organization. Unlike ongoing planning, a specific information systems strategic planning project might be authorized every five years or so, depending on the changes in the industry or company.

In developing the information systems strategic plan, members of the staff look at the overall organization, to anticipate problems rather than react to systems problems as they come up. Several techniques help the organization complete an information systems strategic planning project. A consulting firm is often hired to help with the project. Consultants can offer experience with strategic planning techniques and can train managers and analysts to complete the plan.

Usually managers and staff from all areas of the organization are involved, but the project team is generally led by information systems managers with the assistance of consultants. Systems analysts often become involved in collecting information and interviewing people.

Many documents and existing systems are reviewed. Then the team tries to create a model of the entire organization—to map the business functions it performs. Another model, one that shows the types of data the entire organization creates and uses, is also developed. The team examines all of the locations where business functions are performed and data are created and used. From these models, the team puts together a list of integrated information systems for the organization, called the *application architecture plan.* Then, given the existing systems and other factors, the team outlines the sequence needed to implement the required systems.

Using the list of information systems needed, the team defines the *technology architecture plan*—that is, the types of hardware, software, and communications networks required to implement all of the planned systems. The team must look at trends in technology and make commitments to specific technologies and possibly even technology vendors. The components of the information systems strategic plan are shown in Figure 1-7.

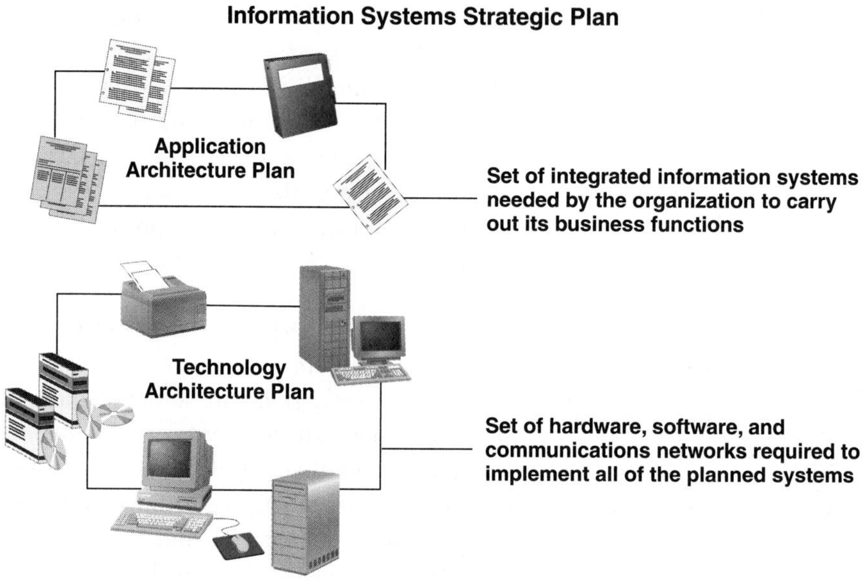

application architecture plan

a description of the integrated information systems that the organization needs to carry out its business functions

technology architecture plan

a description of the hardware, software, and communications networks required to implement planned information systems

FIGURE *1-7*

Components of an information systems strategic plan.

Information Systems Strategic Plan

Application Architecture Plan

Set of integrated information systems needed by the organization to carry out its business functions

Technology Architecture Plan

Set of hardware, software, and communications networks required to implement all of the planned systems

In an ideal world, a comprehensive information systems planning project would solve all of the problems that information systems managers face. Unfortunately, the world continues to change at such a rate that plans must be continually updated. Unplanned information system projects come up all the time, and priorities must be continually evaluated.

Enterprise Resource Planning

An increasing number of organizations are addressing their information systems requirements using an approach called *enterprise resource planning (ERP)*, which commits to using an integrated set of software packages for key information systems. Software vendors such as SAP and PeopleSoft offer comprehensive packages for companies in specific industries. To adopt an ERP solution, the company must carefully study its existing processes and information needs and then determine which ERP vendor provides the best match. ERP systems are so complex that an organization must often commit nearly everyone in the information systems department and throughout the organization to research options. They are also very expensive, in initial cost and in support cost. Much change is involved, for management and for staff. And once the decision is made to adopt an ERP system, it is very difficult to return to the old ways of doing business— and the old systems. An analyst involved in an ERP project can play a key role in the project and will have to draw heavily on technical, business, and people skills.

enterprise resource planning (ERP)

a process in which an organization commits to using an integrated set of software packages for key information systems

ROCKY MOUNTAIN OUTFITTERS AND ITS STRATEGIC INFORMATION SYSTEMS PLAN

To demonstrate the important systems analysis and design techniques in this text, we follow a system development project for a company named Rocky Mountain Outfitters (RMO). RMO is a sports clothing manufacturer and distributor that is about to begin development of a new customer support system. You will encounter RMO customer support system examples in all chapters of this book. For now, try to get a feel for the nature of the business, the approach the company took to define the information systems strategic plan, and the basic objectives of the customer support system that is part of the plan.

Introducing Rocky Mountain Outfitters (RMO)

RMO started in 1978 as the dream of John and Liz Blankens of Park City, Utah. Liz had always been interested in fashion and clothing and had worked her way through college by designing, sewing, and selling winter sports clothes to the local ski shops in Park City. She continued with this side business even after she graduated, and soon it was taking all of her time.

Liz had been dating John Blankens since they met at a fashion merchandising convention. John had worked for several years for a retail department store chain after college and had just completed his MBA. Together they decided to try to expand Liz's business into retailing to reach a larger customer base. Also, around this time they married.

The first step in their expansion involved direct mail-order sales to customers using a small catalog (see Figure 1-8). Liz immediately had to expand the manufacturing operations by adding a designer and production supervisor. As interest in the catalog increased, the Blankenses sought out additional lines of clothing and accessories to sell along with their own product lines. They also opened a retail store in Park City.

FIGURE *1-8*

Early RMO catalog cover (Fall 1978).

By the early 2000s, RMO had grown to become a significant regional sports clothing distributor in the Rocky Mountain and Western states. The states of Arizona, New Mexico, Colorado, Utah, Wyoming, Idaho, Oregon, Washington, and Nevada, and the eastern edge of California had seen tremendous growth in recreation activities. Along with the increased interest in outdoor sports, the market for both winter and summer sports clothes had exploded. Skiing, snowboarding, mountain biking, water skiing, jet skiing, river running, jogging, hiking, ATV biking, camping, mountain climbing, and rappelling had all seen a tremendous increase in interest in these states. Of course, people needed appropriate sports clothes for their activities, so RMO expanded its line of sportswear to respond to this market. It also added a line of casual and active wear to round out its offerings to the expanding market of active people. The current RMO catalog offers an extensive selection (see Figure 1-9).

FIGURE *1-9*

Current RMO catalog cover (Fall 2005).

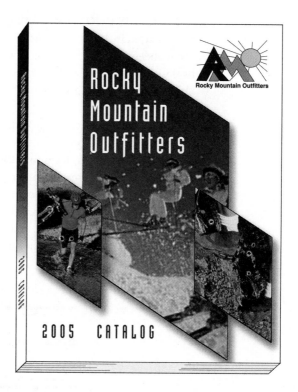

Rocky Mountain Outfitters now employs more than 600 people and produces almost $100 million annually in sales. The mail-order operation is still the major source of revenue, at $60 million. In-store retail sales have remained a modest part of the business, with sales of $2.5 million at the Park City retail store and $5 million at the recently opened Denver store. In the early 1990s, the Blankenses added a phone-order operation that now accounts for $30 million in sales. To the customer, this service was a natural extension, but to RMO, it meant considerable changes in transaction processing systems to handle the phone orders.

RMO Strategic Issues

Rocky Mountain Outfitters was also one of the first sports clothing distributors to provide a Web site featuring its products. The site originally gave RMO a simple Web presence to enhance its image and to allow potential customers to request a copy of the catalog. It also served as a portal for links to all sorts of outdoor sports Web sites. The first RMO Web site enhancement added more specific product information, including weekly specials that could be ordered by phone. Eventually, nearly all product offerings were included in an on-line catalog posted at the Web site. But orders could only be placed by mail or by phone.

John and Liz had considered making a major commitment to business-to-consumer (B2C) e-commerce in the late 1990s. But Liz worried about the risk of sudden and potentially explosive growth. She had seen many small manufacturers and distributors jump into on-line order processing without being able to support the sales properly. Inventory shortages, unreliable service, poorly handled returns, and even occasional double billing had ruined some successful and well-respected brick-and-mortar companies overnight. Liz was determined that RMO would not make the same mistake.

John and Liz could see the potential of e-commerce, but they wanted to be careful to do it right, not just tack it on as an afterthought. The Blankenses had always planned carefully, and they also knew that the role of information technology in their business would continue to grow in strategic importance. Therefore, they decided to think carefully about their entire information technology infrastructure and create a strategic information systems plan. A consulting firm was brought in to help with the strategic information systems planning process.

The consulting firm recommended focusing on two key strategic thrusts:

- Supply chain management
- Customer relationship management

supply chain management (SCM)

a process that seamlessly integrates product development, product acquisition, manufacturing, and inventory management

customer relationship management (CRM)

processes that support marketing, sales, and service operations involving direct and indirect customer interaction

Supply chain management (SCM) concerns processes that seamlessly integrate product development, product acquisition, manufacturing, and inventory management. *Customer relationship management (CRM)* concerns processes that support marketing, sales, and service operations involving direct and indirect customer interaction. Both of these strategies help businesses—especially retailers such as RMO—provide products and services to customers while promoting efficient operations.

The information systems strategic plan includes an application architecture plan, detailing the information systems development projects that the company needed to complete, and a technology architecture plan, detailing the technology infrastructure needed to support the systems. Both components of the plan were based on the supply chain management objectives and the customer relationship management objectives RMO defined for the next five years.

The next section provides some additional background on RMO and summarizes the overall information systems plan it is currently following. In subsequent chapters of this book, we will focus on one of the crucial information systems that is part of the plan—the customer support system.

RMO's Organizational Structure and Locations

Rocky Mountain Outfitters is still managed on a daily basis by John and Liz Blankens. John is president, and Liz is vice president of merchandising and distribution (see Figure 1-10). Other top managers include William McDougal, vice president of marketing and sales, and JoAnn White, vice president of finance and systems. The systems department reports to JoAnn White.

One hundred thirteen employees work in human resources, merchandising, accounting and finance, marketing, and information systems in the corporate offices in Park City, Utah. There are two retail stores: the original Park City store and the newer

FIGURE *1-10*

Rocky Mountain Outfitters' organizational structure.

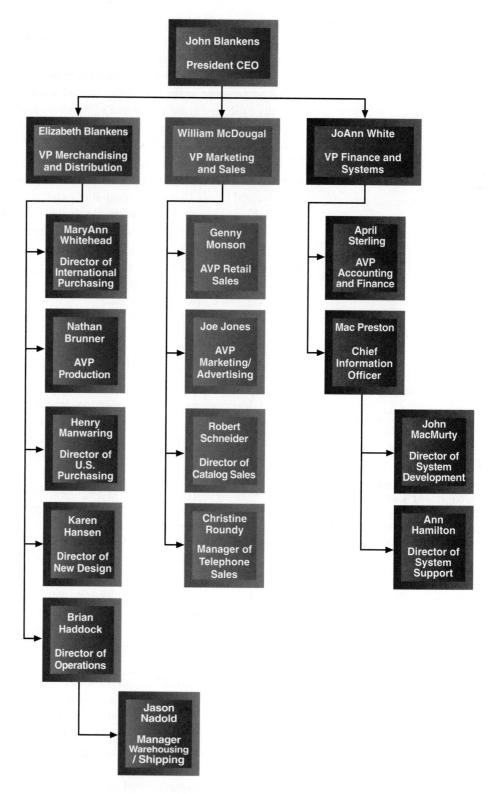

Denver store. Manufacturing facilities are located in Salt Lake City and more recently Portland, Oregon. There are three distribution/warehouse facilities: Salt Lake City, Albuquerque, and Portland. All mail-order processing is done in a facility in Provo, Utah, employing 58 people. The phone-sales center, employing 20, is located in Salt Lake City. Figure 1-11 shows the locations of these facilities.

FIGURE *1-11*

Rocky Mountain Outfitters' locations.

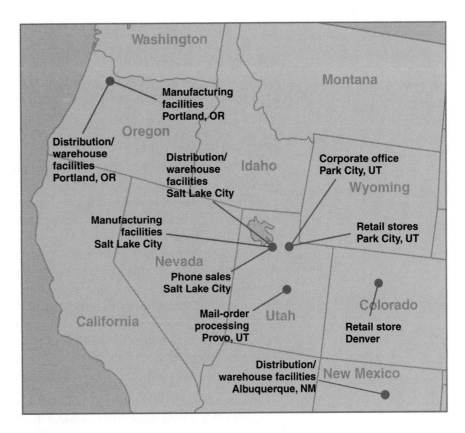

The RMO Information Systems Department

The information systems department is headed by Mac Preston, an assistant vice president with the title chief information officer (CIO), and there are nearly 50 employees in the department (see Figure 1-12). Mac's title of CIO reflects a promotion following the successful completion of the information systems strategic planning project. He is not quite equal to a full vice president, but his position is considered increasingly important to the future of the company. Mac reports to the finance and systems vice president, whose background is in finance and accounting. The information systems department will eventually report directly to the CEO if Mac has success implementing the new strategic information systems plan.

FIGURE *1-12*

RMO information systems department staffing.

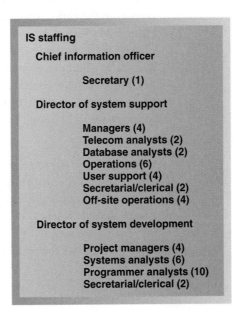

IS staffing

   Chief information officer

         Secretary (1)

   Director of system support

            Managers (4)
            Telecom analysts (2)
            Database analysts (2)
            Operations (6)
            User support (4)
            Secretarial/clerical (2)
            Off-site operations (4)

   Director of system development

            Project managers (4)
            Systems analysts (6)
            Programmer analysts (10)
            Secretarial/clerical (2)

Mac organized information systems into two areas—system support and system development. Ann Hamilton is director of system support. System support involves such functions as telecommunications, database administration, operations, and user support. John MacMurty is director of system development. System development includes 4 project managers, 6 systems analysts, 10 programmer analysts, and a couple of clerical support employees.

Existing RMO Systems

Most of the computer technology and information systems staff at RMO is located at the data center in Park City. A small mainframe computer runs the inventory, mail-order, accounting, and human resource functions by connecting to the distribution and mail-order sites through dedicated telecommunication links. The manufacturing sites have dial-up capability to the mainframe.

Office functions at the home office, distribution sites, and manufacturing sites are supported by local area networks with file servers. Retail stores run a point-of-sale software package with a local server and dial-up batch updates to the inventory system on the mainframe. The phone-sales center has a small local area network with a client-server order-processing application running under Windows. Batch inventory updates are made to the mainframe. The RMO informational Web site is hosted by an Internet service provider (ISP) that maintains the Web content off-site.

The existing information systems and their technology are organized as follows:

- **Merchandising/Distribution.** A mainframe application developed in-house using COBOL/CICS with some DB2 relational database and VSAM files. Implemented 12 years ago.
- **Mail Order.** A mainframe application developed in-house using COBOL. Mail-order clerks in Provo use dedicated terminals. The application is fast and efficient but unsuitable for handling phone orders. Implemented 14 years ago.
- **Phone Order.** A modest Windows application developed using Visual Basic and Oracle as a quick solution to customer demand for phone orders. It is a multiuser file-server design that is not integrated well with merchandising/distribution and has reached capacity. Implemented six years ago.
- **Retail Store Systems.** A retail store package with point-of-sale processing and overnight batch inventory update with the mainframe. Implemented eight years ago.
- **Office Systems.** A local area network with office software, Internet access, and e-mail services in the Park City offices and other sites. Implemented three years ago.
- **Human Resources.** An application developed in-house for payroll and benefits running on the mainframe. Implemented 13 years ago.
- **Accounting/Finance.** A mainframe package from a leading accounting package vendor. Implemented 10 years ago.
- **RMO Informational Web Site.** A static site with catalog information and other links. Implemented and hosted by an ISP three years ago.

The Information Systems Strategic Plan

The information systems strategic plan developed with the help of the consultants includes the technology architecture plan and the application architecture plan. The planning team looked closely at existing systems and at the business objectives of RMO. As initially proposed, supply chain management and customer relationship management provided a vision for the plan. These ideas support the strategic objectives of RMO to build more direct customer relationships and to expand the marketing presence beyond the Western states.

The main features of the plans include the following:

Technology architecture plan

1. Distribute business applications across multiple locations and computer systems, reserving the mainframe for Web server, database, and telecommunications functions to allow incremental and rapid growth in capacity.
2. Move toward conducting strategic business processes via the Internet, first supporting supply chain management, next supporting direct customer ordering on a new, dynamic Web site, and finally supporting additional customer relationship management (CRM) functions that link internal systems and databases.
3. Anticipate the eventual move toward Web-based intranet solutions for business functions such as human resources, accounting, finance, and information management.

Application architecture plan

1. Supply chain management (SCM): implement systems that seamlessly integrate product development, product acquisition, manufacturing, and inventory management in anticipation of rapid sales growth. Custom development with support of consultants.
2. Customer support system (CSS): implement an order-processing and fulfillment system that seamlessly integrates with the supply chain management systems to support the three order-processing requirements: mail order, phone order, and direct customer access via the Web. Custom in-house development.
3. Strategic information management system (SIMS): implement an information system that can extract and analyze supply chain and customer support information for strategic and operational decision making and control. Package solution.
4. Retail store system (RSS): replace the existing retail store system with a system that can integrate with the customer support system. Package solution.
5. Accounting/finance: purchase a package solution, definitely an intranet application, to maximize employee access to financial data for planning and control.
6. Human resources: purchase a package solution, definitely an intranet application, to maximize employee access to human resource (HR) forms, procedures, and benefits information.

   The timetable for implementing the application architecture plan is shown in Figure 1-13. Key components of the supply chain management system, particularly inventory management components, must be defined before the customer support system project can be started. The customer support system project must be started as soon as possible, though, as it is the core system supporting customer relationship management.

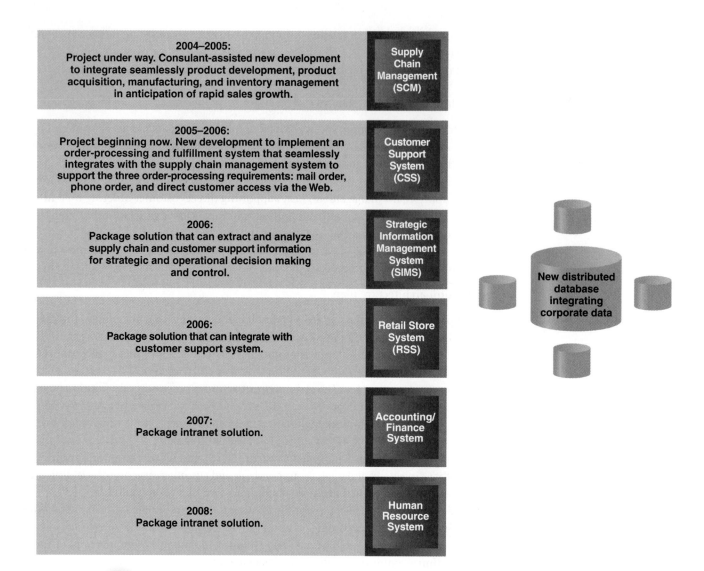

2004–2005: Project under way. Consulant-assisted new development to integrate seamlessly product development, product acquisition, manufacturing, and inventory management in anticipation of rapid sales growth.	**Supply Chain Management (SCM)**
2005–2006: Project beginning now. New development to implement an order-processing and fulfillment system that seamlessly integrates with the supply chain management system to support the three order-processing requirements: mail order, phone order, and direct customer access via the Web.	**Customer Support System (CSS)**
2006: Package solution that can extract and analyze supply chain and customer support information for strategic and operational decision making and control.	**Strategic Information Management System (SIMS)**
2006: Package solution that can integrate with customer support system.	**Retail Store System (RSS)**
2007: Package intranet solution.	**Accounting/ Finance System**
2008: Package intranet solution.	**Human Resource System**

New distributed database integrating corporate data

FIGURE *1-13*

The timetable for RMO's application architecture plan.

The supply chain management (SCM) system and the customer support system (CSS) require custom software development. Custom development is advisable when the system must match very specific company requirements. Consultants have been called in to help define requirements and develop the integration plan for supply chain management. Several leading consulting firms specialize in supply chain management.

The customer support system will probably be developed in-house by the information systems staff, although final decisions about the implementation approach will be deferred until a more complete analysis is completed.

The other systems in the plan will probably be software package solutions selected from among the best-rated software currently available. Package solutions are desirable for standard business systems such as accounting and human resources. The key requirement is that any package must integrate seamlessly with other RMO systems and use intranet technology.

The Customer Support System

The RMO system development project described in this text is the customer support system (CSS). Rocky Mountain Outfitters has always prided itself on its customer orientation. One of the core competencies of RMO has been its ability to develop and maintain customer loyalty. John Blankens has kept himself current on important business concepts, and he is fully aware of the need to develop effective business processes with a customer focus. So even though the customer support system will include all of the required sales and marketing components, he wants everyone to understand that its primary objective is to support RMO's customers. John's long-held view regarding customers has become popular recently with consultants and software vendors, who call the approach customer relationship management (CRM). John smiled approvingly as the consultants explained CRM to his staff.

The application architecture plan detailed some specific objectives for the customer support system. The system should include all functions associated with providing products for the customer, from order entry to arrival of the shipment, such as:

- Customer inquiries/catalog requests
- Order entry
- Order tracking
- Shipping
- Back ordering
- Returns
- Sales analysis

Customers should be able to order by telephone, through the mail, or over the Internet. All catalog items would also be available through a sophisticated RMO Web catalog, and the Web catalog must be consistent with printed catalogs so that customers can browse the printed catalog and then order at the RMO Web site, if they choose. Additionally, customers might find an item in the printed catalog and search for additional information at the Web site.

Order-entry processing needs to support the graphical, self-help style of the Web interface, as well as a streamlined, rapid-response Windows interface required for phone-sales representatives and mail-order clerks. Returns, back orders, and order status also need to be accessible both on the Web and to RMO employees at their desks.

Although some objectives are defined for the system, a complete systems analysis will define the requirements for the system in detail. These objectives only form some guidelines to keep in mind as the project gets under way.

THE ANALYST AS A SYSTEM DEVELOPER (THE HEART OF THE COURSE)

We have discussed many roles that a systems analyst can play in an organization, including strategic planning and helping identify the major information systems projects that the business will pursue. However, the main job of an analyst is working on a specific information systems development project. This text is about planning and executing an information systems project—in other words, working as a system developer. The text is organized around this theme. In this section, we provide an overview of the text—a preview of what system development involves—as exemplified by the development process ahead for Barbara Halifax, who is in charge of the RMO customer support system project that is about to start (see memo).

February 4, 2005

To: John Blankens, President

From: Barbara Halifax, System Development Project Manager

RE: Customer Support System (CSS) Project

John MacMurty suggested I write you a brief note to let you know how excited and proud I am to be in charge of the customer support system (CSS) project. I enjoyed working with you and other RMO staff of the strategic information system plan last year, and I know I have you to thank for putting in a good word for me with Mac Preston (and with John) for this important assignment.

With the SCM system now well under way, we are all ready to begin the CSS project. I have reviewed all of our planning documents and have met regularly with Jack Gracia about the SCM system. Everything I see in the IS trade press concerning integrated supplier and customer systems continues to confirm our assumptions about the need for these projects. My next step is to begin detailed project planning now that it is time to begin the CSS project formally.

Naturally, I'll be writing my regular status reports to John MacMurty, but please let us know if you have any questions or ideas you might want to share with us. I know how important the CSS project is to RMO, and I share your commitment to providing unparalleled customer support with this new system.

Thanks for the opportunity!

BH

cc: Mac Preston, John MacMurty

Part 1: The Modern Systems Analyst

The first part of the text describes the work of the systems analyst. This chapter (Chapter 1) describes the nature of the analyst's work in terms of types of problems solved, the required skills, and the job titles and places where an analyst might work. We hope it is clear so far that the analyst does much more than think about and write programs. The rest of the text is organized around the problem-solving approach we described at the beginning of the chapter.

Not only does the analyst get involved with business problems, but he or she can work on very high-level strategic issues and with people at all levels of the organization relatively early in his or her career. In this chapter we also described Rocky Mountain Outfitters and its strategic information systems plan. The rest of the text will focus on one of the planned new systems—the customer support system—and its development.

Chapter 2 focuses on the variety of approaches available for developing an information system. The system development life cycle (SDLC) is introduced as a technique for managing and controlling a project. A variety of tools, techniques, and methodologies are discussed, including the traditional structured approach and the newer object-oriented approach. System developers should be familiar with the fundamental concepts of both approaches. This text covers both approaches throughout, pointing out where they are similar and where they are different.

Chapter 3 gets to the heart of the system development project by describing how a project is planned and managed. The SDLC provides the structure used for project management. Other project management tools and techniques are also introduced, including feasibility studies, project scheduling, and project staffing. An information systems project is just like any other project in these respects. It is important for an analyst to understand the role of project management. Specific issues also arise when planning an information systems project, and an analyst needs to be familiar with them and the way they relate to the larger context of the activities of project planning for an information systems project.

Part 2: Systems Analysis Tasks

Chapters 4 through 8 cover systems analysis in detail. Chapter 4 discusses techniques for gathering information about the problem that the new system is to solve so that the system requirements can be defined. The various people who are affected by the system (the stakeholders) are also discussed. All of these people need to be interviewed and kept up to date on the status of the project. Techniques such as prototyping and walkthroughs are introduced to help the analyst communicate with everyone involved.

Chapter 5 introduces the concept of models and modeling to record the detailed requirements for the system in a useful form. When discussing an information system, two key concepts are particularly useful: "events" that cause the system to respond, and "things" the system needs to store information about. These two concepts, events and things, are important no matter which approach to system development you are using— either the traditional structured approach or the object-oriented approach. Business events are used to identify system activities in the traditional approach, and use cases in the object-oriented approach. The entity-relationship diagram (ERD) is introduced as a model for showing the things affected in the traditional approach, and the class diagram is introduced as a model of things in the object-oriented approach.

Chapters 6 and 7 continue the discussion of modeling system requirements, at which point the traditional structured approach begins to look different from the object-oriented approach. Chapter 6 covers the traditional approach to requirements, which focuses on processes. Data flow diagrams (DFDs), structured English, and data flow definitions are emphasized. Chapter 7 covers the object-oriented approach to requirements, which focuses on objects and their interactions. Use cases, use case diagrams, and system sequence diagrams are emphasized. System developers should be familiar with both approaches to defining systems requirements, but it is important to recognize that in a given system development project, one approach or the other will be used. They are never used simultaneously.

Chapter 8 demonstrates techniques for generating alternatives for actually implementing the system. Each alternative is described and evaluated carefully for feasibility. Then the best alternative is recommended to management. The final approval of the recommended alternative is a key decision point for the project.

Part 3: Systems Design Tasks

Once one of the alternatives is approved, work on the actual design details begins. Chapters 9 through 15 cover system design issues. Chapter 9 provides an overview of systems design, including the activities completed during the design phase and the general technical environments that are used to implement the system. The three-layer design approach used with both the traditional and object-oriented approaches is introduced. Chapter 10 discusses the traditional approach to system design, showing the types of models used (system flowcharts, structure charts, and pseudocode). Chapters 11 and 12 discuss the object-oriented approach to design, showing the types of models used (sequence diagrams, collaboration diagrams, design class diagrams, and package diagrams). Important design patterns and approaches to evaluating the quality of object-oriented designs are also illustrated.

Chapter 13 describes the issues involved in designing the database for the system, using either a relational database, an object-oriented database, or a hybrid approach that combines relational databases with object technology.

Chapter 14 discusses the user interface to the system, providing an overview of the field of human-computer interaction (HCI) and guidelines for developing user-friendly systems. The chapter covers Windows graphical user interfaces and browser-based interfaces used in Web development. These design concepts apply to both the traditional approach and the object-oriented approach.

Chapter 15 covers the design of system interfaces, system controls, and security. System interfaces include output design of various types of reports that are typically produced on-line and on paper. Information systems controls are discussed, including the importance of ensuring that inputs are accurate and complete and that processing is done correctly. Techniques for protecting the system from unauthorized access are also discussed. These concepts also apply to both the traditional approach and the object-oriented approach.

Part 4: Implementation and Support

Chapter 16 describes the fourth and fifth phases of the SDLC: system implementation and system support. No matter how the system is obtained, a major part of the project is making the system operational and keeping it that way. The analyst's role in implementing the system includes quality control, testing, training users, and making the system operational (conversion). Maintenance and support of the system can continue for many years, involving fixing problems and enhancing the system over time.

Often, a new programmer analyst is involved in maintenance and support of an existing system. Maintenance and support of the system are also the most expensive parts of the project, and decisions made during analysis and design can have a big impact on the ease of maintenance and the overall cost of the system over its lifetime.

This text emphasizes systems analysis and design using a view of the system development process that makes extensive use of iteration and modeling. But you should also be familiar with current trends that focus more explicitly on iteration, risk, and other techniques. The spiral model, extreme programming (XP), the unified process (UP), prototyping, and component-based development are all discussed in Chapter 17.

Implementing a software package instead of developing a system is almost always a viable alternative, and Chapter 18 describes software packages and the comprehensive approach to package solutions referred to as enterprise resource planning (ERP). The appendices cover additional material on project management, project planning, and interviewing.

Summary

A systems analyst is someone who solves business problems using information systems technology. Problem solving means looking into the problem in great detail, understanding everything about the problem, generating several alternatives for solving the problem, and then picking the best solution. Information systems are usually part of the solution, and information systems development is much more than writing programs.

A system is a collection of interrelated components that function together to achieve some outcome. Information systems, like other systems, contain components, and an information systems outcome is the solution to a business problem. Information systems components can be thought of as subsystems that interact or as hardware, software, inputs, outputs, data, people, and procedures. Many different types of systems solve organizational problems, including transaction processing systems, management information systems, executive information systems, decision support systems, communication support systems, and office support systems.

A systems analyst needs broad knowledge and a variety of skills, including technical, business, and people knowledge and skills. Integrity and ethical behavior are crucial to the success of the analyst. Analysts encounter a variety of technologies that often change rapidly. Systems analysis and design work is done by people with a variety of job titles, not only systems analyst but also programmer analyst, systems consultant, systems engineer, and Web developer, among others. Analysts also work for consulting firms, as independent contractors, and for companies that produce software packages.

A systems analyst can become involved in strategic planning by working with executives on special projects, by helping with business process reengineering projects, and by working on company strategic plans. Analysts also assist businesses in their efforts to select and implement enterprise resource planning systems. Sometimes an information systems strategic planning project is conducted for the entire organization, and analysts often are involved. The Rocky Mountain Outfitters information systems strategic planning project described in this chapter is an example.

Usually the systems analyst works on a system development project, one that solves a business problem identified by strategic planning. That is the emphasis in the rest of this text: how the analyst works on a system development project, completing project planning, systems analysis, systems design, systems implementation, and system support activities. The Rocky Mountain Outfitters customer support system project is used to illustrate the system development process.

Key Terms

application architecture plan, p. 16

automation boundary, p. 8

business process reengineering, p. 15

communication support systems, p. 9

customer relationship management (CRM), p. 19

decision support systems (DSS), p. 9

enterprise resource planning (ERP), p. 16

executive information systems (EIS), p. 8

functional decomposition, p. 7

information system, p. 6

information systems strategic plan, p. 15

management information systems (MIS), p. 8

office support systems, p. 9

strategic planning, p. 15

subsystem, p. 6

supersystem, p. 6

supply chain management (SCM), p. 19

system, p. 6

system boundary, p. 8

systems analysis, p. 3

systems analyst, p. 4

systems design, p. 4

techniques, p. 10

technology architecture plan, p. 16

tools, p. 10

transaction processing systems (TPS), p. 8

REVIEW QUESTIONS

1. Give an example of a business problem.
2. What are the main steps followed when solving a problem?
3. Define *system*.
4. Define *information system*.
5. What are the types of information systems found in most organizations?
6. List the six fundamental technologies an analyst needs to understand.
7. List four types of tools the analyst needs to use to develop systems.
8. List five types of techniques used during system development.
9. What are some of the things an analyst needs to understand about businesses and organizations in general?
10. What are some of the things an analyst needs to understand about people?
11. What are some of the types of technology an analyst might encounter?
12. List 10 job titles that involve analysis and design work.
13. How might an analyst become involved with executives and strategic planning relatively early in his or her career?

THINKING CRITICALLY

1. Describe a business problem your university has that you would like to see solved. How can information technology help solve it?
2. Describe how you would go about solving a problem you face. Is the approach taken by a systems analyst, as described in the text, any different?
3. Many different types of information systems were described in this chapter. Give an example of each type of system that might be used by a university.
4. What is the difference between technical skills and business skills? Explain how a computer science graduate might be strong in one area and weak in another. Discuss how the preparation for a CIS or MIS graduate is different from that for a computer science graduate.
5. Explain why an analyst needs to understand how people think, how they learn, how they react to change, how they communicate, and how they work.
6. Who needs greater integrity to be successful, a salesperson or a systems analyst? Or does every working professional need integrity and ethical behavior to be successful? Discuss.
7. Explain why developing an information system requires different skills if the system is a client-server rather than a large-scale centralized mainframe architecture.
8. How might working for a consulting firm for a variety of companies make it difficult for the consultant to understand the business problem a particular company faces? What might be easier for the consultant to understand about a business problem?
9. Explain why a strategic information systems planning project must involve people outside the information systems department. Why would a consulting firm be called in to help organize the project?
10. Explain why a commitment to enterprise resource planning (ERP) would be very difficult to undo once it has been made.

EXPERIENTIAL EXERCISES

1. It is important to understand the nature of the business you work for as an analyst. Contact some information systems developers and ask them about their employers. Do they seem to know a lot about the nature of the business? What types of classes did they take to prepare themselves (for example, banking classes, insurance classes, retail management classes, hospital administration classes, manufacturing technology classes)? Do they plan to take additional courses? If so, which?
2. Think about the type of position you want (working for a specific company, working for a consulting firm, or working for a software package vendor). Do some research on each job by looking at companies' recruiting brochures or Web sites. What do they indicate are the key skills they look for in a new hire? Are there any noticeable differences between consulting firms and the other organizations?
3. You have read an overview of the Rocky Mountain Outfitters' strategic information systems plan, including the technology architecture plan and the application architecture plan. Research system planning at your university. Is there a plan for how information technology will be used over the next few years? If so, describe some of the key provisions of the technology architecture plan and the application architecture plan.

CASE STUDIES

Association for Information Technology Professionals Meeting

"I'll tell you exactly what I look for when I interview a new college grad," Alice Adams volunteered. Alice, a system development manager at a local bank, was talking with several professional acquaintances at a monthly dinner meeting of the Association for Information Technology Professionals (AITP). AITP provides opportunities for information systems professionals to get together occasionally and share experiences. Usually a few dozen professionals from information systems departments at a variety of companies attend the monthly meetings.

"When I interview students, I look for problem-solving skills," continued Alice. "Every student I interview claims to know all about Java and .NET and Dreamweaver and XYZ, or whatever the latest development package is. But I always ask interviewees one thing: 'How do you generally approach solving problems?' And then I want to know if they have even thought much about banks like mine and financial services generally, so I ask, 'What would you say are the greatest problems facing the banking industry these days?'"

Jim Parsons, a database administrator for the local hospital, laughed. "Yes, I know what you mean. It really impresses me if they seem to appreciate how a hospital functions, what the problems are for us—how information technology can help solve some of our problems. It is the ability to see the big picture that really gets my attention."

"Yeah, I'm with you," added Sam Young, the manager of marketing systems for a retail store chain. "I am not that impressed with the specific technical skills an applicant has. I assume they have the aptitude and some skills. I do want to know how well they can communicate. I do want to know how much they know about the nature of our business. I do want to know how interested they are in retail stores and the problems we face."

"Exactly," confirmed Alice.

1. Do you agree with Alice and the others about the importance of problem-solving skills? Industry-specific insight? Communication skills? Discuss.
2. Should you research how a hospital is managed before interviewing for a position with an information systems manager at a hospital? Discuss.
3. In terms of your career, do you think it really makes a difference whether you work for a bank, a hospital, or a retail chain? Or is an information systems job going to be the same no matter where you work? Discuss.

Rethinking Rocky Mountain Outfitters

RMO's strategic information systems plan calls for building a new supply chain management (SCM) system prior to building the customer support system (CSS). John Blankens has stated often that customer orientation is the key to success. If that is so, why not build the CSS first, so customers can immediately benefit from improved customer ordering and fulfillment? Wouldn't that increase sales and profits faster? RMO already has factories that produce many items RMO sells, and RMO has longstanding relationships with suppliers around the globe. The product catalog is well established, and the business has existing customers who appear eager and willing to shop on-line. Why wait? Perhaps John Blankens has made a mistake in planning.

1. What are some of the reasons that RMO decided to build the supply chain management system prior to the customer support system?
2. What are some of the consequences to RMO if it is wrong to wait to build the customer support system?
3. What are some of the consequences to RMO if the owners change their minds and start with the customer support system before building the supply chain management system?

4. What are some other changes that you might make to the RMO strategic information systems plan (both the application architecture plan and the technology architecture plan)? Discuss.

Focusing on Reliable Pharmaceutical Service

The Reliable Pharmaceutical Service is a privately held company incorporated in 1975 in Albuquerque, New Mexico. It provides pharmacy services to health-care delivery organizations that are too small to have their own in-house pharmacy. Reliable grew rapidly in its first decade, and by the late 1980s its clients included two dozen nursing homes, three residential rehabilitation facilities, two small psychiatric hospitals, and four small specialty medical hospitals. In 1990, Reliable expanded its Albuquerque service area to include Santa Fe and started two new service areas in Las Cruces and Gallup.

Reliable accepts pharmacy orders for patients in client facilities and delivers the orders in locked cases every 12 hours. In the Albuquerque and Santa Fe service area, Reliable employs approximately 12 delivery personnel, 20 pharmacist's assistants (PAs), 6 licensed pharmacists, and 10 office and clerical staff. Another 15 employees work in the Las Cruces and Gallup service areas. The management team includes another six people, mainly company owners.

Personnel at each health-care facility submit patient prescription orders by telephone. Many prescriptions are standing orders, which are filled during every delivery cycle until specifically canceled. Orders are logged into a computer as they are received. At the start of each 12-hour shift, the computer generates case manifests for each floor or wing of each client facility. A case manifest identifies each patient and the drugs he or she has been prescribed, including when and how often the drugs should be administered. The shift supervisor assigns the case manifests to pharmacists who, in turn, assign tasks to pharmacy assistants (PAs). Pharmacists supervise and coordinate the PAs' work.

All drugs for a single patient are collected in one plastic drawer of a locking case. Each case is marked with the institution's name, floor number, and wing number (if applicable). Each drawer is marked with the patient's name and room number. Dividers are inserted within a drawer to separate multiple prescriptions for the same patient. When all of the individual components of an order have been assembled, a pharmacist makes a final check of the contents, signs each page of the manifest, and places two copies of the manifest in the bottom of the case, one copy in a file cabinet in the assembly area, and the final copy in a mail basket for billing. When all of the cases have been assembled, they are loaded onto a truck and delivered to the health-care facilities.

Order-entry, billing, and inventory-management procedures are a hodgepodge of manual and computer-assisted methods. Reliable uses a combination of Excel spreadsheets, an Access database, and antiquated custom-developed billing software running on personal computers. Pharmacy assistants use the custom-developed billing software to enter orders received by telephone and to produce case manifests. The system has become increasingly unwieldy as facility contracts and Medicare and Medicaid reimbursement procedures have become more complex. Some costs are billed to the health-care facilities, some to insurance companies, some to Medicare and Medicaid, and some directly to patients. The company that developed and maintained the billing software has gone out of business, and the office staff has had to work around software shortcomings and limitations with cumbersome procedures. Inventory management is done manually.

In 1999 Reliable's revenues leveled off at $40 million and profits plateaued at $5.5 million. By 2003, revenue was declining approximately 4 percent per year, and profit was declining at over 8 percent per year. Several reasons for the decline included the following:

- Price controls in both Medicare and Medicaid reimbursements and contracts with facilities managed by health maintenance organizations (HMOs) and large national health-care companies

- Increasing competition from national retail pharmacy chains such as Walgreens and in-house pharmacies at large local hospitals
- Inefficient operating procedures, which haven't received a comprehensive review or overhaul in almost two decades

Reliable's management team spent most of the last year developing a strategic plan, the key element of which is a major effort to streamline operations to improve service and reduce costs. Management sees this effort as their only hope of surviving in a future dominated by large health-care companies that can dictate price and outsource pharmaceutical services to whomever they choose. They plan a significant expansion into neighboring states after the system is up and running to recoup its costs and increase economies of scale.

Reliable is much smaller than Rocky Mountain Outfitters, the company discussed in this chapter. But the organization still requires an integrated set of information systems to support its operations and management. We will include a case study at the end of each chapter that applies chapter concepts to Reliable Pharmaceutical Service.

1. How many information systems staff members do you think Reliable can reasonably afford to employ? What mix of skills would they require? How flexible would they have to be in terms of the work they do each day?
2. What impact would Web technology have on the way Reliable deploys its systems? Would the Web change the way Reliable does business?
3. Create an application architecture plan and a technology architecture plan for Reliable Pharmaceutical Service to follow for the next five years. What system projects come first in your plan? What system projects come later?

FURTHER RESOURCES

For a comprehensive review of information systems concepts, see:

Effy Oz, *Management Information Systems* (4th ed.). Course Technology, 2004.

Kathy Schwalbe, *Information Technology Project Management* (3rd ed.). Course Technology, 2004.

Ralph M. Stair and George W. Reynolds, *Principles of Information Systems* (6th ed.). Course Technology, 2003.

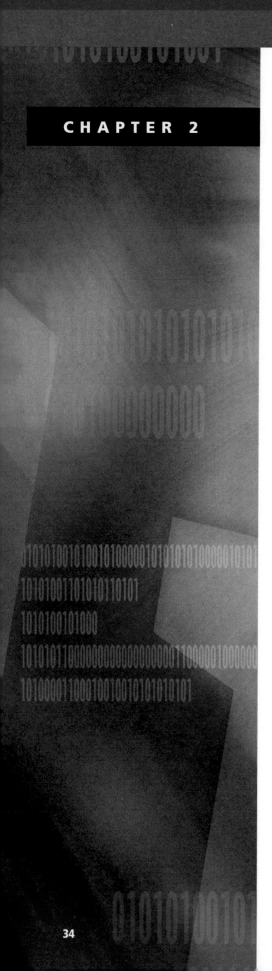

CHAPTER 2

Approaches to System Development

LEARNING OBJECTIVES

After reading this chapter, you should be able to:

- Explain the purpose and various phases of the systems development life cycle (SDLC)

- Explain the differences between a model, a tool, a technique, and a methodology

- Describe the two overall approaches used to develop information systems: the traditional approach and the object-oriented approach

- Describe some of the variations of the systems development life cycle (SDLC)

- Describe the key features of current trends in system development: the spiral model, eXtreme Programming (XP), the Unified Process (UP), and Agile Modeling

- Explain how automated tools are used in system development

CHAPTER OUTLINE

The Systems Development Life Cycle

Methodologies, Models, Tools, and Techniques

Two Approaches to System Development

Systems Development Life Cycle Variations

Current Trends in Development

Tools to Support System Development

Kim, Mary, and Bob, graduating seniors, were discussing their recent interview visits to different companies that recruited computer information system (CIS) majors on their campus. All agreed that they had learned a lot by visiting the companies, but they also all felt somewhat overwhelmed at first.

"At first I wasn't sure I knew what they were talking about," Kim cautiously volunteered. During her on-campus interview, Kim had impressed Ajax Corporation with her knowledge of data modeling. When she visited the Ajax home office data center for the second interview, the interviewers spent quite a lot of time describing the company's system development methodology.

"A few people said to forget everything I learned in school," continued Kim. Ajax Corporation had purchased a complete development methodology called *IM One* from a small consulting firm. Most employees agreed it works fairly well. The people who had worked for Ajax for quite a while thought *IM One* was unique, and they were very proud of it. They had invested a lot of time and money learning and adapting to it.

"Well, that got my attention when they said forget what I learned in school," noted Kim, "but then they started telling me about their SDLC, about iterations, about business events, about data flow diagrams, and about entity-relationship diagrams, and things like that." Kim had recognized that many of the key concepts in the *IM One* methodology were fairly standard models and techniques from the structured approach to system development.

"I know what you mean," said Mary, a very talented programmer who knew just about every new programming language available. "Consolidated Concepts went on and on about things like OMG and UML and UP and some people named Booch, Rumbaugh, and Jacobson. But then it turned out that they were using the object-oriented approach to develop systems, and they liked the fact that I knew Java and VB .NET. No problem once I got past all of the terminology they used. They said they'd send me out for training on Rational Rose, a CASE tool for the object-oriented approach."

Bob had a different story. "A few people said analysis and design were no longer a big deal. I'm thinking, 'Knowing that would have saved me some time in school.'" Bob had visited Pinnacle Manufacturing, which had a small system development group supporting manufacturing and inventory control. "They said they try to just jump in and get to the code as soon as possible. No documentation. No project plan. Then they showed me some books on their desks, and it looked like they had been doing a lot of reading about analysis and design. I could see they were using eXtreme Programming and Agile Modeling techniques and focusing only on best practices needed for their small projects. It turns out they just organize their work differently by looking at risk and writing user stories while building prototypes. I recognized some sketches of class diagrams and sequence diagrams on the boss's whiteboard, so I felt fairly comfortable."

Kim, Mary, and Bob all agreed that there was much to learn in these work environments but also that there are many different terms and points of view used to describe the same key concepts and techniques they learned in school. They were all glad they focused on the fundamentals in their CIS classes and that they had been exposed to a variety of approaches to system development.

OVERVIEW

As the experiences of Kim, Mary, and Bob demonstrate, there are many ways to develop an information system, and doing so is very complex. Project managers rely on a variety of aids to help them with every step of the process. The systems development life cycle (SDLC) introduced in this chapter provides an overall framework for managing the process of system development. But the developer relies on many more concepts for help, including methodologies, models, tools, and techniques. It is very important for you to understand what these concepts are before exploring system development in any detail.

This chapter reviews two main approaches to system development that are currently used to develop business systems: the traditional approach and the object-oriented approach. The traditional approach refers to both structured system development (structured analysis, structured design, and structured programming) and information engineering (IE). The object-oriented approach refers to system development using newer object technologies that require a different approach to analysis, design, and programming.

Traditional and object-oriented approaches use the SDLC as a project management framework, and this chapter describes some important variations of the SDLC.

Additionally, an analyst needs to be familiar with some current trends in system development that may continue to influence analysis and design. Finally, system developers need computer support tools to complete work tasks, including drawing tools and specially designed computer-aided system engineering (CASE) tools. This chapter presents some examples of these software tools. Most of the models, tools, and techniques discussed in this chapter are used during the analysis and design phases of the SDLC.

At Rocky Mountain Outfitters, one of Barbara Halifax's initial jobs as the project manager for the customer support system project is to make decisions about the approach used to develop the system. All of the options described in this chapter are open to her. We will not describe her final decisions, though, because we use the customer support system example throughout this text as we present more details about all approaches.

THE SYSTEMS DEVELOPMENT LIFE CYCLE

project

a planned undertaking that has a beginning and an end, and that produces a desired result or product

Chapter 1 explained that systems analysts solve business problems. For problem-solving work to be productive, it needs to be organized and goal oriented. Analysts achieve these results by organizing the work into projects. A *project* is a planned undertaking that has a beginning and an end and that produces a desired result or product. The term *system development project* describes a planned undertaking that produces a new information system. Some system development projects are very large, requiring thousands of hours of work by many people and spanning several calendar years. In the RMO case study introduced in Chapter 1, the system being developed will be a moderately sized computer-based information system, requiring a moderately sized project lasting less than a year. Many system development projects are smaller, lasting a month or two. For a system development project to be successful, the people developing the system must have a detailed plan to follow. Success depends heavily on having a plan that includes an organized, methodical sequence of tasks and activities that culminate with an information system that is reliable, robust, and efficient.

The development of any new information system normally requires three major sets of activities: analysis activities, design activities, and implementation activities. Analysis activities are those that provide a thorough understanding of the business's information needs and requirements. The focus of analysis is on the business needs, not on any particular computer solution. Design activities are those that define the architecture and structure of a new system to satisfy the business's requirements. During design, analysts begin to specify the details of a computer-system solution. Implementation is the actual construction, testing, and installation of a functioning information system. We call each set of activities a *phase*. Thus, a development project has an analysis phase, a design phase, and an implementation phase.

phase

related system development activities, which are grouped into categories of project planning, analysis, design, implementation, and support

These three phases address the core activities required to develop the information system. But there are two additional sets of activities required for the system development project. At the beginning of any project, there is a project planning phase, consisting of activities that are required to initiate, plan, and obtain approval for the project. This phase is normally short, but it is critical to the overall success of the project. Once the new system is completed and installed, analysts and programmers must perform ongoing activities to maintain and enhance the system over its lifetime. This last phase, called the support phase, obviously lasts much longer than all of the other phases combined. It also is the most expensive. These five phases are shown in Figure 2-1.

systems development life cycle (SDLC)

a project management framework organized into phases and activities

This approach to defining the phases and activities required for a system development project is called the *systems development life cycle (SDLC)*. The activities of every project can be organized into these five phases. The five phases are quite similar to the

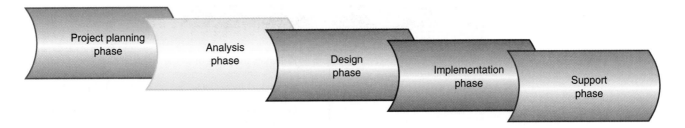

FIGURE *2-1*

Information system development phases.

steps in the general problem-solving approach outlined in Chapter 1. First, the organization recognizes it has a problem to solve (Project Planning). Next, the project team investigates and thoroughly understands the problem and the requirements for a solution (Analysis). Once the problem is understood, a solution is specified in detail (Design). The system that solves the problem is then built and installed (Implementation). As long as the system is being used by the organization, it is maintained and enhanced to make sure it continues to provide the intended benefits (Support).

This classification of the activities of a system development project is a fundamental concept, and it remains the best way to initially learn about system development. However, in today's complex world, there are many variations on the basic five-phase structure illustrated in Figure 2-1. There are also many different approaches to system development, which are also discussed in this chapter. The differences in the various approaches are either in the set of activities composing each phase or in the method of carrying out the activity. But all information systems development requires planning, analyzing the need, designing a solution, and implementing the final system. As you gain an understanding of the objective of each phase and the various alternative activities, you will be able to adjust to any type of system development.

Current approaches to system development also include an additional complicating factor that affects the way the SDLC is used in practice—iteration across phases. Rather than following the SDLC from start to finish, development projects today break the project down into a series of miniprojects. Each miniproject is called an *iteration*, and each iteration involves analysis activities, design activities, and implementation activities. As you learn about the activities and tasks of each phase in this chapter and book, remember that most projects use an SDLC with multiple iterations rather than one sequential SDLC.

All large successful development projects use an organized system development approach based on the SDLC. Small projects also require analysis, design, and implementation, but they may not be as rigorous or as formal or include as much planning. Relating similar activities into phases and defining the steps are what make the SDLC an important concept for development. Without the structure and organization provided by an SDLC, development projects would be at severe risk of missing deadlines, escalating budgets, and producing a low-quality system. As a conceptual framework, the SDLC provides the structure, controls, and checklists needed to ensure successful development.

The Phases of the Systems Development Life Cycle

Understanding information systems development requires identification of the main objectives and primary activities within each of the five phases of the SDLC. The five phases and their objectives are summarized in Figure 2-2.

The Planning Phase

The primary objectives of the *planning phase* are to identify the scope of the new system, ensure that the project is feasible, and develop a schedule, resource plan, and budget for

planning phase

the initial phase of the SDLC, whose objective is to identify the scope of the new system and plan the project

FIGURE *2-2*

SDLC phases and objectives.

SDLC Phase	Objective
Project Planning	To identify the scope of the new system, ensure that the project is feasible, and develop a schedule, resource plan, and budget for the remainder of the project
Analysis	To understand and document in detail the business needs and the processing requirements of the new system
Design	To design the solution system based on the requirements defined and decisions made during analysis
Implementation	To build, test, and install a reliable information system with trained users ready to benefit as expected from use of the system
Support	To keep the system running productively initially and during the many years of the system's lifetime

the remainder of the project. We identify five activities in the project planning phase:

- Define the problem
- Produce the project schedule
- Confirm project feasibility
- Staff the project
- Launch the project

The most important activity of the planning phase is to define precisely the business problem and the scope of the required solution. At this stage in the project, you will not know all of the functions or processes that will be included within the system. However, it is important to identify the major uses of the new system and the business problems that the new system must address.

The two activities of producing the project schedule and staffing the project are clearly closely related. A detailed project schedule listing tasks, activities, and required staff is developed. Fortunately, some excellent methods and tools are available to provide support for this activity, which are explained in the next chapter. Large projects require elaborate schedules with specific, identifiable milestones and control procedures, and a critical part of this phase is identifying the necessary human resources and planning to acquire them at the required times during the project.

The next major element is to confirm that the project is feasible. Many projects are initiated as part of an enterprisewide strategic plan. Within the overall plan, each project must also stand on its own merit. Feasibility analysis investigates economic, organizational, technical, resource, and schedule feasibility. Each of these types of feasibility analysis is explained in more detail in the next chapter.

Finally, the total plan for the project is reviewed with upper management, and the project is initiated. Initiation of the project entails allocating funds, assigning project members, and obtaining other necessary resources such as office and development tools. An official announcement often communicates the project launch.

The Analysis Phase

The primary objective of the *analysis phase* is to understand and document the business needs and the processing requirements of the new system. Analysis is essentially a discovery process. The key words that drive the activities during analysis are *discovery* and *understanding*. Six primary activities are considered part of this phase:

- Gather information
- Define system requirements

analysis phase

the phase of the SDLC whose objective is to understand the user needs and develop requirements

- Build prototypes for discovery of requirements
- Prioritize requirements
- Generate and evaluate alternatives
- Review recommendations with management

Gathering information is a fundamental part of analysis. During this activity, the systems analysts meet with users to learn as much as possible about the *problem domain*—the area of the user's business that needs an information system solution and that is being researched. The analysts obtain information about the problem domain by observing the users as they do their work; by interviewing and asking questions of the users; by reading existing documents about procedures, business rules, and job responsibilities; and by reviewing existing automated systems. In addition to gathering information from the users of the system, the analysts should consult other interested parties. They may include middle management, senior executives, and at times even external customers. Gathering information is the core activity for discovery and understanding.

But it is not sufficient simply to gather information. Analysts must review, analyze, and structure the information obtained so that they can develop an overall understanding of the new system's requirements. This activity is called defining the system requirements, and the primary technique that is used is drawing diagrams to express and model the new system's processing requirements.

One important activity that can help an analyst gather and understand the requirements is to build a prototype of pieces of the new system. Then users can review them. Users often find it easier to express their needs by reviewing working prototypes of alternatives. "A picture is worth a thousand words" is as true in defining system requirements as it is in general, and a prototype is the "picture" that can elicit valuable insights from end users.

As the processing requirements are uncovered, each must be prioritized. There are always more requests for automation support than there is budget or resources to provide it. Thus, the most important needs must be identified and given priority for development. As the analysts prioritize the requirements, they also research various alternatives for implementing the system. Implementation alternatives include building the system in-house, buying a software package, or contracting to a third party to develop and install a new system.

Finally, the team selects and recommends an alternative to upper management. The recommendation recaps the results of the analysis phase activities, and together the team makes firm decisions about an alternative.

The Design Phase

The objective of the *design phase* is to design the solution system based on the requirements defined and decisions made during analysis. High-level design consists of developing an architectural structure for software components, databases, the user interface, and the operating environment. Low-level design entails developing the detailed algorithms and data structures that are required for software development. Seven major activities must be done during the design phase:

- Design and integrate the network
- Design the application architecture
- Design the user interfaces
- Design the system interfaces
- Design and integrate the database
- Prototype for design details
- Design and integrate the system controls

Design activities are closely interrelated and generally are all done with substantial overlap.

The network consists of the computer equipment, network, and operating system platforms that will house the new information system. Many of today's new systems are being installed in network and client-server environments. Design includes configuring these network environments. Sometimes the design is already complete based on an existing operating environment and strategic IT plans. At other times, substantial work must be done to develop an operating environment to provide the level of service the new system requires.

The *application* is the portion of the new information system that satisfies user needs with regard to the problem domain. In other words, the application provides the processing functions for the business requirements. Designing the appropriate computer programs for the application consists of using the diagrams showing the system's requirements that were developed during analysis.

application

the portion of the new information system that satisfies the user's needs in the problem domain

The user interface is a critical component of any new system. During the analysis activities, prototyping may have defined some elements of the user interface. During design, these elements are all combined to yield an integrated user interface consisting of forms, reports, screens, and sequences of interactions.

Most new information systems must also communicate with other, existing systems, so the design of the method and details of these communication links must also be precisely defined. These are called *system interfaces*.

Databases and information files are an integral part of information systems for business. The diagrams of the new system's data storage requirements, developed during analysis, are used to design the database that will support the application portion of the new system. At times, the database for the specific system must also be integrated with information databases of other systems already in use.

During design, it is often necessary to verify the correctness or workability of the proposed design. One important verification method is to build working prototypes of parts of the system to ensure that it will function correctly in the operating environment. In addition, analysts can test and verify alternative design strategies by building prototypes of the new system. Sometimes, if the prototypes are built correctly, they can be saved and used as part of the final system.

Finally, every system must have sufficient controls to protect the integrity of the database and the application program. Because of the highly competitive nature of the global economy and the risks associated with technology and security, every new system must include adequate mechanisms to protect the information and assets of the organization. These controls should be integrated into the new system while it is being designed, not after it has been constructed.

The Implementation Phase

implementation phase

the phase of the SDLC during which the new system is programmed and installed

During the *implementation phase*, the final system is built, tested, and installed. The objective of the activities of this phase is not only to produce a reliable, fully functional information system but also to ensure that the users are all trained and that the organization is ready to benefit as expected from use of the system. All the prior activities must come together during this phase to culminate in an operational system. Five major activities make up the implementation phase:

- Construct software components
- Verify and test
- Convert data
- Train users and document the system
- Install the system

The software can be constructed through various techniques. The conventional approach is to write computer programs using a language such as Visual Basic or Java. Other techniques, based on development tools and existing components, are becoming popular today. The software must also be tested, and the first kind of testing verifies that the system actually works. Additional testing is also required to make sure that the new system meets the needs of the system's users.

During implementation, the analysts may also build additional prototypes. These prototypes are used to verify different implementation strategies and to ensure that the system can handle the volumes of transactions that will exist after it is placed in production.

Almost every new system replaces an existing system, either a completely manual system or an earlier automated system. Normally, the existing information is important and needs to be converted to the format required in the new system. The activity to convert the data often becomes a small project of its own, with analysis, design, and implementation of procedures to clean and convert the data to the new system.

No system is successful unless the users understand it and can use it appropriately. A critical activity during implementation is to train the users on the new system so that they will be productive as soon as possible.

Finally, the actual changeover is the culminating activity in this phase. The new equipment must be in place and functioning, the new computer programs must be installed and working, and the database must be populated and available. The individual pieces of the new system must be up and running before the system can be used for its intended purpose. In today's widely dispersed organizations, the system must frequently be installed in many locations and integrated throughout the organization.

The Support Phase

support phase

the phase of the SDLC whose objective is to keep the system running productively after it is installed

The objective of the *support phase* is to keep the system running productively during the years following its initial installation. The support phase begins only after the new system has been installed and put into production, and it lasts throughout the productive life of the system. The expectation for most business systems is that the system will last for years. During the support phase, upgrades or enhancements may be carried out to expand the system's capabilities, and they will require their own development projects. Three major activities occur during the support phase:

- Maintain the system
- Enhance the system
- Support the users

Every system, especially a new one, contains components that do not function correctly. Software development is complex and difficult, so it is never error-free. Of course, the objective of a well-organized and carefully executed project is to deliver a system that is robust and complete and that gives correct results. However, because of the complexity of software and the impossibility of testing every possible combination of processing requirements, there will always be conditions that have not been fully tested and thus are subject to errors. In addition, business needs and user requirements change over time. Key tasks in maintaining the system include both fixing the errors (also known as *fixing bugs*) and making minor adjustments to processing requirements. Usually a system support team is assigned responsibility for maintaining the system.

Most newly hired programmer analysts begin their careers working on system maintenance projects. Tasks typically completed include changing the information provided in a report, adding an attribute to a table in a database, or changing the design of Windows or browser forms. These changes are requested and approved before the work is assigned, so a change request approval process is always part of the system support phase.

During the productive life of a system, it is also common to make major modifications. At times, government regulations require new data to be maintained or information to be provided. Also, changes in the business environment—new market opportunities, new competition, or new system infrastructure—necessitate major changes to the system. To implement these major system enhancements, the company must approve and initiate an upgrade development project. An upgrade project often results in a new version of the system. During your career, you may have the opportunity to participate in several upgrade projects.

The other major activity during the support phase is to provide assistance to the system users. A *help desk*, consisting of knowledgeable technicians, is a popular method to answer users' questions quickly and help increase their productivity. Training new users and maintaining current documentation are important elements of this activity. As a new systems analyst, you may have the opportunity to conduct training or to staff the help desk to gain experience with user problems and needs. Many newly hired programmer analysts start their careers working at a help desk for part of their workweek.

Scheduling Project Phases

As discussed previously, Figures 2-1 and 2-2 may give the impression that a project always progresses through analysis, then design, and then implementation phases. In fact, during the 1970s and 1980s, analysts tried to make projects proceed from one phase to the next in a sequential fashion. That was called the *waterfall approach*, based on the metaphor that when one phase was finished, the project team dropped down to the next phase. Once you drop over the waterfall to the next phase, there is no going back! The waterfall method required rigid planning and final decision making at each step of the SDLC. The waterfall approach is illustrated in Figure 2-3.

help desk

the availability of support staff to assist users with any technical or processing problem associated with an information system

waterfall approach

an approach to executing an SDLC in which one phase leads (falls) sequentially into the next phase

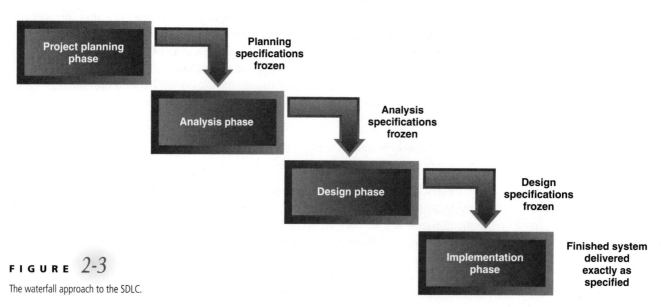

FIGURE *2-3*

The waterfall approach to the SDLC.

Analysts rarely attempt to use a straight waterfall approach. The activities are still classified as planning, analysis, design, and implementation, but many activities of the phases overlap and appear to be done concurrently. Figure 2-4 illustrates how the phases of a typical project might seem to overlap. But, why does overlap occur?

One reason phases overlap is efficiency. At the same time that the team members are analyzing needs, they may be thinking about and designing various forms or reports. To help understand the needs of the users, they may want to design some of the final system. When they do early design, they will frequently throw away part of it. Other com-

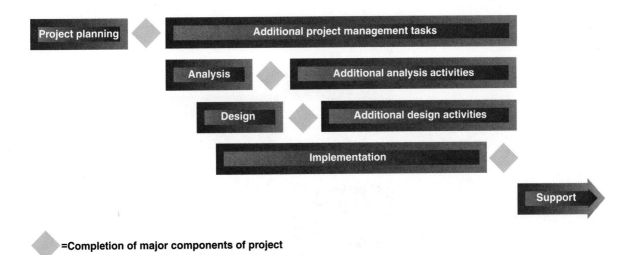

=Completion of major components of project

FIGURE *2-4*

The overlap of system development phases.

ponents may be saved for later inclusion in the final system. In addition, many components of a computer system are interdependent, which necessitates that the analysts do both analysis and some design at the same time.

Then why not overlap all activities completely? The answer is dependency. Some activities naturally depend on the results of prior work. Analysts cannot go very far into design without a basic understanding of the nature of the problem. Thus, some analysis must happen before design. It would also be inefficient for them to write program code before having an overall design structure, because they would have to throw too much away. Experience has proven that the most successful projects have schedules similar to the one shown in Figure 2-4. Such a schedule provides flexibility, yet requires completion of identified milestones on a timely basis.

Understanding Iteration and Project Phases

The overlapping schedule just discussed is best defined and managed using the notion of iteration across phases. Iteration means that work activities are done once, then again, and yet again (they are repeated). With each iteration, the result is refined so that it is closer to what is needed. Iteration assumes that no one gets the right result the first time. With an information system, you need to do some analysis and then some design before you really know whether the system will work and accomplish its goals. Then you do more analysis and design to make improvements (see Figure 2-5). In this view, it is not realistic to complete analysis (define the requirements) before starting work on the design. Similarly, completing the design is very difficult unless you know how the implementation will work (particularly with constantly changing technology). So you complete some design, then some implementation, and the iteration process continues—more analysis, more design, and more implementation. Naturally, the approach to iteration or the amount of iteration depends on the complexity of the project.

A project can be divided into miniprojects, each of which goes through an iteration of analysis, design, and implementation. There are several ways to define each iteration. One approach is to define the key functions that the system must include and then implement those key functions in the first iteration. Once they are completed, the next set of required but less crucial system functions are implemented. Finally, optional or "nice to have" system functions are implemented in the last iteration.

Another approach is to focus on one subsystem at a time. The first subsystem implemented contains the core functionality and data on which the other subsystems depend. Then the next iteration includes an additional subsystem, and so on. Sometimes

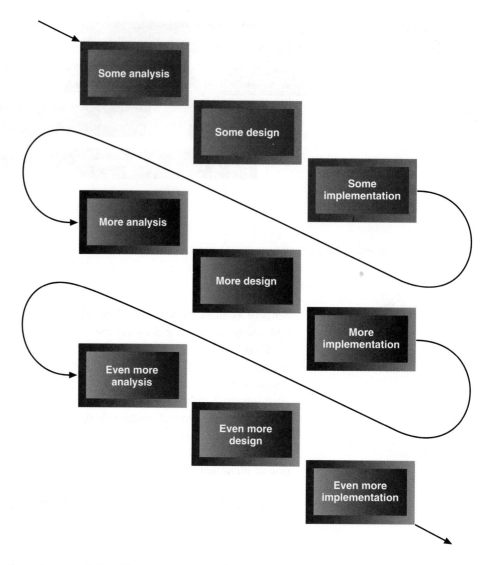

iterations are defined based on the complexity or risk involved (as in the spiral model discussed later in the chapter). Often the most complex or risky parts of the system are handled first because, if you find out early whether they can be handled as planned, you can still change plans if necessary. Other times, the simplest parts are handled first to get as much of the system finished as quickly as possible. How the iterations are defined depends on many factors and might be different with every project you encounter.

A related approach is called *incremental development*. With this approach, you complete parts of the system in one or more iterations and then put the system into operation for users. This approach gets part of the system into the hands of users as early as possible so they can benefit from it. Then you complete a few more iterations to develop another part of the system, integrate it with the first part, and again put it into operation. Finally, you complete the last part and integrate it with the rest.

All approaches to system development can use some amount of iteration. Most development today is done with varying degrees of iteration. The object-oriented approach is always described as highly iterative. Additionally, incremental development can be used with any approach to development, even if each increment is completed using a rigid waterfall approach.

incremental development

a development approach that completes parts of a system in one or more iterations and puts them into operation for users

METHODOLOGIES, MODELS, TOOLS, AND TECHNIQUES

Systems analysts have a variety of aids at their disposal to help them complete activities and tasks in the SDLC. Among them are methodologies, models, tools, and techniques. The following sections discuss each of these aids.

Methodologies

A *system development methodology* provides guidelines to follow for completing every activity in the systems development life cycle, including specific models, tools, and techniques. Some methodologies are homegrown, developed by systems professionals in the company based on their experience. Some methodologies are purchased from consulting firms or other vendors.

Some methodologies (whether homegrown or purchased) contain written documentation that can fill a bookcase. The documentation defines everything the developers might need to produce at any point in the project, including how documentation should look and what reports to management should contain. Other methodologies are much more informal—one document will contain general descriptions of what should be done. Sometimes the methodology that a company adopts is "just follow some sort of methodology," but such freedom of choice is becoming rare. Most people want the methodology to be flexible, though, so that it can be adapted to many different types of projects and systems.

Because a methodology contains instructions about how to use models, tools, and techniques, you must understand what models, tools, and techniques are.

Models

Anytime people need to record or communicate information, in any context, it is very useful to create a model—and a model in information systems development has the same purpose as any other model. A *model* is a representation of an important aspect of the real world. Sometimes the term *abstraction* is used because we abstract (separate out) an aspect of particular importance to us. Consider a model of an airplane. To talk about the aerodynamics of the airplane, it is useful to have a small model that shows the plane's overall shape in three dimensions. Sometimes a drawing showing the cross-sectional details of the wing of the plane is what is needed. In another case, a list of mathematical characteristics of the plane might be necessary to understand how the plane will behave. All of these are models of the same plane.

Some models are physically similar to the real product. Some models are graphical representations of important details. Some models are abstract mathematical notations. Each emphasizes a different type of information. In airplane design, aerospace engineers use lots of different models. Learning to be an aerospace engineer involves learning how to create and use all of the models. It is the same for an information system developer, although models for information systems are not yet as standardized or precise as aerospace models. But system developers are making progress. First, it is important to recognize that the field is very young, and many senior analysts were self-taught. More importantly, though, an information system is much less tangible than an airplane—you can't really see, hold, or touch it. Therefore, the models of the information system can seem much less tangible, too.

What sort of models do developers make of aspects of an information system? Models used in system development include representations of inputs, outputs, processes, data, objects, object interactions, locations, networks, and devices, among other things. Most of the models are graphical models, which are drawn representations that employ agreed-upon symbols and conventions. These are often called *diagrams* and *charts*. You have probably drawn models showing program logic using flowcharts. Much of this text describes how to read and create a variety of models that represent an information system.

Another kind of model important to develop and use is a project-planning model, such as PERT or Gantt charts, which are shown in Chapter 3. These models represent the system development project itself, highlighting its tasks and task completion dates. Another model related to project management is a chart showing all of the people assigned to the project. Figure 2-6 lists some models used in system development.

Some models of system components

Flowchart
Data flow diagram (DFD)
Entity-relationship diagram (ERD)
Structure chart
Use case diagram
Class diagram
Sequence diagram

Some models used to manage the development process

PERT chart
Gantt chart
Organizational hierarchy chart
Financial analysis models – NPV, ROI

Tools

tool

software support that helps create models or other components required in the project

A *tool* in the context of system development is software support that helps create models or other components required in the project. Tools may be simple drawing programs for creating diagrams. They might include a database application that stores information about the project, such as data flow definitions or written descriptions of processes. A project management software tool, such as Microsoft Project (described in Chapter 3), is another example of a tool used to create models. The project management tool creates a model of the project tasks and task dependencies.

Tools have been specifically designed to help system developers. Programmers should be familiar with integrated development environments (IDEs) that include many tools to help with programming tasks—smart editors, context-sensitive help, and debugging tools. Some tools can generate program code for the developer. Some tools reverse-engineer old programs—generating a model from the code so that the developer can figure out what the program does, in case the documentation is missing (or was never done).

CASE tool

a computer-aided system engineering tool designed to help a systems analyst complete development tasks

The most comprehensive tool available for system developers is called a *CASE tool*. *CASE* stands for computer-aided system engineering. CASE tools are described in more detail later in this chapter. Basically, they help the analyst create the important system models, and then the tools automatically check the models for completeness and compatibility with other system models. Finally, the CASE tools can generate program code based on the models. Figure 2-7 lists types of tools used in system development.

F I G U R E *2-7*

Some tools used in system development.

Project management application
Drawing/graphics application
Word processor/text editor
Computer-aided system engineering (CASE) tools
Integrated development environment (IDE)
Database management application
Reverse-engineering tool
Code generator tool

Techniques

technique

a collection of guidelines that help an
analyst complete a system development activity or task

A *technique* in system development is a collection of guidelines that help an analyst complete a system development activity or task. A technique often includes step-by-step instructions for creating a model, or it might include more general advice for collecting information from system users. Some examples include data-modeling techniques, software-testing techniques, user-interviewing techniques, and relational database design techniques.

Sometimes a technique applies to an entire life cycle phase and helps you create several models and other documents. The modern structured analysis technique (discussed later) is an example. Even the strategic system planning techniques discussed in Chapter 1 and project management techniques discussed in Chapter 3 fit this definition. Figure 2-8 lists some techniques commonly used in system development.

F I G U R E *2-8*

Some techniques used
in system development.

Strategic planning techniques
Project management techniques
User interviewing techniques
Data-modeling techniques
Relational database design techniques
Structured analysis technique
Structured design technique
Structured programming technique
Software-testing techniques
Object-oriented analysis and design techniques

How do all these components fit together? A *methodology* includes a collection of *techniques* that are used to complete activities within each phase of the systems development life cycle. The activities include completion of a variety of *models* as well as other documents and deliverables. Like any other professionals, system developers use software *tools* to help them complete their activities. Figure 2-9 shows the relationships among the components of a methodology.

As part of her responsibility as project manager for the new customer support system for Rocky Mountain Outfitters, Barbara Halifax has to make decisions about the methodology to use to develop the system (see Barbara's memo).

FIGURE *2-9*

Relationships among components
of a methodology.

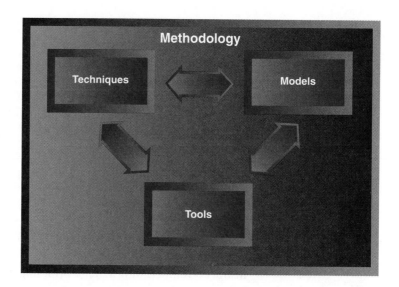

February 14, 2005

To: John MacMurty

From: Barbara Halifax

RE: Customer Support System Planning Update

John, this is just a brief note to let you know I am continuing to work on scheduling and staffing for the project, and as such, we need to make some decisions about the development approach we will use for the project.

In our methodology, we have great latitude in how we schedule phases and activities. I am planning to use an iterative approach and naturally plan to involve users extensively. I have thought about using some concepts from the spiral model we discussed last fall, primarily risk identification. I'm also thinking about incorporating some concepts from the eXtreme Programming approach, such as programming teams, but not until later in the project when programming and testing begin.

More immediate issues involve the choice of development approach, be it traditional or OO. I'm still talking to people about it, and in the early part of analysis, the focus is similar with either approach. I think it is still clear that either approach can be used. Just because we have a Web component doesn't mean we must use OO. On the other hand, we want to get our feet wet with OO and UML at some point. I'm still thinking about the risk and fit for this project. Either way, we plan to use appropriate CASE tools for modeling and for generating some of the code.

That's it for now. Let me know if you have any questions or concerns.

BH

cc: Steven Deerfield, Ming Lee, Jack Garcia

TWO APPROACHES TO SYSTEM DEVELOPMENT

System development is done in many different ways. This diversity can confuse new employees when they go to work as system developers. Sometimes it seems every company that develops information systems has its own methodology. Sometimes different development groups within the same company use different methodologies, and each person in the company may have his or her own way of developing systems.

Yet, as you have already seen in the opening case, there are many common concepts. In virtually all development groups, some variation of the systems development life cycle is used, with phases for project planning, analysis, design, implementation, and support. Additionally, virtually every development group uses models, tools, and techniques that make up an overall system development methodology.

All system developers should be familiar with two very general approaches to system development, because they form the basis of virtually all methodologies: the *traditional approach* and the *object-oriented approach*. This section reviews the major characteristics of both approaches and provides a bit of history.

The Traditional Approach

The traditional approach includes many variations based on techniques used to develop information systems with structured and modular programming. This approach is often referred to as *structured system development*. A refinement to structured development, called *information engineering* (IE), is a popular variation.

Structured System Development

Structured analysis, structured design, and structured programming are the three techniques that make up the *structured approach*. Sometimes these techniques are collectively referred to as the *structured analysis and design technique* (SADT). The structured programming technique, developed in the 1960s, was the first attempt to provide guidelines to improve the quality of computer programs. You certainly learned the basic principles of structured programming in your first programming course. The structured design technique was developed in the 1970s to make it possible to combine separate programs into more complex information systems. The structured analysis technique evolved in the early 1980s to help clarify the requirements for the computer system for the developers before they designed the programs.

Structured Programming

High-quality programs not only produce the correct outputs each time the program runs but also make it easy for other programmers to read and modify the program later. And programs need to be modified all the time. A *structured program* is one that has one beginning and one ending, and each step in the program execution consists of one of three programming constructs:

- A sequence of program statements
- A decision where one set of statements or another set of statements executes
- A repetition of a set of statements

Figure 2-10 shows these three structured programming constructs.

Before these rules were developed, programmers made up programming techniques as they went along, which resulted in some very convoluted programs. Most programmers were happy if the programs ran at all, and they were even happier if the programs produced the right outputs. But following these simple rules made it much easier to read and interpret what a program does.

Another concept related to structured programming is top-down programming. *Top-down programming* divides more complex programs into a hierarchy of program modules (see Figure 2-11). One module at the top of the hierarchy controls program

---

structured approach

system development using structured analysis, structured design, and structured programming techniques

structured program

a program or program module that has one beginning and one ending, and for which each step in the program execution consists of sequence, decision, or repetition constructs

top-down programming

dividing more complex programs into a hierarchy of program modules

Sequence

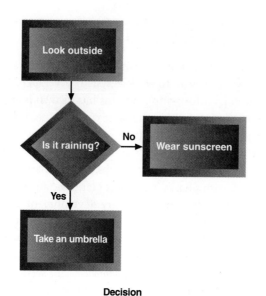

Decision

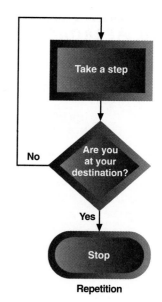

Repetition

FIGURE *2-10*

Three structured programming constructs.

FIGURE *2-11*

Top-down, or modular, programming.

execution by "calling" lower-level modules as required. Sometimes the modules are part of the same program. For example, in COBOL, one main paragraph calls another paragraph using the Perform keyword. In Visual Basic, a statement in an event procedure can call a general procedure. The programmer writes each program module (paragraph or procedure) using the rules of structured programming (one beginning, one end, and sequence, decision, and repetition constructs).

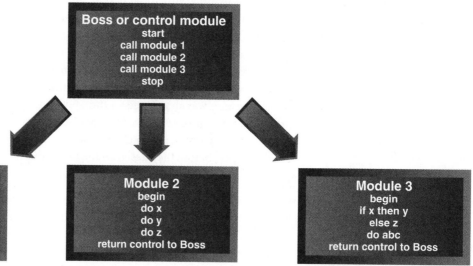

Sometimes separate programs are produced that work together as one "system." Each of these programs follows top-down programming and structured programming rules, but the programs themselves are organized into a hierarchy, as with top-down programming. One program calls other programs. When the hierarchy involves multiple programs, such an arrangement is sometimes called *modular programming*.

Structured Design As information systems continued to become increasingly complex through the 1970s, each system involved many different functions. Each function performed by the system might be made up of dozens of separate programs. The *structured design* technique was developed to provide some guidelines for deciding what the set of programs should be, what each program should accomplish, and how the programs should be organized into a hierarchy. The modules and the arrangement of modules are shown graphically using a model called a *structure chart* (see Figure 2-12).

structured design

a technique providing guidelines for deciding what the set of programs in an IS should be, what each program should accomplish, and how the programs should be organized into a hierarchy

structure chart

a graphical model showing the hierarchy of program modules produced by the structured design technique

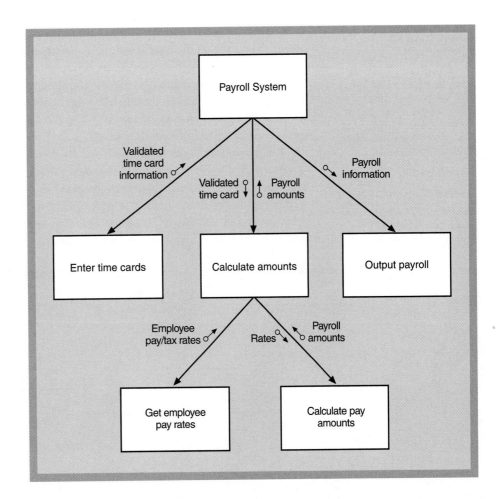

Two main principles of structured design are that program modules should be designed so they are (1) loosely coupled and (2) highly cohesive. *Loosely coupled* means each module is as independent of the other modules as possible, which allows each module to be designed and later modified without interfering with the performance of the other modules. *Highly cohesive* means that each module accomplishes one clear task. That way, it is easier to understand what each module does and to ensure that if changes to the module are required, none will accidentally affect other modules.

The structured design technique defines different degrees of coupling and cohesion and provides a way of evaluating the quality of the design before the programs are actually written. As with structured programming, quality is defined in terms of how easily the design can be understood and modified later when the need arises.

Structured design assumes the designer knows what the system needs to do—what the main system functions are, what the required data are, and what the needed outputs are. Designing the system is obviously much more than designing the organization of the program modules. Therefore, it is important to realize that the structured design technique helps the designer complete part of but not the entire design life cycle phase.

By the 1980s, file and database design techniques were developed to be used along with structured design. Newer versions of structured design assume database management systems are used in the system, and program modules are designed to interact with the database. Additionally, since more nontechnical people were becoming involved with information systems, user-interface design techniques were developed. For example, menus in an interactive system determine which program in the hierarchy gets called. Therefore, a key aspect of user-interface design is done in conjunction with structured design.

Modern Structured Analysis Since the structured design technique requires the designer to know what the system should do, techniques for defining system requirements were developed. System requirements define what the system must do in great detail, but without committing to one specific technology. By deferring decisions about technology, the developers can sharply focus their efforts on what is needed, not on how to do it. If these requirements are not fully and clearly worked out in advance, the designers cannot possibly know what to design.

The *structured analysis* technique helps the developer define what the system needs to do (the processing requirements), what data the system needs to store and use (data requirements), what inputs and outputs are needed, and how the functions work together as a whole to accomplish tasks. The key graphical model of the system requirements used with structured analysis is called the *data flow diagram (DFD)*, and it shows inputs, processes, storage, and outputs, and the way they function together (see Figure 2-13).

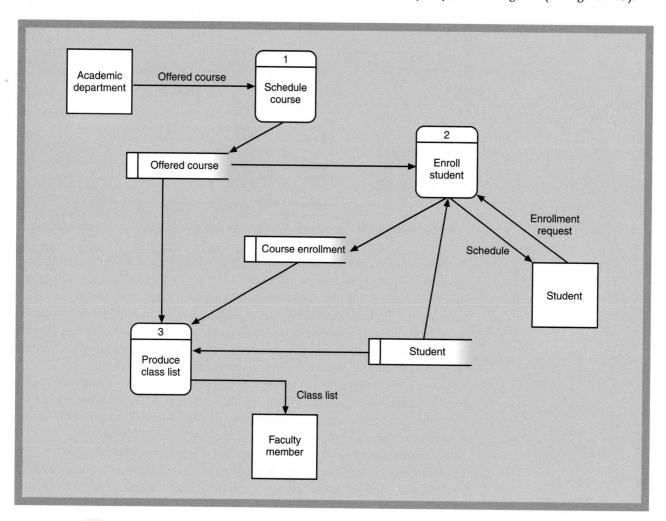

FIGURE *2-13*

A data flow diagram (DFD) created using the structured analysis technique.

The most recent variation of structured analysis defines systems processing requirements by identifying all of the events that will cause the system to react in some way. For example, in an order-entry system, if a customer orders an item, the system must process a new order (a major system activity). Each event leads to a different system activity. The analyst takes each of these activities and creates a data flow diagram showing the processing details, including inputs and outputs.

A model of the needed data is also created based on the types of things about which the system needs to store information (data entities). For example, to process a new order, the system needs to know about the customer, the items wanted, and the details about the order. This model is called an *entity-relationship diagram (ERD)*. The data entities from the entity-relationship diagram correspond to the data storage shown on data flow diagrams. Figure 2-14 shows an example of an entity-relationship diagram. Figure 2-15 illustrates the sequence followed when developing a system using structured analysis, structured design, and structured programming.

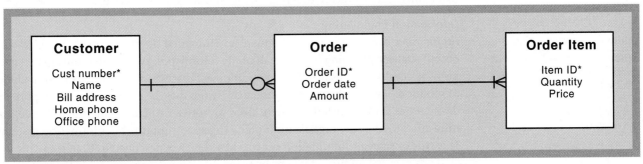

FIGURE 2-14

An entity-relationship diagram (ERD) created using the structured analysis technique.

Weaknesses of the Structured Approach   Because the structured approach to system development evolved over time, many variations can be found in practice. Some people are still following the original versions of structured analysis and structured design they learned years ago, ignoring many improvements. Others picked up bits and pieces of the techniques on the job and never formally studied the details.

Many people have considered the structured approach to be weak because the techniques address only some, but not all, of the activities of analysis and design. Critics desired a more comprehensive and rigorous set of techniques to make system development more like an engineering discipline and less like an art. In addition, many people thought the transition from the data flow diagram (in structured analysis) to the structure chart (in structured design) did not work well in practice. Others thought that data modeling and the entity-relationship diagram were much more important than modeling processes with the data flow diagram. The structured approach, despite its inclusion of data modeling and database design, still made processes rather than data the central focus of the system.

FIGURE 2-15

How structured analysis leads to structured design and to structured programming.

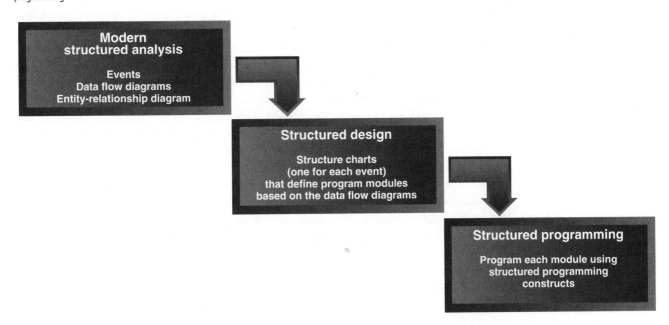

Finally, many people thought that to ensure that systems are comprehensive and co-ordinated, the development of a system should begin only after the organization completed an overall strategic system planning effort. Therefore, they wanted the approach to development to include a strategic system planning technique, both to determine which systems should be built and to provide some initial requirements models that ensured all systems would be compatible. Because of these goals, some developers turned to a refinement of structured development: information engineering.

Information Engineering

Information engineering is a refinement to structured development that begins with overall strategic planning to define all of the information systems that the organization needs to conduct its business (the application architecture plan). The plan also includes a definition of the business functions and activities that the systems need to support, the data entities about which the systems need to store information, and the technological infrastructure that the organization plans to use to support the information systems. This type of strategic information systems planning was described in Chapter 1.

Each new system project begins by using the defined activities and data entities created during strategic systems planning. Then the activities and data are refined as the project progresses. At each step, the project team creates models of the processes, the data, and the ways they are integrated.

The type of data needed to conduct the business changes very little over time, but the processes followed to collect data change frequently. Therefore, the information engineering approach focuses much more on data than the structured approach. Just as the structured approach includes data requirements, information engineering includes processes, too. The processing model of information engineering—the process dependency diagram—is similar to a data flow diagram, but it focuses more on which processes are dependent on other processes and less on data inputs and outputs. Events trigger the processes, as with modern structured analysis.

A final major difference with information engineering is the more complete life cycle support it provides through the use of an integrated CASE tool. The CASE tool helps automate as much of the work as possible. It also forces the analyst to follow the information engineering approach faithfully, sometimes at the expense of flexibility. In many instances, the CASE tool can automatically generate the final program code. CASE tools are also available for the structured approach, but they are often intentionally more flexible, and hence less rigorous.

Information engineering is mainly credited to James Martin, who wrote several books on information engineering and developed CASE tools to support it. By the late 1980s, information engineering was very popular for large mainframe systems. But because they lacked flexibility, the CASE tools that supported information engineering were less useful with smaller desktop applications and client-server applications. By the 1990s, fewer companies were using information engineering exclusively, although many of the concepts and techniques continue to be used, particularly the approach to planning and the emphasis on data modeling.

The information engineering approach refines many of the concepts of the structured approach into a rigorous and comprehensive methodology. Both approaches define information systems requirements, design information systems, and construct information systems by looking at processes, data, and the interaction of the two. This text merges key concepts from these two approaches into one, which we will refer to hereafter as *the traditional approach*. The traditional approach, in one version or another, is still widely used for information system development, although many information systems projects are now using object-oriented technology—which requires a completely different approach.

---

information engineering

a traditional system development methodology thought to be more rigorous and complete than the structured approach, because of its focus on strategic planning, data modeling, and automated tools

---

object-oriented approach

an approach to system development that views an information system as a collection of interacting objects that work together to accomplish tasks

object

a thing in the computer system that can respond to messages

F I G U R E *2-16*

The object-oriented approach to systems (read clockwise starting with user).

The Object-Oriented Approach

An entirely different approach to information systems, the *object-oriented approach*, views an information system as a collection of interacting objects that work together to accomplish tasks (see Figure 2-16). Conceptually, there are no processes or programs; there are no data entities or files. The system consists of objects. An *object* is a thing in the computer system that is capable of responding to messages. This radically different view of a computer system requires a different approach to systems analysis, systems design, and programming.

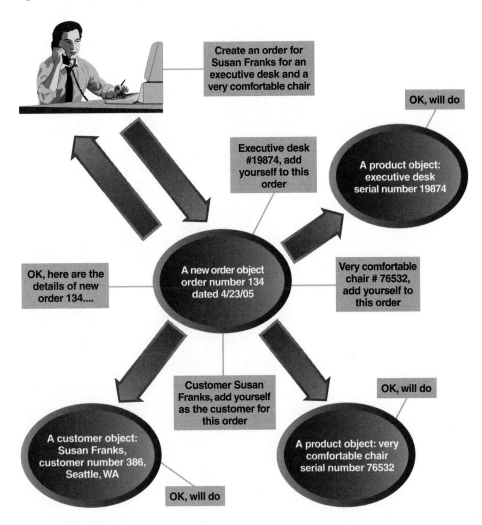

The object-oriented approach began with the development of the Simula programming language in Norway in the 1960s. Simula was used to create computer simulations involving "objects" such as ships, buoys, and tides in fjords. It is very difficult to write procedural programs that simulate ship movement, but a new way of programming simplified the problem. In the 1970s, the Smalltalk language was developed to solve the problem of creating graphical user interfaces (GUIs) that involved "objects" such as pull-down menus, buttons, check boxes, and dialog boxes. Other object-oriented languages include C++ and, more recently, Java and C#. These languages focus on writing definitions of the types of objects needed in a system, and as a result, all parts of a system can be thought of as objects, not just the graphical user interface.

Because the object-oriented approach views information systems as collections of interacting objects, *object-oriented analysis (OOA)* defines all of the types of objects that do the work in the system and shows what user interactions are required to complete

object-oriented analysis (OOA)

defining all of the types of objects that do the work in the system and showing what user interactions are required to complete tasks

tasks. *Object-oriented design (OOD)* defines all of the additional types of objects necessary to communicate with people and devices in the system, shows how the objects interact to complete tasks, and refines the definition of each type of object so it can be implemented with a specific language or environment. *Object-oriented programming (OOP)* consists of writing statements in a programming language to define what each type of object does.

An object is a type of thing—a customer or an employee, as well as a button or a menu. Identifying types of objects means classifying things. Some things, such as customers, exist both outside the system (the real customer) and separately inside the system (a computer representation of a customer). A classification or "class" represents a collection of similar objects; therefore, object-oriented development uses a *class diagram* to show all of the classes of objects in the system (see Figure 2-17). For every class, there may be more specialized subclasses. For example, a savings account and a checking account are two special types of accounts (two subclasses of the class account). Similarly, a pull-down menu and a pop-up menu are two special types of menus. Subclasses exhibit or "inherit" characteristics of the class above them.

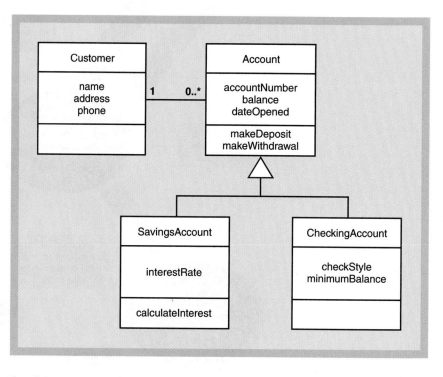

FIGURE *2-17*

A class diagram created during object-oriented analysis.

The object-oriented approach yields several key benefits, among them naturalness and reuse. The approach is natural—or intuitive—for people, because they tend to think about the world in terms of tangible objects. It is less natural to think about complex procedures found in procedural programming languages. Also, because the object-oriented approach involves classes of objects, and many systems in the organization use the same objects, these classes can be used over and over again whenever they are needed. For example, almost all systems use menus, dialog boxes, windows, and buttons, but many systems within the same company also use customer, product, and invoice classes that can be reused. There is less need to "reinvent the wheel" to create an object.

Clearly, the object-oriented approach is quite different from the traditional approach. But in other ways, quite a few traditional concepts are simply repackaged in the object-oriented approach. For this reason, some people find the OO approach difficult to understand at first. Parts 2 and 3 of this book discuss the similarities and differences in detail to help clarify each approach's strengths.

Many systems being developed today combine both traditional and object-oriented technology. Some integrated development environments (IDEs) also combine traditional and object-oriented technology in the same tool—for example, object-oriented programming is used for the user interface, and procedural programming for the rest. Many system projects are also exclusively traditional in analysis and design, and others are exclusively object-oriented, even within the same information systems department. These are some of the reasons that it is important to cover both traditional approaches and newer object-oriented approaches in this text. Everyone should know the basic concepts of each, but your coursework might emphasize one approach over the other.

SYSTEMS DEVELOPMENT LIFE CYCLE VARIATIONS

This text uses a generic systems development life cycle throughout, but it emphasizes that future systems analysts will encounter many variations in use in organizations. This section gives examples of some variations so that you will be familiar with any variation you encounter and comfortable adapting to it. Some of the variations are based on the life cycle phases. Some life cycles differ in their emphasis on the social, as well as technical, subsystems.

Variations of Names for Phases

Figure 2-18 shows some examples of life cycles in use today that use different names for the phases. You may encounter one or several of these variations in your career. In a waterfall approach, the more phases there are, the more rigid cutoffs there are. The first example (an early life cycle example) includes a feasibility study at the beginning, which usually assumes that the project was unplanned rather than part of an overall strategic system planning. System investigation and systems analysis together then make up the analysis phase in this SDLC. The second example is the information engineering life cycle. This life cycle includes the overall strategic planning effort as part of the life cycle

FIGURE *2-18*

Life cycles with different names for phases.

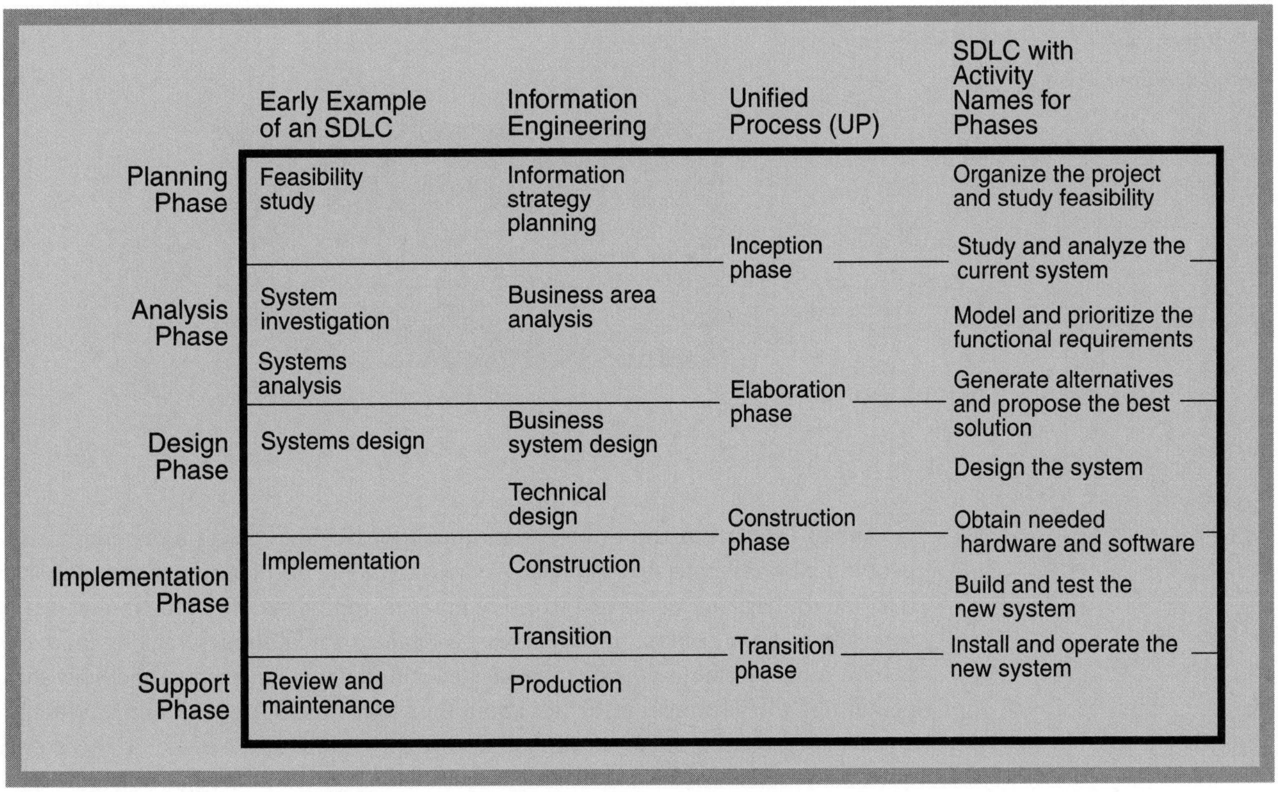

	Early Example of an SDLC	Information Engineering	Unified Process (UP)	SDLC with Activity Names for Phases
Planning Phase	Feasibility study	Information strategy planning	Inception phase	Organize the project and study feasibility
				Study and analyze the current system
Analysis Phase	System investigation	Business area analysis		Model and prioritize the functional requirements
	Systems analysis		Elaboration phase	Generate alternatives and propose the best solution
Design Phase	Systems design	Business system design		Design the system
		Technical design	Construction phase	Obtain needed hardware and software
Implementation Phase	Implementation	Construction		Build and test the new system
		Transition	Transition phase	Install and operate the new system
Support Phase	Review and maintenance	Production		

(the life cycle used in this text includes planning for the project only, although it is assumed that strategic planning has occurred previously). Two design phases are included: business system design and technical design. Finally, construction and transition are two parts of the implementation phase.

The third life cycle model (from the Unified Process) includes only four phases, with less traditional names. You will learn more about the Unified Process later in this chapter. The fourth life cycle model has eight phases, and it uses a different approach to naming the phases—they are named like activities, using the verb-noun form. The generic SDLC model used in this text uses five phases, named in the traditional way: project planning, analysis, design, implementation, and support. The activities in each phase are named using the verb-noun form: gather information, define requirements, prototype for discovery, and so on.

No matter which of these life cycle models you use, the same overall tasks need to be carried out. Phases can overlap and iteration across phases can be used with any of these SDLC examples.

Variations Based on an Emphasis on People

Some life cycles in use reflect differences based on the underlying philosophies of the system development methodologies. Some methodologies recognize more explicitly that information systems are *sociotechnical systems*—that is, they include both social and technical subsystems that must be considered and designed to work well together. Therefore, some life cycles explicitly emphasize these two aspects. Other terms for sociotechnical system development are *user-centered design* and *participatory design*. One example is called Multiview. The life cycle used with Multiview is shown in Figure 2-19. Note that human activity is studied in detail, and analysis and design of sociotechnical aspects of the system are considered in one phase. The design of the human-computer interface as a separate phase also emphasizes the importance of the users and user participation in the development of the system. End users often participate extensively in all aspects of system development.

---

sociotechnical systems

information systems that include both social and technical subsystems designed to work well together

---

FIGURE *2-19*

Phases of the Multiview SDLC.

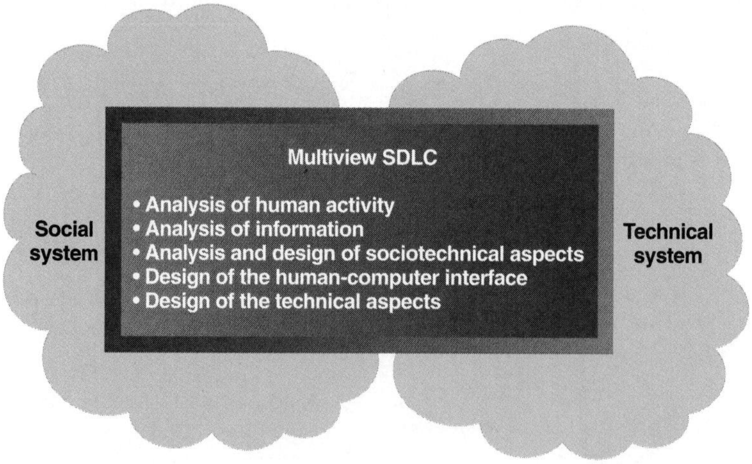

Multiview SDLC

Social system
Technical system

- Analysis of human activity
- Analysis of information
- Analysis and design of sociotechnical aspects
- Design of the human-computer interface
- Design of the technical aspects

The soft systems methodology is another approach that places more emphasis on people. Followers argue that it is relatively easy to model the data and processes in a system, but understanding the real world requires including people in the model. People have different and conflicting objectives, perceptions, and attitudes, making the system behavior unpredictable. One diagram created with the soft systems methodology shows people and their attitudes or beliefs about the system and about each other. When the analyst spends time understanding and modeling human issues, he or she exposes potential problems in a complex system that can be addressed through careful design.

Variations Based on Speed of Development

System developers have always looked for ways to speed the development process. One reason to speed development is the continuing backlog of needed systems. Another reason for speed is the constantly changing technological and business environment. If it takes too long to complete the system, it might be too late to achieve the desired benefits.

One variation of the systems development life cycle that aims to speed up the process radically is called *rapid application development (RAD)*. Some variations of RAD try to speed up the activities in each phase, such as speeding the analysis phase by scheduling intensive meetings of key participants to gather information and make decisions expeditiously. Iterative development is another approach often associated with RAD. It speeds up the process of getting to design and implementation.

Building prototypes of the system during the analysis and design phases of the SDLC also can speed up development, but prototyping is not always done just for RAD. Greater understanding of the system requirements is another objective of prototyping. Prototyping is discussed in Chapter 4 as an information-gathering technique.

Some approaches to RAD are even more radical. They involve creating working prototypes of the proposed system right away so that users can work with and critique them. Then the prototype is expanded into a finished system as soon as users agree on what the system is supposed to do. If not managed carefully, these approaches to RAD can be risky. RAD is discussed in more detail in Chapter 17.

CURRENT TRENDS IN DEVELOPMENT

One thing that never changes in the information systems field is that things are always changing. New tools and techniques are always appearing—sometimes with much publicity and anticipation—and system developers are always looking for new and better ways to work. The techniques and life cycle variations discussed previously are examples of ongoing changes to system development methodologies. A few important current trends in system development are discussed in this section. Any one of these trends could become common and even dominate system development in the future. Or, as with the variations discussed previously, system developers might adapt key concepts or techniques from each of these trends and use them when appropriate.

Risk and the Spiral Model

The spiral model is a highly iterative development approach. Rather than presenting life cycle phases sequentially with some iteration added, the model shows the life cycle as a spiral, starting in the center and working its way around, over and over again, until the project is complete. This model therefore looks very different and sets the tone for the project to be managed differently. Figure 2-20 shows the spiral model graphically.

There are many different ways of implementing the spiral model approach. The example in Figure 2-20 begins with an initial planning phase, as shown in the center of the figure. The purpose of the initial planning phase is to gather just enough information to begin developing an initial prototype. Planning phase activities include a feasibility study, a high-level user requirements survey, generation of implementation alternatives, and choice of an overall design and implementation strategy.

After the initial planning is completed, work begins in earnest on the first prototype (the blue ring in the figure). For each prototype, the development process follows a sequential path through analysis, design, construction, testing, integration with previous prototype components, and planning for the next prototype. When planning for the next prototype is completed, the cycle of activities begins again. Although the figure shows four prototypes, the spiral model approach can be adapted for any number of prototypes.

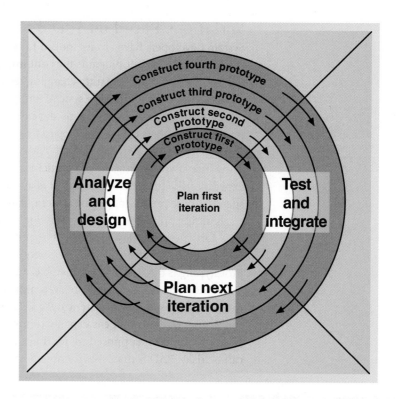

A key concept in the spiral model approach is the focus on *risk*, which is also a core concept in project management, discussed in Chapter 3. Although there are many ways to choose what to focus on in each iteration, the spiral model recommends identifying risk factors that must be studied and mitigated. The first iteration should address the part of the system that appears to have the greatest risk. Sometimes the greatest risk is not one subsystem or one set of system functions; rather, the greatest risk might be the technological feasibility of new technology. If so, the first iteration might focus on a prototype that proves the technology will work as planned. Then the second iteration might begin work on a prototype that addresses risk associated with the system requirements or other issues. Another time, the greatest risk might be user acceptance of change. The first iteration might focus on producing a prototype to show the users that the new system will enrich their working lives.

Although the spiral model approach is similar in many ways to the traditional SDLC when used iteratively, it includes other guidelines and techniques as well. Chapter 17 provides a more detailed description of the spiral model approach.

eXtreme Programming (XP)

eXtreme Programming (XP) is a system development approach recently popularized by Kent Beck. XP adapts techniques from many sources and adds some new ideas. It is sometimes referred to as a "lightweight" system development methodology, meaning it is kept simple and focused on making the development process more efficient for the developer.

The developers begin planning the system project by having the users describe *user stories*. User stories are descriptions of the support the users need from the system—in other words, the required system functionality. The developers document these stories quickly with informal descriptive models. Along with providing the user stories, users describe a set of acceptance tests that will demonstrate that the system provides the required functionality once it is completed.

The developers then plan a series of *releases* for the project, with each release including a working part of the final system, as with incremental development. The project proceeds by working on the first release, which usually takes several iterations to complete. When the first release is completed, the second release is started.

In many ways, XP is much like other iterative and incremental approaches. But XP contains some additional features that make it popular. It requires continuous testing, continuous integration, and heavy user involvement, for example. It also requires that all programming be done by teams, with two programmers working together at one workstation when writing and testing code. This and other features emphasize open and effective communication among team members. A final feature is the firm belief that developers should work no more than 40 hours per week, to prevent burnout but also to demonstrate that system projects can be completed on schedule without overworking the staff if the XP techniques and tools are used for the project. XP is described in more detail in Chapter 17.

The Unified Process (UP)

You learned that some companies obtain complete system development methodologies from consulting firms, either by purchasing rights to the methodology or by contracting for extensive training services from the consulting firm to learn the methodology. The *Unified Process (UP)* is an object-oriented system development methodology offered by Rational Software. Rational Software, which was recently acquired by IBM, is where the three proponents of the Unified Modeling Language (UML) continue to work: Grady Booch, James Rumbaugh, and Ivar Jacobson. The UP is their attempt to define a complete methodology that, in addition to providing several unique features, uses UML for system models. In the UP, the term development *process* is synonymous with development *methodology*.

Although you will learn much about UML because it is a standard modeling notation for the object-oriented (OO) approach, the UP is *not* a standard OO development methodology. UML models described in this text can be used in a variety of ways with any OO development methodology, but because of the stature of Booch, Rumbaugh, and Jacobson, the UP is gaining a lot of attention. Certainly the UP includes many useful and innovative techniques. Booch, Rumbaugh, and Jacobson have written several books about the UP and have endorsed other books about it written by colleagues, so it is possible to learn and to use the UP without purchasing services from Rational.

The UP is designed to reinforce six "best practices" of system development that are common to many system development methodologies:

- Develop iteratively
- Define and manage system requirements
- Use component architectures
- Create visual models
- Verify quality
- Control changes

The UP defines four life cycle phases: inception, elaboration, construction, and transition (seen previously in Figure 2-18). The inception phase defines the scope of the project by specifying *use cases* (which are similar to user stories), as with any development approach. You will learn how to identify use cases and create use case diagrams in Chapters 5 and 7 of this text. The project team also completes a feasibility study to determine whether resources should be invested in the project.

The elaboration phase focuses on several iterations that take part of the system and define the requirements, design the solution, and implement the solution. The team defines the requirements and the design by creating use case diagrams, class diagrams, sequence diagrams, and other UML diagrams. Final cost and benefit estimates are also completed by the end of the elaboration phase.

Unified Process (UP)

an object-oriented system development methodology offered by Rational Software

During the construction phase, you continue to build the system using additional iterations that also include requirements, design, and implementation, possibly creating multiple releases of the system. During the transition phase, you turn the system over to the end users and focus on end-user training, installation, and support.

The four UP phases are different from the traditional SDLC because they do not define generic analysis, design, and implementation phases. Instead, they define the project sequentially by indicating the emphasis of the project team at any point in time. To make iterative development manageable, the UP defines disciplines within each phase. They include business modeling, requirements modeling, design, implementation, testing, deployment, configuration and change management, and project management. Each iteration involves activities from all disciplines. The UP also defines many roles played by developers and many models created during the project. Typical roles include designer, use case specifier, systems analyst, implementer, and architect.

As with any methodology, the UP includes very detailed information about what to do and when to do it for every activity involved in system development. The techniques and models presented in this book are consistent with many of the techniques and models included in the UP, but this book does not focus on the UP exclusively.

Agile Modeling

Many developers prefer the lightweight approach of XP, and many others recognize the need for a comprehensive methodology like the UP. Which approach to use depends on the nature of the project. Many developers using XP want to add more modeling and structure to the project. Many developers using the UP want to streamline the development process.

The latest approach, popularized by Scott Ambler, is referred to as *Agile Modeling*, which encourages developers to combine the best of XP with the best of the UP. It increases the amount of modeling over XP but decreases the formality and documentation required of the UP. Agile Modeling includes the following core practices:

1. **Iterative and Incremental Modeling.** Apply the right models, create several models in parallel, and model in small increments.
2. **Teamwork.** Model with others, get active stakeholder participation, encourage collective ownership, and display models publicly.
3. **Simplicity.** Create simple content, depict models simply, and use the simplest modeling tools.
4. **Validation.** Consider testability, then prove the model is right with code.

As you continue working through this text, you will see that the techniques and models presented can be used with an Agile Modeling approach, whether they are traditional techniques and models or object-oriented techniques and models.

TOOLS TO SUPPORT SYSTEM DEVELOPMENT

No matter which methodology you use, it is important to use automated tools to improve the speed and quality of system development work whenever possible. One type of tool discussed earlier is a CASE tool. CASE tools are specifically designed to help systems analysts complete system development tasks. Analysts use a CASE tool to create models of the system, many of them graphical models. But a CASE tool is much more than a drawing tool.

repository

a database that stores information about the system in a CASE tool, including models, descriptions, and references that link the various models together

CASE Tools

A CASE tool contains a database of information about the project, called a *repository*. The repository stores information about the system, including models, descriptions, and references that link the various models together. The CASE tool can check the models to make sure they are complete and follow the correct diagramming rules. The CASE tool also can check one model against another to make sure they are consistent. If you consider how much time an analyst spends creating models, checking them, revising them, and then making sure they all fit together, it is apparent how much help a CASE tool can provide. Figure 2-21 shows CASE tool capabilities surrounding the repository. If system information is stored in a repository, the development team can use the information in a variety of ways. Every time a team member adds information about the system, it is immediately available for everyone else.

FIGURE *2-21*

A CASE tool repository contains all information about the system.

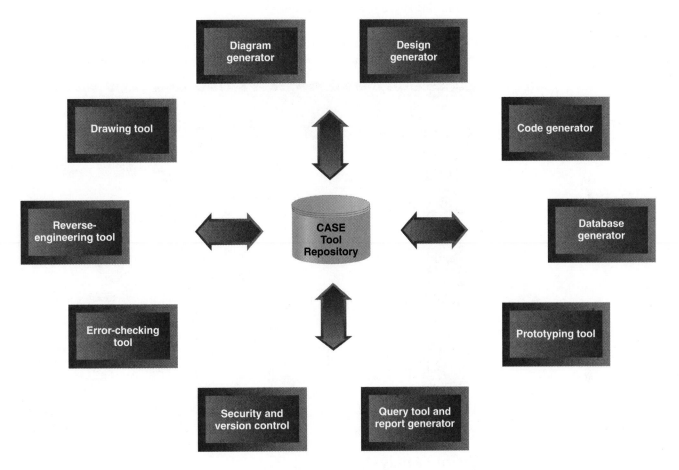

CASE tools are often categorized as upper CASE or lower CASE tools. Upper CASE tools provide support for analysts during the analysis and design phases—for example, creating and checking models and storing information in the repository. Lower CASE tools provide support for implementation, primarily generating programs and database schemas based on specifications in the repository. CASE tools that combine support for the full life cycle are called integrated CASE, or ICASE, tools.

Some CASE tools are designed to be as flexible as possible, allowing analysts to use any development approach they desire. Other CASE tools are designed for very specific methodologies. This section describes some of these tools—those specifically designed for one approach and those designed to be more flexible. Not all of these tools are called CASE tools by their vendors. For many developers and managers, CASE tools failed to meet expectations and fell into disuse. Vendors today refer to their tools as visual modeling tools, integrated application development tools, or round-trip engineering tools.

Microsoft Visio

Although not really a CASE tool, Visio is a drawing tool that analysts use to create just about any system model they might need. Visio comes with a collection of drawing templates that include symbols used in a variety of business and engineering applications. Software and system development templates provide symbols for flowcharts, data flow diagrams, entity-relationship diagrams, UML diagrams, and others found throughout this text. The templates provide a limited repository for storing definitions and descriptions of diagram elements, but Visio does not provide a complete repository for a system development project. Many system developers prefer the flexibility that Visio offers for drawing any diagram needed, however.

Figure 2-22 shows Visio displaying two drawings. One drawing shows several UML diagrams: a class diagram, a use case diagram, and a sequence diagram. The other drawing shows computer and networking symbols that can be used for designing a network.

FIGURE *2-22*

Visio for drawing a variety of diagrams and charts.

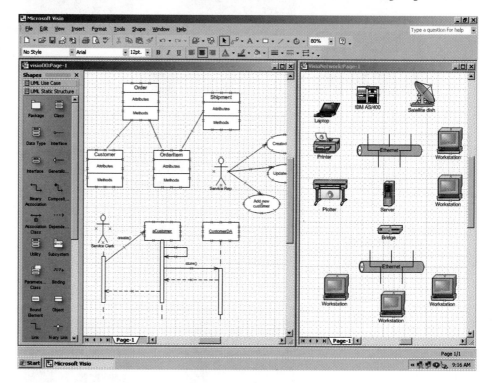

Visible Analyst

Figure 2-23 shows a flexible CASE tool called Visible Analyst from Visible Systems Corporation (www.visible.com). This tool makes it easy to draw typical traditional models, such as data flow diagrams and entity-relationship diagrams, and it also supports object-oriented UML models. Visible Analyst includes a repository for defining system components and provides error-checking and consistency-checking support. Like many vendors, Visible Systems Corporation has stopped using the term *CASE tool*. Visible Analyst is now called an *integrated application development tool*.

Oracle Designer

The Oracle Corporation (www.oracle.com) describes Oracle Designer as a tool set for recording definitions and automating the rapid construction of flexible, graphical, client-server applications. It is integrated with Oracle Developer, a development tool for creating GUI applications with an Oracle relational database. Oracle Designer includes a complete repository, diagramming capabilities, and code-generation capabilities. Oracle Designer is an example of an integrated CASE tool that supports the traditional approach to system development.

FIGURE *2-23*

Visible Analyst showing a data flow diagram and other traditional models.

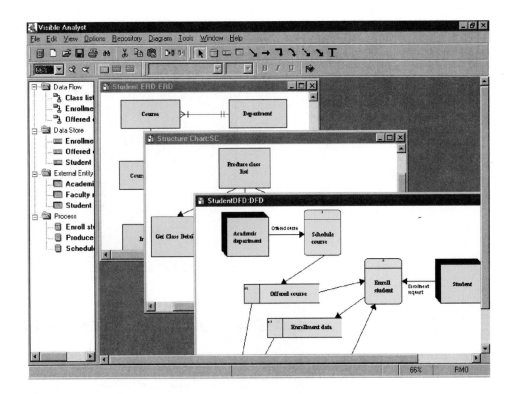

System Modeler, which is used for analysis, includes a process modeler, function-hierarchy diagrammer, data flow diagrammer, and entity-relationship diagrammer. The Design Transformer and Design Editor uses the produced diagrams, along with the detailed definitions in the repository, to create the database and application logic. Figure 2-24 shows the Oracle Designer Front Panel with all of the diagramming options available and a separate screen showing the Entity-Relationship Diagrammer tool being used to create a diagram.

Rational Rose

Rational Rose is a well-known tool from Rational Software, now part of IBM (www.ibm.com), that specifically supports the object-oriented approach. Rational Rose is referred to as a *visual modeling tool* rather than a CASE tool. Rational Rose can be used with the Unified Process (UP) or with any methodology that uses UML diagrams, and because Rational Rose has been so influential, many developers refer to the UP as the "Rational Rose" methodology. The tool provides reverse-engineering and code-generating capability, as well as a repository, and can be integrated with additional tools to provide a complete development environment. Rational Software also has other system modeling tools that are tightly integrated with specific IDEs, including Visual Studio .NET (see below). Figure 2-25 shows Rational Rose displaying some UML models.

Together

A recent advance is the concept of *round-trip engineering*. Because system development can be so iterative, particularly in the object-oriented approach, it is important to synchronize the graphical models (such as class diagrams) continually with generated program code. For example, if the analyst changes the program code, he or she must also update the class diagram. Likewise, changing the class diagram means updating the program code. Unlike ICASE tools that generate code from the graphical models, newer tools automate the process of synchronization in both directions (in a round trip).

FIGURE 2-24

Oracle Designer
(a) Front Panel screen showing requirements modeling, preliminary design
transformer, and design editor options.

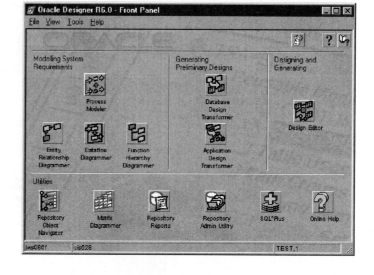

(b) Entity Relationship Diagrammer screen
showing a model.

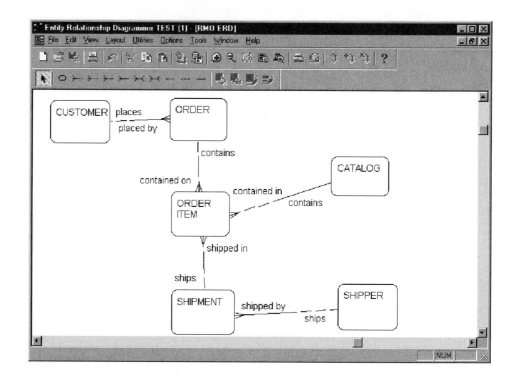

Peter Coad and TogetherSoft, now part of Borland (www.borland.com), pioneered round-trip engineering with the tool named Together. Together uses UML diagrams with several different OO programming languages to provide round-trip engineering support. Figure 2-26 shows Together with a class diagram and synchronized Java code. If the developer prefers to write the code to define the class, the class diagram is updated automatically. If the developer prefers to draw the class diagram first, the code to define the class is updated.

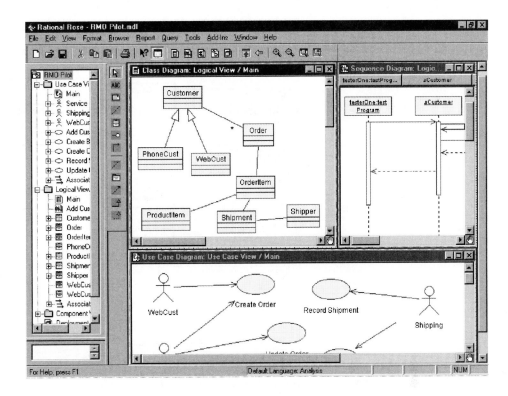

FIGURE *2-26*

The round-trip engineering tool
Together showing a class diagram
with synchronized Java source code.

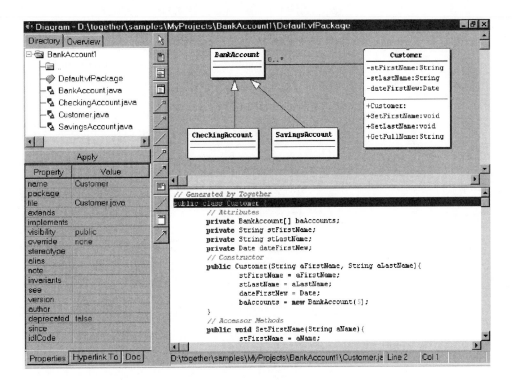

Embarcadero Describe

New products are appearing that include visual modeling and round-trip engineering
features. One example is Embarcadero Describe (www.embarcadero.com). The Enterprise edition of Describe features flexible UML modeling capabilities for analysis and
design, including round-trip engineering with Java (see Figure 2-27). Describe Developer can integrate with several Java development tools.

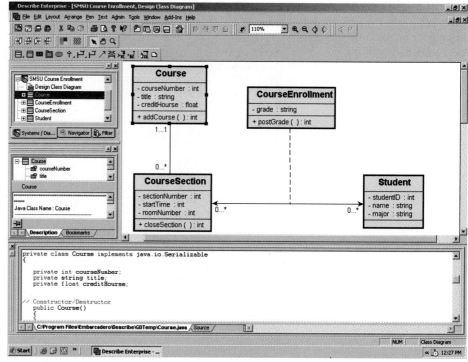

Courtesy of Embarcadero Technologies, Inc.

Rational XDE Professional

The latest approach to system modeling tools combines a full-featured modeling tool
and repository into the integrated development environment (IDE) of another vendor,
where program code is generated and refined. Rational XDE Professional, shown in Figure 2-28, is integrated with the Microsoft Visual Studio .NET IDE, allowing round-trip
engineering and code generation to become part of the IDE used by the developer. The
developer no longer has a CASE tool and a separate IDE. All the support functionality
the developer needs is included in one integrated tool.

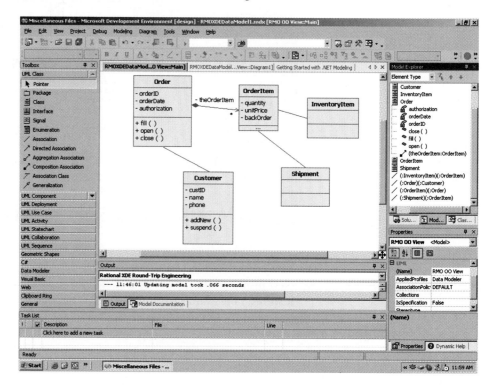

Summary

System development projects are organized around the systems development life cycle (SDLC), and phases of the SDLC include activities that must be completed for any system development project. The SDLC phases are project planning, analysis, design, implementation, and support. System developers learn the SDLC phases and activities sequentially, based on a waterfall approach; in practice, however, the phases overlap and projects contain many iterations of analysis, design, and implementation activities.

There are lots of ways to develop an information system. All development approaches use a systems development life cycle (SDLC) to manage the project, plus models, techniques, and tools that make up a system development methodology. A system development methodology provides guidelines to follow for completing every activity in the SDLC, and there are many different methodologies in use. Most methodologies are based on one of two approaches to information systems development: the traditional approach or the object-oriented approach.

Quite a few variations of the SDLC exist. The original life cycles were waterfall approaches with rigid starting and ending points for each phase. Most variations now use iteration across phases, in which some analysis, then some design, and then some implementation are completed, with the process repeating to refine the system. Some variations, called *participatory* or *user-centered design*, focus on people who use the system. Other variations, called *rapid application development (RAD)*, have the goal of speeding up development.

Some current trends in system development include the spiral model, eXtreme Programming (XP), the Unified Process (UP), and Agile Modeling. These approaches provide innovative insights into best practices in system development and are becoming influential.

Computer-aided system engineering (CASE) tools are special tools designed to help analysts complete development tasks, including modeling and generating program statements directly from the models. This chapter discussed some examples of CASE tools, which are now referred to as integrated application development tools, visual modeling tools, or round-trip engineering tools.

Key Terms

application, p. 40

analysis phase, p. 38

CASE tool, p. 46

class diagram, p. 56

data flow diagram (DFD), p. 52

design phase, p. 39

entity-relationship diagram (ERD), p. 52

help desk, p. 42

implementation phase, p. 40

incremental development, p. 44

information engineering, p. 54

model, p. 45

object, p. 54

object-oriented analysis (OOA), p. 55

object-oriented approach, p. 54

object-oriented design (OOD), p. 55

object-oriented programming (OOP), p. 55

phase, p. 36

planning phase, p. 37

problem domain, p. 39

project, p. 36

repository, p. 62

sociotechnical systems, p. 58

structure chart, p. 50

structured analysis, p. 52

structured approach, p. 49

structured design, p. 50

structured program, p. 49

support phase, p. 41

system development methodology, p. 45

systems development life cycle (SDLC), p. 36

technique, p. 46

tool, p. 46

top-down programming, p. 50

Unified Process (UP), p. 61

waterfall approach, p. 42

REVIEW QUESTIONS

1. What are the five phases of the SDLC?
2. How is the SDLC based on the problem-solving approach described in Chapter 1?
3. What is the objective of each phase of the SDLC? Describe briefly.
4. How is iteration used across phases?
5. What is the difference between a model and a tool?
6. What is the difference between a technique and a methodology?
7. Which of the two approaches to system development was the earliest?
8. Which of the two approaches to system development is the most recent?
9. Which of the traditional approaches focuses on overall strategic systems planning?
10. Which of the traditional approaches is a more complete methodology?
11. What are the three constructs used in structured programming?
12. What graphical model is used with the structured design technique?
13. What graphical model is used with the modern structured analysis technique?
14. What model is the central focus of the information engineering approach?
15. Explain what is meant by a waterfall life cycle approach.
16. What concept suggests repeating activities over and over until you achieve your objective?
17. What concept suggests completing part of the system and putting it into operation before continuing with the rest of the system?
18. What is user-centered and/or participatory design?
19. What is meant by rapid application development (RAD)?
20. What are some features of the spiral model approach to development?
21. What are some of the features of eXtreme Programming (XP)?
22. What are some of the features of the Unified Process (UP)?
23. What are CASE tools? Why are they used?
24. What are some newer terms used to describe CASE tools?

THINKING CRITICALLY

1. Write a one-page paper that distinguishes among the fundamental purposes of the analysis phase, the design phase, and the implementation phase.
2. Describe a system project that might have three subsystems. Discuss how three iterations might be used for the project.
3. Why might it make sense to teach analysis and design phases and activities sequentially, like a waterfall, even though in practice iterations are used in nearly all development projects?
4. List some of the models that architects create to show different aspects of a house they are designing. Explain why several models are needed.
5. What models might an automotive designer use to show different aspects of a car?
6. Sketch the layout of your room at home. Now write a description of the layout of your room. Are these both models of your room? Which is more accurate? More detailed? Easier to follow for someone unfamiliar with your room?
7. Describe a "technique" you use to help you complete the activity "Get to class on time." What are some "tools" you use with the technique?
8. Describe a "technique" you use to make sure you get assignments done on time. What are some "tools" you use with the technique?
9. What are some other techniques you use to help you complete activities in your life?
10. There are at least two approaches to system development, a variety of life cycles, and a long list of techniques and models that are used in some approaches but not in others. Consider why this is so. Discuss these possible reasons, indicating which are the most important: The field is so young; the technology changes so fast; different organizations have such different needs; there are so many different types of systems; and people with widely different backgrounds are developing systems.

EXPERIENTIAL EXERCISES

1. Go to the campus placement office and gather some information on companies that recruit information systems graduates on your campus. Try to find any information about the approach they use to develop systems. Is their SDLC described? Do any mention a CASE tool? Visit the company Web sites and see whether you can find any more information.
2. Visit the Web sites for a few leading information systems consulting firms. Try to find information about the approach they use to develop systems. Are their SDLCs described? Do their sites mention any CASE tools?

A "College Education Completion" Methodology

Like many readers of this book, you are probably a college student working on a degree. Think of completing college as a project—a big project, lasting many years and costing more than you might want to admit. Some students do a better job managing the college completion project than others. Many fail entirely (certainly not you), and most students probably complete college late and way over budget (again, certainly not you).

As with any other project, to be successful, you should follow some sort of "college education completion" methodology. That is, you should follow a comprehensive set of guidelines for completing activities and tasks from the beginning of planning for college through to the successful completion.

1.   What might be the phases of your personal college education completion life cycle?
2.   What are some of the activities of each phase?
3.   What are some techniques you use to help complete the activities? What models might you create during the process of completing college? Differentiate models you create that get you through college from those that help you plan and control the process of completing college.
4.   What are some of the tools you use to help you complete the models?

Factory System Development Project

Sally Jones is assigned to manage a new system development project that will automate some of the work being done in her company's factory. It is fairly clear what is needed: to automate the tracking of the work in progress and the finished goods inventory. What is less clear is the impact of any automated system on the factory workers. Sally has several concerns: How might a new system affect the workers? Will they need a lot of training? Will working with a new system slow down their work or interfere with the way they now work? How receptive will the workers be to the changes the new system will surely bring to the shop floor?

At the same time, Sally recognizes that the factory workers themselves might have some good ideas about what will work and what won't, especially concerning (1) which technology is more likely to survive in the factory environment and (2) what sort of user interface will work best for the workers. Sally doesn't know much about factory operations, although she does understand inventory accounting.

1.   Is the proposed system an accounting system? A factory operations system? Or both?
2.   Which life cycle variations might be appropriate for Sally to consider using?
3.   Which activities of analysis and of design discussed in this chapter should involve factory workers as well as factory management?

Rethinking Rocky Mountain Outfitters

Barbara Halifax wrote her boss that she was still considering many potential approaches to the customer support system development project. She is still completing the project planning phase, so not much time has passed at this point. Consider the training required for the development staff if RMO decides to use an object-oriented approach for the project. How extensive would the training needs be for the RMO staff? What type of training would be required? Is it just about new programming languages, or is it broader than that? How far can the project progress before the decision is made?

Barbara mentions that either approach can be used and that, even though some Web development is involved, the team does not have to use an OO approach. Do you think she is correct? Why or why not? Do some types of projects *require* an OO approach?

Barbara also mentions that she plans to use some iteration and to involve users extensively throughout the project. What life cycle variations are under consideration? What else might she do to speed up the development process? What else might she consider adapting from the spiral model approach, from eXtreme Programming, from the Unified Process, or from Agile Modeling?

Focusing on Reliable Pharmaceutical Service

In Chapter 1, you generated some ideas related to Reliable Pharmaceutical Service's five-year information systems plan. Management has placed a high priority on developing a Web-based application to connect client facilities with Reliable. Before the Web component can be implemented, though, Reliable must automate more of the basic information it handles about patients, healthcare facilities, and prescriptions.

Next, Reliable must develop an initial informational Web site, which will ultimately evolve into an intranet through which Reliable will share information and link its processes closely with its clients and suppliers. One significant requirement of the intranet is compliance with the Health Insurance Portability and Accountability Act of 1996, better known as HIPAA. HIPAA requires healthcare providers and their contractors to protect patient data from unauthorized disclosure. Ensuring compliance with HIPAA will require careful attention to intranet security.

Once basic processes are automated and the intranet Web site is in place, the system will enable clients to add patient information and place orders through the Web. The system should streamline processes for both Reliable and its clients. It should also provide useful query and patient management capabilities to distinguish Reliable's services from those of its competitors, possibly including drug interaction and overdose warnings, automated validation of prescriptions with insurance reimbursement policies, and drug and patient cost data and summaries.

1. One approach to system development that Reliable might take is to start one large project that uses a waterfall approach to the SDLC to thoroughly plan the project, analyze all requirements in detail, design every component, and then implement the entire system, with all phases completed sequentially. What are some of the risks of taking this approach? What planning and management difficulties would this approach entail?

2. Another approach to system development might be to start with the first required component and get it working. Later, other projects could be undertaken to work on the other identified capabilities. What are some of the risks of taking this approach? What planning and management difficulties would this approach entail?

3. A third approach to system development might be to define one large project that will use an iterative approach to the SDLC. Briefly describe what you would include in each iteration. Describe how incremental development might apply to this project. How would an iterative approach decrease project risks compared with the first approach? How might it decrease risks compared with the second approach? What are some risks the iterative approach might add to the project?

FURTHER RESOURCES

Some classic and more recent texts include the following:

Scott W. Ambler, *Agile Modeling: Effective Practices for Extreme Programming and the Unified Process.* Wiley Computer Publishing, 2002.

D. E. Avison and G. Fitzgerald, *Information Systems Development: Methodologies, Techniques and Tools* (2nd ed.). McGraw-Hill, 1995.

Kent Beck, *Extreme Programming Explained: Embrace Change.* Addison-Wesley Publishing Company, 2000.

Tom DeMarco, *Structured Analysis and System Specification*. Prentice Hall, 1978.

C. Gane and T. Sarson, *Structured Systems Analysis: Tools and Techniques*. Prentice Hall, 1979.

Ivar Jacobson et al., *Object-Oriented Software Engineering: A Use Case Driven Approach*. Addison-Wesley, 1992.

Ivar Jacobson, Grady Booch, and James Rumbaugh, *The Rational Unified Process*. Addison-Wesley, 1999.

James Martin, *Information Engineering: A Trilogy* (books 1, 2, and 3). Prentice Hall, 1990.

Steve McConnell, *Rapid Development*. Microsoft Press, 1996.

Meilir Page-Jones, *The Practical Guide to Structured System Design* (2nd ed.). Prentice Hall, 1988.

John Satzinger and Tore Orvik, *The Object-Oriented Approach: Concepts, System Development, and Modeling with UML* (2nd ed.). Course Technology, 2001.

Ed Yourdon, *Modern Structured Analysis*. Prentice Hall, 1989.

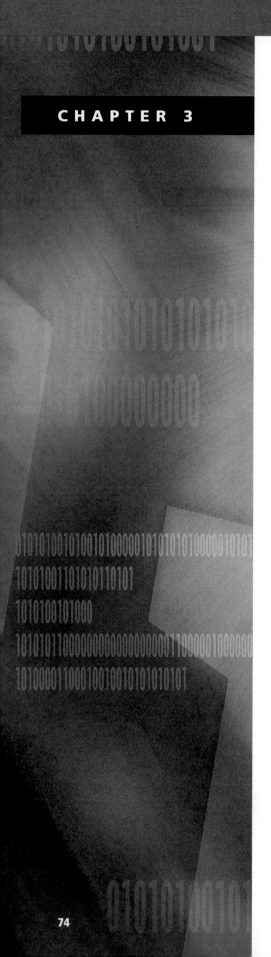

CHAPTER 3

The Analyst as a Project Manager

LEARNING OBJECTIVES

After reading this chapter, you should be able to:

- Explain the elements of project management and the responsibilities of a project manager

- Explain project initiation and the activities in the project planning phase of the SDLC

- Describe how the scope of the new system is determined

- Develop a project schedule using PERT and Gantt charts

- Develop a cost/benefit analysis and assess the feasibility of a proposed project

- Discuss how to staff and launch a project

CHAPTER OUTLINE

Project Management

Project Initiation and the Project Planning Phase

Defining the Problem

Producing the Project Schedule

Confirming Project Feasibility

Staffing and Launching the Project

Recap of Project Planning for RMO

Gary Johnson was just putting the final touches on his presentation for tomorrow's executive review meeting. He was the leader for a small team of three systems analysts who had been working frantically over the last month to plan the development of a new system and to prepare the report. The three had worked closely to gather the information, do the evaluation, and prepare the report and presentation prior to launching the project. He thought back to six weeks before, when he had first received word of this new assignment.

Gary had been the project manager on a successful development project that was in the final stages of testing before going live when he received the assignment to investigate the feasibility of a new system. He spent the first week after being assigned the new project turning over his current management responsibilities to his replacement. The second week he located two other experienced systems analysts to work with him on the new project. It took almost the whole week to revise their schedules so that they could work with him for the month it would take to develop the new project plan. Finally, after two weeks, he and his team were able to begin work in earnest.

The original objective of the new system was to support the sales and administration of retirement account investments for the Blue Sky Family of Mutual Funds. Blue Sky offers many different types of mutual funds to the public, including investments in IRAs (Individual Retirement Accounts). IRAs and Roth IRAs are a special breed of retirement account investments that have very strict information and tax-reporting requirements. Part of the strategic plan for Blue Sky was a project to replace the current information system supporting IRAs. The project was originally scheduled to begin in six months. However, new legislation had just opened a possibility for Blue Sky to offer new types of investment funds that provide savings for college education: state education savings plans and Coverdale educational savings accounts. These accounts also have very strict information and tax-reporting requirements, so they seemed to fit within the scope of the IRA account management system. To maintain its competitive position, Blue Sky needed to begin offering these new investment accounts by next January. Suddenly, the need for the new IRA and educational savings plan system was accelerated. The strategic plan was modified to indicate that development needed to begin immediately.

Even though senior management was almost certain to approve and fund the system, Gary still thoroughly evaluated the project's feasibility. His first activity had been to define the problem precisely and determine exactly what the scope of the project should be. Obviously, he couldn't estimate the size of the project unless he knew what the system needed to include, so he developed a list of business benefits of the new system and a detailed list of functions that the system had to provide. This list of functions, called the *system capabilities*, helped him begin to estimate the cost of developing the system as well as the expected financial benefits. He then began to develop a plan and schedule for the project based on the required system capabilities and some preliminary brainstorming the three team members had done to identify possible system alternatives.

As Gary worked on the schedule, he assigned one of the other analysts to confirm the feasibility of the project and to investigate potential problems that could cause difficulties in the development and deployment of the new system. The third analyst began calculating the costs and benefits to ensure that the new system was still economically feasible. All three team members had to coordinate their efforts because the results were interdependent.

Over the last several days, they had put the results into a final report and prepared for the executive presentation. It always made Gary nervous to present before the executives of the firm, especially for a project that had such high visibility and importance to the company. Although he was nervous, he was also confident. He and the team had done a thorough and professional job. He had already begun planning for his next task, which was to begin staffing the project to investigate the system requirements in much more detail.

OVERVIEW

Chapter 1 described the business environment, with its insatiable need for information systems in today's competitive and rapidly paced global economy. That chapter also discussed the job duties of the systems analyst, including the systems analyst's role in information technology (IT) and IT strategic planning. You also learned about the various types of information systems the analyst might develop and support. Chapter 2 introduced the system development life cycle (SDLC); the methodologies, models, tools, and techniques used to develop systems; and several approaches to system development that are used generally. This chapter begins to narrow the focus to teach the specifics of how an information system is developed within a company. The Rocky Mountain Outfitters (RMO) customer support system project is used as the specific example, and this chapter discusses the project planning phase of the SDLC for RMO. Because of its importance in information systems development and project planning, this chapter introduces the principles of project management. Project management encompasses the skills and techniques that are necessary to succeed in planning and managing the development of a new system. As a knowledgeable worker and problem solver, you will need both technical and management skills to be a contributing member of a system development team. This chapter provides you with the fundamentals of project management, and later chapters will further elaborate on key management principles associated with the various phases of the project.

The second section of the chapter discusses how information systems projects are initiated. Projects are started for two primary reasons. First, a project to develop a new information system may be started because the new system is part of an overall strategic plan, as discussed for RMO in Chapter 1. The second reason that new information system projects are started is to respond to an immediate business need. Such a need usually arises due to some unforeseen information or processing problem within the company.

A major objective of the chapter is also to describe the activities of the project planning phase of the SDLC. The planning process for a new project entails several important steps, such as defining the scope of the project, comparing the estimated costs and anticipated benefits of the new system, and developing a project schedule. The final sections will explain these specific steps and the skills associated with the steps. Because project management, analyzing costs and benefits, and project scheduling are all very large topics, additional information about each of these topics is included in appendices at the end of the book. You are encouraged to review the appendices for more information on these topics.

PROJECT MANAGEMENT

Even though every project team designates one person as the project manager who has primary responsibility for the functioning of the team, all experienced members contribute to the management of the team. This section focuses on project management from the perspective of the project manager, while recognizing that many management tasks are delegated to other team members. The project manager for the RMO customer support system project is Barbara Halifax, but she has one senior systems analyst helping her every step of the way. As the project proceeds, all team members are involved in aspects of managing the project.

There are many different definitions of *management*, and all are helpful in understanding how projects are managed. One common definition is that management is getting things done through other people. Project management is a special type of management. Remember this book's definition of *project*: a planned undertaking that has a beginning and an end and that produces a predetermined result or product. Based on this definition, we extend the definition of management with the following definition of project management: *Project management* is organizing and directing other people to achieve a planned result within a predetermined schedule and budget.

project management

organizing and directing other people to achieve a planned result within a predetermined schedule and budget

Project Success Factors

How important is project management for the success of a system development project? In 1994, the Standish Group began studying system development project success and failure. The surprising initial results indicated that almost 32 percent of all development projects were canceled before they were completed. In addition, more than half of computer system projects cost almost double the original budget. Less than half (about

42 percent) had the same scope and functionality as originally proposed. In fact, many systems were implemented with only a portion of the original requirements satisfied. Depending on company size, completely successful projects (on time, on budget, with full functionality) ranged from only 9 percent to about 16 percent. As of 2000, the percentage of successful system development projects was still a dismal 28 percent, with 72 percent cancelled or completed late, over budget, or with limited functionality. Clearly, system development is a difficult activity requiring very careful planning, control, and execution.

It is interesting to look at the reasons that projects do not fulfill the desired objectives. Some primary reasons why projects fail, or are only partially successful, include the following:

- Incomplete or changing system requirements
- Limited user involvement
- Lack of executive support
- Lack of technical support
- Poor project planning
- Unclear objectives
- Lack of required resources

Even though the studies of failed system development do not specifically identify weak project management as a cause for failure, most of these problems could be corrected with strong project management.

Additional studies of successful projects help to highlight the reasons why projects succeed. Some reasons for success are the following:

- Clear system requirement definitions
- Substantial user involvement
- Support from upper management
- Thorough and detailed project plans
- Realistic work schedules and milestones

The success factors are, in most cases, just the reverse of those for failures. Note that reasons such as "the technology is too complex" do not appear in the lists. What this omission indicates is that projects fail most frequently because project management has failed. Successful projects result from strong project management that ensures that the preceding components are part of the project.

Managing the SDLC

client

the person or group who funds the project

oversight committee

clients and key managers who review and direct the project

user

the person or group of persons who will use the new system

A project manager works with several groups of people. The *client* is the person or group of people who will be paying for the development of the new system—in other words, the customer. So, when we speak of project approval and release of funds, we mean they come from the client. For in-house development, the client can be an executive committee or a particular vice president who is funding the project. For large, mission-critical projects, an *oversight committee* (sometimes called the *steering committee*) may be formed. This committee consists of clients and other key executives who have a vision of the strategic direction of the organization and a strong interest in the project's success. Experience in directing large development projects is helpful but certainly not a prerequisite for a member of the committee. The *users*, on the other hand, are the people who will actually be using the new system. In some cases, the client and user are the same person. Often, however, they are not. The user typically provides information about the detailed functions and operations needed in the new system. The client also provides input on the business framework and strategy, which are important factors that influence the scope and design of a system. In addition, the client approves and oversees the project, along with its funding. Figure 3-1 depicts the various groups of people involved in a development project. Note that the project manager is the focal point of project participants.

FIGURE *3-1*

Participants in a system development project.

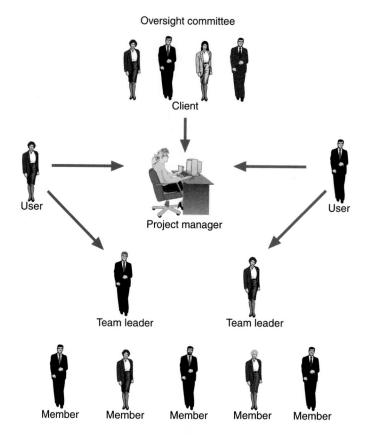

In Chapter 2, you learned that one way to organize the work of a development project is to divide it into phases based on the SDLC. You also saw that these phases help organize and categorize the project activities based on their focus. In other words, analysis phase activities are those that are focused on understanding and developing system requirements. Since the activities of the phases overlap and are completed iteratively, the purpose of breaking the project into phases, activities, and tasks is, not to force sequential steps for the project, but to serve as an organizing structure and provide a checklist of activities and tasks that need to be completed. This checklist ensures that the user needs are well understood and that all necessary components of the final system are designed. It also helps to focus both the users and the analysts on the correct objective for each task. For example, if a meeting is being held to finalize user requirements for a sales commission calculation, then all participants know that design decisions for the database structure should not be finalized during that meeting.

Figure 3-2 (originally shown in Chapter 2) illustrates the overlap of the SDLC phases. It also contains a set of tasks (along the top) called "Additional project management tasks" that spans the breadth of the other phases. This illustration demonstrates that project management tasks are done throughout the entire project—during all the phases. The objective of these project management tasks is to manage all development activities. Figure 3-3 identifies many of the project management tasks that are carried out during the system development project, not including the project management activities of the planning phase. As you look at the detailed project management tasks identified in the figure, you will notice that project management is not a simple process. The chapters throughout this book explain the project management tasks along with the other system development activities.

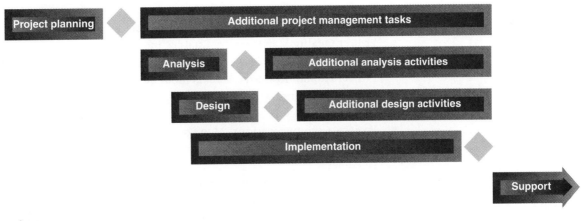

=Completion of major components of project

FIGURE *3-2*

Overlap of SDLC phases with
ongoing project management tasks.

A normal question you may have at this point is, "What makes a successful project manager?" or, "How can I learn to be a project manager?" Good project managers are very valuable to companies. In fact, companies report that one of their most severe shortages is good project managers. A good project manager is worth her weight in gold (and might even earn that much). A first step in becoming an effective project manager is to learn as much about project management concepts and techniques as possible. Appendix A contains a broad explanation of the major project management topics. The explanation of project management principles given in Appendix A is based on the Project Management Body of Knowledge (PMBOK) as developed by the Project Management Institute. Included in this body of knowledge are the principles for the following areas of project management:

FIGURE *3-3*

Project management tasks
corresponding to phases of the SDLC.

- **Project Scope Management**. Defining and controlling the functions that are to be included in the system as well as the scope of the work to be done by the project team
- **Project Time Management**. Building a detailed schedule of all project tasks and then monitoring the progress of the project against defined milestones

PROJECT MANAGEMENT TASKS		
Analysis phase	**Design phase**	**Implementation phase**
-Monitor and control scope	-Monitor and control scope	-Monitor and control scope
-Monitor and control progress	-Monitor and control progress	-Monitor and control progress
-Update schedule	-Monitor and control budget	-Monitor and control budget
-Conduct status reviews	-Conduct status reviews	-Conduct status reviews
-Organize teams	-Coordinate team member training	-Reorganize team assignments
-Provide leadership for teams	-Track open issues	-Coordinate with users/clients
-Coordinate with users/clients	-Encourage/lead team members	-Track testing and quality
-Evaluate risks	-Monitor technical problems	-Take corrective action
-Plan in detail design phase	-Reorganize team assignments	-Coordinate data conversion
-Make presentations	-Monitor subcontractors/vendors	-Conduct system installation
	-Plan in detail implementation phase	-Conduct postimplementation review

- **Project Cost Management.** Calculating the initial cost/benefit analysis and later updates and monitoring expenditures as the project progresses
- **Project Quality Management.** Establishing a total plan for ensuring quality, which includes quality control activities for every phase of the project
- **Project Human Resource Management.** Recruiting and hiring project team members; also training, motivating, team building, and implementing related activities to ensure a happy, productive team
- **Project Communications Management.** Identifying all stakeholders and key communications to each; also establishing all communications mechanisms and schedules
- **Project Risk Management.** Identifying and reviewing throughout the project all potential risks for failure and developing plans to reduce these risks
- **Project Procurement Management.** Developing requests for proposals, evaluating bids, writing contracts, and then monitoring vendor performance

There are many ways that project management tasks can be categorized. These eight knowledge areas form one effective technique to consider the various responsibilities of a project manager and better understand the complexity of project management. In the end, however, a project manager must integrate all of these individual tasks into a cohesive whole. Appendix A discusses each of these knowledge areas in detail, identifies tasks associated with each area, and links these tasks to the respective SDLC phases. Studying Appendix A should provide you with a strong background in the major responsibilities of a project manager.

As you progress in your career, you would be wise to keep a record of project management skills you observe in others as well as those you learn by your own experience. One place to start is with the set of skills described in Chapter 1 for a systems analyst. However, good project management requires more than just those skills. A good project manager knows how to develop a plan, execute it, anticipate problems, and make adjustments for variances. Project management skills *can* be learned. Those who aspire to managing projects are proactive in self-improvement and learn the necessary skills. Build on what you learn in this textbook and continue to practice and hone your project management skills.

PROJECT INITIATION AND THE PROJECT PLANNING PHASE

Information system development projects are initiated for various reasons. Three general driving forces are as follows: (1) to respond to an opportunity, (2) to resolve a problem, and (3) to conform to a directive.

Most companies are continually looking for ways to increase their market share or to open up new markets. One way they create opportunities is through strategic plans, both short term and long term. In many ways, planning is an optimal way to identify new projects. The benefit of this approach is that it provides a more stable and consistent environment in which to develop new systems. As the strategic plans are developed, projects are identified, prioritized, and scheduled. Projects initiated through strategic planning are sometimes described as *top-down* projects.

weighted scoring

a method to prioritize projects based on criteria with unequal weights

To prioritize these projects, companies use a technique called *weighted scoring*. First, the IT strategic planning committee identifies a set of criteria to judge the importance of new projects. Examples of criteria are "opens a new market" or "provides a high net present value." These criteria are weighted for their importance, and each potential project is rated according to the set of criteria. Projects with the highest scores are given priority for initiation.

Projects are also initiated to solve an immediate business problem. These projects attempt to close the gap between what information processing is required to run the business correctly and what is currently in operation. They can be initiated as part of a strategic plan but more commonly are requested by middle managers to resolve some difficulty in company operations. Obviously, senior executives are also aware of internal problems and can initiate projects to solve them. Sometimes these needs are so critical that they are brought to the attention of the strategic planning committee and integrated into the overall business strategy. At other times, an immediate need cannot wait, such as a new sales commission schedule or a new report needed to assess productivity. In these cases, individual managers of business functions will request the initiation of individual development projects.

Finally, projects are initiated to respond to outside directives. One common outside pressure is legislative changes that require new information-gathering and external reporting requirements, such as changes in tax laws and labor laws. For example, regulations in the Health Insurance Portability and Accountability Act (HIPAA) are intended to safeguard patients' medical information. This act affected Reliable Pharmaceutical Service, as discussed in the case at the end of Chapter 2. Legislative changes can also expand or contract the range of services and products that an organization can offer in a market. New regulations and laws can affect the strategic plan, resulting in an expedited need for new systems. We have seen many regulatory changes in the telecommunications industry, with cable TV and telephone companies vying for opportunities to provide cellular services, Internet access, and personalized entertainment.

Several steps normally occur with the initiation of a new project. A project charter is typically identified, which describes the purpose of the new system, the potential start and completion dates, and, very importantly, the key stakeholders and sponsors of the new system.

Whatever the reason for project initiation, it usually requires an initial review to ensure that the benefits outweigh the costs and risks of development. Thus, the first activities of almost every project after it is approved are those that precisely define the business problem, determine the scope of the project, and perform a feasibility analysis, including a cost/benefit analysis. We group all of these initial planning activities together in the project planning phase of the SDLC.

Initiating the Customer Support System for Rocky Mountain Outfitters

As described in Chapter 1, RMO senior executives, with help from an outside consulting firm, had developed a well-thought-out information systems strategic plan. The plan included both a technology architecture component and an application architecture component. Implementation of the plan had begun with the initiation of the supply chain management (SCM) project. The company founders, John and Liz Blankens, understood clearly that to maintain good customer relations, they needed to have systems in place to support the fulfillment of sales as they moved to broader geographical and Internet-based sales. The company would not realize the full benefit of the SCM system until the customer support system (CSS) also came on-line. The SCM system would provide several cost-reduction efficiencies, but RMO expected the real business benefit to come from a dramatic increase in sales from the expanded sales and marketing capabilities of the new CSS.

The supply chain management system was well under way. The project was to be implemented in several increments, since several of RMO's suppliers would also have to upgrade their systems. The first increment was on schedule, the requirements had been finalized, the overall architectural design was firm, and pieces of the new system were expected to be ready early the next year. John Blankens was really excited about the progress and was anxious to get started with the new customer support system. He called a special meeting of the company's executive committee to assess the progress of the current projects and to evaluate the possibility of moving ahead with the new CSS. Prior to the meeting, he asked VP of Finance and Systems JoAnn White to bring a detailed financial analysis of current system budgets and projections of the financial impacts that RMO could expect from beginning the CSS project in the near future. He also invited Chief Information Officer Mac Preston to evaluate the workload of the system development staff and the availability of staff to begin. Several other assignments were given to committee members to consider his proposal carefully.

After a long discussion, the executive committee decided that it was not only feasible to begin the project now, but critical to do so. Other retailers had proven that Internet sales and marketing, if planned and executed correctly, could provide tremendous benefits to a company. Even though there had been several jerks and sputters, e-commerce was here to stay. It was imperative for future viability that RMO, like other brick-and-mortar retailers, also have a strong presence on the Internet.

As a result of the meeting, the committee directed Mac to start the project. First, he met with Director of System Development John MacMurty and asked him to finalize his plans for a project manager and another experienced systems analyst to get the project started. He also asked John to produce a project charter to confirm the decisions made by the executive committee. John began contacting executives to elicit their participation as members of the oversight committee. He understood well that if he could get strong commitment from senior executives in the company, he would ensure good user involvement in the project. One of the key elements of successful projects is to get broad involvement from the users. After a couple of days of discussion with executives throughout the company, the oversight committee was complete. Vice president of marketing and sales William McDougal, who had requested to be the project sponsor since it supported his area directly, was the committee chair. Other members were Robert Schneider, director of catalog sales, Brian Haddock from operations, and, of course, John Blankens and Mac Preston.

The project began with the assignment of Barbara Halifax to the project full time as the project manager, as indicated in the RMO memo in Chapter 1. Barbara has been with RMO for several years. Prior to joining RMO, she worked for the information systems consulting division of one of the large accounting firms. Her experience in consulting gave her broad exposure to many different companies and systems. Senior management in RMO had complete confidence in her abilities to manage the CSS project. Steven Deerfield, a senior systems analyst, was also assigned to the project. Deerfield and Halifax had worked together before and had very compatible work styles. Since this project was a critical component of RMO's long-range strategic plan, two of the very best systems analysts in the company were assigned. Figure 3-4 illustrates the project charter, which documents the preliminary activities to get the project initiated.

As described in Chapter 1, the primary objective of the system is to support RMO's objective of building customer loyalty and of providing all the necessary tools for customer relationship management. The system is to further this objective by supporting all types of customer services—including ordering, returns, and on-line catalogs—for the ongoing telephone sales and a new capability with Internet sales. Customers not only must have access to the on-line catalog of RMO products either via a telephone sales representative or the Internet but must also be able to see their past purchasing history. Managers at RMO would like the system to include several "bells and whistles" to support their vision of RMO customer service.

The following section describes the activities of the project planning phase, using examples from the RMO project. As explained previously, these activities are project management activities to plan, to organize, to schedule, and finally to obtain approval for

FIGURE *3-4*

RMO project charter.

| Project Name: | ***Customer Support System*** |

Project Purpose: To provide increased level of customer support. Should include all customer-related functions from order entry to arrival of the shipment, including customer inquiries/catalog, order entry, order tracking, shipping, back order, returns, and sales analysis.

Anticipated Completion: Within 18 months of project initiation

Approved Budget: Up to $1,500,000

Key Participants:

Participant	Position	Primary responsibilities
Barbara Halifax	Project manager	Manage the entire project
John MacMurty	Director	Supervise project manager Check status weekly Serve on oversight committee
Mac Preston	Chief information officer (CIO)	Serve on oversight committee
William McDougal	Senior VP marketing/sales	Direct project sponsor Approve budget, schedule Serve on oversight committee
Robert Schneider	Director of catalog sales	Serve on oversight committee Provide user support/resources
Brian Haddock	Director of operations	Serve on oversight committee Provide user support/resources
Jason Nadold	Manager of shipping	Provide user support/resources

the project. Note that even though this project seems to have tacit approval from senior management, it must meet the rigorous evaluation criteria of all RMO projects. Even though only two members of the team have been assigned at this point, Barbara and Steve have extensive experience and excellent project management skills.

The Project Planning Phase

The project planning phase of the SDLC, as depicted in Figure 3-5, consists of the activities that are required to get the project organized and started. As discussed in Chapter 2, these activities are:

- Define the problem
- Produce the project schedule
- Confirm project feasibility
- Staff the project
- Launch the project

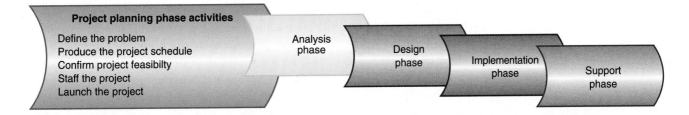

Project planning phase activities

Define the problem
Produce the project schedule
Confirm project feasibilty
Staff the project
Launch the project

Analysis phase

Design phase

Implementation phase

Support phase

FIGURE *3-5*

Activities of the project planning phase.

These activities are essentially project management activities. The project planning phase is usually staffed with only two or three highly experienced systems analysts, one of whom serves as the project manager. The other systems analysts assigned to the team are experienced developers with strong analytical skills, as well as experience in managing and controlling projects. The first team members that are assigned frequently become the core team leaders around which the rest of the team is built. At the successful conclusion of this phase, the project will have begun with resources, schedules, and a budget.

To help you learn about the activities of the project planning phase, the following sections each describe a project planning phase activity and then show how it applies to RMO. Figure 3-6 lists each activity with the key question the project team tries to answer when completing the activity. For example, at the end of the project planning phase, the key question to answer for the *Launch the project* activity becomes, *are we ready to start the project?*

FIGURE 3-6

The activities of the project planning phase and their key questions.

Project planning phase activities	Key questions
Define the problem	Do we understand what we are supposed to be working on?
Produce the project schedule	Can the project be completed on time given the available resources?
Confirm project feasibility	Is it still feasible to begin working on this project?
Staff the project	Are the resources available, trained, and ready to start the project?
Launch the project	Are we ready to start?

DEFINING THE PROBLEM

Carefully defining the problem is one of the most important activities of the project. The objective is to define precisely the business problem to be solved and thereby determine the scope of the new system. This activity defines the target that you want to hit. If the target is ill defined, then all subsequent activities will lack focus. As pointed out earlier, one of the primary causes of project failure is an unclear objective.

The first task within this activity is to review the business needs that originally initiated the project. As with RMO, if the project was initiated as part of the strategic plan, then the planning documents are reviewed. If the project originated from departmental needs, then key users are consulted to help the project team understand the business need. As the needs are identified, the team also develops a detailed list of the expected benefits. We define those as the *business benefits*. The list of business benefits contains the results that the organization anticipates it will accrue from a new system. Business benefits are normally described in terms of the influences that can change the financial statements, either by decreasing costs or increasing revenues.

business benefits

the benefits that accrue to the organization; often measured in monetary terms

The second task in this activity is to identify, at a high level, the expected capabilities of the new system. The objective is to define the scope of the problem in terms of the requirements of the information system that can solve the problem. Although at first defining the expected capabilities may not appear to be defining the problem, it is necessary to understand the scope of the new system and hence of the project.

Members of the development team combine these three components—the problem description, the business benefits, and the system capabilities—to get a *system scope document*. These members (for example, the systems analysts) work with the users and the client to develop this document. Sometimes this document is combined with the project charter; in other cases, it is independent. Figure 3-7 is an example of the system scope document for RMO. Note the differences between the business benefits and the system capabilities. The business benefits focus on the financial benefit to the company. The system capabilities focus on the system itself. The benefits are achieved through the capabilities provided by the system.

system scope document

a document—containing description, business benefits, and system capabilities—to help define the scope of a new system

Problem Description

Catalog sales began in Rocky Mountain Outfitters as a small experiment that soon developed into a rapidly growing division of the company. Support was initially provided by manual procedures with some simple off-the-shelf programs to assist in order taking and fulfillment. By 2004, the growth of catalog sales, including Internet sales, was stretching the abilities of the current system. As a result of a long-term strategic plan, RMO decided to initiate two major system development projects. The first, the supply chain management (SCM) system, was started in 2004 and is progressing on schedule. The second identified system is a customer support system (CSS) to provide sales, marketing, and a full range of customer support functionality. This project is an integral part of the total long-term strategic plan of RMO to continue to grow and maintain its leadership position in the sportswear industry.

Anticipated Business Benefits

The primary business benefit to be obtained from the new system is for RMO to maintain its leadership position in the sportswear industry. More immediate benefits will:

- ◆ Reduce errors caused by manual processing of orders
- ◆ Expedite order fulfillment due to more rapid order processing
- ◆ Maintain or reduce staffing levels in mail-order and phone-order processing
- ◆ Dramatically increase Internet sales through a highly interactive Web site
- ◆ Increase turnover by tracking sales of popular items and slow movers
- ◆ Increase level of customer loyalty through extensive customer support and information

System Capabilities

To obtain the business benefits listed above, the customer support subsystem shall include the following capabilities:

- ◆ Be a high-support system with on-line customer, order, back-order, and returns information
- ◆ Support traditional telephone and mail catalog sales with rapid-entry screens
- ◆ Include Internet customer and catalog sale capability, including purchase and order tracking
- ◆ Maintain adequate database and history information to support market analysis
- ◆ Provide a history of customer transactions for customer query
- ◆ Be able to handle substantial increases in volume (300% or more) without degradation
- ◆ Support 24-hour shipment of new orders
- ◆ Coordinate order shipment from multiple warehouses
- ◆ Maintain history to support analysis of sales and forecasting of market demand

FIGURE *3-7*

System scope document for the RMO customer support system.

proof of concept prototype

a very preliminary prototype built to illustrate that a solution to a business need is feasible

context diagram

a data flow diagram (DFD) showing the scope of a system

At times, especially when the new system is an attempt to push the state of the art, it may be necessary to build a preliminary prototype as a proof of the concept. New solutions, particularly those based on new technology, may not be well accepted or well understood. In that situation, the project team can build a *proof of concept prototype* to illustrate that a solution is possible and feasible. When a proof of concept is necessary, the project scope document will refer to the results of the initial prototype's construction, test, and fitness for purpose. For example, RMO senior management may want the system to automatically suggest complementary accessories for Web customers who purchase items over the Internet. The project team may need to build some prototypes to verify that the request is technologically feasible.

Frequently the project team also develops a diagram to describe the scope of the system in terms of information flowing into and out of the system. This diagram, which is called the *context diagram*, shows the primary users of the system and the information that is exchanged between them and the system. Figure 3-8 is an example of a simplified data flow diagram (DFD) called the context diagram for the RMO customer support system. Since at this point the project team is only documenting the scope of the new system, the diagram includes only major information requirements. In fact, the diagram focuses primarily on output information from the system. Chapter 6 provides a more detailed explanation and example of the RMO context diagram. Chapter 7 explains an object-oriented system overview diagram, called a *use case diagram*.

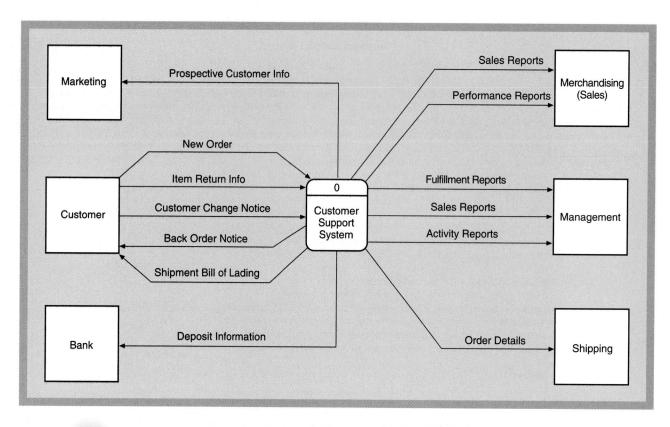

FIGURE *3-8*

Context diagram for the
customer support system.

The box with rounded corners in the middle of the diagram represents the customer support system itself. The boxes around the oval are the entities that provide information to the system or that receive information from the system. The lines with arrowheads are the major inputs to and outputs from the system. This diagram identifies only the major information flows into and out of the system. The objective is to get an overview of a proposed solution and not get involved in the details.

Note that the context diagram is also used during the analysis phase. During the project planning phase, the diagram helps define the scope of the problem. This diagram becomes a starting point for the more detailed investigation done during analysis.

Defining the scope carefully is important for establishing an estimate of the amount of effort required to complete the project. The size or scope of the system determines the amount of effort, which then determines the time and cost of the project. Several techniques can be used to measure the size or scope of a proposed system, although accurate estimates are difficult to achieve. Most approaches count the number of activities or use cases the system is required to support. Some approaches count the number of *function points* that can be identified. One technique, called the COnstructive COst MOdel (COCOMO), attempts to count function points as the number of inputs and outputs, the number of files maintained, the number of updates required, and so on.

The key question to be answered when completing the problem definition activity is, *do we understand what we are supposed to be working on?*

Defining the Problem at RMO

Barbara and Steve, the CSS project team, developed the lists for the system scope document after talking to William McDougal, vice president of marketing and sales, and his assistants. Chapter 4 explains more about interviewing users and eliciting important information. The key point to note here is that it is essential to obtain information from and to involve people who will be using the system and who will obtain the most benefit from it. They provide valuable insights to ensure that the system satisfies the business needs. As noted previously, the most critical element in the success of a system development project is user involvement.

One additional task is required to complete the problem definition activity. The project team conducts a preliminary investigation of alternative solutions to reassess the assumptions the team made when the project was initiated. Since the schedule and budget for the remainder of the project inherently assume a particular approach to developing the system, it is critical to make those implicit assumptions explicit so that all participants understand the constraints on the project schedule and the team can perform an accurate feasibility analysis.

For example, if an "off-the-shelf" program is identified as a possible solution, then part of the schedule during the analysis phase must include tasks to evaluate the program against the needs being researched. If the most viable solution appears to be a new system developed completely in-house, then detailed analysis tasks are planned and scheduled.

While Barbara was finishing the problem definition statement, Steve did some preliminary investigation of possible solutions. He researched the trade magazines, the Internet, and other resources to determine whether sales and customer support systems could be bought and installed rapidly. Although he found several, none seemed to have the exact match of capabilities that RMO needed. He and Barbara, along with William McDougal, had several discussions about how best to proceed. They decided that the best approach was to proceed with the analysis phase of the project before making any final decision about solutions. They would revisit this decision, in much more detail, after the analysis phase activities were completed. For now, Barbara and Steve began developing a schedule, budget, and feasibility statement for the new system.

PRODUCING THE PROJECT SCHEDULE

Before discussing the details of a project schedule, let's clarify three terms: *task*, *activity*, and *phase*. Fundamentally, a *phase* is made up of a group of related activities, and an *activity* is made up of a group of related tasks. A *task*, then, is the smallest piece of work that is identified and scheduled. Activities are also identified, named, and scheduled. For example, suppose that you are scheduling the design phase. Within the design phase, you identify activities such as *Design the user interface*, *Design and integrate the database*, and *Complete the application design*. Within the *Design the user interface* activity, you might identify individual tasks such as *Design the customer entry form* and *Design the order-entry form*. Thus, the phase, activity, and task breakdown provides a three-level hierarchy.

During the project planning phase, it may not be possible to schedule every task in the entire project because it is too early to know all of the tasks that will be necessary. However, one of the requirements of the project planning phase is to provide estimates of the time to complete the project and the total cost of the project. Since one of the major factors in project cost is payment of salaries to the project team, the estimate of the time and labor to complete the project becomes critical. The activity of developing the project schedule is one of the most difficult endeavors of the planning phase, yet it is one of the most important. The development of a project schedule is divided into two main steps:

- Develop a work breakdown structure
- Build a PERT/CPM chart

Developing a Work Breakdown Structure

work breakdown structure (WBS)

the hierarchy of phases, activities, and tasks of a project; one method to estimate and schedule the tasks of a project

A *work breakdown structure (WBS)* is a list of all the individual tasks that are required to complete the project. It is essential in planning and executing the project because it is the foundation for developing the project schedule, for identifying milestones in the schedule, and for managing cost. Figure 3-9 is an example of a work breakdown structure for the planning phase of the RMO project.

Phases, activities, and tasks			Duration in days	Number of resources	Predecessor task
1.0 Project Planning Phase					
	1.1 Define the problem				
		1.1.1 Meet with users	2	2	0.0.0
		1.1.2 Determine scope	1	2	1.1.1
		1.1.3 Write statement of business benefits	1	½	1.1.2
		1.1.4 Write statement of need	1	½	1.1.2
		1.1.5 Define statement of system capabilities	1	½	1.1.2
		1.1.6 Develop context diagram	1	½	1.1.2
	1.2 Produce the project schedule				
		1.2.1 Develop work breakdown schedule	2	2	1.1.3, 1.1.4, 1.1.5, 1.1.6
		1.2.2 Estimate resources, durations, and predecessors	1	2	1.2.1
		1.2.3 Develop PERT chart and Gantt chart	2	2	1.2.2
	1.3 Confirm project feasibility				
		1.3.1 Identify intangible costs and benefits	1	2	1.2.3
		1.3.2 Estimate tangible costs	1	2	1.2.3
		1.3.3 Estimate tangible benefits and do cost/benefit	2	2	1.3.1, 1.3.2
		1.3.4 Evaluate organizational and cultural feasibility	1	1	1.3.3
		1.3.5 Evaluate technical feasibility	2	1	1.3.3
		1.3.6 Evaluate schedule feasibility	1	2	1.3.3
		1.3.7 Evaluate resource availability	1	1	1.3.3
	1.4 Staff the project				
		1.4.1 Develop a project resource plan	1	2	1.3.4, 1.3.5, 1.3.6, 1.3.7
		1.4.2 Identify and request tech staff	1	1	1.4.1
		1.4.3 Meet with users, identify staff	1	1	1.4.1
		1.4.4 Organize project team	1	1	1.4.2, 1.4.3
		1.4.5 Conduct team-building exercises	3	2	1.4.4, 1.5.4
	1.5 Launch the project				
		1.5.1 Prepare presentation materials	1	1	1.3.7
		1.5.2 Make executive presentation	1	1	1.5.1
		1.5.3 Set up project facilities and support resources	3	2	1.5.2
		1.5.4 Conduct official kickoff meeting	1	1	1.4.4, 1.5.3

FIGURE *3-9*

Work breakdown structure for the planning phase of the RMO project.

Notice how this figure is based on the list of activities in the project planning phase of the SDLC. Each activity is further divided into individual tasks to be completed. The WBS identifies a hierarchy, much like an outline for a paper. The project requires a WBS for each phase of the SDLC, and the project planning phase WBS is shown here because we discuss these activities in detail in this chapter. Obviously, the analysis, design, and implementation phase WBSs would be even more important to define, since the project planning phase attempts to schedule the entire project.

Each task has a duration and list of the number of resources required for its completion. As explained in Appendix B, the effort for a task includes the product of the duration, the number of resources, and the percentage of time that the resource is active on that task. In other words, a resource may spend only half time on a two-day task, which results in only one day of effort. Although not shown in this diagram, sometimes three values of the duration are provided: an expected duration, a pessimistic duration, and an optimistic duration. Having various values permits the project manager to develop an optimistic, expected, and pessimistic schedule. However, let's keep our example simple.

It is quite difficult to develop a work breakdown structure from scratch. The project team could meet to brainstorm and try to think of everything that it needs to complete the project. This meeting is a type of bottom-up approach—just brainstorm for every single task and hope you cover everything. Usually, however, one of two techniques is used as a starting point. These techniques are called *standards-based WBS* and *analogy-based WBS*.

A standards-based WBS is built from a standard plan. The example given in Figure 3-9 can be called a standards-based WBS. Based on the standard set of activities defined for the planning phase, the project team expanded each activity into more detailed tasks. The specific methodology and the SDLC in use by the organization play an important role. The key is to have a list of activities and tasks that are predefined for this type of project. Companies that do work for the U.S. government frequently have standard WBS lists for various types of projects. Government contracts frequently require a standard project definition. The project team begins with the standard and then builds on it with the additional tasks that are unique to the particular project.

An analogy-based WBS is similar in form to a standards-based WBS, except that it uses different input as a starting point. In analogy-based WBS, the project team will try to identify another project, an analogous project, that is as similar as possible to the project at hand and is, of course, complete. Then, based on the experience from that previous project, the team will build the new WBS. Companies that do the same types of projects over and over again use this technique. For example, a car producer such as General Motors knows all of the steps required for a project to engineer a new car. Such companies use the same set of procedures that they have successfully used previously.

The major benefit of a good WBS is that it provides the most accurate estimate for the duration and effort required for the project. Time estimates are more accurate if they are developed at a detailed level rather than a "guesstimate" of the major processes. Schedules built from a good WBS are also much easier to monitor and control. The WBS is key to a successful project.

Building a PERT/CPM Chart

PERT stands for *Project Evaluation and Review Technique*, and CPM stands for *Critical Path Method*. Originally they were two distinct techniques, but more recently they have merged into a single scheduling technique. Appendix B details the concepts of a *PERT/CPM* chart and the way to develop one. You learn the way to create one manually and also a little about Microsoft Project, a scheduling software program. Appendix B is available for download at www.course.com via the "Student Downloads" link on the Web page for this book. Almost all live projects use a tool such as MS Project or Primavera to maintain a project schedule. Since every project is a dynamic activity, the schedule changes frequently. Maintaining a schedule by hand would be very time-consuming.

A PERT/CPM is a diagram of all the tasks identified in the WBS, showing the sequence of dependencies of the tasks. An example of a PERT/CPM chart for RMO is given in Figure 3-10. This example, which is only a partial example of the RMO schedule, illustrates the basic ideas of a PERT/CPM chart. Each rectangle represents a single task or a summary activity. Within each rectangle is the name of the task, a unique identifier, the duration, and the beginning and end dates. The task begins at the start of the date listed on the left side of the rectangle and finishes at the end of the date listed on the right side. The connecting lines with arrows indicate the sequence of the tasks. As you look at the diagram, you will note that some tasks are done sequentially, one after the other, and other tasks are done in parallel, or at the same time. The arrows represent task dependencies and indicate the normal sequence of carrying out a project.

By showing which tasks can be done concurrently, a PERT chart also assists in assigning staff. It is always a juggling act to balance the availability and workload of the team members with the dependent and independent tasks.

Building the PERT/CPM chart begins with the list of activities and tasks developed in the WBS. The WBS is analyzed, including the duration and expected resources for each task, to determine the dependencies. For each task, the chart identifies all the immediate

PERT/CPM

a technique for scheduling a project based on individual tasks or activities

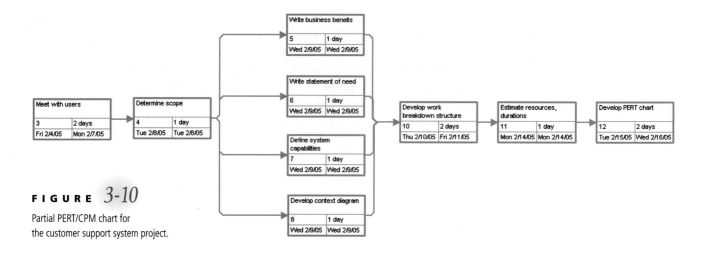

FIGURE *3-10*

Partial PERT/CPM chart for
the customer support system project.

critical path

a sequence of tasks that cannot be
delayed without causing the project to
be completed late

Gantt chart

a bar chart that represents the tasks
and activities of the project schedule

tracking Gantt chart

a type of Gantt chart that indicates the
current date and percentage of work
completed for each task

predecessor tasks and the successor tasks. As you can imagine, completing this sequencing task is quite difficult. Normally for any given project, there are myriad ways to sequence the tasks. For example, should the team design the customer entry form first or the product definition form? Either will work. But decisions must be made concerning the most probable progression of the project. Unfortunately, MS Project, or any other scheduling tool, cannot make such decisions for you. Automation does not help in this step.

The *critical path* is the longest path through the PERT/CPM diagram and contains all the tasks that must be done in the defined sequential order. It is called the critical path because if any task on the critical path is delayed, the entire project will be completed late. Project managers usually monitor the tasks on the critical path very carefully.

One of the benefits of the PERT/CPM technique is that it produces a diagram that makes it easy to see dependencies and the critical path. However, it is not easy to view the project's progress on a PERT chart. Viewing the activities spread out over a calendar requires a different chart called a *Gantt chart*. It is essentially a bar chart, one bar for each task, with the horizontal axis being units of time. A Gantt chart is good for monitoring the progress of the project as it moves along. Figure 3-11 shows a version of a Gantt chart, called a *tracking Gantt chart*, for the same tasks shown in Figures 3-9 and 3-10. The left column shows the names of the tasks. In fact, the list of tasks looks much like the WBS developed earlier. The length of the bars indicates the task durations. In this example of MS Project, the task bars can be color coded to indicate critical path, not started, partially complete, or completed. The red tasks are on the critical path, whereas the blue ones are not on the critical path. Complete tasks, either complete or partially complete, are shown in solid colors. The solid vertical line on February represents today's date. So, the project manager can easily check the status of the tasks on the diagram. Any tasks to the left of the vertical line that are not completed are behind schedule. The tasks to the right that have been completed are ahead of schedule. Tasks that intersect with the vertical line may be either on track, ahead, or behind. Depending on how status is reported, the Gantt chart may or may not indicate the status of those tasks. Most project managers find PERT/CPM charts beneficial while they are developing the schedule, but Gantt charts are most useful once the project begins.

Scheduling the Entire SDLC

The examples shown in Figures 3-9, 3-10, and 3-11 only detail the WBS for the project planning phase. Obviously, the other SDLC phases would also need to be scheduled. The SDLC provides the activities for each phase. Each activity is made up of a list of tasks. A standards-based or analogy-based WBS can be used to provide the detailed list of tasks for each analysis, design, and implementation phase activity. As you learn more about each phase and its activities in this book, you will understand more about the required tasks.

#	Task Name	Duration	Start
1	⊟ Project Planning Phase	22 days	Fri 2/4/05
2	⊟ Define the problem	4 days	Fri 2/4/05
3	Meet with users	2 days	Fri 2/4/05
4	Determine scope	1 day	Tue 2/8/05
5	Write business benefits	1 day	Wed 2/9/05
6	Write statement of need	1 day	Wed 2/9/05
7	Define system capabilities	1 day	Wed 2/9/05
8	Develop context diagram	1 day	Wed 2/9/05
9	⊟ Produce project schedule	5 days	Thu 2/10/05
10	Develop work breakdown structure	2 days	Thu 2/10/05
11	Estimate resources, durations	1 day	Mon 2/14/05
12	Develop PERT chart	2 days	Tue 2/15/05
13	⊟ Confirm project feasibility	5 days	Thu 2/17/05
14	Identify intangible costs/ benefits	1 day	Thu 2/17/05
15	Estimate tangible costs	1 day	Thu 2/17/05
16	Do cost/benefit analysis	2 days	Fri 2/18/05
17	Evaluate organization feasibility	1 day	Tue 2/22/05
18	Evaluate technical feasibility	2 days	Tue 2/22/05
19	Evaluate schedule feasibility	1 day	Tue 2/22/05
20	Evaluate resource availability	1 day	Tue 2/22/05
21	⊟ Staff the project	8 days	Thu 2/24/05
22	Develop a project resource plan	1 day	Thu 2/24/05
23	Identify and request tech staff	1 day	Fri 2/25/05
24	Meet with users, identify staff	1 day	Fri 2/25/05
25	Organize project team	1 day	Mon 2/28/05
26	Conduct team building exercises	3 days	Thu 3/3/05
27	⊟ Launch the project	6 days	Wed 2/23/05
28	Prepare presentation materials	1 day	Wed 2/23/05
29	Make executive presentation	1 day	Thu 2/24/05
30	Set up project facilities	3 days	Fri 2/25/05
31	Conduct official kick-off meeting	1 day	Wed 3/2/05

Tracking Gantt

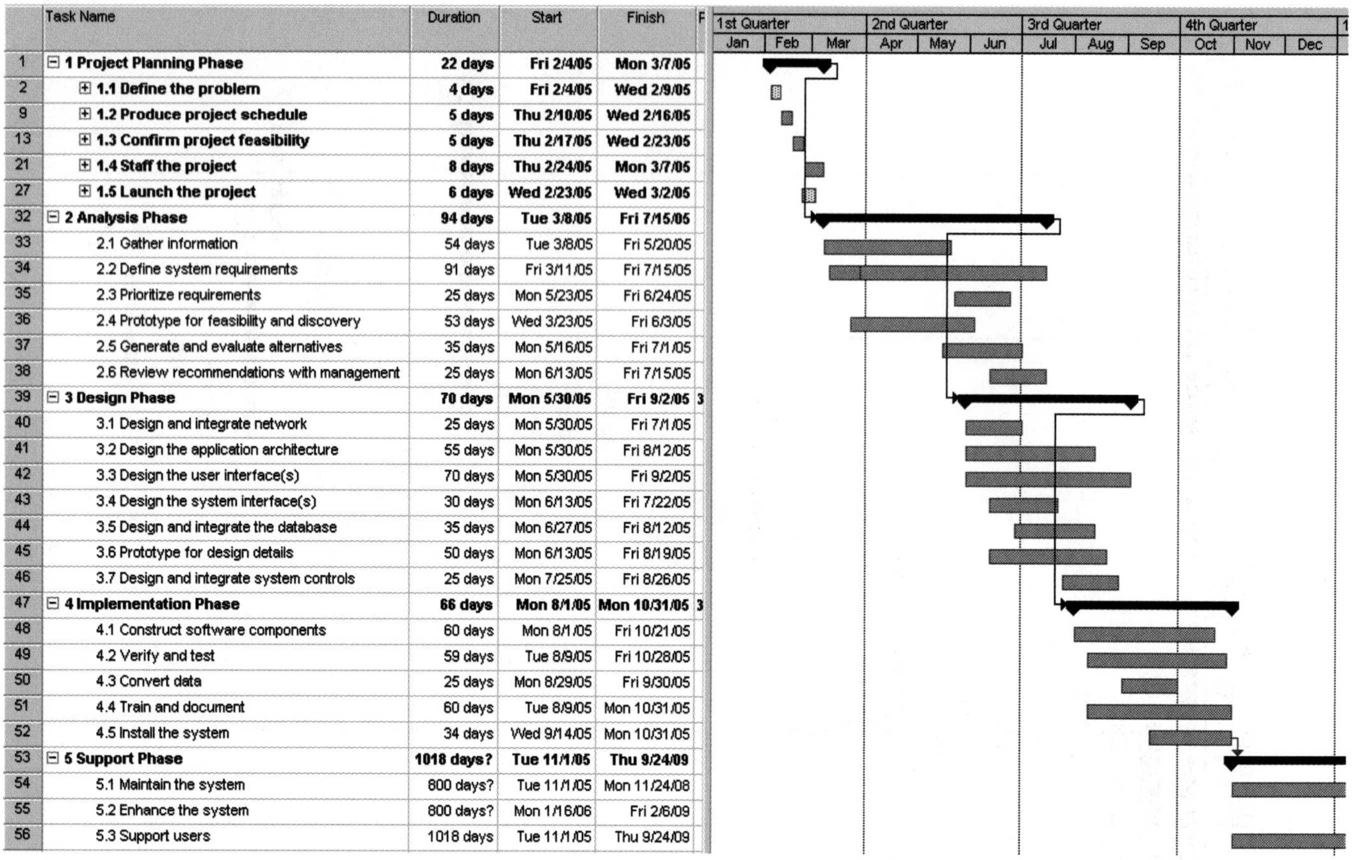

	Task Name	Duration	Start	Finish	
1	⊟ **1 Project Planning Phase**	**22 days**	**Fri 2/4/05**	**Mon 3/7/05**	
2	⊞ **1.1 Define the problem**	**4 days**	**Fri 2/4/05**	**Wed 2/9/05**	
9	⊞ **1.2 Produce project schedule**	**5 days**	**Thu 2/10/05**	**Wed 2/16/05**	
13	⊞ **1.3 Confirm project feasibility**	**5 days**	**Thu 2/17/05**	**Wed 2/23/05**	
21	⊞ **1.4 Staff the project**	**8 days**	**Thu 2/24/05**	**Mon 3/7/05**	
27	⊞ **1.5 Launch the project**	**6 days**	**Wed 2/23/05**	**Wed 3/2/05**	
32	⊟ **2 Analysis Phase**	**94 days**	**Tue 3/8/05**	**Fri 7/15/05**	
33	2.1 Gather information	54 days	Tue 3/8/05	Fri 5/20/05	
34	2.2 Define system requirements	91 days	Fri 3/11/05	Fri 7/15/05	
35	2.3 Prioritize requirements	25 days	Mon 5/23/05	Fri 6/24/05	
36	2.4 Prototype for feasibility and discovery	53 days	Wed 3/23/05	Fri 6/3/05	
37	2.5 Generate and evaluate alternatives	35 days	Mon 5/16/05	Fri 7/1/05	
38	2.6 Review recommendations with management	25 days	Mon 6/13/05	Fri 7/15/05	
39	⊟ **3 Design Phase**	**70 days**	**Mon 5/30/05**	**Fri 9/2/05**	3
40	3.1 Design and integrate network	25 days	Mon 5/30/05	Fri 7/1/05	
41	3.2 Design the application architecture	55 days	Mon 5/30/05	Fri 8/12/05	
42	3.3 Design the user interface(s)	70 days	Mon 5/30/05	Fri 9/2/05	
43	3.4 Design the system interface(s)	30 days	Mon 6/13/05	Fri 7/22/05	
44	3.5 Design and integrate the database	35 days	Mon 6/27/05	Fri 8/12/05	
45	3.6 Prototype for design details	50 days	Mon 6/13/05	Fri 8/19/05	
46	3.7 Design and integrate system controls	25 days	Mon 7/25/05	Fri 8/26/05	
47	⊟ **4 Implementation Phase**	**66 days**	**Mon 8/1/05**	**Mon 10/31/05**	3
48	4.1 Construct software components	60 days	Mon 8/1/05	Fri 10/21/05	
49	4.2 Verify and test	59 days	Tue 8/9/05	Fri 10/28/05	
50	4.3 Convert data	25 days	Mon 8/29/05	Fri 9/30/05	
51	4.4 Train and document	60 days	Tue 8/9/05	Mon 10/31/05	
52	4.5 Install the system	34 days	Wed 9/14/05	Mon 10/31/05	
53	⊟ **5 Support Phase**	**1018 days?**	**Tue 11/1/05**	**Thu 9/24/09**	
54	5.1 Maintain the system	800 days?	Tue 11/1/05	Mon 11/24/08	
55	5.2 Enhance the system	800 days?	Mon 1/16/06	Fri 2/6/09	
56	5.3 Support users	1018 days	Tue 11/1/05	Thu 9/24/09	

FIGURE 3-12

Gantt chart for the complete customer support system project.

If we assume that the project includes overlapping SDLC phases, a Gantt chart showing the entire project at the phase and activity level of detail might look like Figure 3-12. Note that the length of each activity does not imply that the team is working full time on that activity from start to finish. Rather, the activity starts and continues with varying degrees of effort for the duration. All team members get used to multitasking, that is, working on more than one activity or task at the same time. Therefore, this view of the project is not useful for calculating total labor cost, but it does show the completion of the implementation phase by October 31. The elapsed time for the CSS development project is about nine months. After than, the support phase begins.

Recall that the overlapping phases are usually the result of an iterative approach to the SDLC. For planning and scheduling purposes, many project managers use project management software and Gantt charts to plan and track the iterations of the project. Each iteration includes analysis, design, and implementation activities that focus on a portion of the system's functionality. Some analysis activities will be included in every iteration; other activities might only be included in a few. For example, each iteration might include analysis activities: *Gather information* and *Define system requirement;* and design activities: *Design the application architecture, Design the user interfaces*, and *Design and integrate the database.* Similarly, each iteration might include implementation activities: *Construct software components* and *Verify and test.* Other activities from each of these phases might be included in some but not all iterations, depending on the project plan.

A Gantt chart in Figure 3-13 shows how the RMO project might be scheduled with three iterations. The project team does not concern itself with what phase of the SDLC the project is in at any point in time. Rather, the SDLC phases and activities provide the framework for defining project planning and multiple iterations that are scheduled throughout the project.

FIGURE *3-13*

Gantt chart for an iterative approach to the customer support system project.

More detailed information on these scheduling techniques—especially on how to build schedules, including PERT and Gantt charts using MS Project—is given in Appendix B. The key question to be answered when completing this activity is, *can the project be completed on time given the available resources?*

	Task Name	Duration	Start	Finish
1	⊞ 1 Project Planning Phase	22 days	Fri 2/4/05	Mon 3/7/05
32	⊟ 2 Iteration One	44 days	Tue 3/8/05	Fri 5/6/05
33	2.1 Plan iteration	5 days	Tue 3/8/05	Mon 3/14/05
34	2.2 Analysis - gather information	20 days	Tue 3/8/05	Mon 4/4/05
35	2.3 Analysis - define requirements	30 days	Mon 3/14/05	Fri 4/22/05
36	2.4 Design - application architecture	30 days	Mon 3/21/05	Fri 4/29/05
37	2.5 Design - user interface	30 days	Mon 3/21/05	Fri 4/29/05
38	2.6 Design - database	30 days	Mon 3/21/05	Fri 4/29/05
39	2.7 Implement - construct software components	30 days	Mon 3/21/05	Fri 4/29/05
40	2.8 Implement - verify and test	30 days	Mon 3/28/05	Fri 5/6/05
41	2.9 Review with users and management	10 days	Mon 4/25/05	Fri 5/6/05
42	⊟ 3 Iteration Two	45 days	Mon 5/9/05	Fri 7/8/05
43	3.1 Plan iteration	5 days	Mon 5/9/05	Fri 5/13/05
44	3.2 Analysis - gather information	20 days	Mon 5/9/05	Fri 6/3/05
45	3.3 Analysis - define requirements	30 days	Mon 5/16/05	Fri 6/24/05
46	3.4 Design - application architecture	30 days	Mon 5/23/05	Fri 7/1/05
47	3.5 Design - user interface	30 days	Mon 5/23/05	Fri 7/1/05
48	3.6 Design - database	30 days	Mon 5/23/05	Fri 7/1/05
49	3.7 Implement - construct software components	30 days	Mon 5/23/05	Fri 7/1/05
50	3.8 Implement - verify and test	30 days	Mon 5/30/05	Fri 7/8/05
51	3.9 Review with users and management	10 days	Mon 6/27/05	Fri 7/8/05
52	⊟ 4 Iteration Three	45 days	Mon 7/11/05	Fri 9/9/05
53	4.1 Plan iteration	5 days	Mon 7/11/05	Fri 7/15/05
54	4.2 Analysis - gather information	20 days	Mon 7/11/05	Fri 8/5/05
55	4.3 Analysis - define requirements	30 days	Mon 7/18/05	Fri 8/26/05
56	4.4 Design - application architecture	30 days	Mon 7/25/05	Fri 9/2/05
57	4.5 Design - user interface	30 days	Mon 7/25/05	Fri 9/2/05
58	4.6 Design - database	30 days	Mon 7/25/05	Fri 9/2/05
59	4.7 Implement - construct software components	30 days	Mon 7/25/05	Fri 9/2/05
60	4.8 Implement - verify and test	30 days	Mon 8/1/05	Fri 9/9/05
61	4.9 Review with users and management	10 days	Mon 8/29/05	Fri 9/9/05
62	⊟ 5 Project Completion	36 days	Mon 9/12/05	Mon 10/31/05
63	5.1 Convert data	15 days	Mon 9/12/05	Fri 9/30/05
64	5.2 Complete controls and security	30 days	Mon 9/12/05	Fri 10/21/05
65	5.3 Train and document	30 days	Mon 9/12/05	Fri 10/21/05
66	5.4 Install the system	30 days	Tue 9/20/05	Mon 10/31/05
67	⊟ 6 Support Phase	1104 days?	Tue 11/1/05	Fri 1/22/10
68	6.1 Maintain the system	1104 days	Tue 11/1/05	Fri 1/22/10
69	6.2 Enhance the system	1080 days	Mon 12/5/05	Fri 1/22/10
70	6.3 Support users	1005 days?	Tue 11/1/05	Mon 9/7/09

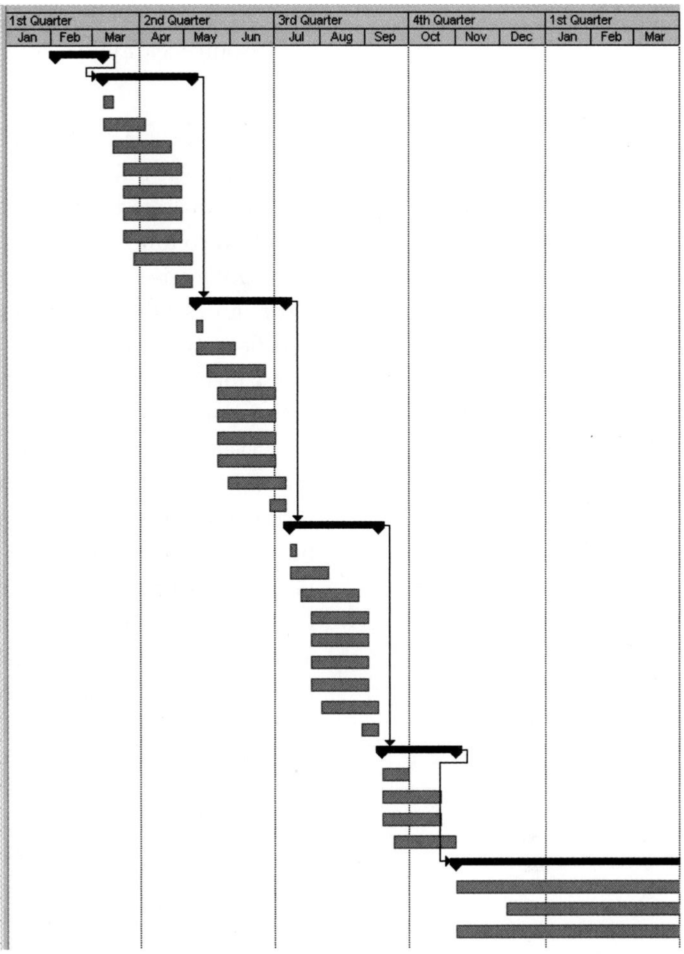

CONFIRMING PROJECT FEASIBILITY

When a project is first initiated, it is assumed that a new system is feasible to develop and install. As more insight is gained during the problem definition activity, it is important to confirm that the project actually is feasible. During project feasibility, the project manager answers questions such as, "Are the expected benefits reasonable?" and "Are the assumed costs realistic?"

The objective in assessing feasibility is to determine whether a development project has a reasonable chance of success. Feasibility analysis essentially identifies all the risks of failure. The project team assesses the original assumptions and identifies other risks that could jeopardize the project's success. The team first identifies those risks and then, if necessary, establishes plans and procedures to ensure that those risks do not interfere with the success of the project. However, if the team suspects there are serious risks that could jeopardize the project, members must discover and evaluate them as soon as possible.

Developing a list of potential risks is fairly difficult, which is another reason that experienced systems analysts are involved in planning. They have encountered and dealt with problems and know where risks are likely. Generally, the team considers five areas of risk for a new system when confirming project feasibility:

- Economic feasibility
- Organizational and cultural feasibility
- Technological feasibility
- Schedule feasibility
- Resource feasibility

Economic Feasibility

Economic feasibility consists of two tests: (1) Is the anticipated value of the benefits greater than projected costs of development? and (2) Does the organization have adequate cash flow to fund the project during the development period? Even though the project may have received initial approval based on the need or strategic plan, final approval usually requires a thorough analysis of the development costs and the anticipated financial benefits. Obviously, the justification for the development of a new system is that it will increase income, either through cost savings or by increased revenues. A determination of the economic feasibility of the project always requires a thorough *cost/benefit analysis*.

Developing a cost/benefit analysis is a three-step process. The first step is to estimate the anticipated development and operational costs. Development costs are those that are incurred during the development of the new system. Operational costs are those that will be incurred after the system is put into production. The second step is to estimate the anticipated financial benefits. Financial benefits are the expected annual savings or increases in revenue derived from the installation of the new system. The third, the cost/benefit analysis step, is calculated based on the detailed estimates of costs and benefits. The most frequent error that inexperienced analysts make during cost/benefit analysis is to try to do the calculations before thoroughly defining costs and benefits. A cost/benefit analysis that does not have thorough and complete supporting detail is valueless.

Development Costs

Although the project manager has final responsibility for estimating the costs of development, senior-level analysts always assist with the calculations. Generally, project costs come in the following categories:

- Salaries and wages
- Equipment and installation
- Software and licenses
- Consulting fees and payments to third parties
- Training
- Facilities
- Utilities and tools
- Support staff
- Travel and miscellaneous

Salaries and wages are calculated based on the staffing requirements for the project. As the project schedule and staffing plans are developed, the estimated cost for salaries and wages can be determined. Figure 3-14 is an example of the estimated cost for salaries for the RMO customer support system. The numbers in this table were calculated by identifying personnel who are needed for the project and the length of time they are to be assigned to the project. As discussed previously, it is not easy to estimate the amount of time a person will be assigned to the project. Some team members work full time on the project; others work on several projects concurrently. If the work breakdown structure (WBS) is detailed and accurate, the salary and wage costs can be more accurately specified.

---

cost/benefit analysis

the analysis to compare costs and benefits to see whether investing in the development of a new system will be beneficial

---

FIGURE *3-14*

Supporting detail for
salaries and wages.

SUPPORTING DETAIL FOR SALARIES AND WAGES	
Team member	**Salary/wage for project**
Project leader	$101,340.00
Senior systems analyst	$90,080.00
Systems analyst	$84,980.00
Programmer analysts	$112,240.00
Programmers	$58,075.00
Systems programmers	$49,285.00
Total salaries and wages	$496,000.00

Each of the other categories of costs requires detailed calculations to determine the estimated costs. The project manager can make detailed estimates of equipment, software licenses, training costs, and so forth. These details are then combined to provide an estimate of the total costs of development. Figure 3-15 is a summary table of all of the costs. Again, each line in the summary table must be supported with details such as those given in Figure 3-14.

FIGURE *3-15*

Summary of development
costs for RMO.

SUMMARY OF DEVELOPMENT COSTS FOR RMO	
Expense category	**Amount**
Salaries/wages	$496,000.00
Equipment/installation	$385,000.00
Training	$78,000.00
Facilities	$57,000.00
Utilities	$152,000.00
Support staff	$38,000.00
Travel/miscellaneous	$112,000.00
Licenses	$18,000.00
Total	$1,336,000.00

Sources of Ongoing Costs of Operations

Once the new system is up and running, normal operating costs are incurred every year. The calculation of the cost and benefit of the new system must also account for these annual operating costs. Generally, analysts do not include the normal costs of running the business in this cost. Only the costs that are directly related to the new system and its maintenance are included. The following list identifies the major categories of costs that might be allocated to the operation of the new system:

- Connectivity
- Equipment maintenance
- Costs to upgrade software licenses
- Computer operations
- Programming support
- Amortization of equipment
- Training and ongoing assistance (the help desk)
- Supplies

Figure 3-16 is a summary of the annual operating costs for RMO. As with the development costs, each entry in the table should be supported with detailed calculations. This figure represents only those costs that are anticipated for the RMO system. Other organizations may have a different set of operating costs.

SUMMARY OF ANNUAL OPERATING COSTS FOR RMO	
Recurring expense	**Amount**
Connectivity	$60,000.00
Equipment maintenance	$40,000.00
Programming	$65,000.00
Help desk	$28,000.00
Amortization	$48,000.00
Total recurring costs	$241,000.00

Sources of Benefits

The project manager and members of the project team can determine most of the development and operational costs. However, the user and the client receive the benefits of the system. Consequently, the client and the user must determine the value of the anticipated benefits. Members of the project team can and do assist, but they should never attempt to determine the value of benefits by themselves.

Benefits usually come from two major sources: decreased costs or increased revenues. Cost savings or decreases in expenses come from increased efficiency in company operations. Areas in which to look for reduced costs include the following:

- Reducing staff by automating manual functions or increasing efficiency
- Maintaining constant staff with increasing volumes of work
- Decreasing operating expenses such as shipping charges for "emergency shipments"
- Reducing error rates through automated editing or validation
- Achieving quicker processing and turnaround of documents or transactions
- Capturing lost discounts on money management
- Reducing bad accounts or bad credit losses
- Reducing inventory or merchandise losses through tighter controls
- Collecting receivables (accounts receivable) more rapidly
- Capturing lost income due to "stock-outs" by implementing better inventory management
- Reducing cost of goods through volume discounts and purchases
- Reducing paperwork cost by implementing electronic data interchange and other automation

This list is just a sampling of the myriad benefits that can accrue. Unlike development costs, there are no "standard" benefits. Each project is different, and the anticipated benefits are different. Figure 3-17 is an example of the benefits that RMO expects from the implementation of the new customer support system.

Financial Calculations

Companies use a combination of methods to measure the overall benefit of the new system. One popular approach is to determine the *net present value (NPV)* of the new system. The two concepts behind net present value are (1) that all benefits and costs are calculated in terms of today's dollars (present value) and (2) that benefits and costs are combined to give a net value. The future stream of benefits and costs are netted together

net present value (NPV)

the present value of dollar benefits
and costs for an investment such as a
new system

FIGURE *3-17*

Sample benefits for RMO.

SAMPLE BENEFITS FOR RMO		
Benefit/cost saving	**Amount**	**Comments**
Increased efficiency in mail-order department	$125,000.00	5 people @ $25,000
Increased efficiency in phone-order department	$25,000.00	1 person @ $25,000
Increased efficiency in warehouse/shipping	$87,000.00	
Increased earnings due to Web presence	$500,000.00	Increasing at 50%/year
Other savings (inventory, supplies, etc.)	$152,000.00	
Total annual benefits	$889,000.00	

and then discounted by a factor for each year in the future. The discount factor is like an interest rate, except it is used to bring future values back to current values. Appendix C provides detailed instructions on how to calculate economic feasibility. You should read Appendix C to ensure you understand the details. Appendix C is available for download at www.course.com via the "Student Downloads" link on the Web page for this book.

Figure 3-18 shows a copy of the RMO net present value calculation done in Appendix C (Figure C-1). In this case, the new system gives an NPV of $3,873,334 over a five-year period using a discount rate of 10 percent.

Another method that organizations use to determine whether an investment will be beneficial is determining the *payback period*. The payback period, sometimes called the *breakeven point*, is the point in time at which the increased cash flow (benefits) exactly pays off the costs of development and operation. Appendix C provides the detailed equations necessary for this calculation. Figure 3-18 illustrates the calculations for the

payback period

the time period where the dollar benefits have offset the dollar costs

breakeven point

the point in time where the dollar benefits have offset the dollar costs

FIGURE *3-18*

Net present value, payback period, and return on investment for RMO.

	RMO cost benefit analysis	Year 0	Year 1	Year 2	Year 3	Year 4	Year 5	Total
1	Value of benefits	$ -	$ 889,000	$ 1,139,000	$ 1,514,000	$ 2,077,000	$ 2,927,000	
2	Discount factor (10%)	1	0.9091	0.8264	0.7513	0.6830	0.6209	
3	Present value of benefits	$ -	$ 808,190	$ 941,270	$ 1,137,468	$ 1,418,591	$ 1,817,374	$6,122,893
4	Development costs	$(1,336,000)						$(1,336,000)
5	Ongoing costs		$ (241,000)	$ (241,000)	$ (241,000)	$ (241,000)	$ (241,000)	
6	Discount factor (10%)	1	0.9091	0.8264	0.7513	0.6830	0.6209	
7	Present value of costs	$ -	$ (219,093)	$ (199,162)	$ (181,063)	$ (164,603)	$ (149,637)	$(913,559)
8	PV of net of benefits and costs	$(1,336,000)	$ 589,097	$ 742,107	$ 956,405	$ 1,253,988	$ 1,667,737	
9	Cumulative NPV	$(1,336,000)	$(746,903)	$ (4,769)	$951,609	$2,205,597	$ 3,873,334	
10	Payback period	2 years + 4796 / (4796 + 951,609) = 2 + .005 or 2 years and 2 days						
11	5-year return on investment	(6,122,893 - (1,336,000 + 913,559)) / (1,336,000 + 913,559) = 172.18%						

payback period. A running accumulated net value is calculated year by year. The year when this value goes positive is the year in which payback occurs. In the RMO example, this payback happens within the third year.

The *return on investment (ROI)* is another evaluation method used by organizations. The objective of the NPV is to determine a specific value based on a predetermined discount rate. The objective of the ROI is to calculate a percentage return (like an interest rate) so that the costs and the benefits are exactly equal over the specified time period. Figure 3-18 shows an ROI calculation for RMO, as developed in Appendix C. The time period can be the expected life of the investment (such as the productive life of the system), or it can be an arbitrary time period.

For RMO, assuming a five-year benefit period, the ROI is 172.18 percent. In other words, the investment in the development costs returned over 172.18 percent on the investment for a period of five years. Obviously, since the system is generating benefits at that point in time, if you assumed that the lifetime was longer, such as 10 years, you would get a much higher ROI.

Intangibles

The previous cost/benefit calculation is dependent on an organization's ability to quantify the costs and the benefits. However, in many instances, an organization cannot measure some costs and benefits and determine a value. If it can estimate a dollar value for a benefit or a cost, the organization treats the value as a *tangible benefit* or cost. Where there is no reliable method of estimating or measuring the value, then it is considered an *intangible benefit*. In some instances, the importance of the intangible benefits far exceed the tangible costs, at least in the opinion of the client, and the client proceeds to develop the system even though the dollar numbers did not indicate a good investment.

Examples of intangible benefits include the following:

- Increased levels of service (in ways that cannot be measured in dollars)
- Increased customer satisfaction (not measurable in dollars)
- Survival (a standard capability common in the industry, or common to many competitors)
- Need to develop in-house expertise (such as with a pilot program with new technology)

Examples of intangible costs include the following:

- Reduced employee morale
- Lost productivity (the organization may not be able to estimate it)
- Lost customers or sales (during some unknown period of time)

Only tangible benefits and costs are used when calculating NPV, payback, and ROI. Even though the intangibles do not enter into the calculations, they should be considered and, in fact, may be the deciding factor on whether the project proceeds or not.

Sources of Funds

As we explained earlier, the project team performs the cost/benefit analysis in conjunction with the development of the project budget. The two components of economic feasibility are concerned with a positive result from the cost/benefit analysis and the source of funds for the system development. Organizations can finance development projects in various ways. Frequently, new information systems are financed using a combination of current cash flows and long-term capital. The project team may not be involved in obtaining the financing for the project. However, the results of the cost/benefit analysis will greatly influence the financing decisions.

---

return on investment (ROI)

a measure of the percentage gain from an investment such as a new system

tangible benefits

benefits that can be measured or estimated in terms of dollars and that accrue to the organization

intangible benefits

benefits that accrue to the organization but which cannot be measured quantitatively or estimated accurately

---

Organizational and Cultural Feasibility

As discussed in Chapter 1, each company has its own culture, and any new system must be accommodated within that culture. There is always the risk that the new system departs so dramatically from existing norms that it cannot be successfully deployed. The analysts involved with feasibility analysis should evaluate organizational and cultural issues to identify potential risks for the new system. Such issues might include the following:

- A current low level of computer competency
- Substantial computer phobia
- A perceived loss of control by staff or management
- Potential shifting of political and organizational power due to the new system
- Fear of change of job responsibilities
- Fear of loss of employment due to increased automation
- Reversal of long-standing work procedures

It is not possible to enumerate all the potential organizational and cultural risks that exist. The project management team needs to be very sensitive to reluctance within the organization to identify and resolve these risks. Again, the question to ask for operational feasibility is, "What items might prevent the effective utilization of the new system and the resulting loss of business benefits?"

After identifying the risks, the project management team can take positive steps to counter them. For example, the team can hold additional training sessions to teach new procedures and provide increased computer skills. Higher levels of user involvement in the development of the new system will tend to increase user enthusiasm and commitment.

Technological Feasibility

Generally, a new system brings new technology into the company. At times the new system stretches the state of the art of the technology. Other projects utilize existing technology but combine it into new, untested configurations. Finally, even existing technology can pose the same challenges as new technology if there is a lack of expertise within the company. If an outside vendor is providing a capability, the client organization usually assumes that it is expert in the area in which it provides service. However, even an outside vendor is subject to the risk that the requested level of technology is too complicated.

The project management team needs to assess very carefully the proposed technological requirements and available expertise. When these risks are identified, the solutions are usually fairly straightforward. The solutions to technological risks include providing additional training, hiring consultants, or hiring more experienced employees. In some cases, the scope and approach of the project may need to be changed to ameliorate technological risk. The important point is that a realistic assessment will identify technological risks early, making it possible to implement corrective measures.

Schedule Feasibility

The development of a project schedule always involves high risk. Every schedule requires many assumptions and estimates without adequate information. For example, the needs, and hence the scope, of the new system are not well known, the estimated time to research and finalize requirements must be estimated, and the availability and capability of team members are questionable.

Another frequent risk in the development of the schedule occurs when upper management indicates that the new system must be deployed within a certain time. Sometimes there is an important business requirement for defining a fixed deadline, such as RMO needing to complete the CSS in time for on-line ordering for the holidays. Similarly, universities require the completion of new systems before key dates in the university schedule. For example, if a new admissions system is not completed before the admissions season, then it might as well wait another full year. In cases like these, schedule feasibility can be the most important feasibility factor to consider.

If the deadline appears arbitrary, the tendency is to build the schedule to show that it can be done. Unfortunately, this practice usually spells disaster. The project team should build the schedule without any preconceived notion of required completion dates. Once the schedule is completed, then comparisons can be done to see whether timetables coincide. If not, then the team can take corrective measures, such as reducing the scope of the project, to increase the probability of the project's on-time completion.

One of the objectives of defining milestones during the project schedule is to permit the project manager to assess the ongoing risk of the schedule's slipping. If the team begins to miss milestones, then the manager can possibly implement corrective measures early. Contingency plans can be developed and carried out to reduce the risk of further slippage.

Allocating adequate personnel with the right experience and expertise to a project is always a problem. Any complex project may incur overruns and schedule extensions. It may be difficult to identify the sources of these risks, but a conscious effort to identify them will at least highlight areas of weakness. Long projects are especially subject to difficulties with resource allocation and to schedule slippage. Solutions can involve contingency plans in case in-house resources are not available.

Resource Feasibility

Finally, the project management team must assess the availability of resources for the project. The primary resource consists of the members of the team. Development projects require the involvement of systems analysts, system technicians, and users. Required people may not be available to the team at the necessary times. An additional risk is that the people who are assigned may not have the necessary skills for the project. Once the team is functioning, team members may have to leave the team. This threat can come either from staff who are transferred internally within the organization if other special projects arise or from qualified team members who are hired away by other organizations. Although the project manager usually does not like to think about these possibilities, skilled people are in short supply and sometimes do leave projects.

The other resources required for a successful project include adequate computer resources, physical facilities, and support staff. Generally, these resources can be made available, but the schedule can be affected if delays in the availability of these resources occur.

Feasibility Analysis

Each of the preceding feasibility analyses has assumed that the project is feasible. In fact, the title of the section is "Confirming Project Feasibility." But, not every project is feasible. For a project to be viable, it must pass all of the feasibility tests. In other words, the team must examine each area of the project carefully and make a determination based on relevant data. If the project is not feasible in any one of the categories, then they must make adjustments. If adjustments cannot improve the situation, then the project should not be initiated. One viable alternative to starting a project that has high risk of failure is simply to do nothing—for now. A project that is not feasible today—for example, due to technical difficulties, high costs, or inadequate expertise—may become feasible in the future. Project managers generally dislike concluding that the project is not feasible and should not be done. The alternative, however, is to begin a project that is destined to fail, harming the company and all involved.

An assessment of each of these five areas of feasibility is an important part of the project planning phase. The key question to be answered when completing this activity is, *is it still feasible to begin working on this project?*

STAFFING AND LAUNCHING THE PROJECT

The responsibility for staffing the project team falls primarily on the project manager. Human resource management, as explained in Appendix A, includes finding the right people with the correct skills and then organizing and managing them throughout the project. The staffing activity consists of five tasks:

- Develop a resource plan for the project
- Identify and request specific technical staff
- Identify and request specific user staff
- Organize the project team into workgroups
- Conduct preliminary training and team-building exercises

Based on the tasks identified in the project schedule, the project manager can develop a detailed resource plan. In fact, the schedule and the resource requirements are usually developed concurrently. If the project manager is using a tool such as Microsoft Project to build the schedule, then the resources required for each task are part of the total schedule. In developing the plan, the project manager recognizes (1) that resources are usually not available as soon as requested and (2) that a period of time is needed for a person to become acquainted with the project.

After developing the plan, the project manager can then identify specific people and request that they become part of the team. Generally, two sources exist for members of the team: (1) technical staff and (2) user staff. Technical staff means the systems analysts, the programmer analysts, the network specialists, and other technicians. Technical staff expect to move from project to project and find change normal. The project manager will meet with the director or vice president of information systems to identify and schedule the necessary resources. In some instances, it may be necessary to hire additional technical staff, so the human resource department may need to become involved. Even though finding technical people for the team is standard procedure, finding and assigning all of the required team members may take some time.

The user staff are people from the user community who are assigned to the team. Sometimes it is difficult to get users assigned to the team full time. Being assigned to a project team is not part of the normal job progression of someone in a user department or group. However, projects do progress more smoothly if a few full-time team members can represent the user community and act as liaisons. Referring back to causes of project failure, it is evident that having users closely associated with and even assigned to the project team will enhance the chances of success.

On small projects, members of the project team may all work together. However, a project team larger than four or five members usually is divided into smaller working groups. Each group will have a group leader who will coordinate the tasks assigned to the group. The responsibility for dividing the team into groups and assigning group leaders falls on the project manager.

Finally, training and team-building exercises are conducted. Training may be done for the project team as a whole when new technology such as a new database or a new programming language is used. In other cases, new team members who are unfamiliar with the tools and techniques that are being used may require individual training. The team should conduct appropriate training for both technical people and users. Team-building exercises are especially important when members have not worked together before. The integration of user members of the team with technical people is an important consideration in developing effective teams and workgroups.

The key question to be answered when completing the staffing activity is, *are the resources available, trained, and ready to start the project?*

After the previous project planning activities are complete, it is time to launch the project. The scope of the new system is defined, the risks have been identified, the project has been found feasible both economically and otherwise, a detailed schedule has been developed, team members have been identified and are ready, and it is now time to start. Two final tasks usually occur at this point. First, the membership of the oversight committee is finalized, and it meets to give final go-ahead for the project, including releasing the necessary funds. Second, the organization makes a formal announcement through its standard communication channels that gives credence to the project and solicits cooperation from all involved parties in the organization. In other words, the project gets the blessing and visible support of the senior executives of the organization. No project should begin without these two events.

The key question to be answered when launching the project is, *are we ready to start?*

RECAP OF PROJECT PLANNING FOR RMO

Barbara and Steve spent the entire month of February putting together the schedule and plans for the CSS. Even though Barbara was the project manager, she and Steve worked together as peers. As a team, they could brainstorm and double-check each other's work. They had worked together before and had an excellent relationship—one based on mutual respect and trust. They could be candid and knew how to work through disagreements as well as how to come to consensus on important issues. Barbara also knew that the work Steve produced was always well thought out and very professionally done. He was a skilled systems analyst and would help make sure that the work done in the planning phase was solid.

The success of the overall project depended heavily on the planning Barbara and Steve did during this phase. The foundation for all other project activities is established during the project planning phase. As Barbara planned for the kickoff meeting to launch the project officially, she reviewed the areas of project management to make sure that she had addressed all of the critical issues.

For project scope management, she developed a list of business benefits, a list of system capabilities, and a context diagram. At this point in the project, the scope definition was still very general. She would make sure the project's scope was precisely defined during the information-gathering activities of the analysis phase.

She and Steve had developed a detailed work breakdown structure and entered the information into MS Project. The schedule was very detailed for the analysis phase, but less so for the design and implementation phases. She would add those details as decisions were made about the implementation approach. She thought that her approach to project time management had been established, and she would have the tools necessary to track the schedule as the project progressed.

The costs and potential benefits had been estimated and used to develop an NPV estimate. She would redo the NPV when she redid the schedule at the end of the analysis phase to ensure that the costs and schedule were within the allowed budget. The other part of cost management was to monitor the costs during the life of the project. MS Project would help her track the costs of each task.

Steve had done a lot of the work to identify and assess risks during the feasibility analysis. Barbara knew that they would both continue to look for risks and assess potential problems during the project. She asked Steve to take time each week to assess the risks and update the list of the highest risks for the project. She felt confident that she would not be blindsided by some unexpected problem.

For project communication and project quality, Barbara established procedures for the project. She set up a central database to post the project's status, decisions, and

February 17, 2005

To: CSS Project Oversight Committee
        William McDougal, Chair

From: Barbara Halifax, Project Manager

Following instructions from John MacMurty, I will send a status report memo to the oversight committee every two weeks during the first two months of the project. Thereafter, I will provide it monthly.

Completed during the last period (two weeks)

Steve Deerfield and I have worked on two major items. First, we have completed a high-level statement of the project scope. As you are aware, we interviewed each of you, as well as other major stakeholders, to determine the business needs and overall scope of the project.

Second, we have developed a project schedule, with a detailed WBS for the planning and analysis phases and a less detailed WBS for the design and implementation phases. I have attached a copy of the project schedule. We anticipate having the system operational before the end of the year.

Plans for the next period (two weeks)

During the next two weeks, we will finalize project feasibility. Primary focus will be on developing a preliminary cost/benefit analysis. As more information is gathered during the analysis phase, this financial analysis will become more precise.

We will also develop a feasibility analysis of the project. Our risk analysis will determine whether the project is feasible and identify areas of high risk. Finally, we will develop a staffing plan, identify team members—both technical staff and users—and begin staffing the project.

Problems, Issues, Open Items -- None

working documents to make sure that all the team members were kept well informed. She established a routine and format for weekly status reports from the team leaders and a status report to the oversight committee. An example of one of her status report memos to the oversight committee follows. These status reports all follow a standard format. In addition to the formal status memos, she would also write more informal memos to John MacMurty. For project quality, internal procedures required that team members and RMO users review all work products. Other quality procedures, such as the test plan, would be established as the project progressed.

She and Steve had identified the other team members they would like to have on the team. John had been especially helpful in finding solid analysts who were available or who would be available soon. In fact, Barbara had already interviewed all of the members who were coming on board. Recognizing the importance of having a team whose members could work together, she had scheduled several days for the team members to get to know each other, to refine their internal working procedures, and to teach them about the tools and techniques that would be used on the project.

All in all, it had been a very hectic but productive month. A lot of work had been done, and a solid foundation had been established for a successful project.

SUMMARY

The focus of this chapter is on project management activities that form the basis of the project planning phase of the SDLC. The chapter covered three major themes: (1) project management, (2) information system project initiation and the project planning phase, and (3) techniques used by the project manager and analysts for completing the project planning activities of the SDLC.

The development of a new system requires an organized, step-by-step approach. We call this approach the systems development life cycle (SDLC), as discussed in Chapter 2. The SDLC defines the phases, activities, and tasks that require attention during the system development project. Project management tasks are involved in the project planning phase at the beginning of the project, but project management tasks continue throughout the project as well.

Project management is the organizing and directing of other people to achieve a planned result within a predetermined schedule and budget. Project management can be divided into eight knowledge areas: scope, time, cost, quality, human resources, communications, risk, and procurement.

Projects are initiated based on information system needs identified and prioritized in strategic plans of the organization. They are also initiated on an ad hoc basis as problems or directives arise. Once a project is initiated, the project planning phase is carried out primarily by the project manager and one or two other senior analysts. Many of the responsibilities of the project manager are carried out via the activities of the project planning phase. This phase consists of five activities: (1) defining the problem, (2) producing the project schedule, (3) confirming project feasibility, (4) staffing the project, and (5) launching the project.

To define the problem, the project manager investigates the problem and the ideas originally defined for a system solution. The scope of the project is established, and an initial system context diagram is used to graphically model the major inputs and outputs. The project schedule is produced by creating a work breakdown structure of phases, activities, and tasks required to complete the project, based on the SDLC. Scheduling is difficult because phases and activities often overlap and several iterations might be used for the project. Scheduling techniques such as PERT/CPM and the critical path are used to investigate scheduling bottlenecks and risks. Ultimately, the project schedule is used as the basis for calculating project labor costs as labor is based on the amount of time spent by project members on project tasks.

Confirming project feasibility requires evaluating risks related to five types of feasibility: economic, technological, schedule, organizational, and resource. Economic feasibility is confirmed using cost/benefit analyses to compare the costs of the project with the expected benefits. Net present value (NPV), payback period, and return on investment (ROI) calculations are used to determine whether the cost/benefit analysis is favorable for the project, although intangible benefits are often important reasons for moving forward with a project.

The project planning phase is completed by a small team, often just the project manager and one or two key analysts. When the project moves on to the analysis phase, additional team members must be identified and assigned to the project. The staffing plan must address team member needs months in the future as well. When the project is ready to be launched, key management personnel and executive sponsors must be notified and involved to ensure project success.

KEY TERMS

breakeven point, p. 97

business benefits, p. 84

client, p. 77

context diagram, p. 85

cost/benefit analysis, p. 94

critical path, p. 90

Gantt chart, p. 90

intangible benefits, p. 98

net present value (NPV), p. 96

oversight committee, p. 77

payback period, p. 97

PERT/CPM, p. 89

project management, p. 76

proof of concept prototype, p. 85

REVIEW QUESTIONS

1. List and explain the activities of the project planning phase.
2. List the seven reasons why projects fail.
3. List the five reasons why projects are successful.
4. What are three reasons projects are initiated?
5. Define *project management*.
6. Explain how information system project management is similar to project management in general.
7. Explain how iterative development makes project scheduling more complex.
8. Describe the five types of feasibility used to evaluate a project.
9. What is the purpose of the cost/benefit analysis used to assess economic feasibility?
10. Explain the difference between tangible and intangible costs and benefits. Which are ignored in cost/benefit analysis?
11. Explain the difference between a PERT chart and a Gantt chart.
12. List at least five possible sources of tangible benefits from the installation of a new system.
13. List at least four sources of development costs.
14. What is meant by the critical path?
15. What is the purpose of a system context diagram?
16. Describe the eight knowledge areas of project management.
17. What activities in the planning phase are specifically focused on project management?

THINKING CRITICALLY

1. Write a short paper that discusses how project management techniques can overcome the reasons for project failure listed at the beginning of the chapter.
2. Given the following narrative, make a list of expected business benefits:

   Especially for You Jewelers is a small jewelry company in a college town. Over the last couple of years, Especially for You has experienced a tremendous increase in its business. However, its financial performance has not kept pace with its growth. The current system, which is partially manual and partially automated, does not track accounts receivables sufficiently, and Especially for You is having difficulty determining why the receivables are so high. In addition, Especially for You runs frequent specials to attract customers. It has no idea whether these specials are profitable or whether the benefit, if there is one, comes from associated sales. Especially for You also wants to increase repeat sales to existing customers, and thus needs to develop a customer database. The jewelry company wants to install a new direct sales and accounting system to help solve these problems.
3. Given the following narrative, make a list of system capabilities:

   The new direct sales and accounting system for Especially for You Jewelers is an important element in the future growth and success of the jewelry company. The direct sales portion of the system needs to track every sale and be able to link to the inventory system for cost data to provide a daily profit and loss report. The customer database needs to be able to produce purchase histories to assist management in prepar-

ing special mailings and special sales to existing customers. Detailed credit balances and aged accounts for each customer would help solve the problem with the high balance of accounts receivables. Special notice letters and credit history reports would help management reduce accounts receivable.

4. Develop a project charter for Especially for You Jewelers based on your work from problems 3 and 4.
5. Build a PERT/CPM chart based on the following list of tasks and dependencies to build and test a screen form for a new system. Identify the critical path.

Task ID	Description	Duration (days)	Predecessor
0	Start	0	—
1	Meet with user	2	0
2	Review existing forms	1	0
3	Identify and specify fields	3	1, 2
4	Build initial prototype	2	3
5	Develop test data (valid data)	4	3
6	Develop error test data	2	5
7	Test prototype	3	4, 6
8	Make final refinements	3	7

6. Suppose that you work in a dentist's office and are asked to develop a system to track patient appointments. How would you start? What would you do first? What kinds of things would you try to find out first? How does your approach compare with what this chapter has described?

EXPERIENTIAL EXERCISES

1. Using Microsoft Project, build a project schedule based on the following scenario. Print out both the PERT chart and the Gantt chart.

   In the table to the right is a list of tasks for a student to have an international experience by attending a university abroad. You can build schedules for several versions of this set of tasks. For the first version, assume that all predecessor tasks must finish before the succeeding task can begin (the simplest version). For a second version, identify several tasks that can begin a few days before the end of the predecessor task. For a third version, modify the second version so that some tasks can begin a few days after the beginning of a predecessor task. Also, insert a few overview tasks such as Application tasks, Preparation tasks, Travel tasks, and Arrival tasks. Be sure to state your assumptions for each version.

2. Build a project plan to show your progress through college. Include the course prerequisite information. If you have access to MS Project or another tool, enter the information in the project management tool.

3. Using information from your organizational behavior classes or other sources, write a one-page paper on what kinds of team-building and training activities might be appropriate as the project team is expanded for the analysis phase.

4. Ask a systems analyst about the SDLC that his or her company uses. If possible, ask the analyst to show you a copy of the project schedule. To what extent is iterative development used?

5. Ask a project manager for his or her opinion on each of the eight project management knowledge areas.

6. Go to the CompTIA (www.compTIA.com) Web site and find the requirements for the project manager exam (IT Project+). Write a one-page summary of the expertise and knowledge required to pass the exam.

Task ID	Description	Duration (days)	Predecessor
1	Obtain forms from the international exchange office	1	None
2	Fill out and send in the foreign university application	3	1
3	Receive approval from the foreign university	21	2
4	Apply for scholarship	3	2
5	Receive notice of approval for scholarship	30	4
6	Arrange financing	5	3, 5
7	Arrange for housing in dormitory	25	6
8	Obtain a passport and the required visa	35	6
9	Send in preregistration forms to the university	2	8
10	Make travel arrangements	1	7, 9
11	Determine clothing requirements and go shopping	10	10
12	Pack and make final arrangements to leave	3	11
13	Travel	1	12
14	Move into the dormitory	1	13
15	Finalize registration for classes and other university paperwork	2	14
16	Begin classes	1	15

Custom Load Trucking

It was time for Stewart Stockton's annual performance review. As Monica Gibbons, an assistant vice president of information systems, prepared for the interview, she reviewed Stewart's assignments over the last year and his performance. Stewart was one of the "up and coming" systems analysts in the company, and she wanted to be sure to give him solid advice on how to advance his career. She knew, for example, that he had a strong desire to become a project manager and accept increasing levels of responsibility. His desire was certainly in agreement with the needs of the company.

Custom Load Trucking (CLT) is a nationwide trucking firm that specializes in the rapid movement of high-technology equipment. With the rapid growth of the communications and computer industries, CLT was feeling more and more pressure from its clients to be able to move its loads more rapidly and precisely. Several new information systems were planned that would enable CLT to schedule and track shipments and trucks almost to the minute. However, trucking was not necessarily a high-interest industry for information systems experts. With the shortage in the job market, CLT had decided not to try to hire project managers for these new projects but to build strong project managers from within the organization.

As Monica reviewed Stewart's record, she found that he had done an excellent job as a team leader on his last project. His last assignment was as a combination team leader/systems analyst on a four-person team. He had been involved in systems analysis, design, and programming, and he also managed the work of the other three team members. He had assisted in the development of the project schedule and had been able to keep his team right on schedule. It also appeared that the quality of his team's work was as good as, if not better than, other teams on the project. She wondered what advice she should give him to help him advance his career. She was also wondering if now was the time to give him his own project.

1. Do you think the decision by CLT to build its own project managers from the existing employee base is a good one? What advice would you give to CLT to make sure that it has strong project management skills in the company?
2. What kind of criteria would you develop for Monica to use to measure whether Stewart (or any other potential project manager) is ready for project management responsibility?
3. How would you structure the job for new project managers to ensure, or at least increase, a high level of success?
4. If you were Monica, what kind of advice would you give to Stewart about managing his career and attaining his immediate goal to become a project manager?

Rethinking Rocky Mountain Outfitters

The chapter identified five areas of project feasibility that need to be evaluated for any new project. However, as indicated, each of these areas of feasibility can also be considered an evaluation of the potential risks of the project. Based on your understanding of Rocky Mountain Outfitters, both from this chapter and the information provided in Chapter 1, build a table that summarizes the risks faced by RMO for this new project. Include four columns titled (1) Project risk, (2) Type of risk, (3) Probability of risk, and (4) Steps to alleviate risk.

Identify as many risks to the project as you can. Type of risk means the category or area of the project feasibility that is at risk. It might help you think about risks in the different categories, for example (1) economic, (2) organizational and cultural, (3) technological, (4) schedule, and (5) resources. The chapter provided a few examples of risk in each of these areas. However, many other risks can cause project failures. Think as broadly as possible and expand the list of potential risks in each area.

Obviously, other kinds of risks are associated with a project of the magnitude of the customer support system. You might want to consider some risks external to the company, such as economic, marketplace, legal environment, and so forth. Other types of internal risks might also be associated with components that are purchased or outsourced, such as development tools, learning curves, poor quality of purchased components, and failure of vendors.

A common risk management technique is to build a table and identify the top 10 risks to the project. Contingency plans can then be built for the top 10 risks. Periodically, the project management team reevaluates the risk list to determine the current top 10 risks. After you build the table, identify which risks you would classify as the top 10 risks.

Focusing on Reliable Pharmaceutical Service

Chapter 2 discussed Reliable Pharmaceutical Service's Web-based application to connect its client nursing homes directly with a new prescription and billing system. You considered both the risks of a sequential, waterfall approach to the SDLC and the risks of an iterative and incremental approach to the SDLC for its development.

1. Now consider the way the project was probably initiated. To what extent is the project the result of (a) an opportunity, (b) a problem, or (c) a directive?
2. Many of the system users (such as employees at health-care facilities) are not Reliable employees. What risks of project failure are associated with the mixed user community? What would you, as a project manager, do to minimize those risks?
3. What are some of the tangible benefits to the project? What are some of the intangible benefits? What are some of the tangible and intangible costs? How would you handle the project's benefits and costs that will accrue to the health-care facilities—would you include tangible benefits and costs to the nursing homes in the cost/benefit analyses? Why or why not?
4. Overall, do you think the approach taken to the project (sequential waterfall versus iterative and incremental) would make a difference in the tangible and intangible costs and benefits? Discuss.
5. Overall, do you think the approach taken to the project would make a difference in minimizing the risks of project failure? Discuss.

FURTHER RESOURCES

Howard Eisner, *Essentials of Project and Systems Engineering Management*. John Wiley & Sons, Inc., 1997.

James P. Lewis, *Team-Based Project Management*. American Management Association, 1998.

Jack R. Meredith and Samuel J. Mantel, Jr., *Project Management, A Managerial Approach* (4th ed.). John Wiley & Sons, Inc., 2000.

Project Management Institute, *A Guide to the Project Management Body of Knowledge*. Project Management Institute, 2000.

Sanjiv Purba, D. Sawh, and B. Shah, *How to Manage a Successful Software Project*. John Wiley & Sons, Inc., 1995.

Felix Redmill, *Software Projects: Evolutionary vs. Big-Bang Delivery*. John Wiley & Sons, Inc., 1997.

Walker Royce, *Software Project Management: A Unified Framework*. Addison-Wesley, Inc., 1998.

Kathy Schwalbe, *Information Technology Project Management* (2nd ed.). Course Technology, Inc., 2002.

Standish Group, "CHAOS," 1994, http://www.pm2go.com/sample_research/chaos_1994_1.asp.

Vijay K. Verma, *Organizing Projects for Success*. Project Management Institute, 1995.

PART 2

Systems
Analysis
Tasks

CHAPTER 4

Beginning the Analysis: Investigating System Requirements

LEARNING OBJECTIVES

After reading this chapter, you should be able to:

- Describe the activities of the systems analysis life cycle phase

- Explain the effect of business process reengineering on activities of the analysis phase

- Describe the difference between functional and nonfunctional system requirements

- Identify and understand the different types of users who will be involved in investigating system requirements

- Describe the kind of information that is required to develop system requirements

- Determine system requirements through review of documentation, interviews, observation, prototypes, questionnaires, vendor research, and joint application design sessions

- Discuss the need for validation of system requirements to ensure accuracy and completeness and the use of a structured walkthrough

CHAPTER OUTLINE

The Analysis Phase in More Detail

Business Process Reengineering and Analysis

System Requirements

Stakeholders—The Source of System Requirements

Techniques for Information Gathering

Validating the Requirements

Joseph Mason, president of Convenience Office Company, was happy that the project was finally getting started. He, along with his vice president of operations and the director of information technology, had developed a new business strategy over the last six months that they thought would really open up some new markets and lock in their current customer base.

Convenience Office Company was a medium-sized distributor of office products to companies in the metropolitan area. Its primary line of business was to provide office supplies, including office furniture, to medium-sized businesses in the area. Its customers included all types of businesses. Companies with substantial office and administrative staffs were prime candidates for the services offered by Convenience, including law firms, accounting and tax offices, securities and trading companies, engineering firms, insurance brokers, insurance claim offices, schools—practically any company with people working at desks.

The office supply business had always been competitive, but with the advent of electronic commerce, it was becoming even more competitive. Many office products are commodities. For example, a pad of paper is the same no matter whom you get it from. So for the company to be successful, Convenience needed to provide excellent service, retain competitive prices, and differentiate itself from competitors. Convenience used several techniques to maintain customer loyalty. One technique was to be a full-service office supply company, offering noncommodity products such as copiers and office furniture. Convenience had also been one of the pioneers in offering on-line Internet services with Web catalogs and Internet ordering.

The new system was the next logical step in Convenience's strategic plan to increase market share locally and to expand into new markets. Basically, the project was to provide standard XML (eXtensible Markup Language) access to all of its systems. Through a standard XML interface, Convenience was beginning to provide B2B (business-to-business) capability for its customers. Not only would this new capability enable Convenience to connect directly to its more advanced customers, but it would also open an opportunity to participate in some of the new B2B exchanges that had begun to appear. Even though participation in a B2B exchange would involve competing at a commodity level, Convenience believed that it could compete successfully and, in fact, would lose market share if it did not move into this new market.

The new project, called the XML Connection Project, was going to be exciting. The new project team had already been identified, and Convenience had put its team together. The team members were now developing the business plan. Hal Turley, the project leader, had been through two other projects as project leader, so he had a good idea how to get started.

One of the initial difficulties facing the project team was identifying users and stakeholders. In most projects, the users were easy to identify. They were the internal employees and users of the new system. During the project, it was also standard procedure to interview the end users, the department heads, and other internal company employees who understood the business processes. These interviews were a necessary and critical part of identifying the system requirements.

This project was very different. It did have all of the normal internal stakeholders; however, since the objective of the system was to interface with external entities, customers, and other companies, the range of potential interested parties was much broader. The participation of these stakeholders was especially critical because the users of the system were not only within Convenience but also from Convenience's existing customer base, potential customers, and other companies. The team needed to consult both technical staff and end-user customers from each of these groups to determine the system interfaces. On a complex B2B e-commerce system, it would be easy to leave out some important users and, consequently, some important requirements.

When Hal met with his team to begin planning requirement definitions, he had three major objectives on the agenda:

- Construct a list of which customers to help develop specifications. Those who were to be involved needed to be as representative as possible of the entire customer base.
- Construct a list of all the important in-house users and technical staff who need to participate in defining system requirements.
- Begin building a list of questions and issues that need to be discussed to define the system requirements.

OVERVIEW

In the previous chapters, you learned that system development consists of four major phases: planning, analysis, design, and implementation. This chapter focuses on the activities of the analysis phase and the skills and detailed tasks that you undertake in this phase. As discussed in Chapter 1, an analyst uses many skills in system development. Two key skills that are needed to perform systems analysis are (1) fact-finding for the investigation of system requirements and (2) modeling of business processes based on the system requirements. Even though the analysis phase includes many other activities, these two skills are fundamental to analysis. In this chapter, you will develop fact-finding and investigation skills. Later chapters will cover modeling.

During the fact-finding and investigation activities, you learn details of business processes and daily operations. In fact, the objective during these activities is to try to become as knowledgeable about how the business operates as the users you interview. Why become an expert? Because only then can you ensure that the system meets the needs of the business. You bring a fresh perspective to the problem and possess a unique set of skills that you can employ to identify new and better ways to accomplish business objectives with information technology. Many current users are so used to the way they have been performing their tasks that they cannot envision better, more advanced ways to achieve results. Your technical knowledge combined with your newly acquired problem domain knowledge can bring unique solutions to business processes—and make a difference in the organization.

An additional benefit to becoming an expert in the problem domain is that you build credibility with the users. Your suggestions will carry more weight because they will meet users' specific needs. During the development of a new system, you will have many suggestions and recommendations about daily business procedures, which usually require major changes in user activities. If you can "walk the walk and talk the talk" of the users' business operations, then they are much more likely to accept your recommendations. Otherwise, you may be viewed as an outsider who really does not understand their problems.

The sections that follow first give an overview of the activities of the analysis phase. They define system requirements and explore the different types of requirements that analysts encounter. Then they explain several techniques analysts use to learn about the business processes and to gather information using both traditional and newer accelerated methods. The chapter ends with a discussion of the need for quality control of the information gathered and the models built during the activities of the analysis phase.

THE ANALYSIS PHASE IN MORE DETAIL

All of the system development approaches described and virtually all of the specific system development methodologies you will encounter in organizations include similar activities in the analysis and design life cycle phases (see Figure 4-1). Naturally, different system development methodologies recommend different techniques for completing these activities. In many cases, they just have different names—the underlying tasks are essentially the same. In some cases, different models might be created to complete an activity. But the activities always involve answering the same key questions.

FIGURE *4-1*

The activities of the analysis phase.

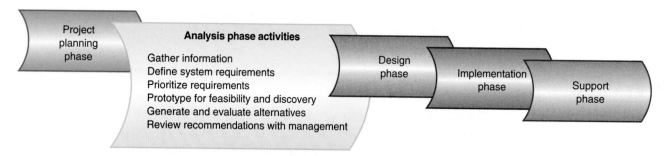

The analysis phase involves defining in great detail what the information system needs to accomplish to provide the organization with the desired benefits. Many alternative ideas should be proposed and the best design solution selected from among them. Later, during systems design, the selected alternative is designed in detail. Six activities must be completed during the analysis phase. These activities are complementary and are usually completed simultaneously. For example, the analyst gathers information continuously and defines requirements based on that information. Both activities are carried out throughout the analysis phase, not just at the beginning.

Gather Information

The analysis phase involves gathering a considerable amount of information. Systems analysts obtain some information from people who will be using the system, either by interviewing them or by watching them work. They obtain other information by reviewing planning documents and policy statements. Documentation from the existing system should also be studied carefully. Analysts can obtain additional information by looking at what other companies (particularly vendors) have done when faced with a similar business need. In short, analysts need to talk to nearly everyone who will use the new system or has used similar systems, and they must read nearly everything available about the existing system.

Beginning analysts often underestimate how much there is to learn about the work the user performs. The analyst must become an expert in the business area the system will support. For example, if you are implementing an order-entry system, you need to become an expert on the way orders are processed (including accounting). If you are implementing a loan-processing system, you need to become an expert on the rules used for approving credit. If you work for a bank, you need to think of yourself as a banker. The most successful analysts become very involved with their organization's main business.

Analysts also need to collect technical information. They try to understand the existing system by identifying and understanding activities of all current and future users, by identifying all present and future locations where work occurs, and by identifying all system interfaces with other systems both inside and outside the organization. Beyond that, analysts need to identify software packages that might be used to satisfy the system requirements. These specifics are discussed later in the chapter.

The key question to be answered when completing this activity is, *do we have all of the information (and insight) we need to define what the system must do?*

Define System Requirements

As all of the necessary information is gathered, it is very important to record it. Some of this information describes technical requirements (for example, facts about needed system performance or expected number of transactions). Other information involves functional requirements—what the system is required to do. Defining functional requirements is not just a matter of writing down facts and figures. Instead, many different types of models are created to help record and communicate what is required.

The modeling process is a learning process for an analyst. As the model is developed, the analyst learns more and more about the system. Modeling continues while information is gathered, and the analyst continually reviews the models with the end users to verify that each model is complete and correct. In addition, the analyst studies each model, adds to it, rearranges it, and then checks how well it fits with other models being created. Just when the analyst is fairly sure the system requirements are fully specified, an additional piece of information surfaces requiring yet more changes, and refinement begins again. Modeling can continue for quite some time, and it does not always have a defined end. The uncertainty involved makes some programmers uncomfortable, but it is unavoidable.

logical model

any model that shows what the system is required to do without committing to any one technology

Two types of system models are developed. A requirements model (or collection of models) is a logical model. A *logical model* shows what the system is required to do in great detail, without committing to any one technology. By being neutral about technology, the development team can focus its efforts first on what is needed, not what form it will take. For example, a model might specify an output of the system as a list of data elements without committing to either paper or on-screen formats. The focus of the model is what information the users need. A *physical model*, on the other hand, shows how the system will actually be implemented. A physical model of the output would include details about format.

physical model

any model that shows how the system will actually be implemented

The difference between logical and physical models is a key concept distinguishing systems analysis and systems design. Generally, systems analysis involves creating detailed logical models, and systems design involves detailed physical models. The design alternatives created during the analysis phase are physical models, but not very detailed.

The specific models created depend on the technique being used for systems analysis. The modern structured analysis technique uses data flow diagrams (DFDs) and entity-relationship diagrams (ERDs). Information engineering uses process dependency diagrams and entity-relationship diagrams. Object-oriented techniques produce class diagrams and use case diagrams. Specific examples of these models are described in detail in Chapters 5, 6, and 7.

The key question to be answered when completing this activity is, *what (in detail) do we need the system to do?*

Prioritize Requirements

Once the system requirements are well understood and detailed models of the requirements are completed, it is important to establish which of the functional and technical requirements are most crucial for the system. Sometimes users suggest additional system functions that are desirable but not essential. However, users and analysts need to ask themselves which functions are truly important and which are fairly important, but not absolutely required. Again, an analyst who understands the organization and the work done by the users will have more insight for answering these questions.

Why prioritize the functions requested by the users? Resources are always limited, and the analyst must always be prepared to justify the scope of the system. Therefore, it is important to know what is absolutely required. Unless the analyst carefully evaluates priorities, system requirements tend to expand as users make more suggestions (a phenomenon called *scope creep*).

The key question to be answered when completing this activity is, *what are the most important things the system must do?*

Prototype for Feasibility and Discovery

Creating prototypes of parts of the new system can be very valuable during systems analysis. The primary purpose for building prototypes during analysis—often called *discovery prototypes*—is to better understand the users' needs. Discovery prototypes are not built with the intent of being fully functional but to check the feasibility of certain approaches to the business need. In many cases, users are trying to improve their business processes or streamline procedures. So, to facilitate the investigation of new business processes, analysts can build prototypes. Using sample screens or reports, analysts discuss with users how the new system can support new processes, and they can demonstrate new business procedures for the new system. Prototypes such as these help users discover requirements they might not have thought about otherwise and get them (and the analysts) thinking creatively "outside of the box."

If the system involves new technology, it is also important early in the project to assess whether the new technology will provide the capabilities to address the business need. Then, the team can be sure that the technology is feasible. Prototypes can prove that the technology will do what it is supposed to do. Also, if the system will include new or innovative technology, the users may need help visualizing the possibilities available from the new technology when defining what they require; prototypes can fill that need.

The prototyping activity during the analysis phase helps answer two key questions: *have we proven that the technology proposed can do what we think we need it to do?* and equally important, *have we built some prototypes to ensure the users fully understand the potential of what the new system can do?*

Generate and Evaluate Alternatives

Many alternatives exist for the final design and implementation of a system. So, it is very important to define carefully and then evaluate all of the possibilities. When requirements are prioritized, the analyst can generate several alternatives by eliminating some of the less important requirements. In addition, technology also raises several alternatives for the system. Beyond those considerations, decisions such as whether to build the system using in-house development staff or a consulting firm affect the outcome. Furthermore, one or more off-the-shelf software packages could possibly satisfy all of the requirements.

Clearly, lots of alternatives are open to the project team, and each needs to be described or modeled at a high (summary) level. Each alternative also has its own costs, benefits, and other characteristics that must be carefully measured and evaluated (as in the feasibility study described in Chapter 3). The best alternative is then chosen. Choosing an alternative is not as easy as it sounds, because costs and benefits are very difficult to measure. And many design details are still uncertain. The analyst is concerned about feasibility of the project overall during the project planning phase, and then the feasibility of each alternative during the analysis phase.

The key question to be answered when completing this activity is, *what is the best way to create the system?*

As discussed previously, each activity has a specific objective, which can be expressed in terms of a question (see Figure 4-2).

FIGURE 4-2

The activities of the analysis phase and their key questions.

Analysis phase activities	Key questions
Gather information	Do we have all of the information (and insight) we need to define what the system must do?
Define system requirements	What (in detail) do we need the system to do?
Prioritize requirements	What are the most important things the system must do?
Prototype for feasibility and discovery	Have we proven that the technology proposed can do what we think we need it to do? Have we built some prototypes to ensure the users fully understand the potential of what the new technology can do?
Generate and evaluate alternatives	What is the best way to create the system?
Review recommendations with management	Should we continue and design and implement the system we propose?

Review Recommendations with Management

All of the preceding activities are done in parallel—gather information, define requirements, prioritize requirements, prototype for feasibility and discovery, and generate and evaluate alternatives. The final analysis phase activity—review recommendations with management—is usually done when all of the other analysis activities are complete or nearly complete. Management should be kept informed of progress through regular project reporting. And the project manager must eventually recommend a solution and obtain a decision from management. Questions the analyst must consider are the following: Should the project continue at all? If the project continues, which alternative is the best choice? Given the recommended alternative, what are the revised budget and schedule for completing the project?

Making a recommendation to senior executives is a major management checkpoint in the project. Every alternative—including cancellation—should be explored. Even though quite a bit of work might already have been invested in the project, it is still possible that the best choice is to cancel the project. Perhaps the benefits are not as great as originally thought. Perhaps the costs are much greater than originally thought. Or, because of the rapidly changing business environment, perhaps the organization's objectives have changed since the project was originally proposed, making it less important to the organization. For any of these reasons, it might be best to recommend that the project be canceled.

If the project is worthwhile, the project team has detailed documentation of the system requirements and a proposed design alternative, so the project manager should be able to produce a more accurate estimate of the budget and schedule for the project. If top managers understand the rationale for continuing the project, then they will probably provide the requested resources. The key point to remember is that continuing the project to the design phase is never automatic. Good project management techniques always require continual reassessment of the feasibility of the project and formal management reviews.

The key question to be answered when completing this activity is, *should we continue with the design and implement the system we propose?*

BUSINESS PROCESS REENGINEERING AND ANALYSIS

Business process reengineering (BPR), which we discussed in Chapter 1, is a movement started in the 1990s that has been the motivation for many new information systems. As the global economy has become more competitive, many companies have initiated a basic rethinking of their internal structure and business procedures. Business process reengineering is a fundamental strategic approach to organizing a company to streamline internal processes and make them as efficient and effective as possible. Previously, the old rule of business procedures was, "If it isn't broken, don't fix it." A newer way of thinking is, "There is always a better way to do it. Let's improve it." Business process reengineering extends this newer approach even further with a more revolutionary idea: "Let's question basic assumptions to find a completely new way to do it that will bring dramatic and profound improvement." Over the last decade, as global competition has become more intense, many companies have found it necessary to rethink completely their most fundamental assumptions about how they do business.

Modern information technology can enable BPR in many ways, including storing and processing large, previously unwieldy databases, providing high-quality information when and where needed, supporting rapid communication among organizational units and functions, and automating decisions previously made by people. For example, many companies have begun strategic thrusts to provide very personalized service to their customers. One of the new buzzwords is customer relationship management (CRM), which we discussed in Chapter 1. Many system vendors are providing specialized software to support CRM. The ability to achieve these goals depends on, first, a high level of automation support for the tremendous amount of detailed information that is generated and, second, very efficient business processes to take advantage of the technology.

One of the classic stories of dramatic improvement is the accounts payable function in the North American division of Ford Motor Company. In the mid-1980s, the accounts payable department employed over 500 people. The original project was to develop an information system to achieve a 20 percent improvement in productivity. However, as the project team and the executives began looking at ways to use automation to improve performance, they found that Mazda, of which Ford owned 20 percent, employed only five people in accounts payable. Even though Ford was a much larger organization, it still had over 100 times as many people performing basically the same function. With a better vision of what was possible, the project team completely redesigned the payables function, utilizing a much higher level of automation. In essence, the accounts payable function was subsumed into a larger purchasing function so that tracking of payables became automated from the time the purchase was made. At the conclusion of the project, Ford was able to carry out its accounts payable functions with a little more than 100 people—a 400 percent improvement.

The decision to approach a system development project as a BPR project has a significant impact on analysis phase activities. The activities of gathering information and defining system requirements are less concerned with detailed operational aspects of the current system and more concerned with exploring new and untried ways of achieving the same goals and satisfying the same business purpose. In a BPR project, an analyst is much more likely to venture beyond the initial scope of the project to pursue information about system requirements and potential improvements and efficiencies. When gathering information and generating and evaluating alternatives, the analyst draws input from a wider variety of internal and external stakeholders. Since business procedures and technology may be more radically altered in a BPR project, prototyping for feasibility and discovery may receive greater emphasis.

SYSTEM REQUIREMENTS

system requirements

specifications that define the functions to be provided by a system

System requirements are all of the capabilities and constraints that the new system must meet. Generally analysts divide system requirements into two categories: functional and nonfunctional requirements. Recall that one of the activities during the planning phase is to identify the system scope. In that activity, the analyst identifies a set of system capabilities. During analysis, the analyst then defines and describes those capabilities in greater detail. In other words, the analyst expands those high-level capabilities into detailed system requirements.

functional requirement

a system requirement that describes an activity or process that the system must perform

Functional requirements are the activities that the system must perform—that is, the business uses to which the system will be applied. They derive directly from the capabilities identified in the planning phase. For example, if you are developing a payroll system, the required business uses might include functions such as "write paychecks," "calculate commission amounts," "calculate payroll taxes," "maintain employee-dependent information," and "report year-end tax deductions to the IRS." The new system must handle all these functions. Identifying and describing all of these business uses requires a substantial amount of time and effort because the list of functions and their relationships can be very complex.

Functional requirements are based on the procedures and rules that the organization uses to run its business. Sometimes they are well documented and easy to identify and describe. An example might be, "All new employees must fill out a W-4 form to enter information about their dependents in the payroll system." Other business rules might be more obtuse or difficult to find. An example from Rocky Mountain Outfitters might be that "an additional 2 percent commission rate is paid to order takers on telephone sales for 'special promotions' that are added to the order." These special promotions are unadvertised specials that are sold by the telephone order clerk—thus the special commission rate. Discovering rules such as this is critical to the final design of the system. If this rule weren't discovered, you might design a system that allows only fixed commission rates and discover much later in the development processing that your design could not accommodate this rule.

Nonfunctional requirements are characteristics of the system other than activities it must perform or support. There are many different types of nonfunctional requirements, including the following:

technical requirement

a system requirement that describes an operational characteristic related to an organization's environment, hardware, and software

- *Technical requirements* describe operational characteristics related to the environment, hardware, and software of the organization. For example, the client components of a new system might be required to operate on portable and desktop PCs running the Windows operating system and using Internet Explorer. The server components might have to be written in Java and communicate with one another using a component interaction standard such as CORBA (Common Object Request Broker Architecture) or SOAP (Simple Object Access Protocol).

performance requirement

a system requirement that describes an operational characteristic related to workload measures, such as throughput and response time

- *Performance requirements* describe operational characteristics related to measures of workload, such as throughput and response time. For example, the client portion of a system might be required to have one-half-second response time on all screens, and the server components might need to support 100 simultaneous client sessions (with the same response time).

usability requirement

a system requirement that describes an operational characteristic related to users, such as the user interface, work procedures, on-line help, and documentation

- *Usability requirements* describe operational characteristics related to users, such as the user interface, related work procedures, on-line help, and documentation. For example, A Web-based interface might be required to follow organizationwide graphic design guidelines such as menu placement and format, color schemes, use of the organization's logo, and required legal disclaimers.

reliability requirement

a system requirement that describes the dependability of a system, such as service outages, incorrect processing, and error detection and recovery

- *Reliability requirements* describe the dependability of a system—how often a system exhibits behaviors such as service outages and incorrect processing and how it detects and recovers from those problems. Reliability requirements are sometimes considered a subset of performance requirements.

security requirement

a system requirement that describes user access to certain functions and the conditions under which access is granted

- *Security requirements* describe which users can perform what system functions under what conditions. For example, access to certain system outputs may be limited to managers at a certain level or employees of a specific department. Some access may be authorized from home and others only from within the organization's local network. Security requirements may also apply to areas such as network communications and data storage. For example, an organization might require encryption of all data transmitted over the Internet and control of all database server access through use of a username and password.

Both functional and nonfunctional system requirements are needed for a complete definition of a new system, and both are investigated and documented during systems analysis. Functional requirements are most often documented in graphical and textual models, as described in Chapters 5 through 7. Nonfunctional requirements are usually documented in narrative descriptions that accompany the models.

STAKEHOLDERS—THE SOURCE OF SYSTEM REQUIREMENTS

stakeholders

all the people who have an interest in the success of a new system

Your primary source of information for system requirements is the various stakeholders of the new system. *Stakeholders* are all people who have an interest in the successful implementation of the system. Generally, we categorize stakeholders into one of three groups: (1) the users, those who actually use the system on a daily basis, (2) the clients, those who pay for and own the system, and (3) the technical staff, the people who must ensure that the system operates in the computing environment of the organization. Figure 4-3 illustrates the various kinds of stakeholders who have an interest in a new system. We have discussed earlier the difference between the users and the clients. During analysis, the analyst also needs to consider the technical staff as well. One of the most important first steps in determining system requirements is to identify these various system stakeholders. In the past, problems have arisen with new systems because only some of the stakeholders were included in the project and the system was built exclusively for them. One of an analyst's first tasks is to identify every type of stakeholder who has an interest in the new system. The second task is to ensure that critical people from each stakeholder category are available to the project as the business experts.

FIGURE 4-3

Stakeholders with an interest in new system development.

Users as Stakeholders

User roles—that is, types of system users—should be identified in two dimensions: horizontally and vertically. By horizontally, we mean that the analyst must look for information flow across business departments or functions. For example, a new inventory system may affect receiving, warehousing, sales, and manufacturing. So, individuals from each of these departments must describe their requirements. The sales department may need to determine when and how to update inventory quantities or to commit inventory at the time of the sale but before it is shipped. Manufacturing may need certain information from the inventory system to assist in scheduling production. So, remembering to include the horizontal dimension in the definition of requirements will ensure that the many different departments, even those that may appear unrelated to the new system, are included.

By vertical dimension, we mean the information needs of clerical staff, of middle management, and of senior executives. Each of these stakeholders has different information requests for the system that must be included in the design. The following sections describe the characteristics and information needs of the various users on the vertical dimension. These same characteristics also apply to each department across the horizontal dimension.

Business Users

Business users are the people who use the system to perform the day-to-day operations of an organization. We often call these operations *transactions*. A ***transaction*** is a piece of work done in an organization, such as "enter an order." In Chapter 1, you learned that a transaction processing system handles these types of business operations. Business users provide information about the daily operations of the business and ways the system must support them.

Information Users

An information user is a person who needs current information from the system. This person may be an operational user or someone else. In some cases, a business may want to make information directly available to customers. However, an information user may not be allowed to enter information on business transactions, just to view specific information. An information user, then, provides an analyst with insight about what kinds of information should be available daily, weekly, monthly, and annually, and about what format is most convenient.

Management Users

Managers are responsible for seeing that the company is performing its daily procedures efficiently and effectively. Consequently, they need statistics and summary information from a system. Management will help an analyst answer the following types of questions:

- What kinds of reports must the system produce?
- What kind of performance statistics must the system maintain?
- What kind of volume information must the system keep, and what volumes of transactions must the new system support?
- Are the controls in the system adequate to prevent errors and fraud?
- How many requests for information will be made and how often?

Executive Users

The top executives of an organization are interested in strategic issues, as well as the daily issues just described. They typically want information from a system so that they can compare overall improvements in resource utilization. They may want the system to interface with other systems to provide strategic information on trends and directions of the industry and the business.

External Users

More and more systems today allow external entities to have direct access to the system. Customers may access the system directly through the Internet. Suppliers may have access to a system to check inventory levels and to initiate billing transactions. These users are more difficult to identify and access since they are not regular members of the organization. However, today they belong to an important group that must be considered in system development.

Client Stakeholders

Although the project team must meet the information processing needs of the users, it also must satisfy the client. Chapter 3 defined the client as the person or group who is providing the funding for the project. In many cases, the client is the same group as the executive users. However, clients may also be a separate group, such as a board of trustees or executives in a parent company. The project team includes the client in its list of important stakeholders because the team must provide periodic status reviews to the client throughout development. The client or a direct representative on a steering or oversight committee also usually approves stages of the project and releases funds.

Technical Stakeholders

Although the technical staff is not a true user group, this group affects many system requirements. The technical staff includes people who establish and maintain the computing environment of the organization. They provide guidance in such areas as programming language, computer platforms, and other equipment. For some projects, the project team includes a member of the technical staff. For other projects, technical personnel are available as needed.

The Stakeholders for Rocky Mountain Outfitters

To demonstrate the different perspectives of stakeholders, let's look at the proposed customer support system for Rocky Mountain Outfitters.

An important part of investigating system requirements is to identify all of the stakeholders. The set of requirements is incomplete if users, clients, external entities, or important technical staff are not consulted as information is being gathered. At RMO, operational users of the new order-processing system include inside sales representatives who take orders over the phone, as well as clerks who process mail orders. They all have different views about what the system should do for them. Sales representatives talk about looking up product information for customers and confirming availability and shipping dates. Mail-order clerks talk about scanning order information into the system to eliminate typing. The warehouse workers who put the shipments together need information about orders that have been shipped, orders to be shipped, and back orders, as well as their normal operational screens that allow them to put orders together into shipments with printed bills of lading.

John and Liz Blankens, as owners, have special interests in reports of the products that have been ordered and shipped. They are interested in watching seasonal trends within and across products. In the sports equipment business, it is critically important to push the trendy items quickly and move on when the trend is past.

The development of the customer support system has been funded in part from internal cash flows. Funds have also been obtained, however, through a special line of credit at the bank. RMO normally has a short-term line of credit for seasonal needs. Since the CSS project is a longer-term investment for a capital good, the Blankenses obtained a different line of financing for it. Their banker is extremely interested in the success of the project, so in this case, the project team even met with the bank's staff to see what special formats of financial information the bank would like the system to maintain.

Finally, since this system will involve new technology—the Internet and distributed systems—very heavy involvement is required by the technical staff. Consequently, many stakeholders will have input into the types of information that can be extracted from the system. Figure 4-4 illustrates, from the upper-level RMO organization chart, people who will be involved. The orange positions indicate the executives and middle managers who will be involved as stakeholders. The project manager will build a list of all users who need to be involved in requirements definition. This organization chart is just the beginning. Other department managers and key employees will also be added.

How did the project team identify which stakeholders to include in the interview schedule? This is always a difficult question. The process begins, however, with an analysis of the scope of the new system. After defining the scope, the team must carefully

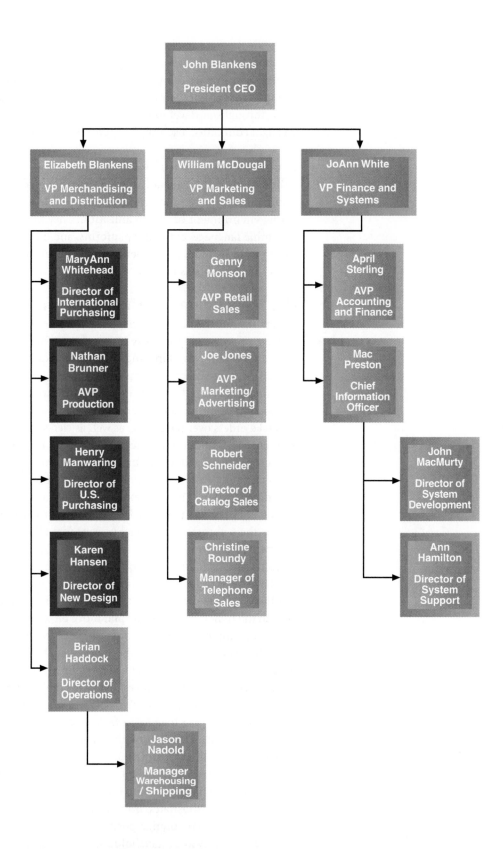

analyze all the people who may require information from the system in any way. At this point, it is better to err on the side of including too many stakeholders rather than missing important sources of requirements. Barbara Halifax sent John MacMurty a memo updating him of her progress in identifying the CSS stakeholders and her upcoming plans for gathering information (see Barbara's memo).

March 10, 2005

To: John MacMurty

From: Barbara Halifax

RE: Customer Support System Update

As we begin the fact-finding activities of the project, I wanted to let you know where we are in the development of an interview schedule and questions. As you can see from the attached organization chart, we have identified the senior managers who should be included in our fact-finding interviews. Please let me know if we have missed anybody who should be included.

We have also worked with the department managers to put together a list of people who will be interviewed in each of the various departments. In some departments, the manager wanted us to have group interviews and have several users participate in a kind of a mini-JAD session. I think this will work well because it will help us get final decisions quickly.

We do have one major question, however. How do we identify external users, such as Web customers, and how do we get them involved? Do you think RMO would be willing to provide some incentive to a couple of customers to help out—maybe a $100 shopping spree or something?

Finally, you should know that we have already collected copies of all the existing forms and reports. We used the existing documents to help us in developing a set of questions to help us get started. Since we want to encourage the users to "think outside the box," we will probably not use the forms in the interviews but only as a source for ideas and questions.

BH

cc: Steven Deerfield, Ming Lee

TECHNIQUES FOR INFORMATION GATHERING

The objective of the analysis phase of system development is to understand the business functions and develop the system requirements. The question that always arises is whether to study and document the existing system or whether to document only the requirements of the new system. When the structured approach, as well as the other approaches explained in Chapter 2, were first developed, systems analysts would first document the existing system and then extrapolate the requirements of the new system from that documentation. In those days, the development of system requirements was a four-step process: (1) identify the physical processes and activities of the existing system, (2) extract the logical business function that was inherent in each existing physical process, (3) develop the logical business functions for the approach to be used in the new system, and (4) define the physical processing requirements of the new system. One disadvantage of this approach was the inordinate amount of time it took. Another problem, frequently with long-term consequences, was that system developers would often simply automate the existing system, in other words "pave the cow paths." As a result, no matter how inefficient the current system was, system developers would simply automate the procedures that were already in place.

As we discussed earlier, in today's competitive world, many companies are using the implementation of new information technology to increase their advantage and are completely redesigning and streamlining internal procedures through business process engineering.

The objective of analysis has not changed; however, the approach to developing system requirements has improved. It is still critical to have a complete, correct set of system requirements, but in today's fast-paced world, there is no time or money to review all the old systems and document all the inefficient procedures. Today, analysts use an accelerated approach by balancing the review of current business functions with the new system requirements. As shown in Figure 4-5, the focus of analysis activities today is to develop a set of logical system requirements for the new system immediately. Analysts review the current system only when they need to understand the business needs, not to define the specific processes of the old system. This focus on the new while sometimes referring to the old is a balancing act for system professionals. They need to understand the business needs in extreme detail (remember, "walk the walk and talk the talk"), but they do not want to get caught up in old, inefficient methods. In fact, in today's development environment, one of the most valuable capabilities that a good system developer can bring is a new perspective to the problem.

FIGURE *4-5*

The relationship between information gathering and model building.

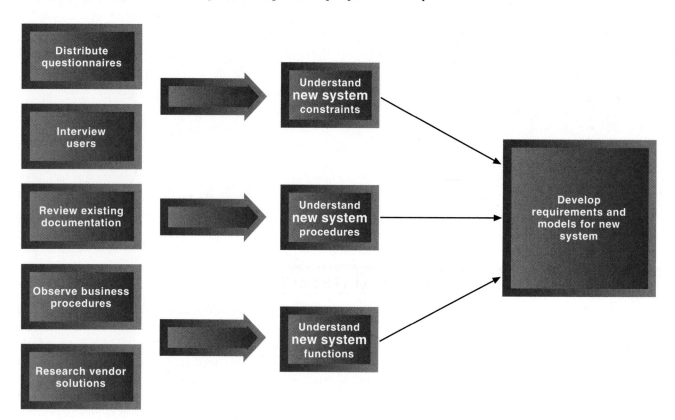

The analysts develop the logical model of the new system as they gather information. The project team creates the physical model (that is, how the system will be built) later as part of systems design. Analysts focus on certain themes and use various techniques to develop the logical model of the system.

Question Themes

The first questions that new systems analysts ask are, "What kind of information do I need to collect? What is a requirement?" Basically, you want to obtain information that will enable you to build the logical model of the new business system. As shown in Figure 4-6, three major themes should guide you as you pursue your investigation.

FIGURE 4-6

Themes for information-gathering questions.

Theme	Questions to users
What are the business operations and processes?	What do you do?
How should those operations be performed?	How do you do it? What steps do you follow?
What information is needed to perform those operations?	What information do you use? What forms or reports do you use?

What Are the Business Processes?

In the first question—What do you do?—the focus is on understanding the business functions. This question is the first step in being able to walk the walk. The analyst must obtain a comprehensive list of all the business processes. In most cases, the users provide answers in terms of the current system, so the analyst must discern carefully which of those functions are fundamental—which will remain and which may possibly be eliminated with an improved system. For example, sales clerks may indicate that the first thing they do when a customer places an order is to check the customer's credit history. In the new system, sales clerks may never need to perform that function; the system might perform the check automatically. The function remains a system requirement, but the method of carrying out the function moves from the clerks to the computer system.

How Is the Business Process Performed?

The second question—How can it be done?— moves the discussion from the current system to the new system. The focus is on how the new system *should* support the function rather than on how it does now. Thus, the first two questions go hand in hand to discover the need and begin to define the system requirement in terms of the new system. The users most frequently talk about the current system, but it is critical for the systems analyst to go beyond the current process. He or she must be able to help the user visualize new and more efficient approaches to performing the business processes made possible by the new technology.

What Information Is Required?

The final question—What information is needed?—elaborates the second question by defining specific information that the new system must provide. The answers to the second and third questions form the basis for the definition of the system requirements. One of the shortcomings of many new systems analysts is that they do not identify all of the required pieces of information. In both this question and the previous one, detail, detail, detail is the watchword. An analyst must understand the nitty-gritty detail in order to develop a correct solution.

Focusing on these three themes helps an analyst ask intelligent, meaningful questions in an investigation. Later, as you learn about models, you will be able to formulate additional meaningful detailed questions to ask.

As you develop skill in asking questions and building models, your problem-solving and analytical skills will increase. Remember, your value as a systems analyst is not that you know how to build a specific model or how to program in a specific language. Your value is in your ability to analyze and solve business information problems—to gather the correct information. Fundamental to that skill is not only how effectively but also how efficiently you can identify and capture these business rules. Effective requirements are complete, comprehensive, and correct. An efficient analyst is one who moves the project ahead rapidly with minimal intrusion on users' time and use of other resources, yet ensures that the information gathered will produce complete, comprehensive, and correct requirements specifications.

The next sections present the various methods of information gathering. All these methods have been proven to be effective, although some are more efficient than others. In most cases, analysts combine methods to increase both their effectiveness and efficiency and provide a comprehensive fact-finding approach. The most widely used methods are the following:

- Review existing reports, forms, and procedure descriptions
- Conduct interviews and discussions with users
- Observe and document business processes
- Build prototypes
- Distribute and collect questionnaires
- Conduct joint application design (JAD) sessions
- Research vendor solutions

Review Existing Reports, Forms, and Procedure Descriptions

This step should probably be the first in fact-finding activities. There are two sources of information for existing procedures and forms. One source is external to the organization—at industrywide professional organizations and at other companies. It may not be easy to obtain information from other companies, but they are a potential source of important information. Sometimes, industry journals and magazines report the findings or "best practices" studies. The project team would be negligent in its duties if its members were not familiar with best practice information. Also, with systems crossing organization boundaries more and more, external sources are an important source of system requirements.

The second source of reports, forms, and procedures is the existing business documents and procedure descriptions within the organization. This internal review serves two purposes. First, it is a good way to get a preliminary understanding of the processes. Often new systems analysts need to learn about the industry or the specific application that they are studying. A preliminary review of existing documentation will bring them up to speed fairly rapidly.

To begin the process, the analysts ask users to provide copies of the forms and reports that they currently are using. They also request copies of procedural manuals and work descriptions. The review of these materials provides an understanding of the business functions. They also form the basis for the development of detailed interview questions.

The second way to use documents and reports is in the interviews themselves. Forms and reports can serve as visual aids for the interview, and the working documents for discussion (see Figure 4-7). Discussion can center on the use of each form, its objective, its distribution, and its information content. The discussion should also include specific business events that initiate the use of the form. Several different business events might require the same form, and specific information about the event and the business process is critical. It is also always helpful to have forms that have been filled out with real information to ensure that the analyst obtains a correct understanding of the fields and data content.

Rocky Mountain Outfitters—Customer Order Form

Name and address of person placing order.
(Please verify your mailing address and make correction below.)
Order Date ___ / ___ / ___

Name

Address   Apt. No

City   State   Zip

Phone: Day ( )_____ Evening ( )_____

Gift Order or Ship To: (Use only if different from address at left.)

Name

Address   Apt. No

City   State   Zip

Gift ☐   Address for this Shipment Only ☐   Permanent Change of Address ☐

Gift Card Message

Delivery Phone ( )_____

Item No.	Description	Style	Color	Size	Sleeve Length	Qty	Monogram	Style	Price Each	Total

Method of Payment

Check/Money Order ☐   Gift Certificate(s) ☐   AMOUNT ENCLOSED $ _____

American Express ☐   MasterCard ☐   VISA ☐   Other ☐

Account Number
☐☐☐☐☐☐☐☐☐☐☐☐☐☐☐☐
MO   YR
___ / ___
Expiration Date

Signature

MERCHANDISE TOTAL _____

Regular FedEx shipping $4.50 per U.S. delivery address _____
(Items are sent within 24 hours for delivery in 2 to 4 days)

Please add $4.50 per each additional U.S. delivery address _____

FedEx Standard Overnight Service _____

Any additional freight charges _____

International Shipping (see shipping information on back) _____

FIGURE 4-7

A sample order form for Rocky Mountain Outfitters.

Reviewing the documentation of existing procedures helps identify business rules that may not come up in the interviews. Written procedures also help discover discrepancies and redundancies in the business processes. However, procedure manuals frequently are not kept up to date, and they commonly include errors. To ensure that the assumptions and business rules that derive from the existing documentation are correct, analysts should review them with the users.

Conduct Interviews and Discussions with Users

Interviewing stakeholders is by far the most effective way to understand business functions and business rules. It is also the most time-consuming and resource-expensive option. In this method, systems analysts meet with individuals or groups of users. A list of detailed questions is prepared, and discussion continues until all the processing requirements are understood and documented by the project team. Obviously, this process may take some time, so it usually requires multiple sessions with each of the users or user groups.

To conduct effective interviews, analysts need to organize in three areas: (1) preparing for the interview, (2) conducting the interview, and (3) following up the interview. Figure 4-8 is a sample checklist that summarizes the major points to be covered and is useful for preparing for and conducting an interview.

Preparing for the Interview

Every successful interview requires preparation. The very first, and most important, step in preparing for an interview is to establish the objective of the interview. In other words, what do you want to accomplish with this interview? Write down the objective so that it is firmly established in your mind. The second step is to determine which users should be involved in the interview. Frequently, the first two steps are so intertwined that both are done together. Even if you don't do anything else to prepare for your

FIGURE *4-8*

A sample checklist to prepare
for user interviews.

Checklist for Conducting an Interview

Before

☐ Establish the objective for the interview
☐ Determine correct user(s) to be involved
☐ Determine project team members to participate
☐ Build a list of questions and issues to be discussed
☐ Review related documents and materials
☐ Set the time and location
☐ Inform all participants of objective, time, and locations

During

☐ Dress appropriately
☐ Arrive on time
☐ Look for exception and error conditions
☐ Probe for details
☐ Take thorough notes
☐ Identify and document unanswered items or open questions

After

☐ Review notes for accuracy, completeness, and understanding
☐ Transfer information to appropriate models and documents
☐ Identify areas needing further clarification
☐ Send thank-you notes if appropriate

interviews, you must at least complete these two steps. The objective and the participants drive everything else in the interview.

The interview participants include both users and project members. Generally, at
least two project members are involved in every interview. The two project members
help each other during the interview and compare notes afterward to ensure accuracy.
The number of users varies depending on the objective of the interview. A small number
of users is generally best when the interview objective is narrow or of a fact-finding nature. In such cases, interviewing more than three users at a time tends to cause unneces-
sarily long discussions. Larger groups are better if the objective is more open-ended,
such as when exploring new process alternatives in a BPR project. Larger groups are often better for generating and evaluating new ideas. However, it can be difficult to manage a large group meeting to ensure high-quality input from all participants. Professional facilitators and formal discovery techniques such as joint application design
(discussed later in this chapter) may be employed if the objective is complex or critical
and the group is large.

The third step is to prepare detailed questions to be used in the interview. Write
down a list of specific questions and prepare notes based on the forms or reports received earlier. Usually you should prepare a list of questions that are consistent with the
objective of the interview. Both open-ended questions and closed-ended questions are
appropriate. Open-ended questions, such as, "How do you do this function?" encourage
discussion and explanation. Closed-ended questions, such as, "How many forms a day
do you process?" are used to get specific facts. Generally, open-ended questions help get
the discussion started and encourage the user to explain all the details of the business
process and the rules.

The last step is to make the final interview arrangements and to communicate those
arrangements to all participants. A specific time and location should be established. If
possible, a quiet location should be chosen to avoid interruptions. Each participant
should know the objective of the meeting and, when appropriate, should have a chance
to preview the questions or materials to be reviewed. Interviews consume a substantial

amount of time, and they can be made more efficient if each participant knows beforehand what is to be accomplished.

Conducting the Interview

New systems analysts are usually quite nervous about conducting interviews. However, remember that, in most cases, the users are excited about getting a better system to help them do their jobs. Practicing good manners usually ensures that the interview will go well. Here are a few guidelines.

Dress appropriately. Dress at least as well as the best-dressed user. In many corporate settings, such as banks or insurance companies with managers present, business suits are appropriate. In factory or manufacturing settings, work dress may be appropriate. The objective in dressing is to project competence and professionalism without intimidating the user.

Arrive on time. If anything, be a little early. If the session is in a conference room, ensure that it is set up appropriately. For a large group or a long session, plan for refreshment breaks.

Limit the time of the interview. Both the preparation and the interview itself affect the time required for the interview. As you set the objective and develop questions, plan for about an hour and a half. If the interview will require more time to cover the questions, it is usually better to break off the discussion and schedule another session. (Other techniques that we discuss later have all-day sessions.) The users have other responsibilities, and the systems analysts can absorb only so much information at one time. It is better to have several shorter interviews than one long marathon. A series of interviews provides an opportunity to absorb the material and to go back to get clarification later. Both the analysts and the users will have better attitudes with several shorter interviews.

Look for exception and error conditions. Look for opportunities to ask "what if" questions. "What if it doesn't arrive? What if the signature is missing? What if the balance is incorrect? What if two order forms are exactly the same?" The essence of good systems analysis is understanding all of the "what ifs." Make a conscious effort to identify all of the exception conditions and ask about them. More than any other skill, the ability to think of the exceptions will strengthen the skill of discovering the detailed business rules. It is a hard skill to teach from a textbook; experience will hone this skill. You will teach yourself this skill by conscientiously practicing it.

Probe for details. In addition to looking for exception conditions, the analyst must probe to ensure a complete understanding of all procedures and rules. One of the most difficult skills to learn as a new systems analyst is to get enough details. Frequently, it is easy to get a general overview of how a process works. But do not be afraid to ask detailed questions until you thoroughly understand how it works and what information is used. You cannot do effective systems analysis by glossing over the details.

Take careful notes. It is a good idea to take handwritten notes. Usually tape recorders make users nervous. Note taking, however, signals that you think the information you are obtaining is important, and the user is complimented. If two analysts conduct each interview, they can compare notes later. Identify and document in your notes any unanswered questions or outstanding issues that were not resolved. A good set of notes provides the basis for building the analysis models as well as establishing a basis for the next interview session.

Figure 4-9 is a sample agenda for an interview session. Obviously, you do not need to conform exactly to a particular agenda. However, as with the interview checklist shown in Figure 4-8, this figure will help prod your memory on issues and items that should be discussed in an interview. Make a copy and use it. As you develop your own style, you can modify the checklist for the way you like to work.

FIGURE 4-9

Sample interview session agenda.

Discussion and Interview Agenda

Setting

Objective of Interview
  Determine processing rules for sales commission rates

Date, Time, and Location
  April 21, 2005, at 9:00 a.m. in William McDougal's office

User Participants (names and titles/positions)
  William McDougal, vice president of marketing and sales, and several of his staff

Project Team Participants
  Mary Ellen Green and Jim Williams

Interview/Discussion

1. Who is eligible for sales commissions?
2. What is the basis for commissions? What rates are paid?
3. How is commission for returns handled?
4. Are there special incentives? Contests? Programs based on time?
5. Is there a variable scale for commissions? Are there quotas?
6. What are the exceptions?

Follow-Up

Important decisions or answers to questions
  See attached write-up on commission policies

Open items not resolved with assignments for solution
  See Item numbers 2 and 3 on open items list

Date and time of next meeting or follow-up session
  April 28, 2005, at 9:00 a.m.

Following up the Interview

Follow-up is an important part of each interview. The first task is to absorb, understand, and document the information that was obtained. Generally, analysts document the details of the interview by constructing models of the business processes and writing textual descriptions of nonfunctional requirements. These tasks should be completed as soon as possible after the interview and the results distributed to the interview participants for validation. If the modeling methods are complex or unfamiliar to the users, the analyst should schedule follow-up meetings to explain and verify the models, as described in the last section of this chapter.

During the interview, you probably asked some "what if" questions that the users could not answer. They are usually policy questions raised by the new system but which management has not considered before. It is extremely important that these questions not get lost or forgotten. Figure 4-10 is a sample form with representative questions from Rocky Mountain Outfitters. If several teams are working, a combined list can be maintained. Other columns that might be added to the list are an explanation of the resolution to the problem and the date resolved.

Finally, make a list of new questions based on areas that need further elaboration or that are missing information. This list will prepare you for the next interview.

OUTSTANDING ISSUES CONTROL TABLE						
ID	Issue Title	Date Identified	Target End Date	Responsible Project Person	User Contact	Comments
1	Partial Shipments	6-12-2005	7-15-2005	Jim Williams	Jason Nadold	Ship partials or wait for full shipment?
2	Return and Commissions	7-01-2005	9-01-2005	Jim Williams	William McDougal	Are commissions recouped on returns?
3	Extra Commissions	7-01-2005	8-01-2005	Mary Ellen Green	William McDougal	How to handle commissions on special promotions?

FIGURE *4-10*

A sample open-items list.

Observe and Document Business Processes

Along with interviews, another extremely useful method of gathering information is to observe users directly at their job sites and to document the processes you observe. This firsthand experience is invaluable to understanding exactly what occurs in business processes.

Observing

The old adage that a picture is worth a thousand words is also true with systems analysis. More than any other activity, observing the business processes in action will help you understand the business functions. However, while observing existing processes, you must also be able to visualize the new system's associated business processes. That is, as you observe the current business processes in order to understand the fundamental business needs, you should never forget that the processes could, and often should, change to be more efficient. Don't get locked into believing there is only one way of performing the process.

You can observe the work in several ways, from a quick walkthrough of the office or plant to doing the work yourself. A quick walkthrough gives a general understanding of the layout of the office, the need for and use of computer equipment, and the general workflow. Spending several hours to observe a user actually doing his or her job provides an understanding of the details of actually using the computer system and carrying out business functions. By being trained as a user and actually doing the job, you can discover the difficulties of learning new procedures, the importance of a system that is easy to use, and the stumbling blocks and bottlenecks of existing procedures and information sources.

It is not necessary to observe all processes at the same level of detail. A quick walkthrough may be sufficient for one process, while another process that is critical or more difficult to understand might require an extended observation period. If you remember that the objective is a complete understanding of the business processes and rules, then you can assess where to spend your time to gain that thorough understanding. As with interviewing, it is usually better if two analysts combine their efforts in observing procedures.

Observation often makes the users nervous, so you need to be as unobtrusive as possible. There are several ways to put users at ease, such as working with a user or observing several users at once. Common sense and sensitivity to the needs and feelings of the users will usually result in a positive experience.

Documenting with Activity Diagrams

As you gather information about business processes, primarily by interviewing the users and by observing the processes, you will need to document your results. One effective way to capture this information is through the use of diagrams. Eventually, you may want to use diagrams to describe the workflows of the new system, but for now, let's just focus on how we would document the current business workflows.

workflow

a sequence of steps to process a business transaction

activity diagram

a type of workflow diagram that describes the user activities and their sequential flow

synchronization bar

a symbol in an activity diagram to control the splitting or uniting of sequential paths

swimlane

a rectangular area on an activity diagram representing the activities done by a single agent

FIGURE *4-11*

Activity diagram symbols.

A *workflow* is the sequence of processing steps that completely handles one business transaction or customer request. Workflows may be simple or complex. Complex workflows may be composed of dozens or hundreds of processing steps and may include participants from different parts of an organization. As an analyst, you may try to depend only on your memory to remember and understand the workflow (a bad idea), you may write it down in a long description, or you can document it with a diagram. The advantages of a simple diagram are that you can visualize it better, and you can review it with the users to make sure it is correct. One of the major benefits of using diagrams and models is that they become a powerful communication mechanism between the project team and the users.

No single method is commonly used to model workflows. Methodologies commonly employed include flowcharts, data flow diagrams, and activity diagrams. Data flow diagrams do a good job of capturing the flow of data within a workflow, but they aren't designed to represent control flows. Flowcharts and activity charts are specifically designed to represent control flow among processing steps, but they don't represent data flow. So, many analysts use a type of workflow diagram called an *activity diagram*. An *activity diagram* is simply a workflow diagram that describes the various user (or system) activities, the person who does each activity, and the sequential flow of these activities.

Figure 4-11 shows the basic symbols used in an activity diagram. The ovals represent the individual activities in a workflow. The connecting arrows represent the sequence between the activities. The black circles are used to denote the beginning and ending of the workflow. The diamond is a decision point at which the flow of the process will either follow one path or the other path. The heavy solid line is a *synchronization bar*, which either splits the path into multiple concurrent paths or recombines concurrent paths. The *swimlane* represents an agent who performs the activities. Since, in a workflow, it is common to have different agents (that is, people) performing different steps of the

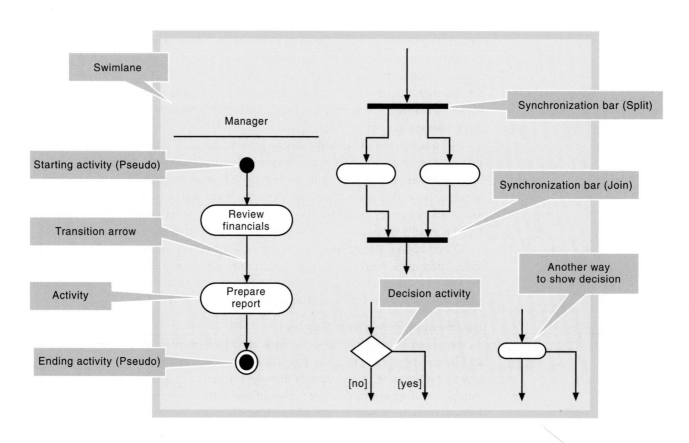

workflow process, the swimlane symbol divides the workflow activities into groups showing which agent performs which activity.

Figure 4-12 is an actual activity diagram for a workflow. This workflow represents a customer requesting a quote from a salesperson. If it is a simple request, the salesperson can enter the data and create the quote. If it is complex, the salesperson requests assistance from a technical expert to generate the quote. In both cases, the computer system calculates the details of the quote.

Suppose in this case that you have interviewed the salesperson and observed the generation of a quote. Looking at Figure 4-12, you can see how the workflow progresses. The customer initiates the first step by requesting a quote. The salesperson performs the next step in the workflow. She writes down the details of the quote request and then decides whether she can do it herself or whether she needs help. If she does not need help, then the salesperson enters the information into the computer system. If the salesperson needs help, then the technical expert performs the next step. The expert reviews the quote request to make sure that the requested components can be integrated into a functioning computer system. The activity of checking the request is fairly complex, and you could in fact break it down into more detailed steps if desired. For now, let's leave the diagram at this level of detail. The expert then enters the information into the system. At this point, the computer system generates the detailed quote. Notice that no matter which path was

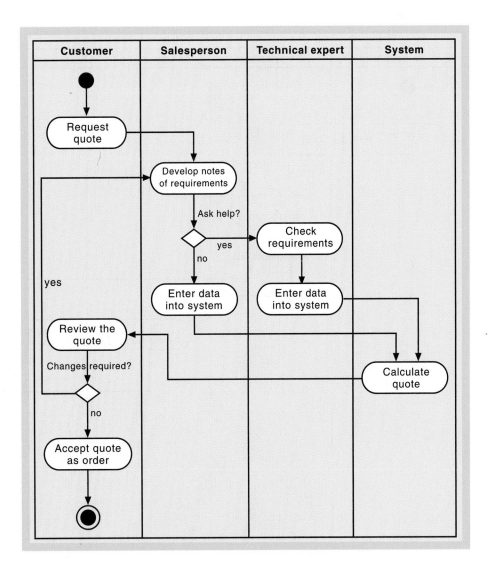

FIGURE *4-12*

A simple activity diagram to demonstrate a workflow.

taken, they both result in this common activity. Finally, the customer reviews the quote and decides whether it needs changes or is acceptable. In this simple case, the customer always does buy something, so this workflow is obviously not completely accurate.

Notice that an activity diagram focuses on the sequence of activities. This diagram is straightforward and quite easy to understand. In fact, one of the strengths of using activity diagrams to document workflows is that users also find them very easy to understand. You can use graphical representations such as this diagram to review your understanding of the particular workflow procedure with the user.

Figure 4-13 illustrates another workflow. This diagram demonstrates some new concepts. Let's assume that the customer from the previous example did want to proceed with an order. This next figure shows the workflow that is required to get the order scheduled for production. The salesperson sends to engineering the printed quote, which has now become an order. This example emphasizes the fact that a document is being transmitted. To indicate that a document is being passed, you place the document symbol at the end of the connecting arrow, and the arrow now becomes a conduit for transmitting a document, not just a flow of activities. After engineering develops the specifications, two concurrent activities happen: purchasing orders the materials, and production writes the program for the automated milling machines. These two activities are completely independent and can occur at the same time. Notice that one synchronization bar splits the path into two concurrent paths, and another synchronization bar reconnects them. Finally, scheduling puts the order on the production schedule.

FIGURE *4-13*

An activity diagram showing concurrent paths.

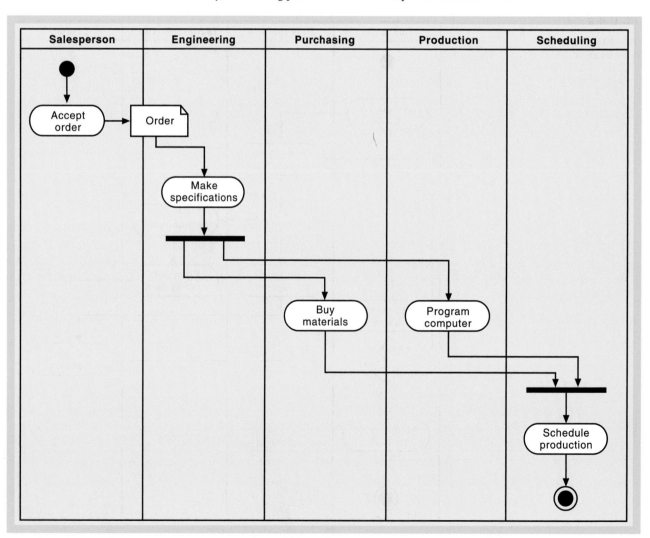

Creating activity diagrams to document workflows is straightforward. The first step is to identify the agents to create the appropriate swimlanes. Next, just follow the various steps of the workflow, and make appropriate ovals for the activities. Connect the activity ovals with arrows to show the flow of the workflow. Here are a couple of simple guidelines:

- Use a decision symbol to represent an either/or situation—one path *or* the other path, but not both. As a shorthand notation, you can merge an activity (using an oval) and a decision (using a diamond) into a single oval with two exit arrows as indicated on the right in Figure 4-11. This notation represents a decision (either/or) activity. Wherever you have an activity that says "verify" or "check," you will probably require a decision—one for the "accept" path and one for the "reject" path. You can merge either/or paths into a common activity (as in *Calculate quote* shown in Figure 4-12) or into other connecting arrows.
- Use synchronization bars for parallel paths—situations in which *both* paths are taken. Include both a beginning and ending synchronization bar. You can also use synchronization bars to represent a loop such as a "do while" programming loop. Put the bar at the beginning of the loop and describe it as "for every." Put another synchronization bar at the end of the loop with the description "end for every."

Build Prototypes

As we discussed briefly in Chapter 2, a *prototype* is an initial, working model of a larger, more complex entity. Prototypes are used for many different purposes, and there are many names to differentiate these uses: throwaway prototypes, discovery prototypes, design prototypes, and evolving prototypes. Each can be used during different phases of the project to test and validate ideas. As already explained, during analysis, prototypes are used to test feasibility and to help identify processing requirements. These prototypes may be in the form of simple screens or report programs. During design, prototypes may be built to test various design and interface alternatives. Even during implementation, prototypes may be built to test the effectiveness and efficiency of different programming techniques.

Prototyping is such a strong tool that you will find it used in almost every development project in some way. As mentioned earlier in the chapter, a discovery prototype is used for a single discovery objective and then discarded once the concept has been tested. For example, if you use a prototype to determine screen formats and processing sequences, once you have finished that definition, you would throw away the prototype. Evolving prototypes, on the other hand, are prototypes that grow and change and may eventually even be used as the final, live system. As you will see later in Chapter 17, one approach to prototyping is to keep modifying the prototype and adding to it until it actually becomes the system that is installed.

The following are characteristics of effective prototypes:

- **Operative.** Generally, a prototype should be a working model, with the emphasis on *working*. A simple start to a prototype, called a *mock-up*, is an electronic form (such as a screen) that shows what it looks like but is not capable of executing an activity. Later, a working prototype will actually execute and provide both "look and feel" characteristics but may lack some functionality.
- **Focused.** To test a specific concept or verify an approach, a prototype should be focused on a single objective. Extraneous execution capability, not part of the specific objective, should be excluded. Although it may be possible to combine several simple prototypes into a larger prototype, the focused objective still applies. Later, the project team can combine prototypes to test the integration of several components.

prototype

a preliminary working model of a larger system

mock-up

an example of a final product that is for viewing only, and not executable

- **Quick.** Tools, such as CASE tools, are needed to develop a prototype so that it can be built and modified quickly. Since a prototype's purpose is to validate an approach, if the approach is wrong, then the tools must be available to modify and test quickly to determine the correct approach.

Integrating prototyping activities in the project is fairly simple. The important point to keep in mind is to have an overall philosophy and purpose for building prototypes and to maintain a consistent focus across all the prototypes that are built.

Distribute and Collect Questionnaires

Questionnaires have a limited and specific use in information gathering. The benefit of a questionnaire is that it enables the project team to collect information from a large number of stakeholders. Even if the stakeholders are widely distributed geographically, they can still help define requirements through questionnaires.

Frequently, the project team can use a questionnaire to obtain preliminary insight on the information needs of the various stakeholders. This preliminary information can then be used to help determine the areas that need further research with document reviews, interviews, and observation. Questionnaires are also helpful to answer quantitative questions such as, "What forms are used to enter new customer information?" and, "On the average, how long does it take to enter one standard order?" Finally, questionnaires can be used to determine the users' opinions about various aspects of the system. Such questions as "On a scale of 1 to 7, how important is it to be able to access a customer's past purchase history?" are often called *closed-ended questions*, because they direct the person answering the question to provide a direct response to only that question. They do not invite discussion or elaboration. The strength of closed-ended questions, however, is that the answers are always limited to the set of choices. The project team can tabulate the answers to determine averages or trends.

Figure 4-14 is a sample questionnaire showing three types of questions. The first part has closed-ended questions to determine quantitative information. The second part consists of opinion questions, in which respondents are asked whether they agree or disagree with the statement. Both types of questions are useful for tabulating and determining quantitative averages. The final part requests an explanation of a procedure or problem. Questions such as these are good as a preliminary investigation to help direct further fact-finding activities.

Questionnaires are not well suited to helping you learn about processes, workflows, or techniques. The questions identified earlier, such as, "How do you do this process?" are best answered using interviews or observation. Questions that encourage discussion and elaboration are called *open-ended questions*. Although a questionnaire can contain a very limited number of open-ended questions, stakeholders frequently do not return questionnaires that contain many open-ended questions.

Conduct Joint Application Design Sessions

Joint application design (JAD) is a technique used to expedite the investigation of system requirements. The normal interview and discussion approach, as explained earlier, requires a substantial amount of time. The analysts first meet with the users, then document the discussion by writing notes and building models, and then review and revise the models. Unresolved issues are placed on an open-items list and may require several additional meetings and reviews to be finalized. This process can extend from several weeks to months, depending on the size of the system and the availability of user and project team resources.

closed-ended questions

questions that have a simple, definitive answer

open-ended questions

questions that require discussion and do not necessarily have a simple, short answer

joint application design (JAD)

a technique to define requirements or design a system in a single session by having all necessary people participate

RMO Questionnaire

This questionnaire is being sent to all telephone-order sales personnel. As you know, RMO is developing a new customer support system for order taking and customer service.

The purpose of this questionnaire is to obtain preliminary information to assist in defining the requirements for the new system. Follow-up discussions will be held to permit everybody to elaborate on the system requirements.

Part I. Answer these questions based on a typical four-hour shift.
1. How many phone calls do you receive?_____
2. How many phone calls are necessary to place an order for a product?_____
3. How many phone calls are for information about RMO products, that is, questions only?_____
4. Estimate how many times during a shift customers request items that are out of stock?_____
5. Of those out-of-stock requests, what percentage of the time does the customer desire to put the item on back order?_____%
6. How many times does a customer try to order from an expired catalog?_____
7. How many times does a customer cancel an order in the middle of the conversation?_____
8. How many times does an order get denied due to bad credit?_____

Part II. Circle the appropriate number on the scale from 1 to 7 based on how strongly you agree or disagree with the statement.

Question	Strongly Agree					Strongly Disagree	
It would help me do my job better to have longer descriptions of products available while talking to a customer.	1	2	3	4	5	6	7
It would help me do my job better if I had the past purchase history of the customer available.	1	2	3	4	5	6	7
I could provide better service to the customer if I had information about accessories that were appropriate for the items ordered.	1	2	3	4	5	6	7
The computer response time is slow and causes difficulties in responding to customer requests.	1	2	3	4	5	6	7

Part III. Please enter your opinions and comments.

Please briefly identify the problems with the current system that you would like to see resolved in a new system.
_____
_____
_____

FIGURE *4-14*

A sample questionnaire.

The objective of JAD is to compress all of these activities into a shorter series of JAD sessions with users and project team members. An individual JAD session may last from a single day to a week. During the session, all of the fact-finding, model-building, policy decisions, and verification activities are completed for a particular aspect of the system. If the system is small, the entire analysis may be completed during the JAD session. The critical factor in a successful JAD session is to have all of the important stakeholders present and available to contribute and make decisions. The actual participants vary depending on the objective of the specific JAD session. Those who are involved may include the following:

■ **The JAD Session Leader.** One of the more important members of the group, the session leader is a person who is experienced or has been trained in understanding group dynamics and in facilitating group discussion. Normally, a JAD session involves quite a few people. Each session has a detailed agenda with specific objectives that must be met, and the discussion must progress toward meeting those objectives. Maintaining focus requires someone with skills and experience to keep people on task tactfully. Often it is tempting to appoint one of the systems analysts as the session leader. However, experience indicates that successful JAD sessions are conducted by someone who is trained to lead group decision making.

- **Users.** Earlier, this chapter identified various classes of users. It is important to have all of the appropriate users in the JAD sessions. Frequently, as requirements are discovered, managers must make policy decisions. If managers are not available in the sessions to make those decisions, then progress is halted. Because of business pressures, it may be difficult for top executives to be present during the entire session. In that case, arrangements should be made for executives to visit the session once or twice a day to become involved in those policy discussions.

- **Technical Staff.** A representative from the technical support staff should also be present in the JAD session. There are always questions and decisions about technical issues that need to be answered. For example, participants may need details of computer and network configurations, operating environments, and security issues.

- **Project Team Members.** Both systems analysts and user experts from the project team should be involved in JAD sessions. These members assist in the discussion, clarify points, control the level of detail needed, build models, document the results, and generally see that the system requirements are defined to the necessary level of detail. The session leader is a facilitator, but often the leader is not the expert on how much detail and definition is required. Members of the project team are the experts on ensuring that the objectives are completely satisfied.

JAD sessions are usually conducted in special rooms with supporting facilities. First, since the process is so intense, it is important to be away from the normal day-to-day interruptions. Sometimes an off-site location may be necessary, or notification that interruptions are not welcome may be posted. On the other hand, it is usually helpful to have telephone access to executives and technical staff who are not involved in the meetings but who may be invited from time to time to finalize policy or technical decisions.

Resources in the JAD session room should include an overhead projector, a black or white board, flip charts, and adequate workspace for the participants. JAD sessions are work sessions, and all the necessary work paraphernalia should be provided.

Recently, JAD sessions have been taking advantage of electronic support to increase their efficiency. Analysis and documentation can be enhanced if participants have personal or laptop computers connected in a network. Then as requirements are documented with narrative descriptions or models, or even as some simple discovery prototypes are built, they can be made available to everybody. Often a CASE tool is provided on the computers to assist in visualization of screen and report layouts and file design. The CASE tool also provides a central repository for all the requirements developed during the session.

Group support systems (GSSs), which also run on the network of computers, allow all participants to post comments (anonymously, if desired) on a common working chat room. This approach helps participants who may be shy in group discussions to become more active and contribute to the group decisions. GSSs also enable the team to store final requirements as decisions are made. Normally JAD sessions are conducted with everyone in the same room. However, GSSs on wider networks provide the opportunity for virtual meetings with participants at geographically dispersed locations.

Figure 4-15 shows an example of a conference room with electronic support. Such a room might be available in larger companies that have development projects in progress

group support system (GSS)
a computer system that enables multiple people to participate with comments at the same time, each on the user's own computer

FIGURE *4-15*

A JAD facility.

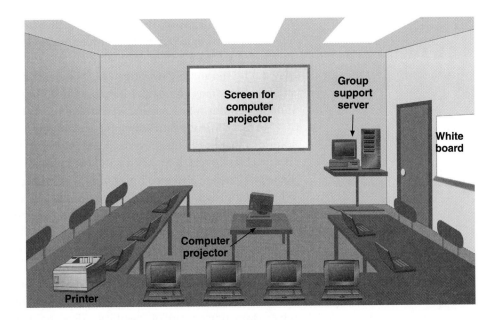

more frequently. The room shown in Figure 4-15 has workstations available to develop model diagrams and prototypes during the JAD sessions. This room could even be quite sophisticated by having computer support for collaborative work (CSCW) software on the computers to facilitate comment and discussion. With CSCW software, certain executives could even participate from remote locations, if necessary.

As stated earlier, one of the dangers of JAD is the risk involved in expediting decisions. Since the objective of JAD is to come to a conclusion quickly on policy decisions and requirements details, sometimes decisions are not optimal. At times, details are inappropriately defined or missed altogether. However, JAD sessions have been largely successful in reducing project development efforts and shortening the schedule.

Research Vendor Solutions

Many of the problems and opportunities that companies want to address with new information systems have already been solved by other companies. In many instances, consulting firms have experience with the same problems, and sometimes software firms already have packaged solutions for a particular business need. Directories, such as Data Sources, also list literally thousands of hardware, software, consulting, and solution developers. As part of the analysis phase, it makes sense to learn about and capitalize on this existing knowledge.

There are three positive contributions and one danger in exploring existing solutions. First, researching alternative solutions will frequently help users generate new ideas of how to better perform their business functions. Seeing how someone else did them, and applying that idea to the culture and structure of the existing organization, will often provide viable alternative solutions for the business need.

Second, some of these solutions are in fact excellent and state of the art. Without this research, the development team may create a system that is obsolete even before it is conceptualized. Companies need solutions that not only solve basic business practices but also are up to date with current competitive practices.

Third, it is often cheaper, and less risky, to buy a solution rather than to build it. If the solution meets the needs of the company and can be purchased, then that is usually a safer, quicker, and less expensive route. There are many ways to buy solutions. Chapter 8 discusses alternative schemes to build and buy. Early in the development project, you want to research other alternatives but not make a final decision until you have investigated all the alternatives.

The danger, or caveat, in this process is that sometimes the users, and even the systems analysts, want to buy one of the alternatives immediately. But if a solution, such as a packaged software system, is purchased too early in the process, the company's needs may not be thoroughly investigated. That is like buying a suit without knowing what size it is. Too many companies have bought a system only to find out later that it only supports half the functions that were needed. Don't fall into this dangerous pit.

The first difficulty in researching vendor alternatives is simply to find out who has solutions that fit the business need. Many of the large software and hardware companies, such as Oracle, IBM, Microsoft, and Computer Associates, have specific solution systems. There are also directories of system solutions—of software, hardware, and developer companies. Data Sources is one of the better ones. You can also search the Internet to find more directories. Sometimes these directories can be found in the library—a company technical library, the city library, or a nearby university library.

Other places to look are in trade journals for the industry. For example, the retail industry has several trade journals. System providers frequently advertise in these journals and at trade shows. Another method is word of mouth from other companies in the industry. Generally, users will have friends who work in competing companies. These people are sometimes aware of specific vendors that have helped solve their own business needs. Although companies compete fiercely on the sales and marketing end, it is not unusual for them to belong to a common trade organization that helps to share knowledge about the industry, including knowledge about system solutions.

Once a list of possible providers has been developed, the next step is to research the details of each solution. It is easy to get the sales and marketing literature, but it is more difficult to get specifics of the system. Techniques that may be useful are (1) technical specifications, (2) a demo or trial system, (3) references of existing clients who would let you observe their system, (4) an on-site visit, and (5) a printout of the screens and reports.

The final step is to review the details of the information received. Depending on the information obtained, it can be reviewed solely by the project team or with the users. In many cases, a review with key users is the most beneficial in understanding and identifying various approaches to addressing the business need. In any event, researching the solutions that have already been developed is an effective early step in understanding the business and identifying possible courses of action.

VALIDATING THE REQUIREMENTS

Now that you have learned about information-gathering techniques and ways to elicit requirements from the users, you need to make sure that the information you gathered is correct. All too frequently, systems analysts think they understand what the users need but have failed to capture some very important subtleties about the business processes. Obviously, correcting such a mistake after the system has been programmed is very expensive. In fact, various studies have indicated that fixing a requirements error later in the development cycle can cost upwards of hundreds of times more than it would have cost to fix it during the requirements definition.

If we compare the development of a new information system to the construction of a house, the requirements determined during analysis are like a house's blueprints and construction designs. The construction of the house is dependent on those blueprints. What if there are errors in the blueprints, such as load-bearing walls that are not strong enough or missing structural supports? If these errors are not discovered until the second story is built, it will be extremely expensive to remove and rebuild walls. So, the blueprints have to be correct. How does an architect ensure that they are correct? By not waiting until the house is being built to "test" the correctness of the blueprints.

System requirements have a similar problem. The design and construction of the system depend on correct requirements. It is too late, and very expensive, if the requirements are "tested" only while the programming is being done. Testing and validation of the system requirements must be done as early as possible.

At this point in your project activities you have collected information about the user requirements. You may have developed some workflow diagrams. In the next chapter, you will learn about building models to describe the system requirements. All of these elements should be thoroughly tested before the actual design and programming begin. When writing a computer program, a programmer must verify the accuracy of the code by conducting various tests. Executing the program on a computer—by entering appropriate input data and observing the output—tests a computer program. Analysts cannot test the requirements that way, so they have to use a different approach.

Various techniques can be used to validate the information from the users and the requirements that are developed from that information. To check internal consistency, analysts build models and verify that they are mathematically consistent. You will learn more about that in the next chapter. One powerful technique, called *structured walkthroughs*, is useful for both validating the requirements against the users needs as well as verifying internal consistency.

A *structured walkthrough*, sometimes just called a *walkthrough*, is a review of the findings from your investigation and of the models built based on those findings. A walkthrough is considered structured because analysts have formalized the review process into a set procedure. The objective of a structured walkthrough is to find errors and problems. Its purpose is to ensure that the blueprint is correct. The fundamental concept is one of documenting the requirements as you understand them and then reviewing them for any errors, omissions, inconsistencies, or problems. A review of the findings can be done informally with colleagues on the project team, but a structured walkthrough must be more formal.

It is important to note one critical point: A structured walkthrough is not a performance review. Managers should be involved only if they were involved in the original fact-finding and thus are required for verification or validation. The review is of an analyst's work and not the person. To help you understand the more structured approach, this section reviews the what, when, who, and how of a structured walkthrough.

One of the major responsibilities of the project manager, as described in Appendix A, is to ensure the quality of the final system. Often during the rush and pressure of a project, systems analysts will think, "My work is good. It does not need to be reviewed." But it is very unwise for a project manager to skip the review. Because of the costly consequences, it just does not make sense to exclude from the project plan specific tasks and procedures to ensure that the requirements are complete and accurate. Omissions such as this will *always* cause problems later in the project. Structured walkthroughs can be performed to validate gathered information; however, they should definitely be performed to validate the specification models (discussed in the next few chapters). This section focuses on the process of a structured walkthrough.

What and When

The first item to be reviewed during a structured walkthrough is the documentation that was developed as part of the analysis phase. It can be a narrative describing a process, a flowchart showing a workflow, or a model diagram documenting an entire procedure. Normally, it is better to conduct several smaller walkthroughs that review three to six pages of documentation than to cover 30 pages of details. Any written work that is a fairly independent package can be reviewed in a walkthrough. It is not uncommon to hold smaller walkthroughs every week or two with members of the project team. The frequency of the walkthroughs is not as critical as the timing—a walkthrough should be scheduled as soon as possible after the documents have been created.

Who

The two main parties involved in walkthroughs are the person or persons who need their work reviewed and the group who reviews it. For verification—that is, internal consistency and correctness—it is best to have other experienced analysts involved in the walkthrough. They look for inconsistencies and problems. For validation—that is, ensuring that the system satisfies all the needs of the various stakeholders—the appropriate stakeholders should be involved. The nature of the work to be reviewed dictates who the reviewers should be. If it is a diagram showing a business process, then the users who supplied the original definition should be involved. If it is a technical specification of design details, then the technical staff should be involved in the review. At times, the reviewers may be members of the project team. In other instances, they are external users or technical staff. Those who can validate the correctness of the work are the people who should be invited.

How

As with an interview, a structured walkthrough requires preparation, execution, and follow-up.

Preparation

The analyst whose work is being reviewed gets the material ready for review. Next, he or she identifies the appropriate participants and provides them copies of the material. Finally, the analyst schedules a time and place for the walkthrough and notifies all participants.

Execution

During the walkthrough, the analyst presents the material point by point. If it is a diagram or flowchart, he or she walks through the flow, explaining each component. One effective technique is to define a sample test case and process it through the defined flow. The reviewers look for inconsistencies or problems and point them out. A librarian, a helper for the presenter, documents the comments made by the reviewers. Presenters should never be their own librarians because they should not be distracted from explaining the documentation. To ensure accuracy, someone else should record the errors, comments, and suggestions.

Corrections and solutions to problems are not made during the walkthrough. At most, some suggested solutions may be provided, but the documentation should not be corrected during the walkthrough. Since presenters are commonly a little nervous, it is unfair to ask them to make wise decisions on the spur of the moment. If a misunderstanding of the user requirements is uncovered, a brief review may be in order. However, if an error is fairly complex, it is better to schedule an additional interview to clarify the misunderstanding. The walkthrough should not get bogged down into a fact-finding session. The reviewer should only provide feedback, and the presenter can integrate it into the material later, when he or she has the entire set of comments and can give the corrections his or her undivided best effort without interruptions or further criticism.

Follow-Up

Follow-up consists of making the required corrections. If the reviewed material has major errors and problems, an additional walkthrough may be necessary. Otherwise, the corrections are made, and the project continues to the next activities.

Figure 4-16 is a sample review form that was used in one of the review sessions at Rocky Mountain Outfitters, for the sales commission rates and rules. Not shown are several attached sheets, including a couple of flowcharts of procedures. The material reviewed in this case is simply a list of business rules for commission rates. The reviewers are senior managers from the user community. Since sales commission business rules are critical, and these managers make the policy decisions on commissions, they are the obvious choices to review the rules as uncovered in discussions and interview sessions.

FIGURE *4-16*

A structured walkthrough evaluation form.

Walkthrough Control Sheet

Project Control Information

Project: *On-line Catalog System, Customer Support Subsystem*

Segment of project being reviewed: *Review of business rules for sales commission rates*

Team leader: *Mary Ellen Green*

Author of work: *Jim Williams*

Walkthrough Details

Date, time, and location
    April 10, 2005. 10:00 a.m. MIS conference room.

Description of materials being reviewed:
    This is a review of the business rules before they are integrated into the diagrams and models. There is a short flowchart attached showing the flow of the commission process. There is another flowchart showing the process to set commission rates. We will also review outstanding issues to ensure that all understand the policy decisions that must be made.

Participating reviewers:
William McDougal, Genny Monson, Robert Schneider

Results of Walkthrough

_____XX__Accept. sign-offs:_____
_____Minor revisions. Description of revisions:

_____Rework and schedule new walkthrough. Description of required rework:
    Excellent and thorough. No rework required.

SUMMARY

The analysis phase can be divided into six primary activities:

- Gather information
- Define system requirements
- Prioritize requirements
- Prototype for feasibility and discovery
- Generate and evaluate alternatives
- Review recommendations with management

Business process reengineering is becoming a widespread method of improving business processes, so it can profoundly affect the analysis phase. It entails a complete redesign of the business processes. Under BPR, new system development is not done merely to automate existing procedures. Instead, the entire process is completely rethought. The objective is to use IT in novel ways to achieve dramatic improvements in efficiencies and levels of service. Because of the special problem-solving, analytical, and modeling skills of systems analysts, they frequently play an important role in BPR efforts.

Generally we divide system requirements into two categories: functional and nonfunctional requirements. The functional requirements are those that explain the basic business functions that the new system must support. Nonfunctional requirements involve the objectives of the system for technology, performance, usability, reliability, and security.

To ferret out the requirements, analysts must work with various stakeholders in the new system. We categorize stakeholders into three groups: (1) the users, those who will actually use the system day to day, (2) the clients, those who pay for and own the system, and (3) the technical staff, the people who must ensure that the system operates within the computing environment of the organization. One of the most important first steps in determining systems requirements is to identify these various system stakeholders.

A fundamental question to investigate system requirements is, "What kind of information do I need?" This chapter provides you with some general guidelines. As you learn more about modeling, you should also understand better what information you need. Three major themes of information should be pursued:

- What are the business processes and operations?
- How are the business processes performed?
- What are the information requirements?

Analysts use seven primary techniques to gather this information, and one technique ensures its correctness. The seven fact-finding techniques are the following:

- Review existing reports, forms, and procedure descriptions
- Conduct interviews and discussions with users
- Observe and document business processes
- Build prototypes
- Distribute and collect questionnaires
- Conduct JAD sessions
- Research vendor solutions

The fundamental idea of a prototype is an initial, working model of a larger, more complex entity. The primary purpose of a prototype is to have a working model that will test a concept or verify an approach. Discovery prototypes are built to define requirements but are then usually discarded or at least not used for the final programming.

Joint application design is a technique used to expedite the investigation of system requirements by holding several marathon sessions with all the critical participants. Discussion results in requirements definition and policy decisions immediately, without the delays of interviewing separate groups and trying to reconcile differences. When done correctly, JAD is a powerful and effective technique.

The review technique to ensure that analysis is accurate and complete is called a *structured walkthrough*. Remember that a structured walkthrough has the objective of reviewing and improving the work. It is not a performance review.

KEY TERMS

activity diagram, p. 134

closed-ended questions, p. 138

functional requirement, p. 119

group support system (GSS), p. 141

joint application design (JAD), p. 139

logical model, p. 115

mock-up, p. 138

open-ended questions, p. 138

performance requirement, p. 120

physical model, p. 116

prototype, p. 137

reliability requirement, p. 120

security requirement, p. 120

stakeholders, p. 120

structured walkthrough, p. 143

swimlane, p. 134

synchronization bar, p. 134

system requirements, p. 119

technical requirement, p. 120

transaction, p. 122

usability requirement, p. 120

workflow, p. 134

REVIEW QUESTIONS

1. What is the difference between functional requirements and nonfunctional requirements?
2. Explain the use of a discovery prototype and an evolutionary prototype.
3. List and describe the three fact-finding themes.
4. What is the objective of a structured walkthrough?
5. Explain the steps in preparing for an interview session.
6. What are the benefits of doing vendor research during information-gathering activities?
7. What categories of stakeholders should you include in fact-finding?
8. What is meant by vertical and horizontal dimensions when determining users to involve?
9. What is JAD? When is it used?
10. What is BPR? What does it have to do with systems analysis?
11. What technique is used to validate user requirements?
12. Describe the open-items list and explain why it is important.
13. What do *correct*, *complete*, and *comprehensive* mean with regard to systems analysis?
14. List and describe the seven information-gathering techniques.
15. What is the purpose of a workflow diagram?
16. Draw and explain the symbols used on a workflow diagram.

THINKING CRITICALLY

1. One of the toughest problems in investigating system requirements is to make sure that they are complete and comprehensive. What things would you do to ensure that you get all of the right information during an interview session?
2. What can you do to ensure that you have included all of the right stakeholders on your list of people to interview? How can you double-check your list?
3. One of the problems you will encounter during your investigation is "scope creep"—that is, user requests for additional features and functions. Scope creep happens because sometimes users have many unsolved problems and the system investigation may be the first time anybody has listened to their needs. How do you keep the system from growing and including new functions that should not be part of the system?
4. It is always difficult to observe users in their jobs. It frequently makes both you and them uncomfortable. What things could you do to ensure that user behavior is not changing because of your visit? How could you make observation more natural?
5. What would you do if you got conflicting answers for the same procedure from two different people you interviewed? What would you do if one was a clerical person and the other was the department manager?

6. You are a team leader of four systems analysts. You have one analyst who has never done a structured walkthrough of his or her work. How would you help the analyst to get started? How would you ensure that the walkthrough was effective?

7. You have been assigned to resolve several issues on the open-items list, and you are having a hard time getting policy decisions from the user contact. How can you encourage the user to finalize these policies?

8. You are going on your first consulting assignment to do systems analysis. Your client does not like to pay to train new, inexperienced analysts. What should you do to appear competent and well prepared? How should you approach the client?

9. In the running case of Rocky Mountain Outfitters, you have set up an interview with Jason Nadold in the shipping department. Your objective is to determine how shipping works and what the information requirements for the new system will be. Make a list of questions, open-ended and closed-ended, that you would use. Include any questions or techniques you would use to ensure you find out about the exceptions.

10. Develop an activity diagram based on the following narrative. Note any ambiguities or questions that you have as you develop the model. If you need to make assumptions, also note them.

The purpose of the Open Access Insurance System is to provide automotive insurance to car owners. Initially, prospective customers fill out an insurance application, which provides information about the customer and his or her vehicles. This information is sent to an agent, who sends it to various insurance companies to get quotes for insurance. When the responses return, the agent then determines the best policy for the type and level of coverage desired and gives the customer a copy of the insurance policy proposal and quote.

11. Develop a workflow diagram based on the following narrative. Note any ambiguities or questions that you have as you develop the model. If you need to make assumptions, also note them.

The purchasing department handles purchase requests from other departments in the company. People in the company who initiate the original purchase request are the "customers" of the purchasing department. A case worker within the purchasing department receives that request and monitors it until it is ordered and received.

Case workers process requests for the purchase of products under $1,500, write a purchase order, and then send it to the approved vendor. Purchase requests over $1,500 must first be sent out for bid from the vendor that supplies the product. When the bids return, the case worker selects one bid. Then, he or she writes a purchase order and sends it to the vendor.

12. Develop an activity diagram based on the following narrative. Note any ambiguities or questions that you have as you develop the model. If you need to make assumptions, also note them.

The shipping department receives all shipments on outstanding purchase orders. When the clerk in the shipping department receives a shipment, he or she finds the outstanding purchase order for those items. The clerk then sends multiple copies of the shipment packing slip. One copy goes to purchasing, and the department updates its records to indicate that the purchase order has been fulfilled. Another copy goes to accounting so that a payment can be made. A third copy goes to the requesting in-house customer so that he or she can receive the shipment.

Once payment is made, the accounting department sends a notification to purchasing. Once the customer receives and accepts the goods, he or she sends notification to purchasing. When purchasing receives these other verifications, it closes the purchase order as fulfilled and paid.

EXPERIENTIAL EXERCISES

1. Conduct a fact-finding interview with someone involved in a procedure that is used in a business or organization. This person could be someone at the university, in a small business in your neighborhood, in the student volunteer office at the university, in a doctor's or dentist's office, in a volunteer organization, or at your local church. Identify a process that is done, such as keeping student records, customer records, or member records. Make a list of questions and conduct the interview. Remember, your objective is to understand that procedure thoroughly—that is, to become an expert on that single procedure.

2. Follow the same instructions as for exercise 1, except make this exercise an observation experience. Either observe the

other person do the work or ask to carry out the procedure yourself. Write down the details of the process you observe.

3. Get a group of your fellow students together and conduct a structured walkthrough of your results from exercise 1 or 2. Using the results of your interview or observation, document the procedure in a flowchart with some narrative. Then, conduct a walkthrough with several colleagues. Or take another assignment, such as Thinking Critically question 9, and walk through your preparation for that assignment. Follow the steps outlined in the text.

4. Research and write a one- to two-page research paper using at least three separate library sources on one of the following topics:

a. Joint application design

b. Prototyping as a discovery mechanism

c. Business process reengineering

d. Computer support for collaborative work (CSCW)

e. Workflow systems

f. Structured walkthrough

5. Using Rocky Mountain Outfitters and the customer support subsystem as your guide, develop a list of all the procedures that may need to be researched. You may want to think about the exercise in the context of your experience with retailers such as L. L. Bean, Lands' End, or Amazon.com. Get some catalogs, check out the Internet marketing done on the retailers' Web sites, and then think about the underlying business procedures that are required to support those sales activities. List the procedures and describe your understanding of each.

CASE STUDIES

John and Jacob, Inc., On-Line Trading System

John and Jacob, Inc., is a regional brokerage firm that has been successful over the last several years. Competition for customers is intense in this industry. The large national firms have very deep pockets, with many services to offer to clients. Severe competition also comes from discount and Internet trading companies. However, John and Jacob has been able to cultivate a substantial customer base from upper-middle income people in the northeastern United States. To maintain a competitive edge with its customers, John and Jacob is in the process of developing a new on-line trading system. The plan for the system identifies many new capabilities that would provide new services to its clients.

Edward Finnigan, the project manager, is in the process of identifying all the groups of people who should be included in the development of the system requirements. He isn't quite sure exactly who should be included. Here are the issues he's considering:

- **Users.** The trading system is to be on-line to each of the company's 30 trading offices. Obviously, the brokers who are going to use the system need to have input, but how should this be done? Edward also isn't sure what approach would be best to ensure that the requirements are complete, yet not require tremendous amounts of time. Including all of the offices would increase enthusiasm and support for the system, but it would take a lot of time. Yet, involving more brokers would bring divergent opinions that would have to be reconciled.

- **Customers.** The trading system will also include confirmations, reports of trades, and customer statements. Web access is also planned, which will enable customers to effect trades and to check their accounts. Consequently, another question Edward has is about involving John and Jacob customers in the development of system requirements. Normally, customers are not asked to participate in the development of systems. However, it would be nice to know how best to serve John and Jacob's customers. He is sensitive to this issue because some of the brokers have indicated to him that many customers do not like the format of their statements from the current system. He would like to involve customers, but he does not know how.

- **Other Stakeholders.** Edward knows he should involve other stakeholders to help define system requirements. He isn't quite sure whom he should contact. Should he go to senior executives? Should he contact middle management? Should he include back-office functions such as accounting and investing? He isn't quite sure how to get organized or how to decide who should be involved.

1. What is the best method for Edward to involve the brokers (users) in development of the new on-line trading system? Should he use a questionnaire? Should he interview the brokers in each of the company's 30 offices, or would one or two brokers representing the entire group be better? How can Edward ensure that the information about requirements is complete, yet not lose too much time doing so?

2. Concerning customer input for the new system, how can Edward involve customers in the process? How can he interest them in participating? What are some ways that Edward can be sure that the customers he does involve are representative of John and Jacob's entire customer group?
3. As Edward considers what other stakeholders he should include, what are some criteria he should use? Develop some guidelines to help him build a list of people to include.

Rethinking Rocky Mountain Outfitters

Barbara Halifax, the project manager for the CSS project, had finished identifying the list of stakeholders in this project. As shown earlier in the chapter, quite a few senior executives would be involved. Most of them would not have major input. Those in Bill McDougal's area would, of course. Not only was he the project sponsor, but all his assistants were excited about this new system and its potential to help the business grow. Barbara had a good working relationship with all of these executives.

Barbara had also identified numerous department managers and senior-level clerks who would be able to provide detailed processing requirements. She had divided her list of stakeholders into two groups. The first group consisted of all those with primary responsibility to help define user requirements. The second group included those who would not have direct use of the system but who would need reports and information from the system. She wanted to make sure the needs of these people were also satisfied.

As an experienced project manager, Barbara had her checklists of things to do. She used a project manager checklist to help her so that she would not forget important tasks. Being a project manager was much too critical, and potentially stressful, to do it "by the seat of your pants."

As she reviewed her list, she noticed several activities that she had not yet considered on the CSS project. She was thinking that before she let her project team start to meet with the users, she ought to consider these items and review them with her team. The items that most caught her attention were the following:

- Develop a communications plan with the user
- Manage user expectations
- Control the scope and avoid scope creep

Based on the concepts you learned in this chapter (you might want to review Appendix A also), what would you do if you were Barbara? Obviously, you want to provide the best possible solution for the company, but you also need to control the project, the scope, and the users so that the system will be successful and be installed in time.

1. Identify the major points you would include in a communications plan at this point in the project.
2. What advice would you give your project team to help it manage the user expectations?
3. What early planning can you do now to ensure that the scope is realistic—to meet the need but within the time and budget allotted?

Focusing on Reliable Pharmaceutical Service

Reliable Pharmaceutical Service plans to develop an extranet that enables its client health-care facilities to order drugs and supplies as if they were ordering from an internal pharmacy. The extranet should enable Reliable's suppliers to function as if they were part of Reliable's internal organization. These views of the final system have significant implications for defining system requirements and for gathering information about those requirements.

1. What information-gathering methods are most appropriate to learn about requirements from Reliable's own management staff and other employees? From client health-care organizations? From suppliers?
2. Should patients in client health-care facilities participate in the information-gathering process? If so, why, and in what ways should they participate?
3. With respect to gathering information from suppliers and clients, how deeply within those organizations should systems analysts look when defining requirements? How might Reliable deal with supplier and client reluctance to provide detailed information on their internal operations?
4. For which user community or communities (internal, supplier, or client) are prototypes likely to be most beneficial? Why?
5. Should Reliable consider adopting business process reengineering (BPR) as a primary approach to defining system requirements? What characteristics of the proposed system and of the development argue for and against BPR?

FURTHER RESOURCES

Vangalur S. Alagar, *Specification of Software Systems*. Springer-Verlag, 1998.

Soren Lauesen, *Software Requirements: Styles and Techniques*, Addison-Wesley, 2002.

Stan Magee, *Guide to Software Engineering Standards and Specifications.* Artech House, 1997.

Suzanne Robertson and James Robertson, *Mastering the Requirements Process.* Addison-Wesley, 2000.

Karl Wiegers, *Software Requirements*, Microsoft Press, 1999.

Jane Wood, *Joint Application Development.* John Wiley & Sons, 1995.

Ralph Young, *Effective Requirements Practices.* Addison-Wesley, 2001.

CHAPTER 5

Modeling System Requirements: Events and Things

LEARNING OBJECTIVES

After reading this chapter, you should be able to:

- Explain the many reasons for creating information system models

- Describe three types of models and list some specific models used for analysis and design

- Explain how events can be used to define system requirements

- Identify and analyze events to which a system responds

- Recognize that events trigger system activities or use cases

- Explain how the concept of things in the system also defines requirements

- Explain the similarities and the differences between data entities and objects

- Identify and analyze data entities and objects needed in the system

- Read, interpret, and create an entity-relationship diagram

- Read, interpret, and create a class diagram

CHAPTER OUTLINE

Models and Modeling

Events and System Requirements

Things and System Requirements

The Entity-Relationship Diagram

The Class Diagram

Where You Are Headed

Waiters on Call is a restaurant meal-delivery service started in 2001 by Sue and Tom Bickford. The Bickfords both worked for restaurants while in college and always dreamed of opening their own restaurant. But unfortunately, the initial investment was always out of reach. The Bickfords noticed that many restaurants offer takeout food, and some restaurants, primarily pizzerias, offer home delivery service. Many people they met, however, seemed to want home delivery service but with a more complete food selection.

Sue and Tom conceived Waiters on Call as the best of both worlds: a restaurant service without the high initial investment. The Bickfords contracted with a variety of well-known restaurants in town to accept orders from customers and to deliver the complete meals. After preparing the meal to order, the restaurant charges Waiters on Call a wholesale price, and the customer pays retail plus a service charge and tip. Waiters on Call started modestly, with only two restaurants and one delivery driver working the dinner shift. Business rapidly expanded, and the Bickfords realized they needed a custom computer system to support their operations. They hired a consultant, Sam Wells, to help them define what sort of system they needed.

"What sort of events happen when you are running your business that make you want to reach for a computer?" asked Sam. "Tell me about what usually goes on."

"Well," answered Sue, "when a customer calls in wanting to order, I need to record it and get the information to the right restaurant. I need to know which driver to ask to pick up the order, so I need drivers to call in and tell me when they are free. Sometimes customers call back wanting to change their orders, so I need to get my hands on the original order and notify the restaurant to make the change."

"Okay, how do you handle the money?" queried Sam.

Tom jumped in, "The drivers get a copy of the bill directly from the restaurant when they pick up the meal. The bill should agree with our calculations. The drivers collect that amount plus a service charge. When drivers report in at closing, we add up the money they have and compare it with the records we have. After all drivers report in, we need to create a deposit slip for the bank for the day's total receipts. At the end of each week, we calculate what we owe each restaurant at the agreed-to wholesale price and send each a statement and check."

"What other information do you need to get from the system?" continued Sam.

"It would be great to have some information at the end of each week about orders by restaurant and orders by area of town—things like that," added Sue. "That would help us decide about advertising and contracts with restaurants. Then we need monthly statements for our accountant."

Sam made some notes and sketched some diagrams as Sue and Tom talked. Then after spending some time thinking about it, he summarized the situation for Waiters on Call. "It sounds to me like you need a system that does some processing when these events occur:

- A customer calls in to place an order.
- A driver is finished with a delivery.
- A customer calls back to change an order.
- A driver reports for work.
- A driver submits the day's receipts.

"Then you need the system to produce information at specific points in time—for example, when it is:

- Time to produce an end-of-day deposit slip
- Time to produce end-of-week restaurant payments
- Time to produce weekly sales reports
- Time to produce monthly financial reports

"Based on the way you have described your business operations, I am assuming you will need a database to store information about these types of things:

- Restaurants
- Menu items
- Customers
- Orders
- Order payments
- Drivers

"Then I suppose you are going to need to maintain information in a database about restaurants and drivers. You'll need to do some processing when:

- You add a new restaurant
- A restaurant changes the menu
- You drop a restaurant
- You hire a new driver
- A driver leaves

"Am I on the right track?"

Sue and Tom quickly agreed that Sam was talking about the system in a way they could understand. They were confident that they had found the right consultant for the job.

The last chapter described the activities of the systems analysis phase of the SDLC and then introduced the many tasks and techniques involved when completing the first analysis activity—gathering information about the system, its stakeholders, and its requirements. An extensive amount of information is required to properly define the system's functional and nonfunctional requirements. This chapter, along with Chapters 6 and 7, presents techniques for documenting the functional requirements by creating a variety of models. These models are created during the analysis phase activity we have named *Define system requirements*, although remember that the analysis phase activities actually are done in parallel and in iterations.

After discussing the types and roles of models, we focus on two key concepts that help define system requirements in both the traditional and the object-oriented approach: events and things. This chapter covers specific models for both the traditional approach and the object-oriented approach, including those based on the Rocky Mountain Outfitters (RMO) customer support system. Keep in mind, though, that in any given system development project, either the traditional approach or the object-oriented approach will be used. However, the two key concepts—events and things—are common to both approaches. Chapter 6 continues the discussion of requirements models for the traditional approach, and Chapter 7 continues the object-oriented approach.

MODELS AND MODELING

An analyst can best describe the requirements for an information system using a collection of models (as we discussed in Chapter 2). Recall that a model is a representation of some aspect of the system being built. Because a system is so complex, an analyst creates a variety of models to encompass the detailed information that he or she collected and digested during the analysis phase (see Figure 5-1). The activity diagram introduced in Chapter 4 is an example of one type of model that focuses on both user and system activities. Also, the analyst uses many different types of models to show the system at different levels of detail (or levels of abstraction), including a high-level overview as well as detailed views of certain aspects of the system. Some models show different parts of the problem and solution; for example, one model might show inputs, another the data stored. Some models show the same problem and solution from different perspectives; one model might show how objects interact from the perspective of outside actors, and another how objects interact in terms of sequencing.

FIGURE *5-1*

An analyst needs a collection of models to understand system requirements.

The Purpose of Models

Some developers think of a model as documentation produced after the analysis and design work is done. But actually, the process of creating a model helps an analyst clarify and refine the design. The analyst learns as he or she completes and then studies parts of the model. Analysts also raise questions while creating a model and answer them as the modeling process continues. New pieces are added; the consequences of changes are evaluated and again questioned. In this respect, the modeling process itself provides direct benefits to the analyst. The technique used to create the model is valuable in itself even if the analyst never shows a particular model to anyone else. But usually models are shared with others as analysis and design progresses.

Another key reason that modeling is important in system development is the complexity of describing information systems. Information systems are very complex, and parts of the systems are intangible. Models of the various parts help simplify the analyst's efforts and focus them on a few aspects of the system at a time. The reason that an analyst uses so many different models is that each relates to different aspects of the system. In fact, some of the models created by the analyst may serve only to integrate these aspects—showing how the other models fit together.

Because of the amount of information gathered and digested and the length of time each analyst spends on a project, analysts need to review the models frequently to help recall details of work previously completed. People can retain only a limited amount of information, so we all need memory aids. Models provide a way of storing information for later use in a form that can be readily digested.

The support for communication is one of the most often cited reasons for creating the models. Given that the analyst learns while working through the modeling process and that the collection of models reduces the complexity of the information system, the models also serve a critical role in supporting communication among project team members and with system users. If one team member is working on models of inputs and outputs, and another team member is working on models of the processes that convert the inputs to outputs, then they need to communicate and coordinate to make sure these models fit together. The second team member needs to see what outputs are desired before modeling the process that creates them. At the same time, both team members need to know what data are stored (the data model) so they know what inputs are needed and what processes are needed to access the required data. Models support essential communication and teamwork among the project team members.

Models also assist in communication with the system users and foster understanding. Typically, an analyst reviews the models with a variety of users to get feedback on the analyst's understanding of the system requirements. Users need to see clear and complete models to comprehend what the analyst is proposing. Additionally, the analyst sometimes works with users to develop the models, so the modeling process helps users better understand the possibilities that the new system can offer. Users also need to communicate among themselves using the models. And the analyst and the users together can use models to relate system capabilities to managers who are responsible for approving the system.

Finally, the requirements models produced by the analyst are used as documentation for future development teams when they maintain or enhance the system. Considering the amount of resources invested in a new system, it is critical for the development team to leave behind a clear record of what was created. An important activity during implementation is to package the documentation accurately, completely, and in a form that future developers can use. Much of the documentation consists of the models created throughout the project. Figure 5-2 summarizes the reasons modeling is important to system development.

Although this book emphasizes models and techniques for creating models, it is important to remember that system projects vary in the number of models required and in their formality. Smaller, simpler system projects will not need models showing every system detail, particularly when the project team has experience with the type of system

Learning from the modeling process

Reducing complexity by abstraction

Remembering all of the details

Communicating with other development team members

Communicating with a variety of users and stakeholders

Documenting what was done for future maintenance/enhancement

being built. Sometimes the key models are created informally in a few hours. Although models are often created using powerful CASE tools and visual modeling tools as discussed in Chapter 2, useful and important models are sometimes drawn quickly over lunch on a paper napkin or in an airport waiting room on the back of an envelope! As with any SDLC activity, an iterative approach is used for creating requirements and design models. The first draft of a model has some but not all details worked out. The next iteration might fill in more details or correct previous misconceptions.

Types of Models

Analysts use many different types of models when developing information systems. The type of model used depends on the nature of the information being represented. Models can be categorized into three general types: mathematical models, descriptive models, and graphical models.

Mathematical Models

A *mathematical model* is a series of formulas that describe technical aspects of a system. Mathematical models are used to represent precise aspects of the system that can be best represented by using formulas or mathematical notation, such as equations that represent network throughput requirements or a function expressing the response time required for a query. These models are examples of *technical* requirements. Additionally, scientific and engineering applications tend to compute results using elaborate mathematical algorithms. The mathematical notation is the most appropriate way to represent these *functional* requirements, and it is also the most natural way for scientific and engineering users to express those requirements. An analyst working on scientific and engineering applications had better be comfortable with math.

But mathematical notation is also sometimes efficient for simpler requirements for business systems. For example, in a payroll application, it is reasonable to model gross pay as regular pay plus overtime pay. A reorder point for inventory, a discount price for a product, or a salary adjustment for a promotion might be modeled with a simple formula.

Descriptive Models

Not all requirements can be precisely defined with mathematics. For these requirements, analysts use *descriptive models*, which can be narrative memos, reports, or lists. Figure 5-3 provides examples of descriptive models for RMO's customer support system. Initial interviews with users might require the analyst to jot down notes in a narrative form,

mathematical model

a series of formulas that describe technical aspects of a system

descriptive model

narrative memos, reports, or lists that describe some aspect of a system such as the description of the phone-order process obtained from phone-order representatives. Sometimes users describe what they do in reports or memos to the analysts. The analyst might convert these narrative descriptions to a graphical modeling notation while compiling all of the information.

FIGURE *5-3*

Some descriptive models.

A narrative description of processing requirements as verbalized by an RMO phone-order representative:

"When customers call in, I first ask if they have ordered by phone with us before, and I try to get them to tell me their customer ID number that they can find on the mailing label on the catalog. Or, if they seem puzzled about the customer number, I need to look them up by name and go through a process of elimination, looking at all of the Smiths in Dayton, for example, until I get the right one. Next, I ask what catalog they are looking at, which sometimes is out of date. If that is the case, then I explain that many items are still offered, but that the prices might be different. Naturally, they point to a page number, which doesn't help me because of the different catalogs, but I get them to tell me the product ID somehow..."

List of inputs for the RMO customer support system:

Item inquiry
New order
Order change request
Order status inquiry
Order fulfillment notice
Back-order notice
Order return notice
Catalog request
Customer account update notice
Promotion package details
Customer charge adjustment
Catalog update details
Special promotion details
New catalog details

Sometimes a narrative description is the best form to use for recording information. Use case descriptions are often written out as one or two short paragraphs of text. More detailed use case descriptions are lists of steps required in processing between the actor and the system. Many useful models of information systems involve simple lists, such as lists of features, inputs, outputs, events, or users. Lists are a form of descriptive or narrative models that are concise, specific, and useful. Figure 5-3 contains a simple list of inputs to the customer support system.

A final example of a descriptive model involves writing a process or procedure in a very precise way, referred to as *structured English* or *pseudocode*. Programmers are familiar with structured English or pseudocode for modeling algorithms that, when followed, always obtain the same result. Therefore, such algorithms are very precise models of processing.

Graphical Models

Probably the most useful models created by the analyst are graphical models. *Graphical models* include diagrams and schematic representations of some aspect of a system. Graphical models make it easy to understand complex relationships that are too difficult

graphical model

diagrams and schematic representations of some aspect of a system to follow when described verbally. Recall the old saying that a picture is worth a thousand words. In system development, a carefully constructed graphical model might be worth a million words!

Some graphical models actually look similar to a real-world part of the system, such as a screen design or a report layout design. But for most of the analyst's work, the graphical models use symbols to represent more abstract things, such as external agents, processes, data, objects, messages, and connections. The key graphical models used for the analysis phase tend to represent the more abstract aspects of a system, since the analysis phase focuses on fairly abstract questions about system requirements without indicating the details of how they will be implemented. The more concrete models of screen designs and report layouts are completed during the systems design phase.

A variety of graphical models are used. Each model highlights (or abstracts) important details of some aspect of the information system. Each type of model should ideally use unique and standardized symbols to represent pieces of information. That way, whoever looks at a model can understand it. However, the number of available symbols is limited—circle, square, rectangle, line, and so on—so be careful when you are first learning the symbols of each model. You will also find variations in the notation used for each type of model in practice. The Unified Modeling Language (UML) now provides diagramming standards for models used in the object-oriented approach. However, diagrams used in the traditional approach are less standardized.

Overview of Models Used in Analysis and Design

The analysis phase activity named *Define system requirements* involves creating a variety of models. They are referred to as *logical models* (as discussed in Chapter 4) because they define in great detail what is required without committing to one specific technology. Analysts create many types of logical models to define system requirements. Figure 5-4 lists some of the more commonly used models. Barbara Halifax currently has her project team working to create requirements models for the customer support system.

Many models are also created during the design phase (see Figure 5-5). Design models are physical models because they show how some aspect of the system will be implemented with specific technology. Some of these models are extensions of requirements models created during systems analysis or derive directly from the requirements models. Some models (for example, a class diagram) are used during analysis and during design.

This chapter and Chapters 6 and 7 describe some of the requirements models in detail. Here we describe events and things, two key concepts common to requirements models in all approaches to development. The models discussed in this chapter include event lists, use cases, event tables, entity-relationship diagrams, and class diagrams. When Barbara Halifax and her team began modeling the new customer support system for Rocky Mountain Outfitters, they identified the events and things the new system needed to accommodate (see Barbara's memo on page 160).

EVENTS AND SYSTEM REQUIREMENTS

event

an occurrence at a specific time and place that can be described and is worth remembering

Now that you know the basic types of models, we turn to the specifics for modeling the functional requirements of an information system. Virtually all approaches to system development begin the modeling process with the concept of an event. An *event* occurs at a specific time and place, can be described, and should be remembered by the system. Events drive or trigger all processing that a system does, so listing them and then analyzing them makes sense when you need to define system requirements.

When defining the requirements for a system, it is useful to begin by asking what events occur that will affect the system being studied. More specifically, what events

FIGURE 5-4

Models created during the analysis phase.

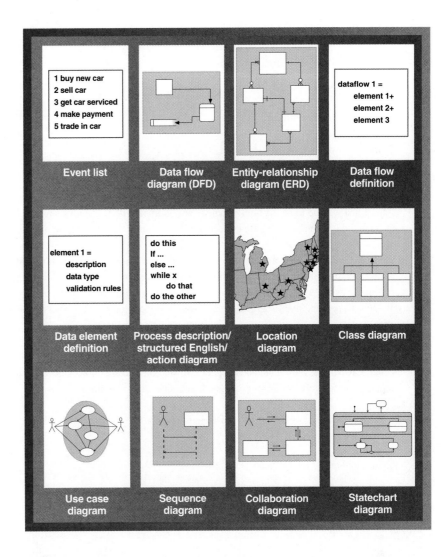

FIGURE 5-5

Some models created during the design phase.

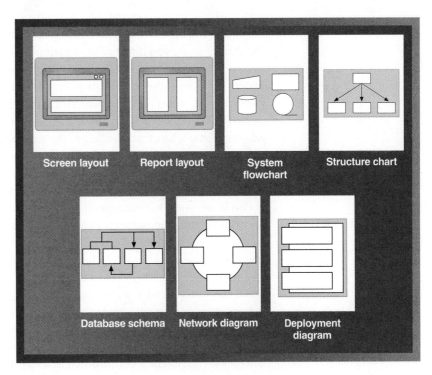

occur that will require the system to respond? By asking about the events that occur that affect the system, you direct your attention to the external environment and look at the system as a black box. This initial view helps keep your focus on a high-level view of the system (looking at the scope) rather than on the inner workings of the system. It also focuses your attention on the system's interfaces to outside people and other systems. End users, those who will actually use the system, can readily describe system needs in terms of events that affect their work. So, the external focus on events is appropriate when working with users. Finally, focusing on events gives you a way to divide (or decompose) the system requirements so you can study each separately. Complex systems need to be broken into manageable units to be understood, and decomposing the system based on events is one way to accomplish this.

Some events important to a charge account processing system for a store are shown in Figure 5-6. The system requirements are decomposed based on six events. A customer triggers three events: makes a charge, pays a bill, or changes address. Three events are triggered inside the system based on time: time to send out monthly statements, time to send late notices, and time to produce end-of-week summary reports. Describing this system in terms of events keeps the focus of the charge account system on the business requirements. Then, to divide up the work among developers, one analyst might focus on the events triggered by people, and another analyst might focus on events triggered internally. The system is decomposed in a way that allows it to be understood in detail.

FIGURE *5-6*

Events affecting a charge
account processing system.

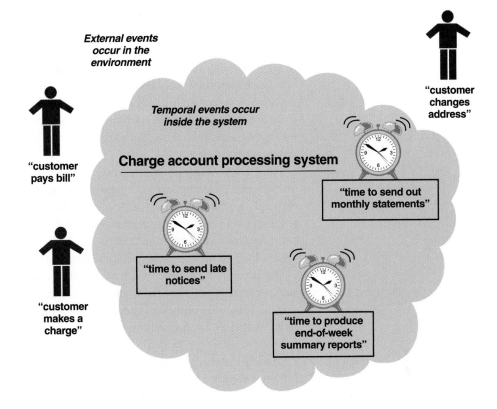

The Background of the Event Concept

The importance of the concept of events for defining system requirements was first emphasized for modern structured analysis as it was adapted to real-time systems in the early 1980s. Real-time systems require the system to react immediately to events in the environment. Early examples of real-time systems included control systems such as manufacturing process control or avionics guidance systems. For example, in process control, if a vat of chemicals is full, then the system needs to turn off the fill valve. The relevant event is "vat is full," and the system needs to respond to that event immediately. In an airplane guidance system, if the plane's altitude drops below 5,000 feet, then the system needs to turn on the low-altitude alarm. Business systems analysts did not have to create similar types of applications, so real-time system extensions to structured analysis did not receive much emphasis in business systems analysis and design at the time.

However, information systems have become much more interactive. Now most information systems being developed are so interactive that they can be thought of as real-time systems (in fact, people expect a real-time response to almost everything). Now the traditional approach, including structured analysis and information engineering, uses the event concept. More important, the object-oriented approach has also embraced the event concept. No matter which approach you use, a key analysis step is listing and analyzing the events to which the system must respond.

Types of Events

There are three types of events to consider: external events, temporal events, and state events. The analyst begins by trying to identify and list as many of these events as possible, refining the list while talking with system users.

External Events

An *external event* is an event that occurs outside the system, usually initiated by an external agent or actor. An external agent (or actor) is a person or organizational unit that supplies or receives data from the system. To identify the key external events, the analyst

---

external event

an event that occurs outside the system, usually initiated by an external agent or actor first tries to identify all of the external agents that might want something from the system. A classic example of an external agent is a customer. The customer may want to place an order for one or more products. This event is of fundamental importance to an order-processing system like the one needed by Rocky Mountain Outfitters. But other events are associated with a customer. Sometimes a customer wants to return an ordered product, or a customer needs to pay the invoice for an order. External events such as these are the types that the analyst looks for because they begin to define what the system needs to be able to do. They are events that lead to important transactions that the system must process.

When describing external events, it is important to name the event so that the external agent is clearly defined. The description should also include the action that the external agent wants to pursue. So, the event *Customer places an order* describes the external agent (a customer) and the action that the customer wants to take (to place an order for some products) that directly affects the system. Again, if the system is an order-processing system, the system needs to process the order for the customer.

Important external events can also result from the wants and needs of people or organizational units inside the company—for example, management requests for information. A typical event in an order-processing system might be *Management checks order status*. Perhaps managers want to follow up on an order for a key customer, and the system must routinely provide that information.

Another type of external event occurs when external entities provide new information that the system simply needs to store for later use. For example, a regular customer reports a change in address, phone, or employer. Usually one event for each type of external agent can be described to handle updates to data, such as *Customer updates account information*. Figure 5-7 provides a checklist to help in identifying external events.

FIGURE *5-7*

External event checklist.

External Events to Look for Include:
√ External agent wants something resulting in a transaction
√ External agent wants some information
√ Data changed need to be updated
√ Management wants some information

Temporal Events

temporal event

an event that occurs as a result of reaching a point in time

A second type of event is a *temporal event*, an event that occurs as a result of reaching a point in time. Many information systems produce outputs at defined intervals, such as payroll systems that produce a paycheck every two weeks (or each month). Sometimes the outputs are reports that management wants to receive regularly, such as performance reports or exception reports. These events are different from external events in that the system should automatically produce the required output without being told to do so. In other words, no external agent or actor is making demands, but the system is supposed to generate needed information or other outputs when they are needed.

The analyst begins identifying temporal events by asking about the specific deadlines that the system must accommodate. What outputs are produced at that deadline? What other processing might be required at that deadline? The analyst usually identifies these events by defining what the system needs to produce at that time. The payroll example discussed previously might be named *Time to produce biweekly payroll*. The event defining the need for a monthly summary report might be named *Time to produce monthly sales summary report*. Figure 5-8 provides a checklist to use in identifying temporal events.

FIGURE *5-8*

Temporal event checklist.

Temporal Events to Look for Include:
√ Internal outputs needed
    √ Management reports (summary or exception)
    √ Operational reports (detailed transactions)
    √ Internal statements and documents (including payroll)
√ External outputs needed
    √ Statements, status reports, bills, reminders

Temporal events do not have to occur on a fixed date. They can occur after a defined period of time has elapsed. For example, a bill might be given to a customer when a sale has occurred. If the bill has not been paid within 15 days, the system might send a late notice. The temporal event, *Time to send late notice*, might be defined as a point 15 days after the billing date.

State Events

state event

an event that occurs when something happens inside the system that triggers the need for processing

A third type of event is a *state event*, an event that occurs when something happens inside the system that triggers the need for processing. For example, if the sale of a product results in an adjustment to an inventory record and the inventory in stock drops below a reorder point, it is necessary to reorder. The state event might be named *Reorder point reached*. Often state events occur as a consequence of external events. Sometimes they are similar to temporal events, except the point in time cannot be defined. The reorder event might be named *Time to reorder inventory*, which sounds like a temporal event.

Identifying Events

It is not always easy to define the events that affect a system. But some guidelines can help an analyst think through the process.

Events versus Prior Conditions and Responses

It is sometimes difficult to distinguish between an event and part of a sequence of prior conditions that leads up to the event. Consider an example of a customer buying a shirt from a retail store (see Figure 5-9). From the customer's perspective, this purchase involves a long sequence of events. The first event might be that a customer wants to get dressed. Then the customer wants to wear a striped shirt. Next, his striped shirt appears to be worn out. Then the customer decides to drive to the mall. Then he decides to go into Sears. Then he tries on a striped shirt. Then the customer decides to leave Sears and go to Wal-Mart to try on a shirt. Finally, the customer wants to purchase the shirt. The analyst has to think through such a sequence to arrive at the point where an event directly affects the system. In this case, the system is not affected until the customer is in the store, has a shirt in hand ready to purchase, and says, "I want to buy this shirt."

In other situations, it is not easy to distinguish between an external event and the system's response. For example, when the customer buys the shirt, the system requests a credit card number, and the customer supplies the credit card. Is the act of supplying the credit card an event? In this case, no. It is part of the interaction that occurs while completing the original transaction.

The way to determine whether an occurrence is an event or part of the interaction following the event is by asking whether any long pauses or intervals occur—that is, can the system transaction be completed without interruption? Or is the system at rest again waiting for the next transaction? Once the customer wants to buy the shirt, the process continues until the transaction is complete. There are no significant stops once the transaction begins. Once the transaction is complete, the system is at rest, waiting for the next transaction to begin.

FIGURE 5-9

Sequence of actions that lead up to
only one event affecting the system.

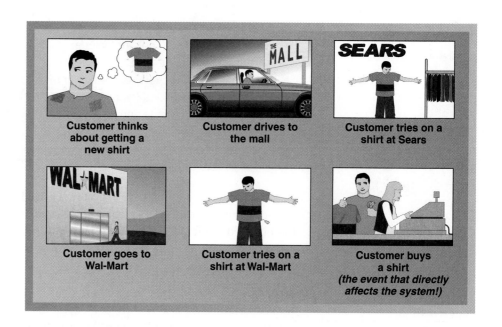

On the other hand, separate events occur when the customer buys the shirt using his store credit card account. When the customer later pays the bill at the end of the month, is the processing part of the interaction involving the purchase? In this case, no. The system records the transaction and then does other things. It does not halt all processes to wait for the payment. A separate event occurs later that results in sending the customer a bill (this is a temporal event: *Time to send monthly bills*). Eventually, another external event occurs (*Customer pays the bill*).

The Sequence of Events: Tracing a Transaction's Life Cycle

It is often useful in identifying events to trace the sequence of events that might occur for a specific external agent or actor. In the case of Rocky Mountain Outfitters' new customer support system, the analyst might think through all of the possible transactions that might result from one new customer (see Figure 5-10). First, the customer wants a catalog or asks for some information about item availability, resulting in a name and address being added to the database. Next, the customer might want to place an order. Perhaps he or she will want to change the order, correcting the size of the shirt, for example, or buy another shirt. Next, the customer might want to check the status of an order to find out the shipping date. Perhaps the customer has moved and wants an address change recorded for future catalog mailings. Finally, the customer might want to return an item. Thinking through this type of sequence can help identify events.

Technology-Dependent Events and System Controls

Sometimes the analyst is concerned about events that are important to the system but do not directly concern users or transactions. Such events typically involve design choices or system controls. During analysis, the analyst should temporarily ignore these events. They are important for the design phase, however.

Some examples of events that affect design issues include external events that involve actually using the physical system, such as logging on. Although important to the final operation of the system, such a detail of implementation should be deferred. At this stage, the analyst should focus only on the functional requirements—the work that the system needs to complete. A logical model does not need to indicate how the system is actually implemented, so the model should omit the implementation details.

Most of these events involve *system controls*, which are checks or safety procedures put in place to protect the integrity of the system. Logging on to a system is required be-

system controls

checks or safety procedures put in place to protect the integrity of the system

Customer requests
a catalog

Customer wants to
check item availability

Customer places
an order

Customer changes or
cancels an order

Customer wants to
check order status

Customer updates
account information

Customer returns
the item

FIGURE *5-10*

The sequence of "transactions" for one specific customer resulting in many events.

perfect technology assumption

the assumption that events should be included during analysis only if the system would be required to respond under perfect conditions

FIGURE *5-11*

Events deferred until the design phase.

cause of system security controls, for example. Other controls protect the integrity of the database, such as backing up the data every day. Both of these controls are important to the system, and they will certainly be added to the system during design. But spending time on these controls during analysis only adds to the requirements model details that the users are not typically very concerned about (they trust information services to take care of such details).

One technique used to help decide which events apply to controls is to assume that technology is perfect. The *perfect technology assumption* states that events should be included during analysis only if the system would be required to respond under perfect conditions—that is, with equipment never breaking down, capacity for processing and storage being unlimited, and people operating the system being completely honest and never making mistakes. By pretending that technology is perfect, analysts can eliminate events like *Time to back up the database* because they can assume that the disk will never crash. Again, during design, the project team adds these controls because technology is obviously not perfect. Figure 5-11 lists some examples of events that can be deferred until the design phase.

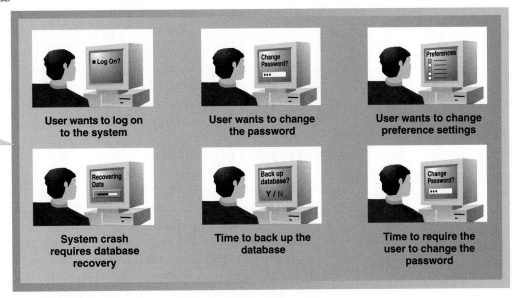

Don't worry much about these until the design phase

User wants to log on
to the system

User wants to change
the password

User wants to change
preference settings

System crash
requires database
recovery

Time to back up the
database

Time to require the
user to change the
password

Events in the Rocky Mountain Outfitters Case

The Rocky Mountain Outfitters customer support system involves a variety of events, many of them similar to those just discussed. A list of the external events is shown in Figure 5-12. Some of the most important external events involve customers: *Customer wants to check item availability, Customer places an order, Customer changes or cancels an order*. Other external events involve RMO departments: *Shipping fulfills order, Marketing wants to send promotional material to customers, Merchandising updates catalog*. The analyst can develop this list of external events by looking at all of the people and organizational units that want the system to do something for them.

FIGURE *5-12*

External events for the RMO customer support system.

Customer wants to check item availability
Customer places an order
Customer changes or cancels order
Customer or management wants to check order status
Shipping fulfills order
Shipping identifies back order
Customer returns item (defective, changed mind, full or partial returns)
Prospective customer requests catalog
Customer updates account information
Marketing wants to send promotional materials to customers
Management adjusts customer charges (correct errors, make concessions)
Merchandising updates catalog (add, change, delete, change prices)
Merchandising creates special product promotion
Merchandising creates new catalog

The customer support system also includes quite a few temporal events, shown in Figure 5-13. Many of these produce periodic reports for organizational units: *Time to produce order summary reports, Time to produce fulfillment summary reports, Time to produce catalog activity reports*. The analyst can develop the list of temporal events by looking for all of the regular reports and statements that the system must produce at certain times.

FIGURE *5-13*

Temporal events for the RMO customer support system.

Time to produce order summary reports
Time to produce transaction summary reports
Time to produce fulfillment summary reports
Time to produce prospective customer activity reports
Time to produce customer adjustment/concession reports
Time to produce catalog activity reports

Looking at Each Event

While developing the list of events, the analyst should note for later use any additional information about each event. This information is entered in an event table. An *event table* includes rows and columns, representing events and their details, respectively. Each row in the event table records information about one event. Each column in the table represents a key piece of information about that event. The information about an event *Customer wants to check item availability* is shown in Figure 5-14.

First, for each event, how does the system know the event has occurred? A signal that tells the system an event has occurred is called the *trigger*. For an external event, the trig-

event table

a table that lists events in rows and key pieces of information about each event in columns

trigger

a signal that tells the system that an event has occurred, either the arrival of data needing processing or a point in time

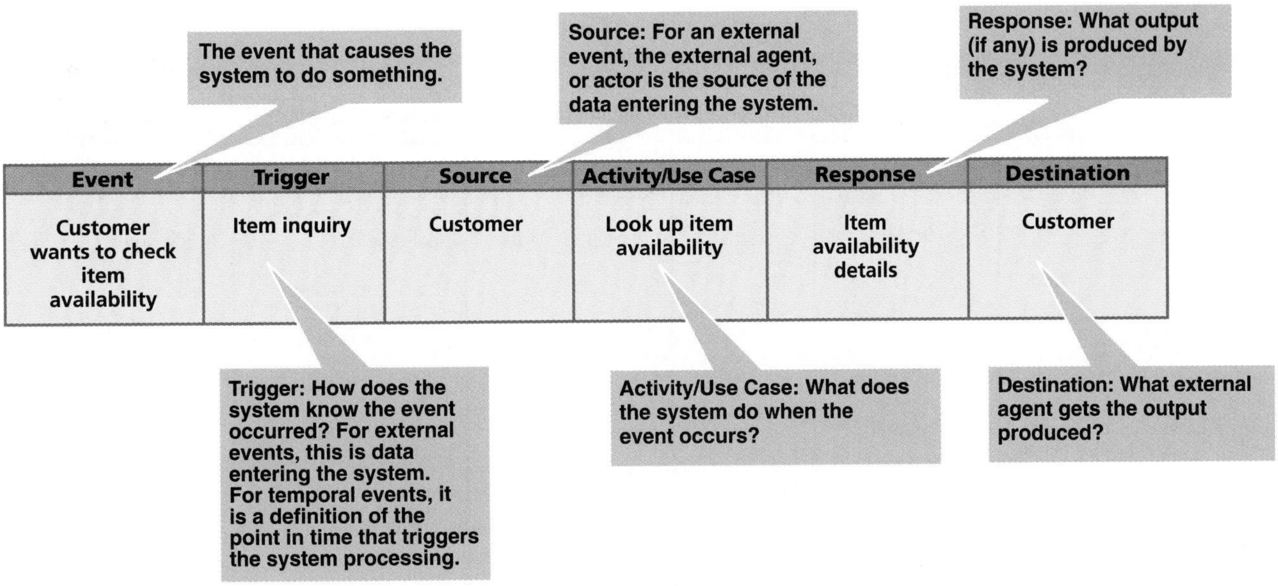

The event that causes the system to do something.

Source: For an external event, the external agent, or actor is the source of the data entering the system.

Response: What output (if any) is produced by the system?

Event	Trigger	Source	Activity/Use Case	Response	Destination
Customer wants to check item availability	Item inquiry	Customer	Look up item availability	Item availability details	Customer

Trigger: How does the system know the event occurred? For external events, this is data entering the system. For temporal events, it is a definition of the point in time that triggers the system processing.

Activity/Use Case: What does the system do when the event occurs?

Destination: What external agent gets the output produced?

FIGURE *5-14*

Information about each event in an event table.

source

an external agent or actor that supplies data to the system

activity

behavior that the system performs when an event occurs (similar to a use case)

use case

a series of actions that a system performs that result in a defined outcome (similar to an activity)

response

an output, produced by the system, that goes to a destination

destination

an external agent or actor that receives data from the system

ger is the arrival of data that the system must process. For example, when a customer places an order, the new order details are provided as input. The *source* of the data is also important to know. In this case, the source of the new order details is the customer—an external agent or actor. For a temporal event, the trigger is a point in time. For example, at the end of each business day, the system knows it is time to produce transaction summary reports.

Next, what does the system do when the event occurs? What the system does (the reaction to the event) is called an *activity* or *use case*. When a customer places an order, the system carries out the activity *Create a new order*. When it is time to produce transaction summary reports, the system carries out the activity *Produce transaction summary reports*. The term *use case* comes from the object-oriented approach. A use case is very similar to an activity. A use case is defined as a series of actions a system performs that result in a defined outcome. Another way to think about a use case is as a *case* where the system is *used* for some purpose, such as to *Create a new order*. Many analysts are finding the concept of a use case helpful when thinking through and modeling system requirements whether they are using a traditional or an object-oriented approach. Much has been written recently about use cases, and you will learn how to describe use cases in detail in Chapter 7.

Finally, what response from the system does the activity or use case produce? A *response* is an output from the system. When the system produces transaction summary reports, those reports are the outputs. One activity can generate several responses. For example, when the system creates a new order, an order confirmation goes to the customer, order details go to shipping, and a record of the transaction goes to the bank. The *destination* is the place where any response (output) is sent, again an external agent or actor. Sometimes an activity generates no response at all. For example, if the customer wants to update account information, the information is recorded in the database, but no output needs to be produced. Recording information in the database is part of the activity or use case.

The list of events—together with the trigger, source, activity or use case, response(s), and destination(s) for each event—can be placed in an event table so that the analyst can keep track of them for later use. An event table is a convenient way to record key information about the requirements for the information system. The event table for the RMO customer support system is shown in Figure 5-15. The event table will later be used in Chapters 6 and 7 to develop requirements models for both the traditional approach and the object-oriented approach.

FIGURE 5-15 The complete event table for the RMO customer support system.

CUSTOMER SUPPORT SYSTEM EVENT TABLE

Event	Trigger	Source	Activity/Use Case	Response	Destination
1. Customer wants to check item availability	Item inquiry	Customer	Look up item availability	Item availability details	Customer
2. Customer places an order	New Order	Customer	Create new order	Real-time link	Credit bureau
				Order confirmation	Customer
				Order details	Shipping
				Transaction	Bank
3. Customer changes or cancels order	Order change request	Customer	Update order	Change confirmation	Customer
				Order change details	Shipping
				Transaction	Bank
4. Time to produce order summary reports	"End of week, month, quarter and year"		Produce order summary reports	Order summary reports	Management
5. Time to produce transaction summary reports	"End of day"		Produce transaction summary reports	Transaction summary reports	Accounting
6. Customer or management wants to check order status	Order status inquiry	Customer or management	Look up order status	Order status details	Customer or management
7. Shipping fulfills order	Order fulfillment notice	Shipping	Record order fulfillment		
8. Shipping identifies back order	Back-order notice	Shipping	Record back order	Back-order notification	Customer
9. Customer returns item	Order return notice	Customer	Create order return	Return confirmation	Customer
				Transaction	Bank
10. Time to produce fulfillment summary reports	"End of week month, quarter, and year"		Produce fulfillment summary reports	Fulfillment summary reports	Management
11. Prospective customer requests catalog	Catalog request	Prospective customer	Provide catalog info	Catalog	Prospective customer
12. Time to produce prospective customer activity reports	"End of month"		Produce prospective customer activity reports	Prospective customer activity reports	Marketing
13. Customer updates account information	Customer account update notice	Customer	Update customer account		
14. Marketing wants to send promotional materials to customers	Promotion package details	Marketing	Distribute promotional package	Promotional package	Customer and prospective customer
15. Management adjusts customer charges	Customer charge adjustment	Management	Create customer charge adjustment	Charge adjustment notification	Customer
				Transaction	Bank

(continued)

Event	Trigger	Source	Activity/Use Case	Response	Destination
16. Time to produce customer adjustment/ concession reports	"End of month"		Produce customer adjustment reports	Customer adjustment reports	Management
17. Merchandising updates catalog	Catalog update details	Merchandising	Update catalog		
18. Merchandising creates special product promotion	Special promotion details	Merchandising	Create special promotion		
19. Merchandising creates new catalog	New catalog details	Merchandising	Create new catalog	Catalog	Customer and prospective customer
20. Time to produce catalog activity reports	"End of month"		Produce catalog activity reports	Catalog activity reports	Merchandising

THINGS AND SYSTEM REQUIREMENTS

Another key concept used to define system requirements involves understanding and modeling things about which the system needs to store information. To the users, these items are the things they deal with when they do their work—products, orders, invoices, and customers—that need to be part of the system. They are often referred to as things in the problem domain of the system. For example, an information system needs to store information about customers and products, so it is important for the analyst to identify lots of information about them. Often these things are similar to the external agents or actors that interact with the system. For example, a customer external agent places an order, but the system also needs to store information about the customer. In other cases, these things are distinct from external agents. For example, there is no external agent named *product*, but the system needs to store information about products.

In the traditional approach to development, these things make up the data about which the system stores information. The type of data that need to be stored is definitely a key aspect of the requirements for any information system. In the object-oriented approach, these things are the objects that interact in the system. No matter which approach to developing an information system you use, identifying and understanding these things are both key initial steps.

Types of Things

As with events, an analyst should ask the users to discuss the types of things that they work with routinely. The analyst can ask about several types of things to help identify them. Many things are tangible and therefore more easily identified, but others are intangible. Different types of things are important to different users, so it is important to include information from all types of users.

Figure 5-16 shows some types of things to consider. Tangible things are often the most obvious, such as an airplane, book, or vehicle. In the Rocky Mountain Outfitters case, a catalog and an item in the catalog are tangible things of importance. Another common type of thing in an information system is a role played by a person, such as employee, customer, doctor, or patient. A customer is obviously a very important role a person plays in the Rocky Mountain Outfitters case.

FIGURE 5-16

Types of things.

Tangible things	Roles played	Organizational units	Devices	Incidents, events, or interactions	Sites/ locations
airplane book vehicle document worksheet	employee customer doctor patient end user system administrator	division department section task force workgroup	sensor timer controller printer disk drive keyboard display window mouse menu button	flight service call logon logoff contract purchase order payment	warehouse branch office factory retail store desktop

Other types of things can include organizational units, such as a division, department, or workgroup. Similarly, sites or locations might be important in a particular system, such as a warehouse, a store, or a branch office. Finally, information about an incident or interaction of importance can be considered a thing—information about an order, a service call, a contract, or an airplane flight. An order, a shipment, and a return are important incidents in the Rocky Mountain Outfitters case. Sometimes these incidents are thought of as relationships between things. For example, an order is a relationship between a customer and an item of inventory. Initially, the analyst might simply list all of these as things and then make adjustments that might be required by different approaches to analysis and design.

The analyst identifies these types of things by thinking about each event in the event list and asking questions. For example, for each event, what types of things are affected that the system needs to know about and store information about? When a customer places an order, the system needs to store information about the customer, the items ordered, and the details about the order itself, such as the date and payment terms.

Procedure for Developing an Initial List of Things

The general guidelines just discussed reveal that analysts can use many sources of information to develop an initial list of things about which the system needs to store information. A useful procedure to follow is to begin by listing all of the *nouns* that users mention when talking about the system. Consider the events, the activities or use cases, the external agents or actors, and the triggers and responses from the event table as potential things, for example. Then add to the list any additional nouns that appear in information about the existing system or that come up in discussions with stakeholders.

Step One   Using the event table and information about each event, identify all of the nouns.

For the RMO customer support system, the nouns include RMO, customer, product item, order, confirmation, transaction, shipping, bank, change request, summary report, management, transaction report, accounting, back order, back-order notification, return, return confirmation, fulfillment reports, prospective customer, catalog, marketing, customer account, promotional materials, charge adjustment, catalog details, merchandising, and catalog activity reports.

Step Two Using other information from existing systems, current procedures, and current reports or forms, add items or categories of information needed.

For the RMO customer support system, these items might include more detailed information, such as price, size, color, style, season, inventory quantity, payment method, shipping address, and so forth. Some of these items might be additional categories, and some might be more specific pieces of information about things you have already identified (called *attributes*).

Step Three Refine the list and record assumptions or issues to explore.

As this list of nouns builds, it will be necessary to refine it. Ask these questions about each noun to try to decide whether you should *include* it:

- Is it a unique thing the system needs to know about?
- Is it inside the scope of the system I am working on?
- Does the system need to remember more than one of these items?

Ask these questions about each noun to decide whether you should *exclude* it:

- Is it really a synonym for some other thing I have identified?
- Is it really just an output of the system produced from other information I have identified?
- Is it really just an input that results in recording some other information I have identified?

Ask these questions about each noun to decide whether you should *research* it:

- Is it likely to be a specific piece of information (attribute) about some other thing I have identified?
- Is it something that I might need if assumptions change?

Figure 5-17 lists some of the nouns from the RMO customer support system event table and other sources, with some notes about each one.

FIGURE *5-17*

Partial list of "things" based on nouns for RMO.

Identified Noun	Notes on Including Noun as a Thing to Store
Accounting	We know who they are. No need to store it.
Back order	A special type of order? Or a value of order status? Research.
Back-order information	An output that can be produced from other information.
Bank	Only one of them. No need to store.
Catalog	Yes, need to recall them, for different seasons and years. Include.
Catalog activity reports	An output that can be produced from other information. Not stored.
Catalog details	Same as catalog? Or the same as product items in the catalog? Research.
Change request	An input resulting in remembering changes to an order.
Charge adjustment	An input resulting in a transaction.
Color	One piece of information about a product item.
Confirmation	An output produced from other information. Not stored.
Credit card information	Part of an order? Or part of customer information? Research.
Customer	Yes, a key thing with lots of details required. Include.

(continued)

Identified Noun	Notes on Including Noun as a Thing to Store
Customer account	Possibly required if an RMO payment plan is included. Research.
Fulfillment reports	An output produced from information about shipments. Not stored.
Inventory quantity	One piece of information about a product item. Research.
Product item	Yes, what RMO includes in a catalog and sells. Include.
Management	We know who they are. No need to store.
Marketing	We know who they are. No need to store.
Merchandising	We know who they are. No need to store.
Order	Yes, a key system responsibility. Include.
Payment method	Part of an order. Research.
Price	Part of a product item. Research.
Promotional materials	An output? Or documents stored outside the scope? Research.
Prospective customer	Possibly same as customer. Research.
Return	Yes, the opposite of an order. Include.
Return confirmation	An output produced from information about a return. Not stored.
RMO	There is only one of these! No need to store.
Season	Part of a catalog? Or is there more to it? Research.
Shipment	Yes, a key thing to track. Include.
Shipper	Yes, they vary and we need to track the order. Include.
Shipping	Our department. No need to store.
Shipping address	Part of customer? Or order? Or shipment? Research.
Size	Part of a product item. Research.
Style	Part of a product item. Research.
Summary report	An output produced from other information. Not stored.
Transaction	Yes, each one is important and must be remembered. Include.
Transaction report	An output produced from transaction information. Not stored.

Relationships among Things

After recording and refining the list of things, the analyst researches and records additional information. Many important relationships among things are important to the system. A *relationship* is a naturally occurring association among specific things, such as an order is placed by a customer and an employee works in a department (see Figure 5-18). *Is placed by* and *works in* are two relationships that naturally occur between specific things. Information systems need to store information about employees and about departments, but equally important is storing information about the specific relationships—John works in the accounting department, and Mary works in the marketing department, for example. Similarly, it is quite important to store the fact that Order 1043 for a shirt was placed by John Smith.

Relationships between things apply in two directions. For example, *a customer places an order* describes the relationship in one direction. Similarly, *an order is placed by a cus-*

FIGURE *5-18*

Relationships naturally
occur among things.

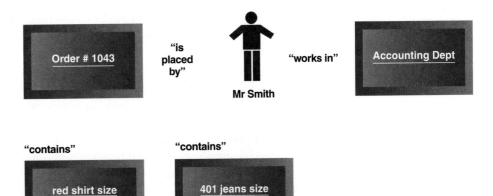

tomer describes the relationship in the other direction. It is important to understand the relationship in both directions because sometimes it might seem more important for the system to record the relationship in one direction than in the other. For example, Rocky Mountain Outfitters definitely needs to know what items a customer ordered so the shipment can be prepared. However, it might not be apparent initially that the company needs to know all of the customers who have ordered a particular item. What if the company needs to notify all customers who ordered a defective or recalled product? Knowing this information would be very important, but the operational users might not immediately recognize that fact.

It is also important to understand the nature of each relationship in terms of the number of associations for each thing. For example, a customer might place many different orders, but an order is placed by only one customer. The number of associations that occur is referred to as the *cardinality* of the relationship. Cardinality can be one to one or one to many. Again, cardinality is established for each direction of the relationship. The term *multiplicity* is used to refer to the number of associations in the object-oriented approach. Figure 5-19 lists examples of cardinality/multiplicity associated with an order.

cardinality

the number of associations that occur among specific things, such as a customer places many orders and an employee works in one department

multiplicity

a synonym for cardinality (used with the object-oriented approach)

FIGURE *5-19*

Cardinality/multiplicity of relationships.

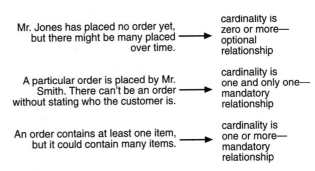

Sometimes it is important to describe not just the cardinality but also the range of possible values of the cardinality (the minimum and maximum cardinality). For example, a particular customer might not ever place an order. In this case, there are zero associations. Alternatively, the customer may place one order, meaning one association exists. Finally, the customer might place two, three, or even more orders. The relationship for a customer placing an order can have a range of zero, one, or more, usually indicated as zero or more. The zero is the minimum cardinality, and "more" is the maximum cardinality. These terms are referred to as "cardinality constraints."

In some cases, at least one association is required (a mandatory as opposed to optional relationship). For example, the system might not record any information about a customer until the customer places an order. Therefore, the cardinality would read "customer places one or more orders."

A one-to-one relationship can also be refined to include minimum and maximum cardinality. For example, the order is placed by one customer—it is impossible to have an order if there is no customer. Therefore, one is the minimum cardinality, making the relationship mandatory. Since there cannot be more than one customer for each order, one is also the maximum cardinality. Sometimes such a relationship is read as "an order must be placed by one and only one customer."

The relationships described here are between two different types of things—for example, a customer and an order. These are called *binary relationships*. Sometimes a relationship is between two things of the same type. For example, the relationship *is married to* is between two different people. This type of relationship is called a *unary relationship* (sometimes called a *recursive relationship*). Another example of a unary relationship is an organizational hierarchy where one organizational unit reports to another organizational unit—the packing department reports to shipping, which reports to distribution, which reports to marketing.

A relationship can also be among three different types of things, called a *ternary relationship*, or any number of different types of things, called an n-*ary relationship*. One particular order, for example, might be associated with a specific customer plus a specific sales representative, requiring a ternary relationship.

Storing information about the relationships is just as important as storing information about the specific things. It is important to have information on the name and address of each customer, but it is equally important (or perhaps more so) to know what items each customer has ordered.

Attributes of Things

Most information systems store and use specific pieces of information about each thing, as discussed in Figure 5-17 earlier. The specific pieces of information are called *attributes*. For example, a customer has a name, a phone number, a credit limit, and so on. Each of these details is an attribute. The analyst needs to identify the attributes of each thing that the system needs to store. One attribute may be used to identify a specific thing, such as a Social Security number for an employee or an order number for a purchase. The attribute that uniquely identifies the thing is called an *identifier*, or *key*. Sometimes the identifier is already established (a Social Security number, vehicle ID number, or product ID number). Sometimes the system needs to assign a specific identifier (an invoice number or transaction number).

A system may need to remember many similar attributes. For example, a customer has several names—a first name, a middle name, a last name, and possibly a nickname. A *compound attribute* is an attribute that contains a collection of related attributes, so an analyst may choose one compound attribute to represent all of these names, perhaps naming it *Customer full name*. A customer might also have several phone numbers—a home phone number, office phone number, fax phone number, and cellular phone number. The analyst might start out by describing the most important attributes but later add to the list. Attribute lists can get quite long. Some examples of attributes of a customer and the values of attributes for specific customers are shown in Figure 5-20.

Data Entities and Objects

When describing things important to a system, so far we have mainly used examples of things the system needs to store information about. In the traditional approach to system development, these things are called *data entities*. The data entities, the relationship between data entities, and the attributes of data entities are modeled using an entity-

---

binary relationships

relationships between two different types of things, such as a customer and an order

unary (recursive) relationship

a relationship among two things of the same type, such as one person being married to another person

ternary relationship

a relationship among three different types of things

n-ary relationship

a relationship among *n* (any number of) different types of things

attribute

one piece of specific information about a thing

identifier (key)

an attribute that uniquely identifies a thing

compound attribute

an attribute that contains a collection of related attributes

data entities

the things the system needs to store information about in the traditional approach to information systems

FIGURE 5-20

Attributes and values.

All customers have these attributes:	Each customer has a value for each attribute:		
Customer ID	101	102	103
First name	John	Mary	Bill
Last name	Smith	Jones	Casper
Home phone	555-9182	423-1298	874-1297
Work phone	555-3425	423-3419	874-8546

relationship diagram (ERD). Computer processes interact with the data entities, creating them, updating attribute values, and associating one with another. The entity-relationship diagram (described in detail later) is a requirements model that is later used to create the database design model, usually for a relational database. The entity-relationship diagram is used with the structured approach and the information engineering approach to system development.

The other way to think about things is as objects that interact in the system. Objects in the work environment of the user (sometimes called *objects in the problem domain*) in the object-oriented approach are often similar to the data entities in the traditional approach. The main difference is that objects do the work in the system, they do not just store information. In other words, objects have behaviors as well as attributes. This simple distinction has profound effects on the way the system is viewed and constructed. At the early stages of modeling requirements, the object-oriented approach is quite similar to the traditional approaches. Figure 5-21 compares the traditional and object-oriented views of data entities and objects.

FIGURE 5-21

Data entities compared with objects.

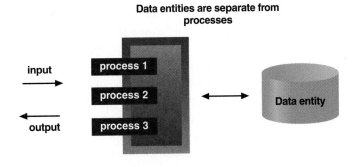

Data entities are separate from processes

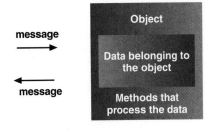

Objects encapsulate data and the methods that process the data into one unit

With the object-oriented approach, each specific thing is an object (John, Mary, Bill), and the type of thing is called a *class* (in this case, Customer). The word *class* is used because all of the objects are classified as one type of thing. The classes, associations among classes, and the attributes of classes are modeled using a class diagram.

class

the type or classification to which all similar objects belong

Additionally, the class diagram shows some of the behaviors of objects of the class, called *methods*.

Methods of a class are the behaviors all objects in the class are capable of doing. A behavior is an action that the object processes itself. Instead of an outside process updating data values, the object updates its own values when asked to do so. To ask the object to do something, another object sends it a message. One object can send another object a message, or an object can send the user a message. The information system overall becomes a collection of interacting objects, as described in Chapter 2.

Because each object contains values for attributes and methods for operating on those attributes (plus other behaviors), an object is said to be *encapsulated* (covered or protected)—a self-contained unit. We will return to the discussion of the class diagram later in this chapter.

THE ENTITY-RELATIONSHIP DIAGRAM

The traditional approach to system development (the structured techniques and information engineering approaches, as described in Chapter 2) places a great deal of emphasis on data storage requirements for a new system. Data storage requirements include the data entities, their attributes, and the relationships among the data entities. As just discussed, the model used to define the data storage requirements is called the *entity-relationship diagram* (ERD).

Examples of ERD Notation

On the entity-relationship diagram, rectangles represent data entities, and the lines connecting the rectangles show the relationships among data entities. Figure 5-22 shows an example of a simplified entity-relationship diagram with two data entities, Customer and Order. Each Customer can place many Orders, and each Order is placed by one Customer. The cardinality is one to many in one direction and one to one in the other direction. The "crow's feet" symbol on the line next to the Order data entity indicates "many" orders. But other symbols on the relationship line also represent the minimum and maximum cardinality constraints. See Figure 5-23 for an explanation of relationship symbols. The model in Figure 5-22 actually says that a Customer places a minimum of zero and a maximum of many Orders. Reading the other direction, the model says an Order is placed by at least one and only one Customer. This notation can express precise details about the system. The constraints reflect the business policies that management has defined, and the analyst must discover what these policies are. The analyst does not determine that two customers cannot share one order; management does.

FIGURE *5-22*

A simple entity-relationship diagram.

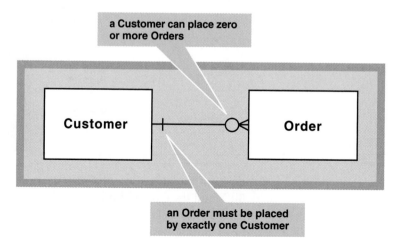

a Customer can place zero or more Orders

Customer | | ○< Order

an Order must be placed by exactly one Customer

FIGURE *5-23*

Cardinality symbols of relationships.

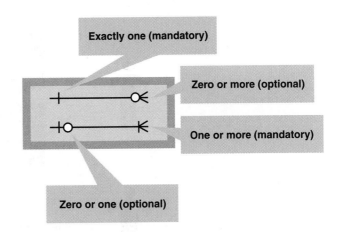

Figure 5-24 shows the model expanded to include the order items (one or more specific items included on the order). Each order contains a minimum of one and a maximum of many items (there could not be an order if it did not contain at least one item). For example, an order might include a shirt, a pair of shoes, and a belt, and each of these items is associated with the order. This example also shows some of the attributes of each data entity: A customer has a customer number, a name, a billing address, and several phone numbers. Each order has an order ID, order date, and so on. Each order item has an item ID, quantity, and price. The attributes of the data entity are listed below the name, with the key identifier listed first.

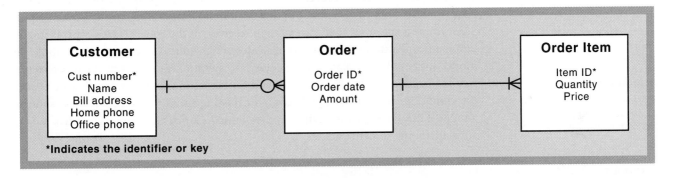

FIGURE *5-24*

An expanded ERD with attributes shown.

Figure 5-25 shows how the actual data in some transactions might look. John is a customer who has placed two orders. The first order, placed on February 4, was for two shirts and one belt. The second order, placed on March 29, was for one pair of boots and two pairs of sandals. Mary is a customer who has not yet placed an order. Recall that a customer might place zero or more orders. Therefore, Mary is not associated with any orders. Finally, Sara placed an order on March 30 for three pairs of sandals.

While working on the model, the analyst often refines the ERD. One example of refinement is analyzing many-to-many relationships. Figure 5-26 shows an example of a many-to-many relationship. At a university, courses are offered as course sections, and a student enrolls in many course sections. Each course section contains many students. Therefore, the relationship between course section and student is many to many. There are situations where many-to-many relationships occur naturally, and they can be modeled as shown, with "crow's feet" on both ends of the relationship. If a relational database is designed from an ERD with a many-to-many relationship, a separate table containing keys from both sides of the relationship is created because relational databases cannot directly implement many-to-many relationships. Chapter 13 discusses relational databases in detail.

FIGURE *5-25*

Customers, orders, and order items consistent with the expanded ERD.

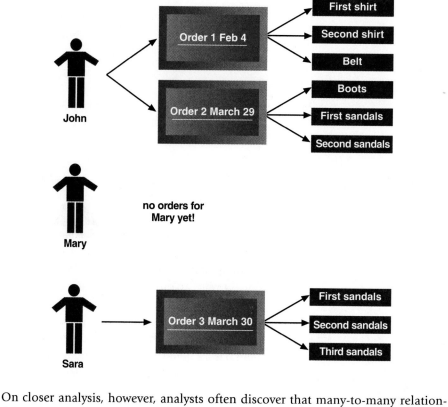

On closer analysis, however, analysts often discover that many-to-many relationships involve additional data that must be stored. For example, in the ERD in Figure 5-26, where is the grade that each student receives for the course stored? This is important data, and although the model indicates which course section a student took, the model does not have a place for the grade. The solution is to add a data entity to represent the relationship between student and course section, sometimes called an *associative entity*. The associative entity is given the missing attribute. Figure 5-27 shows the expanded ERD with an associative entity named Course Enrollment, which has an attribute for the student's grade.

associative entity

a data entity that represents a many-to-many relationship between two other data entities

FIGURE *5-26*

A university course enrollment ERD with a many-to-many relationship.

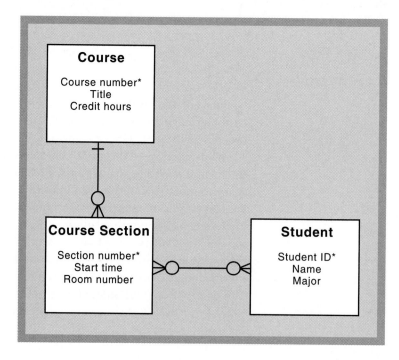

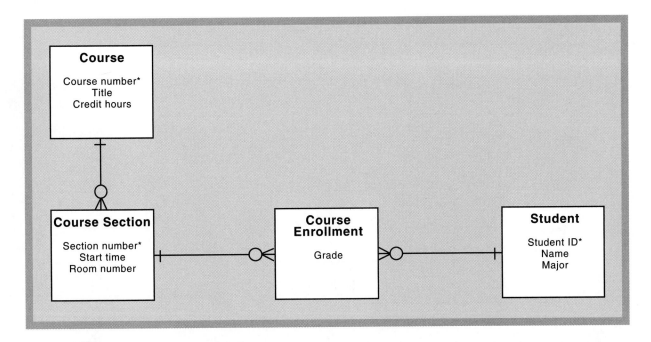

FIGURE *5-27*

A refined university course enrollment
ERD with an associative entity.

Reading the relationships in Figure 5-27 from left to right, the ERD says that one course section has many course enrollments, each with its own grade, and each course enrollment applies to one specific student. Reading from right to left, it says one student has many course enrollments, each with its own grade, and each course enrollment applies to one specific course section. A database implemented using this model will be able to produce grade lists showing all students and their grades in each course section, as well as grade transcripts showing all grades earned by each student.

Other refinements are made to the ERD during the modeling process. One major refinement process that applies to designing relational databases, called *normalization*, is discussed in Chapter 13.

The Rocky Mountain Outfitters ERD

The Rocky Mountain Outfitters entity-relationship diagram is a variation of the customer and order example already described. Most of the data entities are from the list of things developed in Figure 5-17. Figure 5-28 shows a fairly complete version of the model but without the attributes.

Each customer can place zero or more orders. Each order can have one or more order items, meaning the order might be for one shirt and two sweaters. Each order item is for a specific inventory item, meaning a specific size and color of shirt. Although the diagram does not show such an attribute, an inventory item should have an attribute for quantity on hand of that size and color. Since there are many colors and sizes (each with its own quantity), each inventory item is associated with a product item that describes the item generically (vendor, gender, description).

An earlier version of the model showed that each product item is contained in one or more catalogs, and each catalog contains one or more product items, a many-to-many relationship. Therefore, this model adds an *associative entity* named Catalog Product between Catalog and Product Item because the relationship has some attributes that need to be remembered, specifically the regular and special prices. Each catalog can list a different price for the same product item (ski pants might be cheaper in the spring catalog).

The entity-relationship diagram for Rocky Mountain Outfitters also has information about shipments. Since this diagram includes requirements for orders, not retail sales, each order item is part of a shipment. A shipment may contain many order items. Each shipment is shipped by one shipper.

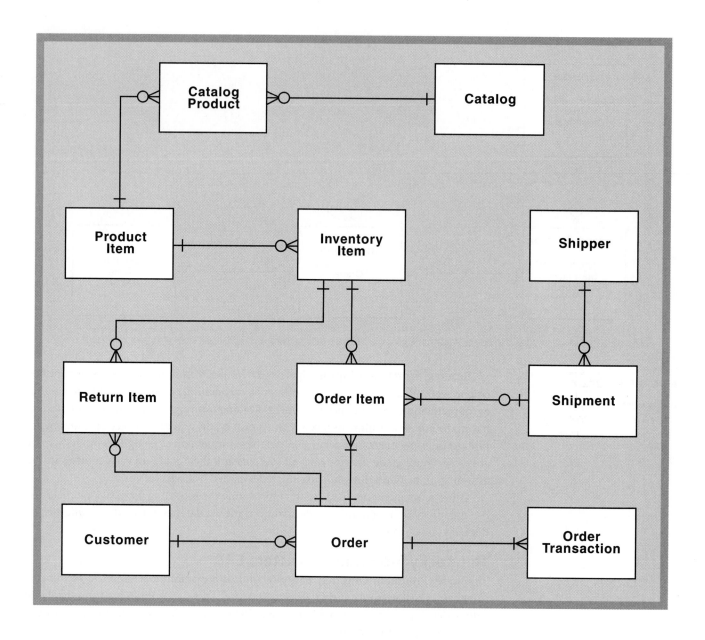

FIGURE *5-28*

Rocky Mountain Outfitters customer support system entity-relationship diagram (ERD) without attributes.

The ERD shown in Figure 5-28 contains a lot of very specific information about the requirements for the system. Be sure that you can trace through all of the relationships shown and try to describe a specific example of each data entity involved in one specific order. Try listing the key attributes for each data entity to check your understanding. Draw a sketch similar to that shown in Figure 5-25 to show some actual data this ERD describes. You can check your understanding by looking ahead to the class diagram in the next section and to the relational database design in Chapter 13.

Once it is developed, a model like this entity-relationship diagram needs to be walked through carefully, as you would walk through the logic of a program. Being able to walk through and "debug" any model is a very important skill in system development, as discussed in Chapter 4.

To test your understanding of the diagram, consider whether one order might have items shipped by different shippers. Is it possible, given the requirements shown in this diagram? The answer is yes. Operationally, some order items might be back ordered, so when they are finally shipped, they are part of a different shipment. A different shipper could handle this shipment.

Other requirements shown in the model include one or more order transactions for each order. An order transaction is a record of a payment or a refund for the order. One order transaction is created when the customer initially pays for the order. Later, though, the customer might add another item to the order, generating an additional charge. This involves a second order transaction. Finally, the customer might return an item, requiring a refund and a third order transaction.

THE CLASS DIAGRAM

The object-oriented approach also emphasizes understanding the things involved in a system. As discussed previously, this approach models classes of objects instead of data entities. The classes of objects have attributes and associations, just like the data entities. Cardinality (called *multiplicity* in the object-oriented approach) also applies among classes. The main difference, as already discussed, is that the objects do the actual processing in the system as well as storing information. The processing done (the behavior of the object) is possible because objects have methods as well as attributes. The sets of requirements models for the traditional approaches and object-oriented approaches eventually become quite different because of the object behavior. The design models are definitely very different.

More Complex Issues about Classes of Objects

Some other issues about things come up more frequently with the object-oriented approach than with the traditional approach, although the issues are not exclusively object-oriented. These issues are two additional ways that people structure their understanding of things in the real world: generalization/specialization hierarchies and whole-part hierarchies. This section discusses these concepts first before discussing the class diagram used to model things with the object-oriented approach.

Generalization/Specialization

generalization/specialization hierarchies

hierarchies that structure or rank classes from the more general superclass to the more specialized subclasses; sometimes called inheritance hierarchies

Generalization/specialization hierarchies are based on the idea that people classify things in terms of similarities and differences. Generalizations are judgments that group similar types of things; for example, there are many types of motor vehicles—cars, trucks, and tractors. All motor vehicles share certain general characteristics, so a motor vehicle is a more general class. Specializations are judgments that categorize different types of things—for example, special types of cars include sports cars, sedans, and sport utility vehicles. These types of cars are similar in some ways, yet different in other ways. Therefore, a sports car is a special type of car.

A generalization/specialization hierarchy is used to structure or rank these things from the more general to the more special. As discussed previously, classification refers to defining classes of things. Each class of thing in the hierarchy might have a more general class above it, called a *superclass*. At the same time, a class might have a more specialized class below it, called a *subclass*. In Figure 5-29, a car has three subclasses and one superclass (Motor Vehicle).

We mentioned that people structure their understanding by using generalization/specialization hierarchies. That is, people learn by refining the classifications they make about some field of knowledge. A knowledgeable banker can talk at length about special types of loans and deposit accounts. A knowledgeable merchandiser like John Blankens at Rocky Mountain Outfitters can talk at length about special types of outdoor activities and clothes. Therefore, when asking users about their work, the analyst is trying to understand the knowledge the user has about the work, which the analyst can represent by constructing generalization/specialization hierarchies. At some level, the

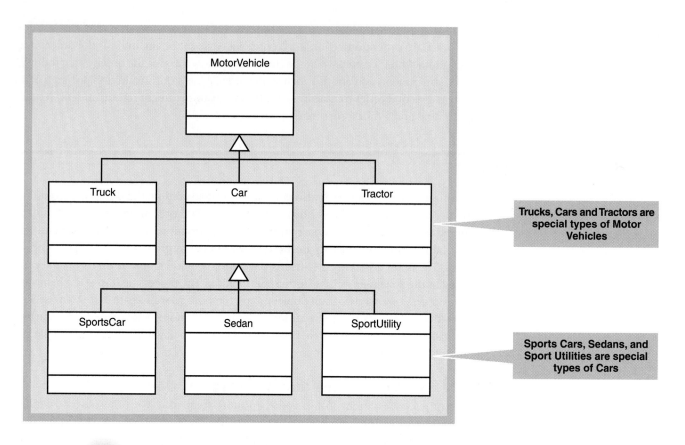

FIGURE 5-29

A generalization/specialization
hierarchy for motor vehicles.

motivation for the new customer support system at RMO started with John's recognition
that Rocky Mountain Outfitters might handle many special types of orders with a new
system (Web orders, telephone orders, and mail orders). These special types of orders
are shown in Figure 5-30.

FIGURE 5-30

A generalization/specialization
hierarchy for orders.

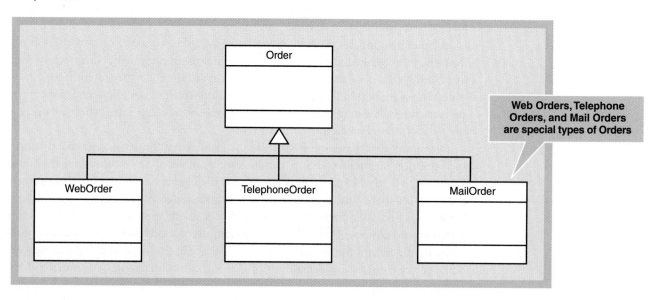

inheritance

a concept that allows subclasses to share characteristics of their superclasses

whole-part hierarchies

hierarchies that structure classes according to their associated components

aggregation

whole-part relationship between an object and its parts

composition

whole-part relationship in which the parts cannot be dissociated from the object

FIGURE *5-31*

Whole-part (aggregation) relationships between a computer and its parts.

Inheritance allows subclasses to share characteristics of their superclasses. Returning to Figure 5-29, a car is everything any other motor vehicle is but also something special. A sports car is everything any other car is plus something special. In this way, the subclass "inherits" characteristics. In the object-oriented approach, inheritance is a key concept that is possible because of generalization/specialization hierarchies. Sometimes these hierarchies are referred to as *inheritance hierarchies*.

Whole-Part Hierarchies

Another way that people structure information about things is by defining them in terms of their parts. For example, learning about a computer system might involve recognizing that the computer is actually a collection of parts—processor, main memory, keyboard, disk storage, and monitor. A keyboard is not a special type of computer; it is part of a computer. Yet, it is also something entirely separate in its own right. *Whole-part hierarchies* capture the relationships that people make when they learn to make associations between an object and its components.

There are two types of whole-part hierarchies: aggregation and composition. The term *aggregation* is used to describe a form of association that specifies a whole-part relationship between the aggregate (whole) and its components (parts) where the parts can exist separately. Figure 5-31 demonstrates the concept of aggregation in a computer system, showing the diamond symbol to represent aggregation. The term *composition* is used to describe whole-part relationships that are even stronger, where the parts, once associated, can no longer exist separately. The diamond symbol is filled in to represent composition.

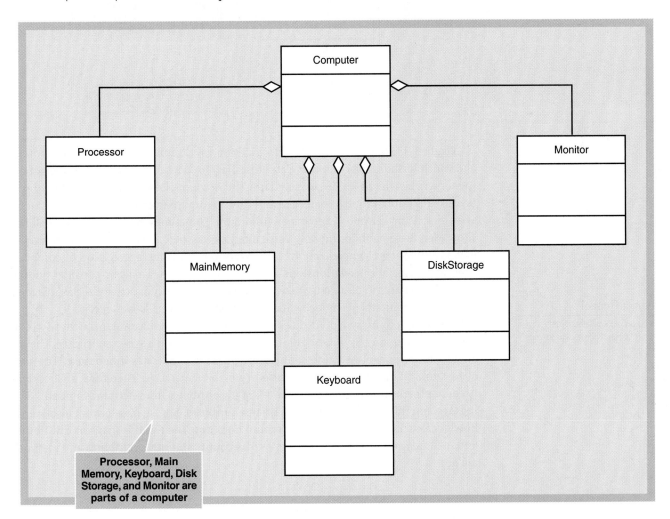

Whole-part hierarchies, both aggregation and composition, serve mainly to allow the analyst to express subtle distinctions about associations among classes. As with any association relationship, cardinality/multiplicity can apply, such as when a computer has one or more disk storage devices.

Examples of Class Diagram Notation

As mentioned previously, the class diagram is the model used to show classes of objects for a system. The notation is based on the Unified Modeling Language (UML), which has become the standard for models used with object-oriented system development. This section introduces the notation for the class diagram. This notation has been used previously in Figures 5-29, 5-30, and 5-31.

Figure 5-32 shows an example of a symbol for one class: Customer. The class symbol is a rectangle with three sections. The top section contains the name of the class, the middle section lists the attributes of the class, and the bottom section lists the important methods of the class. Methods are not always shown in the class symbol if they are fairly standard. It can usually be assumed that a new customer can be added, deleted, changed, and connected to an account.

FIGURE *5-32*

The class symbol with three sections for name, attributes, and methods.

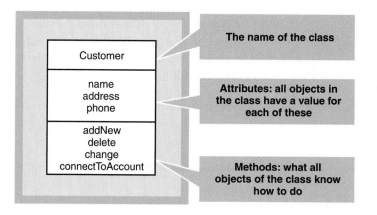

Figure 5-33 shows part of a class diagram for a system that maintains bank accounts and which includes the Customer class. This diagram does not show the methods of the Customer class because they are standard. The Account class lists a couple of methods because they are unique to bank accounts and central to the processing of the system. They include makeDeposit and makeWithdrawal. Sometimes the class diagram will not show even these methods initially. Most analysts wait before worrying about object behavior until classes, generalization/specialization hierarchies, aggregation or composition, and association relationships are worked out. Many analysts consider adding methods to classes to be a design activity rather than an analysis activity. A class diagram created as a requirements model without methods is called a *domain model*.

The bank account system includes a generalization/specialization hierarchy: Account is the superclass, and SavingsAccount and CheckingAccount are two subclasses. The triangle symbol drawn on a line connecting the classes indicates inheritance. The subclasses inherit attributes and behaviors from the superclass. Therefore, a CheckingAccount inherits the two methods plus all of the attributes from Account. Similarly, SavingsAccount inherits the same methods and attributes. But SavingsAccount also knows how to calculate interest; CheckingAccount does not. The result is that some attributes and methods are common to both types of accounts, but other attributes and methods are not.

domain model

a class diagram without methods, which is created as a requirements model

FIGURE *5-33*

A bank account system class diagram.

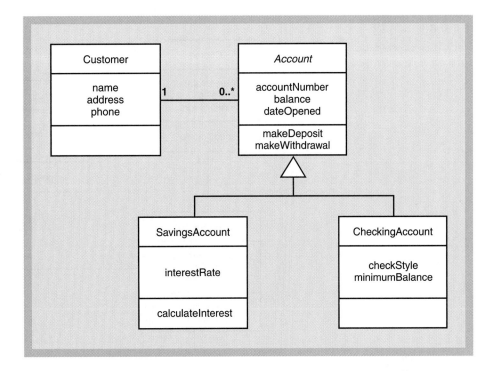

In this example, inheritance means that when a SavingsAccount object is created (or instantiated), it will require values for four attributes, but a CheckingAccount object will require values for five attributes. The CheckingAccount object can be asked to make a deposit, as can the SavingsAccount object. The SavingsAccount object can be asked to calculate interest, but the CheckingAccount object cannot. Each object, or instance, maintains information and can be asked to invoke one of its methods.

The Customer class and the Account class can be associated, just as they would be in an entity-relationship diagram. Each customer can have zero or more accounts. Note that the diagram indicates the minimum and maximum multiplicity on the line connecting the classes. (Recall that the term *multiplicity* is used as a synonym for cardinality on the class diagram.) The asterisk means "many," so the multiplicity between Customer and Account is a minimum of zero and a maximum of many (0..* usually just shown as *). To indicate a mandatory relationship, the diagram indicates a minimum of 1 and a maximum of many (1..*). Similarly, each account is owned by one and only one customer in this example (1). The diagram can indicate optional one-to-one associations as a minimum of zero and a maximum of 1 (0..1).

Note also that the SavingsAccount and CheckingAccount classes inherit the association with Customer, so a customer is actually associated with a checking account or savings account, as appropriate. In fact, the bank does not offer anything like the simple Account class; the class exists only to allow special types of accounts to inherit attributes, methods, and associations. The Account class is an *abstract class*, a class that cannot be instantiated. On the class diagram, an abstract class has its name in italics, as in Figure 5-33. CheckingAccount, SavingsAccount, and Customer are examples of *concrete classes* that can be instantiated.

UML also provides a notation for showing classes that are really association relationships between other classes, such as in the example of the university course enrollment ERD that needed to store the student's grade (Figure 5-27). It is called an *association class* (as opposed to an *associative entity*). The university course enrollment class diagram is shown in Figure 5-34. The multiplicity of the association between CourseSection and Student is many to many. A dashed line connects the association line to the association class named CourseEnrollment that has the grade attribute.

abstract class

a class that cannot be instantiated (no objects can be created), existing only to allow subclasses to inherit its attributes, methods, and associations

concrete class

a class that can be instantiated (objects can be created)

association class

a class that represents a many-to-many relationship between two other classes

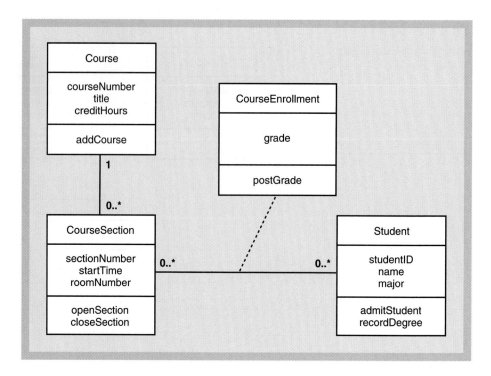

The class diagram in Figure 5-34 includes some methods to reinforce the idea that the classes have attributes and methods. The class diagram shows more about the requirements than the ERD. Course has a method to add a course. Course sections can be opened for enrollment and later closed. A student can be admitted and later receive a degree. Once the semester is over, a course enrollment can post a grade. No generalization/specialization is shown in this example, but there could be special types of students (undergraduate and graduate) and special types of courses (credit and noncredit) in a university enrollment system.

The Rocky Mountain Outfitters Case Class Diagram

The class diagram for Rocky Mountain Outfitters is shown in Figure 5-35. As you can see, it is very similar to the entity-relationship diagram shown previously in Figure 5-28. The main attributes of all classes are shown, although as with the ERD, attributes can be left off the class diagram when presenting an overview of the model.

A generalization/specialization hierarchy is included to show that an order can be any one of three types—Web order, telephone order, and mail order—as discussed previously. Note that all types of orders share the attributes listed for Order, but each special type of order has some additional attributes. Order is an abstract class (the name is in italics), since any order must be one of the three special types.

The other classes and associations among classes are similar to the RMO entity-relationship diagram. CatalogProduct is an association class attached to the association between Catalog and ProductItem. Multiplicity for association relationships is indicated with both minimums and maximums. No whole-part associations (aggregation or composition) are shown, although it might be argued that an OrderTransaction is part of an Order or that a ProductItem is part of a Catalog. It does not make much difference in this example because whole-part and association relationships are similar when they are implemented. Many analysts choose not to indicate aggregation or composition on class diagrams for business systems.

Note that this diagram shows no methods. The initial class diagram developed during the systems analysis phase of the project would not include methods and—as mentioned earlier—is called the *domain model*. As behaviors of objects are further developed

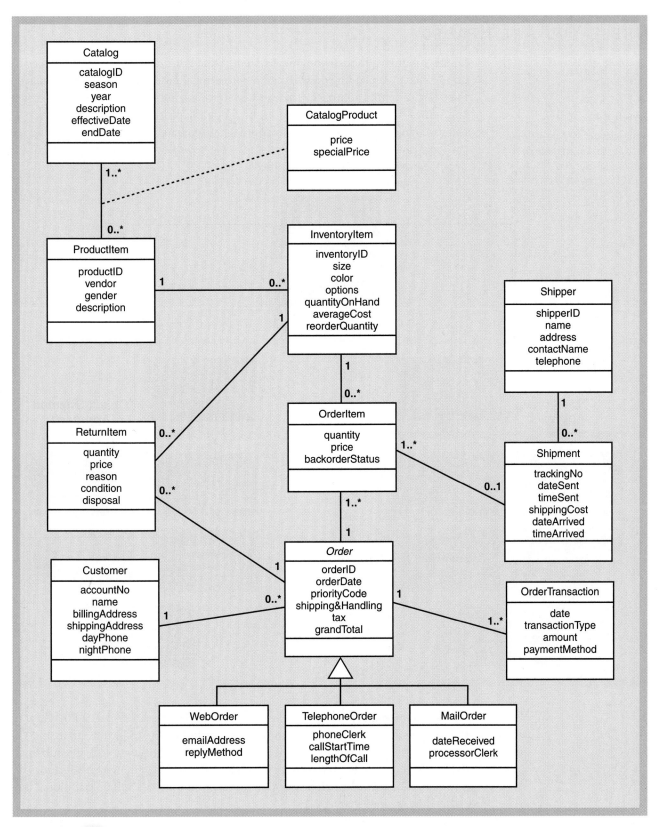

FIGURE *5-35*

Rocky Mountain Outfitters class diagram.

during design, methods are added to the diagram. For now, remember that objects of each class have behaviors as well as attributes. For example, remember that the Shipper class knows how to create a new shipper object, delete a shipper object, change a name or address, and connect to a specific shipment. All of the other classes have these standard capabilities also.

The requirements models for a new system created during the analysis phase become quite different depending on whether the project team uses the traditional approach or the object-oriented approach. The two key concepts discussed in this chapter—events and things—are the starting places in the modeling process for both approaches. The next two chapters discuss these two approaches separately, in both cases starting with the same preliminary information. Figure 5-36 shows how the two approaches diverge once the events and things are identified.

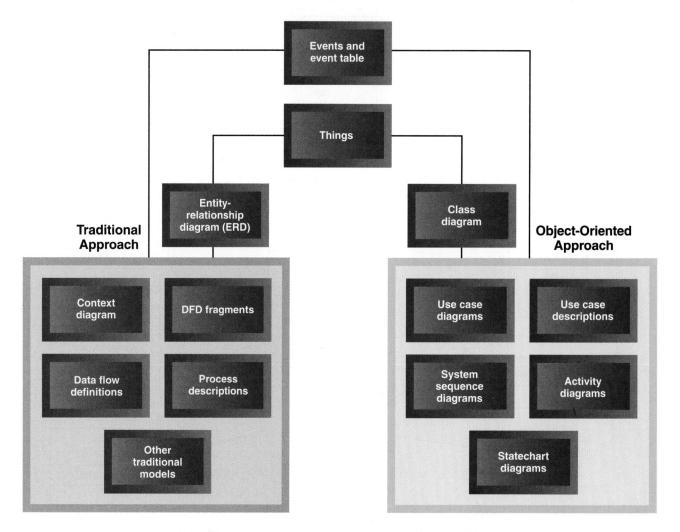

FIGURE *5-36*

Requirements models for the traditional approach and the object-oriented approach.

The traditional approach takes the event table and creates a set of data flow diagrams (DFDs) based on the information in the table, including the context diagram and DFD fragments. The entity-relationship diagram (ERD) defines the data storage requirements that are included in the DFDs. Other information about the requirements includes data flow definitions and process descriptions. These and some additional traditional models are discussed in Chapter 6.

The object-oriented approach takes the event table and creates use case diagrams and a set of use case descriptions. Use cases and the class diagram are used to create additional models of system requirements including activity diagrams, system sequence diagrams, and statechart diagrams. These models are discussed in Chapter 7.

SUMMARY

This chapter is the first of three chapters that present techniques for modeling a system's functional requirements, highlighting the tasks that are completed during the analysis phase activity named *Define system requirements*. Models of various aspects of the system are created to document the requirements. Events and things in the user's work environment are key concepts common to all approaches to system development. The traditional approach uses entity-relationship diagrams (ERD), and the object-oriented approach uses class diagrams as key models.

Models are useful for many reasons, including furthering the learning process that occurs while creating them, reducing complexity, documenting the many details that need to be remembered, communicating with team members and users, and documenting system requirements for future use during system support. Many types of models are used, including mathematical models, descriptive models, and graphical models. Many of the models described in this text are graphical models, some used during analysis and some used during design.

A key early step in the modeling process is to identify and list the events that require a response from the system. An event is something that can be described that occurs at a specific time and place and is worth remembering. External events occur outside the system, usually triggered by someone who interacts with the system. Temporal events occur at a defined point in time, such as the end of a work day or the end of every month. State events occur based on an internal system change. Information about each event is recorded in an event table, which lists the event, the trigger for the event, the source of the trigger, the activity or use case that the system must carry out, the response produced as system output, and the destination for the response.

The other key concept involves the things users deal with when doing their work that the system needs to remember, such as products, orders, invoices, and customers. There are many naturally occurring relationships among things the user works with: A customer places an order, and an order requires an invoice. Cardinality (or multiplicity) of a relationship refers to the number of associations involved in a relationship: A customer might place many orders, and each order is placed by one customer. Attributes are specific pieces of information about a thing, such as a name and an address for a customer. The traditional approach models these things as data entities that represent data that are stored. The object-oriented approach models these things as objects belonging to a class, which have attributes as well as behaviors (called *methods*). Although the two approaches to development use the same basic concept to identify these data entities or objects, the way objects are used in the object-oriented approach is quite different from the traditional approach because of the object behavior.

The traditional approach uses the entity-relationship diagram to show data entities, attributes of data entities, and relationships. The object-oriented approach uses the class diagram to show the classes, attributes, and methods of the class, and associations among classes. Two additional concepts are used in class diagrams (although they are sometimes used in entity-relationship diagrams, too): generalization/specialization hierarchies, which allow inheritance from a superclass to a subclass, and whole-part hierarchies, which allow a collection of objects to be associated as a whole and its parts.

The next two chapters discuss requirements models produced by the traditional approach and the object-oriented approach separately.

KEY TERMS

abstract class, p. 185

activity, p. 167

aggregation, p. 183

association class, p. 185

associative entity, p. 179

attribute, p. 174

binary relationships, p. 174

cardinality, p. 173

class, p. 175

composition, p. 183

compound attribute, p. 174

concrete class, p. 185

data entities, p. 175

descriptive model, p. 156

destination, p. 167

domain model, p. 184

encapsulation, p. 175

event, p. 158

event table, p. 166

external event, p. 161

generalization/specialization hierarchies, p. 181

graphical model, p. 157

identifier (key), p. 174

inheritance, p. 183

mathematical model, p. 156

methods, p. 175

multiplicity, p. 173

n-ary relationship, p. 174

perfect technology assumption, p. 165

relationship, p. 171

response, p. 167

source, p. 167

state event, p. 163

system controls, p. 164

temporal event, p. 162

ternary relationship, p. 174

trigger, p. 166

unary (recursive) relationship, p. 174

use case, p. 167

whole-part hierarchies, p. 183

REVIEW QUESTIONS

1. What are some of the reasons for creating models during system development?

2. What are three types of models?

3. What are the two key concepts used to begin defining system requirements?

4. What is an event?

5. What are the three types of events?

6. Which type of event results in data entering the system?

7. Which type of event occurs at a defined point in time?

8. Which type of event does not result in data entering the system but always results in an output?

9. What type of event would be named *Employee quits job*?

10. What type of event would be named *Time to produce paychecks*?

11. What are some examples of system controls?

12. What does the perfect technology assumption state?

13. What are the columns in an event table?

14. What is a trigger? A source? An activity or use case? A response? A destination?

15. What is a "thing" called in models used in the traditional approach?

16. What is a "thing" called in the object-oriented approach?

17. What is a relationship?

18. What is cardinality of a relationship (also called *multiplicity*)?

19. Describe how an entity-relationship diagram shows the minimum and maximum cardinality.

20. What are unary, binary, and *n*-ary relationships?

21. What are attributes and compound attributes?

22. What is an associative entity?

23. What are the symbols shown in an entity-relationship diagram?

24. What are the symbols shown in a class diagram?

25. What is encapsulated along with the values of attributes in an object?

26. What is a generalization/specialization hierarchy?

27. From what type of class do subclasses inherit?

28. What are two types of whole-part hierarchies?

29. What three pieces of information about a class are put in the three parts of the class symbol?

30. What does the triangle symbol indicate on a line connecting classes on the class diagram?

31. What is the difference between an abstract and a concrete class?

32. How is an association class shown on a class diagram?

33. What type of classes are shown in a domain model?

THINKING CRITICALLY

1. Provide an example of each of the three types of models that might apply to designing a car, to designing a house, and to designing an information system.

2. Explain why requirements models are logical models rather than physical models.

3. Explain why the event concept did not initially receive much attention in business systems analysis.

4. Review the external event checklist in Figure 5-7, and think about a university course registration system. What is an example of an event of each type in the checklist? Name each event using the guidelines for naming an external event.

5. Review the temporal event checklist in Figure 5-8. Would a student grade report be an internal or external output? Would a class list for the instructor be an internal or external output? What are some other internal and external outputs for a course registration system? Using the guidelines for naming temporal events, what would you name the events that trigger these outputs?

6. In a course registration system, for the event *Student registers for classes*, create an event table entry listing the event, trigger, source, activity, response(s), and destination(s). For the event *Time to produce grade reports*, create another event table entry.

7. Consider the following sequence of actions taken by a customer at a bank. Which action is the event the analyst should define for a bank account transaction-processing system? (1) Kevin gets a check from Grandma for his birthday. (2) Kevin wants a car. (3) Kevin decides to save his money. (4) Kevin goes to the bank. (5) Kevin waits in line. (6) Kevin makes a deposit in his savings account. (7) Kevin grabs the deposit receipt. (8) Kevin asks for a brochure on auto loans.

8. Consider the perfect technology assumption, which states that events should be included during analysis only if the system would be required to respond under perfect conditions. Could any of the events in the event table for Rocky Mountain Outfitters be eliminated based on this assumption? Explain. Why are events such as *User logs on to system* and *Time to back up the data* required only under imperfect conditions?

9. Draw an entity-relationship diagram, including minimum and maximum cardinality for the following: The system stores information about two things: cars and owners. A car has attributes for make, model, and year. The owner has attributes for name and address. Assume that a car must be owned by one owner, and an owner can own many cars, but an owner might not own any cars (perhaps she just sold them all, but you still want a record of her in the system).

10. Draw a class diagram for the cars and owners described in exercise 9 but include subclasses for sports car, sedan, and minivan with appropriate attributes.

11. Consider the entity-relationship diagram shown in Figure 5-27, the refined ERD showing course enrollment with an associative entity. Does this model allow a student to enroll in more than one course section at a time? Does the model allow a course section to contain more than one student? Does the model allow a student to enroll in several sections of the same course and get a grade for each enrollment? Does the model store information about all grades earned by all students in all sections?

12. Again consider the entity-relationship diagram shown in Figure 5-27. Add the following to the diagram and list any assumptions you had to make. A faculty member usually teaches many course sections, but some semesters a faculty member may not teach any. Each course section must have at least one faculty member teaching it, but sometimes teams teach course sections. Furthermore, to make sure that all course sections are similar, one faculty member is assigned as course coordinator to oversee the course, and each faculty member can be coordinator of many courses.

13. If the entity-relationship diagram you drew in exercise 12 showed a many-to-many relationship between faculty member and course section, a further look at the relationship might reveal the need to store some additional information. What might this information include? (Hint: Does the instructor have specific office hours for each course section? Do you give an instructor some sort of evaluation for each course section?) Expand the ERD to allow the system to store this additional information.

14. Draw a class diagram for the course enrollment system completed in exercise 13. Be sure to use the correct notation for association classes (see Figure 5-34).

15. Consider a system that needs to store information about computers in a computer lab at a university, such as the features and location of each computer. What are the things that might be included in a model? What are some of the relationships among these things? What are some of the attributes of these things? Draw an entity-relationship model for this system.

16. Draw a class diagram for the computer lab system described in exercise 15.

17. Refer to the class diagram for bank accounts in Figure 5-33. Expand the model to show that there are special types of customers—personal and commercial. All customers have a name and mailing address. Commercial customers have additional attributes for credit rating, contact person, and contact person phone. Personal customers have attributes for home phone and work phone. Additionally, expand the model to show that the bank has multiple branches, and each account is serviced by one branch. Naturally, each branch has many accounts.

18. Consider the class diagram for Rocky Mountain Outfitters shown in Figure 5-35. If a Web order is created, how many attributes does it have? If a telephone order is created, how many attributes does it have? If an existing customer places a phone order for one item, how many new objects are created overall for this transaction?

19. A product item for RMO is not the same as an inventory item. A product item is something like a men's leather hunting jacket supplied by Leather 'R' Us. An inventory item is a specific size and color of the jacket—like a size medium brown leather hunting jacket. If RMO adds a new jacket to its catalog, and six sizes and three colors are available in inventory, how many objects need to be added overall?

20. Consider the following class diagram showing college, department, and faculty members.
    a. What kind of relationships are shown in the model?
    b. How many attributes does a "faculty member" have? Which (if any) have been inherited from another class?
    c. If you add information about one college, one department, and four faculty members, how many objects do you add to the system?

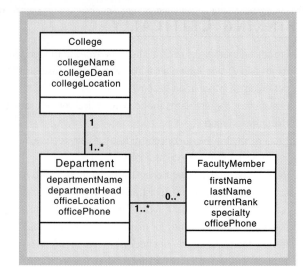

d. Can a faculty member work in more than one department at the same time? Explain.
e. Can a faculty member work in two departments at the same time, where one department is in the college of business and the other department is in the college of arts and sciences? Explain.

EXPERIENTIAL EXERCISES

1. Set up a meeting with a librarian. During your meeting, ask the librarian to describe the situations that come up in the library to which the book checkout system needs to respond. List these external events. Now ask about points in time, or deadlines, that require the system to produce a statement, notice, report, or other output. List these temporal events. Does it seem natural for the librarian to describe the system in this way? Similarly, ask the librarian to describe the things about which the system needs to store information. See whether you can get the librarian to list the important attributes and describe relationships among things. Does it seem natural for the librarian to describe these things? Create either an ERD or a class diagram based on what you learn.

2. Visit a restaurant or the college foodservice and talk to a server (or talk with a friend who is a food server). Ask about the external events, temporal events, and data entities or objects, as you did in exercise 1. What are the events for order

processing at a restaurant? Complete an event table and either an ERD or class diagram.

3. Review the procedures for course registration at your university and talk with the staff in advising, in registration, and in your major department. Think about the sequence that goes on over an entire semester. What are the events that students trigger? What are the events that your major department triggers? What are the temporal events that result in information going to students? What are the temporal events that result in information going to instructors or departments?

4. Again review information about your own university. Create generalization/specialization hierarchies using the class diagram notation for (1) types of faculty, (2) types of students, (3) types of courses, (4) types of financial aid, and (5) types of housing. Include attributes for the superclass and the subclasses in each case.

CASE STUDIES

The Spring Breaks 'R' Us Travel Service Booking System

Spring Breaks 'R' Us Travel Service (SBRU) books spring-break trips at resorts for college students. During the fall, resorts submit availability information to SBRU indicating rooms, room capacity, and room rates for each week of the spring-break season. Each resort offers bookings for a different number of weeks each season, and rooms have different rates depending on the week. Usually, the resorts make a variety of rooms with different capacities available so students can book the right size room. Couples can book a two-person room, for example, and four people can book a room for four.

In December, SBRU generates a list of resorts, available weeks, and room rates that is distributed to college campus representatives all over the country. When a group of students submits a reservation request for a week at a particular resort, SBRU assigns the students to a room with sufficient capacity and sends each student a confirmation notice. When the cutoff date for a week arrives, SBRU sends each resort a list of students booked in each room for the following week. When the students arrive at the resort, they pay the resort directly for the room. Resorts send commission checks directly to the SBRU accounting system, which is separate from the booking system. When spring break is over, students return to their schools and hit the books.

1. To what events must the SBRU booking system respond? Create a complete event table listing the event, trigger, source, activity, response, and destination for each event. Be sure to consider only the events that trigger processing in the booking system, not the SBRU accounting system or the systems operated by the resorts.
2. List the data entities (or classes) that are mentioned. List the attributes of each data entity (or class). List the relationships among data entities (or classes).
3. Which classes might be refined into a generalization/specialization hierarchy? List the superclass and any subclasses for each of them.

The Real Estate Multiple Listing Service System

The Real Estate Multiple Listing Service system supplies information that local real estate agents use to help them sell houses to their customers. During the month, agents list houses for sale (listings) by contracting with homeowners. The agent works for a real estate office, which sends information on the listing to the multiple listing service. Therefore, any agent in the community can get information on the listing.

Information on a listing includes the address, year built, square feet, number of bedrooms, number of bathrooms, owner name, owner phone number, asking price, and status code. At any time during the month, an agent might directly request information on listings that match customer requirements, so the agent contacts the multiple listing service with the request. Information on the house, on the agent who listed the house, and on the real estate office the agent works for is provided. For example, an agent might want to call the listing agent to ask additional questions or call the homeowner directly to make an appointment to show the house. Twice each month (on the 15th and 30th), the multiple listing service produces a listing book that contains information on all listings. These books are sent to all of the real estate agents. Many real estate agents want the books (which are easier to flip through), so they are provided even though the information is often out of date. Sometimes agents and owners decide to change information about a listing, such as reducing the price, correcting previous information on the house, or indicating that the house is sold. The real estate office sends in these change requests to the multiple listing service when the agent asks the office to do so.

1. To what events must the multiple listing service system respond? Create a complete event table listing the event, trigger, source, activity, response, and destination for each event.
2. Draw an entity-relationship diagram to represent the data storage requirements for the multiple listing service system, including the attributes mentioned. Does your model include data entities for offer, buyer, and closing? If so, reconsider. Include information that the multiple listing service needs to store, which might be different from information the real estate office needs to store.
3. Draw a class diagram that corresponds to the ERD but shows that different types of listings have different attributes. The description in the case assumes all listings are for single-family houses. What about multifamily listings or commercial property listings?

The State Patrol Ticket Processing System

The purpose of the State Patrol ticket processing system is to record driver violations, to keep records of the fines paid by drivers when they plead guilty or are found guilty of moving violations by the courts, and to notify the court that a warrant for arrest should be issued when such fines are not paid in a timely manner. A separate State Patrol system records accidents and verification of financial responsibility (insurance). Yet a third system produces driving record reports from the ticket and accident records for insurance companies. Finally, a fourth system issues, renews, or suspends driver's licenses. These four systems are obviously integrated in that they share access to the same database, but otherwise, they are operated separately by different departments of the State Patrol. State Patrol operations (what the officers do) are entirely separate.

The portion of the database used with the ticket processing system involves driver data, ticket data, officer data, and court data. Driver data, officer data, and court data are used by the system. The system creates and maintains ticket data. Driver attributes include license number, name, address, date of birth, date licensed, and so on. Ticket attributes include ticket number (each is unique and preprinted on each sheet of the officer's ticket book), location, ticket type, ticket date, ticket time, plea, trial date, verdict, fine amount, and date paid. Court and officer data include the name and address of each, respectively. Each driver may have zero or more tickets, and each ticket applies to only one driver. Officers write quite a few tickets.

When an officer gives a ticket to a driver, a copy of the ticket is turned in and entered into the system. A new ticket record is created, and relationships to the correct driver, officer, and court are established in the database. If the driver pleads guilty, he or she mails in the fine in a preprinted envelope with the ticket number on it. In some cases, the driver claims innocence and wants a court date. When the envelope is returned without a check and the trial request box has an "X" in it, the system notes the plea on the ticket record, looks up driver, ticket, and officer information, and sends a ticket details report to the appropriate court. A trial date questionnaire form is also produced at the same time and is mailed to the driver. The instructions on the questionnaire tell the driver to fill in convenient dates and mail the questionnaire directly to the court. Upon receiving this information, the court schedules a trial date and notifies the driver of the date and time.

When the trial is completed, the court sends the verdict to the ticketing system. The verdict and trial date are recorded for the ticket. If the verdict is innocent, the system that produces driving record reports for insurance companies will ignore the ticket. If the verdict is guilty, the court gives the driver another envelope with the ticket number on it for mailing in the fine.

If the driver fails to pay the fine within the required period, the ticket-processing system produces a warrant request notice and sends it to the court. This happens if the driver does not return the original envelope within two weeks or does not return the court-supplied envelope within two weeks of the trial date. What happens then is in the hands of the court. Sometimes the court requests that the driver's license be suspended, and the system that processes drivers' licenses handles the suspension.

1. To what events must the ticket processing system respond? Create a complete event table listing the event, trigger, source, activity, response, and destination for each event.
2. Draw an entity-relationship diagram to represent the data storage requirements for the ticket processing system, including the attributes mentioned. Explain why it is important to understand how the system is integrated with other State Patrol systems.
3. Draw a class diagram that corresponds to the ERD but assumes there are different types of drivers. Classifications of types of drivers vary by state. Some states have restricted licenses for minors, for example, and special licenses for commercial vehicle operators. Research your state's requirements, and create a generalization/specialization hierarchy for the class Driver, showing the different attributes each special type of driver might have. Consider the same issues for types of tickets. Include some special types of tickets in a generalization/specialization hierarchy in the class diagram.

Rethinking Rocky Mountain Outfitters

When listing nouns and making some decisions about the initial list of things (see Figure 5-17), the RMO team decided to research Customer Account as a possible data entity or class if the system included an RMO payment plan (similar to a company charge account plan). Many retail store chains have their own charge accounts for the convenience of the customer—to increase sales to the customer and to better track customer purchase behavior.

Consider the implications to the system if management decided to incorporate an RMO charge account and payment plan as part of the customer support system.

1. Discuss the implications that such a change would have on the scope of the project. How might this new capability change the list of stakeholders the team would involve when collecting information and defining the requirements? Would the change have any effect on other RMO systems or system projects planned or under way? Would the change have any effect on the project plan originally developed by Barbara Halifax? In other words, is this a minor change or a major change?
2. What events need to be added to the event table? Complete the event table entries for these additional events. What activities or use cases for existing events might be changed because of a charge account and payment plan? Explain.
3. What are some additional things and relationships among things that the system would be required to store because of the charge account and payment plan? Modify the entity-relationship diagram and the class diagram to reflect these charges.

Focusing on Reliable Pharmaceutical Service

In Chapter 1 you learned about the background and prescription-processing operations for Reliable Pharmaceutical Service. As discussed in this chapter, defining the requirements for the new system starts by taking the information gathered about the needed system and then focusing on the events that require system processing and on the things about which the system needs to store information. The full system would involve many events and things. In this chapter's case exercise, we focus on only a subset of events for the system and a subset of data entities or classes. Later chapters' exercises will add to the scope and complexity of the requirements for Reliable.

1. Create an event table that lists information about system requirements based on the following specific system processing: When a nursing home needs to fill prescriptions for its patients, it provides order details to Reliable. Reliable immediately records information about the order and prescriptions. Prescription orders come in from all of Reliable's nursing-home clients throughout the day. At the start of each 12-hour shift, Reliable prepares a case manifest, detailing all recent orders, which is given to one of the pharmacists. When the pharmacist has assembled the orders for each client, the pharmacist records the order fulfillment. (Review the Reliable case description at the end of Chapter 1 for more details.) In addition, the system needs to add or update patient information, add or update drug inventory information, produce purchase orders to replenish the drug inventory, record inventory adjustments, and generate various management reports. For now, ignore any billing, payments, or insurance processing.

2. Create an entity-relationship diagram that shows the data storage requirements for the following portion of the system: Add a few attributes to each data entity and show minimum and maximum cardinality. To process the prescription order, Reliable needs to know about the patients, the nursing home, and the nursing-home unit where each patient resides. Each nursing home has at least one, but possibly many, units. A patient is assigned to a specific unit. An order consists of one or more prescriptions, each for one specific drug and for one specific patient. An order, therefore, consists of prescriptions for more than one patient. Careful tracking and record keeping is obviously crucial. Additionally, each patient has many prescriptions. One pharmacist fills each order.

3. Create a class diagram for the object-oriented approach that shows the same requirements as described in step 2. Include a few attributes for each class and show minimum and maximum multiplicity. Be sure to identify any association classes and use the correct notation.

4. How important is it to understand that each order includes prescriptions for more than one patient? Is this the type of information that is difficult to sort out at first? Did you see the implications initially, or did you have to work through the model until it made sense to you? Discuss.

FURTHER RESOURCES

Some classic and more recent texts include the following:

Events and event analysis:

Stephen McMenamin and John Palmer, *Essential Systems Analysis*. Prentice Hall, 1984.

Ed Yourdon, *Modern Structured Analysis*. Prentice Hall, 1989.

Data modeling, entity-relationship diagrams, and database management:

Peter Rob and Carlos Coronel, *Database Systems: Design, Implementation, and Management* (5th ed.). Course Technology, 2002.

Objects, object behavior, and the class diagram:

Grady Booch, Ivar Jacobson, and James Rumbaugh, *The Unified Modeling Language User Guide*. Addison-Wesley, 1999.

Peter Coad, David North, and Mark Mayfield, *Object Models, Strategies, Patterns, and Applications* (2nd ed.). Prentice Hall, 1997.

CHAPTER 6

The Traditional Approach to Requirements

LEARNING OBJECTIVES

After reading this chapter, you should be able to:

- Explain how the traditional approach and the object-oriented approach differ when an event occurs

- List the components of a traditional system and the symbols representing them on a data flow diagram

- Describe how data flow diagrams can show the system at various levels of abstraction

- Develop data flow diagrams, data element definitions, data store definitions, and process descriptions

- Develop tables to show the distribution of processing and data access across system locations

- Read and interpret Information Engineering models that can be incorporated within traditional structured analysis

CHAPTER OUTLINE

Traditional and Object-Oriented Views of Activities

Data Flow Diagrams

Documentation of DFD Components

Information Engineering Models

Locations and Communication through Networks

Arturo Romero and Lei Xu were meeting to review a first draft of several data flow diagrams for San Diego Periodicals' new advertising billing system. Arturo was the analyst assigned to define the new system requirements. Lei was the manager in charge of advertising accounts—she knew virtually all there was to know about how the current system operated. The two had met several times before. Their most recent meeting (just last week) reviewed details of the current system events, ad request processing, and process participants. Arturo left that meeting with pages of notes and sample forms and reports from the current system. Arturo had telephoned Lei several times since that meeting to ask additional questions.

Arturo began the review by saying, "The materials and information that you gave me last week took me quite awhile to absorb, but I think I was able to understand and document all of the system functions that we discussed. The purpose of this meeting is to ensure that the processing requirements that I've written down are complete and accurate. Let's start with a few of the diagrams that I've created."

Arturo laid out three diagrams on the table. Lei looked at them briefly and said, "I've never seen diagrams like this before—they look like blueprints for playing a game of marbles. And I thought the entity-relationship diagrams were strange!"

Arturo replied, "I expect this review will go slowly since this is your first look at this style of documentation. I'll explain how to interpret the diagrams as we go along. Ask as many questions as you like. The quality of our work depends on your understanding the diagrams, so don't be shy."

Arturo continued, "The pictures are called data flow diagrams. They divide your system into processing functions represented by the rectangles with rounded-off corners. The arrows show data movement among processes and between processes and files." Lei pointed to a square on one of the diagrams and said, "I assume that this is a company purchasing ad space?"

Arturo replied, "Yes, the squares represent people or organizations that supply inputs or expect output data from the system."

Lei said, "I think I can get the hang of this. I recognize most of the names that you've used for the processes and data. I'm not sure what these other symbols are—they're named for things that we store in our manual files and database, but they don't seem to correspond exactly to our system."

"They don't," Arturo replied. "They're entities from the entity-relationship diagram that we developed a couple of weeks ago. But let's skip over those for the moment. Why don't we walk through the processing sequence for booking an ad, and we'll discuss the entities as we get to them?"

Arturo and Lei continued reviewing the DFDs, and the next thing they knew, over an hour had passed. Several pages of Arturo's notepad had been filled, and 25 corrections and comments were noted in red on the data flow diagrams. Lei said, "My brain feels completely drained. I don't think that I can do any more of this today."

Arturo replied, "You've given me plenty of things to work on, so let's call it quits for now. Can we meet for two hours at nine o'clock on Thursday?"

Lei replied, "Yes, I'm free then. So, will you be bringing more data flow diagrams, or do you have something even weirder up your sleeve?"

Arturo smiled and said, "The toughest stuff is behind us, but you should expect a few more surprises."

OVERVIEW

Chapter 5 described two key concepts associated with modeling system requirements in both the traditional and the object-oriented (OO) approaches to information systems development: events and things. In this chapter, the focus turns to what the system does when an event occurs: activities and interactions.

This chapter describes the traditional structured approach to representing activities and interactions. We describe and present the diagrams and other models of the traditional approach, and we provide examples from the Rocky Mountain Outfitters customer support system to show how each model is related. Also, we briefly discuss how the traditional and Information Engineering approaches and their models can be used together to describe a system. Chapter 7 describes details of the OO approach to representing activities and interactions.

Modeling activities and interactions is a difficult process with either the traditional or OO approach. Building models is a challenging and time-consuming task. Activities and interactions must be specified in exacting detail. Analysts and users must jointly evaluate model completeness, correctness, and quality. As illustrated in the accompanying RMO progress memo, coordinating the efforts of project participants and building a consensus about detailed system requirements are complex project management activities (see memo).

April 14, 2005

To: John MacMurty

From: Barbara Halifax

RE: Customer support system update

John, I just wanted to provide a status report on the analysis phase. We've completed new system documentation for customer orders and most of the management and accounting reports. So far we've developed several dozen data flow diagrams and about 100 pages of supporting data definitions and process specifications. I anticipate that we'll triple that number before the analysis phase is completed.

Working out how to handle returns and back orders has been problematic. As you know, our existing procedures leave much to be desired, so we started with a blank sheet of paper. But the clean break with the past can't eliminate the inherent complexity in processing returns and back orders. We're bogged down in the details, and I've had trouble getting users from marketing, shipping, and IS to agree on procedures that are efficient for all, especially the customer.

Scheduling time with users to review analysis models is an ongoing problem. It seems that some of the users thought that their work was nearly finished when we completed the initial data models. It's been a rude shock to them to find that the intensity level and time demands are increasing.

If you think it's necessary, would you have a word with Jason and Genny and let them know the importance of rapid user feedback on our analysis models? We're on track with the orginal schedule, but it won't take many missed meetings with users to put us behind schedule. We need commitment from everyone to meet the analysis phase deadlines.

Thanks and take care.

BH

cc: Ming Lee, Jack Garcia

Traditional and Object-Oriented Views of Activities

The traditional and OO approaches to system development differ in how a system's response to an event is modeled and implemented. The traditional approach views a system as a collection of processes, some performed by people and some performed by computers. Traditional computer processes are much like procedural computer programs—they contain instructions that execute in a sequence. When the process executes, it interacts with stored data, reading data values and then writing other data values back to the data file. The process might also interact with people, such as when an instruction asks the user to input a value or it displays information to the user on the computer screen. The traditional approach to systems, then, involves processes, stored data, inputs, and outputs. When modeling what the system does in response to an event, the traditional approach includes processing models that emphasize these system features.

In contrast, the OO approach views a system as a collection of interacting objects. The objects are the things discussed in Chapter 5. Objects are capable of behaviors (called *methods*) that allow them to interact with each other and with people using the system. One object asks another object to do something by sending it a message. There are no conventional computer processes or data files per se. Objects carry out the activities and remember the data values. When modeling what the system does in response to an event, the OO approach includes models that show objects, their behavior, and their interactions with other objects.

Figure 6-1 summarizes the differences between traditional and OO approaches to systems. Because of these differences, the traditional and OO approaches to requirements employ different models, as summarized in Figure 6-2. The remainder of this chapter will explore the traditional models on the left side of Figure 6-2.

FIGURE *6-1*

Traditional versus OO approaches.

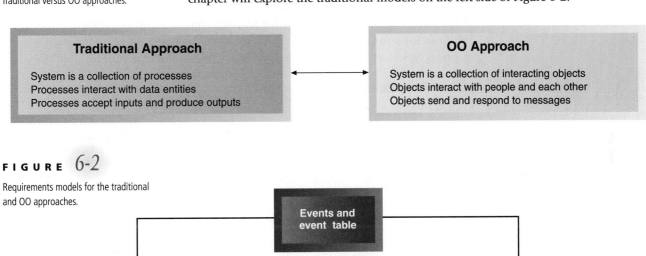

Traditional Approach

System is a collection of processes
Processes interact with data entities
Processes accept inputs and produce outputs

OO Approach

System is a collection of interacting objects
Objects interact with people and each other
Objects send and respond to messages

FIGURE *6-2*

Requirements models for the traditional and OO approaches.

DATA FLOW DIAGRAMS

The traditional approach to information system development describes activities as processes carried out by people or computers. A graphical model that has proven to be quite valuable for modeling processes is the data flow diagram. There are other process models, such as the process dependency diagram used in the Information Engineering approach and the workflow diagrams used with business process reengineering, but the data flow diagram is the most commonly used process model.

A *data flow diagram (DFD)* is a graphical system model that shows all of the main requirements for an information system in one diagram: inputs and outputs, processes, and data storage. Everyone working on a development project can see all aspects of the system working together at once with the DFD. That is one reason for its popularity. The DFD is also easy to read because it is a graphical model and because there are only five symbols to learn (see Figure 6-3). End users, management, and all information systems workers typically can read and interpret the DFD with minimal training.

FIGURE *6-3*

Data flow diagram symbols.

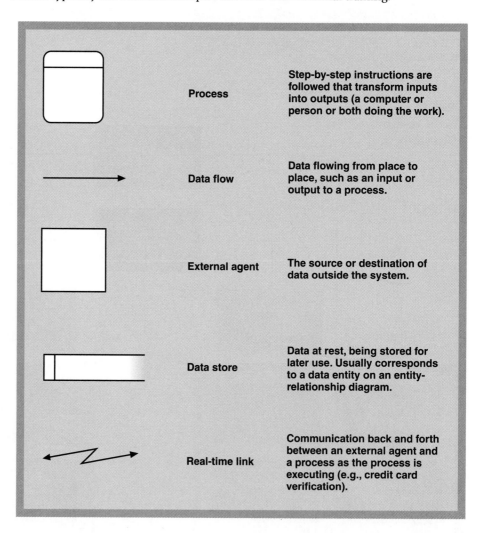

	Process	Step-by-step instructions are followed that transform inputs into outputs (a computer or person or both doing the work).
	Data flow	Data flowing from place to place, such as an input or output to a process.
	External agent	The source or destination of data outside the system.
	Data store	Data at rest, being stored for later use. Usually corresponds to a data entity on an entity-relationship diagram.
	Real-time link	Communication back and forth between an external agent and a process as the process is executing (e.g., credit card verification).

Figure 6-4 shows an example of a data flow diagram representing a portion of the Rocky Mountain Outfitters (RMO) customer service system. The square is an *external agent*, Customer, the source and destination for some data outside the system. The rectangle with rounded corners is a *process* named *Look up item availability* that can also be referred to by its number, 1. A process defines rules for transforming inputs to outputs.

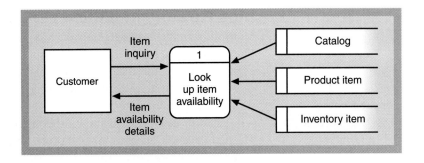

data flow

an arrow on a DFD that represents data movement among processes, data stores, and external agents

data store

a place where data are held pending future access by one or more processes

The lines with arrows are *data flows*. Figure 6-4 shows two data flows between Customer and process 1: a process input named Item inquiry and a process output named Item availability details. The final symbol—the flat, open-ended rectangle—is a *data store*. Each data store represents a file or part of a database that stores information about a data entity. In this example, data flows (lines with arrows) point from the data stores to the process, meaning that the process looks up information in the data stores named Catalog, Product item, and Inventory item.

You might recognize that the process in Figure 6-4 corresponds to an activity in the event table for RMO shown in Chapter 5 (see Figure 5-15). The event was *Customer wants to check item availability*, the trigger was Item inquiry, the source was Customer, the response was Item availability details, and the destination for the response was Customer. Therefore, the data flow diagram shows the system activity in response to this one event in graphical form.

But another piece of information on the DFD is not in the event table: the data stores containing information about an item's availability. Each data store in Figure 6-4 represents a data entity from the entity-relationship diagram (ERD) shown in Chapter 5 (see Figure 5-28). The process on the DFD uses information that we provided by including these data entities and their attributes in the ERD for the system. Therefore, the data flow diagram integrates processing triggered by events with the data entities modeled using the ERD. Figure 6-5 summarizes the correspondences among components of the DFD, events described in the event table, and entities defined in the ERD.

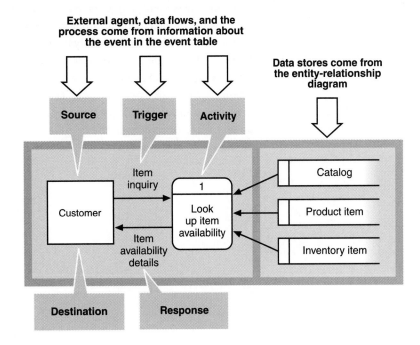

Data Flow Diagrams and Levels of Abstraction

Many different types of data flow diagrams are produced to show system requirements. The example just described is a DFD fragment, showing one process in response to one event. Other data flow diagrams show the processing at either a higher level (a more general view of the system) or at a lower level (a more detailed view of one process). These differing views of the system (high level versus low level) are called *levels of abstraction*.

Data flow diagrams can show either higher-level or lower-level views of the system. The high-level processes on one DFD can be decomposed into separate lower-level, detailed DFDs. Processes on the detailed DFDs can also be decomposed into additional diagrams to provide multiple levels of abstraction.

Figure 6-6 shows how DFDs at each level of detail provide additional information about one process at the next higher level. The topmost DFD shows the most abstract representation of the course registration system as a single process. The middle DFD shows internal details of a context diagram process. The bottom DFD shows internal details of process 1 in the middle DFD. Each DFD abstraction level is described further in the following sections.

Context Diagram

A *context diagram* is a DFD that describes the most abstract view of a system. All external agents and all data flows into and out of the system are shown in one diagram, with the entire system represented as one process. The topmost DFD in Figure 6-6 is a context diagram for a simple university course registration system that interacts with three external agents: Academic department, Student, and Faculty member. Academic departments supply information on offered courses, students request enrollment in offered courses, and faculty members receive class lists when the registration period is complete.

A context diagram clearly shows the system boundary. The system scope is defined by what is represented within the single process and what is represented as external agents. External agents that supply or receive data from the system are outside the system scope, and everything else is inside the system scope. The context diagram does not usually show data stores because all of the system's data stores are considered to be within the system scope (that is, part of the internal implementation of the process that represents the system). However, data stores may be shown when they are shared by the system being modeled and another system.

The context diagram is usually created in parallel with the event table described in Chapter 5. Each trigger for an external event becomes an input data flow, and the source becomes an external agent. Each response becomes an output data flow, and the destination becomes an external agent. Triggers for temporal events are not data flows, so there are no input data flows for temporal events. Note that the context diagram DFD can be created directly from the event table. The two models provide alternative views of the same system requirements information.

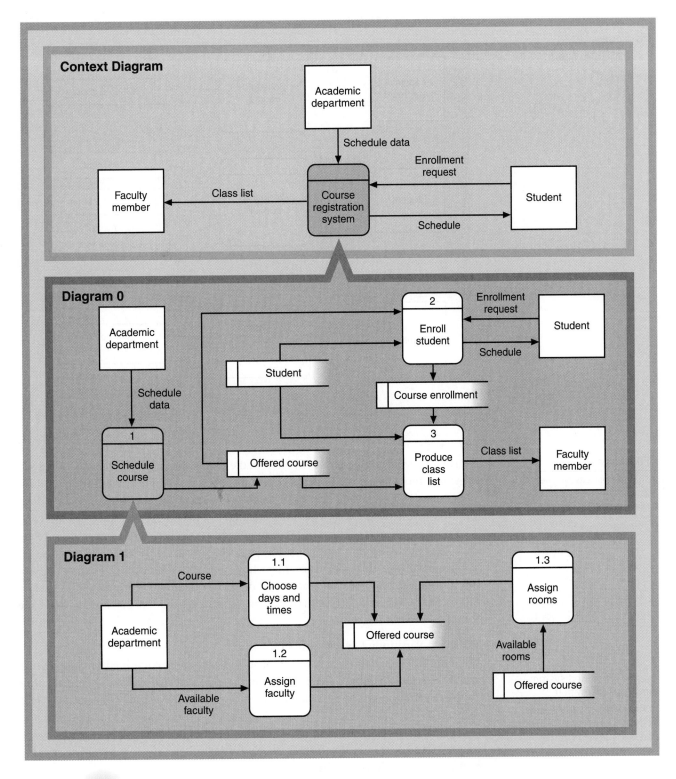

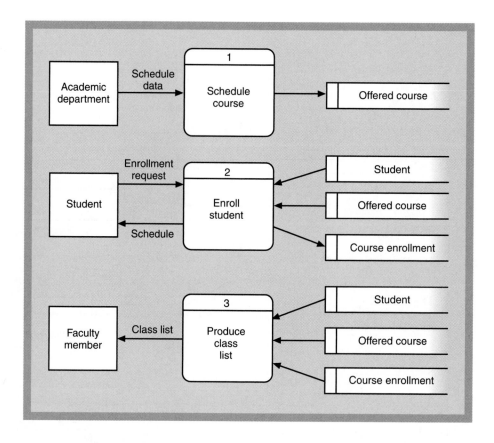

FIGURE *6-7*

DFD fragments for the course
registration system.

DFD Fragments

A *DFD fragment* is created for each event in the event table. Each DFD fragment is a self-contained model showing how the system responds to a single event. The analyst usually creates DFD fragments one at a time, focusing attention on each part of the system. The DFD fragments are drawn after the event table and context diagram are complete.

Figure 6-7 shows the three DFD fragments for the simple course registration system. Each DFD fragment represents all processing for an event within a single process symbol. The fragments show details of interactions among the process, external agents, and internal data stores. The data stores used on a DFD fragment represent entities on the ERD. Each DFD fragment shows only those data stores that are actually needed to respond to the event.

The Event-Partitioned System Model

All of the DFD fragments for a system or subsystem can be combined on a single DFD called the *event-partitioned system model*, or *diagram 0*. Figure 6-8 shows how the three course registration system DFD fragments shown in Figure 6-7 are combined to create diagram 0.

Diagram 0 is used primarily as a presentation tool. It summarizes an entire system or subsystem in greater detail than does a context diagram. However, analysts often avoid developing diagram 0 because:

- The information content duplicates the set of DFD fragments.
- The diagram is often complex and unwieldy, particularly for large systems that respond to many events.

As we'll discuss later in the chapter, redundancy and complexity are two DFD characteristics that analysts should avoid whenever possible.

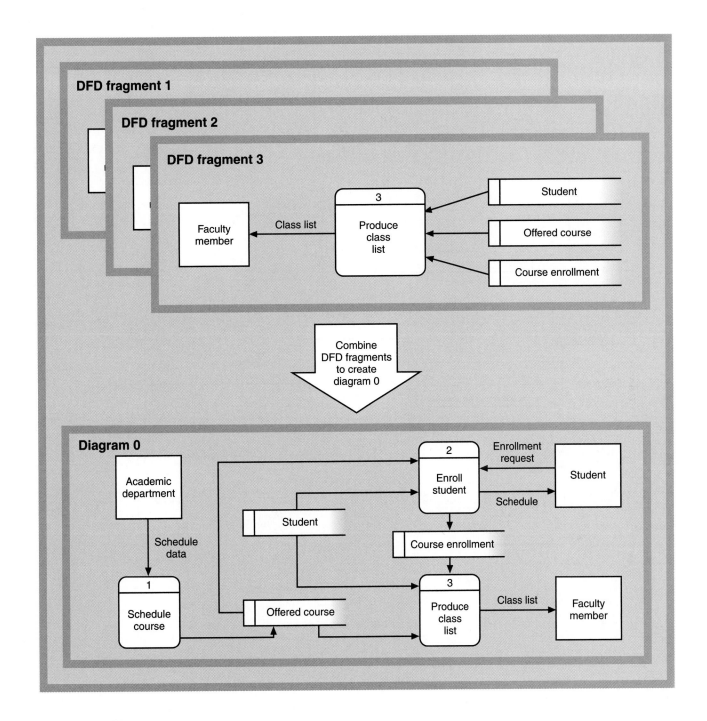

FIGURE *6-8*

Combining DFD fragments to create the event-partitioned system model for the course registration system.

RMO Data Flow Diagrams

Figure 6-9 shows a context diagram for the Rocky Mountain Outfitters customer support system. Normally, data flows and external agents on the context diagram are taken directly from the event table as discussed previously, but since the RMO customer support system responds to 20 events, this figure combines data flows for some events for simplicity. In a smaller system example with 10 to 15 events, you should include all data flows on the context diagram.

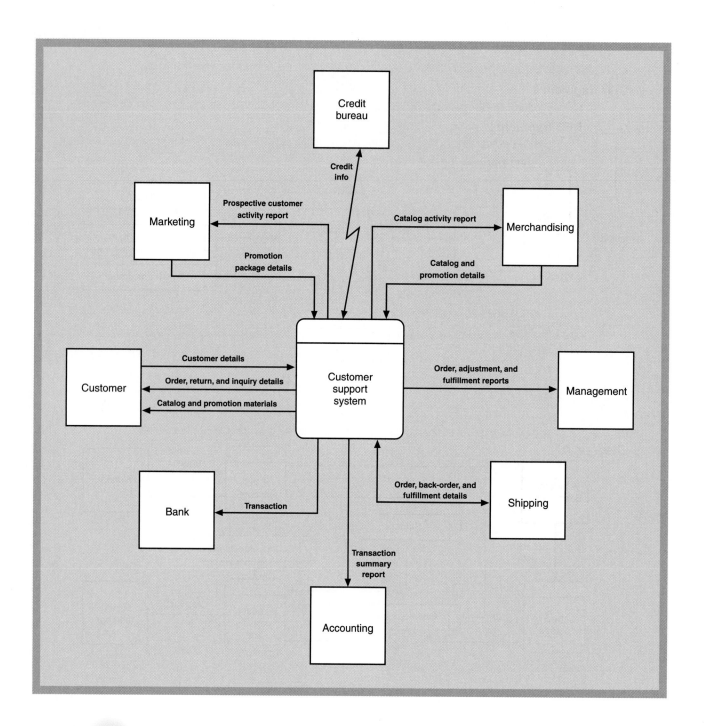

FIGURE 6-9

A context diagram for the RMO
customer support system.

When a system responds to many events, it is commonly divided into subsystems, and a context diagram is created for each subsystem. Figure 6-10 divides the RMO customer support system into subsystems based on event similarities, including interactions with external agents, interactions with data stores, and similarities in required processing. Figure 6-11 shows the context diagram for the order-entry subsystem. Note that all data flows from the event table for this subsystem are shown on the DFD.

FIGURE *6-10*

RMO subsystems and events
for each subsystem.

Order-entry subsystem

    Customer wants to check item availability
    Customer places an order
    Customer changes or cancels an order
    Time to produce order summary reports
    Time to produce transaction summary reports

Order fulfillment subsystem

    Customer or management wants to check order
      status
    Shipping fulfills an order
    Shipping identifies a back order
    Customer returns the item (the item is defective,
      the customer has changed his mind, full or
      partial returns)
    Time to produce fulfillment summary reports

Customer maintenance subsystem

    Prospective customer requests a catalog
    Time to produce prospective customer activity reports
    Customer updates account information
    Marketing wants to send promotional materials
      to customers
    Management adjusts customer charges (correct
      errors, make concessions)
    Time to produce customer adjustment/concession reports

Catalog maintenance subsystem

    Merchandising updates the catalog (add, change,
      delete, change prices)
    Merchandising creates a special product promotion
    Merchandising creates a new catalog
    Time to produce catalog activity reports

FIGURE *6-11*

A context diagram for the RMO
order-entry subsystem.

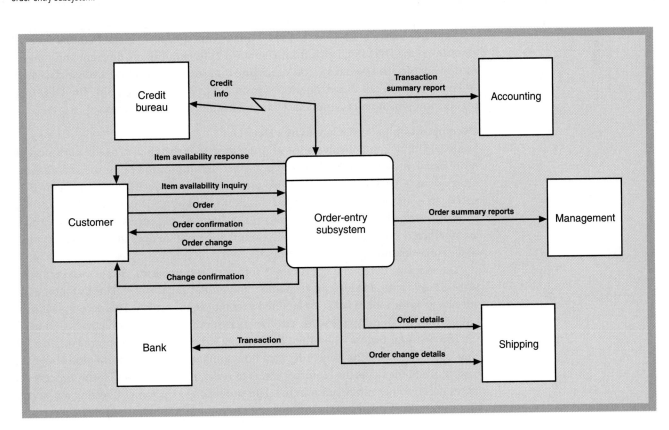

FIGURE *6-12*

DFD fragments for the RMO
order-entry subsystem.

Figure 6-12 shows the DFD fragments for the RMO order-entry subsystem. Note that there are five DFD fragments, one for each order-entry subsystem event listed in Figure 6-11.

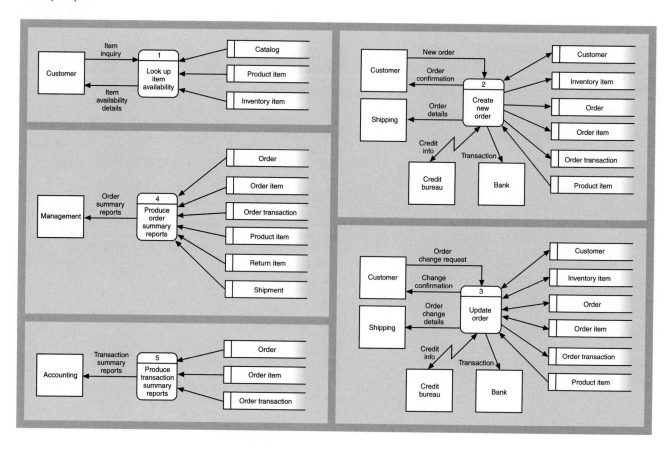

Similarly, Figure 6-13 shows the RMO order-entry subsystem diagram 0, the result of combining the DFD fragments from Figure 6-12. To simplify the diagram and make it more readable, the seven data stores in Figure 6-12 are collapsed into a single data store in Figure 6-13. Recall that diagram 0 is just used as a presentation aid. The DFD fragments show which processes interact with which individual data stores.

Decomposition to See One Activity's Detail

Some DFD fragments involve a lot of processing that the analyst needs to explore in more detail. As with any modeling step, further decomposition helps the analyst learn more about the requirements while also producing needed documentation. Figure 6-14 shows an example of a more detailed diagram for RMO DFD fragment 2, *Create new order*. It is named diagram 2 since it shows the "insides" of process 2. The subprocesses are numbered 2.1, 2.2, 2.3, and 2.4. The numbering system does not necessarily imply sequence of subprocess execution, though.

The diagram decomposes process 2 into four subprocesses: *Record customer information, Record order, Process order transaction,* and *Produce confirmation.* These subproc-esses are viewed as the four major steps required to complete the activity. This decomposition is just one way to divide up the work. Another analyst might arrive at a different solution.

The first step begins when the customer provides the information making up the New order data flow. The New order data flow contains all of the information about the customer and the items the customer wants to order. If the customer is new, process 2.1 stores the customer information in the data store named Customer (creating a new customer record or updating existing customer information as required). Remember that

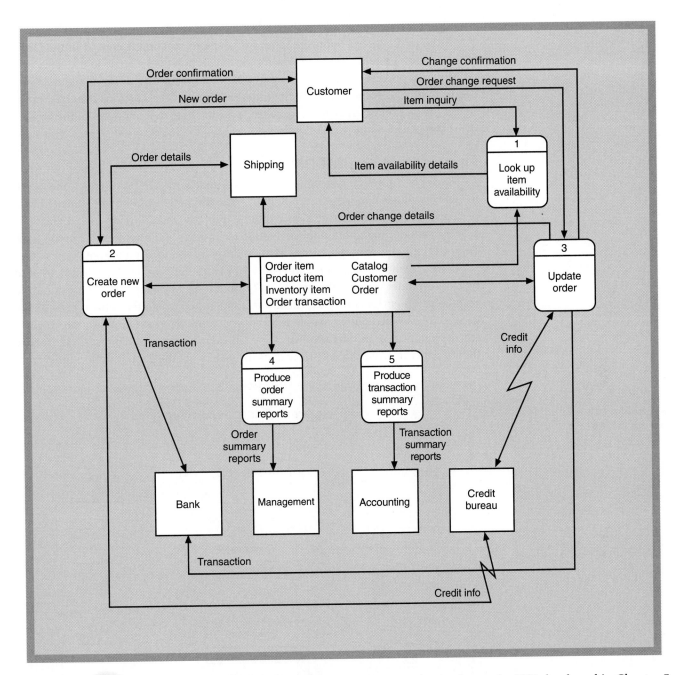

FIGURE *6-13*

An event-partitioned model
of the order-entry subsystem.

the data store represents the customer data entity on the ERD developed in Chapter 5 (see Figure 5-28). Process 2.1 then sends the rest of the information about the order, a data flow named Order details, on to process 2.2.

Process 2.2 takes the Order details data flow and creates a new order record by adding data to the Order data store. Then for each item ordered, the stock on hand and the current price are looked up in the Product item and Inventory item data stores. If adequate stock is on hand, an order item record is created for that item, and the stock on hand for the inventory item record is changed. If three items are ordered, one order record is created and three order item records are created.

Process 2.2 adds up the total amount due for the order (price times quantity for each item) and sends the data flow named Transaction details to process 2.3 to record the transaction. Transaction details include the order number, amount, and credit card information. Process 2.3 needs a real-time link to a credit bureau to get a credit authorization for the customer's credit card. This needs to be a real-time link rather than a data flow because data need to flow back and forth rapidly while the process is executing. If the credit card is approved, a record of the transaction is created in the Order transaction data store, and a data flow for the transaction goes directly to the bank.

The final process produces the order confirmation for the customer and the order details that go to shipping. Using the order number, process 2.4 looks up data on the Order, the Customer, and each Order item (plus the item description from the Product item) and produces the required outputs.

Physical and Logical DFDs

A DFD can be a physical system model, a logical system model, or a blend of the two. If the DFD is a logical model, then it assumes that the system will be implemented with perfect technology, as described in Chapter 5. If the DFD is a physical model, then one or more assumptions about implementation technology will be embedded in the DFD. These assumptions may take many forms and may be very difficult to spot.

Consider whether diagram 2 in Figure 6-14 is a logical model. First, is it clear what type of computer is doing the processing? Could it be a desktop system? A centralized mainframe system? A networked client-server system? Or, could the entire process as just described be carried out by people without any computer at all? Similarly, are the data stores sequential computer files? Are they tables in a relational database? Or are they files of paper in a file cabinet? How does the system get the data flow New order from

FIGURE *6-14*

A detailed diagram for *Create new order* (diagram 2).

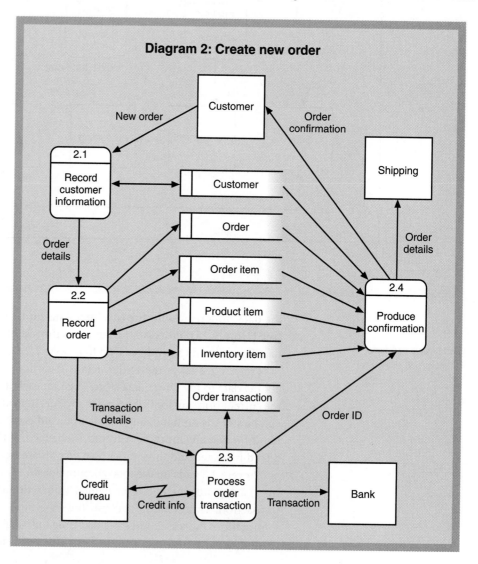

the customer so it can be processed? By clicking check boxes and list boxes in a Windows application? Or on a Web page? Or by manually filling out a form that a clerk types into the system? Or by talking to the clerk over the phone? Or by talking to a speech recognition program over the phone?

All of the alternatives described are possible, and if the model is a logical model, you should not be able to tell how the system is implemented. At the same time, the processing requirements (what must go on) should be fairly detailed, down to indicating what attribute values are needed. The model could be even more detailed and still be a logical model.

Now consider whether the DFD in Figure 6-15 is a physical system model by comparing it with diagram 1 in Figure 6-6. A number of elements indicate assumptions about implementation technology, including:

- Technology-specific processes
- Actor-specific process names
- Technology- or actor-specific process orders
- Redundant processes, data flows, and files

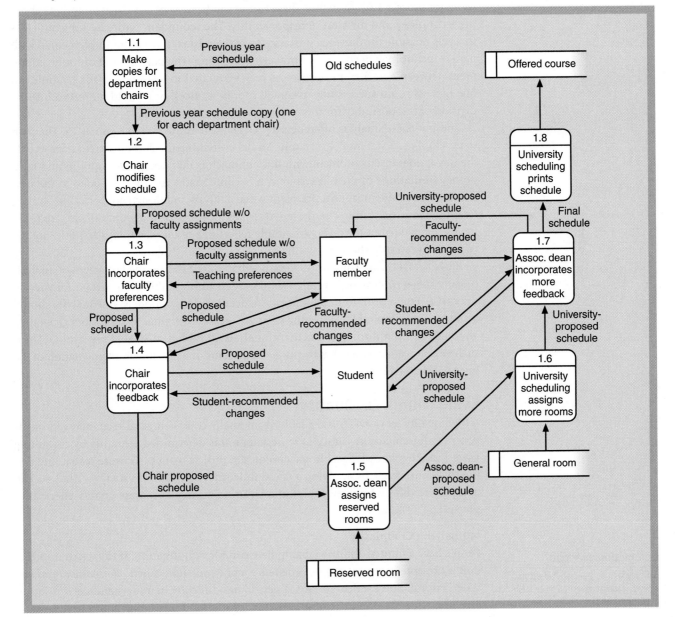

The most obvious technology assumption is embedded in the name of process 1.1. Making copies is an inherently manual task, which implies that the data store Old schedules and the data flows into and out of process 1.1 are paper. It is possible that the data store and flows are electronic, but if so, the question arises why a process would be needed to make electronic copies.

Many of the process names include actors in the system. References to Chair, Assoc. dean, and University scheduling all indicate that a particular individual or department performs a process. The sequential flow of data among the processes is a by-product of the person or department that carries out each process. One can imagine alternate implementations with fewer processes, different process orders, or different assignment of processes to individuals and departments. The DFD clearly models one very specific set of decisions about process ordering and responsibility.

The DFD also includes processes with similar or redundant processing logic. For example, faculty input is accepted early, but faculty members later perform error checking twice (the data flows from processes 1.4 and 1.7). Also, rooms are assigned at two different times from two different data stores (Reserved room for process 1.5, and General room for process 1.6). As before, these features indicate very specific assumptions about the technology and division of responsibility. The redundant error checking indicates that it is possible for previous processes to make mistakes. A system implemented with perfect technology needs no internal error checking. The partitioning of room assignment between two files and processes may be related to technology (for example, no one process could successfully assign all rooms at once), or it could indicate a historic division of responsibility for room assignment.

Inexperienced analysts often develop DFDs such as the one in Figure 6-15. The path to developing such a model is simple: Model everything the current system does exactly the way it does it. The problem with this approach is that design assumptions and technology limitations of the old system can become inadvertently embedded in the new system. This problem is most prevalent when analysis and design are performed by different persons or teams. The designer(s) may not realize that some of the "requirements" embedded in the DFDs are simply reflections of the way things are now, not the way they necessarily should be in the future.

Physical DFDs are sometimes developed and used during the last stages of analysis or early stages of design. They are useful models for describing alternate implementations of a system prior to developing more detailed design models. But analysts should avoid creating physical DFDs during all analysis phase activities, except when generating alternatives. Even during that activity, analysts should clearly label physical DFDs as such so that readers know that the model represents one possible implementation of the logical system requirements.

Evaluating DFD Quality

A high-quality set of DFDs is readable, is internally consistent, and accurately represents system requirements. Accuracy of representation is determined primarily by consulting users and other knowledgeable stakeholders. A project team can ensure readability and internal consistency by applying a few simple rules to DFD construction. Analysts can apply these rules while developing the DFDs or during a separate quality check after preparing DFD drafts.

Minimizing Complexity

People have a limited ability to manipulate complex information. If too much information is presented at once, people experience a phenomenon called *information overload*. When information overload occurs, a person has difficulty in understanding. The key to avoiding information overload is to divide information into small and relatively inde-

information overload

difficulty in understanding that occurs when a reader receives too much information at one time

pendent subsets. Each subset should contain a comprehensible amount of information that people can examine and understand in isolation.

A layered set of DFDs is an example of dividing a large set of information into small, independent subsets. Each DFD can be examined in isolation. The reader can find additional detail about a specific process by moving down to the next level, or find information about how a DFD relates to other DFDs by examining the next higher-level DFD.

An analyst can avoid information overload within any single DFD by following two simple rules of DFD construction:

- 7 ± 2
- Interface minimization

rule of 7 ± 2

the rule of model design that limits the number of model components or connections among components to no more than nine

The *rule of 7 ± 2* (also known as Miller's Number) derives from psychology research, which shows that the number of information "chunks" that a person can remember and manipulate at one time varies between five and nine. A larger number of chunks causes information overload. Information chunks may be many things, including names, words in a list, digits, or components of a picture.

Some applications of the rule of 7 ± 2 to DFDs include:

- A single DFD should have no more than 7 ± 2 processes.
- No more than 7 ± 2 data flows should enter or leave a process, data store, or data element on a single DFD.

These rules are general guidelines, not unbreakable laws. DFDs that violate these rules may still be readable, but violations should be considered a warning of potential problems.

Minimization of interfaces is directly related to the rule of 7 ± 2. An interface is a connection to some other part of a problem or description. As with information chunks, the number of connections that a person can remember and manipulate is limited, so the number of connections should be kept to a minimum. Processes on a DFD represent chunks of business or processing logic. They are related to other processes, entities, and data stores by data flows. A single process with a large number of interfaces (data flows) may be too complex to understand. This complexity may show up directly on a process decomposition as a violation of the rule of 7 ± 2. An analyst can usually correct the problem by dividing the process into two or more subprocesses, each of which should have fewer interfaces.

minimization of interfaces

a principle of model design that seeks simplicity by limiting the number of connections among model components

Pairs or groups of processes with a large number of data flows between them are another violation of the interface minimization rule. Such a condition usually indicates a poor partitioning of processing tasks among the processes. The way to fix the problem is to reallocate the processing tasks so that fewer interfaces are required. The best division of work among processes is the simplest, and the simplest division is one that requires the fewest interfaces among processes.

Ensuring Data Flow Consistency

An analyst can often detect errors and omissions in a set of DFDs by looking for specific types of inconsistency. Three common and easily identifiable consistency errors are:

- Differences in data flow content between a process and its process decomposition
- Data outflows without corresponding data inflows
- Data inflows without corresponding outflows

balancing

equivalence of data content between data flows entering and leaving a process and data flows entering and leaving a process decomposition DFD

A process decomposition shows the internal details of a higher-level process in a more detailed form. In most cases, the data content of flows to and from a process at one DFD level should be equivalent to the content of data flows to and from all processes in a decomposition. This equivalency is called *balancing*, and the higher-level DFD and the process decomposition DFD are said to be "in balance."

Note the use of the term *data content* in the previous paragraph. Data flow names can vary among DFD levels for a number of reasons, including decomposition of one combined data flow into several smaller flows. Thus, the analyst must be careful to look at the *components* of data flows, not just data flow names. For this reason, detailed analysis of balancing should not be undertaken until data flows have been fully defined.

Unbalanced DFDs may be acceptable when the imbalance is due to data flows that were ignored at the higher levels. For example, diagram 0 for a large system usually ignores details of error handling, such as when an item is ordered but is later determined to be out of stock and discontinued by its manufacturer. A process called *Fulfill order* on diagram 0 would not have any data flows associated with this condition. In the process decomposition of *Fulfill order*, the analyst might add a process and data flows to handle discontinued items.

Another type of DFD inconsistency can occur between the data inflows and outflows of a single process or data store. By definition, a process transforms data inflows into data outflows. In a logical DFD, data should not be needlessly passed into a process. The following consistency rules can be derived from these facts:

- All data that flow into a process must flow out of the process or be used to generate data that flow out of the process.
- All data that flow out of a process must have flowed into the process or have been generated from data that flowed into the process.

Figure 6-16 shows an example that violates the first rule. Compare Figure 6-16 with the first DFD fragment in Figure 6-12, and note the difference in the data inflows to the process. Looking up item availability requires only information to identify the item and access to corresponding data stores. In Figure 6-16, excess data input (an entire order) flows into the process, and the process accesses more data stores than needed to generate the data outflow Item availability details. A process such as the one shown in Figure 6-16 is sometimes called a **black hole** because some or all of the data that enter never leave.

black hole

a process or data store with a data input that is never used to produce a data output

FIGURE *6-16*

A process with unnecessary data input—a black hole.

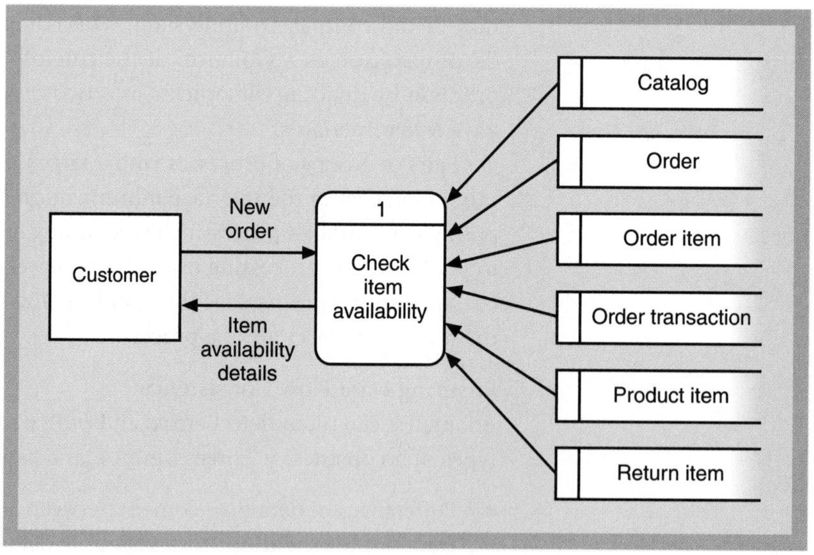

Figure 6-17 shows an example that violates the second rule. Compare Figure 6-17 with the bottom DFD fragment in Figure 6-7, and note the difference in the data inflows to the process. In Figure 6-17, insufficient data enter the process to produce the data output. Required data inputs from the Offered course and Course enrollment are missing. A process such as the one shown in Figure 6-17 is sometimes called a **miracle** because data emerge from the process without any apparent source.

miracle

a process or data store with a data element that is created out of nothing

FIGURE *6-17*

A process with an impossible
data output—a miracle.

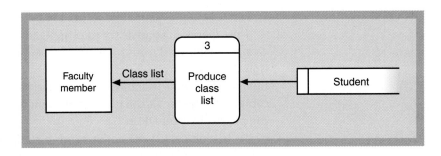

Analysts sometimes can spot black holes and miracles simply by examining the DFD. In other cases, close examination of the data dictionary or process descriptions is required. In Figure 6-18, data elements A, B, and C flow into the process but do not flow out. Data element A is used to determine what formula to apply to recompute the value of X, so that data element is a necessary input. However, data elements B and C play no role in generating process output and thus should be eliminated as unnecessary inflows.

FIGURE *6-18*

A process with unnecessary data input.

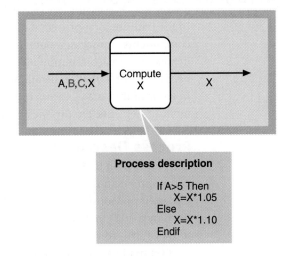

In Figure 6-19, data elements A, B, and Y flow out of the process. Data element A flows into the process. Data element Y is computed by an algorithm based on data element A. However, data element B does not flow into the process and is not computed by internal processing logic. Thus, data element B indicates either an error in the data outflow (B should be eliminated) or an omission in the internal processing logic (the rule that determines B is missing).

FIGURE *6-19*

A process with an impossible
data output.

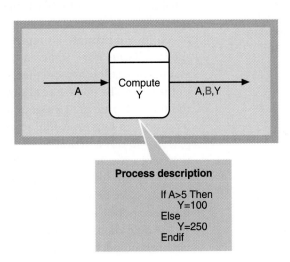

Note that both consistency rules apply to data stores as well as processes. Any data element that is read from a data store must have been previously written to that data store. Similarly, any data element that is written to a data store eventually must be read from the data store. Examining the consistency of data flows to and from a data store is complicated by the fact that a data element may flow into and out of a data store on completely different DFDs.

Evaluating data flow consistency is a straightforward but tedious process. Fortunately, most CASE tools automatically perform data flow consistency checking. But those CASE tools place rigorous requirements on the analyst to specify the internal logic of processes precisely. Without precise process descriptions, it is impossible for the CASE tool (or a human being) to know what data elements are used as input or generated as output by internal processing logic.

DOCUMENTATION OF DFD COMPONENTS

In the traditional approach, data flow diagrams show all three types of internal system components—processes, data flows, and data stores—on one diagram, but additional details about each component need to be described. First, each lowest-level process needs to be described in detail. Additionally, the analyst needs to define each data flow in terms of the data elements it contains. Data stores also need to be defined in terms of the data elements. Finally, the analyst also needs to define each data element.

Process Descriptions

Each process on a DFD must be defined formally. There are several options for process definition, including one that has already been discussed—process decomposition. As discussed previously, in a process decomposition, a higher-level process is formally defined by a DFD that contains lower-level processes. These lower-level processes may in turn be further decomposed into even lower-level DFDs.

Eventually a point is reached at which a process doesn't need to be defined further by a DFD. This point occurs when a process becomes so simple that it can be described adequately by other methods—structured English, decision tables, or decision trees. With each method, the process is described as an algorithm, and an analyst chooses the most appropriate presentation format by determining which is most compact, readable, and unambiguous. In most cases, structured English is the preferred method.

Structured English uses brief statements to describe a process very carefully. Structured English looks a bit like programming statements, but without references to computer concepts. Rules of structured programming are followed, and indentation is used for clarity. For example, a simple set of instructions for processing ballots after a vote is shown in Figure 6-20. Some statements are simply instructions. Other statements repeat instructions. Still other statements direct the program to execute one set of instructions or the other. The procedure always starts at the top and ends at the bottom. Therefore, the rules of structured programming apply. Note, though, that a process described by structured English is not necessarily a computer program—it might be done by a person—so it is a logical model. It is unambiguous, so anyone following the instructions will arrive at the same result.

An example of a process description for Rocky Mountain Outfitters is shown in Figure 6-21. Note how the process description provides more specific details about what the process does. If one process description method becomes too complex, then the analyst should choose another. Excess length (for example, more than 20 lines) or multiple levels of indentation (indicating complex decision logic) indicate that a structured

structured English

a method of writing process specifications that combines structured programming techniques with narrative English

FIGURE 6-20

A structured English example.

```
Process Ballots Procedure

Collect all ballots
Place all ballots in a stack
Set Yes count and No count to zero
Repeat for each ballot in the stack
    If Yes is checked then
        Add one to Yes count
    Else
        Add one to No count
    Endif
    Place ballot on counted ballot stack
Endrepeat
If Yes count is greater than No count then
    Declare Yes the winner
Else
    Declare No the winner
Endif
Store the counted ballot stack in a safe place
End Process Ballots Procedure
```

FIGURE 6-21

RMO process 2.1 *(Record customer information)* and its structured English process description.

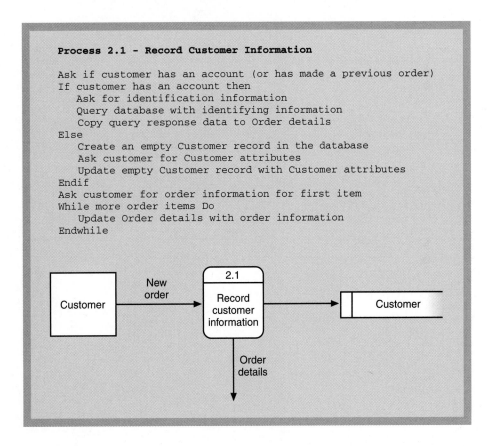

```
Process 2.1 - Record Customer Information

Ask if customer has an account (or has made a previous order)
If customer has an account then
    Ask for identification information
    Query database with identifying information
    Copy query response data to Order details
Else
    Create an empty Customer record in the database
    Ask customer for Customer attributes
    Update empty Customer record with Customer attributes
Endif
Ask customer for order information for first item
While more order items Do
    Update Order details with order information
Endwhile
```

English description may be too complex. An analyst can sometimes address excess indentation by converting the description to an equivalent decision table or decision tree. In other cases, a process decomposition may be required.

Structured English is well suited to describing processes with many sequential processing steps and relatively simple control logic (such as a single loop or an *if-then-else* statement). Structured English is not well suited for describing processes with the following characteristics:

- Complex decision logic
- Few (or no) sequential processing steps

Decision logic is complex when multiple decision variables and a large number of possible combinations of those variables need to be considered. When a process with complex decision logic is described with structured English, the result is typically a long and difficult-to-read description. For example, consider the structured English description for calculating shipping costs shown in Figure 6-22. Note that the description is relatively long and consists mostly of control structures (*if*, *else*, and *endif* statements).

FIGURE 6-22

A structured English process description for determining delivery charges.

```
If YTD purchases > $250 then
    If number of items ordered < 4 then
        If delivery date is next day then
            delivery charge is $25
        Endif
        If delivery date is second day then
            delivery charge is $10
        Endif
        If delivery date is seventh day then
            delivery charge is $1.50 per item
        Endif
    Else
        If delivery date is next day then
            delivery charge is $6 per item
        Endif
        If delivery date is second day then
            delivery charge is $2.50 per item
        Endif
        If delivery date is seventh day then
            delivery charge is zero (free)
        Endif
    Endif
Else
    If number of items ordered < 4 then
        If delivery date is next day then
            delivery charge is $35
        Endif
        If delivery date is second day then
            delivery charge is $15
        Endif
        If delivery date is seventh day then
            delivery charge is $10
        Endif
    Else
        If delivery date is next day then
            delivery charge is $7.50 per item
        Endif
        If delivery date is second day then
            delivery charge is $3.50 per item
        Endif
        If delivery date is seventh day then
            delivery charge is $2.50 per item
        Endif
    Endif
Endif
```

decision table

a tabular representation of processing logic containing decision variables, decision variable values, and actions or formulas

decision tree

a graphical description of process logic that uses lines organized like branches of a tree

Decision tables and *decision trees* can summarize complex decision logic more concisely than structured English. Figures 6-23 and 6-24 show a decision table and decision tree that represent the same logic as the structured English example in Figure 6-22. Both incorporate decision logic into the structure of the table or tree to make the descriptions more readable than their structured English equivalent. The decision table is more compact, but the decision tree is easier to read. Sometimes an analyst needs to describe a process all three ways before deciding which approach describes a particular process best.

The following steps are used to construct a decision table:

FIGURE *6-23* A decision table for calculating shipping charges.

YTD purchases > $250	YES						NO					
Number of Items (N)	N≤3			N≥4			N≤3			N≥4		
Delivery Day	Next	2nd	7th	Next	2nd	7th	Next	2nd	7th	Next	2nd	7th
Shipping Charge ($)	25	10	N*1.50	N*6.00	N*2.50	Free	35	15	10	N*7.50	N*3.50	N*2.50

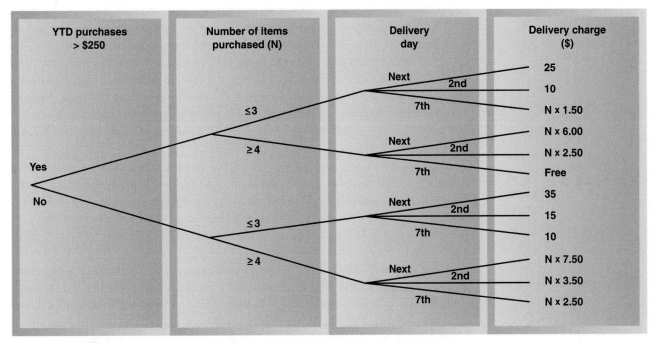

FIGURE *6-24*

A decision tree for calculating shipping charges.

1. Identify each decision variable and its allowable values (or value ranges).
2. Compute the number of decision variable combinations as the product of the number of values (or value ranges) of each decision variable.
3. Construct a table with one more columns than the number of decision variable combinations computed in step 2 (the extra column is for decision variable names and process action or computation descriptions). The table should have a row for each decision variable and a row for each process action or computation.
4. Assign the decision variable with the fewest values (or value ranges) to the first row of the table. Put the decision variable name in the first column. Divide the remaining columns into sets of columns for each decision variable value (or value range).
5. Choose the next decision variable with the fewest values (or value ranges) for the second row. Put the variable's name in the first column. Compute the number of column groups as the product of the number of values (or value ranges) of this variable and all the variables above it in the table. Divide the remaining columns into the computed number of groups, and insert values (or value ranges) in a regular pattern.
6. Continue inserting rows as instructed in step 5 until all decision variables have been included in the table.
7. Add a row for each calculation or action. For each calculation cell, insert the appropriate constant value or formula for the combination of decision variable values that appear above the cell in the same column. For each action cell, place a check mark in the cell if that action is performed when the decision variables have the values shown in the column above the cell.

Now let's follow these steps to show how the decision table in Figure 6-23 was constructed. There are three decision variables: year to date (YTD) purchases, number of items ordered, and delivery day. YTD purchases has two relevant ranges: less than $250, and greater than or equal to $250. Note that decision variable ranges must be mutually exclusive and collectively exhaustive. Number of items ordered also has two relevant ranges: less than or equal to three, and greater than or equal to four. Delivery day has three possible values: next day, second day, and seventh day. There are $2 \times 2 \times 3 = 12$ combinations of values, so there are 13 columns in the table to allow for a decision variable name, the formula, and action names.

Both YTD purchases and number of items have two relevant value ranges, so either can occupy the first row. We chose YTD purchases. It has two value ranges, so we created two groups of $12 \div 2 = 6$ columns, and labeled one for each possible value. The next row is for number of items. It has two ranges, so we need four groups of three columns—that is, $12 \div 2$ value ranges for number of items $\div 2$ value ranges for YTD purchases. We insert the value ranges for number of items into the column groups in a regular pattern, as shown in the sample figure. The delivery day is now inserted into the table. Since it is the last decision variable, we don't need to group any columns beneath it. We simply insert the values of the delivery date into individual columns in a regular pattern, as we did for the other decision variables.

The final step is to insert the row containing formulas and values for the shipping charges. Each cell contains a value or formula for the combination of decision variable values in the columns above. For example, shipping is free for customers with YTD purchases greater than $250, orders of more than three items, and seventh-day delivery. The shipping charge is $35 for customers with YTD purchases less than $250, an order of three items or fewer, and next-day delivery.

If the decision table is used to represent a process that implements one or more actions—instead of value calculations, as in the previous example—then the table must contain a row for each action. Cells in these rows are checkmarked to indicate which actions are performed under which conditions. Figure 6-25 shows a simple example of this type of table. Two action rows are included, and the action is performed if a check mark appears in the cell immediately below the decision variable values. For example, if the customer is new and the shipment contains an item back-ordered more than 25 days, then the shipment is expedited and the detailed return instructions are included in the container. If the customer isn't new and the order contains no items back-ordered more than 25 days, then neither action is taken.

FIGURE 6-25

A simple decision table with multiple action rows.

New customer	Yes		No	
Item back order ≥ 25 days	Yes	No	Yes	No
Include detailed return Instructions	✓	✓		
Expedite delivery	✓		✓	

You can construct a decision tree using almost the same steps as listed previously for constructing a decision table. The primary difference is that rows in a decision table are columns in a decision tree, and vice versa. To see this for yourself, draw an imaginary line through the table in Figure 6-23 from the top-left to bottom-right. Then flip the table along the imaginary line and compare the structure of the flipped table to the decision tree in Figure 6-24. The only other significant difference between a table and a tree is that a tree uses labeled branches instead of grouped columns to represent decision variable values.

Data Flow Definitions

data flow definition

a textual description of a data flow's content and internal structure

A data flow is a collection of data elements, so *data flow definitions* list all the elements. For example, a simplified New order data flow (to process 2.1 in Figure 6-14) consists of a customer name, credit card number, and list of catalog item numbers and quantities. Some of these elements are actually structures of other elements, such as a customer name consisting of first name, middle initial, and last name. The system stores most of these data elements, so they coincide with the attributes of data entities included in the ERD.

Sometimes data flow definitions contain a more complex structure. In the New order example, each data flow consists of many catalog items and quantities (a repeating group). It is important to document this structure. The notations for data flow definitions vary. One approach is simply to list the data elements, as shown in Figure 6-26. The elements that can have many values are indicated. Another approach uses an algebraic notation such as that shown in Figure 6-27. The data flow "equals" or "consists of" one element plus another element, and so on. Groups of elements that can have many values are enclosed in curly braces. This example shows New order "equals" the customer name "plus" customer address "plus" credit card information "plus" "one or more" inventory item number and quantity. In this example, the customer name can be defined separately as a structure of elements.

FIGURE *6-26*

Data flow definitions simply listing elements.

```
Customer-Name
Customer-Address
Credit-Card-Information
Item-Number
Quantity
```

FIGURE *6-27*

Algebraic notation for data flow definition (New-Order).

$$\text{New-Order} = \text{Customer-Name} + \text{Customer-Address} + \text{Credit-Card-Information} + {}^{n}_{1}\{ \text{ Item-Number} + \text{Quantity } \}$$

Figures 6-28 and 6-29 show a complex report and its corresponding data flow definition. The structure of the report is a repeating group of products with an embedded repeating group of inventory items. The data flow definition captures this structure by embedding the item repeating group within one set of curly braces and the product repeating group within the outermost set of curly braces.

Data Store Definitions

Since a data store on the DFD represents a data entity on the ERD, no separate definition is typically needed (except perhaps a note referring the reader to the ERD). If data stores are not linked to an ERD, then the analyst simply defines the data store as a collection of elements (possibly with a structure) in the same way that data flows are defined.

Data Element Definitions

Data element definitions describe a data type, such as string, integer, floating point, or Boolean. Each element should also be described to indicate specifically what it represents. Sometimes these descriptions are very specific. A date of sale might be defined as the date the payment for the order was received. Alternately, the date of sale might be the date an order is placed. Sometimes different departments in the same company have different definitions for the same element, so it is very important for the analyst to confirm exactly what the element means to users.

Other parts of a data element definition vary depending on the type of data. A length is usually defined for a string. For example, a middle initial might be one character maximum, but how long should a first name be? Numeric values usually have a minimum

Rocky Mountain Outfitters — Products and Items

ID	Name	Season	Category	Supplier	Unit Price	Special	Special Price	Discontinued
RM0125	Outdoor Field	Spr/Fall	Mens C	8201	$39.00	▦	$0.00	No
Description Outdoor Nylon Jacket with Lining								

Size	Color	Style	Units in Stock	Reorder Level	Units on Order
Large	Blue		1500	150	
Large	Green		1500	150	
Large	Red		1500	150	
Large	Yellow		1500	150	
Medium	Blue		1500	150	
Medium	Green		1500	150	
Medium	Red		1500	150	
Medium	Yellow		1500	150	
Small	Blue		1500	150	
Small	Green		1500	150	
Small	Red		1500	150	
Small	Yellow		1500	150	
Xlarge	Blue		1500	150	
Xlarge	Green		1500	150	
Xlarge	Red		1500	150	
Xlarge	Yellow		1500	150	

ID	Name	Season	Category	Supplier	Unit Price	Special	Special Price	Discontinued
RM0125	Hiking Walkers	All	Footwear	7993	$49.95	▦	$0.00	No
Description Hiking Walkers with Patterned Tread Durable Uppers								

Size	Color	Style	Units in Stock	Record Level	Units on Order
10	Brown		1000	100	
10	Tan		1000	100	
11	Brown		1000	100	
11	Tan		1000	100	
12	Brown		1000	100	
12	Tan		1000	100	
13	Brown		1000	100	
13	Tan		1000	100	
7	Brown		1000	100	
7	Tan		1000	100	
8	Brown		1000	100	
8	Tan		1000	100	
9	Brown		1000	100	
9	Tan		1000	100	

FIGURE *6-28*

A sample report produced by the RMO customer support system.

data dictionary

a repository for definitions of data flows, data elements, and data stores

and maximum value that can be defined as a valid range. Sometimes specific values are allowed for the element, such as valid codes. If the element is a code, it is important to define the valid codes and their meaning. For example, code A might mean ship immediately, code B might mean hold for one day, and code C might mean hold shipment pending confirmation. Some sample data element definitions are shown in Figure 6-30.

Analysts need to maintain a central store of all these definitions as a project reference and to ensure consistency. A *data dictionary* is a repository for definitions of data flows, data stores, and data elements. A data dictionary may be a simple loose-leaf notebook or word processing file, in smaller development projects. In larger projects, a project management or CASE tool usually holds the data dictionary. The data dictionary may also hold process descriptions.

FIGURE *6-29*

A data flow definition for the RMO products and items report.

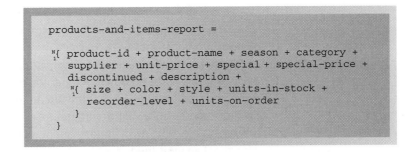

```
products-and-items-report =

N{ product-id + product-name + season + category +
1     supplier + unit-price + special + special-price +
      discontinued + description +
    N{ size + color + style + units-in-stock +
    1     recorder-level + units-on-order
      }
   }
```

FIGURE *6-30*

Data element definitions.

```
units-in-stock =
   a positive integer

supplier =
   a four digit numeric code

unit-price =
   a positive real number accurate to two decimal places,
   always in U.S. dollars

description =
   a text field containing a maximum of 50 printable characters

special =
   a coded field with one of the following values
   0: item is not "on special"
   1: item is "on special"
```

DFD Summary

Figure 6-31 shows each of the components of a traditional analysis model—an entity-relationship diagram, data flow diagrams, process definitions, and data definitions. The four components form an interlocking set of specifications for most system requirements. The data flow diagram provides the highest-level view of the system, summarizing processes, external agents, data stores, and the flow of data among them. Each of the other components describes some aspect of the data flow diagram in greater detail.

FIGURE *6-31*

The components of a traditional systems analysis model.

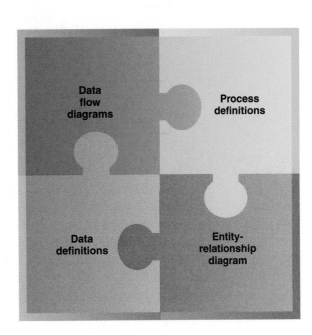

The models described thus far were developed in the 1970s and 1980s as part of the traditional structured analysis methodology (see Yourdon 1989 in the "Further Resources" section). They were designed to document completely the logical requirements of a system. However, some analysts choose to augment the structured models with models borrowed from other methodologies. Such models may be used to describe information not captured by the structured models. Or they may be used to present similar information in a slightly different form. The remainder of this chapter describes some of these "borrowed" models and ways they can be used to augment the traditional structured analysis models.

INFORMATION ENGINEERING MODELS

Information Engineering (IE)

a system development methodology that focuses on strategic planning, data modeling, and automated tools, and is thought to be more rigorous and complete than the structured approach

Information Engineering (IE) was developed by James Martin in the early 1980s. At that time, IE and the structured system development methodology were considered competitors, although they have some features in common. As both methodologies matured, many analysts sought to combine the best features of both. This section provides an overview of the IE methodology and describes some IE analysis models and ways they can be used with the structured analysis models covered earlier in this chapter.

The IE Systems Development Life Cycle

Figure 6-32 shows the phases of the IE systems development life cycle (SDLC). The first phase of the IE SDLC concentrates on modeling the entire enterprise and producing coordinated information systems plans. The enterprise is modeled in terms of its strategic goals and plans, information technology infrastructure, stored data requirements, and major functional areas. Overlaps between the data and functional models of the enterprise are used to identify specific business areas for further analysis in the second IE SDLC phase.

F I G U R E *6-32*

Information Engineering systems development life cycle phases.

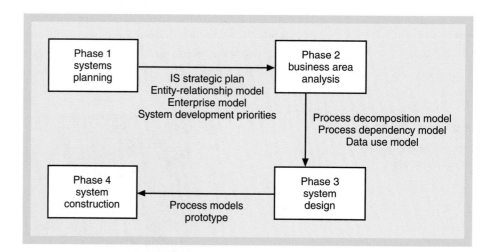

Two key features of IE that are apparent in the first phase of its SDLC are its focus on the business as a whole and its focus on stored data. IE concentrates first on describing the enterprise and its environment in terms of strategies, plans, goals, objectives, and organization structure. Only then does it define requirements for information technology infrastructure and information-processing applications. The focus on stored data follows from a key assumption that business data are an organizational resource that is much less subject to change than business processes. Internal and external data needs are assumed to drive processing requirements and process structure, not vice versa.

The second phase of the IE SDLC describes the processing requirements of each business area in the enterprise. Each business area is decomposed into a hierarchy of related processes. A process decomposition model describes the hierarchy, and a process dependency model describes interrelationships among processes. A set of data use models specifically describes relationships between the process and data models.

The third phase of the IE SDLC develops the detailed design of business processes. IE uses a number of different techniques and models to describe processes. The methodology stresses user input to process design and to process model validation. Developers develop graphical process models. Users then validate those models.

The final phase of the IE SDLC is system construction. The IE methodology assumes that CASE tools and code generators are used during the construction phase. Process models and prototypes developed in the previous phase are used to generate system components automatically. Code generators take data and process models developed with CASE tools as inputs and generate application programs and database creation commands as outputs. Historically, IE has targeted procedural programming languages such as COBOL and C and relational database management systems. Later generations of IE CASE tools and code generators have targeted OO environments.

IE and Structured Development Compared

Figure 6-33 compares features of IE and traditional structured methodologies. The most significant difference between the methodologies is their scope. IE is designed to address enterprisewide information-processing requirements. Its first phase looks at the enterprise as a whole and describes data and information processing within the context of organizational structure, goals, and strategies.

	Information Engineering	Structured methodology
For what system size is it best suited?	enterprise	any size
Does it explicity incorporate strategic planning?	yes	no
How tightly integrated are its techniques and models?	highly integrated	moderately integrated
What is the central focus of analysis?	data	processes
Can techniques from other methodologies be easily incorporated?	no	yes

FIGURE *6-33*

A comparison of the Information Engineering and structured methodologies.

In comparison, the structured methodology makes no assumptions about either the scope of its application or about planning activities that precede structured analysis. Structured development can be applied to entire enterprises, relatively small systems within an enterprise, or anything in between. Structured analysis does not explicitly require enterprise-level planning and description activities. Thus, its models do not directly incorporate information about organizational structure, enterprise strategies and goals, or existing or proposed information technology infrastructure.

Another significant difference between the methodologies is how tightly integrated their tools and techniques are. IE is a tightly integrated methodology in which the outputs of each phase completely specify inputs required in the next phase. The methodology is complete and all encompassing. Thus, it is difficult (and generally unnecessary) to incorporate techniques and models from other methodologies within IE. In contrast, the structured methodology is a general set of techniques and models that are only moderately integrated. There are some gaps in the methodology; thus, it is possible (and usually desirable) to incorporate techniques and models from other methodologies such as IE.

An ERD for the entire enterprise is developed during the first IE phase. This model provides information that is used extensively in later IE modeling and design activities. Structured analysis also develops and uses an ERD, although it plays a less central role in the structured methodology than in the IE methodology. Because both methods develop an ERD, it is possible to incorporate into the structured methodology other IE models that depend on the ERD. The following sections describe some of the IE models that are useful additions to the structured analysis models described earlier in the chapter.

Process Decomposition and Dependency Models

IE takes a different approach to modeling processes and data flows than does structured analysis. IE process models show three types of information:

- Decomposition of processes into other processes
- Dependency relationships among processes
- Internal processing logic

Analysts first model business areas by decomposing them into subsidiary processes. Figure 6-34 shows an example of a *process decomposition diagram* for the RMO customer support system. This diagram shows the decomposition of the system (or business area) into the major subsystems described earlier in Figure 6-10. The order-entry subsystem is further decomposed into processes that are in turn decomposed further. Processes that are further decomposed on a different diagram or page are indicated by three dots in the upper-left corner.

Compare the information content of Figure 6-34 with earlier examples of RMO DFDs (for example, Figures 6-9 and 6-14). There are several similarities between process representations used with IE and structured development. One obvious similarity is the symbol used to represent a process. Both methodologies use an oval, though it is oriented differently on the page.

Another similarity is the hierarchical relationship among processes at different levels of abstraction. Structured DFDs show these relationships indirectly by placing process decompositions in different DFDs and using a hierarchical numbering system to tie the various DFDs together. The process dependency diagram directly represents hierarchical relationships among process levels on a single page. Both methods show the same process decomposition information in different ways.

Figure 6-35 shows a *process dependency diagram* for *Create new order*. This diagram corresponds to the *Create new order* process decomposition DFD shown in Figure 6-14. The large arrow labeled *Order received* represents the event that triggers the first process on the diagram. The lines and arrows connecting processes indicate dependency among processes. The process *Produce confirmation* depends on the process *Process order transaction*, which implies that *Produce confirmation* cannot begin until *Process order transaction* is completed.

The difference between a dependency and a data flow is subtle but important. Process dependency diagrams do not directly show data flows among processes (note the lack of names or labels on the lines connecting processes). *Produce confirmation* depends on *Process order transaction* because it makes no sense to confirm an order that hasn't been finalized (such as an order for which credit has not been approved). Even if one process creates data that another process uses, a process dependency diagram never directly shows the flow of that data among processes.

Figure 6-36 shows a process dependency diagram with added data flows. Process dependencies are shown in black, and data flows are shown in red. IE calls such a diagram a data flow diagram even though it is quite different from a structured DFD. The differences between Figures 6-14 and 6-36 are striking. Although the number and ordering of processes are the same, there are significant differences in information content. The

process decomposition diagram

a model that represents the hierarchical relationship among processes at different levels of abstraction

process dependency diagram

a model that describes the ordering of processes and their interaction with stored entities

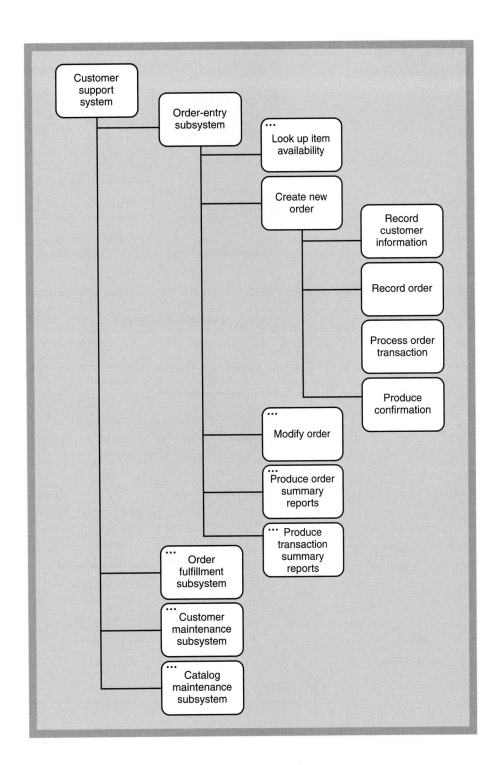

structured DFD explicitly shows data flows to and from processes. These flows (and the entities) are ignored on the process dependency diagram. The structured DFD routes data directly from one process to another, but the process dependency diagram routes all data through the data stores. The reason is that during the business area analysis phase, IE assumes that processes read and write data to and from the database.

The differences between the diagrams demonstrate a different approach to representing a logical (rather than physical) model of processing. IE purposely avoids showing data flows among entities and processes because they might introduce unnecessary physical assumptions about process implementation. IE assumes that the "when" and

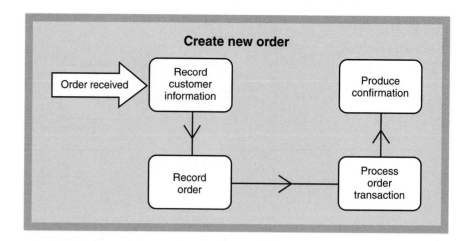

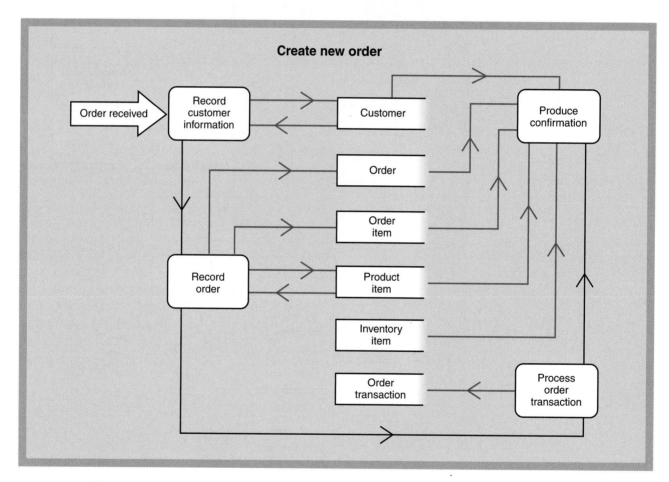

"where" of data flows are design decisions. Data sources, destinations, and routing are not explicitly modeled until the systems design phase.

In contrast, structured analysis requires representing flows to and from entities and processes. Data flow among processes can be forced through data stores, but this is neither required nor encouraged by structured analysis rules. The IE approach makes it difficult to create a model that is not logical. The structured approach provides a general set of modeling tools that can be applied to logical or physical models. It is up to the analyst to decide what restrictions (if any) are needed when developing logical DFDs.

Both process decomposition and process dependency diagrams can be useful additions to the structured methodology. Process decomposition diagrams are useful for graphically showing the hierarchical relationships among a large number of processes. Process dependency diagrams can be a useful alternative for examining and describing the logical essence of process relationships and data flows.

The primary danger of using either model in addition to traditional structured models is redundancy. The structured models are designed to be as nonredundant as possible, thus making them easier to modify. The IE models represent information that is also represented on structured DFDs. Thus, any change to a structured model requires a change to a corresponding IE model. This requirement makes implementing a change more cumbersome and opens up the possibility of inconsistency among the models.

LOCATIONS AND COMMUNICATION THROUGH NETWORKS

Because structured systems analysis concentrates on logical modeling, physical issues such as processing locations and networks are sometimes ignored during analysis. However, a great deal of information about process, data, and user distribution is needed during the early stages of design. Examples include:

- Number of locations of users
- Processing and data access requirements of users at specific locations
- Volume and timing of processing and data access requests

This information is needed to make initial design decisions such as the distribution of computer systems, application software, and database components. It is also needed to determine required network capacity among user and processing locations. The Information Engineering development methodology specifies several tables that describe the relationships among entities on the ERD and other information system and enterprise components. These tables describe the sources and uses of data within the enterprise. Examples include tables that describe which processes initially capture information about an entity and which processes later use that data. IE develops these tables early in the SDLC and uses them to guide later decisions about process design and structure.

Many analysts gather and summarize location information early in the analysis phase of the traditional SDLC. Gathering the information during analysis enables analysts to make better decisions during the last two analysis phase activities—*Generate and evaluate alternatives* and *Review recommendations with management*. Location information is also useful in many design phase activities, including *Design and integrate the network*, *Design the application architecture*, and *Design and integrate the database*.

The first step in gathering location information is to identify and describe the locations where work is being or will be performed. Possible locations include business offices, warehouses, and manufacturing facilities, and less obvious locations such as customer or supplier offices, employee homes, hotel rooms, and automobiles. All of these locations should be listed, and a *location diagram* should be drawn to summarize the locations graphically. A location diagram for Rocky Mountain Outfitters is shown in Figure 6-37. The location diagram shows the analyst what network connections might be required, but it also has the added benefit of reminding everyone that users at all locations should be consulted about the system.

The next step is to list the functions that are performed by users at each location. Using the event table, the analyst can list where each activity is performed. Figure 6-38 shows an *activity-location matrix* that summarizes this information. Each row is a system activity, and each column represents a location. Many activities are performed at multiple locations.

Recall that Rocky Mountain Outfitters also has a system project under way for the inventory management system. The inventory management system will involve many ac-

location diagram

a diagram or map that identifies all of the processing locations of a system

activity-location matrix

a table that describes the relationship between processes and the locations in which they are performed

FIGURE *6-37*

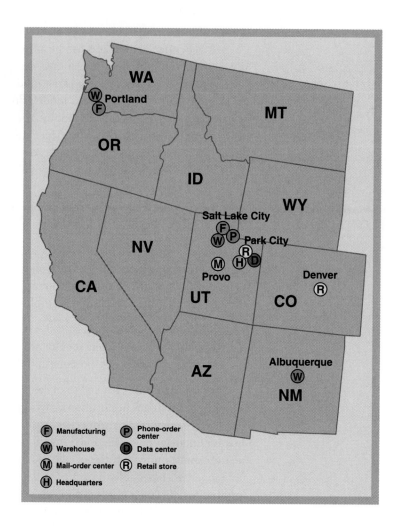

activity-data matrix

a table that describes stored data entities, the locations from which they are accessed, and the nature of the accesses

tivities at the manufacturing facilities, but the customer support system will not. Additionally, RMO has a plan for integrating the system at the retail stores with the inventory management system, but not with the customer support system. Therefore, these locations are not shown on the activity-location matrix.

Other matrices can be created to highlight access requirements. One approach is to list activities and data entities (or classes of objects) in an *activity-data matrix*. This matrix shows which activities require access to the data or objects. This information can be found on the DFD fragments for the traditional approach and on the sequence diagrams for the OO approach. In either approach, creating a matrix to summarize this information can be useful.

ACTIVITY	LOCATION				
	Corporate offices (Park City)	Distribution warehouses (Salt Lake City, Albuquerque, Portland)	Mail-order (Provo)	Phone sales (Salt Lake City)	Customer direct interaction (Anticipated)
Look up item availability	X	X	X	X	X
Create new order			X	X	X
Update order				X	X
Look up order status	X	X		X	X
Record order fulfillment		X			
Record back order		X			
Create order return		X			
Provide catalog info			X	X	X
Update customer account	X		X	X	X
Distribute promotional package	X				
Create customer charge adjustment	X				
Update catalog	X				
Create special promotion	X				
Create new catalog	X				

FIGURE *6-38*

Activity-location matrix for the Rocky Mountain Outfitters customer support system.

CRUD

acronym of create, read, update, and delete

Figure 6-39 shows an activity-data matrix for Rocky Mountain Outfitters. The cells of the matrix show additional information to clarify what the activity does to the data. The letter *C* means the activity creates new data, *R* means the activity reads data, *U* means the activity updates data, and *D* means the activity might delete data. The acronym *CRUD* (create, read, update, and delete) is often used to describe this type of matrix.

FIGURE 6-39

Rocky Mountain Outfitters
activity-data matrix.

ACTIVITIES / **DATA ENTRIES**

ACTIVITIES	Catalog	Customer	Inventory item	Order	Order item	Order transaction	Package	Product item	Return item	Shipment	Shipper
Look up item availability			R				R	R			R
Create new order		CRU	RU	C	C	C	R	R		C	R
Update order		RU	RU	RUD	RUD	RUD	R	R		CRUD	R
Look up order status		R		R	R	R				R	
Record order fulfillment					RU					RU	
Record back order					RU					CRU	
Create order return		CRU		RU		C			C		
Provide catalog info	R		R				R	R			
Update customer account		CRUD									
Distribute promotional package	R	R	R				R	R			
Create customer charge adjustment		RU				CRUD					
Update catalog	RU		R				RU	R			
Create special promotion	R		R				R	R			
Create new catalog	C		R				CRU	R			

C = Creates new data, R = Reads existing data, U = Updates existing data, D = Deletes existing data

SUMMARY

Data flow diagrams (DFDs) are used in combination with the event table and entity-relationship diagram (ERD) to model system requirements. DFDs model a system as a set of processes, data flows, external agents, and data stores. DFDs are relatively easy to read because they graphically represent key features of the system using a small set of symbols. Because there are many features to be represented, many types of DFDs are developed, including context diagrams, DFD fragments, subsystem DFDs, event-partitioned DFDs, and process decomposition DFDs.

Each process, data flow, and data store requires a detailed definition. Analysts may define processes in a number of ways, including a structured English process specification, a decision table, a decision tree, or a process decomposition DFD. Process decomposition DFDs are used when internal process complexity is too great to allow the creation of a readable, one-page definition by any other means. Data flows are defined in terms of their component data elements and their internal structure. Data elements may be further defined in terms of their type and allowable content. Data stores correspond to entities on the ERD, and thus require no additional definition.

Models borrowed from other methodologies may supplement data flow diagrams to provide additional information or an alternate view of requirements. Information Engineering (IE) models are a useful complement to traditional analysis models. Particularly useful models include the process decomposition diagram, process dependency diagram, location diagram, activity-location matrix, and activity-data (or CRUD) matrix. The process decomposition model provides a useful summary of how processes on multiple DFD levels are related to one another. The process dependency diagram provides a view of processing details that emphasizes interaction with stored entities instead of data flow among processes.

The location diagram, activity-location matrix, and activity-data matrix describe important information about system locations. The location diagram summarizes geographic locations where the system is to be used. The activity-location matrix describes which processes are implemented at which locations. The activity-data matrix summarizes where and how each data store is used.

We've now covered all of the models that are used to document system requirements in the traditional approach to systems analysis. Chapter 7 will cover the models used to document system requirements in the OO approach to systems analysis. Chapter 8 will cover the transition from systems analysis to systems design.

KEY TERMS

activity-data matrix, p. 233

activity-location matrix, p. 233

balancing, p. 216

black hole, p. 216

context diagram, p. 204

CRUD, p. 233

data dictionary, p. 225

data flow, p. 203

data flow definition, p. 223

data flow diagram (DFD), p. 202

data store, p. 203

decision table, p. 220

decision tree, p. 220

DFD fragment, p. 206

event-partitioned system model, or diagram 0, p. 206

external agent, p. 202

Information Engineering (IE), p. 226

information overload, p. 215

level of abstraction, p. 204

location diagram, p. 232

minimization of interfaces, p. 215

miracle, p. 217

process, p. 202

process decomposition diagram, p. 228

process dependency diagram, p. 230

rule of 7 ± 2, p. 215

structured English, p. 219

REVIEW QUESTIONS

1. List at least three different types of DFDs. What is each diagram type used to represent?
2. List the five component parts (symbols) of a DFD. Briefly describe what each symbol represents.
3. How does an analyst determine whether a person or organization should be represented on a DFD as an external agent or by one or more processes?
4. Processes on an event-partitioned DFD can be described by a detailed DFD or a process specification. How does an analyst determine which is the most appropriate form of description?
5. Describe how each column of an event table is represented on a DFD (that is, what symbols are used?).
6. How are entities from the ERD represented on a DFD? How are relationships from the ERD represented on a DFD?
7. What features may be present on a physical DFD that should never be present on a logical DFD?
8. What DFD characteristics does an analyst examine when evaluating DFD quality?
9. What is a black hole? What is a miracle? How can each be detected?
10. Why might an analyst describe a process with a decision table or tree instead of structured English?
11. List IE process models that are useful supplements to the traditional analysis-phase modeling tools (such as DFDs, data definitions, and process specifications). Describe the advantages and disadvantages of each IE model compared with a similar traditional model.
12. What are the disadvantages of using both traditional and IE process models to describe the same system?
13. List and briefly describe the life cycle phases of Information Engineering.
14. What is an activity-location matrix? How is it related to DFDs?
15. What is an activity-data matrix? How is it related to DFDs and the ERD?

THINKING CRITICALLY

1. Assume that you are preparing a DFD to describe the process of creating, approving, and closing a mortgage loan by a mortgage broker. Should the broker be represented as an external agent or by one or more processes? Why? What about the closing agent, the credit bureau, and the bank that issues the mortgage note?
2. Examine the course registration system described in Figure 6-6. Are there any other processes that would be required to implement a fully functioning system? Hint: Black holes and miracles may indicate processing steps that were left out of the DFD.
3. Develop a process dependency diagram with data flows for process 3 *(Update order)* in Figure 6-12. What similarities are there between your diagram and the one shown in Figure 6-13? Try to redraw all three diagrams to eliminate redundancies. Are your revised diagrams more or less readable than the original diagrams?
4. Assume that the transaction summary report for the RMO order-entry subsystem (see process 5 in Figure 6-12) contains a listing of every order that was created during a date range entered by the user. The report title page contains the report name, the date range, and the date and time the report was prepared. For each order, the report lists the order number, order date, order total, and form of payment. Within each order, the report lists all order items and returns, including item number, quantity ordered (or returned), and price. Report totals include the sum of all order totals, average order total, average item price, and average return price. Write a data flow definition entry for the report, and write a process specification for the process that produces the report.
5. Create a process dependency diagram for all the processes shown in Figure 6-6.
6. Create a process decomposition diagram (with data flows) for process 1 in Figure 6-6.
7. Create an activity-data (CRUD) matrix for the course registration system in Figure 6-6.

EXPERIENTIAL EXERCISES

1. Develop a physical DFD that models the process of grocery shopping, from the time you write down a shopping list until the time you store purchased groceries in your home. Construct your DFD as a linear sequence of processes. Now develop a logical DFD to describe the same scenario. Try to develop a diagram that is equally valid as a logical description of the way you currently buy groceries and as a logical description of ways you might buy groceries without ever leaving your home.

2. Consider the admissions requirements for a degree program, major, or concentration at your school. Look up the requirements in the school catalog and rewrite them in structured English. Develop an equivalent decision table and/or decision tree. Which is easier to understand? Why?

3. Get a copy of your school transcript. Write a data definition that describes its contents. Write data element definitions for the fields Grade, Credits, and Degree.

4. Define process 2 in Figure 6-7 as it is implemented at your school. Use whatever combination of process decomposition and process specification is appropriate. If you develop any process decomposition DFDs, then be sure to define all data flows.

CASE STUDIES

The Real Estate Multiple Listing Service System

Refer to the description of the Real Estate Multiple Listing Service system in the Chapter 5 case studies. Use the event list and ERD for that system as a starting point for the following exercises:

1. Draw a context DFD.
2. Draw an event-partitioned DFD.
3. Draw any required process decomposition DFDs.

State Patrol Ticket Processing System

Refer to the description of the State Patrol ticket processing system in the Chapter 5 case studies. Use the event list and ERD for that system as a starting point for the following exercises:

1. Draw a context DFD.
2. Draw an event-partitioned DFD.
3. Draw any required process decomposition DFDs.
4. Create data flow definitions for any data flows that are fully described in the written system description.

Rethinking Rocky Mountain Outfitters

This chapter contains many DFDs describing the RMO order-entry subsystem but no DFDs describing the RMO order fulfillment subsystem, customer maintenance subsystem, or catalog maintenance subsystem (see the subsystem event lists in Figure 6-10). Review the RMO event table (Figure 5-15) and ERD (Figure 5-28) and perform the following tasks:

1. Develop DFD fragments for all of the events not documented in Figure 6-12.
2. Develop a single DFD that shows processing for all events, using one process for each subsystem and showing all needed data stores. To simplify the diagram, place all external agents along the outer edge, and duplicate them as necessary to minimize long or crossing data flows. Place all data stores in the middle of the diagram.
3. Develop a data flow definition for the RMO customer order form in Figure 6-40.

FIGURE 6-40

RMO catalog order form.

Rocky Mountain Outfitters—Customer Order Form

Rocky Mountain Outfitters

Name and address of person placing order.
(Please verify your mailing address and make correction below.)
Order date ___/___/___

Name

Address Apt. no.

City State Zip

Phone: Day ( )________ Evening ( )________

Gift Order or Ship To: (Use only if different from adress at left.)

Name

Address Apt. No

City State Zip

Gift ☐ Address for this Shipment Only ☐ Permanent Change of Address ☐

Gift Card Message

Delivery Phone ( )________

Item no.	Description	Style	Color	Size	Sleeve Length	Qty	Monogram	Style	Price Each	Total

MERCHANDISE TOTAL ________

Total Monogramming charges ($5.00 per line per item) ________

Sales tax on merchandise delivered in Colorado and Utah ________

Regular FedEx shipping $4.50 per U.S. delivery address
(Items are sent within 24 hours for delivery in 2 to 4 days) $4.50

Please add $4.50 per each additional U.S. delivery address ________

FedEx Standard Overnight Service
Add $6.00 for delivery in 2 business days to confirm U.S. address* ________

Any additional freight charges ________

International Shipping (see shipping information on back) ________

Method of Payment

Check/Money Order ☐ Gift Certificate(s) ☐ AMOUNT ENCLOSED $ ________

American Express ☐ MasterCard ☐ VISA ☐ Other ☐

Account Number

MO YR
___/___
Expiration Date

Signature ________________

Reliable PHARMACEUTICALS

Focusing on Reliable Pharmaceutical Service

Continue your modeling efforts for the Reliable Pharmaceutical Service case by performing the following tasks:

1. Create a context diagram for the Reliable Pharmaceutical case based on the system description in Chapter 1 and the event table that you developed in Chapter 5.
2. Create DFD fragments for each event from the event table and ERD that you developed in Chapter 5.
3. Create an event-partitioned model (diagram 0) by combining the DFD fragments you created for question 2.
4. Create a logical DFD showing the processing details for the event *Time to generate orders (shipments)* based on the description in Chapter 1. Pay careful attention to modeling data movement and processing, not the movement and processing of physical goods (e.g., drugs). Create any process descriptions and data definitions needed to fully specify system requirements.
5. Consider the problem of modeling the billing procedures briefly described in Chapter 1. Should a physical DFD of billing procedures be developed? Why, or why not?

FURTHER RESOURCES

J. Martin, *Information Engineering: Book I Introduction*. Prentice Hall, 1988.

J. Martin, *Information Engineering: Book II Planning and Analysis*. Prentice Hall, 1989.

Stephen M. McMenamin and John F. Palmer, *Essential Systems Analysis*. Yourdon Press, 1984.

G. A. Miller, "The magical number seven, plus or minus two: Some limits on our capacity for processing information," *Psychological Review*, volume 63 (1956), pp. 81–97.

Edward Yourdon, *Modern Structured Analysis*. Yourdon Press, 1989.

CHAPTER 7

The Object-Oriented Approach to Requirements

LEARNING OBJECTIVES

After reading this chapter, you should be able to:

- Develop use case diagrams

- Write use case and scenario descriptions

- Develop activity diagrams and system sequence diagrams

- Refine and enhance the domain model class diagram

- Explain how UML diagrams work together to define functional requirements for the object-oriented approach

CHAPTER OUTLINE

The Unified Modeling Language and the Object Management Group

Object-Oriented Requirements

The System Activities—A Use Case/Scenario View

Identifying Inputs and Outputs—The System Sequence Diagram

Problem Domain Modeling—The Domain Model Class Diagram

Integrating Object-Oriented Models

Electronics Unlimited is a warehousing distributor that buys electronic equipment from various suppliers and sells it to retailers throughout the United States and Canada. It has operations and warehouses in Los Angeles, Houston, Baltimore, Atlanta, New York, Denver, and Minneapolis. Its customers range from large nationwide retailers, such as Target, to medium-sized independent electronics stores.

Many of the larger retailers are moving toward integrated supply chains. Information systems used to be focused on processing internal data; however, today these retail chains want suppliers to become part of a totally integrated supply chain system. In other words, the systems need to communicate between companies to make the supply chain more efficient.

To maintain its position as a leading wholesale distributor, Electronics Unlimited has to convert its system to link with both its suppliers—the manufacturers of the electronic equipment—and its customers—the retailers. It is developing a completely new system utilizing object-oriented techniques to provide these links. Object-oriented techniques facilitate system-to-system interfaces by using predefined components and objects to accelerate the development process. Fortunately, many of the system development staff have recently begun learning about object-oriented development and are eager to apply the techniques and models to a system development project.

William Jones is explaining object-oriented development to the group of systems analysts who are being trained in this approach. "We're developing most of our new systems using object-oriented principles. The complexity of the new system, along with its interactivity, makes the object-oriented approach a natural way to develop requirements. It takes a little different thought process than you may be used to, but the object-oriented models track very closely with the new object-oriented programming languages."

William continued, "This way of thinking about a system in terms of objects is very interesting. It also is consistent with the object-oriented programming techniques you learned in your programming classes. You probably first learned to think about objects when you developed screens for the user interface. All of the controls on the screen, such as buttons, text boxes, and drop-down boxes, are objects. Each has its own set of trigger events that activate its program functions.

"Now you just extend that same thought process so that you think of things like purchase orders and employees as objects, too. We can call them *problem domain* or sometimes *business objects* to differentiate them from screen objects such as windows and buttons. During analysis, we have to find out all of the trigger events and methods associated with each business object."

"How do we do that?" one of the analysts asked.

"You continue with your fact-finding activities and build a scenario for each business process. The way the business objects interact with each other in the scenario determines how you identify the initiating activity. We refer to those activities as the *messages* between objects. The tricky part is that you need to think in terms of objects instead of just processes. Sometimes it helps me to pretend I am an object. I will say, 'I am a purchase order object. What functions and services are other objects going to ask me to do?' Once you get the hang of it, it works very well, and it is enlightening to see how the system requirements unfold as you develop the diagrams."

OVERVIEW

The basic objective of requirements definition is understanding—understanding users' needs, understanding how the business processes are carried out, and understanding how the system will be used to support those business processes. As we indicated in Chapter 2, system developers use a set of tools and techniques to discover and understand the requirements for a new system. This activity is a key part of the systems analysis phase of the systems development life cycle. In object-oriented development, the set of analysis activities is more specifically referred to as *object-oriented analysis* (OOA). The first step in the process for developing this understanding requires the fact-finding skills you learned in Chapter 4. Fact-finding activities are also called *discovery activities*, and obviously discovery must precede understanding. In this chapter, you learn to take discovery to the next level—to build understanding.

Chapter 5 introduced the concepts of models and modeling activities as a way to define and document system requirements. The models introduced in Chapter 5 focus on two primary aspects of functional requirements: the events and the things involved in users' work. As you learned, events are happenings in the business's environment to which the system must respond. Those events are identified and documented in an event table. The new system must be able to respond to business events by carrying out system activities, which are called *use cases*.

A new system also needs to record and store information about things involved in the business processes. In a manual system, the information would be recorded on paper and stored in a filing cabinet. In an automated system, the information is stored in electronic files or a database. The information storage requirements of a system are documented either with entity-relationship diagrams (ERDs) in the traditional approach or with class diagrams in the object-oriented approach.

In this chapter you learn how to understand and define the requirements for a new system using object-oriented analysis models and techniques. You should be aware that the line between object-oriented analysis and object-oriented design is somewhat fuzzy because the models that are built to define requirements during analysis are refined and extended to produce a systems design. Recall that we mentioned the object-oriented approach uses an iterative approach to development, which identifies some of the requirements, then does some preliminary design and implementation, then iterates again and again through requirements, design, and implementation. So, even though we do not focus here on the iterative nature of requirements definition, it is a normal part of the object-oriented approach. Chapters 11 and 12 extend the requirements into a complete object-oriented design that can serve as the foundation for programming the new system.

THE UNIFIED MODELING LANGUAGE AND THE OBJECT MANAGEMENT GROUP

The object-oriented (OO) modeling notation that we present in this textbook is the Unified Modeling Language (UML). We have mentioned UML and its importance to the object-oriented approach in prior chapters. UML is the successor to the modeling techniques found in Grady Booch's Object Technology, James Rumbaugh's Object Modeling Technique (OMT), Ivar Jacobson's Object Oriented Software Engineering, and several other methods. In 1995, a preliminary version of UML was presented to the Object-Oriented Programming, Systems, Languages, and Applications (OOPSLA) conference. By January 1997, UML had gone through several iterations and reviews, incorporating public feedback and revision by the primary authors. In January 1997, UML was presented to the Object Management Group (OMG) in response to its request for a standard modeling technique.

The OMG is a consortium of over 800 software vendors, developers, and organizations that have combined their efforts to develop and propagate uniformity in object-oriented systems. Established in 1989, OMG's mission is to promote the theory and practice of object technology for the development of distributed computing systems. The goal is to provide a common architectural framework for object-oriented applications based on widely available interface specifications. Since January 1997, many revisions to UML have been developed and submitted to the OMG for consideration. The OMG remains the organization that will approve any changes to the standards for OO modeling. Therefore, the UML standards will continue to evolve, but they will remain standardized for the benefit of system developers (and students!). More details about UML and the OMG can be found on the OMG Web site at http://www.omg.org.

OBJECT-ORIENTED REQUIREMENTS

As discussed in Chapter 5, one of the great benefits of using models to document requirements is that it helps you, as the system developer, to think clearly and carefully about the details of the processing and information needs of the stakeholders. As you read this chapter and work the exercises associated with it, you should pay careful attention to how the models require you to search out and understand the user needs. Because of the benefit derived from developing models, object-oriented system requirements are specified and documented through the process of building models.

As shown in Figure 7-1, the system development process starts with the identification of events and things. Events are the business processes that a new system must address, and things are the problem domain objects that are involved in the business process. The problem domain objects are important in both the development of the new system itself as well as the design of the database. New developers frequently ask which to define first, the processes or the classes of objects. In reality, both aspects are closely related and are usually defined together. Experienced developers often move back and forth between identifying classes and business processes, and they make several passes before completing a set of requirements. Do not be discouraged if you find yourself changing your diagrams and models as you work to define requirements.

FIGURE *7-1*

Requirements diagrams for traditional and object-oriented models.

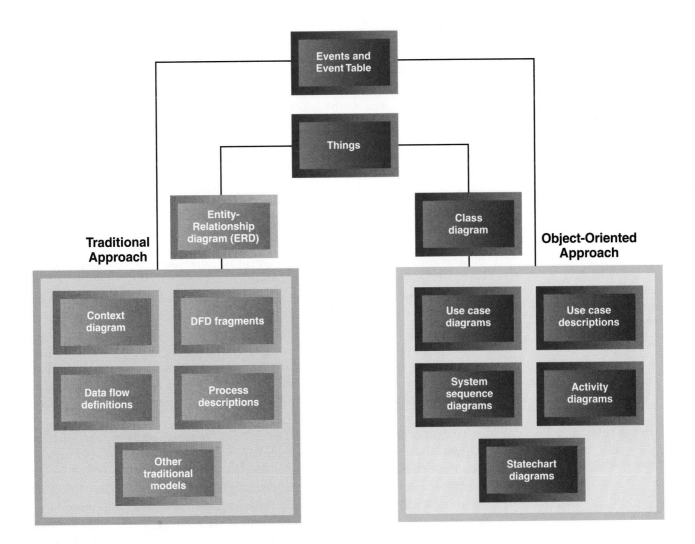

The object-oriented approach requires several interrelated models to create a complete set of specifications. Even though it may seem complex at first to have so many different types of diagrams, as you use them, you will learn to appreciate how they all fit together like a puzzle to produce a complete specification. Essentially, the object-oriented approach "divides and conquers" complex systems. Each model describes a different aspect of the system, so you only focus on one aspect at a time. But you must learn all the different models and the way they fit together. Later, at the end of the chapter, we discuss how all of the diagrams unite to form a complete view of a system's functional requirements. As a beginner with UML you should concentrate now on learning each new model and understanding its role in specifying the total system.

This chapter focuses on a collection of models that can be used to capture system requirements based on use cases with the object-oriented approach. Four models—use case diagrams, use case descriptions, activity diagrams (discussed in Chapter 4), and system sequence diagrams—are used to describe the system use cases from various points of view. System sequence diagrams are a type of interaction diagram. One other type of diagram—the class diagram (discussed in Chapter 5)—is used to define the classes of objects in the problem domain. In many cases, analysts use all the models to completely define the system requirements. However, sometimes only two or three models may be required to specify the requirements accurately.

The purpose of a *use case diagram* is to identify the "uses," or use cases, of the new system—in other words, to identify how the system will be used. The use case diagram can be derived directly from the event table from the column titled "Activity/Use Case." A use case diagram is a convenient way to document the system activities. Sometimes a single, comprehensive diagram is used to identify all use cases for an entire system. At other times, a set of smaller use case diagrams are used.

Each use case must also be described in detail. One way is to write out a narrative description of the steps that the user and the system do together to complete the use case. Each use case can also be defined using diagrams. As you learned in Chapter 4, activity diagrams can be used to describe any business processes done by people in an organization. However, they are also used to describe processes that include both manual and automated system activities, so they can also be used to define a use case.

System sequence diagrams (SSDs) are used to define the inputs and outputs and the sequence of interactions between the user and the system for a use case. They are used in conjunction with detailed descriptions or with activity diagrams. In a sequence diagram, these information flows in and out of a system are called *messages*. The users are identified, and the detailed messages are described.

You learned about classes of objects and the class diagram in Chapter 5. Class diagrams are used to identify the real world "things" that determine the structure of the programming classes (and also the database structure). In the object-oriented view of systems, everything is considered an object. In Chapter 5, we explained that the objects identified belonged to problem domain classes and that these classes consisted of both concrete things, such as customers, and more abstract things, such as orders or airplane flights. Constructing a class diagram helps identify information about the real-world objects that will be part of the new system.

One other diagram identified in Figure 7-1 is the statechart diagram. A *statechart diagram*, or more simply just a statechart, describes the collection of states of each object. Some objects that are identified in the class diagram have status conditions that need to be tracked, and the processes allowed for that object depend on its status. A customer order may have several important status conditions that control the processing of that order—for example, an order that is not complete should not be shipped. A statechart identifies these status conditions and specifies the processes allowed. Statecharts are also used during design to identify various states of the system itself and allowable

use case diagram

a diagram to show the various user roles and how those roles use the system

system sequence diagram

a diagram showing the sequence of messages between an external actor and the system during a use case or scenario

message

the communication between objects within a use case

statechart diagram

a diagram showing the life of an object in states and transitions

events that can be processed. So, statecharts can be considered either as an analysis tool or a design tool. We defer discussion of statecharts until Chapter 12, where we discuss advanced topics of object-oriented analysis and design.

THE SYSTEM ACTIVITIES—A USE CASE/SCENARIO VIEW

The objective of use case analysis is to identify and define all of the business processes that the system must support. Analysts define the use cases at two levels—an overview level and a detailed level. The event table and the use case diagrams provide an overview of all the use cases for a system. Detailed information about each use case is described with a use case description, an activity diagram, and a system sequence diagram, or a combination of these models.

Use Cases and Actors

A use case is an activity the system carries out, usually in response to a request by a user of the system. You can think of a use case as a situation in which the system must accomplish some goal of a user. For example, consider the RMO system. One of the processes that the RMO system must perform is to process new customer orders. So, one use case for this system is *Create new order*. Notice that the focus is on the automated system—on the activities that the *system* must perform to create an order.

Implied in all use cases is a person who uses the system. In UML, that person is called an *actor*. An actor is always outside the automation boundary of the system but may be part of the manual portion of the system. In this respect, an actor is not always the same as the source of the event in the event table. A source of an event is the initiating person, such as a customer, and is always external to the system, including the manual system. In contrast, an actor in use case analysis is the person who is actually interacting with the computer system itself. By defining actors that way—as those who interact with the system—we can more precisely define the exact interactions to which the automated system must respond. This tighter focus helps define the specific requirements of the automated system itself—to refine them as we move from the event table to the use case details. One way to help identify actors at the right level of detail is to assume that actors must have hands. Thinking of actors as having hands encourages us to define actors as those who actually touch the automated system. But remember that some actors are not people. They can also be other systems or other devices that receive services from the system.

Another way to think of an actor is as a role. For example, in the RMO case, the use case *Create new order* might involve an order clerk talking to the customer on the phone. Or, the customer might be the actor if the customer places the order directly, through the Internet. One final way to think about an actor and a use case is that a use case is a goal that the actor wants to achieve. One way to state this goal is to say, "The order clerk uses the system to create a new order." Notice that in this sentence both the actor—the *order clerk*—and the use case—*Create a new order*—are identified. In fact, stating the use cases in sentence form is a good technique to understand the relationship between use cases and actors.

The Use Case Diagram

Figure 7-2 shows how a use case is documented in a use case diagram. A simple stick figure is used to represent an actor. The stick figure is given a name that characterizes the role the actor is playing. The use case itself is symbolized by an oval with the name of the use case inside. The connecting lines between actors and use cases indicate which actors utilize which use cases. Although hands are not part of the standard UML notation, the actor in this figure is drawn with hands to help you remember that this actor must have direct access to the automated system.

FIGURE *7-2*

A simple use case with an actor.

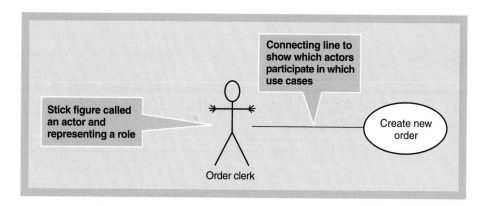

A use case diagram is a graphical model that summarizes the information about the actors and use cases. To do use case analysis, a system developer looks at the system as a whole and tries to identify all of its major uses.

Automation Boundary and Organization

Figure 7-3 expands the use case given in Figure 7-2 to include additional use cases and additional actors. In this instance, both the order clerk and the customer are allowed to access the system directly. As indicated by the relationship lines, each actor can use every use case. A boundary line is also drawn around the entire set of use cases. This boundary is the automation boundary. It denotes the boundary between the environment, where the actors reside, and the internal components of the computer system.

FIGURE *7-3*

A use case diagram of the Order-entry subsystem for RMO, showing a system boundary.

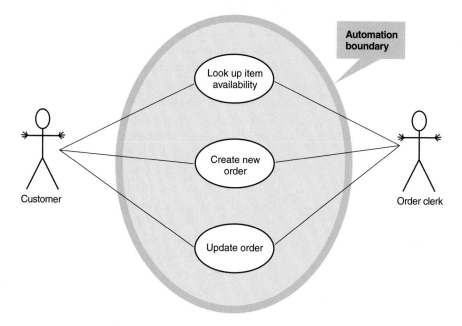

There are various ways to organize the use cases to depict different points of view. One is to organize the use cases by subsystem. Figure 7-3 is a partial use case diagram for one such subsystem, namely, the Order-entry subsystem. Figure 7-4 expands that figure to show all the subsystems for the customer support system of RMO with some of the associated use cases. Another way is to include all use cases that involve a specific actor. Figure 7-5 shows all the use cases involving the Customer actor. This diagram is useful for showing all of the activities that are accessible through the Internet. An analyst chooses to draw use case diagrams based on the needs of the project team. If the plan is to meet with the marketing department managers to discuss all use cases involving direct customer interaction, then the use case diagram in Figure 7-5 will be very useful in the meeting.

FIGURE *7-4*

A use case diagram of the customer
support system (by subsystem).

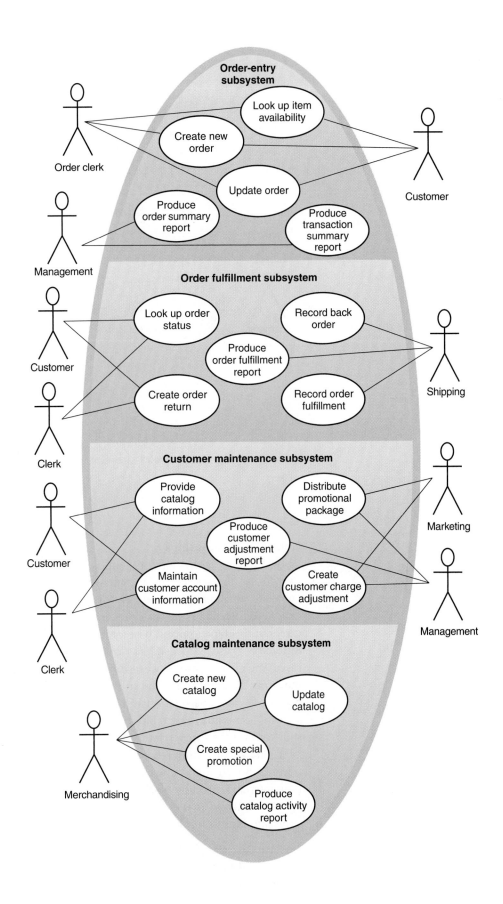

FIGURE *7-5*

All use cases involving Customer.

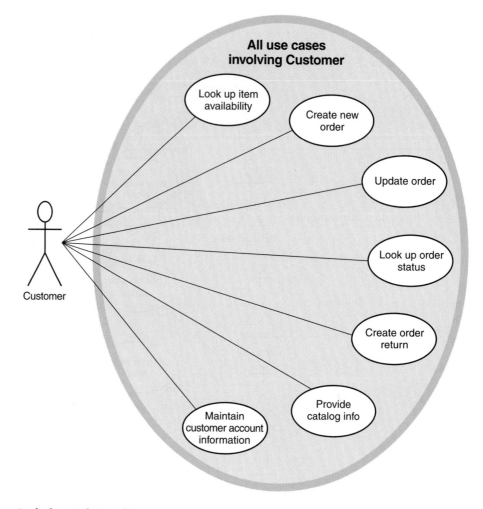

All use cases
involving Customer

«Includes» Relationships

Frequently during the development of a use case diagram, it is reasonable for one use case to use the services of a common subroutine. For example, two of the Order-entry subsystem use cases are *Create new order* and *Update order*. Each of these use cases may need to validate the customer account. A common subroutine may be defined to carry out this function, and it becomes an additional use case. Figure 7-6 shows the additional use case, named *Validate customer account*, which is used by both the other use cases. The relationship between these use cases is denoted by the connecting line with the arrow. The direction of the arrow indicates which use case is included as a part of the major use case. The relationship is read *Create new order «includes» Validate customer account.* Sometimes this relationship is referred to as the «includes» relationship, or sometimes as the «uses» relationship.

Figure 7-6 also shows that *Look up item availability* can be part of an «includes» relationship. So, an analyst can define two types of «includes» use cases: one that is a common internal subroutine, such as *Validate customer account,* and is not directly referenced by an external actor and one that is directly referenced by external actors. *Look up item availability* is an example of the latter.

The Use Case Diagram Compared with the Event Table

As indicated earlier, the event table and the use case diagram contain much of the same information. One of the questions you may be asking yourself is, "If they are so similar, do I need to develop both models?" In fact, for any given project, you may not develop both models. Some analysts prefer to start by listing use cases rather than events, and they move directly to the use case diagram (as discussed in the following section). The

FIGURE *7-6*

An example of the Order-entry
subsystem with «includes» use cases.

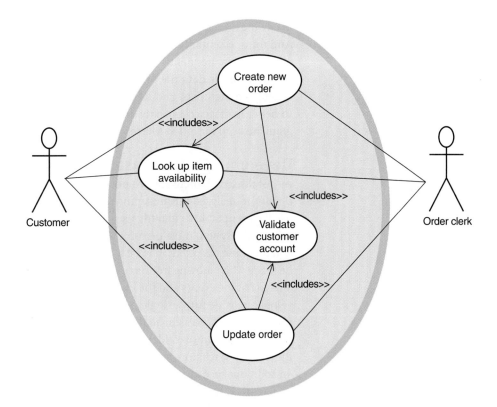

event table can be used as the foundation for either traditional structured development or object-oriented development. As a well-educated and skilled professional, you should be proficient with both techniques.

However, some differences do exist between the two models. First, the point of view of each is slightly different. An event table always focuses on the business processes. It does so by identifying business events and external, initiating sources for those events. These external sources are the ones that cause the business event to be initiated, and they can be somewhat removed from the automated system. On the other hand, a use case diagram emphasizes the automated system. Because it is concerned only with the automated system, the actors actually have contact with the automated system and may not necessarily be the original initiators of the business event.

Another difference between the two models can be seen when identifying temporal and state events. Since use cases are usually initiated by external actors, temporal and state events are often overlooked if the analyst does not carefully identify all events. This is a deficiency of use case modeling if use cases are defined too narrowly. As discussed in Chapter 14, on-line system menus typically include menu options representing each temporal event from the event table so that such an event may be triggered by a user as well as being a purely temporal event. Therefore, we recommend including a use case for each temporal and state event to assure these requirements are not overlooked.

It is important to remember that the analyst will be completing the event table and the use case diagrams concurrently. The analyst will also continually refine and update events and use cases. The refinements that occur usually involve adjustments to balance the scope of each use case. For example, during the development of the event table two events *Add new customer* and *Update customer information* may have been identified. From the system's point of view, the use case for both business events is almost the same because they both involve updating the customer file. A single use case could be defined to support both business events. The use case could be named *Maintain customer account information*. It is common to define a single use case to support multiple business events if the following three criteria are met: First, essentially the same processing is occurring inside the automated system. Second, essentially the same information is being updated. And third, essentially the same information is input and output from the events. These conditions are frequently met for business events that perform basic file maintenance on a single, simple data file or table. Sometimes a single event triggers very complex processing requirements, and it makes sense to divide the system activity into two use cases to better manage complexity. In all of these situations, the event table and use case diagram are both modified to keep the models synchronized.

Developing a Use Case Diagram

As just discussed, there are two possible starting points to develop a use case diagram. If a developer analyzed business processes and constructed an event table, then he or she will use the event table to identify use cases. Each event is analyzed in detail to determine the process the system must perform to support that event, the actors who initiate the event, and other use cases that may be invoked because of the event. Refinements should always be made when moving from one model to a more detailed model, so a careful review and analysis of the events and the event table are important. After additional analysis, the developer may identify a single event as a use case, combine several events to form a single use case if the processing required seems similar, or identify multiple use cases if the processing seems complex. Identification of multiple use cases usually occurs when they have the «includes» relationship and two use cases are factored out of one large use case or when an additional use case is defined based on a common subroutine, as discussed previously.

Figure 7-4, which showed the customer support subsystems, was developed using this approach. You will note that most of the use cases defined in the figure come directly from the event table given in Figure 5-15. In fact, the name of the use cases in Figure 7-4 come from the description provided in the Activity/Use Case column of the event table. There are a couple of exceptions to this pattern. Since temporal events normally also can be initiated manually, we have used the option of identifying an external actor for each temporal use case. The other exception is with event number 13, *Customer updates account information*. In this instance, the use case definition is expanded to include all scenarios having to do with maintaining customer information. In this instance, the use case is titled *Maintain customer account information* to denote that it will include additions, updates, and deletions. These are examples of where the use case diagram could initiate a refining update to the event table.

If an event table has not been created, the other starting point to develop a use case diagram is to begin by identifying the actors and the functions they perform with the system. To do so, you must remember two preconditions. First, you must make the system boundary an automated system so that the actors you identify actually contact the system—that is, have hands. Second, you must assume perfect technology. Be sure that the use cases are based on business events and not technical activities like logging on to the system. Given those preconditions, you can develop the use case diagram in two steps, which are done in iteration.

1. Identify the actors of the system. Note that actors are actually roles played by users. Instead of listing the actors as Bob, Mary, or Mr. Hendricks, you should identify the specific roles that these people play. Remember that the same person may play various roles as he or she uses the system. Those roles become such titles as order clerk, department manager, auditor, and so forth. It is important to be comprehensive and to identify every possible role that will use the system. Other systems may also be actors of a system.

2. Once the actor roles have been identified, the next step is to develop the list of goals those roles have in the use of the automated system. A goal is a task performed by an actor to accomplish some business function that adds value to the business. Goals are such tasks as "process a sale," "accept a return," or "ship an order." Goals

are units of work that can be identified and described. At the completion of the goal, the data of the system should be stable for some time.

These two steps are performed in brainstorming sessions with project team members and users. There is no magical way to find or identify use cases. Even though the focus is on the automated system, a thorough analysis of the business processes is required to understand all ways that actors will need to use the system.

Another important technique that you should use when identifying events for the event table or when developing the use case diagram directly is called *CRUD analysis*, which compares the identified use cases with the domain model class diagram. Analysts do CRUD analysis after making an initial use case diagram to double-check their work. Recall that CRUD stands for Create, Read (or Report), Update, and Delete. CRUD analysis was first introduced in Chapter 6, and it is a technique associated with Information Engineering (IE). A CRUD analysis requires that every class in the class diagram have sufficient use cases to support creating new object instances, reading or reporting on those objects, updating those objects, and in many cases deleting object instances. The use case may not be named *create or update*, but the underlying process should add a new instance or update an existing instance. For example, a use case named *Record payment* does not explicitly indicate that a new payment object is created, but a detailed description of the use case will indicate that a new payment is created. The use case *Create new order* might create OrderItem objects and update InventoryItem objects. In other cases, many of the use cases are named beginning with the word *maintain* to cover routine additions, updates, reads, and deletions.

To do a CRUD analysis, simply look at every class in the domain model class diagram and be sure that there are use cases somewhere in the use case diagram to support all the CRUD functions that are appropriate for the application. Keep in mind, though, that with integrated systems, one system might be responsible for creating objects and another system might only update them. The CRUD analysis approach provides a cross-check, not a final solution, and it also provides an opportunity to confirm important system integration requirements that otherwise might not be obvious.

Use Case Detailed Descriptions

As indicated earlier, creating a use case diagram is only one part of use case analysis. The use case diagram helps identify the various processes that are performed by the users and that must be supported by the new system. Careful system development requires that we go to a much lower level of detail. To create a comprehensive, robust system that truly meets users' needs, we must understand all of the detailed steps. Internally, a use case includes a whole sequence of steps to complete a business process. For example, frequently several variations of the business steps exist within a single use case. The use case *Create new order* will have a separate flow of activities depending on which actor invokes the use case. The processes for an order clerk creating a new order over the telephone may be quite different from the processes for a customer creating an order over the Internet. Each flow of activities is a valid sequence for the *Create new order* use case. These different flows of activities are called *scenarios*, or sometimes *use case instances*. Thus, a scenario is a unique set of internal activities within a use case and represents a unique path through the use case.

A use case can be elaborated using various diagrams and descriptions. One of the more useful diagramming techniques for documenting a use case is an activity diagram, discussed later. Activity diagrams were first introduced in Chapter 4. Many analysts prefer to write narrative descriptions of use cases, which can be written at various levels of detail depending on the need. Typically, use case descriptions are written at three separate levels of detail: brief description, intermediate description, and fully developed

scenario, or use case instance

a particular sequence of steps within a use case; a use case may have several different scenarios

description. Written descriptions and activity diagrams can be used in any combination, depending on an analyst's needs.

Brief Description

A brief description can be used for very simple use cases, especially when the system to be developed is also a small, well-understood application. A simple use case would normally have a single scenario and very few, if any, exception conditions. A brief description used in conjunction with an activity diagram provides a solid description for a simple use case. Figure 7-7 provides a brief description of the *Create new order* use case. Generally, a use case such as *Create new order* is complex enough that either an intermediate or fully developed description is developed. We illustrate those descriptions next.

FIGURE *7-7*

Brief description of *Create new order* use case.

Create new order description

When the customer calls to order, the order clerk and system verify customer information, create a new order, add items to the order, verify payment, create the order transaction, and finalize the order.

Intermediate Description

The intermediate-level use case description expands the brief description to include the internal flow of activities for the use case. If there are multiple scenarios, then each flow of activities is described individually. Exception conditions can be documented, if they are needed. Figures 7-8 and 7-9 show intermediate descriptions that document the two scenarios of *Order clerk creates telephone order* and *Customer creates Web order*. These two scenarios were identified earlier as separate work flows for the *Create new order* use case. Notice that each describes what the user and the system require to carry out the processing for the scenario. Exception conditions are also listed. Each step is identified with a number to make it easier to read. In many ways this description is a version of structured English, which can include sequence, decision, and repetition blocks.

FIGURE *7-8*

Intermediate description of the telephone order scenario for *Create new* order.

Flow of activities for scenario of *Order Clerk creates telephone order*

Main Flow:

1. Customer calls RMO and gets order clerk.
2. Order clerk verfies customer information. If a new customer, invoke *Maintain customer account information* use case to add a new customer.
3. Clerk initiates the creation of a new order.
4. Customer requests an item be added to the order.
5. Clerk verifies the item and adds it to the order.
6. Repeat steps 4 and 5 until all items are added to the order.
7. Customer indicates end of order; clerk enters end of order; system computes totals.
8. Customer submits payment; clerk enters amount; system verifies payment.
9. System finalizes order.

Exception Conditions:

1. If an item is not in stock, then customer can
   a. choose not to purchase item, or
   b request item be added as a back-ordered item.
2. If customer payment is rejected due to bad-credit verification, then
   a. order is cancelled, or
   b. order is put on hold until check is received.

FIGURE *7-9*

Intermediate description of the Web order scenario for *Create new order*.

Fully Developed Description

The fully developed description is the most formal method for documenting a use case. Even though it takes a little more work to define all the components at this level, it is the preferred method of describing the internal flow of activities for a use case. One of the major difficulties that software developers have is that developers often struggle to obtain a deep understanding of the users' needs. But if you create a fully developed use case description, you increase the probability that you thoroughly understand the business processes and the ways the system must support them. Figure 7-10 is an example of a fully developed use case description of the telephone order scenario of the *Create new order* use case, and Figure 7-11 shows the Web order scenario for the same use case.

Figures 7-10 and 7-11 can also serve as a standard template for documenting a fully developed description for other scenarios and use cases. The first and second compartments are used to identify the use cases and scenarios within use cases, if needed, that are being documented. In larger or more formal projects, a unique identifier can also be added for the use case, with an extension identifying the particular scenario. Sometimes the name of the system developer who produced the form is also added.

The third compartment identifies the trigger that initiates the use case. This trigger is the same as that described in the event table, as explained in Chapter 5. There are two points of view from which to describe a trigger. One way is to identify the business event that initiates the process. For example, in Figure 7-10 the *Create new order* process is started by the customer telephoning RMO. This point of view is centered on the outside world. A second point of view of a trigger is as an activity that causes the automated system to first recognize that the use case has begun. From the second viewpoint, the trigger could be described as "The clerk enters a request for a new order." Since at this point in the development of the new system the objective is to understand the business need, the better viewpoint is the first—identifying the external event that initiates the entire process.

The fourth compartment is a brief description of the use case or scenario. Analysts may just duplicate the brief description they constructed earlier here. The fifth compartment identifies the actor or actors. This compartment duplicates some of the information that is contained in the use case diagram itself. The sixth compartment identifies

Use Case Name:	Create new order	
Scenario:	Create new telephone order	
Triggering Event:	Customer telephones RMO to purchase items from the catalog.	
Brief Description:	When customer calls to order, the order clerk and system verify customer information, create a new order, add items to the order, verify payment, create the order transaction, and finalize the order.	
Actors:	Telephone sales clerk.	
Related Use Cases:	Includes: *Check item availability*.	
Stakeholders:	Sales department: to provide primary definition. Shipping department: to verify information content is adequate for fulfillment. Marketing department: to collect customer statistics for studies of buying patterns.	
Preconditions:	Customer must exist. Catalog, Products, and Inventory items must exist for requested items.	
Postconditions:	Order and order line items must be created. Order transaction must be created for the order payment. Inventory items must have the quantity on hand updated. The order must be related (associated) to a customer.	
Flow of Events:	**Actor**	**System**
	1. Sales clerk answers telephone and connects to a customer.	
	2. Clerk verifies customer information.	
	3. Clerk initiates the creation of a new order.	3.1 Create a new order.
	4. Customer requests an item be added to the order.	
	5. Clerk verifies the item (*Check item availability* use case).	5.1 Display item information.
	6. Clerk adds item to the order.	6.1 Add create an order item.
	7. Repeat steps 4, 5, and 6 until all items are added to the order.	
	8. Customer indicates end of order; clerk enters end of order.	8.1 Complete order. 8.2 Compute totals.
	9. Customer submits payment; clerk enters amount.	9.1 Verify payment. 9.2 Create order transaction. 9.3 Finalize order.

Exception Conditions:	2.1 If customer does not exist, then the clerk pauses this use case and invokes *Maintain customer information* use case.
	
	2.2 If customer has a credit hold, then clerk transfers the customer to a customer service representative.
	
	4.1 If an item is not in stock, then customer can
	    a. choose not to purchase item, or
	    b. request item be added as a back-ordered item.
	
	9.1 If customer payment is rejected due to bad-credit verification, then
	    a. order is cancelled, or
	    b. order is put on hold until check is received.	

FIGURE *7-10*

Fully developed description of the telephone order scenario for *Create new* order.

Use Case Name:	Create new order	
Scenario:	Create new Web order	
Triggering Event:	Customer logs on to the RMO Web site and requests to purchase an item.	
Brief Description:	Customer logs in and requests the new order form. The customer searches the catalog on-line and purchases items from the catalog. The system adds the purchased items to the order. At the end the customer enters credit-card information.	
Actors:	Customer.	
Related Use Cases:	Includes: *Register new customer, Check item availability.*	
Stakeholders:	Sales department: to provide primary definition. Shipping department: to verify information content is adequate for fulfillment. Marketing department: to collect customer statistics for studies of buying patterns.	
Preconditions:	Catalog, Products, and Inventory items must exist for requested items.	
Postconditions:	Order and order line items must be created. Order transaction must be created for the order payment. Inventory items must have the quantity on hand updated. The order must be related (associated) to a customer.	
Flow of Events:	**Actor**	**System**
	1. Customer connects to the RMO home page and then links to the order page. 2. If this is a new customer, then customer links to the customer account page and adds the appropriate information to establish a customer account. 2a. If existing customer, customer logs in. 3. Customer searches catalog. 4. When customer finds the correct item, he/she requests it be added to the order. 5. Repeat steps 3 and 4. 6. Customer requests end of order. 7. Customer makes any changes. 8. Customer requests payment screen. 9. Customer enters payment information.	2.1 Create new customer record. 2a.1. Validate customer account. 2.2. Create a new shopping cart order; display order form with catalog frame. 3.1 Display products from catalog based on searches and selections. 4.1 Add item to shopping cart order. 6.1 Display shopping cart items, with totals and amounts due; edit and submit buttons. 8.1 Display payment details screen. 9.1 Accept payment, finalize order, send confirmation e-mail.

Exception Conditions:	4.1 If an item is not in stock, then customer can
	
	   a. choose not to purchase item, or
	
	   b. request item be added as a back-ordered item.
	
	8.1 If customer payment is rejected due to bad-credit verification, then
	
	   a. order is cancelled, or
	
	   b. order is put on hold until check is received.	

FIGURE *7-11*

Fully developed description of the Web order scenario for *Create new* order.

other use cases and the way they are related to this use case, for example, the «includes» relationship we discussed earlier. Other, more advanced relationships can also be identified. The important point, however, is to add a cross-reference to other use cases to understand all aspects of the users' requirements.

The stakeholders compartment identifies interested parties—other than specific actors. They may be users who do not actually invoke the use case but who have an interest in results produced from the use case. For example, in Figures 7-10 and 7-11 no one in the marketing department actually creates new orders, but they do perform statistical analysis of the orders that were entered. So, marketers have an interest in the data that are captured and stored from the *Create new order* use case. Considering all stakeholders is an important step for system developers so that they ensure they have understood all requirements.

The next two compartments provide critical information about the state of the system before and after the use case executes, called *preconditions* and *postconditions*. *Preconditions* state what conditions must be true before a use case begins. In other words, it identifies what the state of the system must be for the use case to begin, including what objects must already exist, what information must be available, and even what the condition of the actor is prior to beginning the use case.

A *postcondition* identifies what must be true upon completion of the use case. The same items that are used to describe the precondition should be included in the statement of the postcondition. For example, during the processing of a use case that updates various financial accounts, some accounts will be out of balance. So, a postcondition for that use case would be that the updates should be complete for all accounts and that they should all be in balance.

The final two compartments in the template describe the detailed flow of activities of the use case. In this instance we have shown a two-column version, identifying the steps performed by the actor and the responses required by the system. The item numbering helps identify the sequence of the steps. Some developers prefer the one-column version, as shown at the intermediate level. Alternative activities and exception conditions are described in the final compartment. The numbering of exception conditions also helps tie the exceptions with specific steps in the use case description.

Activity Diagram Description

The other way to document a use case scenario is with an activity diagram. In Chapter 4, you learned about activity diagrams as a form of workflow diagram. You learned that an activity diagram is an easily understood diagram to document the workflows of the business processes. Activity diagrams are a standard UML diagram. In this instance, activity diagrams are an effective technique to document the flow of activities for each use case scenario.

Figures 7-12 and 7-13 are the activity diagrams that document the same two scenarios as shown in Figures 7-10 and 7-11. In Figure 7-12 the customer interacts with the order clerk, who in turn uses the system. Since the purpose of a use case is to specify the interaction of an actor (with hands) with the system, the figure includes swimlanes for the Order Clerk and the Computer System. However, to aid in understanding the total flow of activities for the scenario, the Customer—the one who initiates the steps—is also included. Note that the Customer swimlane is an optional addition in Figure 7-12 that simply aids in understanding the total workflow. In Figure 7-13 the customer is the actor who interacts with the computer system, so only two swimlanes are required to describe the steps in the scenario.

An activity diagram can be used to support any level of use case descriptions. As you can see, activity diagrams are very similar to the two-column description in the fully developed description. The benefit of creating an activity diagram is that it is more visual and can help both the user and the developer work together to fully document the use case.

---

precondition

a set of criteria that must be true prior to the initiation of a use case

postcondition

a set of criteria that must be true upon completion of the execution of a use case

---

FIGURE *7-12*

Activity diagram of the
telephone order scenario.

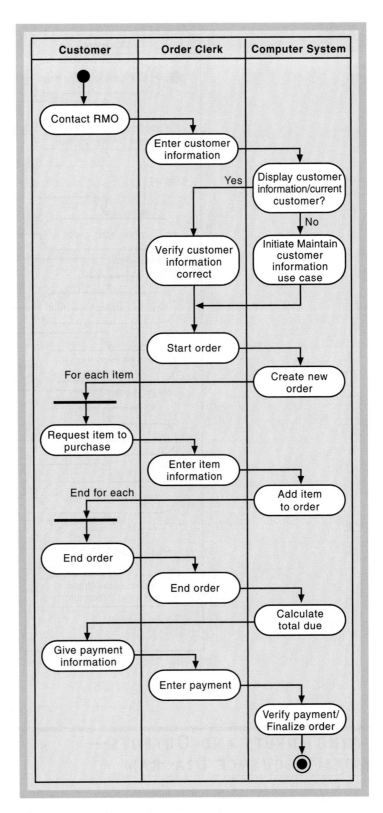

As a quick glance at Figures 7-12 and 7-13 demonstrates, the two scenarios of the
Create new order use case are quite different. Even though the scenarios carry out the
same basic function, the set of screens and options on the screens may be quite different
for each. Activity diagrams are also helpful in developing system sequence diagrams, as
explained in the next section.

FIGURE *7-13*

Activity diagram of the
Web order scenario.

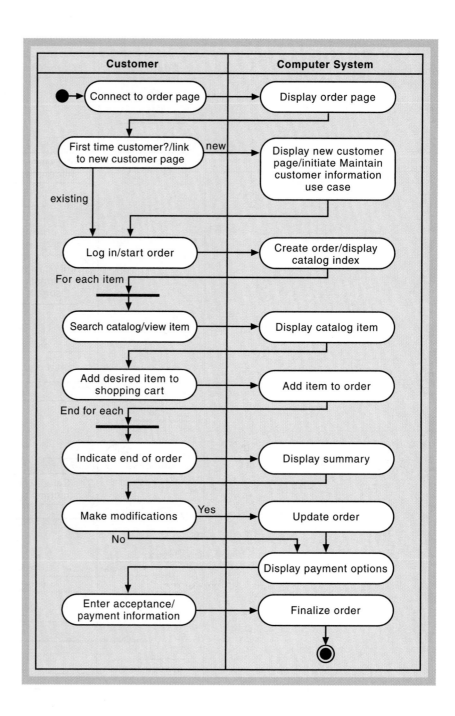

IDENTIFYING INPUTS AND OUTPUTS— THE SYSTEM SEQUENCE DIAGRAM

interaction diagram

either a collaboration diagram or a sequence diagram that shows the interactions between objects

In the object-oriented approach, the flow of information is achieved through sending messages either to and from actors or back and forth between internal objects. A system sequence diagram (SSD) is used to describe this flow of information into and out of the automated system. So, an SSD documents the inputs and the outputs and identifies the interaction between actors and the system. An SSD is a type of *interaction diagram*. In the following sections, and in industry practice, we often use the terms *interaction* and *message* interchangeably.

FIGURE *7-14*

Sample system sequence
diagram (SSD).

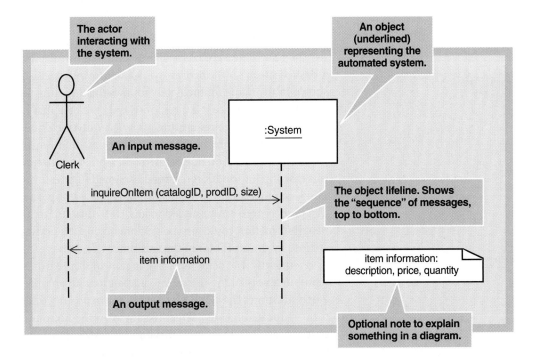

SSD Notation

Figure 7-14 shows a generic SSD. As with a use case diagram, the stick figure represents an actor—a person (or role) that interacts with the system. In a use case diagram, the actor "uses" the system, but the emphasis in an SSD is on how the actor "interacts" with the system by entering input data and receiving output data. The idea is the same with both diagrams; the level of detail is different.

The box labeled :System is an object that represents the entire automated system. In SSDs, and all interaction diagrams, instead of using class notation, analysts use object notation. Object notation indicates that the box refers to an individual object and not the class of all similar objects. The notation is simply a rectangle with the name of the object underlined. The colon before the underlined class name is a frequently used, but optional, part of the object notation. In an interaction diagram, the messages are sent and received by individual objects, not by a class. In an SSD, the only object included is one representing the entire system.

lifeline, or object lifeline

the vertical line under an object on a sequence diagram to show the passage of time for the object

Underneath the actor and the :System are vertical dashed lines called *lifelines*. A *lifeline*, or *object lifeline*, is simply the extension of that object, either actor or object, throughout the duration of the SSD. The arrows between the lifelines represent the messages that are sent or received by the actor or the system. Each arrow has an origin and a destination. The origin of the message is the actor or object that sends it, as indicated by the lifeline at the arrow's tail. Similarly, the destination actor or object of a message is indicated by the lifeline that is touched by the arrowhead. The purpose of lifelines is to indicate the sequence of the messages sent and received by the actor and object. The sequence of messages is read from top to bottom in the diagram.

A message is labeled to describe both the message's purpose and any input data being sent. The syntax of the message label has several options; the simplest forms are shown in Figure 7-14. Remember that the arrows are used to represent both a message and input data. But what is meant by the term *message* here? In a sequence diagram, a message is considered to be an action that is invoked on the destination object, much like a command. Notice in Figure 7-14 that the input message is called inquireOnItem. The clerk is sending a request, or a message to the system, to find an item. The input data that is sent with the message is contained within the parentheses, and in this case it is data to identify the particular item. The syntax is simply the name of the message followed by the input parameters in parentheses. This form of syntax is attached to a solid arrow.

The return message has a slightly different format and meaning. Notice the arrow is a dashed arrow. A dashed arrow is used to indicate a response or an answer and, as shown in the figure, immediately follows the initiating message. The format of the label is also different. Since it is a response, only the data that is sent on the response is noted. There is no message requesting a service, only the data being returned. In this case, a valid response might be a list of all the information returned—such as description, price, and quantity of an item. However, an abbreviated version is also satisfactory. In this case the information returned is named item information. Additional documentation is required to show the details. In Figure 7-14 this additional information is shown as a note. A note can be added to any UML diagram to add explanations. The details of item information could also be documented in supporting narratives or even simply referenced by the attributes in the Customer class.

Frequently, the same message is sent multiple times. For example, when entering items on an order, the message to add an item to an order is sent multiple times. Figure 7-15a illustrates the notation to show this repeating operation. The message and its return are located inside a larger rectangle. At the bottom of the large rectangle is the descriptive text to control the behavior of the messages within it. The asterisk (*), indicates that the messages in the box repeat. The asterisk is the UML notation for repeat or loop on a message. Recall that the asterisk (*) on a class diagram means multiplicity, or many. A benefit of UML is that the same symbol has the same meaning from diagram to diagram. An asterisk indicates to repeat something many times or is associated with many instances. The square brackets and text inside them is called a *true/false condition* for the messages in the rectangle. It is interpreted to mean that the messages in the rectangle repeat as long as the true/false condition evaluates to true. Square brackets mean the same thing in other diagrams, such as statecharts, which are discussed in Chapter 12.

Figure 7-15b shows an alternative, abbreviated notation for several of the concepts just discussed. First, a message and the returned data can be shown in one step. Note that the return data is identified as a return value on the left side of an assignment operator—the := sign. This alternative simply shows a value that is returned. Second, the true/false condition is placed on the message itself. Note that in this example, the true/false condition is used for the control of the loop. True/false conditions are also used to evaluate any type of test that determines whether a message is sent. For example, [credit card payment] might be used to control whether a message is sent to the system to verify a credit-card number. Finally, the asterisk is also placed on the message itself. So, for simple repeating messages, the alternate notation is shorter. However, if several messages are included within the repeat or there are multiple messages, each with its own true/false condition, then the more detailed notation is more explicit and precise.

The complete notation for a message is the following:

* [true/false condition] return-value := message-name (parameter-list)

Any part of the message may be omitted. In brief, the notation components are the following:

- Asterisk (*) indicates repeat or looping of the message.
- Brackets [ ] indicate a true/false condition. It is a test for that message only. If it evaluates to true, the message is sent. If it evaluates to false, the message is not sent.
- Message-name is the description of the requested service. It is omitted on dashed-line return messages, which only show the return data parameters.
- Parameter-list (with parentheses on initiating messages and without parentheses on return messages) shows the data that is passed with the message.

FIGURE *7-15*

Repeating message.
(a) Detailed notation.
(b) Alternate notation.

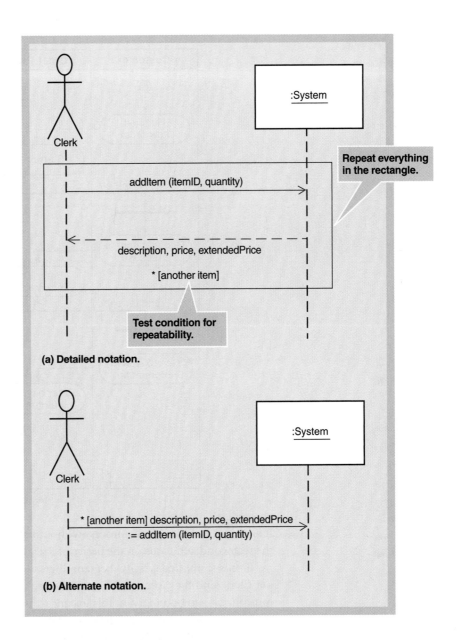

(a) Detailed notation.

(b) Alternate notation.

■ Return-value on the same line as the message (requires :=) is used to describe data being returned from the destination object to the source object in response to the message.

Developing a System Sequence Diagram

An SSD is normally used in conjunction with the use case descriptions to help document the details of a single use case or scenario within a use case. To develop an SSD, you will need to have a detailed description of the use case, either in the fully developed form as shown in Figures 7-10 or 7-11 or as activity diagrams as shown in Figures 7-12 and 7-13. These two models identify the series of activities within a use case, but they do not explicitly identify the inputs and outputs. An SSD will provide this explicit identification of inputs and outputs. One advantage of using activity diagrams is that it is easy to identify when an input or output occurs. Inputs and outputs occur whenever an arrow in an activity diagram goes from an external actor to the computer system. Figure 7-16 is a simplified version of Figure 7-12 for the telephone order scenario of the RMO *Create new order* use case. Obviously, the simplified version has many things missing, but

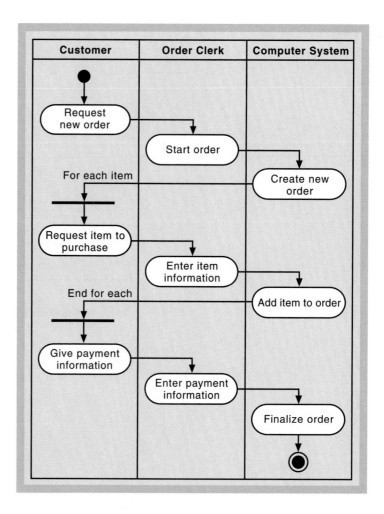

it allows us to focus on the process without having to consider all of the complexity of the real world and focus on the basics of SSD development.

In this simplified activity diagram, there are three swimlanes: the Customer, the Order Clerk, and the Computer System. Before beginning the SSD, you must first determine the system boundary. In this instance, the system boundary coincides with the vertical line between the Order Clerk swimlane and the Computer System swimlane. Since the purpose of the SSD is to describe the inputs and outputs to the automated computer system, in the SSD only the Order Clerk and the Computer System will be included. It is not wrong to include both actors in the SSD, but it is more focused to show only the system and the actor who sends the inputs and receives the outputs.

The development of an SSD based on a activity diagram can be divided into four steps:

■ **Identify the input messages.** In Figure 7-16 there are three locations with a workflow arrow crossing the boundary line between the clerk and the system. At each location that the workflow crosses the automation boundary, input data is required; therefore, a message is needed.

■ **Describe the message from the external actor to the system using the message notation described earlier.** In most cases, you will need a message name that describes the service requested from the system and the input parameters being passed. Figure 7-17, the SSD for the *Create new order* use case, illustrates the three messages. Notice that the names of the messages reflect the services that the actor is requesting of the

FIGURE *7-17*

An SSD of the simplified telephone
order scenario for the *Create new
order* use case.

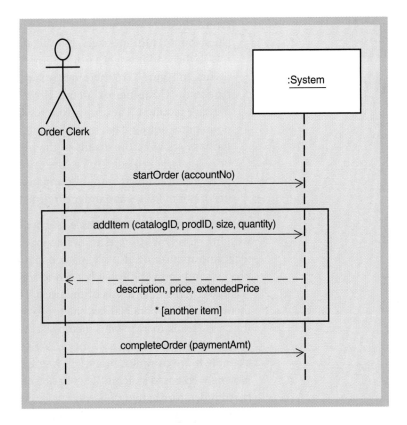

system, startOrder, addItem, and completeOrder. Other names could also have been
used. For example, instead of addItem, the name could be enterItemInformation.

The other information required is the parameter list for each message. Determining exactly which data items must be passed in is more difficult. In fact, developers frequently find that determining the data parameters requires several iterations before a
correct, complete list is obtained. The important principle for identifying data parameters is to base it on the class diagram. In other words, the appropriate attributes from
the classes are listed as parameters. Looking at the attributes, along with an understanding of what the system needs to do will help you find the right attributes.

In the example of the first message, startOrder, the precondition for this use
case states that a customer should exist. A postcondition is that the order must be
connected to the customer. So, for this simplified version of the use case, the first
message passes in the accountNo, which is the identifier in the customer class. (Refer to the RMO class diagram shown later in Figure 7-20 to find the attributes for
each class.) Other than the accountNo, no other parameters are needed for the system to locate the existing customer details.

In the second message, addItem, parameters are needed to identify the item
from the catalog and the quantity to be purchased. The parameters catalogID,
prodID, and size are used to describe the inventory item that will be added to the
order. The quantity field, of course, simply identifies how many.

The third message, based on the activity diagram, enters the payment amount.
This parameter corresponds to the amount attribute in the OrderTransaction class.

- **Identify and add any special conditions on the input messages, including iteration and true/false conditions.** In this instance, we have the iteration box and the
true/false condition associated with it shown in square brackets.
- **Identify and add the output return messages.** Remember, there are two options to
show return information, either as a return value on the message itself, or a separate

return message with a dashed-line arrow. The activity diagram can provide some clues about return messages, but there is no standard rule that when a transition arrow in the workflow goes from the system to an external actor that an output always occurs. In Figure 7-16 there are two arrows going from the Computer System swimlane to the Customer swimlane. However, in Figure 7-17, there is only one output message required. The arrow from the Create new order activity in Figure 7-16 does not require output data. In this instance the only output identified is on the middle message showing the details of the item added to the order—the description, the price, and the extended price (the price times quantity). The other messages could possibly have shown output information such as customer name and address for the first input message, and order confirmation for the third one.

Remember that the objective is discovery and understanding, so you should be working closely with users to define exactly how the workflow proceeds and exactly what information needs to be passed in and is provided as output. This is an iterative process, and you will probably need to refine these diagrams several times before they accurately reflect the needs of the users. During Rocky Mountain Outfitters' development project, Barbara Halifax, the project manager, has reviewed many diagrams with the users (see Barbara's status memo).

Let's now develop an SSD for the Web scenario of *Create new order*. Not only is this example more complex, but it will highlight how to develop the requirements for deploying Web-based systems. Refer back to Figure 7-13 for the activity diagram of a Web-based order. Notice that this workflow is fairly complex.

Figure 7-18 is the completed SSD for the Web-based scenario. In Figure 7-13, the workflow crosses the automated system boundary from the Customer to the Computer System eight times, some of which are optional flows. The first message, with its response message, begins the use case by requesting the new order page (requestNewOrder). The system does not need input data to perform the processes requested by these two messages, so no input parameters are required. The next input message is a request for the new customer page (newCustomerPage). On this message, there is a true/false condition to test whether this is a new customer. Thus, the message only fires if the new customer condition evaluates to true. Since the objective of a sequence diagram is only to show the messages and not to show processing logic, there is no message to show the branching out to another use case; a simple note is added to remind the developers about that jump.

The third message just allows the user to actually start an order (beginOrder). The message shows that the customer account number is an input parameter. When the user interface is actually developed, this information may already be in the system since it may be on the screen from adding a new customer. However, by showing it as an input parameter, the developers will know that it has to be available, either from the user or captured from another page.

The next process is one of adding items to the order. The activity diagram in Figure 7-13 shows a loop to add items. That is captured by the iteration box. However, one of the activities in the workflow says *Search catalog/view item*. Even though a loop is not explicitly shown, a search normally implies a loop of some type. So, on the input message to view a product on Figure 7-18, an asterisk has been added for iteration. The two asterisks, one inside an iteration box, creates a nested loop condition. Note that on these two messages, the return-value method is used to return data. The remaining messages and responses follow the activity diagram.

These first sections of the chapter have explained the set of models that are used in object-oriented development to specify the processing aspects of the new system. The use case diagram provides an overview of all of the events that must be supported. The scenario descriptions, as provided by written narratives or activity diagrams, give the de-

April 14, 2005

To: John MacMurty

From: Barbara Halifax, Project Manager

RE: Customer Support System status

John, here is the status report for our work over the last two weeks and the plans for work over the next period. I have also attached a copy of our schedule, highlighting which tasks are completed. You will note that we are nearly on schedule. A couple of the tasks took longer than expected due to delays in getting final decisions from the sales department.

Completed during the last period (two weeks)

We made progress in the last two weeks with the development of fully developed use case descriptions, activity diagrams, and system sequence diagrams for the use cases that had been defined earlier. As of today, we have completed the system sequence diagrams for all of the use cases in this iteration, with four exceptions. During recent meetings with the users, four new use cases were defined. We decided to include two of those in this iteration but delay the other use cases to a later iteration. Detailed documentation for the delayed use cases will also be done later.

Also, as we were reviewing the diagrams with the users in our structured walkthrough sessions, we discovered that some of our classes were missing critical attributes. It was a real eye-opener for some of the newer analysts to see the importance of quality controls and structured walkthroughs of our work.

Plans for the next period (two weeks)

During the next period, we will begin looking at some design issues for this iteration, including some preliminary database designs. We will also begin investigating implementation alternatives and network requirements.

Problems, issues, open items

There are no major problems at this point. We do have about 20 items on the Outstanding Items Log, but none of them is holding up the project. In your oversight committee meeting, you might just emphasize to the department heads the importance of getting those items resolved as soon as possible.

BH

cc: Steven Deerfield, Ming Lee, Jack Garcia

tails of the internal steps within each use case. Precondition and postcondition statements help define the context for the use case—that is, what must exist before and after processing. Finally, the system sequence diagram describes the inputs and outputs that occur within a use case. Together, these models provide a comprehensive description of the processing requirements for the system and give the foundation for system design.

Now that the use cases have been explained, let's review the structural aspects of the new system, which are based on the class diagram.

FIGURE *7-18*

An SSD of the Web order scenario
for the *Create new order* use case.

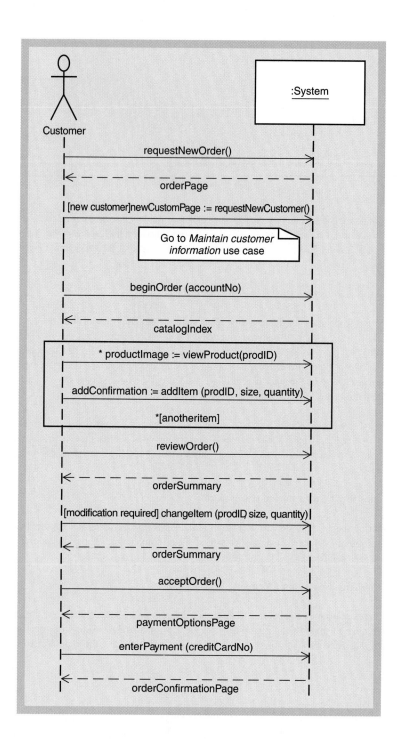

PROBLEM DOMAIN MODELING—THE DOMAIN MODEL CLASS DIAGRAM

Your first exposure to a class diagram was provided in Chapter 5 in the discussion of things involved in the users' work. Since things are in reality objects, and objects are the basis of the object-oriented approach, the class diagram is usually considered the most important of all the UML models. In fact, the class diagram is the focal point of object-oriented development. Every other UML model must be consistent with the class diagram. Analysts always spend considerable time developing, reviewing, and correcting the class diagram. The initial domain model class diagram is just the beginning. You will continue to enhance the class diagram as you move from analysis to design to implementation, adding more details. As you develop other UML models, you should always go back to the class diagram to be sure that it is still correct and to add details.

Recall also that class diagrams can be used for many purposes. The term *class diagram* has both a broad and a narrow definition. In the broad definition, a class diagram is a UML diagram that depicts classes. It can be used to discuss any set of classes, such as defining problem domain classes during analysis or defining the classes making up the user interface during system design. Sometimes class diagrams are created to show design details for problem domain classes. At other times they are used to distinguish problem domain classes from other types of classes (operating environments, controllers, user interfaces, and databases). Therefore, a collection of class diagrams is created during a development project.

However, when defining system requirements, developers focus only on problem domain classes. The problem domain classes form the foundation for later design of the software classes and also drive the definition of all other classes. The class diagram created to define these system requirements is called the *domain model class diagram,* or simply the *domain model,* because it shows the classes that are part of the problem domain of the user. The RMO class diagram developed in Chapter 5 is a domain model. The domain model class diagram does not show methods, focusing instead on the users' real-world objects. Chapter 5 described classes and the class diagram more broadly to emphasize how the OO approach differs from the traditional approach. The purpose of this section is to review the notation and use of the domain model class diagram for describing requirements.

The domain model class diagram serves two critically important purposes. First, it describes the basic structure of the classes that must be implemented with object-oriented programming. Second, it is used as a conceptual data model—to describe the classes that will be used for database definition. Database definition is explained in Chapter 13.

Chapter 5 identified three compartments in the documentation for a class. The three compartments contain the class name, the class attributes, and the class methods. For requirements definition, only the top two compartments are used. The class name and the class attributes are derived directly from the real-world objects that are being described. The third compartment—class methods—are a design construct and are not used yet. Real-world objects do not have methods, but when we design software classes for the new system, we do add methods. In fact, it is common in UML notation to use class rectangles with only two compartments in domain models because it emphasizes that a particular class diagram is a requirements model and not a design model.

In Chapter 5 a class diagram example for a course registration system was described to illustrate class diagram notation. Figure 7-19 shows an expanded version of the example that illustrates additional UML class notation and some of the key concepts that are utilized for a domain model class diagram. The rectangles contain two compartments, one for the class name, which is capitalized, and one for the attribute names, whose first words are not capitalized. Some additional notation on attributes is also shown. When doing domain modeling, it is often important to identify the primary key field or fields. In UML, properties of attributes are shown with curly braces as shown by the courseNumber {key} in the Course class. The attribute numberOfSections is underlined to indicate that it is a class-level attribute. Usually, an attribute has a unique value for each object. However, a class-level attribute has one value that applies to all objects of the class. In Java it is implemented as a static attribute, and in VB .NET it is implemented as a shared attribute. In this example, the system needs to store one value representing the number of course sections that are offered. Each time an additional section is added, the value is increased by one.

domain model class diagram, or domain model

a class diagram that focuses on the user's problem domain classes, describing the basic structure of classes and the conceptual data model

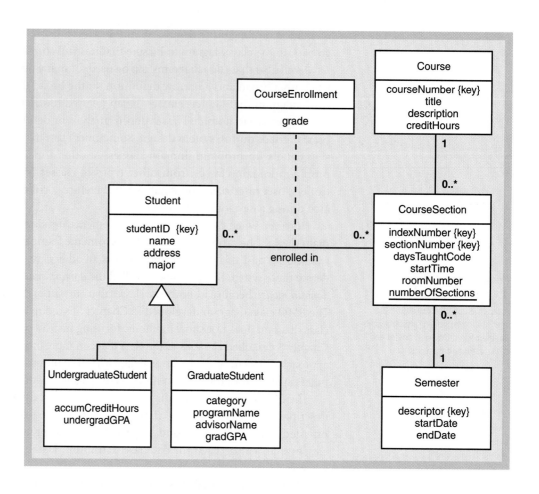

An association relationship between classes is indicated by the connecting lines. In UML, association names are optional. The association between CourseSection and Student is labeled *enrolled in*. The complete name of the association is *Student enrolled in CourseSection*. Also, in this case, the association has a multiplicity (cardinality) of many-to-many. Whenever a many-to-many association exists, an association class may potentially exist. However, an association class is only possible when three criteria are met. First, the association must have a many-to-many multiplicity. Second, the association must have an attribute that applies only to the association itself and not to either connected class. For example, the attribute grade applies to the student enrolled in a particular course section. It does not just apply to the student or to the course section. Third, since each object in the association class represents a single associative link between two objects in the connected classes, the key for the association class objects is always the combination of the keys of the connected classes. If some other attribute is also needed within the key, then the association class has been defined incorrectly.

Figure 7-19 also extends the example from Chapter 5 to show generalization/specialization from Student to UndergraduateStudent and GraduateStudent. For example, every object in UndergraduateStudent is also a Student. The specialization classes inherit all of the attributes and associations of the generalization class. Again, this domain model does not include methods, but it does show how the domain model can be expanded to show more detail as the requirements are understood in greater detail.

A class diagram for the RMO customer support system was shown in Figure 5-35 in Chapter 5. We duplicate that figure here in Figure 7-20. The formatting is changed slightly to emphasize that we are using this diagram as a domain model. Key properties have also been added for some classes.

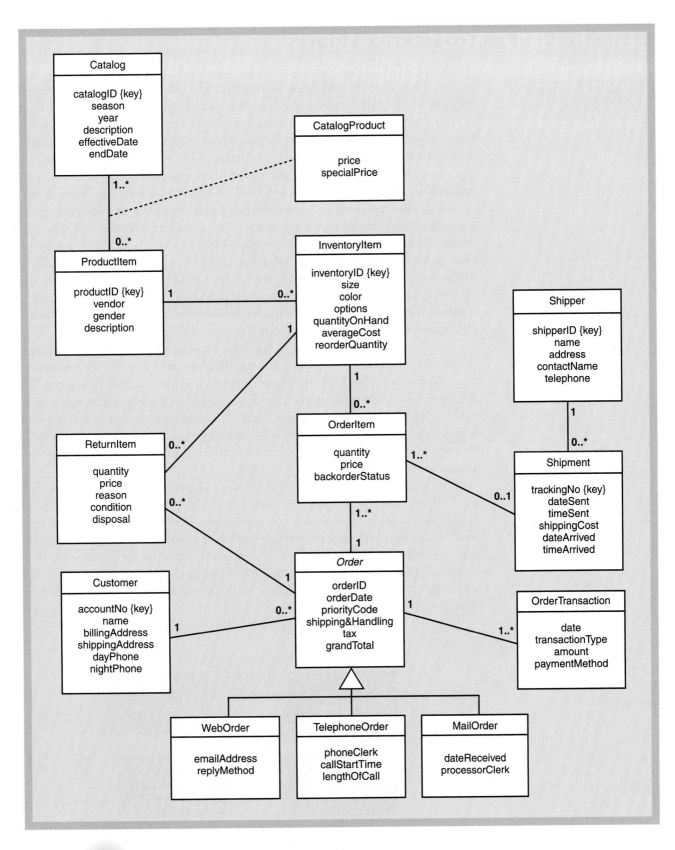

FIGURE *7-20*

RMO domain model class diagram.

INTEGRATING OBJECT-ORIENTED MODELS

The diagrams described in this chapter allow analysts to completely specify the system requirements. If you were developing a system using a waterfall systems development life cycle, then you would develop the complete set of diagrams to represent all system requirements before continuing with design. However, because you are using an iterative approach, you would only construct the diagrams that are necessary for a given iteration. A complete use case diagram would be important to get an idea of the total scope of the new system. But the supporting details included in use case descriptions, activity diagrams, and system sequence diagrams need only be done for use cases in the specific iteration.

The domain model class diagram is a special case. Much like the entire use case diagram, the domain model class diagram should be as complete as possible for the entire system, as shown for RMO. The number of problem domain classes for the system provides an additional indicator of the total scope of the system. Refinement and actual implementation of many classes will wait for later iterations, but the domain model should be fairly complete. The domain model is necessary to identify all of the domain classes that are required in the new system. Although we do not focus on database design in this chapter, the domain model is also used to design the database.

Throughout the chapter, you have seen how the construction of a diagram depends on information provided by another diagram. You have also seen that the development of a new diagram often helps refine and correct a previous diagram. You should also have noted that the development of detailed diagrams is critical to gain a thorough understanding of the user requirements. Figure 7-21 illustrates the primary relationships between the requirements models for OO development. The dependencies generally flow from top to bottom, but some arrows have two heads to illustrate that influence goes in both directions.

Note that the use case diagram and the class diagram are the primary models on which others draw information. You should develop those two diagrams as completely as possible. Earlier in the chapter it was noted that a CRUD analysis between those two models will help ensure that they are as complete as possible. The detailed descriptions, either in narrative format or in activity diagrams, are important for development of system sequence diagrams. As the development of the system progresses, and especially as you begin doing detailed system design, you will find that the relationship between these models strengthens.

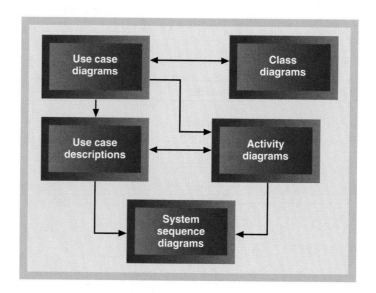

FIGURE *7-21*

Relationships between OO requirements models.

SUMMARY

The object-oriented approach has a complete set of diagrams that together document the user's needs and define the system requirements. These requirements are specified using the following models:

- domain model class diagrams
- use case diagrams
- use case detailed models, either descriptive format or activity diagram
- system sequence diagrams (SSDs)

A use case diagram documents the various ways that the system can be used. It can be developed independently or in conjunction with the event table where one event triggers one use case. A use case consists of actors, use cases, and connecting lines. A use case identifies a single function that the system supports. An actor represents a role of someone or something that uses the system. The connecting lines indicate which actors invoke which use cases. Use cases can also invoke other use cases as a common subroutine. This type of connection between use cases is called the «includes» relationship.

The internal activities of a use case are first described by an internal flow of activities. It is possible to have several different internal flows, which represent different scenarios of the same use case. Thus, a use case may have several scenarios. These details are documented either in descriptive format or with activity diagrams to describe the workflows. Preconditions and postconditions can be defined for each use case to assist in understanding other effects on the system of executing a use case.

Another diagram that provides more details of the processing requirements of a use case is a system sequence diagram, or SSD. An SSD documents the inputs and outputs of the system. The scope of each SSD is usually a use case or a scenario within a use case. The components of an SSD are the actor—the same actor identified in the use case—and the system. The system is treated as a black box, in that the internal processing is not addressed. Messages, which represent the inputs, are sent from the actor to the system. Output messages are returned from the system to the actor. The sequence of messages is indicated by the top-to-bottom sequence of messages.

The domain model class diagram continues to be refined when defining requirements. Often the class symbol includes just two sections to indicate the diagram represents a domain model, leaving out the methods. This is because the classes in the domain model represent the real-world objects involved in the users' work; they do not represent software classes. Additional class diagram notation can be used, such as labeling association relationships, indicating class attributes, and identifying properties.

KEY TERMS

domain model class diagram, or domain model, p. 267
interaction diagram, p. 258
lifeline, or object lifeline, p. 259
message, p. 244
postcondition, p. 256
precondition, p. 256

scenario, or use case instance, p. 251
statechart diagram, p. 244
system sequence diagram, p. 244
true/false condition, p. 260
use case diagram, p. 244

REVIEW QUESTIONS

1. What is the OMG?
2. What is UML? What type of modeling is it used for?
3. What are the two primary uses of a class diagram?
4. Why is the class diagram at the requirements stage of the development cycle called the *problem domain class diagram*?
5. How does UML denote a key field in a problem domain class diagram?
6. What is an association class?
7. What does it mean when an attribute is underlined in a class?
8. What three criteria must be true in order for an association class to be defined?
9. What are the two basic parts of a use case model? What is its purpose or objective?
10. What is the difference between a use case description and an activity diagram?
11. What is meant by a scenario? What is another term used to describe a scenario?
12. Why do actors have "hands"?
13. What is the «includes» relationship used for?
14. What is the difference in the focus on the boundary condition of a use case diagram and an event table?
15. With regard to a use case, what is an activity diagram used for?
16. What is a precondition? What is a postcondition? Why are they important in use case development?
17. What is the purpose of a system sequence diagram? What symbols are used in a system sequence diagram?
18. What are the steps required to develop a system sequence diagram?
19. What is meant by CRUD analysis? How is it done?
20. Identify the models explained in this chapter and their relationship to each other.

THINKING CRITICALLY

1. To review your skills in developing a class diagram, develop a domain model class diagram, including associations and multiplicities, based on the following narrative.

   This case is a simplified (initial draft) of a new system for the University Library. Of course, the library system must keep track of books. Information is maintained about both book titles and the individual book copies. Book titles maintain information about title, author, publisher, and catalog number. Individual copies maintain copy number, edition, publication year, ISBN, book status (whether it is on the shelf or loaned out), and date due back in.

   The library also keeps track of patrons to the library. Since it is a university library, there are several types of patrons, each with different privileges. There are faculty patrons, graduate student patrons, and undergraduate student patrons. Basic information about all patrons is name, address, and telephone number. For faculty patrons, additional information is office address and telephone number. For graduate students, information such as graduate program and advisor information is maintained. For undergraduate students, program and total credit hours are maintained.

   The library also keeps information about library loans. A library loan is a somewhat abstract object. A loan occurs when a patron approaches the circulation desk with a stack of books to check out. Over time a patron can have many loans. A loan can have many physical books associated with it. (And a physical book can be on many loans over a period of time. Information about past loans is kept in the data-base.) So, in this case, an association class should probably be created for loaned books.

   If a book is checked out that a patron wants, he/she can put that title on reserve. This is another class that does not represent a concrete object. Each reservation is for only one title and one patron. Information such as date reserved, priority, and date fulfilled is maintained. When it is fulfilled, the system associates it with the loan on which it was checked out.

2. Develop a use case diagram for the university library system.

   Part a. Based on the following descriptions, build a preliminary use case diagram.

   Patrons have access to the library information to search for book titles and to see whether a book is available. A patron can also reserve a title if all copies are checked out. When patrons bring books to the circulation desk, a clerk checks out the books on a loan. Clerks also check books in. When books are dropped in the return slot, they check in the books. Stocking clerks keep track of the arrival of new books.

   The managers in the library have their own activities. They will print out reports of book titles by category. They also like to see (on-line) all overdue books. When books get damaged or destroyed, they will delete information about book copies. Managers also like to see what books are on reserve.

   Part b. Given the class diagram you developed in question 1, do a CRUD analysis and list any new use cases you discover.

Or, if you change the name of any use cases, indicate that also. In this case, patron information can be accessed and downloaded from another university database.

3. To review your skills in developing a class diagram, develop a class diagram, including associations and multiplicities, based on the following narrative.

A clinic with three dentists and several dental hygienists needed a system to help administer patient records. This system does not keep any medical records. It only processes patient administration.

Each patient has a record with his/her name, date of birth, gender, date of first visit, and date of last visit. Patient records are grouped together under a household. A household has attributes such as name of head of household, address, and telephone number. Each household is also associated with an insurance carrier record. The insurance carrier record contains name of insurance company, address, billing contact person, and telephone number.

In the clinic, each dental staff person also has a record that tracks who works with a patient (dentist, dental hygienist, x-ray technician). Since the system focuses on patient administration records, only minimal information is kept about each dental staff person, such as name, address, and telephone number. Information is maintained about each office visit, such as date, insurance copay amount (amount paid by the patient), paid code, and amount actually paid. Each visit is for a single patient, but, of course, a patient will have many office visits in the system. During each visit, more than one dental staff person may be involved in the visit by doing a procedure. For example, the x-ray technician, dentist, and dental hygienist may all be involved on a single visit. In fact, some dentists are specialists in such things as crown work, and even multiple dentists may be involved with a patient. For each *staff person does procedure in a visit* combination (many-to-many), detailed information is kept about the procedure. This information includes type of procedure, description, tooth involved, the copay amount, the total charge, the amount paid, and the amount the insurance company denied.

Finally, the system also keeps track of invoices. There are two types of invoices: invoices to insurance companies and invoices to heads of household. Both types of invoices are fairly similar, listing each visit, the procedures involved, the patient copay amount, and the total due. Obviously, the totals for the insurance company are different from the patient amounts owed. Even though an invoice is a report (printed out), it also maintains some information such as date sent, total amount, amount already paid, amount due and also the total received, date received, and total denied. (Insurance companies do not always pay all they are billed.)

4. Develop a use case diagram for the dental clinic.

Part a. Develop a use case diagram based on the following narrative.

The receptionist keeps track of patient and head-of-household information. He/she will enter information about the patients and head of household. He/she will also keep track of office visits by the patients. Patient information is also entered and maintained by the office business manager. In addition, the business manager maintains the information about the dental staff persons.

The business manager also prints the invoices. Patient invoices are printed monthly and sent to the head of household. Insurance invoices are printed weekly. When the invoices are printed, the business manager double-checks a few invoices against information in the system to make sure it is being aggregated correctly. She also enters the payment information when it is received.

Each dental staff person is responsible for entering information about the dental procedures that he/she performs.

The business manager also prints an overdue invoice report showing heads of household who are behind on their payments. Sometimes dentists like to see a list of the procedures they performed during a week or month, and they can request that report.

Part b. Expand the use case diagram you have developed based on a CRUD analysis of the class diagram you developed in the previous problem.

5. Interpret and explain the use case diagram in Figure 7-22. Explain the various roles of those using the system and the functions that each role requires. Explain the relationships and the ways the use cases are related to each other.

6. Given the following narrative, do the following:
   a. Develop an activity diagram for each scenario, and
   b. Complete a fully developed use case description for each scenario.

Quality Building Supply has two kinds of customers: contractors and the general public. Sales to each are slightly different.

When a contractor buys materials, he/she takes them to the contractor's checkout desk. The clerk enters the contractor's name into the system. The system displays the contractor's information, including his/her current credit standing.

The clerk then opens up a new ticket (sale) for the contractor. Next, the clerk scans in each item to be purchased. The system finds the price of the item and adds the item to the ticket. At the end of the purchase, the clerk indicates the end of the sale. The system compares the total amount against the contractor's current credit limit, and if it is acceptable, finalizes the sale. The system creates an electronic ticket for the items, and the contractor's credit limit is reduced by the amount of the sale. Some contractors like to keep a record of their purchases, so they request that the ticket details be printed out. Others aren't interested in a printout.

A sale to the general public is simply entered into the cash register, and a paper ticket is printed as the items are

FIGURE 7-22

A use case diagram for the
inventory system.

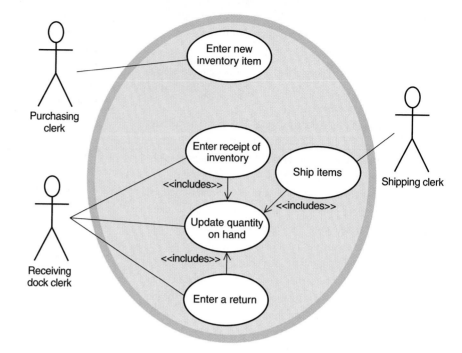

identified. Payment can be by cash, check, or credit card. The
clerk must enter the type of payment to ensure that the cash
register balances at the end of the shift. For credit-card payments, the system prints out a credit-card voucher that the
customer must sign.

7. Given the following narrative, develop either an activity diagram or a fully developed description for a use case of *Add a
new vehicle to an existing policy* in a car insurance system.

   A customer calls an insurance clerk at the insurance
company and gives his/her policy number. The clerk enters
this information, and the system displays the basic insurance
policy. The clerk then checks the information to make sure
the premiums are current and the policy is in force.

   The customer gives the make, model, year, and vehicle
identification number (VIN) of the car to be added. The clerk
enters this information, and the system validates that the
given data is valid. Next, the customer selects the types of
coverage desired and the amount of each. The clerk enters
the information, and the system records each and validates
the requested amount against the policy limits. After all of
the coverages have been entered, the system validates the total coverage against all other ranges, including other cars on
the policy.

   Finally, the customer must identify all drivers and the
percentage of time they drive the car. If a new driver is to be
added, then another use case, *Add new driver*, is invoked.

   At the end of the process, the system updates the policy, calculates a new premium amount, and prints the updated policy statement to be mailed out to the policy owner.

8. Given the following list of classes and relationships for the
previous car insurance system, list the preconditions and the
postconditions for the *Add a new vehicle to an existing policy* use case.

   Classes in the system:

   • Policy
   • InsuredPerson
   • InsuredVehicle
   • Coverage
   • StandardCoverage (list standard insurance coverages with
     prices by rating category)
   • StandardVehicle (lists all types of vehicles ever made)

   Relationships in the system:

   • Policy has InsuredPersons (one to many)
   • Policy has InsuredVehicles (one to many)
   • Vehicle has Coverages (one to many)
   • Coverage is a type of StandardCoverage
   • Vehicle is a StandardVehicle

9. Develop a system sequence diagram based on the narrative
and your activity diagram for problem number 6.

10. Develop a system sequence diagram based on the narrative
or your activity diagram for problem number 7.

EXPERIENTIAL EXERCISES

1. The functionality required by Rocky Mountain Outfitters' customer support system is also found in several real-world companies. Based on your experience with on-line shopping and shopping carts, build a use case diagram of functions that can be done by a Web customer (similar to Figure 7-5). Web sites that you might refer to include L.L. Bean (http://www.llbean.com/), Lands' End (http://www.landsend.com/), Amazon.com (http://www.amazon.com/), and Barnes and Noble Booksellers (http://www.barnesandnoble.com/).

2. Develop fully developed use case descriptions for each of the use cases you defined in exercise 1.

3. Based on the flow of activities you developed in exercise 2, develop system sequence diagrams for those same use cases and scenarios. Add preconditions and postconditions to each use case.

4. Analyze the information requirements of the Web site from exercise 1. Doing a reverse CRUD analysis (going from the use case diagram to the domain model class diagram) will help you identify classes. Develop a domain model class diagram.

5. Locate a company in your area that does software development. A company that is either a consulting company or has a large staff of information systems professionals will tend to be more rigorous in its approach to software development. Set up an interview. Determine the development approaches that they use. Many companies still use traditional structured techniques combined with some object-oriented development. In other companies, some projects are structured, while other projects are object-oriented. Find out what kinds of modeling the company does for requirements specification. Compare your findings with the techniques taught in this chapter.

6. IBM Rational is a company that is a wholly own subsidiary of IBM. The authors of UML have also been executives in Rational Rose. Consequently, IBM Rational was an early leader in developing CASE tools to support UML and object-oriented modeling. You can download an evaluation copy of IBM Rational's UML CASE (IBM Rational XDE Modeler) tool and use the tool to draw the RMO diagrams. This will give you experience with a widely used industry tool. Alternatively, your college or university can enroll in the Seed program and provide copies of the tools in its laboratories. The URL is http://www.rational.com/.

CASE STUDIES

The Real Estate Multiple Listing Service System

Refer to the description of the Real Estate Multiple Listing Service system in the Chapter 5 case studies. Using the event list and ERD for that system as a starting point, develop the following object-oriented models:

1. Convert your ERD to a domain class diagram.
2. Develop a use case diagram.
3. Develop a fully developed use case description or an activity diagram for each use case.
4. Develop a system sequence diagram for each use case.

The State Patrol Ticket Processing System

Refer to the description of the State Patrol ticket processing system in the Chapter 5 case studies. Using the event list and ERD for that system as a starting point, develop the following object-oriented models:

1. Convert your ERD to a class diagram.
2. Develop a use case diagram.
3. Develop fully developed use case descriptions for two of the primary use cases, such as *Recording a traffic ticket* and *Scheduling a court date*.
4. Develop system sequence diagrams for those same use cases.

The DownTown Videos Rental System

DownTown Videos is a chain of 11 video stores scattered throughout a major metropolitan area in the Midwest. The chain started with a single store several years ago and has grown to its present size. Paul Lowes, the owner of the chain, knows that to compete with the national chains will require a state-of-the-art movie rental system. You have been asked to develop the system requirements for the new system.

Each store has a stock of movies and video games for rent. For this first iteration, just focus on the movies. It is important to keep track of each movie title to know and to identify its category (classical, drama, comedy, and so on), its rental type (new release, standard), movie rating, and other general information such as movie producer, release date, and cost. In addition to tracking each title, the business must track individual copies to note their purchase date, their condition, their type (VHS or DVD) and their rental status. User functions must be provided to maintain this inventory information.

Customers, the lifeblood of the business, are also tracked. DownTown considers each household to be a customer, so special mailings and promotions are offered to each household. For any given customer, several people may be authorized to rent videos and games. The primary contact for each customer can also establish rental parameters for other members of the household. For example, if a parent wants to limit a child's rental authorization to only PG and PG-13 movies, the system will track that.

Each time a movie is rented, the system must keep track of which copies of which movies are rented; the rental date and time and the return date and time; and the household and person renting the movie. Each rental is considered to be open until all of the movies and games have been returned. Customers pay for rentals when checking out videos at the store.

For this case, develop the following diagrams:

1. A class diagram
2. A use case diagram. Analyze user functions. Also do a CRUD analysis based on the class diagram.
3. An activity diagram for each of the use cases having to do with renting and checking in movies and for the use cases to maintain customer and family member information.
4. A system sequence diagram for each of the use cases from problem 3.

TheEyesHaveIt.com Book Exchange

TheEyesHaveIt.com Book Exchange is a type of e-business exchange that does business entirely on the Internet. The company acts as a clearinghouse for both buyers and sellers of used books.

For a person to offer books for sale, he/she must register with EyesHaveIt. The person must provide a current physical address and telephone number, as well as a current e-mail address. The system will then maintain an open account for this person. Access to the system as a seller is through a secure, authenticated portal.

A seller can list books on the system through a special Internet form. Information required includes all of the pertinent information about the book, its category, its general condition, and the asking price. A seller may list as many books as desired. The system maintains an index of all books in the system so that buyers can use the search engine to search for books. The search engine allows searches by title, author, category, and keyword.

People wanting to buy books come to the site and search for the books they want. When they decide to buy, they must open an account with a credit card to pay for the books. The system maintains all of this information on secure servers.

When a request to purchase is made, along with the payment, TheEyesHaveIt.com sends an e-mail notice to the seller of the book that was chosen. It also marks the book as sold. The system maintains an open order until it receives notice that the books have been shipped. Once the seller receives notice that a listed book has been sold, he/she must notify the buyer, via e-mail within 48 hours that the purchase is noted. Shipment of the order must be made within 24 hours after the seller sends the notification e-mail. The seller sends a notification to both the buyer and TheEyesHaveIt.com when the shipment is made.

After receiving notice of shipment, TheEyesHaveIt.com maintains the order in a shipped status. At the end of each month, a check is mailed to each seller for the book orders that are have been in a shipped status for 30 days. The 30-day waiting period is to allow the buyer to notify TheEyesHaveIt.com if the shipment does not arrive for some reason, or if the book is not in the same condition as advertised.

The buyers can, if they desire, enter a service code for the seller. The service code is an indication of how well the seller is servicing book purchases. Some sellers are very active and use TheEyesHaveIt.com as a major outlet for selling books. So, a service code is an important indicator to potential buyers.

For this case, develop the following diagrams:

1. A class diagram
2. A use case diagram
3. A fully developed description of the use cases for two use cases such as *Add a seller* and *Record a book order*.
4. A system sequence diagram for each of the two use cases in problem 3.

Rethinking Rocky Mountain Outfitters

The event table for RMO is given in Figure 5-15. Based on this event table and the class diagram shown in Figure 7-20, the use case diagram in Figure 7-4 was developed. The chapter illustrates detailed models (activity and system sequence diagrams) for *Create new order*.

Using the information provided in the RMO case descriptions and the figures in the book (Figure 5-15, 7-4, and 7-20), develop a fully developed use case description and system sequence diagram for each of the following Customer actor use cases: (1) *Update order* and (2) *Create order return*. Now do the same for both of the Shipping actor use cases.

Focusing on Reliable Pharmaceutical Service

Previous chapters have described the activities and processes of Reliable Pharmaceutical Service. Use the previous descriptions, particularly the basic description in Chapter 1 and the detailed descriptions from Chapter 5, as well as the following, additional description of the case, to develop object-oriented requirements models.

Company processes (for use case development):

There are several points in the order-fulfillment process where information must be recorded in the system. Obviously, new orders must be recorded. Case manifests must be printed at the start of each shift. In fact, since a prescription itself may take a fairly long time to be completely used, as in the case of long-standing prescriptions, each time a medication is sent out (prescription fulfillment), information must be entered into the system, noting the quantity of medication that was sent and which pharmacist filled the prescription for that shift.

As explained in Chapter 5, basic information about all of the patients, the nursing homes, the staff, the insurance companies, and so forth must also be recorded in the system.

Information requirements (for class diagram requirements):

Reliable needs to know about the patients, the nursing home, and the nursing-home unit where each patient resides. Each nursing home has at least one but possibly many units. A patient is assigned to a specific unit.

Prescriptions are rather complex entities. They contain basic information, such as ID number, original date of order, drug, unit of dosage (pill, teaspoon, suppository), size of dosage (milligrams, number of teaspoons), frequency or period of dosage (daily, twice a day, every other day, every 4 hours), and special considerations (take with food, take before meals). In addition, there are several types of prescriptions, each with unique characteristics. Some orders are for a single one-time-only prescription. Some orders are for a certain number of dosages (pills). Some orders are for a time period (start date, end date). Information about the prescription order must be maintained. An order occurs when the nursing home phones in the needed prescriptions. Since prescriptions may last for an extended period of time, a prescription is a separate entity from the order itself. The system records which employee accepted and entered the original order.

The system also has basic data about all drugs. Each drug has generic information, such as name, chemical, and manufacturer. However, more detailed information for each type of dosage, such as size of each pill, is also kept. A single drug may have many different dosage sizes and types.

In addition, information about the fulfillment of orders must also be maintained. For example, on a prescription for a number of pills, each time a pill or a number of pills is dispensed, the system must keep a record of that fact. A record is also maintained of which pharmacist or assistant fulfilled the order. Assume all prescriptions are dispensed only as needed for a 12-hour shift.

Basic data is kept about prescription payers, such as name, address, and contact person. For this first iteration, do not worry about billing or payments. Those capabilities may be added in a later iteration.

Based on your previous work, the cases from prior chapters, and the description here, do the following:

1. Refine and extend your class diagram developed in Chapter 5 as necessary.
2. Develop a use case diagram. Base it directly on the event table you created for Chapter 5. Be sure to include a CRUD analysis with your class diagram from question 1 and discuss what additional use cases might be needed based on your CRUD analysis.
3. Develop an activity diagram for each use case related to entering new orders, creating case manifests, and fulfilling orders. You should have at least three activity diagrams. Write a fully developed use case description for each of these use cases.
4. Develop a system sequence diagram for each use case you developed in question 3.

FURTHER RESOURCES

Grady Booch, James Rumbaugh, and Ivar Jacobson. *The Unified Modeling Language User Guide.* Addison-Wesley, 1999.

E. Reed Doke, J.W. Satzinger, and S.R. Williams. *Object-oriented Application Development Using Java.* Course Technology, 2002.

Hans-Erik Eriksson and Magnus Penker. *UML Toolkit.* John Wiley & Sons, 1998.

Martin Fowler. *UML Distilled: Applying the Standard Object Modeling Language.* Addison-Wesley, 1997.

Ivar Jacobson, Grady Booch, and James Rumbaugh. *The Unified Software Development Process.* Addison-Wesley, 1999.

Philippe Kruchten. *The Rational Unified Process, An Introduction.* Addison-Wesley, 2000.

Craig Larman. *Applying UML and Patterns: An Introduction to Object-oriented Analysis and Design and the Unified Process* 2nd Edition. Prentice-Hall, 2002.

James Rumbaugh, Ivar Jacobsen, Grady Booch. *The Unified Modeling Language Reference Manual.* Addison-Wesley, 1999.

CHAPTER 8

Evaluating Alternatives for Requirements, Environment, and Implementation

LEARNING OBJECTIVES

After reading the chapter, you should be able to:

- Prioritize the system requirements based on the desired scope and level of automation for the new system

- Describe the strategic decisions that integrate the application deployment environment and the design approach for the new system

- Determine alternative approaches for system development

- Evaluate and select a development approach based on the needs and resources of the organization

- Describe key elements of a request for proposal and evaluate vendors' proposals for outsourced alternatives

- Develop a professional presentation of findings to management

CHAPTER OUTLINE

Project Management Perspective

Deciding on Scope and Level of Automation

Defining the Application Deployment Environment

Choosing Implementation Alternatives

Contracting with Vendors

Presenting the Results and Making the Decisions

Robert Holmes wasn't exactly sure how to proceed with his project. He had six proposals from software vendors to develop an Internet-based ordering system for his company, Tropic Fish Tales. He and his project team had to figure out some way to make a meaningful comparison among the proposals to determine which alternative best fit the needs of the company. Then he had to make a presentation of his analysis and recommendations.

The problem was that none of the six proposals was the same. He and his team had spent a tremendous amount of time developing a request for proposal (RFP) that they had sent to several firms providing custom solutions. They had worked hard on the RFP to make sure that it contained a very precise definition of business requirements. Even with this well-designed RFP, none of the six proposals looked the same. He was going to have to devise a method to do a fair comparison among the proposals. Otherwise, how would he know which solution was the best for Tropic Fish Tales?

His company had made an early decision to develop an RFP and obtain outside assistance with the development. The project appeared to be pretty large, and the information system staff was quite small and inexperienced. The least-expensive solution was from a company that had a standard off-the-shelf ordering system. The advantage was that it would be quick and fairly inexpensive to install and get working. However, the disadvantage was that it did not quite fit all of the requirements. Robert wasn't sure how important the missing functionality was to his company. The system could be made to work with some modifications to work procedures and forms.

At the other end of the spectrum was a proposal for a completely new state-of-the-art system for Internet sales, with electronic interfaces to suppliers and shippers. This system was a complete electronic commerce solution with fully automated support. The proposal also indicated that substantial transaction, customer, and order history information would be retained and available in real time. The system also contained automated inventory management functions. Although the system had more capability than the company really needed, it would certainly bring Tropic Fish Tales to the forefront of high-technology solutions. He wondered whether the company could afford the price, however, which was about three times the cost of the low-cost solution.

The other proposals ranged in between the two extremes. One company proposed to develop a system from the ground up, working very closely with Robert's firm to ensure that the system fit the requirements perfectly. Another company had a base system that it proposed to modify. The base system was for a different industry and was not currently Internet based, so substantial modifications would be necessary. One solution ran only on UNIX machines. Even though the system appeared to have most of the desired functionality, it would take some work to modify it for a Novell network, which is the current environment for the company.

Robert was scheduled to meet Bill Williams, the director of information systems, later in the day. He hoped Bill would have some suggestions about how to address this problem.

OVERVIEW

As we discussed in previous chapters, the six major activities of the analysis phase of system development are the following:

- Gather information
- Define system requirements
- Prototype for feasibility and discovery
- Prioritize requirements
- Generate and evaluate alternatives
- Review recommendations with management

You have already learned about fact finding, defining system requirements, and prototyping activities of the analysis phase. You learned how to define system requirements using either a traditional approach or an object-oriented approach. This chapter explains the last three activities of the analysis phase—that is, the transitional activities that refocus the project from discovery and analysis to solutions and design. These final activities are pivotal in the project; they set the direction for the design and implementation of the solution system.

We first discuss the project management orientation that underlies all three activities. Recall from Chapter 3 that one of the major responsibilities of the project manager is to define and control the scope of the new system. The objective of prioritizing the requirements is to define the scope of the system precisely, and the scope directly affects the project cost and schedule, which are also the project manager's responsibility. Evaluating implementation alternatives guides the rest of the project. The outcome of these final activities determines the detailed schedule for the final phases of the project.

Next, we discuss evaluating and prioritizing the system requirements. It is normal during analysis to uncover many more requirements and needs than can reasonably be included within the system, so the development team must categorize and prioritize the requirements to determine what to include. Frequently, two or three alternative combinations of requirements will be developed, along with their required resources, and then an oversight committee of executives, users, and technical managers decides which approach is most viable. This chapter discusses various strategies for prioritizing requirements and selecting a scope and level of automation.

Then we discuss the various alternatives for the production environment, including alternatives for the hardware configuration and operating systems. The existing or planned environment for the new system is a critical consideration—that is, what hardware, system software, networks, and standards will support the new system? The chapter contains a brief overview of choices and constraints for the deployment and development environments. We also demonstrate the important points to consider in the environment by discussing them in the context of Rocky Mountain Outfitters and its new system.

Next, we discuss the alternatives for design and implementation. The focus is on the various options for actually building and installing the system. Once the system scope is determined and a decision on the environment has been made, then several alternative methods of development are reviewed. These alternatives can range from building the new system completely from scratch to buying a system from someone else to outsourcing the entire development and daily operation. We review the most popular alternatives and discuss steps used to make a selection. Included within this discussion are instructions on how to develop and use a request for proposal (RFP).

Although these three activities are described as the final activities for the analysis phase, in most cases they are done in parallel with all other analysis phase activities. For example, when requirements are being developed, it may make sense to prioritize them as they are identified. As the development team gains an understanding of the requirements, the team is also normally deciding on the operating system. These types of decisions require careful consideration and ongoing evaluation. We describe them as final and pivotal activities, even though most of the actual work may have been done concurrently with the fact-finding and requirements-development tasks.

Finally, the chapter discusses the need to organize these results in a professional format and present them to upper management. This point in the project is a logical milestone at which to assess the user requirements, update the feasibility analyses, and obtain a commitment—and funding—for the remainder of the project.

PROJECT MANAGEMENT PERSPECTIVE

In Chapter 3 we saw that the activities of the planning phase required the project manager's heavy involvement. The analysis phase activities described in this chapter are no different in that regard. But in addition to a management component, they also have a critical technical requirement. Thus, both the project manager and senior technical members of the project team must work together to complete these activities successfully.

System development projects come in all sizes and complexities and vary in the level of their formality. One effective technique for managing large or complex projects is to develop decision metrics—systems to measure the alternatives and to evaluate them based on their relative scores. Other projects are smaller and less complex and can use more informal techniques with fewer metrics. In this chapter, we discuss several techniques to help in evaluating alternatives.

Appendix A identifies eight areas of project management: scope, time, cost, quality, human resources, communications, risk, and procurement. The activities discussed in this chapter relate to seven of those areas. Let's discuss some of the project management tasks related to the analysis phase activities of prioritizing requirements, evaluating alternatives, and reviewing recommendations with senior management.

A project's scope is directly affected by the priorities established for the system requirements. While prioritizing requirements, the project manager precisely defines the functions that will be included in the project and sets a baseline, which he or she can use to control and direct the rest of the project. A firm list of functions that users and project staff have agreed to can control the scope of the project and keep it manageable. If no firm decisions are ever made about what should be in the new system, then it is almost impossible for the project manager to control the size of the project.

The schedule, which is part of project time management, is further developed at this time, as decisions are made about the scope, the environment, and implementation. In fact, in many projects, the schedule is not completed until these decisions are made. For example, if the team decides to purchase components for the new system or to hire outside programmers, then the project schedule must reflect those decisions.

Project cost management involves both estimating the project costs and controlling them. Costs and schedule profoundly affect decisions regarding a project's scope, environment, and implementation. Frequently, a project manager must recalculate the cost/benefit ratio to confirm a project's financial feasibility. On many projects, a go/no-go decision is made at this stage of analysis—when the project manager recalculates costs and benefits.

Presenting findings to the oversight committee is a key responsibility of a project manager. Project communications management involves collecting and explaining all of the key decisions, feasibility analyses, risks, benefits, schedules, and costs to the stakeholders who are funding the project.

As the team makes decisions, particularly technical decisions about environment and implementation, the project manager must determine and evaluate the various risks associated with each alternative. A complete risk analysis and feasibility assessment is done for each of the alternatives being considered. Since key project decisions are being made, it is important for the project manager to conduct a thorough risk analysis.

As implementation alternatives are evaluated, the project manager begins activities associated with procurement management. Vendors must be identified and evaluated. Requests for proposals are developed, and proposals are evaluated. Contract negotiations may even begin. An effective project manager must have good procurement skills to ensure that reliable and professional vendor relationships are established and good purchase decisions are made.

Finally, even though specific tasks may not be directly associated with project quality management, it should be obvious that quality is the objective of all activities.

Project management runs throughout a project's lifetime, but the two times when project management tasks are most prevalent are during the initial planning phase and during the later evaluation activities of the analysis phase. The skills of the project manager are most evident as critical decisions and project directions are established.

The decisions affecting requirements, environment, and implementation approach are made together because they are interdependent. In the following sections, we treat each topic separately, but in reality they are all intertwined. First, we address the requirements and project scope.

DECIDING ON SCOPE AND LEVEL OF AUTOMATION

Prioritizing requirements includes tasks to define both the scope and the level of automation for the new system. Scope and level of automation are two very closely related aspects of the new application system. The scope of the system defines which business functions will be included in the system. For example, in the current Rocky Mountain Outfitters (RMO) point-of-sale system, the scope includes handling mail and telephone sales, but not Internet sales. The level of automation is how much computer support exists for the functions that are included. In the new system, a very low level of automation for telephone sales would be to require telephone clerks to use printed catalogs at their desks to verify customer requests. The system would then support only simple data entry of the order information. A higher level of automation for telephone sales would be to have the catalog and customer information on-line so that telephone clerks get automated entry and verification of inventory items and customer name and address information.

Controlling a Project's Scope

One common problem with development projects is scope creep. As the name implies, the development team may receive requests to add new system functions after the requirements have been defined and decisions finalized. One way to help control this problem is by formalizing the process to identify, categorize, and prioritize the functions that will be included within the new system so that everyone agrees to and signs off on system functions. In Chapter 5 you learned that the event table describes all of the business events that the system must support. Continuing to use the event table to control which business functions will be supported by the new system is an effective technique to control the project's scope.

During analysis, users usually request many more business functions than the schedule and budget can allow. The team needs to decide which functions are critical and must be included and which can be deferred until later. A common approach to determine the scope is to list each requested function and rate its importance, using such categories as "mandatory," "important," and "desirable." Determining the priority of each function is usually done in conjunction with a description of the level of automation for each function.

Determining the Level of Automation

The level of automation describes the support the system will provide for each function. For most functions of an application system, at least three levels of automation can be defined: low, middle, and high. At the lowest level, the computer system only provides simple record keeping. Data input screens allow employees to capture information and insert it into a database. Simple field edits and validation of input data are also included. For example, a low level of automation for an order-entry function has a data-entry screen to enter customer and order information. The system date may be used for the order date. The user manually enters each line item for the order. The system may or may not automatically calculate the price. Usually, stock on hand and anticipated shipment dates cannot be verified. At the end of order entry, the information is stored in the database and the function is concluded.

Analysts also define a middle-range level of automation for each function, which may be a single midrange point or various midrange alternatives. Usually, the midrange alternative is a combination of features from the high-level and the low-level automation alternatives. They try to make their best guess of what is necessary and what is justified at the current stage of technology and within the budget.

A high level of automation occurs when the system takes over, as much as possible, the processing of a function. Usually, it is more difficult for an analyst to define high-end automation than low-end automation because low-end automation is basically an automated version of a current manual procedure. However, generating a high-level automation alternative requires brainstorming and thinking "outside the box" to create new processes and procedures. Chapter 4 discussed some ideas associated with business process reengineering (BPR), which attempts to get people to rethink completely the way that business functions are performed to achieve radical improvements in processing speeds and levels of service. Successful reengineering of business functions depends on computer systems that provide high levels of automation support.

Figure 8-1 is a table that contains both scoping and level of automation information for each function of the RMO customer support system. The figure contains all of the business events from the original event table (see Figure 5-15) as well as seven new functions that were identified during systems analysis. The objective of this table is to identify all of the potential events and functions that the new system needs to perform. Each business function is prioritized as mandatory, important, or desirable. Users and clients prioritize the functions based on the needs of the business and the objectives of the new system. For example, if one objective of the system is to increase customer support, then functions that allow RMO to respond to customer requests will be mandatory functions, at least at some level of automation.

FIGURE *8-1*

RMO's CSS functions with priorities and three levels of automation.

Functions (Expanded from Event List)	Priority (Mandatory, Important, Desirable)	Low-End Automation	Medium (Most Probable) Automation	High-End Automation (Medium Level plus ...when + Appears)
Check item availability	Important	Periodic listing of quantity on hand	Real-time. Internal and Web	+ Sales prompting
Place Order	Mandatory	Clerk data entry	Clerk real-time and customer via Web	+ Promotion prompting and stock-out alternatives
Change or cancel order	Important	Clerk overnight	Clerk real-time and customer via Web for 24 hours	Clerk real-time and customer via Web up to shipment
Check order status	Important	Clerk overnight	Clerk real-time and customer via Web	+ Automatic notification
Fulfill order	Mandatory	Print pull list and shipping label	Pull list, shipping label, real-time update	Automated warehouse Real-time update
Create back order	Important	Clerk data entry	Real-time	+ System automatic and notify supplier
Return item	Important	Clerk data entry	Real-time, clerk update restock, and customer	Automatic inventory and account update
Mail catalog	Mandatory	Print labels	Personalize cover letter	+ Personalize throughout
Correct customer account	Important	Data entry	Real-time	+ Automatic from activity
Send promotional material	Important	Print labels	Personalized cover page	Personalized based on buying history

(continued)

Functions (Expanded from Event List)	Priority (Mandatory, Important, Desirable)	Low-End Automation	Medium (Most Probable) Automation	High-End Automation (Medium Level plus ...when + Appears)
Adjust customer charges	Mandatory	Data entry	Real-time update	+ Automatic from activity
Update catalog	Mandatory	Data entry	Real-time	+ Automatic suggestions from sales history
Create promotional materials	Important	Data entry	Real-time	Recommendations based on sales history
Create new catalog	Mandatory	Record keeping of products, prices, etc.	Record keeping of products, prices, pictures layouts	Digital scan and page layout
System Reports				
Produce order summary reports	Important	Printed on request	On-line view and real-time	Data visualization tools
Produce activity reports	Important	Printed on request	On-line view and real-time	Data visualization tools
Produce transaction summary reports	Important	Printed on request	On-line view and real-time	Data visualization tools
Produce customer adjustment report	Important	Printed on request	On-line view and real-time	Data visualization tools
Produce fulfillment reports	Important	Printed on request	On-line view and real-time	Data visualization tools
Produce catalog activity reports	Important	Printed on request	On-line view and real-time	Data visualization tools
Newly Identified Events				
Maintain customer purchase history	Important	Archive files with summary reports	Archive, printed promotional notices	Automatic, real-time for sales prompting
Provide ongoing feed to manufacturing	Desirable	Printed reports	Daily update	Real-time and trend analysis
Provide EDI feed to suppliers from sales data	Desirable	Printed reports and history	Daily update	Real-time and trend analysis
Tie in to shipper system	Desirable	No link	Daily update and e-mail notification to customer	Automatic feed and shipment tracking via Web link
Perform data warehousing and conduct data analysis	Desirable		Trend analysis	Trend analysis, data visualization tools
Prompt automated sales	Desirable		Based on promotions	Based on sales promotions and history
Conduct expanded sales analysis with DSS	Desirable		Printed reports	Data visualization tools

The table also includes various levels of automation for each function. The analysts do not attempt to describe every characteristic of each level of automation in the table. Supporting descriptions will describe each cell in more detail. The table provides an overview of the functions, the priority of each, and the various methods to implement each function at the different levels of automation.

Let's take the order-entry function within RMO as an example. We start by identifying the best customer service possible. We ask the question, "Why does a customer need to be able to order exactly what she wants at the most convenient time?" Also, "What does a customer want to know about her order at the time she has finished ordering?" (Normally, we would try to reengineer the entire process and would also ask questions such as, "When, or how soon, does the customer want to receive the items she has ordered?" But let's limit our discussion only to the order-entry component of the process.)

To answer the order-entry questions, RMO staff decide that they want to provide their on-line customers all the benefits of catalog shopping: convenience, the ability to order anytime of day or night, no crowds, the ability to order from home, wide selection, simplicity of ordering, and privacy. RMO also wants to provide, as much as it is feasible, the benefits of store shopping: being able to inspect the items; trying them on and comparing sizes, colors, and patterns; having items and related accessories near each other; examining several products together for match and compatibility; and so forth.

Given the stated desires, a high-end system might have the following characteristics:

- Customers can access the catalog on-line, with full-color, three-dimensional pictures. For more technical products, the catalog should include detailed descriptions and diagrams showing their construction and other details. This service can be provided for Internet customers through the Web. For telephone customers, the catalog can be provided, through the phone line, directly to a television set.
- The catalog is also interactive and allows the customer to combine several items with graphical imaging that displays them together (for example, showing a shirt, jacket, and shorts on a simulated person).
- The user interface to the catalog and order system is either voice activated or keypad activated.
- The system should make suggestions of related items that customers may need or desire to purchase at the same time.
- The system should verify that all items are in stock and establish a firm time when shipment will occur. (The fulfillment portion of the system should support shipment within 24 hours or less or, even better, guarantee same-day delivery.)
- Items not in stock should be immediately ordered from the manufacturer or other supply source (the system will immediately send the transaction to the other systems), enabling RMO to ensure delivery to the customer at a future date.
- Payment is verified on-line, just as in a store.
- The customer can see a history of all prior orders and can check the status of any individual order either with the telephone or on the Web.

Interestingly, all of these capabilities can be supported with current technology. The question, of course, is whether RMO can justify the cost at this time. In any event, we have defined high-end automation for the order-entry portion of the new system.

Selecting Alternatives

Once the identified functions have been prioritized and the levels of automation have been analyzed, then the project team reviews all the alternatives. Preliminary decisions may have been made based on individual need or importance, but the entire group of alternatives is normally evaluated together. This will provide a more global, or "big picture," view of the proposed system. In recent years, companies have been building new systems to gain competitive advantage in the marketplace. In addition, more and more companies are entering into e-commerce ventures on both the supply and delivery sides. By establishing more global and strategic criteria, companies can make better long-term decisions for their new systems. The following list identifies some of the key criteria that are used:

- **Strategic plan.** The initial decision to develop a new information system is frequently an outgrowth of a long-term strategic plan. As discussed previously, strategic planning occurs both for the long-term organizational strategy and for information technology to support organizational plans. As decisions are made concerning individual capabilities of the new system, the strategic plan is frequently used as a global measuring rod. For example, if an organization's long-term goal is to develop a supply chain system with automatic interfaces between itself and its suppliers, then the system must be designed to support these interfaces even though they may not be implemented in a first phase.

■ **Economic feasibility.** Obviously, higher levels of automation require substantially more funds to implement. Frequently, development teams generate several groups of capabilities and levels of automation and then project costs to develop those different packages. With more detailed information about requirements and assessments of the difficulty of developing certain capabilities, a more accurate cost/benefit analysis can be generated.

■ **Schedule and resource feasibility.** Including more advanced features in a system not only costs more but also lengthens the schedule. One effective method to minimize the immediate impact on a project is to plan for future system upgrades. All commercial software developers work this way, and it is a viable alternative for in-house development. A new system often has less capability than the organization ultimately desires. But as users gain experience with the new system and information systems staff learn from past experience, together they can enhance the system until they achieve the desired level of automation.

■ **Technological feasibility.** Not only must project teams review the technical feasibility of desired alternatives, but they must also carefully consider whether the organization has in-house expertise to develop and implement the system. Frequently, organizations hire additional staff or contract with outside resources to obtain technical expertise. Usually it is more prudent to select alternatives that do not require pushing the state of the art. Yet, some companies with substantial funds and broad access to resources do so. For cutting-edge projects, a detailed risk analysis is critical.

■ **Operational, organizational, and cultural feasibility.** Changes in business processes also involve risk—risk that must be managed. Broader scope and higher levels of automation usually require organizations to reengineer their business functions and manual processes. Benefits can be substantial and dramatic; however, users need support for the change to maintain morale and commitment to the new system. Typically, information systems staff underestimate the difficulties of changing people's work procedures and job activities. For that reason, it is usually a good idea to involve people who have been trained in organizational behavior to assist in managing changes.

Evaluating Alternatives for RMO

Rocky Mountain Outfitters is in the preliminary stages of selecting functions and automation levels for the customer support system. A final decision depends on the alternatives the development team chooses for implementation. We explore those alternatives next.

Based on preliminary budget and resource availability, the project team at RMO determined that it is possible to include all functions that were categorized as either mandatory or important in the table in Figure 8-1. For each of those functions, the team does a detailed analysis for the desired level of automation. Fundamental decisions about level of automation affect several functions at the same time. For example, three levels of automation affect *Check item availability, Place order, Change or cancel order*, and so forth. The three basic alternatives listed in Figure 8-1 are (1) data entry of information with overnight processing, (2) real-time entry for both employee clerks and for customers via the Web, and (3) the same system as the medium level with added sales prompting based on promotions and even personal customer purchase history. In fact, these three alternatives could even be divided into more alternatives, such as Web versus no Web and sales prompting for promotions but not purchase history. A fundamental decision on the automation level will then need to be consistent across the listed functions.

Figure 8-2 lists the functions and shows by shading which functions are to be included and at what level of automation. Since some basic decisions affect several functions, it was not necessary to do a detailed calculation of the cost/benefit, operational impacts, technical feasibility, and resource impacts for each function. The low level of automation was not acceptable to RMO management. Most of the current systems already provided that level of automation. As can be seen from the table, RMO selected the medium or most probable level of automation for most functions. The high-end automation would have required a substantial jump not only in software but also in hardware processing power. For example, high-end support for placing an order would require both clerks and the Web-based system to encourage the customer to purchase other items based on sales promotions and past purchases. This capability would require a very large on-line database with very fast access speeds. The cost for that capability is not within the budget for RMO at this time.

FIGURE *8-2*

Preliminary selection of alternative functions and level of automation for RMO (selections are shaded).

Functions (Expanded from Event List)	Priority (Mandatory, Important, Desirable)	Low-End Automation	Medium (Most Probable) Automation	High-End Automation (Medium Level plus ...when + Appears
Check item availability	Important	Periodic listing of quantity on hand	Real-time. Internal and Web	+ Sales prompting
Place order	Mandatory	Clerk data entry	Clerk real-time and customer via Web	+ Promotion prompting and stock-out alternatives
Change or cancel order	Important	Clerk overnight	Clerk real-time and customer via Web for 24 hours	Clerk real-time & customer via Web up to shipment
Check order status	Important	Clerk overnignt	Clerk real-time and customer via Web	+ Automatic notification
Fulfill order	Mandatory	Print pull list and shipping label	Pull list, shipping label, real-time update	Automated warehouse. Real-time update
Create back order	Important	Clerk data entry	Real-time	+ System automatic and notify supplier
Return item	Important	Clerk data entry	Real-time, clerk update restock, and customer	Automatic inventory and account update
Mail catalog	Mandatory	Print labels	Personalize cover letter	+ Personalize throughout
Correct customer account	Important	Data entry	Real-time	+ Automatic from activity
Send promotional material	Important	Print labels	Personalized cover page	Personalized based on buying history
Adjust customer charges	Mandatory	Data entry	Real-time update	+ Automatic from activity
Update catalog	Mandatory	Data entry	Real-time	+ Automatic suggestions from sales history
Create promotional materials	Important	Data entry	Real-time	Recommendations on sales history
Create new catalog	Mandatory	Record keeping of products, prices, etc.	Record keeping of products, prices, pictures, layouts	Digital scan and page layout

Functions (Expanded from Event List)	Priority (Mandatory, Important, Desirable)	Low-End Automation	Medium (Most Probable) Automation	High-End Automation (Medium Level plus ...when + Appears)
System Reports				
Produce order summary reports	Important	Printed on request	On-line view and real-time	Data visualization tools
Produce activity reports	Important	Printed on request	On-line view and real-time	Data visualization tools
Produce transaction summary reports	Important	Printed on request	On-line view and real-time	Data visualization tools
Produce customer adjustment report	Important	Printed on request	On-line view and real-time	Data visualization tools
Produce fulfillment reports	Important	Printed on request	On-line view and real-time	Data visualization tools
Produce catalog activity reports	Important	Printed on request	On-line view and real-time	Data visualization tools
Newly Identified Events				
Maintain customer purchase history	Important	Archive files with summary reports	Archive, printed promotional notices	Automatic, real-time for sales prompting
Provide ongoing feed to manufacturing	Desirable	Printed reports	Daily update	Real-time and trend analysis
Provide EDI feed to suppliers from sales data	Desirable	Printed reports and history	Daily update	Real-time and trend analysis
Tie in to shipper system	Desirable	No link	Daily update and e-mail notification to customer	Automatic feed and shipment tracking via Web link
Perform data warehousing and conduct data analysis	Desirable		Trend analysis	Trend analysis, data visualization tools
Prompt automated sales	Desirable		Based on promotions	Based on sales promotions and history
Conduct expanded sales analysis with DSS	Desirable		Printed reports	Data visualization tools

Of the seven newly identified functions, RMO management decided to include three in the project. The first addition is to prepare for the high-end support by including the function to maintain customer history and use it to develop special promotions. Feasibility analysis indicates that this alternative does not require substantial increases in cost or length of the project schedule.

The second addition is a more rapid update of inventory levels to the manufacturing facilities. The cost/benefit analysis of this alternative indicates an immediate return by a reduction of back orders and stock-outs.

The third addition is a subsystem to provide more sophisticated analysis of sales trends. This subsystem will utilize the database of sales orders and time series data based on customer histories. Thus, it builds on the data being included for individual customers. It was difficult for the project team to calculate a precise cost/benefit ratio for this new capability, but the sales manager convinced the oversight committee that the capability was critical to the future competitiveness of RMO. So, resources will be dedicated to add this capability. This subsystem is somewhat independent, so to minimize its impact on the schedule, the project team will implement the subsystem several months after the rest of the system.

Defining the Application Deployment Environment

application deployment environment

the configuration of computer equipment, system software, and networks for the new system

One of the primary considerations in developing a new information system is the application deployment environment. The *application deployment environment* is the configuration of computer hardware, system software, and networks in which the new application software will operate. An important part of any project is ensuring that the application deployment environment is defined and well matched to application requirements. At this life cycle stage, the analyst's goal is to define the environment in sufficient detail to be able to choose from among competing alternatives and to provide sufficient information for design to begin. Additional details are added as design proceeds.

Hardware, System Software, and Networks

In the early years of computer applications, there was only one application type and one deployment environment: a batch-mode application executing on a centralized mainframe using files stored on disk or tape, with off-line data-entry devices such as keypunch machines. As computing technology has matured, the range of application types has grown to include:

- Stand-alone applications on mini- and personal computers
- On-line interactive applications
- Distributed applications
- Web-based applications

Just as the number of application types has proliferated, so has the variety of hardware, system software, and networks that support them. Computers now range in size from handheld devices to large supercomputers. In addition, analysts are faced with many choices in supporting software such as operating systems (for example, UNIX and Windows), database management systems (for example, Oracle and DB2), component infrastructure software and standards (for example, CORBA and .NET), and Web server software (for example, Internet Information Server and Apache). Modern application software relies on a complex infrastructure that includes client and server hardware, supporting system software, computer networks, and the standards that enable them to operate smoothly together.

When choosing or defining the deployment environment, analysts are concerned with several important characteristics, including:

- **Compatibility with system requirements.** Requirements such as user locations, speed of access and update, security, and transaction volume have a significant impact on environmental requirements. For example, high-volume transaction processing systems such as credit-card payment-processing systems require secure high-speed networks, powerful servers, and compatible operating systems and database management systems (DBMSs).
- **Compatibility among hardware and system software.** Although hardware and system software compatibility has generally improved over time, there are still significant considerations. For example, since Oracle and Sun Microsystems are frequent partners in software and standards development, it is no surprise that the Oracle DBMS performs well on Sun servers running Solaris (Sun's version of UNIX). Similarly, Microsoft operating systems and database management systems are well suited to computers using Intel processors. Ensuring good compatibility of hardware and system software simplifies a system's installation and configuration, improves performance, and minimizes long-term operating costs.
- **Required interfaces to external systems.** Modern applications often interact with external systems operated by entities such as credit-reporting agencies, customers,

suppliers, and the government. Implementing external interfaces may require a certain system software and, less frequently, specific hardware. For example, a credit-reporting agency might provide services via Web-based XML requests or a published CORBA-compliant component. An application that interacts with the credit-reporting system must support one or both of those interfaces and include whatever system software is compatible with the interfaces.

- **Conformity with the IT strategic plan and architecture plans.** Because there are so many choices in hardware and system software, organizations find it difficult and expensive to support many different types. Most medium- and large-scale organizations have strategic application and technology architecture plans that focus their efforts on a limited set of hardware and software alternatives. For example, an organization might choose to emphasize a standard platform consisting of UNIX, Oracle, CORBA-compliant distributed systems, and compatible hardware from Sun Microsystems and Hewlett-Packard. Although that environment might not be best for every application type, sticking to it whenever possible will minimize the total cost of infrastructure maintenance and maximize the long-term compatibility among systems for that particular organization.

- **Cost and schedule.** Deployment environment alternatives may vary in their impact on project cost and schedule. Typically, environment choices that match the IT strategic plan and existing systems are the fastest and least expensive to acquire, configure, and support.

In sum, the analyst must define an application deployment environment that enables the application to meet stated requirements, fits within the organization's IT plans, and can be acquired and configured within acceptable limits of budget and schedule.

Development Tools

development environment
the programming languages, CASE tools, and other software used to develop application software

Analysts must also consider and select development tools. The *development environment* consists of the programming language(s), CASE (computer-assisted software engineering) tool(s), and other software used to develop application software. The specific deployment environment usually limits development environment choices. For example, choosing a deployment environment based on Microsoft .NET limits the set of compatible development tools to those provided by Microsoft (for example, Visual Studio .NET) and a relatively small number of third-party vendors. System software choices will also be limited to those most compatible with the deployment and development environment (for example, Microsoft server operating systems, Internet Information Services, and SQLServer, for a .NET application).

Normally, companies have a preferred language for system development, and their analysts are familiar with its features. However, as technology changes, newer languages frequently provide additional capabilities. Analysts can choose from numerous development languages—from structured languages such as COBOL to object-oriented languages such as Smalltalk, C++, and Java to Web-based languages such as JavaScript and PHP. Using a new language does require additional commitment and funding to provide the development team with necessary training.

If a company has invested heavily in a CASE tool, then all new development may be restricted to the tool's methodology, since a CASE tool usually dictates the implementation language and methodology. Some CASE tools also generate database descriptions and distributed code modules for only one database management system and distributed software standard.

Even if the organization hasn't invested in a CASE tool, the choice of development tools, such as compilers, debuggers, and integrated development environments, is usually limited by the target operating system, database management system, and component or

Web service standards. For example, a deployment environment consisting of UNIX, Oracle, and CORBA would usually lead developers to choose the Java programming language and a tool suite such as Oracle JDeveloper, Sun ONE Studio, or IBM WebSphere.

Many corporations have committed to a particular database management system, and it can limit tool selection also. Most DBMS vendors also supply a compatible set of development tools, which can substantially accelerate the development of some application types compared with development tools not optimized to a particular DBMS. Examples include Microsoft Access and Visual Basic, Microsoft SQLServer and Visual Studio .NET, and Oracle Application Server and JDeveloper.

In sum, application deployment environment choices, particularly the operating system, DBMS, and distributed software standard, tend to limit development tool choices. Thus, an analyst should consider the deployment and development environments together when determining their fit to a particular application.

The Environment at Rocky Mountain Outfitters

The systems environment at RMO had been built piecemeal over the life of the company to support the business functions at the various locations. Currently, there are two major manufacturing plants, which provide products for three warehouses. The warehouses also stock items from other manufacturers. RMO's manufacturing facilities, warehouses, retail stores, the mail-order center, the phone center, and the data center are currently networked.

The Current Environment

Figure 8-3 illustrates the current computer environment at RMO. Existing technology consists of a mainframe computer located at the home office in Park City, with dedicated lines connected to the three warehouse distribution sites in Salt Lake City, Portland, and Albuquerque. A dedicated line also connects the mainframe to the mail-order center in Provo, Utah. Both mail-order and distribution functions are linked directly to the mainframe to allow real-time connection of terminals. The communication technology is based on high-volume mainframe transaction technology. The merchandising and distribution system is a mainframe application written in COBOL and using a combination of IBM's DB2 database system and VSAM files. The system was written by RMO staff with assistance from outside technical consultants.

Dial-up telephone lines are used to communicate with the manufacturing sites in Salt Lake City and Portland. Each manufacturing facility also has its own LAN system to

FIGURE *8-3*

The existing processing environment at RMO.

Location and facility	Equipment	Connection
Park City—Data Center	Mainframe	
Park City—Retail	Client-server	Daily dial-up
Salt Lake City—Manufacturing	Local LAN	Daily dial-up
Salt Lake City—Warehouse	Midrange computer	Dedicated line to Data Center
Salt Lake City—Phone Order	Client-server	Daily dial-up
Provo—Mail Order Center	Client-server	Dedicated line to Data Center
Portland—Warehouse	Midrange computer	Dedicated line to Data Center
Portland—Manufacturing	Local LAN	Daily dial-up
Denver—Retail	Client-server	Daily dial-up
Albuquerque—Warehouse	Midrange computer	Dedicated line to Data Center

support specific manufacturing information. Updates to the central inventory system are done in a batch mode daily via the dial-up connection.

The retail stores have local client-server retail systems that collect sales and financial information through the cash registers. This information is also forwarded to the central accounting and financial systems residing on the mainframe. The transmittal is done in batch mode daily.

The phone-order system, in Salt Lake City, is a fairly small Windows application running in a client-server environment. It was built by RMO staff as an independent application and is not well integrated with the rest of the inventory and distribution systems. Information is forwarded daily in batches to the system in Park City.

Other applications, such as human resources and general accounting, are also mainframe systems running in Park City.

The Proposed Environment

Many of the decisions associated with the target environment are made during strategic planning, which establishes long-term directions for an organization. In other situations, the strategic plan is modified as new systems are developed to use the latest technological advancements. In RMO's case, many technical decisions were made during the initial phases of the supply chain management (SCM) project that is well under way. Since the new customer support system (CSS) must integrate seamlessly with the SCM, technical decisions must be consistent with prior decisions as well as the long-term technology plan.

Since environment decisions are corporatewide strategic decisions, RMO convened a meeting to discuss the technology alternatives and to make decisions. Attendees consisted of Mac Preston, chief information officer; John MacMurty, director of system development; and Barbara Halifax, project manager. Additional technical staff were also included in the meeting to provide details as needed.

To ensure that all participants were aware of potential alternatives, Barbara presented and reviewed the information shown in Figure 8-4. This figure identifies potential implementation alternatives and was similar to the one used to make decisions for the

FIGURE *8-4*

Processing environment alternatives.

Alternative	Description
1. Move all functionality to be browser based (intranet/Internet)	Make both the internal and external applications Web-based with browser interface. This solution would provide a consistent interface and facilitate e-commerce growth.
2. Use internal LAN/WAN technology	Internal transactions cans be faster. The database would not need an interface to the Web. Put only the catalog on the Web.
3. Use a mix of alternatives 1 and 2	Use the Web for customer interactions, but use internal LAN/WAN technology for back-end processing and an interface to SCM and other internal systems.
4. Use the mainframe as the central database server	Support for high-volume transactions. It will serve as a centralized database for all systems. It provides high security, control, and consistency.
5. Use a distributed database on multiple servers	Distributed data provides rapid response and load leveling. Growth can be done incrementally. Updating is more complex.
6. Use complete OO components such as Common Object Request Broker Architecture (COBRA) objects	This solution would make seamless interfaces between applications—SCM, CSS, and other systems—with object brokers. It would position RMO for future OO migration. This solution requires middleware integration software.
7. Use OO for the user interface with a back-end relational database	Use Visual Basic or Java to develop the applications. Use DB2 or relational Oracle for database processing. This solution would be very efficient for high volumes.
8. Use OO for the user interface plus CORBA objects for communication between systems	This alternative would position RMO for a move to a complete OO environment. Middleware software is required for integration of systems.

SCM project. The alternatives are listed by type of technology and degree of centralization. The first three alternatives considered are whether to:

- Move to Internet technology
- Use internal LAN/WAN technology
- Use a mix of the two options

The next two alternatives focus on the equipment—whether to:

- Use a mainframe central processor
- Use distributed client-server processors

Finally, the location and type of database are considered. The decision is whether to use more traditional relational database technology or to move to more advanced object-oriented databases. Any decisions made for the CSS would need to be consistent with prior decisions for the SCM.

RMO wants its system to be state of the art, but it also does not want to have a high-risk project and attempt new technology that is not yet proven. Figure 8-5 lists the major components of the strategic direction for RMO.

Current, well-tested technology can provide client-server processing on a rack of multiple processors to support high-volume Internet transactions. Microsoft's Internet Server will provide Internet support. The existing DB2 database on the mainframe is a very viable option to provide efficient back-end processing. The database will require redesign and must be rebuilt for the new system, but the fundamental processing environment is solid.

All of the COBOL applications will be replaced with new systems that will be written using Java, Visual Basic, VBScript, and PHP as appropriate.

In this approach, the mainframe will remain as the central database server. The other two tiers will be application servers. The users will have individual client personal computers that are connected to the application servers. Barbara Halifax attached the table shown in Figure 8-5 to her biweekly status report to John MacMurty (see the accompanying memo). She also identified the other open issues that needed to be decided. Even though the operating environment decisions are critical to the progress of the project, other important issues also still need to be addressed.

FIGURE *8-5*

Strategic directions for the processing environment at RMO.

Issue	Direction(s)
Required interfaces to other systems	1. Automatic feed to SCM system 2. Interface to feed the accounting general ledger 3. Interface to provide automatic feed to external systems—credit card verification and package shipping 4. Potential move to XML for a common interface language
Equipment configuration	1. Servers with multiple CPU configuration for front-end applications 2. Database support provided with mainframe central processor
Operating system	1. Windows Server 2003 front-end servers 2. MVS for mainframe
Network configuration	1. Windows network 2. IIS for Web servers
Language environment	1. Visual Basic, Java, and PHP for application and Web development
Database environment	1. Maintain DB2 database on mainframe 2. Reevaluate long-term strategy for the OO database
CASE tools	1. Use only diagramming portions; various alternatives available

April 28, 2005

To: John MacMurty

From: Barbara Halifax

RE: Customer Support System status

I appreciate your participation and support in last week's meeting to decide on the application deployment environment and agree with the decisions we made. They are consistent with the directions we had assumed in our preliminary plans. Here's our current status:

Completed during the last period

We have spent a lot of time finalizing the scope of the new system. We have been reviewing with each department head the list of business events (from the event table) for that department. We have also reviewed the data model to ensure that we have identified all the necessary attributes. I feel confident that we now have good agreement on the scope and requirements for the new system.

Much of my time during this last week was spent preparing for our meeting on the deployment environment.

Plans for next period

As you know, we still have one more critical meeting coming up next week—to make the final go/no-go decision. We already have the system specifications defined,which we will present. We will complete the reevaluation of all feasibility issues for the schedule, the cost/benefit, and other risks for both the project itself and the deployment of the new system. We already addressed technological feasibility in the last meeting. I have reviewed our major findings with you and other members of the oversight committee, so there should be no surprises. The meeting will allow us to formalize and finalize the approval and funding for the rest of the project.

BH

cc: Steven Deerfield, Ming Lee, Jack Garcia

CHOOSING IMPLEMENTATION ALTERNATIVES

So far, we have described the analysis and fact-finding activities in the development project. As the project team makes decisions about the scope, level of automation, and processing environment, it also makes related decisions about the actual approach to designing, programming, and installing the system. There are numerous ways to implement a solution. For example, if an application is fairly standard, perhaps the organization could just buy a computer program or system to support it. Even for more sophisticated systems, other companies may have already developed standard systems that can be purchased. If purchasing is not an option, then the organization may decide to build the system in-house, and even then there are various alternatives. Outside programmers can be contracted for a range of services or specific technical expertise. The point is that at this stage the organization must plan the rest of the project, and there are a multitude of options.

Figure 8-6 presents some variations on implementing a system. The left axis represents the build-versus-buy options. The bottom axis shows the alternatives of develop-

FIGURE *8-6*

Implementation alternatives.

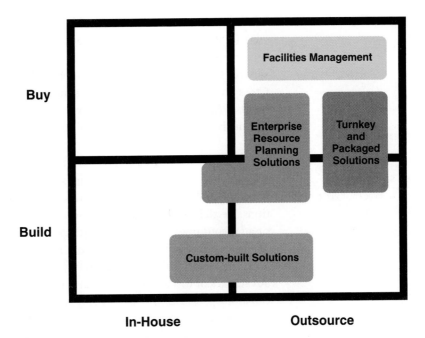

ing the system in-house versus outsourcing the project. Each axis represents a continuum. For example, an entire system can be bought, or the entire solution can be built. But in between those extremes are systems in which portions are purchased and portions are built. In other words, a basic solution may be purchased, but it may require modification or programming of some components to interface with existing systems. Similarly, many options exist for all or part of the solution to be developed in-house or outsourced.

The shapes between the two axes show various general approaches to obtaining a system. Facilities management occurs when the entire system, including development and operation, is contracted to another company. Below that is packaged software or a turnkey system. Although slightly different approaches—packaged software is shrink-wrapped, off-the-shelf, while a turnkey system is a customized package—both usually require some modification to fit the existing environment. Thus, these options usually have a "build" component. Enterprise resource planning (ERP) solutions begin with a standard system, but they require substantial integration with a company's business processes. ERP solutions are integrated so tightly with the entire organization and all its systems that the implementation frequently requires a substantial effort. Custom-built systems require substantial programming by either in-house staff or outsourced consultants and programmers. Each alternative is explained in more detail in the following paragraphs.

Facilities Management

Facilities management is the outsourcing of the entire data processing and information support capability for an entire organization. Facilities management is not an actual development technique or implementation alternative. Instead, it is the result of an organization's strategic decision to move all system development, implementation, and operation to an outside provider. For example, a bank may hire a facilities management firm to provide all of its data processing capability. The computers, software systems, networks—even the technical staff—all belong to the outside firm. The bank in essence has opted to let another firm become its information systems department.

Outsourcing of all IS functions is a long-term, strategic decision. It applies to an entire organization and not just a single development project. So, even though we discuss it as one of the alternatives for implementation, this decision is not typically made by any project team. It is usually a top executive decision. Normally, a facilities management contract

facilities management

the outsourcing of all data processing and information technology to an outside vendor

between an organization and a provider is a multimillion-dollar contract that covers services for 8 to 10 years. Electronic Data Systems (EDS), a multibillion-dollar company, is one company that obtains the majority of its revenues by providing facilities management services to many industries. EDS supports the banking, health insurance (such as Blue Cross and Blue Shield), grocery, insurance, and retailing industries, as well as governments. EDS can provide high-quality facilities management services in these various industries by employing a staff of highly experienced industry specialists.

Packaged, Turnkey Software, and ERP Systems

Packaged software comprises software systems that are purchased to support a particular application. A strict definition implies that the software is used as is, with no modifications. We all have packaged software on our personal computers, such as a word processor or an accounting/general ledger package. We buy the software components without the source code but with documentation, install it, and use it. We don't modify it or try to add new capabilities. We use it exactly as it comes, with only the built-in options. The advantages of this software are that it works well and is inexpensive for the amount of capability provided. It is also usually well documented, relatively error free, and stable.

Packaged software has its place in the overall scheme of an organization's IS strategy. First, many packages can become part of a larger project. For example, a standard reporting system package may provide reporting capabilities to users. Generally, whenever possible, companies try to find packaged software to perform those standard functions.

A *turnkey system* is provided by an outside company as a complete solution, including hardware and software, and the organization only needs to turn it on. In most cases, the outside vendor specializes in a particular industry and its application software. Literally hundreds of firms, many of them small to medium-sized, specialize in systems for particular needs. These turnkey system firms advertise in trade journals for an industry. A few examples are legal systems for law firms, video systems for video stores, patient record systems for dentists and doctors, point-of-sale systems for small retail firms, construction management systems for construction firms, library systems for libraries, and so forth. The list is almost endless.

One critical problem with turnkey systems is that they often do not exactly meet the needs of an organization, and the organization frequently has the onerous task of modifying the way it does business to conform to the computer system. Some turnkey system vendors will modify their systems to suit particular customers. An organization normally purchases the base system, a certain number of customized changes, and a service agreement. The vendor firm analyzes the unique requirements of the organization and makes those changes to the program code. The service agreement can range from simple input form and report modifications to more extensive modifications over a period of months or years. In some cases, only executable code is provided; in others, both executable and source code are provided so that the organization can also make its own modifications. Sometimes the vendor firm makes all modifications; other times the purchasing organization may have programmers work with the vendor's project team to reduce the cost of customization and to gain experience on the new system. Again, numerous combinations are possible, and this method is very popular for obtaining software for small and medium-sized applications that are somewhat, but not completely, standard.

In the past, turnkey systems were used only for specialized systems within an organization. However, recently, several large firms have introduced this approach for enterprisewide systems. These systems, called *enterprise resource planning (ERP)* systems, support all operational functions of an entire organization. Companies such as SAP and PeopleSoft have had good success introducing ERP systems into organizations. Obviously, when the support is enterprisewide, the deployment is a major undertaking. Many of these projects take longer than a year to install and cost millions of dollars.

packaged software

software that is already built and can be purchased as a package

turnkey system

a complete system solution, including software and hardware, that can be turned over to the purchasing organization

The advantage of ERP systems is that a new system can usually be obtained at a much lower cost and risk than through in-house development. The cost is lower because 60 to 80 percent of the application already exists in the base system. Risk is lower because the base system is usually well developed and tested. In addition, other organizations are already using it, so it has a track record of success.

The disadvantage is that the ERP system may not do exactly what the organization needs, even after the system has been customized. Frequently, a gap exists between the exact needs of the organization and the functionality the system provides. The company then must modify its internal processes and train its users to conform to the new system. ERP systems are discussed in more detail in Chapter 18.

Custom-Built Software Systems

Custom-built software systems are those that are developed partly or completely by an outside organization and tailored to the exact needs of an organization. The new system is developed from scratch, based on the systems development life cycle. In some cases, the project team is staffed entirely by a consulting firm; in others, the project team is a combination of in-house staff and outside consultants.

The advantage of custom development is that an organization purchases a tremendous amount of experience and expertise to build a new system. Usually, the consulting firm has developed similar systems in the past and has extensive domain knowledge for a particular industry and application. It also will have a large pool of very experienced staff to solve complex technical problems. In addition, a large, experienced staff can be brought to the project rapidly to meet schedules and deadlines. Outsourcing and contract development are the fastest-growing segments of the IS industry.

The major disadvantage of custom development, of course, is the cost. Not only is the organization paying for the development of a new system, but it is paying for it in hourly wages for consultants. Typically, organizations opt for custom development when they do not have in-house expertise or have very aggressive schedules that must be met. Normally, the anticipated return on investment for the new system must be quite high to justify the cost of this approach. Custom systems are usually large systems with very high transaction volumes. One example is a health-care system with millions of claims to process. When the system can reduce the cost of processing a claim by one or two dollars, the total savings reach millions of dollars quickly due to the high volume.

Most large and medium-sized companies have an in-house information systems development staff. In fact, you may find excellent employment opportunities as a member of the development staff in such companies. One of the main problems with in-house development, particularly in medium-sized firms, is that a portion of a project may require special technical expertise beyond employees' experience. As a result, one alternative is to use company employees to manage and staff the project but to hire special consultants to assist in areas where extra expertise is required. That way, the organization can maintain control and ensure progress but still obtain assistance when needed.

The advantages of this approach are primarily control of the project and knowledge of the project team. Company staff also have a better understanding of the internal culture of the organization and the specific processing needs of various business groups. One other major benefit is that the organization can build internal expertise by developing the system in-house.

The major disadvantage is that the in-house staff may not recognize when they need assistance. At times, the "not invented here" syndrome—the notion that "if we did not think of it or develop it, it is no good"—complicates development because perfectly good, reasonably priced solutions are not utilized. Sometimes the technical problems are more complex than anticipated, and in-house people do not recognize the need to obtain expert assistance.

Selecting an Implementation Alternative

At times, selecting an implementation alternative is straightforward. At other times, deciding among alternatives can be difficult, especially when outside providers are included. For example, one solution may have some of the required functions, but not all. Another solution may have the requisite functions but may only run on an undesirable platform and operating system. Some solutions may provide a quick, inexpensive solution for existing problems but may be limited for future growth; others offer long-term capabilities but are very expensive and take a long time to develop. One vendor may propose a turnkey system, another custom development, another a turnkey system with a particular database management system and platform, and yet another a joint development project. The problem in selecting is the proverbial comparison of apples and oranges. Frequently, there is very little in common among the solutions proposed by outside vendors because each vendor proposes a system that fits its own strength. The systems analyst must establish a set of common criteria to compare the alternatives with as much consistency as possible.

Identifying Criteria

To begin selection, you must identify the criteria that you will use to compare the various alternatives. You will use these criteria to compare all viable alternatives, although differences among alternatives may make some criteria more or less applicable to those proposals. In particular, there are usually some differences in criteria or evaluation methods in comparing packaged and turnkey with custom-built systems. For example, packaged and turnkey systems typically have an existing base of users who can be queried regarding system functionality, reliability, and other important characteristics. For custom-built systems, it may be more important to ascertain the technical skills of the vendor staff. For alternatives that combine purchase of existing solutions with substantial customization or new development, both criteria would merit careful scrutiny.

Different criteria and evaluation methods may also be applied to alternatives presented by outside vendors and those presented by an internal IS department. However, there is an inherent danger of bias toward internal providers when applying different criteria for internally generated alternatives. In theory, criteria such as "vendor reliability" and long-term costs should be evaluated similarly for both internal and external providers. But as a practical matter, some criteria are often ignored or given less emphasis due to the perception of lower risk and greater control over internal IS staff and departments. The project manager and oversight committee must carefully examine selection criteria and measurement methods to ensure fair and complete comparison of internal and external alternatives.

There are three major areas to consider in selecting an implementation alternative:

- General requirements
- Technical requirements
- Functional requirements

General requirements include considerations that are important but not directly associated with the computer system itself. The first major component of general requirements is the feasibility assessment, which was discussed earlier in the context of selecting the scope and level of automation. Each of the implementation alternatives under consideration must meet the requirements for cost, technology, operations, and schedule defined in the feasibility analysis. The following list identifies several criteria that can be included in this section:

- The performance record of the provider
- Level of technical support from the provider

- Availability of experienced staff
- Development cost
- Expected value of benefits
- Length of time (schedule) until deployment
- Impact on internal resources
- Requirements for internal expertise
- Organizational impacts (retraining, skill levels)
- Expected cost of data conversion
- Warranties and support services (from outside vendors)

Obviously, some criteria are more important to the organization than others. For example, in the preceding list, we may want to purchase only from a very reputable, stable, and experienced provider. So, the performance record of the provider is extremely important. On the other hand, we may have some leeway in the schedule, so a very short deployment schedule may not be critical. The relative importance of each item in the list can be weighted with a numbering scale. Figure 8-7 provides a sample table of general criteria and weighting factors for RMO. That table uses a five-point weighting scale. Criteria that are more important are given a higher number, such as a five or maybe a four. Those that are less important are assigned lower numbers. The extended score is the weight times the raw score for each category.

FIGURE *8-7*

A matrix showing a partial list of general requirements.

General requirements criteria	Weight (5=high, 1=low)	Alternative 1 In-house		Alternative 2 Package #1 + modify		Alternative 3 Package #2 + modify		Alternative 4 Custom development	
		Raw	Extended	Raw	Extended	Raw	Extended	Raw	Extended
Availability of experienced staff	4	3	12	3	12	3	12	5	20
Developmental cost	3	5	15	5	15	3	9	1	3
Expected value of benefits	5	5	25	3	15	4	20	3	15
Length of time until deployment	4	2	8	5	20	4	16	2	8
Low impact on internal resources	2	2	4	4	8	5	10	4	8
Requirements for internal expertise	2	2	4	4	8	5	10	4	8
Minimal organizational impacts	3	4	12	3	9	4	12	4	12
Performance record of the provider	5	5	25	4	20	4	20	4	20
Level of technical support provided	4	5	20	3	12	3	12	3	12
Warranties and support services provided	4	5	20	4	16	4	16	4	16
Total			**145**		**135**		**137**		**122**

The four alternatives along the top of the table represent various implementation options. The first alternative is to develop the system in-house. The second and third alternatives are different turnkey systems that start with a basic package and modify it. The last option is to contract with a consulting firm to develop a completely new system from the ground up. The four alternatives are for illustration only—to show the various weighting values possible.

Functional requirements represent the functions that must be included within the system. These requirements are developed during the analysis phase, identified in the event table, and described in the data flow diagrams or use case diagrams. Each project has a unique set of functional requirements based on the needs of the system.

Figure 8-8 illustrates a partial list of functional requirements for the RMO customer support system. The weighting technique is the same as is used for general requirements.

FIGURE *8-8*

A matrix showing a partial list of functional requirements.

Functional requirements criteria	Weight (5=high, 1=low)	Alternative 1 In-house		Alternative 2 Package #1 + modify		Alternative 3 Package #2 + modify		Alternative 4 Custom development	
		Raw	Extended	Raw	Extended	Raw	Extended	Raw	Extended
Make inquiry on items	4	5	20	4	16	5	20	5	20
Create customer order	5	5	25	5	25	5	25	5	25
Change order	4	5	20	5	20	5	20	5	20
Make inquiry on orders	4	5	20	5	20	4	16	5	20
Package order	5	5	25	5	25	5	25	5	25
Ship order	5	5	25	5	25	5	25	5	25
Create back order	4	5	20	5	20	5	20	5	20
Accept return	4	5	20	5	20	4	16	5	20
Correct customer account	4	5	20	3	12	4	16	5	20
Update catalog	5	5	25	2	10	3	15	5	25
Create special promotions	3	5	15	0	0	2	6	5	15
Initiate a promotion mailing	3	5	15	0	0	2	6	5	15
Create sales summaries	3	5	15	3	9	3	9	5	15
Create order summaries	2	5	10	3	6	3	6	5	10
Create shipment summaries	2	5	10	2	4	5	10	5	10
Total			**285**		**212**		**235**		**285**

In addition to the functional and general requirements, each new system normally has a set of technical requirements that must be met. Technical requirements are also system constraints—the constraints under which the system must operate. This category includes all other requirements that are placed on the system, its method of operation, its performance, its utility, and so forth. The following list indicates some of the items that should be considered under technical requirements:

- Robustness (the software does not crash)
- Programming errors (the software calculates correctly)
- Quality of code (maintainability)
- Documentation (user and system, on-line and written)
- Ease of installation
- Flexibility (the software makes it easy to adjust to new functionality and new environments)
- Structure (maintainable, easy to understand)
- User-friendliness (natural and intuitive use)
- Performance (response time)
- Scalability (ability to handle large volumes)
- Compatibility with operating environment (hardware, operating system)

FIGURE *8-9*

A matrix showing a partial list of technical requirements.

Technical requirements criteria	Weight (5=high, 1=low)	Alternative 1 In-house		Alternative 2 Package #1 + modify		Alternative 3 Package #2 + modify		Alternative 4 Custom development	
		Raw	Extended	Raw	Extended	Raw	Extended	Raw	Extended
Robustness	5	?	*18	3	15	4	20	?	*18
Programming errors	4	?	*16	4	16	4	16	?	*16
Quality of code	4	?	*18	4	16	5	20	?	*18
Documentation	3	5	15	3	9	4	12	4	12
Easy installation	3	5	15	5	15	4	12	4	12
Flexibility	3	4	12	3	9	4	12	5	15
Structure	3	4	12	4	12	4	12	4	12
User-friendliness	4	5	20	3	12	4	16	5	20
Total			**126**		**104**		**120**		**123**

Figure 8-9 shows possible weighting factors and scores for technical requirements. For alternatives that are already built, such as packages or ERP systems, scores can usually be derived. However, for custom-built alternatives, such as in-house projects, these points become objectives for the new system. In other words, since nothing is built yet, these items cannot be measured or evaluated. However, they do become criteria for the construction of the new system. In Figure 8-9, to make balanced comparisons between the alternatives, we have assigned values to "build" alternatives that are the averages of the "buy" alternatives (the values are marked by asterisks).

Probably the most difficult part of this exercise is establishing the weighting factors. The client, system users, and the project team should all have a voice in establishing the weighting factors. Consideration must be given not only to the relative importance of each criteria within each area—general, functional, or technical—but also to the balance among all major areas. In other words, the rating team must ensure that the relative weight of general requirements compared with functional requirements truly represents the desires of the client.

Making the Selection

Once requirements have been considered and rated, each alternative can then be evaluated with a raw score based on how well it meets the criteria. Ranges for raw scores generally vary from a simple three-point scale to a more finely ratcheted six-point scale. For example, a three-point scale could contain these ratings: Fully Satisfy (2), Partially Satisfy (1), and Not Satisfy (0). A six-point scale could represent these ratings: Superior (5), Excellent (4), Good (3), Fair (2), Poor (1), and Disqualify (0). To calculate a weighted score in each criterion for each alternative, staff would multiply the raw score by the weighting factor. An overall score, which is the sum of the individual criteria scores, determines a ranking among the various alternatives for this category.

RMO decided to undertake most of the CSS development with in-house staff. As seen in Figure 8-7, the first three alternatives rank very close together on general requirements, with in-house development having a slight advantage. In Figure 8-8, alternatives 1 and 4 are approximately equal and better than alternatives 2 and 3. In Figure 8-9, which shows technical requirements, alternatives 1, 3, and 4 are very close. So, overall, in-house development does provide a slight advantage. RMO's in-house systems analysts and technical staff had proven several times in the past that they could handle development of complex systems. In addition, this approach would enable RMO to continue to build in-house expertise. But although the approach selected was to do development in-house, the information systems group was not averse to hiring specialists when needed.

RMO has sufficient in-house expertise to develop the networks and the database portions of the system. The Web-based development will also be done in-house, but RMO will probably have to hire several specialists. Since RMO wanted to keep the expertise within the company, hiring some new specialists seemed to be a viable method.

Some of the integration issues, such as integrating new client-server architecture with mainframe systems, could possibly become quite complicated. Some very experienced consultants were available, and RMO anticipated that it would need to retain a couple to oversee this portion of the project.

RMO staff is now ready to proceed with the CSS project. A review of the feasibility constraints identified no serious problems. The project is still on schedule and within budget, and the attitude within the company is very positive. With the availability of these additional resources, the project also appears to be technically feasible.

CONTRACTING WITH VENDORS

For RMO's customer support system, in-house development was the chosen alternative for implementation. But to get the information needed to evaluate all the other alternatives, the CSS team sent out a formal request for proposal to each prospective vendor.

Generating a Request for Proposal

As just mentioned, a *request for proposal (RFP)* is a formal document sent to vendors. Its basic purpose is to state requirements and solicit proposals from vendors to meet those requirements. Use of RFPs is almost universal in government contracts and fairly common in private industry. The project manager has primary responsibility for developing the RFP and evaluating submitted proposals.

Often, particularly in governmental purchasing, an RFP is a legal document. Vendors rely on information and procedures specified in the RFP. That is, they invest resources in responding with the expectation that certain procedures will be followed consistently and completely. Thus, an RFP is often considered to be a contractual offer, and a vendor's response represents an acceptance of that offer.

A good RFP includes a detailed explanation of the information needs of an organization and the processing requirements that must be fulfilled. Chapter 4 defined system requirements as consisting of functional and technical requirements. A good RFP will provide detailed explanations of both these types of system requirements. If the early project assumptions indicated that the most viable option for the new system would be to purchase a turnkey solution or outsource for custom development, then the analysis activities are geared toward developing an RFP. When the outside firm is selected, it will then ensure that its staff obtains in-depth knowledge about the problem domain before beginning detailed design or customization.

To develop and distribute a good RFP, the purchasing organization must do an in-depth analysis. This work is not usually a problem for firms that have in-house information systems staff. However, for firms that do not have information systems staff, determining processing requirements can be a problem. They may also have difficulty generating a meaningful RFP or evaluating the various purchase alternatives. Smaller, unsophisticated firms tend to ignore this problem and simply try to make the best decision they can, often in ignorance. A wiser approach is to hire an independent consultant, one who will not be involved in the development, to help establish the selection criteria and decide on a vendor. The same criteria for choosing a final vendor should be used in selecting this independent consultant.

Figure 8-10 shows an outline of a generic RFP. Obviously, each RFP must be tailored to the specific needs of the organization and the requirements of the project. The first part of the RFP, comprising items I and II, provides background information on the company and

FIGURE *8-10*

A sample RFP table of contents.

Request for Proposal
Table of Contents

I.   Introduction and background
     A. Background on company
     B. Overview of industry/business

II. Overview of need
     A. Description of business need
     B. Expected business benefits
     C. Overview of system requirements

III. Description of technical requirements
     A. Operating environment
     B. Performance requirements
     C. Integration, interfaces, and compatibility
     D. Hardware specifications
     E. Expansion and growth requirements
     F. Maintainability requirements

IV. Description of functional requirements
     A. Specification of primary functions
     B. Specification of information outputs
     C. Specification of the user interface
     D. Identification of optional functions and enhancements

V.   Description of general requirements
     A. Maintenance and support
     B. Documentation and training
     C. Future releases
     D. Other contractual requirements

VI. Requested provider and project information
     A. Request for statement of work and project schedule
     B. Request for reference list of provider
     C. Request for project personnel information

VII. Details for submitting the proposal
     A. Time requirements
     B. Format requirements

VIII. Evaluation criteria and process
     A. Expected timetable of evaluation
     B. Method of evaluation of technical, functional, and general requirements

the need for a new system. Next, items III through V describe in detail all of the requirements that the new system must meet. In this example, we have divided the requirements into technical, functional, and general requirements. Section VI requests information on the provider's background and experience. The final two sections, VII and VIII, indicate how the proposal should be submitted and how it will be evaluated.

The RFP should clearly state the procedural requirements for submitting a valid proposal. When possible, the organization should include an outline of a valid proposal, along with a statement of the contents of each section. In addition, the RFP should clearly state deadlines for questions, proposal delivery, and other important events.

The requirements statement constitutes the majority of the RFP. The body of the RFP can formalize and state the guidelines previously described. Requirements should be separated into those that are absolute (essential) and those that are optional or subject to negotiation. This categorization is a more formal version of the prioritization discussed earlier. The RFP also should state explicitly the evaluation criteria—for example, the categories that are to be evaluated as well as the weighting factors.

Benchmarking and Choosing a Vendor

One method to evaluate the quality of a vendor's system is either to observe it in use or to install it on a trial basis and test it out. However productive this approach may be, it is also expensive and difficult. The format of the data, the forms, and even the platform may be alien to existing configurations. If the system is complex, then people must be trained to use it. Staff must also be available to do testing. Although this approach can be quite expensive, it may be less expensive than making a bad decision.

benchmark

an evaluation of a system against some standard

Some applications are amenable to a more rigorous evaluation called a *benchmark*. A benchmark is a performance evaluation of application software (or test programs) using actual hardware and systems software under realistic processing conditions. In years past, benchmarking was often difficult to perform because of the expense of the hardware and software configurations and the length and cost of installation. Currently, these problems are less severe because hardware is cheaper, installation procedures have been streamlined, and competition among vendors is fierce.

Another way to observe a system in use is to visit another company. Sometimes the vendor will have demonstration versions already installed in its own facilities. Potential purchasers are permitted to go to the vendor's plant and test the system. Vendors also have previous clients who are using the system. It is almost always a good idea to visit these previous clients. When prospective clients make a site visit, they are not permitted to test the system themselves with their own data, but they will be permitted to observe the other company using the system. They can also talk to previous clients about their experience with the system and the vendor. It is always a good idea to get references and to talk with other companies that have done business with the vendor.

Frequently, companies want to add capabilities to the software as the business environment changes. To do so, they must make the enhancements themselves or have their vendor supply upgrades and fixes. Organizations should ascertain the level of ongoing research and development being done by the vendor. How compatible are these new capabilities to a particular system after it has been customized? Is there an active users group that can suggest new enhancements and modifications to the vendor? The company is making a major investment, and it is important that the investment be considered in the long term. The largest investment in any new system is the long-term cost of maintenance. A good vendor will help its clients leverage their maintenance dollars by providing upgrade support and new capabilities based on feedback from existing clients.

Developing a Contract

Once a final decision is made on which proposal provides the best solution and value, then a contract is written. Contract development and negotiation are usually a team effort involving the project manager, legal counsel, and frequently other senior executives. The project manager's involvement is essential to ensure that the contract meets the needs of the project and that important performance and termination clauses are included.

Contracts can be divided into several different types, which shift the risk either to the purchaser or the vendor. Fixed-dollar contracts put most of the risk on the vendor. The advantage to the purchasing company is that the vendor assumes the burden of project delays and overruns. However, the vendor usually sets a high price to compensate for the risk. Cost-plus-percentage contracts put the risk on the purchaser. In fact, cost-plus-percentage contracts encourage the vendor to spend more since its income is directly proportional to the costs of the project. A middle ground, with both sharing the risk, is a cost-plus-fixed-fee or cost-plus-incentive contract. In this case, both the purchaser and the vendor benefit if the project finishes as quickly as possible. With a fixed fee, the profit margin for the vendor is high if the project progresses quickly. If the project drags on, however, the fixed fee results in a lower profit margin.

PRESENTING THE RESULTS AND MAKING THE DECISIONS

The results of the investigation and analysis activities described in this chapter are normally summarized in a written report and presented orally to executives. The intended audience is the executive oversight committee, which has decision-making and funding responsibility for the project. The objective of the documentation and presentation is to provide the necessary background so that informed decisions can be made.

The responsibility of the project team, including both technical and user members, is to do the detailed investigation and calculations to enable an informed analysis of all of the alternatives. However, the final decision of which alternative, or mix of alternatives, is chosen rests with the executive oversight committee. This committee not only controls the budget and provides the funding but also is responsible for the overall strategic direction of the company.

One of the more difficult tasks for the project team is to compile, organize, and present the alternatives and critical issues in a way that is easy to understand yet accurate and complete. The executive oversight committee usually consists of people who are, first and foremost, business executives. They generally are not technical experts, yet they need to make decisions that affect the entire organization. So, presenting findings is one of the most difficult tasks that the project team will have. It requires careful consideration to find the right balance of detail. At one extreme is so much technical detail that the oversight committee cannot understand or follow the logic and becomes lost or bored. At the other end are recommendations without sufficient supporting detail or logic.

The formality of the presentation varies from organization to organization. Some companies require very formal written reports and oral presentations. Other organizations require nothing written and only informal discussion between the client (the person funding the project) and the project team leader. Smaller organizations tend to be less formal, and large corporations typically have standard policies and procedures for approval.

The format of the document and report varies considerably, depending on the desires of the audience. A detailed description of how to develop this presentation is provided in Appendix D. (Appendix D is available for download at www.course.com via the "Student Downloads" link on the Web page for this book.) Generally, the written documentation will follow the same format as the presentation.

SUMMARY

The activities explained in this chapter are primarily project manager responsibilities. The focus of the project changes at this point from one of discovering the requirements to that of developing a solution system. So, the activities described are pivotal for the project to change emphasis. Obviously, the project manager takes primary responsibility for this change in direction. These activities involve seven of the eight project management knowledge areas that are described in Appendix A.

One of the important activities in the analysis phase is to prioritize the system requirements based on the scope and level of automation desired. The scope of the new system determines which functions it will support. The level of automation is a measure of how automated the selected functions will be. Highly automated functions have sophisticated computer systems such as expert systems to help carry out the business functions.

The application deployment environment is the configuration of computer hardware, systems software, and networks in which the new system must operate. It determines the constraints that are imposed on the system development alternatives. The analyst must define an environment, or multiple environmental choices, that match application requirements and the organization's strategic application and technology architecture plans.

Another activity that is done in conjunction with prioritizing the requirements is determining what alternatives are possible for developing the solution and then selecting one of those alternatives. Implementation alternatives include such options as building the system in-house, buying a packaged or turnkey solution, or contracting with a developer to build it (outsourcing). When outsourcing is anticipated, a request for proposal (RFP) is developed and sent out. The RFPs are then evaluated for how well they match the requirements. Selecting from the various alternatives should be a careful process. The evaluation includes consideration of such factors as the match of the proposed system to the functional and technical requirements and the reputation and performance record of the submitting vendor.

One of the final tasks in the analysis phase is developing recommendations and presenting them to management. After the analysis phase, a more knowledgeable decision can be made about the direction, cost, feasibility, and approach of the rest of the project. The systems analyst documents the results of the analysis phase and presents them in a logical fashion that is focused toward the executives who make funding decisions.

KEY TERMS

application deployment environment, p. 291

benchmark, p. 306

development environment, p. 292

facilities management, p. 297

packaged software, p. 298

request for proposal (RFP), p. 304

turnkey system, p. 298

REVIEW QUESTIONS

1. What is meant by the *application deployment environment*? Why is it important in the consideration of a development approach?

2. List and briefly describe the characteristics that an analyst examines when choosing or defining the deployment environment.

3. Describe the relationship between the application deployment and development environments.

4. Explain the fundamentals of facilities management.

5. What is the difference between scope and level of automation?

6. What is meant by the build-versus-buy decision?

7. Define a *packaged solution*. Explain what is entailed in the packaged solution approach.

8. What is meant by *ERP*? How does an ERP approach affect acquiring a new solution?

9. What does *outsourcing* mean? How does it affect a project?
10. Define *benchmark*. Why is it useful in selecting a new system?
11. What is an *RFP*? Why is it developed at the end of the analysis phase instead of at the beginning?
12. What is the difference between general requirements, technical requirements, and functional requirements?

THINKING CRITICALLY

1. What are the advantages of purchasing a packaged solution? What are the disadvantages or dangers?
2. What are the advantages of building a solution from the ground up? What are the disadvantages?
3. What are the advantages to outsourcing a development project? What are the disadvantages?
4. Discuss the importance of developing a formal technique and specific criteria for evaluation alternatives.
5. Given the following narrative, identify the functions to be included within the scope of the system. Also identify several levels of automation for each function. The purpose of this question is to give you an opportunity to think creatively, especially to identify high-level automation alternatives for the various functions.
   Conference Coordinators (CC) assists organizations or corporations in coordinating and organizing conferences and meetings. It provides such services as designing and printing brochures, handling registration of attendees, fielding questions from attendees, securing meeting spaces and hotel rooms, and planning extracurricular activities. CC gets its business in two ways: by following up on leads that a company is going to be holding a conference and by having the company contact CC directly. When a contact is made, the client is asked for basic information about the desired event: city, dates, anticipated number of attendees, price range, and external activities desired. From this information, CC prepares a bid. CC likes to keep its turnaround time on bids to under five working days. Each project is assigned to a project manager, who will gather information from the support staff to prepare the bid. If necessary, he or she may also request information from the visitors' center for the desired city.
6. What are important points that determine weighting factors for the functional requirements listed in the system requirements for a proposed system?
7. List the important points that determine weighting factors in the general and technical requirements for a proposed system.
8. Given the following matrix of various technical requirements, develop your own weighting factors for an inventory management system at a small plumbing supplier. Justify your weights. Extend the raw scores to the Extended column and calculate the totals. Which would you choose? Justify your selection: Did you go strictly by the numbers, or are there other factors you might consider? How do you handle a number that is not given: give it an average of the others, pick the best of the others, guess a value, or assign a zero? (Raw numbers use a six-point scale.)

Category	Weight Build in-house	Alternative 1 Buy turnkey		Alternative 2 Buy package		Alternative 3	
		Raw	Extended	Raw	Extended	Raw	Extended
Robustness		5		3		3	
Programming errors		?		4		4	
Quality of code		?		4		5	
Documentation		4		4		3	
Easy installation		5		5		4	
Flexibility		5		4		3	
User-friendliness		5		5		5	
Total							

EXPERIENTIAL EXERCISES

1. Assume that the deployment environment for a high-volume payment processing system consists of the following:
   - DB2 DBMS running under the OS/390 operating system on an IBM S/390 mainframe
   - WebSphere application server running under the Z/OS operating system on an IBM zSeries 900 mainframe
   - CORBA-compliant component-based application software written in Java that will be executed by other internal and external systems

   Investigate possible development environments for this deployment environment. Describe their advantages and disadvantages and recommend a specific set of development tools.
2. Set up an interview with an organization that uses information systems. Ask for an example of an RFP for a software system. Identify the parts of the RFP. Compare them with the recommended components discussed in this chapter.
3. From a news article or Internet information, find an example of a company that is installing an ERP package (SAP, PeopleSoft, or other company). If possible, get a copy of the overall project plan and analyze the various activities. Compare them with a standard SDLC. Find out the total budget for the project.
4. Develop an RFP for RMO to be sent out to various vendors.
5. Develop a recommended implementation approach for RMO. Also develop a presentation of your recommendation to upper management.
6. Look through some trade magazines (*Software*, *CIO*, *Datamation*, *Infoweek*, and so on) to find examples of companies that have done an evaluation of vendors. Describe their methods and comment on their strengths and weaknesses.

CASE STUDIES

Tropic Fish Tales' RFPs

Now that you have read and studied the chapter, review the opening case on Tropic Fish Tales. Your job is to provide specific advice for Robert Holmes or Bill Williams on how to evaluate the various RFPs.

Assuming that you can build some matrices that measure relative strengths among the proposals, comment on the applicability of doing an evaluation based strictly on the numbers. In other words, assume that Robert and Bill were able to create criteria and weights to measure the benefit to the company of the different alternatives.

1. Do you think it would be possible to sum up the resulting values and make a decision based only on the numbers? Support your answer.
2. What factors, other than those in the matrix of weighted criteria, might Robert and Bill need to consider in making a decision? Can these other factors influence the decision as strongly as the quantified criteria?
3. What if the values of several alternatives are very close? What other factors might Robert and Bill need to consider?

The Real Estate Multiple Listing Service System

Consider the requirements of the multiple listing service system developed in Chapters 5, 6, and 7. Assume that you're the project manager and that you work for a consulting firm hired by the multiple listing service to perform only the survey and analysis phase activities.

1. Assume that system users and owners have indicated a strong desire for a system that can be accessed "anytime, anywhere." Discuss the implications of their desire for the system scope. Given the preferences of the system users and owners, should you prepare a table similar to Figure 8-2? Why or why not?
2. Discuss the implications of the anytime, anywhere requirement for the application deployment environment. What type(s) of hardware, network, and software architecture will be required to fulfill that requirement?

3. Investigate the availability of packaged and turnkey systems for multiple listing services. Search the Internet and real estate trade magazines and Web sites. Discuss the pros and cons of choosing a packaged or turnkey system.
4. Develop an RFP outline that covers packaged, turnkey, and custom-developed systems. What are the difficulties of writing one RFP that covers all three scenarios? Who should be involved in evaluating RFP responses?

Rethinking Rocky Mountain Outfitters

Various application deployment environments would actually be acceptable for RMO's strategic plan. The staff's current thinking was to move more toward a Microsoft solution, using the latest version of Microsoft Server with Microsoft's IIS as the Web server. However, Linux with Apache servers offers another large installed base of servers. Considering that RMO could also take that approach, do the following:

1. Describe a viable configuration using Apache/Linux.
2. Compare the relative market penetration of Microsoft and Apache/Linux (a good starting place is http://news.netcraft.com).

The database issue is another potential controversy for RMO. The current decision is to keep the mainframe and run DB2, a very efficient relational database. However, another alternative would be to implement an Oracle database. Oracle is also very strong in the marketplace. Given these two alternatives, do the following:

3. Compare the relative market penetration of these two solutions.
4. List the strengths and weaknesses of each approach, that of the DB2 mainframe approach and that of Oracle running on some type of multiple processor server computer.

Focusing on Reliable Pharmaceutical Service

Assume that Reliable has completed a thorough analysis of system requirements (part of which you worked on as case exercises in Chapters 4 through 7). Management is now confronted with the task of choosing a system scope and implementation approach. To summarize the alternatives, you have prepared the following table, which divides the requirements into functional subsets, estimates the duration of design and implementation for each function if software is custom-built, and categorizes the risk for each function based on software complexity, technology maturity, and certainty about requirements.

Function	Project Duration	Risk
Inventory and purchasing	9 months	moderate
Order fulfillment (manual data entry)	6 months	low
Web-based order entry	9 months	high
Prescription warning	12 months	high
Billing	18 months	high

Top executives have evaluated the table and determined that all of the functions are high-priority needs. The project is critical to restoring profitability and maintaining market share. Reliable is well behind the technology curve for its industry, and it needs to modernize to reduce costs and to provide expected levels of service. Unfortunately, overlap and dependency among the functions makes it difficult to consider implementing only a subset of the functions. Executives would prefer to implement all functions in a single project, but they consider the combined project duration for all functions to be much too long.

Significant parts of the proposed system, such as inventory, purchasing, and prescription warning, are similar to systems used by retail pharmacies and in-house pharmacies in large hospitals and health maintenance organizations. But some significant differences exist in Reliable's requirements for order entry, product delivery, and billing. There are a handful of large vendors and several dozen smaller vendors that specialize in pharmacy systems.

Assume that management has identified the following options for proceeding with the system development or acquisition:

- Contract with a vendor to modify a packaged prescription software system to suit Reliable's needs
- Contract with a vendor to purchase the generic parts of a prescription system and extend the system to address Reliable's unique needs with custom-built software
- Contract with a system development firm to custom-build a system, possibly making use of some off-the-shelf components for inventory management and prescription warning

Reliable's executives have assigned you the following tasks:

1. Develop an RFP outline that addresses the options identified by the executives. List and briefly describe each general, technical, and functional requirement.
2. Assume that you have already developed a complete set (over 100 printed pages) of analysis documents using either the traditional or object-oriented approach. Should those be included in the RFP? Why or why not?
3. Develop matrices (similar to Figures 8-7, 8-8, 8-9) for evaluating RFP responses.
4. Develop a list of vendors to whom the RFP should be sent.

FURTHER RESOURCES

Scott E. Donaldson and Stanley G. Siegel, *Cultivating Successful Software Development: A Practitioner's View.* Prentice Hall, 1997.

Ralph L. Kliem and Irwin S. Ludin, *Project Management Practitioner's Handbook.* American Management Association, 1998.

Sanjiv Purba, David Sawh, and Bharat Shah, *How to Manage a Successful Software Project, Methodologies, Techniques, Tools.* John Wiley & Sons, 1995.

John J. Rakos, *Software Project Management for Small to Medium Sized Projects.* Prentice Hall, 1990.

Kathy Schwalbe, *Information Technology Project Management* (2nd ed.) Course Technology, 2002.

Neal Whitten, *Managing Software Development Projects: Formula for Success.* John Wiley & Sons, 1995.

System Design Tasks

CHAPTER 9

Moving to Design

LEARNING OBJECTIVES

After reading this you chapter, you should be able to:

- Discuss the issues related to managing and coordinating the design phase of the SDLC

- Explain the major components and levels of design

- Describe each design phase activity

- Describe common deployment environments and matching application architectures

- Develop a simple network diagram and estimate communication capacity requirements

CHAPTER OUTLINE

Understanding the Elements of Design

Design Phase Activities

Project Management: Coordinating the Project

Deployment Environment

Application Architecture

Network Design

James Schultz is a summer intern with Fairchild Pharmaceuticals. He has been assigned to an ongoing development project for a production scheduling and control system. The project was nearing the end of the analysis phase when James joined the team two weeks ago.

James works for Carla Sanchez, the chief analyst and project manager. For the past two weeks, James has been shadowing Carla as she completed final analysis phase tasks. He helped Carla prepare presentations to the project oversight committee and system users. He has absorbed a lot of information about the project in a very short time, but the details and overall project direction haven't yet formed a complete and coherent picture in his mind. Yesterday, the oversight committee signed off on the work to date, so Carla asked James to stop by her office first thing this morning to discuss his assignment for the next project phase.

James knocked on Carla's open door and asked, "Is this a good time or should I come back later?"

"I have time now," she said. "Come in and have a seat. Let's start by reviewing the results of yesterday's meetings, and I'll answer any questions you have. Then we'll narrow down your tasks for the next few weeks of the project."

James said, "I have two questions that I think are related. The first is, which implementation details have been decided and which haven't? The discussions with the users and oversight committee left me with the impression that the decision to go with a full-blown Web-based system had already been made. Yet none of the supporting infrastructure for a Web-based system currently exists, or does it?"

Carla replied, "A few elements are in place, but most of it will need to be designed and acquired. But let's hear your other question before we get into that."

James continued, "Well, you've sort of anticipated my next question, which is what do we need to do next? There seem to be several important tasks that need to be started now, such as choosing system software to support Web services, determining what changes will be needed to the company network, and designing the database. But I suspect that I've left out a few important pieces. Also, the tasks and decisions are so interdependent that I don't know which should be tackled first."

Carla smiled before she replied, "Well, you really were awake during all of the meetings! You should take some pride in knowing that you're confused about exactly the right things at this point in the project. The transition from analysis to design is an important but uncertain step in all projects, and this one is no exception. It's hard to move from a detailed knowledge of what the user wants and needs to a precise blueprint of a system that will satisfy those wants and needs. As you've correctly observed, there are many important decisions that need to be made very quickly, and they overlap. They're also heavily constrained by available time, budget, and existing systems, skills, and infrastructure."

Looking a bit relieved, James replied, "So what's up first, and where do I fit in?"

Carla replied, "The generic name for the next step is architectural design. It's where we'll finalize all of the big-picture decisions, such as what hardware will support the new system, what operating systems we'll use, how we'll store and access data, and what languages and tools we'll use. Some of these issues were briefly addressed at the start of the project, and some decisions were implicit in the choice of deployment environment and automation scope approved by the oversight committee yesterday. What we need to do now is to lay all of them on the table, make sure they're compatible with one another and with existing systems and capabilities, and parcel out the detailed tasks associated with each.

Carla continued, "I spent yesterday afternoon dividing the work into major categories, including hardware and operating systems, Web support services, database design, application software design, and user interface design. I summarized the choices made so far and the remaining decisions we need to discuss. Key players will meet as a group for the rest of the week to discuss options in each area and develop the system architecture. For example, we'll decide whether to extend our existing database to support the new system or develop a new database with a new DBMS. By the end of the week, we'll have made all of the critical architectural decisions, ensured that the pieces all fit together, and developed plans to tackle each area with personnel assignments and time lines. From that point forward, work in each area can proceed in parallel. Professor Chen told me that you've done an independent study in Web services support software, right?"

"Yes," James replied. "I did a comparative study of infrastructure requirements and communication protocols for Web services using COM+, CORBA, and SOAP. I did an in-depth technology review of each and visited two sites using each technology to see how they worked in practice."

"Good," said Carla. "That knowledge will come in handy, since we need to decide whether to base the new system on SOAP and, if so, what supporting infrastructure and development tools to use. I think that you'll learn a lot by working with me for another week or two as we hammer out the architectural design. Once we get the detailed design tasks rolling, we'll choose one for you that suits your interests and abilities. There'll be plenty of interesting tasks from which to choose and more than enough work to keep you busy for the next month or two."

OVERVIEW

Chapter 8 described the activities that complete the analysis phase and begin the transition from analysis to design. This chapter completes that transition and discusses issues related to the design of the new system. While analysis focuses on what the system should do—that is, the requirements—design is oriented toward how the system will be built—on defining structural components. Such activities as defining the scope and prioritizing the requirements are clearly analysis tasks. However, defining the application deployment environment and determining levels of automation are tasks begun during analysis but completed in more detail during design. Thus, we began the transition to design in the last chapter, and we move completely into design considerations in this chapter.

This chapter is the first of seven chapters discussing design. In this chapter, we briefly describe all design phase activities and discuss the first activity in more detail. Later chapters explore other design phase activities using both traditional and object-oriented models and techniques.

UNDERSTANDING THE ELEMENTS OF DESIGN

Systems design is the process of describing, organizing, and structuring the components of a system at both the architectural level and a detailed level with a view toward constructing the proposed system. Systems design is like a set of blueprints used to build a house. The blueprints are organized by the different components of the house and describe the rooms, the stories, the walls, the windows, the doors, the wiring, the plumbing, and all other details. We do the same organizing in systems design, although the components we are describing are the components of the new system. We design and specify various components of the solution.

To understand the various elements of systems design, we must consider two questions:

- What are the components that require systems design?
- What are the inputs to and outputs of the design process?

Major Components and Levels of Design

To do design, analysts first partition the entire system into its major components because an information system is much too complex to design all at once. Figure 9-1 depicts how these various components fit together. The icons on the figure are pieces of hardware, and inside the hardware are the software components. The cloud represents the entire system, and the various icons show the parts of the system that must work together to make the system functional. Information systems professionals must ensure that they develop a total solution for the users—they have not done their job if they haven't provided an integrated, complete solution.

As we will see in an upcoming section, the activities of the design phase of the SDLC support this partitioning of the final system into design components. Basically, each design phase activity is focused around designing one of the identified components shown in Figure 9-1.

A second important idea underlying systems design is that of the different levels of design. During analysis, we first identified the scope of the problem before we tried to understand all of the details. We called this step top-down analysis. Analysis, as it was presented, included both top-down activities (for example, scope first, then details) and bottom-up activities (for example, DFD fragments first, then the middle-level diagram). The same ideas apply during design.

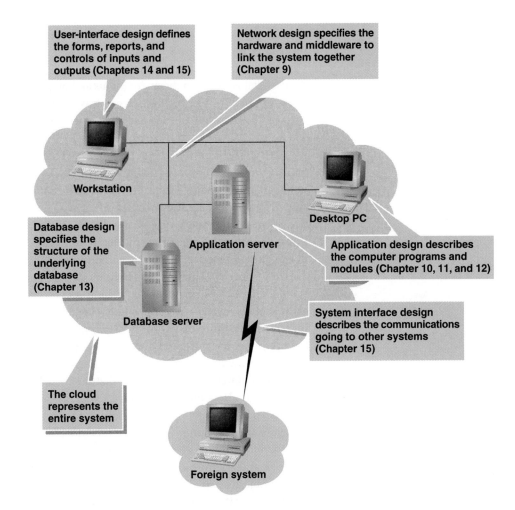

FIGURE *9-1*

System components requiring
systems design.

User-interface design defines
the forms, reports, and
controls of inputs and
outputs (Chapters 14 and 15)

Network design specifies the
hardware and middleware to
link the system together
(Chapter 9)

Workstation

Desktop PC

Database design
specifies the
structure of the
underlying
database
(Chapter 13)

Application server

Application design describes
the computer programs and
modules (Chapter 10, 11, and 12)

Database server

System interface design
describes the communications
going to other systems
(Chapter 15)

The cloud
represents the
entire system

Foreign system

architectural design

broad design of the overall system
structure; also called *general design* or
conceptual design

detail design

low-level design that includes the design of specific program details

As you begin working in industry, you will find that various names are given to the design at the highest level, including architectural design, general design, and conceptual design. We will use the term *architectural design*. During architectural design, you first determine the overall structure and form of the solution before trying to design the details. Designing the details is usually called *detail design*. It is not so important at this point to distinguish which activities are architectural design and which are detail design. Neither is it important to identify which models or documents belong to architectural design or to detail design. What is important is to recognize that design should proceed in a top-down fashion. Let's review the implications of this approach for each of the design components identified in Figure 9-1.

For the entire system, the analysts first identify the overall application deployment environment. They determine the overall architectural requirements and structure of the network before specifying the details of the routers, firewalls, servers, workstations, and other components. This approach was introduced in Chapter 8 and will be expanded in this chapter.

For the application software, the first steps are to identify the various subsystems and their relationships to the network, the database, and the user-interface components. Part of that early design is the automation system boundary. The system boundary identifies which functions are included within the automated system and which are manual procedures. Notice that we began this process by identifying the level of automation, which was explained in Chapter 8.

For the database component, the first steps are to identify the type of database to be used and the database management system. Some details of the record structures and the data fields may have been identified, but the final design decisions will depend on the architecture.

For the user interface, the first steps are to identify the general form and structure of the user dialog based on the major inputs and outputs. The project team also describes the relationship of the user-interface elements with the application software and the hardware equipment. Afterward, the detailed window and report layouts can be developed.

Moving from Analysis to Design

During the activities of the analysis phase, we built documents and models. For traditional analysis, models such as the event table, data flow diagrams, and entity-relationship diagrams were built. For object-oriented analysis, we also used the event table and developed other models such as class diagrams, use case diagrams, and use case descriptions. Regardless of the approach, the input to the design phase activities is the set of documents and models that were built during earlier phases.

During analysis, analysts also built models to represent the real world and to understand the desired business processes and the information used in those processes. Basically, analysis involves decomposition—breaking a complex problem with complicated information requirements into smaller, more understandable components. Analysts then organize, structure, and document the problem domain knowledge by building requirements models. Analysis and modeling require substantial user involvement to explain the requirements and to verify that the models are accurate.

Design is also a model-building activity. Analysts use the information gathered during analysis—the requirements models—and convert that information into models that represent the solution system. Design is much more oriented toward technical issues and therefore requires less user involvement and more involvement by other systems professionals. Figure 9-2 illustrates this flow from analysis to design, highlighting the distinct objectives of each phase.

The original definition of design indicates that design involves describing, organizing, and structuring the system solution. The output of the design activities is a set of diagrams and documents that achieves this objective. These diagrams model and document various aspects of the solution system. As with the analysis models, some components are similar for structured and OO approaches, but other components are very different.

FIGURE *9-2*

Analysis objectives to design objectives.

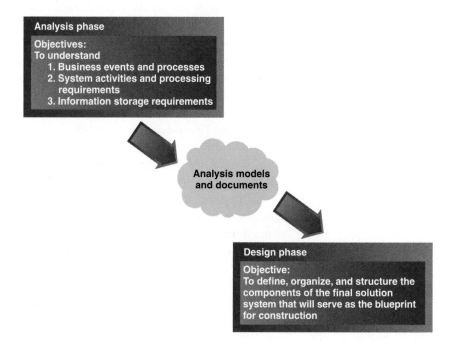

FIGURE *9-3*

Traditional structured and object-oriented models.

Figure 9-3 duplicates the information about traditional and OO requirements models originally shown in Figure 5-36 and extends it with the design models for both traditional structured design and object-oriented design. As noted in the figure, the models developed during the analysis phase feed directly into the models built for design—the

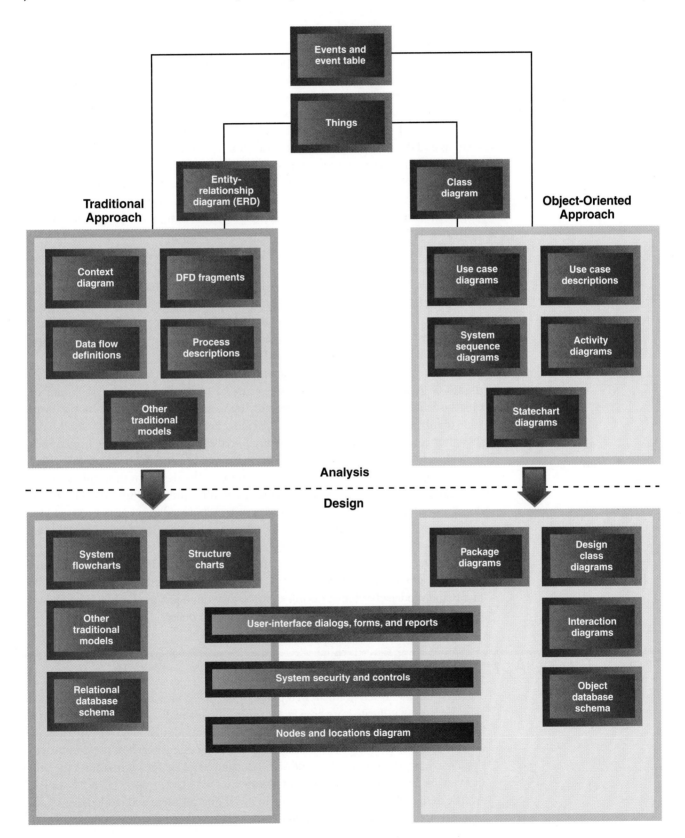

traditional analysis models feed the traditional design models, and the object-oriented analysis models feed the object-oriented design models. Note also that several design models (shown in green and spanning the two sides of the figure) are common to both approaches.

For database design, the traditional approach usually uses a relational database model. The object-oriented technique can design either a relational database model or a newer object-oriented database model. For user-interface design, both techniques include the design of the human-computer dialog, forms, and reports. Both database and user-interface design share many of the same techniques, whether a structured approach or an object-oriented approach is used.

For application architecture design, however, traditional structured techniques and object-oriented techniques do differ substantially. Structured techniques, including analysis and design models, have been used for many years to describe the structure and organization of systems written using the input-process-output model of software. These models are well suited to describing business applications, most of which rely on databases or files and do not require sophisticated real-time processing. These models were originally developed to support application software design and programming using COBOL and BASIC programming languages. They are equally well suited to programming in other languages, such as C, FORTRAN, Pascal, and other business-oriented programming languages.

Object-oriented techniques are newer techniques that have become widely used since the late 1980s. They are well suited to real-time, interactive, and event-driven software such as operating systems that require multitasking capabilities. Object-oriented development is rapidly becoming the preferred approach for developing business applications, which are usually interactive and event driven.

A frequently asked question is, can structured techniques and object-oriented techniques be mixed? In other words, is it possible to do structured analysis and then object-oriented design of the application or vice versa? In some situations it may be possible to mix and match, such as when designing and implementing the interface using OO after completing traditional structured analysis. But generally, such mixing and matching does not work well for application design, because the basic philosophies of the two approaches are so fundamentally different. The design of the application software using a traditional approach provides an architectural structure based on the top-down procedural functions of the system. A system designed using object-oriented techniques has an architectural structure based on the set of interacting objects for each use case.

Whatever approach has been used, once analysts have addressed the major components of a system, have considered its architectural design, and have in hand the documents and models developed during analysis, they can begin to consider how to design the system. We turn next to the activities of the design phase.

DESIGN PHASE ACTIVITIES

The activities identified in the design phase of the SDLC provide an overview of the design process. As indicated previously, these activities provide the design for each of the components illustrated in Figure 9-1. More details about design processes are explained later in this chapter and in subsequent chapters as we discuss each of the design activities. First, let's get an overview of the SDLC design phase. Figure 9-4 identifies the activities that are associated with the design phase.

The design phase involves specifying in detail how a system will work using a particular technology. Some of the design details will have been developed during systems

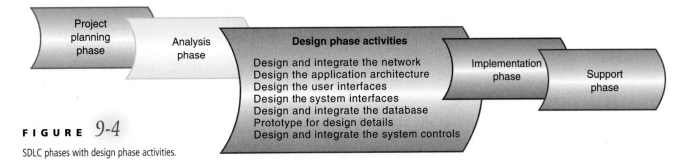

FIGURE *9-4*

SDLC phases with design phase activities.

analysis, when the alternatives were described. But much more detail is required. Sometimes systems design work is done in parallel with the analysis phase, and usually the activities within systems design are done in parallel. For example, the database is typically designed at the same time the user interface is designed. An iterative approach to the SDLC is also commonly used, so when we talk about the design phase activities, sometimes the activities are completed for one iteration and then later for another iteration. As with the activities in the analysis phase, each activity of the design phase can be summarized with a question, as shown in Figure 9-5.

FIGURE *9-5*

Design phase activities and key questions.

Design phase activity	Key question
Design and integrate the network	*Have we specified in detail how the various parts of the system will communicate with each other throughout the organization?*
Design the application architecture	*Have we specified in detail how each system activity is actually carried out by the people and computers?*
Design the user interface(s)	*Have we specified in detail how all users will interact with the system?*
Design the system interface(s)	*Have we specified in detail how the system will work with all other systems inside and outside our organization?*
Design and integrate the database	*Have we specified in detail how and where the system will store all of the information needed by the organization?*
Prototype for design details	*Have we created prototypes to ensure all detailed design decisions have been fully understood?*
Design and integrate the system controls	*Have we specified in detail how we can be sure that the system operates correctly and the data maintained by the system are safe and secure?*

Each of the activities develops a specific portion of the final set of design documents. Just as a set of building blueprints has several different documents, a systems design package also consists of several different sets of documents that specify the entire system. In addition, just as the blueprints must all be consistent and integrated to describe the same physical building, the various systems design documents also must be consistent and integrated to provide a comprehensive set of specifications for the complete system.

Design and Integrate the Network

Sometimes a new system is implemented along with a new network. If this is the case, then the network needs to be designed. More often, though, network specialists have established the network based on an overall strategic plan. The systems design alternative chosen was one that fit the existing network plan. So rather than designing a network, the project team typically must integrate the system into an existing network.

Important technical issues arise when making the system operate over a network, such as reliability, security, throughput, and synchronization. Again, specialists are often brought in to help with the technical details. The requirements developed during systems analysis specify what work goes on at what locations, so these locations need to be connected. Technical requirements (as opposed to functional requirements) often have to do with communication via networks.

Later in this chapter, we highlight critical issues in network design and planning. The key question to be answered when completing the "Design and integrate the network" activity is, *have we specified in detail how the various parts of the system will communicate with each other throughout the organization?*

Design the Application Architecture

Designing the application architecture involves specifying in detail how all system activities will actually be carried out. These activities are described during systems analysis in great detail as logical models, without indicating what specific technology would be used. Once a specific design alternative is chosen, the detailed computer processing—the physical models—can be designed. A key decision is to define the automation boundary, discussed in Chapter 1, which separates the manual work done by people from the automated work done by computer. Models created include physical data flow diagrams, structure charts, interaction diagrams, and other physical models.

The approach to application design and the design models created vary depending on the development and deployment environments. If the programming language is Visual Basic, for example, the type and nature of the models developed will be different than if the language were COBOL. If client-server architecture is used, the models used are different than with a centralized architecture. If object-oriented technology is used, the models definitely are quite different. Additionally, some activities are carried out by people rather than computers, so manual procedures need to be designed.

The key question to be answered when completing the "Design the application architecture" activity is, *have we specified in detail how each system activity is actually carried out by the people and computers?*

Design the User Interfaces

A critical aspect of the information system is the quality of the user interface. The design of the user interface defines how the user will interact with the system. To most users, the interface is a graphical user interface with windows, dialog boxes, and mouse interaction. Increasingly, it can include sound, video, and voice commands. Users' capabilities and needs differ widely; each user interacts with the system in different ways. Additionally, different approaches to the interface might be needed for different parts of the system. Therefore, there are many user interfaces to consider. And as information systems become increasingly interactive and accessible, the user interface is becoming a larger part of the system.

Analysts should remember that to the user of the system, the user interface *is* the system. The user interface is more than just the screens—it is everything the user comes into contact with while using the system, conceptually, perceptually, and physically. So, the user interface is not just an add-on to the system. It is something that needs to be considered throughout the development process.

The nature of the user interface begins to emerge very early in the development process, when requirements are being defined. The specification of the tasks the users complete begins to define the user interface. Then when alternatives are being defined, a key aspect of each alternative is its type of user interface. The activity of designing the user interface in detail, however, occurs during systems design.

Sometimes specialists in user-interface design are brought in to help with the project. These specialists might be called *interface designers*, *usability consultants*, or *human factors engineers*. The visual programming environments now available make it easy for developers to create graphical user interfaces for applications. But it is still very difficult to make a graphical user interface friendly or intuitive.

The processes associated with user-interface design are discussed in Chapter 14. The key question to be answered when completing the "Design the user interfaces" activity is, *have we specified in detail how all users will interact with the system?*

Design the System Interfaces

No system exists in a vacuum. A new information system will affect many other information systems. Sometimes one system provides information that is later used by another system. Other times systems exchange information continuously as they run. The component that enables systems to share information is the system interface, and each system interface needs to be designed in detail.

From the very beginning of system design, analysts must ensure that all of these systems work together well. Some system interfaces link internal organizational systems, so the analyst may have information available about the other systems. In some cases, the new system needs to interface with a system outside the organization—for example, at a supplier's site or customer's home. In other cases, the new system needs to interface with a package application that the organization has purchased and installed. System interfaces can become quite complex, particularly with so many types of technology available today. Often, an organization needs people with very specialized technical skills to work on these interfaces.

System interface design is discussed in more detail in Chapter 15. The key question to be answered when completing the "Design the system interfaces" activity is, *have we specified in detail how the system will work with all other systems inside and outside our organization?*

Design and Integrate the Database

Designing the database for the system is another key design activity. The data model (a logical model) created during systems analysis is used to create the physical model of the database. Sometimes the database is a collection of traditional computer files. More often, it is a relational database consisting of dozens or even hundreds of tables. Sometimes files and relational databases are used in the same system. Sometimes object-oriented databases might be used instead of relational databases.

Analysts must consider many important technical issues when designing the database. Many of the technical (as opposed to functional) requirements defined during systems analysis concern database performance needs (such as response times). Much of the design work might involve performance tuning to make sure the system actually works fast enough. Another key aspect of designing the database is making sure that new databases are properly integrated with existing databases.

A detailed discussion of database design is presented in Chapter 13. The key question to be answered when completing the "Design and integrate the database" activity is, *have we specified in detail how and where the system will store all of the information needed by the organization?*

Prototype for Design Details

During the design phase, it is important to continue creating and evaluating prototypes. Often associated with interface design, prototyping can also be used to confirm design choices about the database, network architecture, controls, or even programming environments being used. Therefore, when analysts consider all of the design activities, they think about how prototypes might be used to help understand a variety of design decisions. It is also important to recognize that rapid application development (RAD) approaches develop prototypes during design that evolve into the finished system. In those cases, the prototype is the system.

The key question to be answered when completing the "Prototype for design details" activity is, *have we created prototypes to ensure that all detailed design decisions have been fully understood?*

Design and Integrate the System Controls

A final design activity involves ensuring that the system has adequate safeguards to protect organizational assets. These safeguards are referred to as system controls. This activity is not listed last because it is less important than the others. On the contrary, it is a crucial activity. It is listed last because controls have to be considered for all other design activities—user interface, system interface, application architecture, database, and network design.

User-interface controls limit access to the system to authorized users. System interface controls ensure that other systems cause no harm to this system. Application controls ensure that transactions recorded and other work done by the system are done correctly. Database controls ensure that data are protected from unauthorized access and from accidental loss due to software or hardware failure. Finally, and of increasing importance, network controls ensure that communication through networks is protected. All of these controls need to be designed into the system, based on the existing technology. Specialists are often brought in to work on some controls, and all system controls need to be thoroughly tested.

Control issues are addressed in several chapters but more explicitly in Chapter 15. The key question to be answered when completing the "Design and integrate the systems controls" activity is, *have we specified in detail how we can be sure that the system operates correctly and the data maintained by the system are safe and secure?*

PROJECT MANAGEMENT: COORDINATING THE PROJECT

The initiation of design activities is a pivotal point in the development project. The focus changes from discovery to solution development, and the whole tenor of the project changes. Coordinating all of the ongoing activities is challenging for even the best project managers because myriad details and tasks must be handled to keep the project on track. Figure A-4 of Appendix A identifies many project management tasks that are required during the design phase. Most of these tasks involve monitoring progress and coordinating the ongoing work.

Even though analysis for an iteration is essentially complete at this point, some analysis tasks remain. Every new system has a multitude of business rules that must be integrated into it. For example, a set of business rules concerning sales commissions includes when and how commissions are calculated, what happens to commissions on merchandise returns, when commissions are paid, how the commission schedule varies to encourage sales of high-margin items and sale items, and so forth. All of these business rules must be defined to develop the commission programs properly. However, what if management is still making decisions about these business rules? You would not want to hold up the entire project for a few of these decisions. On the other hand, you want to make sure that they do not fall through the cracks.

In addition, as the project team members, including the users, better understand the potential capabilities of the new system, they may ask to modify the business definition to provide higher levels of automation and support. This request is good for the company, which will benefit from a better system. However, the request is bad for the project because it will increase the scope and delay the project. How does the project manager control the scope of the project while being sensitive to these additional requests that may be critical to the company?

Design activities also require substantial coordination. The project begins to fragment as an increasing number of design issues are addressed. Frequently, the system is divided into subsystems, and each subsystem has unique design requirements. The project team may also be divided into smaller teams to focus on the various subsystems and on other design issues. Some technical issues—such as network configuration, database design, distributed processing needs, and communications capability—are common to all subsystems. Other issues—such as response time and specialized input equipment—may be limited to specific subsystems. Coordinating and integrating all technical issues for all subsystems, along with any middleware software, are fundamental to the smooth functioning of the overall system. Ultimately, the success of a development project depends on how well all of the work of the design teams is coordinated.

Two other miniprojects may also be initiated at this point: a data conversion project and a test case development project. We will explain more about the details of these projects in Chapter 16. We mention them here simply to note that they insert additional complexity to the management of the project.

Finally, activities of the implementation phase, such as programming, also begin around this time. In fact, design and programming for an iteration frequently are conducted concurrently. Programming can begin as soon as design decisions are made. So, in addition to the various groups working on design issues, groups of programmers and programmer/analysts will probably be added to the project team.

Control and coordination of these various activities can be complicated further by the fact that people involved in the project may work at different locations. Project communications management, which is essential to successful coordination, becomes many times more complicated as people are added to the team.

Given these complexities, let's discuss several project management tools and techniques to help in the coordination of the project.

Coordinating Project Teams

The fundamental tool to coordinate the various project teams' activities is the project schedule. As the activities of the design phase begin, the project manager must update the schedule by identifying and estimating all tasks associated with design and implementation, as well as any outstanding tasks associated with ongoing requirements definition. The project schedule usually must be reworked substantially to ensure that the project remains organized.

During the analysis phase, the project manager and an assistant often do project management tasks, but when the project expands and several teams are formed, management of the project becomes more complicated. Frequently, a committee composed of the leaders of the key design and implementation teams assumes more responsibility for coordinating and controlling aspects of project management. Weekly, and sometimes daily, status meetings are held. If this group includes people at remote locations, teleconferencing support may be required.

The Project Team at RMO

As the customer support system project moves forward into design at RMO, the project team has been enhanced with the addition of new team members. Consistent with the earlier discussion, RMO initiated two new subprojects at this time, one for data conversion and one for the system and acceptance test plans. To integrate new people into the team, Barbara Halifax reorganized the structure of the project team. Those who had been on the team throughout the analysis phase are now key players in getting the new team members up to speed. The accompanying RMO memo highlights some of the changes going on in the project at this point.

Coordinating Information

As design moves forward, the development teams begin to generate a tremendous amount of detailed information about the system. Modules, classes, data fields, data structures, forms, reports, methods, subroutines, and tables are all being defined in substantial detail. Tremendous coordination is needed to keep track of all of these pieces of information. But two kinds of tools can help in this process.

May 5, 2005

To: John MacMurty

From: Barbara Halifax

Re: Design of the Customer Support System

John, this has been a very busy period! Here is a quick status report on completed and upcoming activities.

Completed during the last period
After receiving approval from the oversight committee to move ahead with design, I have been finalizing the project schedule to carry us through design and implementation. We have added five more people to the project team: two users, one senior technical design person to help us with the system design, and two programmer/analysts.

In addition to our main team, which is now focusing on design, I have organized two small teams, one to begin scoping out the data-conversion activities and another to begin developing acceptance test cases. The new user team members have been assigned to these subteams, one to each.

Plans for the next period
We will move ahead rapidly to complete our architectural design. Once we have that design finished, we will finalize the database design, continue with the user-interface design, and initiate low-level application design.

I have also initiated some control procedures. To stabilize the system's scope, all new requirements changes will now be tracked via the Change Request Log and require committee approval. We will also begin internal team status meetings to coordinate our design activities.

Problems, issues, and open items
The only major problem at this point is that there are several items on the Open Items List that need to be finalized—both technical items that my staff needs to resolve and user requirements definitions that need decisions. We're making progress.

BH

cc. Steven Deerfield, Ming Lee

The most common and widespread technique to record and track project information is to use a CASE tool. Most CASE tools have a central repository to capture information. In Chapter 2, you learned about CASE tools, and Figure 2-21 illustrated the various component tools that make up a comprehensive CASE system. A major element in a CASE tool system is the central repository of information. The central repository not only records all design information but also is normally configured so that all teams can view project information to facilitate communication among the teams of a project. Figure 9-6 illustrates the various information components that may exist within a CASE data repository.

FIGURE *9-6*

System development information stored in the CASE repository.

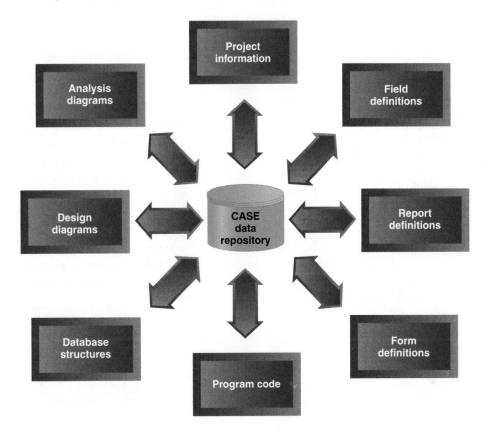

Other electronic tools are also available to help with team communication and information coordination. These tools and techniques, often referred to as computer support for collaborative work, not only record final design information but also assist in team collaboration. Often during the development process, several people need to work together in the development of the design, so they need to discuss and dynamically update the working documents or diagrams. One collaborative tool that is frequently used is Lotus Notes. Other software programs allow figures and diagrams to be updated with tracking and version information so that the evolution of the result can be documented.

One especially difficult part of development projects is tracking open items and unresolved issues. We mention this issue here, not because it requires a new technique, but because the problem is pervasive and all good project managers develop techniques to track these items. One simple method is to have an open items control log. In Chapter 4, we presented the idea of an open-items control list and an example in Figure 4-10. It is a sequential list of all open items with information to track responsibilities and resolution of the open item, as shown by the column headings. Frequently, closed items are shaded for easy identification. This technique is effective not only during analysis but also throughout the entire project.

Now that you have an overview of the design phase and major issues of project management, we turn to the tasks associated with architectural and network design. First, we discuss technical aspects of the deployment environment that directly affect architectural design. Then, we describe common approaches to architectural design and the software and network design issues associated with each approach.

DEPLOYMENT ENVIRONMENT

In Chapter 8, you learned that defining the deployment environment is an activity that bridges analysis and design. The deployment environment consists of the hardware, system software, and networking environment in which the system will operate. In this section, we describe common deployment environments in detail, and in the next section we'll explore related design patterns and architectures for application software.

Single-Computer and Multitier Architecture

As its name implies, *single-computer architecture* employs a single computer system and its directly attached peripheral devices, as shown in Figure 9-7a. It can be a stand-alone PC application, but in this context we are discussing large mainframe applications where users interact with the system via input/output devices of limited functionality that are directly connected to the computer. Single-computer architecture usually requires all system users to be located near the computer. The primary advantage of single-computer architecture is its simplicity. Information systems deployed on a single-computer system are relatively easy to design, build, operate, and maintain.

The capacity limits of a single computer may make single-computer architecture impractical or unusable for large information systems. Many systems are so large that even

FIGURE *9-7*

Single-, clustered, and multicomputer architectures.

IBM S/390 G5

(a) Single-computer architecture

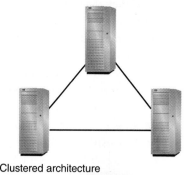

(b) Clustered architecture

IBM zSeries 800   IBM FAStT600   IBM pSeries 690

(c) Multicomputer architecture

multitier architecture

architecture that distributes application-related software or processing load across multiple computer systems

clustered architecture

a group of computers of the same type that share processing load and act as a single large computer system

multicomputer architecture

a group of dissimilar computers that share processing load through specialization of function

the largest mainframe computer cannot perform all the required processing, data storage, and data retrieval tasks. Such systems require another architectural approach.

Multitier architecture employs multiple computer systems in a cooperative effort to meet information-processing needs. Multitier architecture can be further subdivided into two types:

- *Clustered architecture*, shown in Figure 9-7b, employs a group (or cluster) of computers, usually from the same manufacturer and model family. Programs are allocated to the least utilized computer when they execute so that the processing load can be balanced across all machines. In effect, a cluster acts as a single large computer system. Clustered computer systems are normally located near one another so that they can be connected with short high-capacity communication links.
- *Multicomputer architecture*, shown in Figure 9-7c, also employs multiple computer systems. But hardware and operating systems are not required to be as similar as in a clustered architecture. A suite of application or system programs and data resources is exclusively assigned to each computer system. Each computer system is optimized to the role that it will play in the combined system, such as database or application server.

Centralized and Distributed Architecture

centralized architecture

architecture that locates all computing resources in a central location

The term *centralized architecture* describes deployment of all computer systems in a single location. Centralized architecture is generally used for large-scale processing applications, including both batch and real-time applications. Such applications are common in industries such as banking, insurance, and catalog sales. Information systems in such industries often have the following characteristics:

- Some input transactions do not need to be processed in real time (for example, out-of-state checks delivered in large nightly batches from central bank clearinghouses).
- On-line data-entry personnel can be centrally located (for example, a centrally located group of telephone order takers can serve geographically dispersed customers).
- The system produces a large amount of periodic outputs (for example, monthly credit-card statements mailed to customers).
- A high volume of transactions occurs between high-speed computers (for example, business-to-business processing for supply chain management).

Any application that has two or three of these characteristics is a viable candidate for implementation on a centralized mainframe. Current trends in conducting e-business have instilled new life into centralized mainframe computing because of the transaction volumes of many business-to-business (B2B) processes.

Centralized computer systems are seldom used as the sole hardware platform for an information system. Most systems have some transaction inputs that must be accepted from geographically dispersed locations and processed in real time—for example, a cash withdrawal from an ATM. Most systems also have some outputs that are requested from and delivered to remote locations in real time—for example, an insurance policy inquiry by a state motor vehicle department. Thus, centralized computer systems are typically used to implement one or more subsystems within a larger information system that includes on-line, batch, and geographically dispersed components.

Components of a modern information system are typically distributed across many computer systems and geographic locations. For example, corporate financial data might be stored on a centralized mainframe computer. Minicomputers in regional offices might be used to generate accounting and other reports periodically based on data

F I G U R E *9-8*

A possible network
configuration for RMO.

stored on the mainframe. Personal computers in many locations might be used to access and view periodic reports as well as to directly update the central database. Such an approach to distributing components across computer systems and locations is generically called *distributed architecture*. Distributed architecture relies on communication networks to connect geographically dispersed computer hardware components.

Computer Networks

A *computer network* is a set of transmission lines, specialized hardware, and communication protocols that enable communication among different users and computer systems. Computer networks are divided into two classes depending on the distance they span. A *local area network (LAN)* is typically less than one kilometer long and connects computers within a single building or floor. The term *wide area network (WAN)* can describe any network over one kilometer, though the term typically implies much greater distances spanning cities, countries, continents, or the entire globe.

Figure 9-8 shows a possible computer network for RMO. A single LAN serves each geographic location, and all LANs are connected by a WAN. Users and computers in a single location communicate via their LAN. Communication among geographically dispersed sites uses the LANs at both sites and the WAN. A *router* connects each LAN to the WAN. A router scans messages on the LAN and copies them to the WAN if they are addressed to a user or computer on another LAN. The router also scans messages on the WAN and copies them to the LAN if they are addressed to a local user or computer.

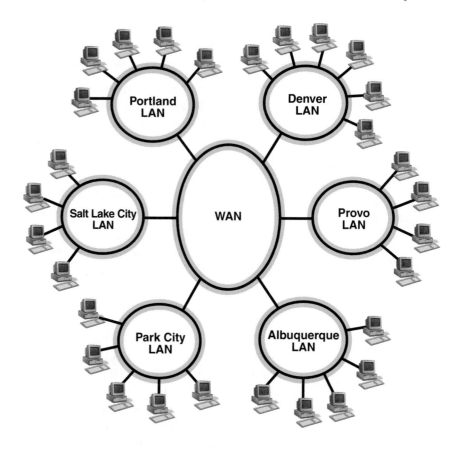

Technologies such as Ethernet are typically used to implement LANs. They provide low to moderate amounts of message-carrying capacity at relatively low cost. WAN technologies such as asynchronous transmission mode are more complex and expensive, though they typically provide higher message-carrying capacity and greater reliability.

WANs may be constructed using purchased equipment and leased long-distance transmission lines. WAN setup and operation may also be subcontracted from a long-distance telecommunications vendor such as AT&T or Sprint.

Computer networks provide a generic communication capability among computer systems and users. This generic capability can support many services, including direct communications (such as telephone service and video conferencing), message-based communications (such as e-mail), and resource sharing (such as access to electronic documents, application programs, and databases). A single network can simultaneously support multiple services with appropriate hardware and sufficient transmission capacity.

There are many ways to distribute information system resources across a computer network. Users, application programs, and databases can be placed on the same computer system, on different computer systems on the same LAN, or on different computer systems on different LANs. Application programs and databases can also be subdivided and each distributed separately.

The Internet, Intranets, and Extranets

The *Internet* is a global collection of networks that are interconnected using a common low-level networking standard—TCP/IP (Transmission Control Protocol/Internet Protocol). The *World Wide Web (WWW)*, also called simply the *Web*, is a collection of resources (programs, files, and services) that can be accessed over the Internet by a number of standard protocols, including the following:

- Formatted and linked document protocols, such as HyperText Markup Language (HTML), eXtensible Markup Language (XML), and HyperText Transfer Protocol (HTTP)
- Executable program standards, including Java, JavaScript, and Visual Basic Script (VBScript)
- Distributed software and Web-service standards, including Distributed Computing Environment (DCE), Common Object Request Broker Architecture (CORBA), and Simple Object Access Protocol (SOAP)

The Internet is the infrastructure on which the Web is based. In other words, resources of the Web are delivered to users over the Internet.

An *intranet* is a private network that uses Internet protocols but is accessible only by a limited set of internal users (usually members of the same organization or workgroup). The term also describes a set of privately accessible resources that are organized and delivered via one or more Web protocols over a network that supports TCP/IP. An intranet uses the same protocols as the Internet and Web but restricts resource access to a limited set of users. Access can be restricted in various ways, including unadvertised resource names, firewalls, and user/group account names and passwords.

An *extranet* is an intranet that has been extended to include directly related business users outside the organization (such as suppliers, large customers, and strategic partners). An extranet allows separate organizations to exchange information and coordinate their activities, thus forming a *virtual organization*. One widely used method of implementing an extranet is through a *virtual private network (VPN)*. A private network is a network that is secure and accessible only to members of an organization (or virtual organization). Historically, implementing a private network required an organization to own and operate its own network lines or leased dedicated telephone lines. A VPN sends encrypted messages through public Internet service providers.

Internet

a global collection of networks that use the same networking protocol—TCP/IP

World Wide Web (WWW), or Web

a collection of resources such as files and programs that can be accessed over the Internet using standard protocols

intranet

a private network that is accessible to a limited number of users, but which uses the same TCP/IP protocol as the Internet

extranet

an intranet that has been extended outside the organization to facilitate the flow of information

virtual organization

a loosely coupled group of people and resources that work together as though they were an organization

virtual private network (VPN)

a network with security and controlled access for a private group but built on top of a public network such as the Internet

---

APPLICATION ARCHITECTURE

Simple deployment environments, such as a single centralized computer with video display terminals, can be matched to relatively simple application architectures. More complex distributed and multitier hardware and network architectures require more complex software architectures. This section describes common examples of application architecture for distributed and multitier deployment environments and the design issues and decisions associated with each.

Client-Server Architecture

server

a process, module, object, or computer that provides services over a network

client

a process, module, object, or computer that requests services from one or more servers

Client-server architecture divides programs into two types: client and server. A *server* manages one or more information system resources or provides a well-defined service. A *client* communicates with a server to request resources or services, and the server responds to those requests.

Client-server architecture is a general model of software organization and behavior that can be implemented in many different ways. A typical example is the interaction between a client application program executing on a workstation and a database management system (DBMS) executing on a larger computer system (see Figure 9-9). The application program sends database access requests to the database management system via a network. The DBMS accesses data on behalf of the application and returns a response such as the results of a search operation or the success or failure result of an update operation.

FIGURE *9-9*

Client-server architecture with a shared database.

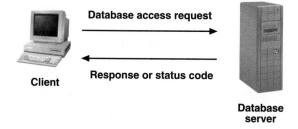

The architectural issues to be addressed when designing client-server software are:

- Decomposing the application into client and server programs, modules, or objects
- Determining which clients and servers will execute on which computer systems
- Describing the communication protocols and networks that connect clients and servers

The key to decomposing the application into clients and servers is identifying resources or services that can be centrally managed by independent software units. Examples of centrally managed services include security authentication and authorization, credit verification, and scheduling. In each case, a service provides a set of well-defined processes such as retrieval, update, and approval based on a data store that is hidden from the client, as shown in Figure 9-10.

FIGURE *9-10*

Interaction among client, server, and a service-related data store.

Client and server software can execute on any computer system. But the most typical arrangement is to place server software on separate server computer systems and to distribute client software to computer systems "close" to end-users, such as desktop workstations. Figure 9-11 shows a typical arrangement for an order-processing application. Credit verification, delivery scheduling, and database server processes execute on a centrally located minicomputer or mainframe, and users execute multiple copies of the client software on workstations.

FIGURE *9-11*

Interaction among multiple clients and a single server.

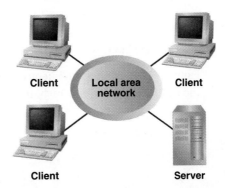

Client and server communicate via well-defined communication protocols over a physical network. In Figure 9-11, the network is a LAN, and an appropriate low-level network protocol such as TCP/IP provides basic communication services. But the system designer must also specify higher-level protocols, or languages, by which client and server exchange service requests, responses, and data. In some cases, such as communication with a DBMS, standard protocols and software may be employed, such as Structured Query Language (SQL) via an Open Database Connectivity (ODBC) database connection. But in other cases, the designer must define the exact format and content of valid messages and responses. If a service is provided by an external organization (for example, credit verification), then the external organization will have already designed an appropriate protocol, and the application designer must ensure that clients adhere to the protocol.

The primary advantage of client-server architecture is deployment flexibility. Client-server architecture arose as an approach to distributing software across networked computers. It provides the inherent advantages of a networked environment including:

■ **Location flexibility.** The ability to "move" system components without "disturbing" other system components, in response to changing organizational parameters such as size and physical location
■ **Scalability.** The ability to increase system capacity by upgrading or changing the hardware on which key software components execute
■ **Maintainability.** The ability to update the internal implementation of one part of a system without needing to change other parts (for example, the credit verification software can be rewritten or replaced as long as the new software uses the existing client-server protocol)

The primary disadvantages of client-server architecture are the additional complexity introduced by the client-server protocols and the potential performance, security, and reliability issues that arise from communication over networks. A centralized application executing as one large program on a single computer needs no client-server protocols, and all communication within the application occurs within the relatively secure, reliable, and efficient confines of a single machine.

For most organizations, the flexibility advantages of client-server far outweigh the disadvantages. As a result, client-server architecture and its newer variants have become the dominant architecture for the vast majority of modern software.

three-layer architecture

a client-server architecture that divides an application into the view layer, business logic layer, and data layer

data layer

the part of three-layer architecture that interacts with the database

business logic layer

the part of three-layer architecture that contains the programs that implement the business rules of the application

view layer

the part of three-layer architecture that contains the user interface

Three-Layer Client-Server Architecture

A widely applied variant of client-server architecture, called *three-layer architecture*, divides application software into a set of client and server processes independent of hardware or locations. All layers might reside on one processor, or three or more layers might be distributed across many processors. In other words, the layers might reside on one or more tiers. The most common set of layers includes:

- The *data layer*, which manages stored data, usually in one or more databases
- The *business logic layer*, which implements the rules and procedures of business processing
- The *view layer*, which accepts user input and formats and displays processing results

Figure 9-12 illustrates the interaction of the three layers. The view layer acts as a client of the business logic layer, which, in turn, acts as a client of the data layer.

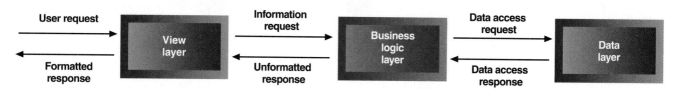

FIGURE *9-12*

Three-layer architecture.

Like earlier forms of client-server architecture, three-layer architecture is inherently flexible. Interactions among the layers are always requests or responses, which makes the layers relatively independent of one another. It doesn't matter where other layers are implemented or on what type of computer or operating system they execute. The only interlayer dependencies are a common language for requests and responses and a reliable network with sufficient communication capacity.

Multiple layers can execute on the same computer, or each layer can operate on a separate computer. Complex layers can be split across two or more computers. System capacity can be increased by splitting layer functions across computers or by load sharing across redundant computers. In the event of a malfunction, redundancy improves system reliability if the server load can be shifted from one computer to another. In sum, three-layer architecture provides the flexibility needed by modern organizations to deploy and redeploy information-processing resources in response to rapidly changing conditions.

Three-layer architecture is currently a widely applied architectural design pattern with both the traditional approach and the object-oriented approach. As with other forms of client-server architecture, the key design tasks are decomposing the application into layers, clients, and servers, distributing the "pieces" across hardware platforms, and defining the physical network and protocols.

The business logic layer is the core of the application software and is constructed according to the requirements models developed during analysis, as described in Chapters 5 through 7. For example, in the traditional approach, all of the business logic defined for system activities within the RMO data flow diagrams would be implemented as functions or procedures in the business logic layer. The window or browser forms making up the view layer would not contain much procedural code. In the object-oriented approach, the classes of objects in the RMO class diagram (see Figure 5-35) would be implemented within the business logic layer as classes of objects that interact to complete user tasks. In either case, the business logic layer is a server for the view layer and is a client of the data layer. However, the business logic layer may itself be decomposed into multiple clients and services. Three-layer architecture is usually implemented with object-oriented techniques and tools, as described in Chapter 11, though it is also implemented with traditional design techniques and programming languages, as described in Chapter 10. In this respect, three-layer architecture is a prominent architectural design pattern that applies to both traditional and OO approaches.

In this text, Chapters 10 and 11 describe the how view and data layer software is designed with traditional and OO approaches. Chapter 13 describes the details of the database that is accessed by the data access layer. Chapter 14 describes user-interface design techniques and guidelines that are independent of the software that implements the view layer, such as the arrangement of interface elements on a video display and the dialog between user and computer that supports a specific application task.

Middleware

middleware

computer software that implements communication protocols on the network and helps different systems communicate

Client-server and three-layer architecture relies on special programs to enable communication between the various layers. Software that implements this communication interface is usually called *middleware*. Middleware connects parts of an application and enables requests and data to pass between them. There are various methods to implement the middleware functions. Some common types of middleware include teleprocessing monitors, transaction processing monitors, and object request brokers (ORBs). Each type of middleware has its own set of protocols to facilitate communication between the various components of an information system.

When specifying the protocols to be used for client-server or interlayer communication, the designer usually relies on standard frameworks and protocols incorporated into middleware. For example, interactions with DBMSs usually employ standard protocols such as ODBC and SQL with supporting software obtained from the DBMS vendor or a third party. Third-party service providers such as credit bureaus and electronic purchasing or bidding services usually employ a standard Web protocol such as HTTP or XML. Industry-specific protocols have been developed in many industries such a health care and banking.

Complex OO software distributed across multiple layers and hardware platforms relies on an ORB based on a distributed object interface standard such as CORBA. Distributed non-OO software relies on different middleware products based on standards such as DCE or Microsoft's COM+. Web-based applications rely on Web-oriented protocols such as Microsoft's .NET and Sun's J2EE and specific middleware products that implement and support those protocols. Support protocols and software for distributed objects, components, and Web-based applications are discussed in more detail in Chapter 16.

Internet and Web-based Application Architecture

The Web is a complex example of client-server architecture. Web resources are managed by server processes that can execute on dedicated server computers or on multipurpose computer systems. Clients are programs that send requests to servers using one or more of the standard Web resource request protocols. Web protocols define valid resource formats and a standard means of requesting resources and services. Any program (not just a Web browser) can use Web protocols. Thus, Weblike capabilities can be embedded in ordinary application programs.

Internet and Web technologies present an attractive alternative for implementing information systems. For example, consider the problem of data entry and access by an RMO buyer when purchasing items from its suppliers. Buyers are typically on the road for several months a year, often for weeks at a time. A traveling buyer needs some means of remotely interacting with RMO's supply chain management (SCM) system to record purchasing agreements and query inventory status.

One way of providing these capabilities would be to design custom application software and a private network to connect to that software. The primary portion of the system could be installed on a server at RMO. The client portion of the application—for data entry—would then be installed on the buyers' laptop computers. A buyer would then connect to the system from remote locations to gain access to the application server, make queries to the database, and enter data.

Another alternative for implementing remote access for buyers would be to construct an application that uses a Web browser interface. The application would execute on a Web server, communicate with a Web browser using HTML or XML, and be accessible from any computer with an Internet connection. Buyers could use a Web browser on their laptop computer and connect to the application via an Internet service provider wherever they're currently located. Buyers could also access the application from any other computer with Internet access (for example, a computer in a vendor's office, hotel business suite, or copy center such as Kinko's).

Flexibility is the key to the Internet alternative. Implementing the application via the Internet greatly expands the application's accessibility and also eliminates the need to install custom client software on buyers' laptop computers. With Internet technology, client software can be updated by simply updating the version stored on the Web server. The application is relatively cheap to develop and deploy because existing Web standards and networking resources are employed. Custom software and private access via modems require more complex development and maintenance of a greater number of customized resources.

Implementing an application via the Web, an intranet, or an extranet has a number of advantages over traditional client-server applications, including the following:

- **Accessibility.** Web browsers and Internet connections are nearly ubiquitous. Internet, intranet, and extranet applications are accessible to a large number of potential users (including customers, suppliers, and off-site employees).
- **Low-cost communication.** The high-capacity WANs that form the Internet backbone were funded primarily by governments. Traffic on the backbone networks travels free of charge to the user, at least for the present. Connections between private LANs and the Internet can be purchased from a variety of private Internet service providers at relatively low cost. In essence, a company can use the Internet as a low-cost WAN.
- **Widely implemented standards.** Web standards are well known, and many computing professionals are already trained in their use. Server, client, and application development software is widely available and relatively cheap.

Information resource delivery via an intranet or extranet enjoys all of the advantages of Web delivery since intranets and extranets use Web standards. In many ways, intranets, extranets, and the Web represent the logical evolution of client-server computing into an off-the-shelf technology. Organizations that had shied away from client-server computing because of the costs and required learning curve can now enjoy client-server benefits at substantially reduced complexity and cost.

Of course, there are negative aspects of application delivery via the Internet and Web technologies, including the following:

- **Security.** Web servers are a well-defined target for security breaches because Web standards are open and widely known. Wide-scale interconnection of networks and the use of Internet and Web standards makes servers accessible to a global pool of hackers.
- **Reliability.** Internet protocols do not guarantee a minimum level of network throughput or even that a message will ever be received by its intended recipient. Standards have been proposed to address these shortcomings, but they have yet to be widely adopted.
- **Throughput.** The data transfer capacity of many home users is limited by analog modems to under 56 kilobits per second. Internet service providers and backbone WANs can become overloaded during high-traffic periods, resulting in slow response time for all users and long delays when accessing large resources.

- **Volatile standards.** Web standards change rapidly. Client software is updated every few months. Developers of widely used applications are faced with a dilemma: use the latest standards to increase functionality or use older standards to ensure greater compatibility with older user software.

The primary disadvantages to RMO of implementing the customer order application via the Internet are security, performance, and reliability. If a buyer can access the system via the Web, then so can anyone else. Access to sensitive parts of the system can be restricted by a number of means, including user accounts and passwords. But the risk of a security breach will always be present. Performance and reliability are limited by the buyer's Internet connection point and the available Internet capacity between that connection and the application server. Unreliable or overloaded local Internet connections can render the application unusable. RMO has no control over these factors.

The key architectural design issues for Web-based applications are similar to those for other client-server architectures: defining client and server processes or objects, distributing them across hardware platforms, and connecting them with appropriate networks, middleware, and protocols. However, for Web-based applications, the choices for middleware and protocols tend to be much more limited than for other forms of client-server architecture.

Now that we've discussed common approaches to application architecture, we'll turn our attention to designing the networking infrastructure that connects parts of a modern information system.

NETWORK DESIGN

Networks are used throughout organizations today. As a result, many new development projects involve network design. Network planning and design are critical issues that must be dealt with early in the design phase for any multitiered system. The key design issues are:

- Integrating network needs of the new system with existing network infrastructure
- Describing the processing activity and network connectivity at each system location
- Describing the communication protocols and middleware that connect layers
- Ensuring that sufficient network capacity is available

Network Integration

Modern organizations rely on networks to support many different applications. Thus, the majority of new systems must be integrated into existing networks without disrupting existing applications. Network design and management are highly technical tasks, and most organizations have permanent in-house staff, contractors, or consultants to handle network administration.

The analyst for a new project begins network design by consulting with the organization's network administrators to determine whether the existing network can accommodate the new system. In some cases, the existing network capacity is sufficient, and only minimal changes are required, such as adding connections for new servers or modifying routing and firewall configuration to enable new application layers to communicate.

Planning for more extensive changes—such as significant capacity expansion, new communication protocols, or modified security protocols—is much more complex. Typically, the network administrator assumes the responsibility of acquiring new capacity and making any configuration changes to support the new system since he or she understands the existing network and the way other network-dependent applications operate. The analyst's role for the new system in these cases is to provide the network administrator with sufficient information and time to enable system development, testing, and deployment.

Network Description

Location-related information gathered during analysis may have been documented using location diagrams (such as Figure 6-37), activity-location matrices (such as Figure 6-38), and activity-data matrices (such as Figure 6-39). During network design, the analyst expands the information content of these documents to include processing locations, communication protocols, middleware, and communication capacity.

There are many different ways to describe the network infrastructure for a specific application. Figure 9-13 shows a *network diagram* that describes how application layers are distributed across locations and computer systems for the RMO customer support system. The diagram summarizes key architectural decisions from Figure 8-5 and combines them with specific assumptions about where application software will execute, where servers and workstations will be located, and how network resources will be organized.

> **network diagram**
>
> a model that shows how application layers are distributed across locations and computer systems

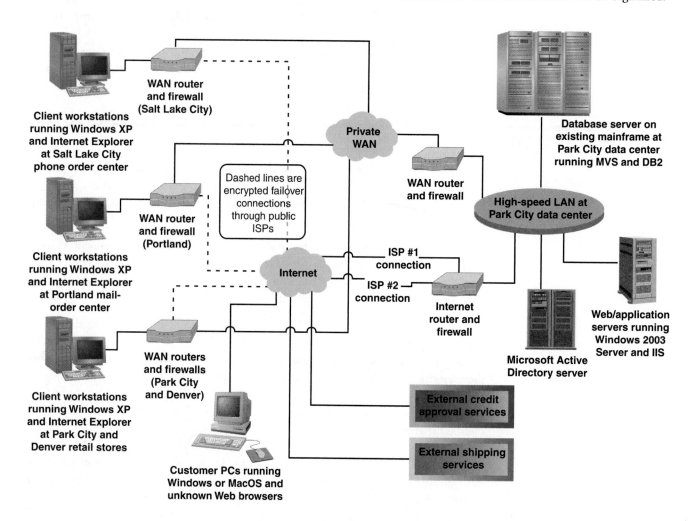

The diagram embodies specific assumptions about server locations, which would be decided in consultation with network administrators. The Web/application servers could have been distributed outside the Salt Lake City data center, which might have improved system response time and reduced data communication capacity requirements on the private WAN. However, distributing the servers would also entail duplication of server administration at multiple locations, which would increase operational complexity and cost. Decisions such as server locations, communication routes, and network security options are determined both by application requirements and organizationwide policies.

Communication Protocols and Middleware

The network diagram is also a starting point for specifying protocol and middleware requirements. For example, the private WAN connections must support protocols required to process Microsoft Active Directory logins and queries. If the WAN fails, messages are routed through encrypted (VPN) connections over the Internet, so those connections must support the same protocols as the private WAN. All clients must be able to send HTTP requests and receive active content such as HTML forms and embedded scripts. Application servers must be able to communicate with credit verification and shipping services via the Internet. Firewalls and routers must be configured to support all interactions among the workstations, customer PCs, Web/application servers, the Active Directory server, and external credit and shipping services. The Park City data center LAN must support at least one protocol for transmitting database queries and responses among the mainframe and Web/application servers.

Network Capacity

Information from the activity-location and activity-data matrices is the starting point for estimating communication capacity requirements for the various LAN, WAN, and Internet connections. Figure 9-14 reproduces data from the RMO activity-data matrix (see Figure 6-39) covering two activities (*Look up item availability* and *Create new order*) and three data entities (Customer, Inventory Item, and Order). Similar tables would be required for all combinations of activity, data entity, and location. Figure 9-14 includes estimates of data size per access type and the average and peak number of access per minute or hour.

Data size per access type is an educated guess at this point in the system design since none of the software layers, interlayer communication dialogs, or databases have yet been designed. Once those components have been designed in more detail or implemented, analysts can refine their estimates or actually sample and measure real data transmissions. Actual data transmission capacity will include communication protocols in addition to raw data.

	DATA ENTITIES		
Activities and Locations	**Customer**	**Inventory Item**	**Order**
Look up item availability (Salt Lake City phone order center)		R (125 bytes, 25/min average, 250/min peak)	
Look up item availability (Park City retail store)		R (125 bytes, 5/hr average, 15/hr peak)	
Look up item availability (Denver retail store)		R (125 bytes, 5/hr average, 15/hr peak)	
Create new order (Salt Lake City phone order center)	C (500 bytes, 2/min average, 10/min peak) R (500 bytes, 8/min average, 80/min peak) U (500 bytes, 2/min average, 10/min peak)	R (60 bytes, 30/min average, 300/min peak) U (60 bytes, 30/min average, 300/min peak)	C (200 bytes, 10/min average, 100/min peak)
Create new order (Portland mail-order center)	C (500 bytes, 1/min average, 10/min peak) R (500 bytes, 4/min average, 40/min peak) U (500 bytes, 1/min average, 10/min peak)	R (60 bytes, 15/min average, 150/min peak) U (60 bytes, 15/min average, 150/min peak)	C (200 bytes, 5/min average, 50/min peak)
Create new order (Park City retail store)	C (500 bytes, 1/hr average, 5/hr peak) R (500 bytes, 4/hr average, 20/hr peak) U (500 bytes, 1/hr average, 5/hr peak)	R (60 bytes, 15/hr average, 75/hr peak) U (60 bytes, 15/hr average, 75/hr peak)	C (200 bytes, 5/hr average, 25/hr peak)
Create new order (Denver retail store)	C (500 bytes, 1/hr average, 5/hr peak) R (500 bytes, 4/hr average, 20/hr peak) U (500 bytes, 1/hr average, 5/hr peak)	R (60 bytes, 15/hr average, 75/hr peak) U (60 bytes, 15/hr average, 75/hr peak)	C (200 bytes, 5/hr average, 25/hr peak)

C = Creates new data,   R = Reads existing data,   U = Updates existing data,   D = Deletes existing data

FIGURE *9-14*

Partial activity-data matrix for RMO customer support system updated with data access size and volume.

SUMMARY

Systems design is the process of organizing and structuring the components of a system to allow the construction (that is, programming) of the new system. The design phase of a project consists of activities that relate to the design of the various components of the new system. The components that need to be designed include the application architecture, the user interfaces, the system interfaces, the database, the network, and system controls. Prototyping may be required to fully specify any part or all of the design.

The inputs to the design activities are the diagrams, or models, that were built during analysis. The outputs of the design are also a set of diagrams, or models, that describe the architecture of the new system and the detailed logic within the various programming components. The inputs, design activities, and outputs are different depending on whether a structured approach or an object-oriented approach is used.

Project management is critical when design activities are initiated. The tenor of the project changes at this point. During design, technical staff are involved. The project manager has to pay particular attention to the schedule, staff, and design activities. Sometimes programming is begun, which requires the addition of programming staff. Other side projects for data conversion and test-data development are also begun at this time.

Designing the application architecture can be subdivided into architectural and detail design. Architectural design adapts the application to the deployment environment, including hardware, software, and networks. Modern application software is usually deployed in a distributed multi-computer environment and is organized according to client-server architecture or a variant such as three-layer architecture. Architectural design decisions include decomposing the application into clients, servers, or layers, distributing software across hardware platforms, and specifying required protocols, middleware, and networks.

Architectural design decisions can be documented in a network diagram. The network diagram describes the organization of computing and network resources and specifies details such as what the required protocols are and which application software and middleware executes on which computer systems. Required network capacity can be determined by expanding the activity-location and activity-data matrices to include estimates of message size and volume.

KEY TERMS

architectural design, p. 317
business logic layer, p. 334
centralized architecture, p. 328
client, p. 332
clustered architecture, p. 328
computer network, p. 330
data layer, p. 334
detail design, p. 317
distributed architecture, p. 330
extranet, p. 331
interface designers, p. 323
Internet, p. 331
intranet, p. 331
local area network (LAN), p. 330

middleware, p. 335
multicomputer architecture, p. 328
multitier architecture, p. 328
network diagram, p. 338
router, p. 330
server, p. 332
single-computer architecture, p. 328
three-layer architecture, p. 334
view layer, p. 334
virtual organization, p. 331
virtual private network (VPN), p. 331
wide area network (WAN), p. 330
World Wide Web (WWW), or Web, p. 331

REVIEW QUESTIONS

1. What is the primary objective of systems design?
2. What is the difference between analysis and design? List the activities in the design phase of the SDLC.
3. Why is project management so critical during the design phase? What tools can a project manager use during the design phase?
4. Explain the difference between centralized architecture and distributed architecture.
5. Explain the difference between clustered architecture and multicomputer architecture in a centralized system.
6. How are the Internet, intranets, and extranets similar? How are they different?
7. Describe client-server architecture and list the key architecture design issues that must be addressed when developing a client-server information system.
8. List and briefly describe the function of each layer in three-layer architecture.
9. What role does middleware play?
10. Describe the process of network design.
11. What roles do systems analysts and network administrators play in network design?
12. What is a network diagram? What information does it convey and where does the analyst gather that information?
13. How does the analyst generate estimates of required communication capacity? What analysis phase models are used as input?

THINKING CRITICALLY

1. Discuss the evolution of client-server computing from file server to multilayer applications to Web-based applications. What has been the driving force causing this evolution? Where do you think network computing will be in the next five years? Ten years?
2. Assume that the deployment environment for a high-volume payment processing system consists of the following (these assumptions are from the scenario presented in Chapter 8's first Experiential Exercise):
   - DB2 DBMS running under the OS/390 operating system on an IBM S/390 mainframe
   - WebSphere application server running under the Z/OS operating system on an IBM zSeries 900 mainframe
   - CORBA-compliant component-based application software written in Java that will be executed by other internal and external systems
   What are the key architectural design decisions that must be made for the system? When should the decisions be made and who should make them? Outline the subsequent design tasks that should occur after the key architectural design decisions are made. To what extent can the subsequent steps be performed in parallel?
3. Develop a network diagram that supports the architectural design decisions in your answer to number 2.

EXPERIENTIAL EXERCISES

1. Set up a meeting with the chief analysts of a medium- or large-scale development project and discuss the transition from analysis to design for that project. How and when were key architectural decisions such as automation boundary, network design, and supporting infrastructure made? Who made the decisions? Were the early architectural decisions modified later in the project? If so, how and why?
2. Find an example of an application system that is browser based and uses TCP/IP standards. Explain how it works, showing sample screens and reports. List each middleware component and describe its function. List each protocol employed and identify the standard family or families to which the protocol belongs.
3. Examine the RMO network diagram in Figure 9-13 and note the connections to external service providers for credit verification and shipping services. Identify at least three companies that can provide each service. Investigate their on-line Web-based service capabilities and describe the protocols used by clients to interact with their services.

CASE STUDIES

The Real Estate Multiple Listing Service System

In Chapter 8 you were asked to discuss the implications of the "anytime, anywhere" requirement for the application deployment environment and to describe the type(s) of hardware, network, and software architecture required to fulfill that requirement. Assume that you addressed that question by specifying a three-layer architecture using ordinary PCs running Web browsers to implement the view layer. Draw a network diagram that represents your chosen solution.

Today's computer-based real estate listings typically include graphical data, such as still and moving pictures in addition to text descriptions of properties. What is the impact of such data on data communication requirements within your network design, assuming 10 listing accesses per hour? 100 listing accesses per hour? 1,000 listing accesses per hour?

Rethinking Rocky Mountain Outfitters

In Chapter 8, you were asked to consider an alternative deployment scenario for RMO based on Apache Web servers running under Linux and an Oracle database server. Modify the network diagram in Figure 9-13 to reflect the alternative deployment scenario. What changes, if any, are required for the client workstations and customer PCs? What changes, if any, are required in middleware and communication protocols? Will there be any change in the estimates of required data-communication capacity among client workstations and servers located at the Park City data center? Why or why not?

Focusing on Reliable Pharmaceutical Service

Assume the same facts as presented in the Chapter 8 Reliable Pharmaceutical case. Also assume that you are the project manager for the selected vendor's development team. Your company, RxTechSys, develops and markets software to retail and hospital pharmacies and has decided to take on the Reliable project to expand potential market share. RxTechSys and Reliable will jointly develop the new software. RxTechSys will then market the finished product to other companies and pay a royalty to Reliable for each sale.

RxTechSys has been in the pharmacy software business for 20 years. The latest version of the software is a Web-based application built on the Microsoft .NET platform. Major functions such as inventory control, purchasing, billing, and prescription warning are implemented as separate .NET Web services. As part of the team that prepared the response to Reliable's RFP, you determined that RxTechSys's current system can be adapted to Reliable's needs as follows:

- Existing browser-based prescription entry can be modified to handle data input from multiple customer locations over a VPN. This is a significant modification due to expanded data content and greater security requirements.
- Order fulfillment software will have to be written from scratch.
- Billing software will require significant modification since your current system assumes that all patients have their health care managed by a single institution, with possible third-party reimbursements through Medicaid/Medicare.
- Other parts of your existing system can be used with little or no modification.

Reliable has provided you with a complete set of object-oriented analysis models, the quality of which you approved during contract negotiations. Your task is to move the project forward through design and implementation.

Reliable has assigned an operational manager with some computer experience to your team full-time, and she is authorized to assign other Reliable personnel to your project as needed. You have been assigned a full-time staff of four developers, two of whom have substantial design experience and all of whom participated in developing the most recent version of RxPharmSys software.

Develop a design plan and schedule covering the next 4 to 6 weeks (your expected project duration is 10 months). What design decisions must be made within the next two weeks? Who should make them? How will design and development proceed thereafter—what tasks must be performed and in what order? How will you manage and control the project?

FURTHER RESOURCES

Robert Orfali, Dan Harkey, and Jeri Edwards, *Client/Server Survival Guide,* third edition. Wiley, 1999.

CHAPTER 10

The Traditional Approach to Design

LEARNING OBJECTIVES

After reading this chapter, you should be able to:

- Develop a system flowchart

- Develop a structure chart using transaction analysis and transform analysis

- Write pseudocode for structured modules

CHAPTER OUTLINE

Bernard closed the office door and spoke to his officemate, Stana, in an exasperated voice. "I don't understand why Jim insists that I update these system flowcharts and structure charts. We should throw this all out and start from scratch with object-oriented (OO) design. I drew a few of these traditional diagrams in school, but we spent most of our time with object-oriented diagrams and techniques. I feel as if I'm being asked to fix a computer with a hammer and saw!"

Bernard, a recent MIS college graduate, was a new employee at Theatre Systems, Inc., which sells and supports financial reporting software for small and medium-sized U.S. theatre chains. He was hired just as an upgrade project was moving from analysis to design. Although the company's software had been updated regularly with improvements and new features, it lacked some modern capabilities, such as a Web-based interface and scalable multilayer architecture.

Stana, who had worked for the company for almost four years, replied, "Well, you need to remember two things. First, many of the IS staff here don't fully understand OO analysis and design techniques. Version 1 of our software and its analysis and design documentation was developed in the early 1980s. All of the upgrades since then have been incremental, so there's been no need to burn the old design models and start from scratch. Significant chunks of the system are unchanged in over a decade.

"Second, there's a question of suitability of the tools to the task. If our goal was to develop and implement an entirely new system that could scale from the smallest mom-and-pop theatre to the largest nationwide theatre chain and be deployed and redeployed at whim, then we'd almost certainly be using the latest distributed software technologies, OO programming languages, and the OO analysis and design tools that best match them. We'd also throw away most of our existing source code and redevelop the entire system from scratch. But our current project calls for grafting some Web browser front-end interfaces onto a system written in C

with as little change as possible to the existing code. Structured design models work very well for the existing C programs and functions."

"So how do I represent Web interfaces and client-server interactions with structured techniques?" Bernard asked.

Stana replied, "The trick is to think of the Web server as a container for application software programs that communicate with the Web browser over a real-time link—the Internet or an intranet. In structured design, the primary software units are programs and modules. So in the current system, the modules are C functions, which are packaged into a small number of complex programs that do many things, with all-encompassing menu-based front ends. One of your most important tasks for this upgrade is to decompose those large programs into smaller ones and move functions that implement the existing user interface out of the C code and into Web-page code. The remaining functions are application logic that you can package into small programs that can be called from Web-server scripts. Each of those small programs should be one structure chart and one box on a system flowchart. You should be able to cut and paste from the existing structure charts to create rough drafts as a starting point."

Bernard looked relieved at first, but then confusion and concern crossed his face. "Jim is going to review my work at the end of the week. I'm worried that I'll make some huge mistakes and that he'll think he made a mistake in hiring me. Would you look over some of my work before I meet with him?"

Stana gave Bernard a reassuring smile. "Jim put you in this office with me. And even though I'm assigned to other project tasks, he asked me to help you when you needed it. Software development only succeeds when everyone works as a team. People who *don't* ask for help are the ones that get fired. So why don't you spend the rest of the morning designing the entry and verification modules for the snack bar receipts, and we'll sit down after lunch and go over them?"

OVERVIEW

This chapter describes the traditional approach to designing software. The chapter begins with an overview of the structured models, model development process, and related terminology. We then describe how data flow diagrams are annotated with automation boundary information. Next, we explore how information from analysis phase models is transformed into design models using system flowcharts, structure charts, and module pseudocode. Then we discuss how traditional software design is integrated with other design phase activities. The chapter concludes with an examination of how the traditional approach is applied to designing a three-layer architecture.

As described in the opening case, traditional software design and structured design models are relatively old. They are commonly used with systems developed using procedural programming languages and are well suited to describing systems with both batch and on-line components. Most new systems are developed with object-oriented programming languages, so traditional system design models are decreasing in popularity. However, as illustrated in the case, many older systems in use today were designed and documented using traditional methods and models. Also, traditional design concepts such as coupling, cohesion, and top-down partitioning underlie both traditional and object-oriented design methods, so it is important to understand those concepts. Finally, traditional models are sometimes adapted to newer software development methods and paradigms such as multilayered software. So, analysts should be knowledgeable about the traditional approach to design.

THE STRUCTURED APPROACH TO DESIGNING THE APPLICATION ARCHITECTURE

The application architecture consists of the application software programs that will carry out the functions of the system. Application design must be done in conjunction with the design of both the database and the user interface. However, we focus exclusively on the design of the computer software itself here for ease of understanding.

You may already have written business programs using a third-generation language such as Visual Basic, C, COBOL, or Pascal. Third-generation programming languages are organized around *modules* that are arranged in a hierarchy like a tree. The top module is often called the *boss module* or the *main module*. The middle-level modules are *control modules*, and the *leaf modules* (those at the ends of the branches) are the detailed modules that contain most of the algorithms and logic for the program. A module, then, is a small section of program code that carries out a single function. A *computer program* is a set of modules that are compiled into a single executable entity. The design of a computer program is specified with a structure chart, which will be discussed in detail later in this section.

In large systems, a single program usually cannot perform all of the required functions. Sometimes one program is written to perform on-line activities, and another program carries out periodic functions that are executed once a day. Other programs may have specialized functions, such as backing up the data or producing year-end financial reports. All these individually executable entities, or programs, compose the entire system.

Both the structure of the overall system and any individual subsystems are documented using a *system flowchart*. The system flowchart identifies each program, along with the data it accesses. The system flowchart also shows the relationships among the various programs, subsystems, and their files or databases. It documents the architectural structure of the overall system. We describe how to design a system flowchart later in this chapter.

Finally, the project team must also design the internal logic of individual modules. The internal algorithms that compose the logic of the modules are usually documented using *pseudocode*. If you have taken programming classes, you probably had to write algorithms in pseudocode before you actually coded your programs. Pseudocode is very much like the structured English described in Chapter 6. Pseudocode is simply structured-programming-like statements that describe the logic of a module.

In general, analysts use a top-down approach for design. The inputs for the development of the design models and documents are the data flow diagrams and their detailed documentation, in structured English, and the detailed data flow definitions. Figure 10-1 illustrates the process. Analysts use an intermediate form of the DFD called the *DFD with automation system boundary*, which divides the computerized portions of the system from the manual portions. So, this diagram determines which portions of the system need to be included in the design. This enhanced DFD is actually used as the source for the design models, as shown in the figure.

module

an identifiable component of a computer program that performs a defined function

computer program

an executable entity made up of a set of modules

system flowchart

a diagram that describes the overall flow of control between computer programs in a system

pseudocode

structured-programming-like statements that describe the logic of a module

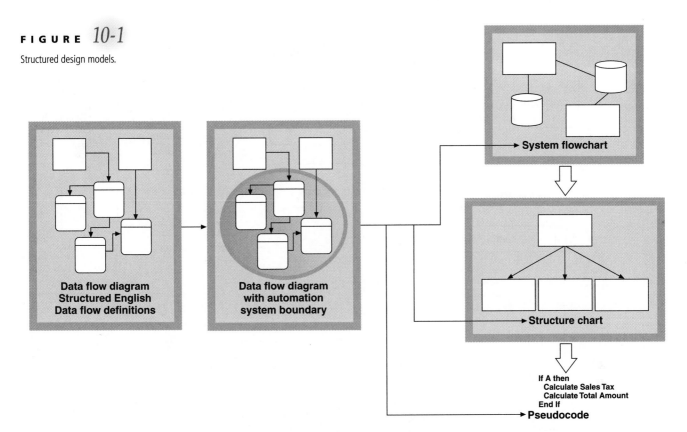

The following sections of the chapter trace the sequence of activities shown in Figure 10-1. First, the automation system boundary is discussed. Then we explain the development of the system flowchart. Next, the approach to the design of the structure chart is elaborated. Finally, we describe the form and method of writing pseudocode.

THE AUTOMATION SYSTEM BOUNDARY

The automation system boundary partitions the data flow diagram processes into manual processes and those that are to be included in the computer system. During the analysis phase, we looked at the business events and all of the processes that described those events. At that time, we did not try to distinguish between manual and automated processes, but to develop the computer system's design, we must identify the processes that will be automated.

Figure 10-2 illustrates a typical data flow diagram with the automation system boundary added. This figure shows both the system boundary, which identifies the entire automated system, and program boundary lines, which partition the DFD into separate programs. This diagram is the first step in design, and it determines what the programs are and what processes are included within those programs. In this context, we define a program as a separate executing entity.

Processes can either be inside or outside the system boundary. Processes that are outside are manual processes, such as sorting and inspecting paper documents, entering customer orders, or visually inspecting incoming shipments. The processes that are inside the boundary may be carried out in on-line or batch modes. On-line processes are usually active every day during working hours. Batch processes may be activated each night or only periodically. In some cases, the system boundary goes right through a process, which indicates that the process is a mid- or high-level process and is exploded on a more detailed diagram. Some of the processes in the exploded detail will be inside and some outside the system boundary.

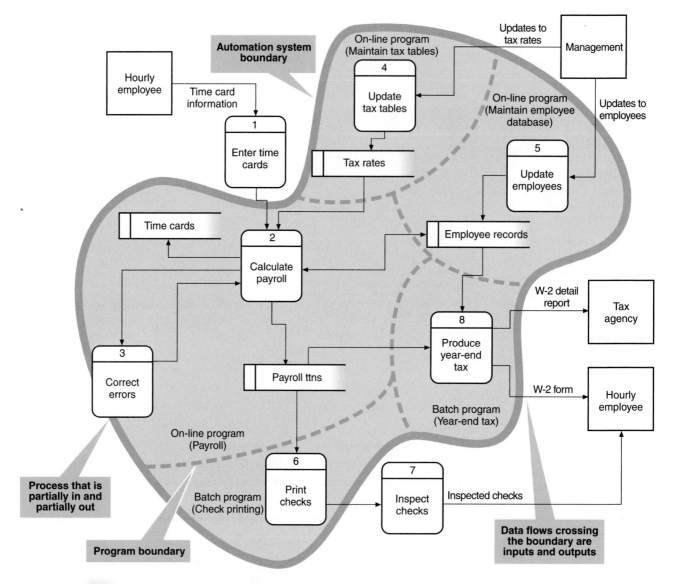

Diagram labels:

- **Automation system boundary**
- Hourly employee
- Time card information
- On-line program (Maintain tax tables)
- 4 — Update tax tables
- Updates to tax rates
- Management
- On-line program (Maintain employee database)
- Updates to employees
- 1 — Enter time cards
- Tax rates
- 5 — Update employees
- Time cards
- 2 — Calculate payroll
- Employee records
- 3 — Correct errors
- 8 — Produce year-end tax
- W-2 detail report
- Tax agency
- Payroll ttns
- **Process that is partially in and partially out**
- On-line program (Payroll)
- Batch program (Year-end tax)
- W-2 form
- Hourly employee
- 6 — Print checks
- 7 — Inspect checks
- Inspected checks
- Batch program (Check printing)
- **Program boundary**
- **Data flows crossing the boundary are inputs and outputs**

FIGURE *10-2*

The data flow diagram with an automation system boundary.

Data flows are found inside, outside, or crossing the system boundary and the program boundaries. The data flows that cross the system boundary are particularly important; they represent inputs and outputs of the system. In other words, the design of the program interfaces, including both the user interface and transmittals to other systems, is defined by the boundary-crossing data flows. In the final system, these data flows will be forms or reports in the user interface, or files or telecommunication transmittals between systems. Data flows that cross the boundaries between programs represent program-to-program communication. In the final system, these data flows will also be files or telecommunication transmittals between programs.

Figure 10-2 is the high-level data flow diagram showing all of the major processes for a payroll program. The system boundary can also be drawn on each data flow fragment to show more detail about which processes are internal or external and which low-level data flows cross the boundary.

THE SYSTEM FLOWCHART

The system flowchart is the representation of various computer programs, files, databases, and associated manual procedures that make up a complete system. Processes are grouped into programs and subsystems based on similarities such as shared timing (for example, a process performed monthly), shared access to stored data (for example, all processes that update employee data), and shared users (for example, processes that produce reports for the marketing department). The programs and subsystems thus created have complex interdependencies, including flow of data, flow of control, and interaction with permanent data stores. The system flowchart is frequently constructed during the analysis activities. For example, the subsystems of the RMO customer support system were defined during the analysis phase (see Chapter 6), and the set of events allocated to each subsystem makes up the program modules. We also saw this division in the definition of various use cases (Chapter 7).

A system flowchart graphically describes the organization of the subsystems into automated and manual components, showing the flow of data and control among them. System flowcharts are used primarily to describe large information systems consisting of distinct subsystems and dozens or more programs. They are also used to describe systems that perform batch processing (for example, systems used to process bank transactions, payroll checks, and utility bills). A common characteristic of such systems is the division of processing into discrete steps (such as validation of input transactions, update of a master file with transaction data, and production of periodic reports) with a fixed execution sequence. Many batch systems also make extensive use of files in addition to or instead of databases.

System flowcharts first came into widespread use to document the processing and data flow between programs that processed information through batch transaction files. Frequently, in these systems, one program would produce a file of all the daily transactions. Then another program would process the transactions and update a master file. Yet another program would be used to produce the various reports required from the system.

Today's newer systems perform much processing in real time, as each transaction is entered. These systems also usually make updates to a relational database system instead of a master file. Centralized database management systems also include many of the processes that were previously done by individual programs. System flowcharts also may be drawn for these newer systems, although the diagrams tend to be much simpler because the processing is more centralized to a single program or subsystem. But since the systems developed these days are generally much more complex overall, you will still see system flowcharts used to represent how all of the pieces fit together.

Many business systems today also have both real-time and batch components. For example, today your credit-card purchases are at least verified in real time and may even be posted in real time. However, monthly account statements and customer payments typically are processed in batches. A system flowchart is used to describe the overall organization of this type of system and show the relationship between the real-time components and the batch processing.

Figure 10-3 illustrates the most common symbols that are used in system flowcharts. These symbols are fairly common throughout the industry, although from time to time you will see variations.

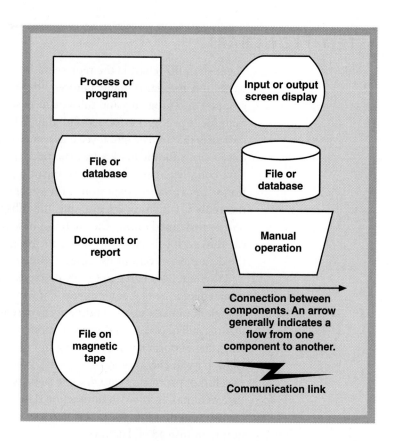

Figure 10-4 is a system flowchart for the payroll system shown on the DFD in Figure 10-2. Note that the system flowchart identifies the files, programs, and manual processes of the total system. We have added physical implementation descriptions by identifying the file media, disk or tape. Frequently, we also include additional system functions and files such as backups and history files. Even though the information shown in Figures 10-2 and 10-4 is very similar, the focus of the diagrams is different. The system flowchart focuses on the implementation of physical objects such as executable programs, files, and documents.

Figure 10-4 shows that the payroll program has four inputs and produces three outputs. The inputs are the time cards, the tax rate table file, the employee database, and corrections. Outputs produced are an error report, a payroll transaction file, and an updated payroll history file. Other programs (that is, independent executables) are the two programs to maintain the tax rate tables and the employee database, and the other two programs to write checks and to produce year-end income tax reports.

Figure 10-5 is an example of a system flowchart for Rocky Mountain Outfitters. The four main programs correspond to the subsystems identified in the list of subsystems in Figure 6-10. As in the example of the order-entry subsystem shown in Figure 6-12, each subsystem will include DFD fragments for each of the events in the subsystem. In the system flowchart, the individual data stores have been converted into database files. As you can see, the creation of the system flowchart requires the architectural design of the major program steps, the architectural design of the databases, the identification of the major interfaces, and the identification of the primary outputs.

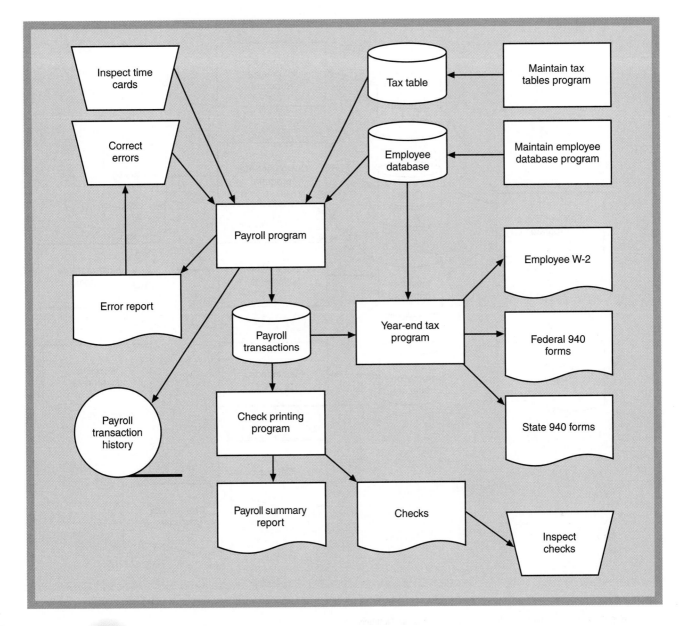

FIGURE *10-4*   A sample system flowchart for a payroll system.

Figure 10-5 also shows one additional subsystem that is not identified in Figure 6-10. During the scoping and level of automation review discussed in Chapter 8, RMO decided that it needed a higher level of automation for a couple of sales analysis reports. In this instance, instead of adding the reports to an existing subsystem, the project team defined a new subsystem.

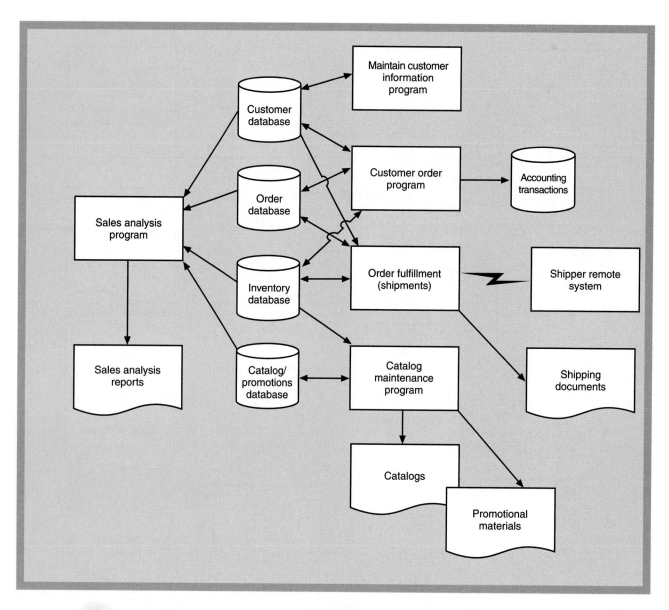

FIGURE *10-5* The system flowchart for Rocky Mountain Outfitters.

THE STRUCTURE CHART

The primary objective of structured design is to create a top-down decomposition of the functions to be performed by a given program in a new system. Each independent program shown in the system flowchart performs a set of functions. Using structure chart techniques always provides a hierarchical organization of program functions. First, this section explains what a structure chart is and how to interpret one. We explain how each structure chart is related to the DFDs created during systems analysis, and then we show how to use a detailed data flow diagram to develop one.

A *structure chart* hierarchy describes the functions and the subfunctions of each part of a system. For example, the program may have a function called *Calculate pay amounts*. Some subfunctions are *Calculate base amount*, *Calculate overtime amount*, and *Calculate taxes*. In a structure chart, these functions are drawn as a rectangle. Each rectangle represents a module.

A module is the basic component of a structure chart and is used to identify a function. Figure 10-6 shows a simple structure chart for the payroll example. Modules, as

structure chart

a hierarchical diagram showing the relationships between the modules of a computer program

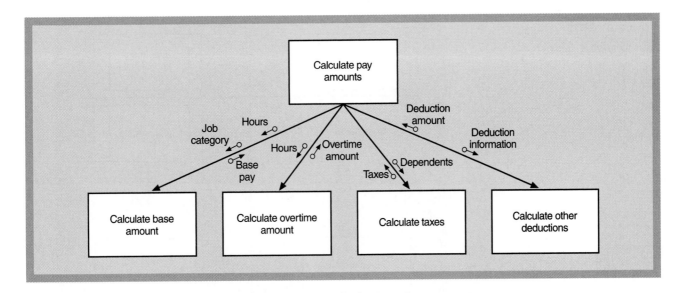

FIGURE *10-6*

A simple structure chart for
the *Calculate pay amounts* module.

shown by the rectangles, are relatively simple and independent components. Higher-level modules are control modules that control the flow of execution. Lower-level modules are "worker bee" modules and contain the program logic to actually perform the functions. In Figure 10-6, for example, the *Calculate pay amounts* module may do nothing more than call the lower-level modules in the correct sequence to carry out the logic of calculating payroll.

Notice how a structure chart provides a simple and direct organization to a computer program whose purpose is to calculate payroll amounts. As you have learned in your programming class, modular programming is a well-accepted method to write computer programs that are easy to understand and maintain. Breaking a complex program into small modules makes initial programming and maintenance programming easier. The development of a structure chart is based on rules and guidelines. The key points are that the program is a hierarchy and that the modules are well formed with high internal cohesiveness and minimal data coupling. Later, we describe in more detail the characteristics of good modules.

The lines connecting the modules indicate the calling structure from the higher-level modules to the lower-level modules. The little arrows next to the lines show the data that are passed between modules and represent the inputs and outputs of each module. At the structure chart level, we are not yet concerned with what is happening inside the module. We only want to know that somehow the module does the function indicated by its name, using the input data and producing the output data.

Figure 10-7 shows the common symbols used to draw structure charts. The rectangle represents a module. In a structure chart, a module can represent something as simple as a block of code, such as a paragraph or section in a COBOL program. In other languages, a module typically represents a function, procedure, or subroutine. Examples of modules as program fragments include subroutines (as in FORTRAN and BASIC), paragraphs or subprograms (as in COBOL), procedures (as in Pascal), and functions (as in FORTRAN, C, and C++). A module also can be a separately compiled entity such as a complete C program. The rectangle with the double bars is simply an existing module or a module that is used in several places. Use of the double bar notation is optional.

Part c of Figure 10-7 shows a call from a higher-level module to the lower-level module. A *program call* occurs when one module invokes a lower-level module to perform a needed service or calculation. The implementation of a program call varies among programming languages. For example, a program call can be implemented as a function call in C or C++, a procedure call in Pascal, or a subroutine call in FORTRAN. In each case,

program call

the transfer of control from a module
to a subordinate module to perform
a requested service

FIGURE *10-7*

Structure chart symbols.

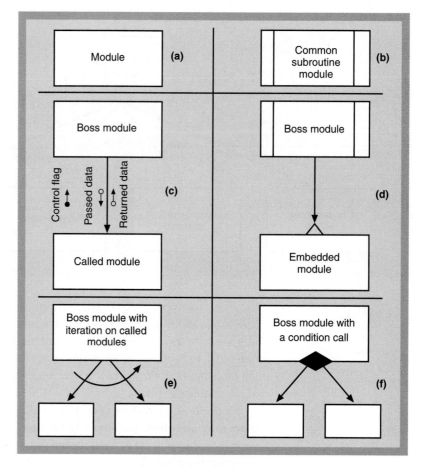

the program passes control to the called module, the called module executes a series of program statements, the called module passes control back to the calling module, and execution resumes with the statement or instruction immediately following the call.

Figure 10-7c also indicates how data are passed between modules. The arrows with the open circle, called *data couples*, represent data being passed into and out of the module. A data couple can be an individual data item (such as a customer account number) or a higher-level data structure (such as an array, record, or other data structure). The type of coupling used at each level of the structure chart usually follows the principle of layering of detail. That is, coupling between modules near the top of a structure chart typically consists of data structures representing high-level aggregations of data. Coupling between modules at the bottom of the structure chart typically consists of single data items, flags, and relatively small data structures.

The arrow with the darkened circle is a control couple flag. A flag is purely internal information that is used between modules to indicate some result. Flags originating from lower-level modules often indicate a result, such as a record passing a validation test. Another common use is to indicate that the end of a file was reached.

Figure 10-7d illustrates a lower-level module that is broken out on the structure chart but that in all probability will be subsumed into the calling module for programming. This documentation technique primarily ensures that the function performed by the module is highlighted. Figure 10-7e and f show two alternatives for program calls. In 10-7e we show the notation used to indicate iteration through several modules. In 10-7f we show conditional calling of low-level modules—that is, the program calls modules only when certain conditions exist.

Figure 10-8 is a more complete view of the Payroll program, including the original *Calculate pay amounts* function from Figure 10-6. Notice that the entire structure chart

data couples

the individual data items that are passed between modules in a program call

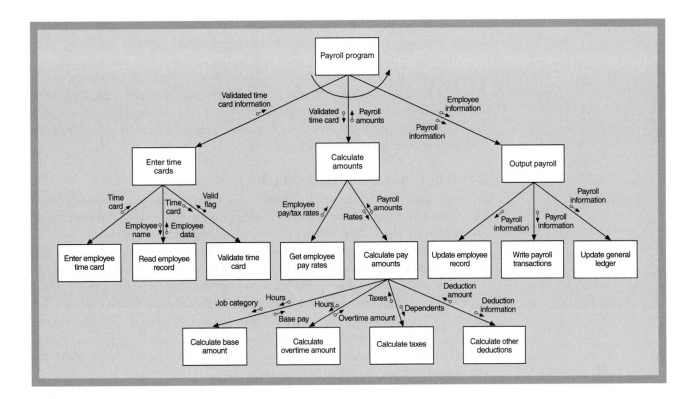

FIGURE *10-8*

A structure chart for the entire Payroll program.

shown in Figure 10-8 is based on the system activities following the temporal event *Time to produce payroll.* During systems analysis for the payroll system, the analyst would have identified this event as one that occurs at the end of every week for hourly employees. Many other events would have been identified as well.

A basic idea of structured programming is that each module only has to do a very specific function. The module at the very top of the tree is the boss module. Its functions are to call the modules on the next tier, pass information to them, and receive information back. The function of each middle-level module is to control the processing of the modules below it. Each has control logic and any error-handling logic that is not handled by the lower-level module. The modules at the extremities, or the leaves, contain the actual algorithms to carry out the functions of the program. This approach to programming separates program control logic from business algorithm logic and makes programming much easier.

The arrows from the higher-level modules to the lower-level modules indicate the program call. The direction of the call is always from left to right. Notice that the structure chart maintains a strict hierarchy in the calling structure. A lower module never calls a higher module. The curved arrow immediately below the boss module indicates a loop across all three calls. In other words, the main module will have an internal loop that includes calls to all three lower-level modules within the same loop.

In the example, you can see the flow of information downward and back up. Usually, a higher-level module will request a service from a lower-level module and pass the necessary input information. The lower-level module then returns the requested information or some control information, as a flag, to notify the higher-level module of the successful completion of the task. Looking at the *Enter time cards* subhierarchy, you can observe that the employee time card information is passed to the boss. Then the employee name is passed to the next module, which reads some employee information and passes it back up. Finally, employee data and time card information are passed to the rightmost module, which validates the time card. This module passes upward a control flag, indicating success or failure of the validation. If the validation fails, then the program sends error messages and goes into its error-handling routines. We have not shown all the complexities required in a real program, especially the error-handling modules.

Included within the structure chart will be the modules that access data from the outside world. It is important that the design of these modules be consistent with the design of the user interface, the interface to other systems, and the database design. The structure charts that have been developed should also be consistent with the system flowchart. If changes were made during this design activity, then the project team should update the system flowchart accordingly.

Developing a Structure Chart

Structure charts create a hierarchy of modules for a program. A structure chart looks like a tree with a root module and branches. A subtree is simply a branch that has been separated from the overall tree. When the subtree is placed back in the larger tree, the root of the subtree becomes just another branch in the overall tree. Why is this important? The structure chart can be developed in pieces and combined for the final diagram.

Figure 10-5 showed the system flowchart for RMO's customer support system. Each major program corresponds to a subsystem in the event-partitioned diagram. Each program will have its own structure chart. However, as can be seen in Figure 6-10, each program—that is, subsystem—consists of several events. Each event corresponds to a process on the event-partitioned DFD, and each process will be detailed in a DFD fragment based on the event from the event table.

There are two methods to develop structure charts: transaction analysis and transform analysis. *Transaction analysis* uses as input the system flowchart and the event table to develop the top level of the tree—that is, the main program boss module and the first level of called modules. *Transform analysis* uses as input the DFD fragments to develop the subtrees, one for each event in the program. Each subtree root module corresponds to the first-level branch of the main program structure chart. We discuss each method in turn.

transaction analysis

the development of a structure chart based on a DFD that describes the processing for several types of transactions

transform analysis

the development of a structure chart based on a DFD that describes the input-process-output data flow

Transaction Analysis

In transaction analysis, the first step is to examine the system flowchart and identify each major program, such as the Customer order program in Figure 10-5. In Figure 10-9, we duplicate Figure 6-13, the event-partitioned DFD for the order-entry subsystem. This DFD shows the five processes derived from the five events of this subsystem. In this subsystem, the five primary processes are five different transactions that must be supported. These transactions are *Look up item availability, Create new order, Update order, Produce order summary reports,* and *Produce transaction summary reports.*

Figure 10-10 shows the structure chart based on transaction analysis for this program. As already mentioned, transaction analysis is the process of identifying each separate transaction that must be supported by the program and constructing a branch for each one. In essence, this program, at least at the highest level, is simply a module to display a screen for the user to enter a transaction choice and then to invoke the appropriate module to process the transaction. This diagram does not show the additional detail below each of the transaction modules. Each of the transaction modules, which are named after the transactions, will be the main boss module for a subtree to process the transaction. Each subtree will be developed based on the DFD fragment for that event and will be developed utilizing transform analysis.

This structure chart also has very few data couples. Essentially, the only information passed is the transaction selection from the *Get transaction choice* module. That information is used by the control module to select the correct processing module. The subtree beneath the processing module will display the appropriate screens to accept and pass the detailed information required.

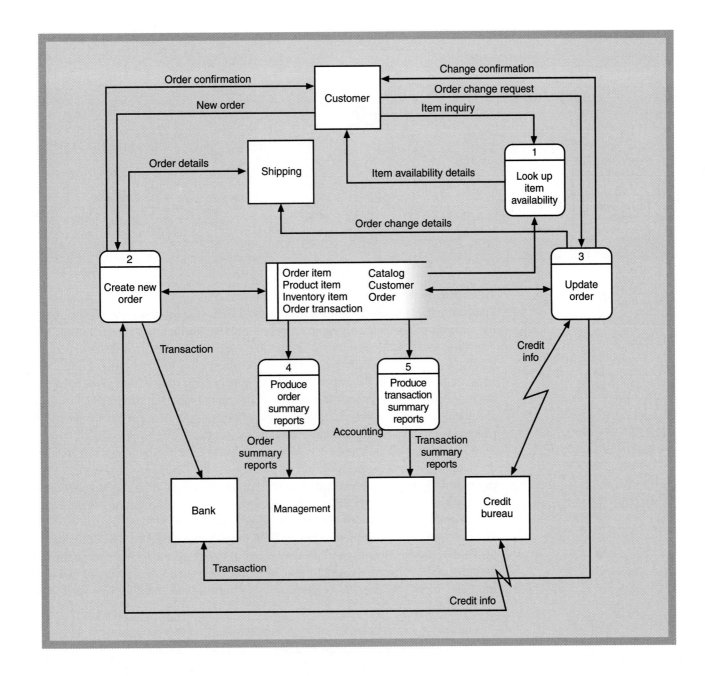

FIGURE *10-9*

Event-partitioned DFD for the
order-entry subsystem.

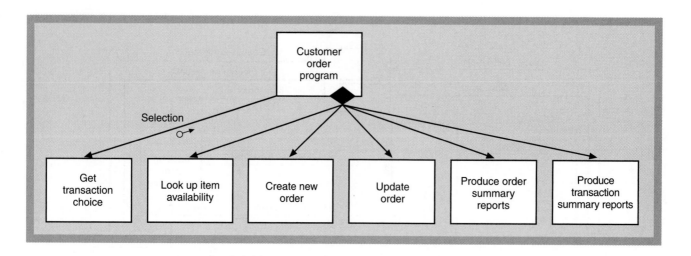

FIGURE *10-10*

High-level structure chart for
the Customer order program.

Transform Analysis

Transform analysis is based on the idea that the computer program "transforms" input data into output information. Structure charts developed with transform analysis usually have three major subtrees: an input subtree to get the data, a calculate subtree to perform the logic, and an output subtree to write the results. Figure 10-8 is a good example of a structure chart that was developed using transform analysis, for the process of transforming time card inputs into payroll outputs following one event. Note that a DFD fragment usually follows this pattern of input-process-output, and the structure chart converts the processing on the DFD fragment to a top-down structure of program modules.

Sometimes DFD fragments are decomposed into detailed diagrams. The detailed diagrams provide more detail than can be used for the structure chart. Figures 10-11 through 10-14 provide an example of transform analysis from Rocky Mountain Outfitters. Figure 10-11 shows the DFD fragment created for the *Create new order* event. Figure 10-12 contains the exploded view for that event, which is used for transform analysis. The structure chart is developed directly from the data flow diagram. The fundamental idea is that the detailed diagram processes from the data flow diagram become the leaf modules in the structure charts. The mid-level processes—that is, the processes that were exploded to derive the low-level processes—become the intermediate-level boss

FIGURE *10-11*

The *Create new order* DFD fragment.

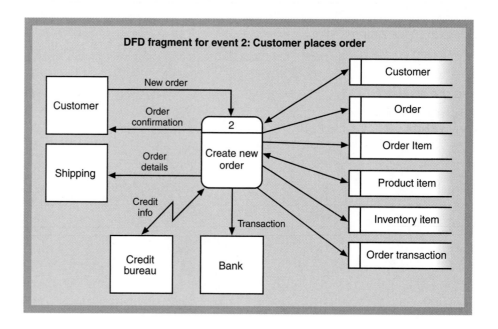

FIGURE *10-12*

Exploded view of the
Create new order DFD.

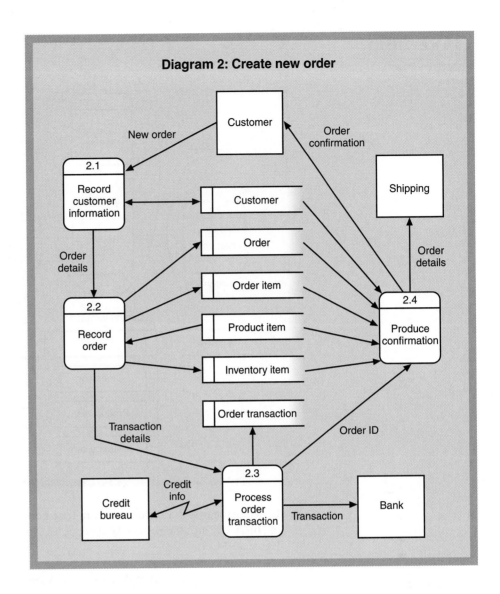

Diagram 2: Create new order

modules on the structure chart. Thus, the hierarchy of the structure chart directly reflects the organization of the set of nested, or leveled, data flow diagrams. Additional boss modules may need to be developed to provide the correct structure to the structure chart.

As stated previously, the general form of a structure chart developed with transform analysis is input-process-output. The method to develop a structure chart from a data flow diagram fragment consists of the following steps:

1. Determine the primary information flow. This flow is the main stream of data that are transformed from some input form to the output form.
2. Find the process that represents the most fundamental change from an input stream to an output stream (see Figure 10-13). The input data stream is called the *afferent data flow*. The output data stream is called the *efferent data flow*. The center process is called the *central transform*.
3. Redraw the data flow diagram with the input to the left and the output to the right. The central transform process goes in the middle. If this diagram is an exploded-view data flow diagram, add the parent process to the diagram. You can omit nonprimary data flows to simplify the drawing. An example of this redrawn data flow diagram is given in Figure 10-13.

afferent data flow

the incoming data flow from a sequential set of DFD processes

efferent data flow

the outgoing data flow in a sequential set of DFD processes

central transform

set of DFD processes that are located between the input and output processes

FIGURE *10-13*

Rearranged view of the
Create new order DFD.

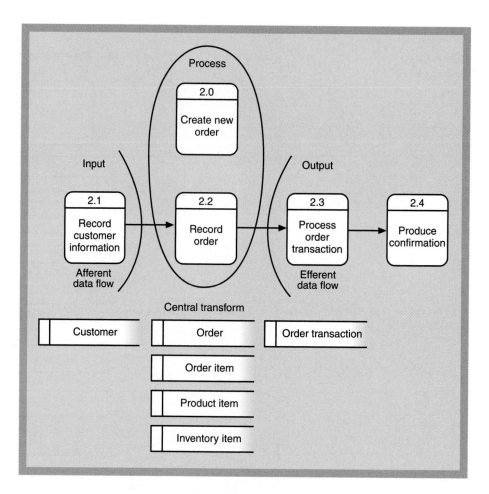

4. Generate the first-draft structure chart, based on the redrawn data flow, with the calling hierarchy and the required data couples. An example of this diagram is given in Figure 10-14.

FIGURE *10-14*

First draft of the structure chart.

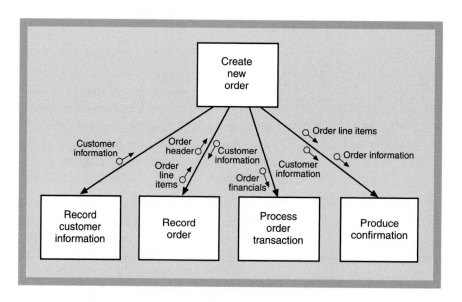

5. Add other modules as necessary to get input data via the user-interface screens, read from and write to the data stores, and write output data or reports. Usually, these modules are lower-level, or utility, modules. Add the appropriate data couples based on the data flows to and from these data stores.

6. Using any structured English or decision table documentation as a basis, add other required intermodule relationships such as looping and decision symbols.
7. Make the final refinements to the structure chart based on the quality-control concepts discussed in the following section.

Through step 4, as can be seen in Figure 10-14, the organization of the structure chart very closely mirrors the data flow diagram from which it derives. In step 5 we begin to enhance the first-draft structure chart with additional modules to provide modules to read and write data. Frequently, there are no corresponding processes on the data flow diagram. Thus, at this point we depend less on the data flow diagram information and more on the requirements of a good design. Figure 10-15 illustrates the structure chart for the next step—step 5.

FIGURE *10-15*

The structure chart for the *Create new order* program.

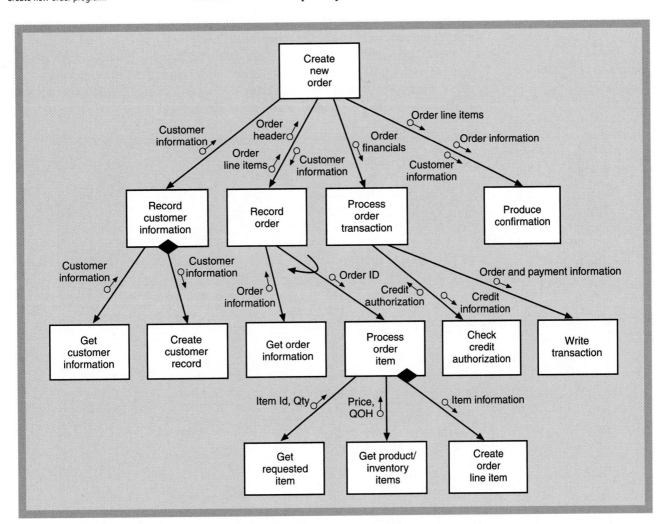

Comparing Figure 10-14 with 10-15, notice that Figure 10-14 indicates that all the input information comes from the far-left module, *Get customer information*. In Figure 10-15, observe that the accessing of information has been distributed across other branches of the structure chart. Customer information is retrieved through the far-left branch, but additional customer information about the order is retrieved in the second branch of the chart. Even though this organization is not exactly true to the data flow diagram, it is a more logical organization of the structure chart. The addition of these data access modules is truly a design process—the creation of new components based on systems design principles.

In addition to distributing the access of customer input data, the structure chart in Figure 10-15 has other data access modules to retrieve product and inventory information. This type of data retrieval corresponds to the data flows on the data flow diagram between the processes and the data stores. During design, we must explicitly identify the modules that actually read from and write to the data stores. The module *Get product/inventory items* is added to the structure chart to provide the retrieval of the product information. As the additional modules are added to the structure chart, then the data couples are defined more precisely to reflect this more detailed design structure.

In Figure 10-15, we have also added the symbols concerning looping and optional calls. The black diamond indicates that the call to create a customer record is optional and in fact is required only if the customer is a new customer. The general form of the structure chart has the inputs to the left and the outputs to the right. The black diamond on the call to *Create order line item* indicates a situation where an item is out of stock, so the call to *Create order line item* is conditional.

The high-level boss module, *Create new order*, and its tree of modules can be plugged into the transaction structure chart in Figure 10-10. Figure 10-16 illustrates the process of combining the top-level structure chart, developed using transaction analysis, with the lower-level subtrees, developed with transform analysis.

FIGURE *10-16*

Combination of structure charts
(data couple labels are not shown).

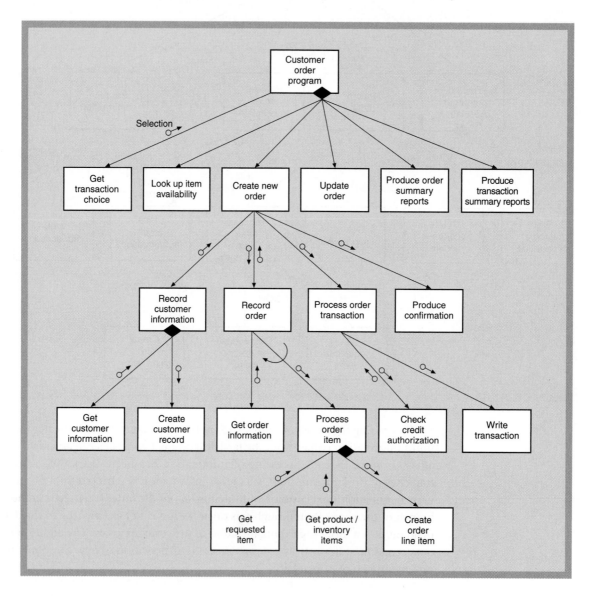

Evaluating the Quality of a Structure Chart

The process of developing structure charts from DFDs can become rather involved. Rules and guidelines can be used to test the quality of the final structure chart, however. Two measures of quality are *module coupling* and *module cohesion*. Generally, it is desirable to have highly cohesive and loosely coupled modules.

module coupling

the manner in which modules relate to each other; the preferred method is data coupling

module cohesion

a measure of the internal strength of a module

The principle of coupling is a measure of how a module is connected to the other modules in the program. The objective is to make modules as independent as possible because a module that is independent can execute in almost any environment. An independent module has a well-defined interface of several predefined data fields, and it passes back a well-defined result in predefined data fields. The module does not need to know who invoked it and, in fact, can be invoked by any module that conforms to the input and output data structure. The best coupling is through simple data coupling. In other words, the module is called, and a specific set of data items is passed. The module performs its function and returns the output data items. This kind of module can be reused in any structure chart that needs the specific function performed.

Cohesion refers to the degree to which all of the code within a module contributes to implementing one well-defined task. Modules with high cohesion implement a single function. All of the instructions within the module are part of that function, and all are required for the function. Modules with low (or poor) cohesion implement multiple or loosely related functions.

Note that the amount of coupling and the specific data items being passed are good indicators of the degree of module cohesion. Modules that implement a single task tend to have relatively low coupling because all of their internal code acts on the same data item(s). Modules with poor cohesion tend to have high coupling because loosely related tasks typically are performed on different data items. Thus, a module with low cohesion generally has several unrelated data items passed by its superior.

A flag passed down the structure chart is also an indicator of poor cohesion in the lower-level module. Flags passed into a module are used typically to select the part of the recipient module's code that will be executed. Part a of Figure 10-17 shows an example of poor cohesion. A project team can improve cohesion by partitioning the module into separate modules, one for each value of the flag, as shown in part b of Figure 10-17. The code of the superior module is programmed to use the flag to decide which of the partitioned subordinate modules to call.

FIGURE *10-17*

Examples of module cohesion.

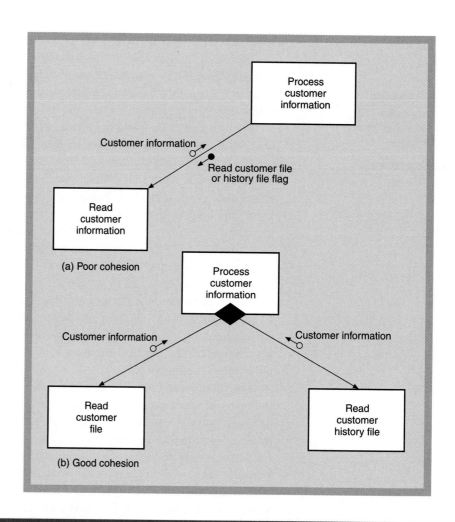

(a) Poor cohesion

(b) Good cohesion

MODULE ALGORITHM DESIGN: PSEUDOCODE

The previous two models, the system flowchart and the structure chart, provide the overall structure of the system and the structure within each program. The next requirement of design is to describe the internal logic within each module. Three common methods are used to describe module logic: flowcharts, structured English, and pseudocode. All three methods are equivalent in their ability to describe logic. Flowcharting is a visual method that uses boxes and lines to describe the flow of logic in a program. In the early days of computing, flowcharting was used almost exclusively. Today, however, versions of pseudocode and structured English have replaced flowcharting. You learned about structured English in Chapter 6. Pseudocode is a variation of structured English that is closer to a programming language. Frequently, analysts write pseudocode using statements that are very similar to the target language. If they are writing to COBOL, they use COBOL-like syntax. If they are writing in Visual Basic or C, they use a syntax that mirrors those languages.

Figure 10-18 shows a simple example of the logic of the payroll system. Pseudocode statements for the *Payroll program, Calculate amounts, Calculate pay amounts,* and *Calculate taxes* modules are shown. This figure shows examples of each of the three types of control statements used in structured programming: sequence, a sequence of executable statements; decision, if-then-else logic; and iteration, do-until or do-while.

```
Payroll program
DoUntil No more time cards
     Call Enter time cards
     Call Calculate amounts
     Call Output payroll
End Until

Calculate amounts
     Call Get employee pay rates
     Call Calculate pay amounts

Calculate pay amounts

Call Calculate base amount
If (HoursWorked > 40) Then
     Call Calculate overtime amount
End If
Call Calculate taxes
If (SavingsDeduction=yes) or (MedicalDeduction=yes) or (UnitedWay=yes) Then
     Call Calculate other deductions
End if

Calculate taxes

Get Tax Rates based on Number Dependents, Payrate
Calculate Income Tax = PeriodPayAmount * IncomeTaxRate
If YTD Pay < FICA MaximumAmount Then
     Calculate EmpFICA = PeriodBasePay * FICAEmployeeRate
     Calculate CorpFICA = PeriodBasePay * FICACorpRate
End If
If StateTax Required Then
     Get StateTaxRate based on State, NumberDependents, Payrate
     CalculateStateTax = PeriodPayAmount * StateTaxRate
End If
If StateTaxRequired Then
     Get StateTaxRate based on State, NumberDependents, Payrate
     Calculate StateTax = PeriodPayAmount * StateTaxRate
End If
If OvertimePay > 0 Then
     Calculate OvertimeIncomeTax = PeriodOvertimePay * IncomeTaxRate
     Add to OvertimeIncomeTax to Incometax
     If YTDPay < FICAMaximum Amount Then
          Calculate EmpOvertimeFICA = PeriodOvertimePay * FICAEmployeeRate
          Calculate CorpOvertimeFICA = PeriodOvertimePay * FICACorpRate
     End If
     If StateTaxRequired Then
          Calculate StateOvertimeTax = PeriodOvertimePay * StateTaxRate
     End If
End If
```

FIGURE *10-18*

Pseudocode for the *Calculate pay amounts* hierarchy.

INTEGRATING STRUCTURED APPLICATION DESIGN WITH OTHER DESIGN TASKS

So far, you've learned how to develop a structure chart based on the information in a data flow diagram. The primary focus was on capturing the information in the structural relationship between the processes. The structure chart developed from either transaction analysis or transform analysis will be correct, but it may not be complete. Before the structure chart can be considered complete, it usually must be modified or enhanced to integrate the design of the user interface, the database, and the network, as Barbara Halifax discusses in the accompanying RMO memo. Since user-interface and database design are discussed in later chapters, we will only briefly discuss in this section and the next section on three-layer design the types of changes that need to be made.

May 19, 2005

To: John MacMurty

From: Barbara Halifax

Re: Software design modifications for the Customer Support System

John, we're behind a few days in finalizing all of the design details, but I think we can make up the time as we move into implementation. It took longer than expected to develop dialogs and other user-interface specifications. Since those results didn't match up well with our initial expectations, we need to go back to the software design and make some modifications, which I expect to be completed by the end of the week.

Also, during a walkthrough with Ann, we discovered some inconsistencies and potential network bottlenecks in our original plans for distributing software modules across layers and computer systems. We've made appropriate modifications to the application architecture and identified some changes that will be needed in the network, particularly in our firewall and router configurations. Ann has scheduled those updates for the middle of June, which is well in advance of our first round of software testing activities.

Let me know if you have any questions.

BH

cc: Ann Hamilton

The user interface consists of a set of input forms, output forms, and reports. Interactive user interfaces are usually based on a dialog between the user and the system and include a series of input and output forms. Every form must be displayed and the data retrieved somewhere in a module in the structure chart. As these forms are developed, the structure chart needs to be evaluated from three aspects:

- Are additional user-interface modules needed?
- Does the pseudocode in the interface modules need to be modified?
- Are additional data couples needed to pass data?

In earlier chapters, you learned that the entity-relationship diagram (ERD) must be consistent with the data stores found on the data flow diagrams. There is not necessarily a one-to-one correspondence between data stores and database tables, but the information on every data store must be somewhere in the database. In addition, every database table and field must be represented by a data store somewhere. During design, the project team performs this same type of analysis and makes appropriate changes to the structure chart.

The same three aspects—modules, pseudocode, and data couples—need to be evaluated for the database. If a database management system is being used, then a common interface is usually provided. The designer can make the database accessible either by calling a database interface module or by embedding SQL (Structured Query Language) statements within the pseudocode.

Finally, the structure charts and system flowcharts must be checked for correspondence to the existing or planned network architecture. Since architectural design normally precedes or runs concurrently with detailed software design, system flowcharts and structure charts will normally be developed with the proper assumptions about network architecture. However, changes may be made and new issues may be uncovered as additional details are added to the design. Thus, an important final step in detailed design is to reevaluate its correspondence to the network architecture, particularly with respect to required protocols, capacity, and security.

The linear nature of a textbook makes it necessary to present details of design activities in a fixed sequence. However, in real development projects, the order in which design activities are performed varies greatly. Some projects assign detailed design tasks such as software, user interface, system interface, and database design to multiple teams operating in parallel. If an iterative approach to the SDLC is used, detailed design tasks are completed for each iteration. Other projects may follow a more linear sequence due to lack of personnel or specific project characteristics. In projects that follow a relatively linear order, early detailed design decisions of all types must be reevaluated after later design tasks are completed.

THREE-LAYER DESIGN

Chapter 9 described three-layer design and its division of application software into the view, business logic, and data access layers. Structure charts and system flowcharts predate the development of three-layer architecture by at least a decade. Still, they can be used to describe design decisions and software structuring based on three-layer architecture.

Figure 10-19 shows a system flowchart for the RMO Customer order program. The flowchart divides processing according to a three-layer architecture of view layer, business logic layer, and data layer, as described in Chapter 9. Each layer communicates using well-defined protocols as noted on the flowchart, which enables the layers to be located on different machines, if desired. As described in Chapter 9, the choice of protocols, such as HTTPS and SQL, and the choice of deployment environments, such as Internet Explorer,

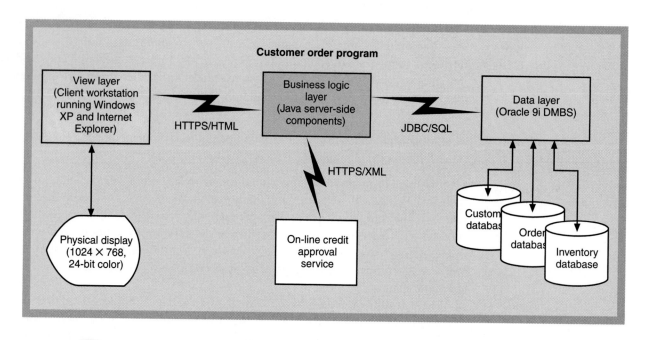

FIGURE *10-19*

A system flowchart showing three-layer architecture for the *Customer order* program.

Java components, and an Oracle database management system (DBMS) are architectural design decisions made early in the design phase. Annotating a system flowchart with specifics of the deployment environment is one way of documenting the decision and communicating important constraints to other project participants.

However, a system flowchart doesn't necessarily describe where software layers execute. In Figure 10-19, the view layer is described as executing on a client workstation, but no locations are given for the business logic and data layers. They could also execute on the client workstation (an unusual arrangement), both could execute on a single server, or each could execute on a separate server or cluster of servers. Unless specific location information is included, a system flowchart should be assumed to describe the distribution of processing functions across programs or groups of modules, not across computer systems.

Describing software structure with structure charts under a three-layer architecture can be quite different from the examples shown earlier in the chapter, depending on the deployment environment and development tools. As a point of comparison, consider the structure charts in Figures 10-15 and 10-16. Structure charts were developed in an era when application programs were functionally all encompassing. That is, they included modules to handle all operational aspects of the application task, including input/output (view layer), business logic, and interaction with data stored in files or databases (data layer). Figure 10-16 describes just such a program, as might be implemented in a traditional programming language such as C or COBOL and deployed in a centralized hardware architecture.

Figure 10-20 shows an alternate structure chart for a three-layer architecture for creating an order, based on a form-oriented dialog that will be discussed in Chapter 14. View layer modules are shown in yellow, business logic layer modules in red, and data layer modules in green. All view layer modules are shown, but only the business logic modules called by the Customer form are shown. Other view layer modules would require similar business logic layer modules. The data layer is composed of a DBMS, and business logic layer modules include code to generate appropriate database access commands and process responses.

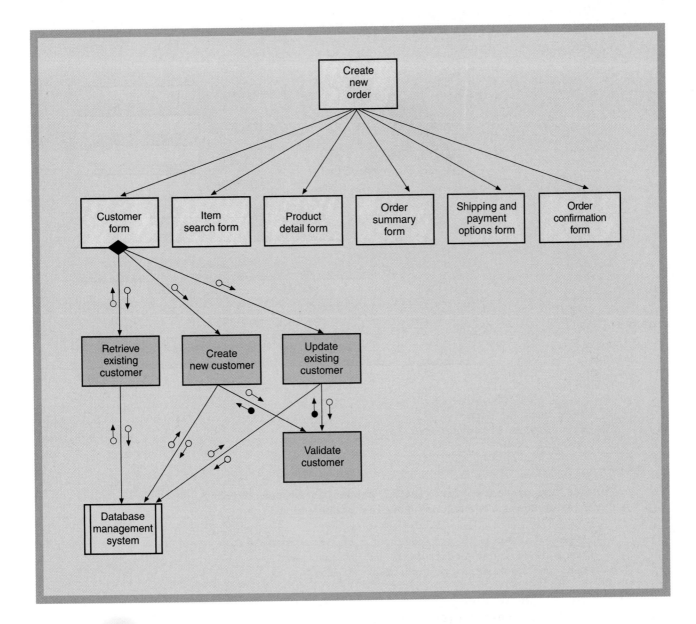

FIGURE *10-20*

A structure chart showing three-layer architecture for the activity *Create new order.*

Figure 10-21 shows the RMO Customer form used to find, add, or update customer data as depicted on the structure chart. Figure 10-22 shows how some of the code attached to the Customer form might be implemented in the Visual Basic programming language to represent the view layer. The event procedure btnSearch_Click() executes when the user clicks the Search button. The CustomerID number typed into the form is passed to one of the functions in a code module representing the business logic layer. The view layer does not include the details of the customer search.

Figure 10-23 shows a template for the business logic layer function RetrieveExisting-Customer() that is called by the btnSearch_Click() event procedure. It indicates the code insertion point for database retrieval statements. Note that this function handles retrieval of existing customer data but not data entry for new customer data or database update of new or existing customer data. Additional functions or procedures shown on the structure chart below the Customer form would also be included in the business layer to handle this functionality. The example demonstrates a clean division of program code between the view and business logic layers, even though they might execute within the same program on a single machine.

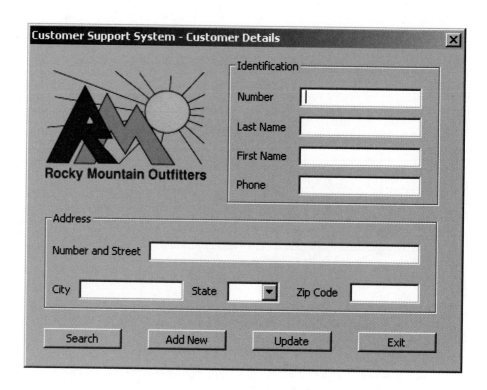

<figure>**FIGURE** *10-21*

A simple RMO form to find, add, or update Customer data.</figure>

```
Public Class CustomerForm
    Inherits System.Windows.Forms.Form

    Windows Form Designer generated code

    Private Sub btnSearch_Click(ByVal sender As System.Object, _
    ByVal e As System.EventArgs) Handles btnSearch.Click

        'when btnSearch is clicked, call the code module
        'in the Business Logic Layer using customerID
        'and get back array of customer details

        Dim custID As String
        Dim customerDetails(6) As String
        custID = txtCustomerID.Text
        Try
            customerDetails = OrderEntryModule.RetrieveExistingCustomer(custID)
            txtLastName.Text = customerDetails(0)
            txtFirstName.Text = customerDetails(1)
            txtStreetAddress.Text = customerDetails(2)
            txtCity.Text = customerDetails(3)
            txtState.Text = customerDetails(4)
            txtZipCode.Text = customerDetails(5)
            txtPhone.Text = customerDetails(6)
        Catch
            MessageBox.Show("Customer not found")
        End Try

    End Sub
```

FIGURE *10-22*

Visual Basic code for the form shown in Figure 10-21.

```
Public Class OrderEntryModule

    'This module contains functions and procedures that are
    'called from event procedures on forms

    'It represents part of the Business Logic Layer for shared
    'order entry processing code.

    Public Shared Function RetrieveExistingCustomer(ByVal anID As String) As String()

        'This function queries the database based on the
        'customer ID to search for an existing customer
        'and then it returns the customer details

        Dim customerDetails(6) As String

        'Insert needed code here

        Return customerDetails

    End Function

    'Continue with other order entry functions and procedures
    '. . .

End Class
```

FIGURE *10-23*

A Visual Basic code template
to search for an existing Customer.

When layers are distributed across multiple computer systems, programs are more specialized and numerous than in centralized architectures. Rather than combining user interface, business logic, and data access modules within a single program and structure chart, distributed three-layer architecture employs multiple programs. Some layers, such as the view layer, may not even be programs in the traditional sense, such as when a Web-based HTTP user interface is used. The layers that are written in traditional or OO programming languages are usually separate programs. For example, in a Web-based system, the top two layers of the structure chart in Figure 10-20—the yellow modules—would be implemented as a set of Web pages. The third-level modules *Retrieve existing customer*, *Create new customer*, and *Update existing customer*—the red modules—would probably be separate programs, stored on an application or Web server and executed via calls from a Web page. Thus, the single structure chart in Figure 10-20 would be decomposed into several smaller structure charts.

Note that independent programs executing on different computer systems can't communicate using function or procedure parameters represented as data couples on a system flowchart. Instead, modules in different layers communicate over real-time links using well-defined protocols. That form of communication is represented on the system flowchart, as shown in Figure 10-19. The exact format and content of messages passed among layers must be specified within module pseudocode or elsewhere.

We've now covered architectural and detailed software design using traditional models and methods. In the next two chapters, we'll cover those same design tasks using object-oriented tools and techniques. After that, we'll turn our attention to the remaining design activities—database, user interface, and system interface design.

SUMMARY

For the traditional structured approach to systems design, the primary input is the data flow diagram. The data flow diagram is first enhanced by the addition of a system boundary. The designer sketches the system boundary to show the overall system. He or she also sketches the boundary on the DFD fragments to show program boundaries at a lower level.

The designer describes processes within each DFD boundary using one or more structure charts. The designer develops structure charts using transaction analysis, transform analysis, or both. Transaction analysis is appropriate for the upper structure chart levels of a system that processes multiple input or transaction types. Transform analysis is appropriate for designing programs that transform a single transaction from its input form to an output. Structure charts may also be based on three-layer architecture, in which case modules will be clearly identified by layer and the structure chart may be decomposed into smaller structure charts if layers will execute on multiple computer systems.

A structured design may also include system flowcharts and module pseudocode. System flowcharts show the movement of data among programs, files, and manual processing steps, thus providing an overall view of an entire system. System flowcharts can also describe the interaction between layers of a multilayered system. Module pseudocode describes the internal logic of a structure chart module.

KEY TERMS

afferent data flow, p. 361

central transform, p. 361

computer program, p. 348

data couples, p. 356

efferent data flow, p. 361

module, p. 348

module cohesion, p. 364

module coupling, p. 364

program call, p. 355

pseudocode, p. 348

structure chart, p. 354

system flowchart, p. 348

transaction analysis, p. 358

transform analysis, p. 358

REVIEW QUESTIONS

1. Explain the relationship and differences between a module and a program.
2. What is the purpose of the automation system boundary? How do you develop one?
3. What is a system flowchart used for?
4. What symbols are used on a system flowchart?
5. What is the purpose of a structure chart?
6. What are the symbols used on a structure chart?
7. Explain *transaction analysis.*

8. Explain *transform analysis.* What is meant by the term *central transform*?
9. What is the difference between afferent and efferent data flow?
10. Explain module coupling and module cohesion. Why are these concepts important?
11. Describe how structure charts for three-layer architecture are different from those for all-encompassing programs that execute on a single computer system.

THINKING CRITICALLY

1. Given the data flow diagram shown in Figure 10-24, do the following: (a) draw a system boundary; (b) divide the DFD into program components such as real-time, monthly, daily, periodic, and so forth; and (c) draw a system flowchart based on the division into program components.

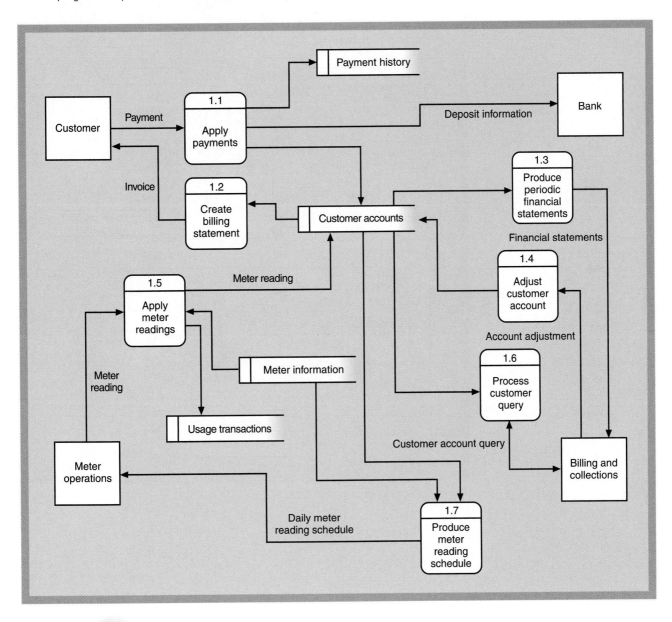

FIGURE *10-24*

Electric company customer billing.

2. Given the data flow diagram shown in Figure 10-25, and using transaction analysis, develop a structure chart.

FIGURE *10-25*

Student registration program.

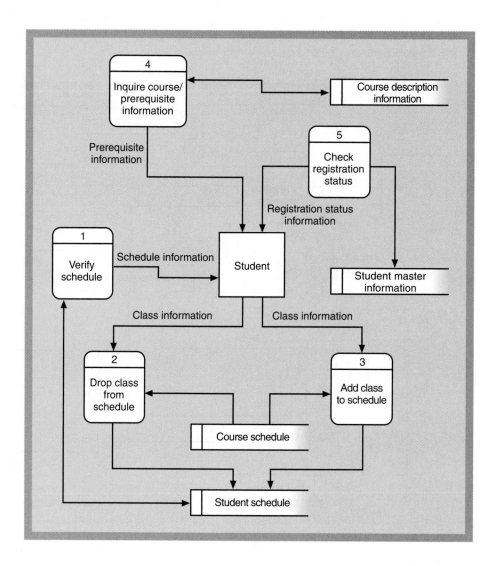

3. Given the data flow diagram shown in Figure 10-26, and using transform analysis, develop a structure chart.

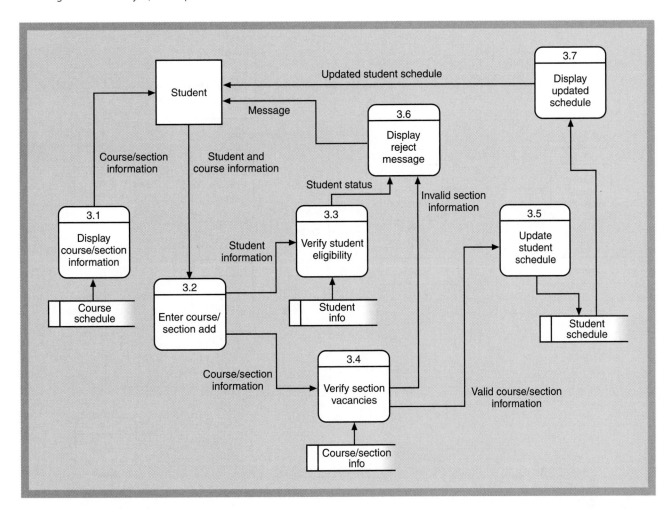

FIGURE *10-26*

Explosion of *Add class to schedule.*

4. Integrate the structure charts from problems 2 and 3 into a single structure chart.

5. Given the data flow diagram shown in Figure 10-27, and using transform analysis, develop a structure chart.

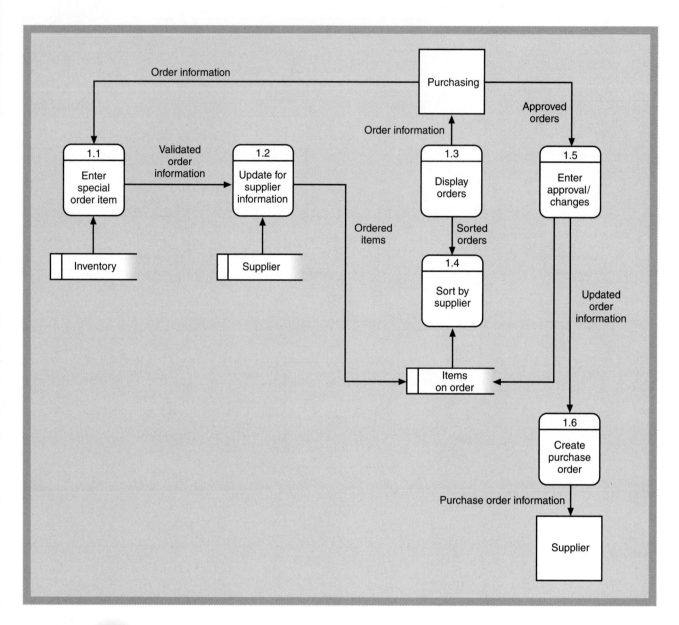

FIGURE *10-27*

Special-order purchasing.

6. Finish developing business logic layer modules for the view layer modules in Figure 10-20.

EXPERIENTIAL EXERCISES

1. Discuss the hierarchical nature of traditional structured design. What kinds of systems and architectures are naturally more inclined to a hierarchy?
2. Find an example of a business system written in COBOL or BASIC that has a hierarchical structure and was probably developed using traditional methods. Search the Internet for COBOL, Microfocus COBOL, or Visual Basic.
3. Find a local company that is doing development using traditional structured techniques. Set up an interview with an IS staff member. Gather as much information as you can about the company's systems. Review the company's techniques and SDLC methodology.

CASE STUDIES

The Real Estate Multiple Listing Service System

Refer to the description of the Real Estate Multiple Listing Service system in the case studies of Chapter 5 and the DFDs you developed in the case studies for Chapter 6. Develop a structure chart for the system. Follow the steps indicated in this chapter, including any additional modules required for accessing data.

Rethinking Rocky Mountain Outfitters

Review the decisions about the deployment environment and design for the Rocky Mountain Outfitters customer support system as described in Chapters 8 and 9 (see pages 311 and 343) and the related traditional design models in this chapter. Specifically for this system, what are the comparative advantages and disadvantages of software design with traditional methods and models compared with object-oriented methods and models?

Focusing on Reliable Pharmaceutical Service

Based on the description of the Reliable Pharmaceutical Service system in Chapters 5 and 6 and the DFDs you developed for Chapter 6, develop a system flowchart and structure charts for the system. Assume that the system will be designed and deployed according to three-layer architecture.

FURTHER RESOURCES

Tom DeMarco, *Structured Analysis and System Specification.* Yourdon Press, 1979.

Meilir Page-Jones, *The Practical Guide to Structured Systems Design*, second edition, Yourdon Press, 1988.

Edward Yourdon, *Modern Structured Analysis.* Yourdon Press, 1989.

Edward Yourdon and Larry L. Constantine, *Structured Design: Fundamentals of a Discipline of Computer Program and Systems Design*, Prentice Hall, 1979.

CHAPTER 11

The Object-Oriented Approach to Design: Use Case Realization

LEARNING OBJECTIVES

After reading this chapter, you should be able to:

- Explain the purpose and objectives of object-oriented design

- Develop design class diagrams

- Develop interaction diagrams based on the principles of object responsibility and use case controllers

- Develop detailed sequence diagrams as the core process in systems design

- Develop collaboration diagrams as part of systems design

- Document the architectural design using package diagrams

CHAPTER OUTLINE

Object-Oriented Design—The Bridge between Analysis and Programming

Design Classes and Design Class Diagrams

Interaction Diagrams—Realizing Use Cases and Defining Methods

Designing with Sequence Diagrams

Designing with Collaboration Diagrams

Updating the Design Class Diagram

Package Diagrams—Structuring the Major Components

Implementation Issues for Three-Layer Design

Even though there had been some hiccups at the beginning of the project, things seemed to be under control now. Bill Santora, who was the project leader responsible for the development of an integrated customer account system at New Capital Bank, had just finished a technical review of the new system's first-cut design with the review committee. This first-cut design focused on six core use cases, which had been chosen as the most fundamental to the business and were going to be implemented in the first iteration.

New Capital Bank had been using object-oriented languages for quite a while, but it had been slower to adopt object-oriented analysis and design techniques. Bill had been involved in some early pilot projects that had used the Unified Process (UP) and the Unified Modeling Language (UML) to develop systems using object-oriented techniques. However, this development project was his first large-scale project that would be entirely object oriented.

As Bill collected his presentation materials, his supervisor, Mary Garcia, spoke. "Your technical review went very well Bill. The committee found only a few minor items that need to be fixed. Even though I am not completely current on the new object-oriented techniques, it was easy for me to understand what you presented and how these core functions will work. I still find it hard to believe that you will have these six pieces implemented in the next couple of weeks."

"Wait a minute," Bill laughed. "It won't be ready for the users then. Getting these six core functions coded and running doesn't mean that we are almost done. This project is still going to take a year to complete."

"Yes, I know. But it is nice that we will have something to show after only two months. Not only do I feel more confident in this project, but the users love to see things developing."

"I know. Remember how much grief I got when I originally laid out this plan based on the UP? Since the UP is an iterative approach, it was more difficult to detail the project schedule for the later iterations. So I had a hard time convincing everybody that the project schedule was not too risky. The upside is that since each iteration is only six weeks long, we have something to show right at the beginning. You don't know how relieved I am that the design passed the review! The team has done a lot of work to make sure the design was solid, and we all felt confident. It is good to get confirmation, though. And we really will have some basic pieces of the new system working in two more weeks."

"Well, building it incrementally makes a lot of sense and certainly seems to be working. I especially liked the detailed sequence diagrams you showed for each use case. It was terrific how you showed that the three-layer design supported each use case. Even though I do not consider myself an advanced object-oriented technician, I could understand how each use case was implemented. I think you wowed everybody when you demonstrated how you could use the same basic design to support both our internal bank tellers and a Web portal for our customers. Congratulations."

Bill's response reflected Mary's enthusiasm. "How about the design class diagrams? Don't they give a nice overview of the classes and the methods? We use them extensively as a focus for discussion on the team. They really help the programmers write good, solid code."

"By the way, have you scheduled a review with the users?" Mary asked.

"No. We worked intimately with the users on developing the use cases and creating the fully developed use case descriptions. We also worked with the users to develop prototype screens for all of these use cases. So, rather than try to explain the details of the design models to the users, we are just going to code these use cases next. After all, we will have something to show them in a couple of weeks. Then we will have another round of meetings with users to let them verify our work and to begin work on the next iteration. We need their involvement as we develop the use case descriptions for the next set of use cases."

"I am anxious to see the first pieces run. It just makes so much sense to be able to test these core functions during the rest of the project. Let me congratulate you again," Mary said as she and Bill headed off to lunch together.

OVERVIEW

Recall from Part 2 that object-oriented analysis consists of addressing two objectives—discovery and understanding. You learned that discovery is made up of fact-finding activities such as interviewing system users, but that understanding develops from taking the information gleaned from user interviews and constructing a set of interrelated and comprehensive models. In Chapter 5, you learned how to identify problem domain classes and use cases, which provide the foundation for understanding system requirements. In Chapter 7, you learned how to complete the object-oriented analysis requirements models by extending each use case with a detailed description, an activity diagram, and a system sequence diagram (SSD) and by continuing to refine the domain model class diagram. Model building is an essential part of understanding user needs and the way they influence the proposed system. However, remember that the objective of the analysis models is, not to describe the new system, but to understand, in precise terms, the system requirements.

Chapter 9 then introduced you to the concepts of architectural and detailed design. You learned that systems design focuses on many system components—network design, application design, database design, user-interface design, and system interface design. The primary focus of this chapter and the next is on how to develop detailed object-oriented design models, which the programmers then use to code the system. The two most important models that must be developed are design class diagrams and interaction diagrams (sequence diagrams and collaboration diagrams). You will learn how to develop design class diagrams for each layer of three-layer design: domain layer, view layer, and data access layer. Design class diagrams extend the domain model that was developed during analysis. Interaction diagrams extend the system sequence diagrams, also developed during analysis. We also discuss how to associate classes into package diagrams to show relationships and dependencies. Finally, we discuss some principles of good design and ways to apply them.

OBJECT-ORIENTED DESIGN—THE BRIDGE BETWEEN ANALYSIS AND PROGRAMMING

If we compare the development of a new software system with building a house, we might say that analysis is like the architect's initial sketches and conceptual drawings. Those drawings depict what the homeowner wants for rooms, floors, layout, location on the lot, and so forth. However, those drawings and sketches are inadequate for a contractor to build the house. The contractor needs much more detail. So, the architect takes the initial drawings and builds a set of detailed plans, called *blueprints*, which show exact measurements for walls, floors, and ceilings; specifications for plumbing and electrical wiring; and even the specific materials to be used. These detailed blueprints are similar to the design models that software developers build.

The analogy can be extended even further. While the sketches are being drawn, the homeowner is very involved in dictating his or her desires, but during development of the detailed blueprints, the homeowner is less involved. He or she does review and verify the specifications in the blueprints, but development of the wiring diagrams and plumbing layout in the walls is done by an expert, the architect, to meet the buyer's desires and to direct the contractor later. Similarly, in software development the detailed design specifications are primarily done by the software design experts, with only occasional input by the users—primarily for verification of the design.

So what is object-oriented design? It is the process by which a set of detailed object-oriented design models are built, which the programmers will later use to write code and test the new system. Systems design is essentially a bridge between a user's requirements and programming for the new system. Just as a builder would never try to build anything larger than a doghouse or a shed without a set of blueprints, a system developer would never try to develop a large system without a set of design models. Sometimes students who are building personal Web pages or small systems for course assignments think that design models are unnecessary. Just remember that blueprints may not be necessary for a doghouse, but they are critical for something more complex, such as a home. Designing the application software, which is covered in this chapter, is just one part of OO design. As you learned in Chapter 9, user-interface design, network design, controls and security design, and database design also require design tasks and design models.

Overview of Object-Oriented Programs

Before going further, let's quickly review how an object-oriented program works. An object-oriented program consists of a set of objects that cooperate to accomplish a result. Each object contains program logic and any necessary attributes in a single unit. These objects work by sending each other messages and collaborating to support the functions of the main program.

Figure 11-1 depicts how an object-oriented program works. The program includes an Input window object that displays a form in which to enter student identification and other information. After the student ID is entered, the Input window object sends a message (message number 2) to the Student class to tell it to create a new Student object in the program and also to go to the database and get the student information and put it in the Student object (message 3). Once that is done, the new Student object also sends that information back to the Input window object to display it on the screen. The registration clerk then enters the personal information updates (message 4), and another sequence of messages is sent to update the Student object in the program and the student information in the database.

FIGURE *11-1*

Object-oriented event-driven program flow.

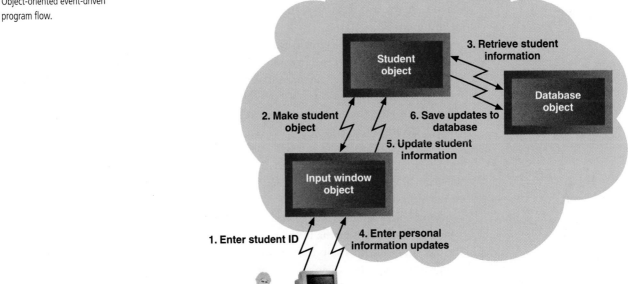

One common question about object-oriented programs is, who is in charge? In a structured program, it is obvious who the boss module is and who controls the computing. In an object-oriented program, it is not so obvious. In fact, no one may be in charge. Yes, there is one program that gets things started, but once the program is executing, no particular module or object has to be in charge.

Let's compare traditional structured programs with object-oriented programs through an analogy—a computer analogy, but one with which you may be familiar. A mainframe computer is a large computer with a massive amount of computing capability. A mainframe computer may be connected to thousands of individual work terminals, and it controls them all. No individual terminal does work unless the main computer directly instructs it to do so. The mainframe also does all the database access and execution. This system is much like a traditional structured program.

In contrast, a network of personal computers consists of many individual computers connected by network cables. Each computer has its own computing capability, but if you are working at one computer, you can communicate with another personal computer through a message and ask for assistance. For example, some of the individual computers on the network have large disk drives and special databases that you can access. They are called *file servers*. Other resources, such as printers, are also on the network. Typically, in most PC networks you have encountered, there is no single, coherent purpose for all of the individual computers. But sometimes it is desirable to have many computers work together for a specific purpose. The way they can work together is again by sending messages to each other to fulfill the overall computing objective. This system works in much the same way that an object-oriented system or program is designed to.

An object-oriented system consists of sets of computing objects. Each object has encapsulated within itself data and program logic. Analysts define the structure of the program logic and data fields by defining a class. The class definition describes the structure, or a template of what an executing object looks like. The object itself does not come into existence until the program begins to execute. We call that *instantiation* of the class, or making an instance (an object), based on the template provided by the class definition.

The objective of object-oriented design is to identify and specify all of the objects that must work together to carry out each use case, such as user-interface objects, problem domain objects, and database access objects. As was discussed in Chapter 9, this is called a *multilayer design*, since we can logically divide the objects into three layers. Besides identifying the classes, another design objective is to specify the classes with enough detail so that a programmer can understand how the objects work together to carry out the use cases and can write object-oriented code for each class. For a student registration system, the partial class definitions in a programming language might look like those in Figures 11-2 a and b. First, we show a class definition in Java and then some in Visual Basic .NET.

As an object-oriented system designer, you will need to provide enough detail so that a programmer can write the initial class definitions and then add considerable detail to the code. As you will see in the following sections, the primary components of the OO design are design class diagrams, interaction diagrams, and, for some classes, statechart diagrams (see Chapter 12 for a detailed discussion of statecharts). For example, a design class specification helps define the attributes and the methods. You will learn the detailed notation for design classes later in the chapter, but as a preliminary example, observe Figure 11-3. You should be able to see where the attributes and method signatures come from for the sample code shown in Figure 11-2. Notice that the class name, the attributes, and the method names are derived from the design class notation. Of course, in the design class we took some liberties by abbreviating first name and last name to simply *name* and by combining all the components of an address into one field called *address*. If there is any question whether a programmer will know that he or she should break these shortened names out into the detailed fields, then the designer should not take these shortcuts. Other code that needs to be added to the class definition can be derived from the other design models, including the interaction diagrams and the statechart diagrams. As we discuss design classes, be sure to refer back on the code snippets shown in Figure 11-2 to help you see the connection between design and programming.

We will also describe other models and diagrams as part of OO design. These other models, such as package diagrams and deployment diagrams, are helpful as supporting

instantiation

creation of an object based on the template provided by the class definition

```
public class Student
{
      //attributes
      private int studentID;
      private String firstName;
      private String lastName;
      private String street;
      private String city;
      private String state;
      private String zipcode;
      private Date dateAdmitted;
      private float numberCredits;
      private String lastActiveSemester;
      private float lastActiveSemesterGPA;
      private float gradePointAverage;
      private String major;

      //constructors
      public Student (String inFirstName, String inLastName, String inStreet,
           String inCity, String inState, String inZip, date inDate)
      {
           firstName = inFirstName;
           lastName = inLastName;
           . . .
      }
      public Student (int inStudentID)
      {
           //read database to get values
      }

      //get and set methods
      public String getFullName ( )
      {
           return firstName + " " + lastName;
      }
      public void setFirstName (String inFirstName)
      {
           firstName = inFirstName
      }
      public float getGPA ( )
      {
           return gradePointAverage;
      }
      //and so on

      //processing methods
      public void updateGPA ( )
      {
           //access course records and update lastActiveSemester and
           //to-date credits and GPA
      }
{
```

FIGURE *11-2*

Example of class definitions for a
student registration system.
(a) Student class definition in Java.

```
Public Class Student

    'attributes
    Private studentID As Integer
    Private firstName As String
    Private lastName As String
    Private street As String
    Private city As String
    Private state As String
    Private zipcode As String
    Private dateAdmitted As Date
    Private numberCredits As Single
    Private lastActiveSemester As String
    Private lastActiveSemesterGPA As Single
    Private gradePointAverage As Single
    Private major As String

    'constructor methods
    Public Sub New(ByVal inFirstName As String, ByVal inLastName As String,
        ByVal inStreet As String, ByVal inCity As String, ByVal inState As String,
        ByVal inZip As String, ByVal inDate As Date)
        firstName = inFirstName
        lastName = inLastName
        . . .
    End Sub

    Public Sub New(ByVal inStudentID)
        'read database to get values
    End Sub

    'get and set accessor methods
    Public Function GetFullName() As String
        Dim info As String
        info = firstName & " " & lastName
        Return info
    End Function

    Public Function SetFirstName (ByVal inName As String)
        firstName = inFirstName
    End Function

    Public Function Get GPA() As Single
        Return gradePointAverage
    End Function

    'Processing Methods
    Public Function UpdateGPA()
        'read the database and update last semester
        'and to date credits and GPA
    End Function
End Class
```

FIGURE *11-2*

Example of class definitions for a
student registration system.
(b) Student class definition in Visual Basic.

FIGURE *11-3*

Simplified design class for Student class.

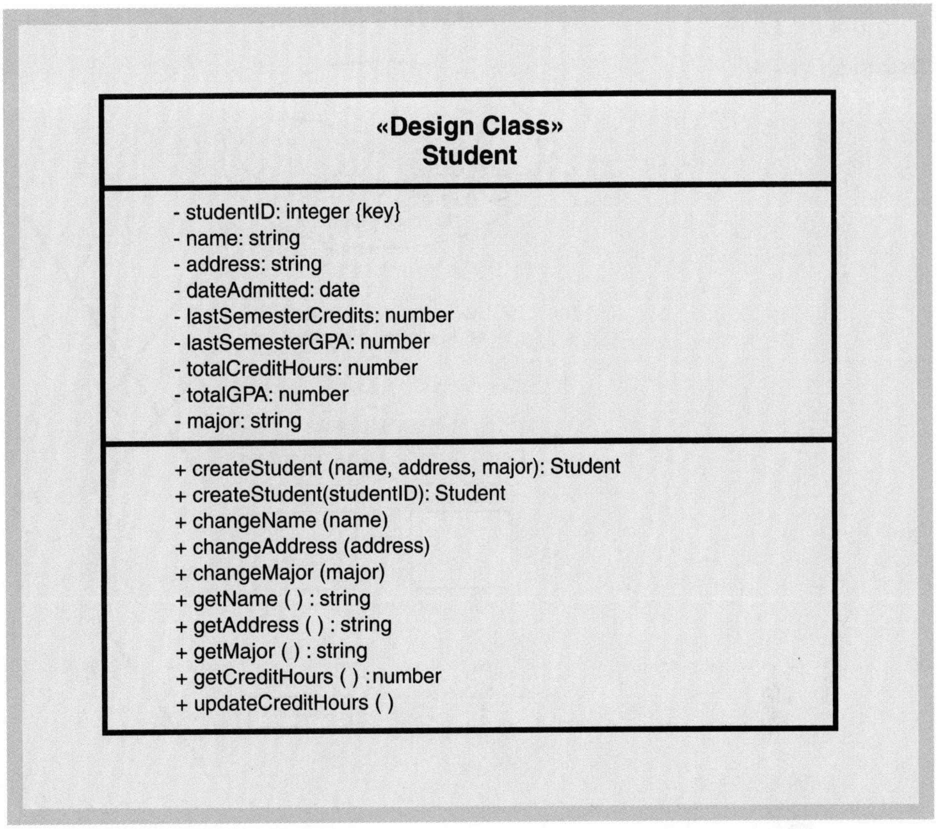

```
                    «Design Class»
                       Student

      - studentID: integer {key}
      - name: string
      - address: string
      - dateAdmitted: date
      - lastSemesterCredits: number
      - lastSemesterGPA: number
      - totalCreditHours: number
      - totalGPA: number
      - major: string

      + createStudent (name, address, major): Student
      + createStudent(studentID): Student
      + changeName (name)
      + changeAddress (address)
      + changeMajor (major)
      + getName ( ) : string
      + getAddress ( ) : string
      + getMajor ( ) : string
      + getCreditHours ( ) :number
      + updateCreditHours ( )
```

documentation. Remember, however, that the primary OO design models are design class diagrams and interaction diagrams, with an occasional statechart diagram where necessary.

Object-Oriented Design Processes and Models

We focus on three object-oriented design models in this chapter—design class diagrams, interaction diagrams, and package diagrams. Figure 11-4 illustrates which analysis models are directly used to develop which design models. The models on the left side—use case diagrams, use case descriptions and activity diagrams, domain model class diagrams, and system sequence diagrams—are those that were developed during analysis. Those on the right side—design class diagrams, interaction diagrams, and package diagrams—will be developed during design. As you might infer from the number of arrows pointing to them in the figure, interaction diagrams are the core diagrams in design, and you will learn how to develop interaction diagrams in this chapter.

The other major design model, which you will also learn to develop in this chapter, is the design class diagram. Its main purpose is to document and describe the programming classes that will be built for the new system. Design class diagrams are the end result of the design process. They describe the set of object-oriented classes needed for programming, navigation between the classes, attribute names and properties, and method names and properties. As shown in Figure 11-4, this information is derived from two sources: the domain model class diagram and the interaction diagrams. In fact, notice that the arrow between design class diagrams and interaction diagrams points two ways. The two-headed arrow indicates that some information in the design class diagram is needed to complete interaction diagrams and that information in the interaction diagrams is used to develop design class diagrams.

The process of designing requires several steps, involving iterations. First, a preliminary version, or first-cut model, of the design class diagrams is created. Some basic

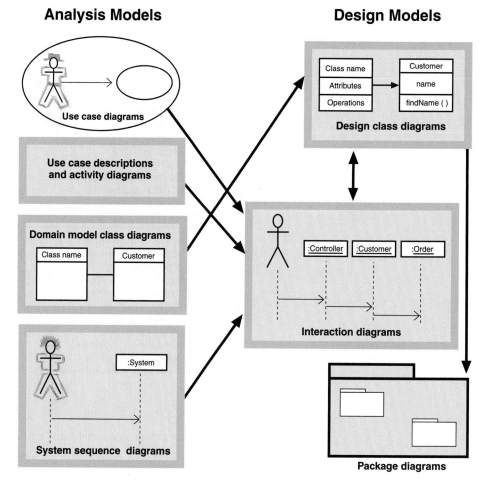

FIGURE *11-4*

Design models with their respective input models.

information, such as attribute names, must be obtained from the first-cut model to develop the interaction diagrams.

The second step in design is to develop interaction diagrams—resulting in one for each use case or scenario. Developing an interaction diagram is a multistep process of determining which objects work together and how they work together. Development of the interaction diagrams is the heart of object-oriented systems design. As shown in Figure 11-4, input models for interactions diagrams are use case diagrams, use case descriptions, system sequence diagrams, and design class diagrams. The end result of the development of these design models is called *realization of use cases*. In this instance, the term *realization* is the specification of the detailed processing that the system must do to carry out the use case—in other words, to make a set of software blueprints. Notice that just as object-oriented analysis was driven by use cases, so also is object-oriented design. That is, design is done on a use case–by–use case basis.

The third step in OO design is to return to the design class diagram and develop method names based on information developed during the design of the interaction diagrams. The navigation visibility and attribute information is also updated in this iteration of the design class diagram.

The final design step is to partition the design class diagram into related functions using package diagrams. There are several ways that a system might be partitioned. One way is by subsystem. Another way is by layers. In Chapter 9, you learned about multilayer architectures and multiple tiers. In this chapter, you will learn how to partition design class diagrams into packages to represent the multiple layers in a multiple-tier system. We focus on a basic multilayer design consisting of the view layer (user-interface classes), the domain layer (problem domain classes from the domain model class dia-

realization of use cases

specification of all detailed system processing for each use case

gram), and the data layer (database access classes). Note that several terms are used to denote the domain layer. Synonymous terms include *domain layer*, *problem domain layer*, and *business logic layer*. Package diagrams provide an architectural, high-level view of the final system.

DESIGN CLASSES AND DESIGN CLASS DIAGRAMS

As shown in Figure 11-4, the design class diagrams and the detailed interaction diagrams use each other as inputs for design and are developed in parallel. A first iteration of the design class diagram is done based on the domain model and on engineering design principles. The preliminary design class diagram is then used to help develop interaction diagrams. As design decisions are made during development of the interaction diagrams, the results are used to refine the design class diagram.

As we mentioned previously, the design class diagram is an extension of the domain model class diagram developed during OO analysis. The domain model class diagram shows a set of problem domain classes and their association relationships. During analysis, since it is a discovery process, analysts generally do not worry too much about the details of the attributes or the methods. However, in object-oriented programming, the attributes of a class must be declared as public or private, and each attribute must also be defined by its type, such as character or numeric. During design, it is important to elaborate these details, as well as to define parameters that are passed to the methods and return values from methods. Sometimes, analysts also define the internal logic of each method at this point. Thus, the design class diagram for design is a more detailed version of the domain model class diagram. We complete it by integrating information from interaction diagrams and other models.

As developers build the design class diagrams, they will add many more classes than were originally defined in the domain model. Since the objective of analysis was to understand the business need, the focus was only on specifying the classes that defined the problem domain. However, to build a complete object-oriented system, many other design classes must be identified and specified. Referring back to Figure 11-1, the Input window objects and Database objects are examples of these additional classes that must be defined. As these classes are defined, we usually document them on various class diagrams. The classes in a system can be partitioned into various distinct class diagrams, such as user-interface classes. At times, we may also develop distinct class diagrams by subsystem. The point is that class diagramming is a tool that can be used in different ways. We now turn to design class diagram notation and the design principles used in the first iteration of developing the design class diagram.

Design Class Symbols

Unified Modeling Language (UML) does not specifically distinguish between design class notation and domain model notation. However, practical differences occur simply because the objective of design modeling is distinct from domain modeling. Domain modeling shows things in the users' work environment and the naturally occurring associations among them. The classes are not specifically software classes. Once we start a design class diagram, though, we are specifically defining software classes. Since many different types of design classes will be identified during the design process, UML has a special notation, called a *stereotype*, that allows designers to designate a special type of class. A *stereotype* is simply a way to categorize a model element as a certain type. A stereotype extends the basic definition of a model element by indicating that it has some special characteristic that we want to highlight. The notation for a stereotype is the name of the type placed within printer's guillemets, like this: «control». You were first

stereotype

a way of categorizing a model element by its characteristics, indicated by guillemets (« »)

exposed to a stereotype when you were developing a use case diagram. You learned that connecting lines between actors and use cases indicated a relationship and that a certain type of relationship existed between use cases, called the «includes» relationship. That was a stereotype because it categorized the relationship of one use case as including another use case.

There are four types of design classes that are considered to be standard: an entity class, a control class, a boundary class, and a data access class. Figure 11-5 shows the notation used to identify these four stereotypes. Two types of notation can be used for design classes. The class rectangles on the left show the full symbols. Notice that the stereotypes are placed above the name in the name compartment. The circular symbols on the right are shorthand notation for these stereotypes, called *icons*. We will use the stereotype icons from time to time, but in most cases we prefer the full notation.

FIGURE *11-5*

Standard stereotypes found in design models.

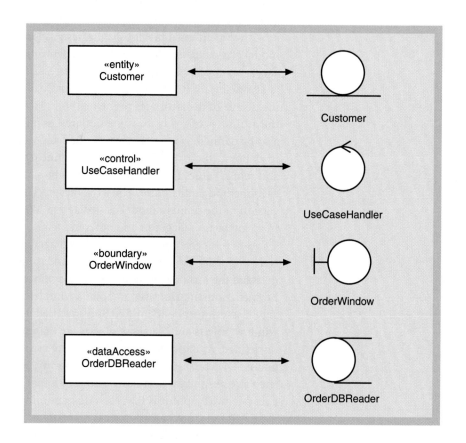

An *entity class* is the design identifier for a problem domain class. In other words, it comes from the domain model. These objects are normally passive, in that they wait for business events to occur before they do anything, and they are also usually persistent classes. A *persistent class* is one that exists after the program quits. In other words, the data must persist after the system is shut down. Obviously, the way to do that is to write it out to a file or database.

A *boundary class* is a class that is specifically designed to live on the system's automation boundary. In a desktop system these classes would be the windows classes and all the other classes associated with the user interface.

A *control class* is a class that mediates between the boundary classes and the entity classes. In other words, its responsibility is to catch the messages from the boundary class objects and send them to the correct entity class objects. It acts as a kind of switchboard, or controller, between the view layer and the domain layer.

entity class

design identifier for a problem domain class

persistent class

an entity class that exists after a system is shut down

boundary class

a class that exists on a system's automation boundary, such as an input window

control class

a class that mediates between boundary classes and entity classes, acting as a switchboard between the view layer and domain layer

data access class

a class that is used to retrieve data from a database

A *data access class* is a class that is used to retrieve data from and send data to a database. Rather than insert database access logic, including SQL statements, into the entity class methods, a separate layer of classes to access the database is often included in the design.

Design Class Notation

Figure 11-6 shows the details within a design class symbol. The name compartment includes the class name and the stereotype information. The lower two compartments contain more details about the attributes and the methods.

FIGURE *11-6*

Internal symbols used to define a design class.

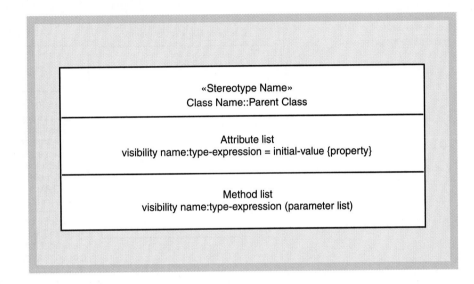

«Stereotype Name»
Class Name::Parent Class

Attribute list
visibility name:type-expression = initial-value {property}

Method list
visibility name:type-expression (parameter list)

The format that analysts use to define each attribute includes the following:

visibility

notation of whether an attribute can be directly accessed by another object; indicated by plus or minus signs

- Attribute visibility. *Visibility* denotes whether other objects can directly access the attribute. (A + sign indicates an attribute is visible, or is public, and a – sign means that it is not visible, or is private.)
- Attribute name
- Type-expression (such as character, string, integer, number, currency, or date)
- Initial-value
- Property (within curly braces), such as {key}

The third compartment contains the method signature information. A *method signature* shows all of the information needed to invoke (or call) the method. It shows the format of the message that must be sent, which consists of the following:

method signature

notation that shows all of the information needed to invoke, or call, the method

- Method visibility
- Method name
- Type-expression (the type of the return parameter from the method)
- Method parameter list (incoming arguments)

In object-oriented programming, analysts use the entire signature to identify a method. Some OO languages allow multiple methods to have the same name as long as they have different parameter lists or return types. In those languages, both the method name and the parameter-list are used to invoke the correct method. For example, suppose that we want to be able to find a customer record either by the customer ID number or by the customer name. We could identify two methods, each with the same name, such as getCustomer (customerID) and getCustomer (customerName). When a method, such as getCustomer, has the same name but different parameter lists, we say that method is an *overloaded method*. To know which method to invoke, the run-time environment must also note what parameters are included—whether a number (customerID) or a text field (customerName) was entered.

Chapter 7 illustrated several classes from the domain model for a student registration system. Figure 11-7 shows an enhanced domain model Student class and the design class diagram Student class for comparison.

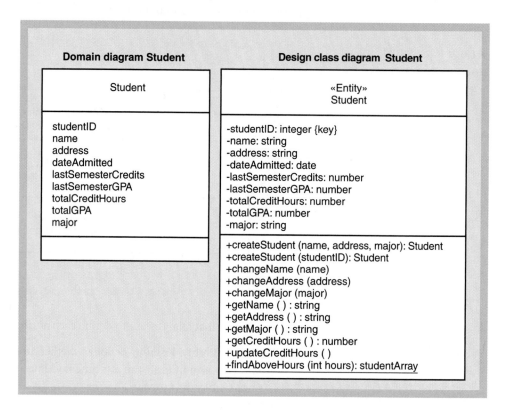

The domain model attribute list contains all attributes discovered during analysis activities. The design class diagram includes more information on attribute types, initial values, and properties. It can also include a stereotype for clarification, and the Student class is shown as an «entity» stereotype. In the design class diagram Student class, the third compartment contains the method signatures for the class. Most of the method signatures will be developed during the design of the interaction diagrams. Remember that UML is meant to be a general object-oriented notation technique and not specific to any one language. So, the notation will not be exactly the same as programming method notation.

For example, for those of you who have programming experience, the constructor notation we use is createStudent: Student (name, address, major). Remember that the constructor is the method that makes a new object for that class. In many programming languages, the constructor is given the same name as the class. However, in this situation we use a create statement to follow more closely the message names used in interaction diagrams. The figure also illustrates another constructor. In the second method line, only the student ID is passed in. This implies that a student with an ID exists, and the constructor itself must fill in the information about the student. This usually requires access to a database to get values for the fields.

The method called findAboveHours (int hours): studentArray, as denoted with an underlined name, is a special kind of method. Remember in the object-oriented ap-

class-level method

a method that is associated with a class instead of with objects of the class

proach that a class is a template to create individual objects or instances. Most of the methods apply to each instance of the class. However, frequently analysts need to look through all of the instances at once. That type of method is called a *class-level method*, which is denoted by an underline. In VB .NET it is a *shared* method and in Java it is a *static* method. This kind of method is one that is executed by the class instead of a specific object of the class. Because these methods are used at the class level, they do not depend on the existence of a particular object and, if necessary, can access data across all objects. In this example, the findAboveHours method looks through all the instances of this class and returns those having total hours greater than the input parameter.

Some Fundamental Design Principles

Now that you understand how an object-oriented program works and you know the notation for a design class, let's review several basic principles that will guide design decisions. We will mention these principles of good object-oriented design throughout the chapter as we discuss the steps of object-oriented design. The following basic principles are important to all parts of object-oriented design.

Encapsulation and Information Hiding

encapsulation

a design principle of objects in which both data and program logic are included within a single self-contained unit

Encapsulation is the design concept that each object is a self-contained unit that includes both data and program logic. Each object internally carries its own data and provides a set of methods that access the data. Each object also provides a set of services that are invoked by calling the object's methods. One of the benefits of this approach to design is that a software developer can design the system in a building-block fashion. Nearly all engineering disciplines have standard units that serve as building blocks and that can be combined into a final design. Encapsulated objects are the software equivalent of building blocks.

object reuse

a design principle in which a set of standard objects can be used over and over again within a system

Programmers also depend heavily on the benefits of encapsulation to support the idea of *object reuse*. Every object-oriented language comes with a set of standard objects that are used over and over again throughout a system. These standard sets of objects provide basic services that are used many times in the same system—and sometimes even in multiple systems. One frequent application of reuse is in the design of the user interface either for desktop or Web applications. Designers often reuse the same classes for developing windows and window components such as buttons, menus, icons, and so forth. Problem domain classes can also be reused.

information hiding

a design principle in which data associated with an object is not visible to the outside world but methods are provided to access or change the data

Related to encapsulation is the concept of *information hiding*. Information hiding dictates that the data associated with an object is not visible to the outside world. In other words, the object's attributes are private. A set of methods is provided to access the data and to modify or change it. Although this principle is primarily a programming concept and is most beneficial for programming and testing, several important design principles are based on it. The linkage or coupling between objects in a system is better if access to data attributes is done via a standard interface of method names, as explained in the next section.

Navigation Visibility

navigation visibility

a design principle in which one object is able to view and interact with another object

As stated earlier, an object-oriented system is a set of interacting objects. The interaction diagrams developed during design document what interactions occur between which objects. However, for one object to interact with another object by sending a message, the first object must be visible to the second object. *Navigation visibility*, in this context, refers to the ability of one object to be able to view and interact with another object. In programming jargon, invoking a method on an object frequently requires dot notation to invoke the correct method on the correct object. Sometimes developers refer to navigation visibility as just *navigation* or just *visibility*.

As a designer, you must always be aware of navigation visibility. Interactions between objects can only be accomplished with navigation visibility. One of the responsibilities of a design is to specify which classes have navigation visibility to other classes. Navigation visibility can be either one way or two way. For example, a Customer object may be able to view an Order object. That means that the Customer object knows which orders a customer has placed. In programming terms, the Customer class has a variable, or array of variables, that point to the Order object or objects for that customer. If navigation is two way, then each Order object will also have a variable that refers to the Customer object. If the navigation is not two way, then Order objects will not have a variable to point to the Customer object. In a design class diagram, navigation visibility is identified by an arrow between the classes, where the arrow points to the visible class.

Figure 11-8 shows one-way navigation visibility between the Customer class and the Order class. Notice that there is a variable called myOrder in the Customer class. This variable holds a value to refer to an order instance. Frequently, the reference variable myOrder is not explicitly shown in the design class. The navigation arrow indicates that visibility is required. We have included the variable in this example to emphasize the concept.

FIGURE *11-8*

Navigation visibility between Customer and Order.

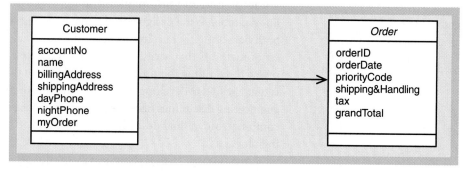

Coupling

Coupling is a general term that is derived from navigation visibility. In the previous example where Customer had navigation visibility to Order, one could also say that Customer and Order are coupled, or linked. Now, extend this same idea of visibility throughout all the classes in the entire system. *Coupling* is a qualitative measure of how closely the classes in a design class diagram are linked. A simple way to think about coupling is the number of navigation arrows on the design class diagram. Low coupling for the system is usually better than high coupling. In other words, fewer navigation visibility arrows indicates that a system is easier to understand and maintain.

We say that coupling is a qualitative measure because no specific number measures coupling in a system. A designer must develop a feel for coupling—to recognize when there is too much coupling or to know what is a reasonable amount of coupling. Coupling is evaluated as a design progresses use case by use case. Generally, if each use case design has a reasonable level of coupling, the entire system will, too.

Refer back to Figure 11-1 and observe the flow of messages between the objects. Obviously, objects that send messages to each other must have navigation visibility and thus are coupled. For the Input window object to send a message to the Student object, it must have navigation visibility to it. So, the Input window object is coupled to the Student object. But notice that the Input window object is not connected to the Database object, so those objects are not coupled. If we designed the system so that the Input window object accessed the Database object, then the overall coupling for this use case would increase—that is, there would be more connections. Is that good or bad? In this simple example, it might not be a problem. But for a system with even 10 or more classes, undisciplined connections with navigation visibility can cause very high levels of coupling, making the system more complex.

coupling

a qualitative measure of how closely the classes in a design class diagram are linked

So why is high coupling bad? It is bad primarily because it adds unnecessary complexity to a system, making it very hard to maintain. A change in one class ripples throughout the entire system. So, experienced analysts make every effort to simplify coupling and reduce ripple effects in the design of a new system.

Cohesion and Separation of Responsibilities

cohesion

a qualitative measure of the consistency of functions within a single class

Cohesion refers to the consistency of the functions within a single class. *Cohesion* is a qualitative measure of the focus or unity of purpose within a single class. For example, in Figure 11-1, you would expect the Student class to have methods, that is, functions, to enter student information such as student identification number or name. That would represent a unity of purpose and a highly cohesive class. But what if that same object also had methods to make classroom assignments or assign professors to courses? The cohesiveness of the class would be reduced.

Classes with low cohesion have several negative effects. First, they are hard to maintain. Since they do many different functions, they tend to be overly sensitive to changes within the system, suffering from ripple effects. Second, it is hard to reuse such classes. Since they have many different—and often unrelated—functions, it usually does not make sense to reuse them in other contexts. For example, a button class that processes button clicks can easily be reused. However, a button class that processes button clicks and user logons has very limited reusability. A final drawback is that classes that are not cohesive are usually difficult to understand. Frequently, their functions are intertwined and their logic is complex.

Although there is no firm metric to measure cohesiveness, we can think about classes as having very low, low, medium, or high cohesion. Remember, high cohesion is the most desirable. An example of very low cohesion would be a class that has responsibility for services in different functional areas, such as a class that accesses both the Internet and a database. These two types of activities are very different and accomplish different purposes. To put them together in one class causes very low cohesion.

An example of low cohesion would be a class that has different responsibilities but in related functional areas, perhaps a class that does all database access for every table in the database. It would be better to have different classes to access customer information, order information, and inventory information. Although the functions are the same— that is, they access the database—the types of data passed and retrieved are very different. So, a class that is connected to the entire database is not as reusable as one that is only connected to the customer table.

An example of medium cohesion would be a class that has closely related responsibilities, such as a single class that maintains customer information and customer account information. Two highly cohesive classes could be defined, one for customer information, such as name and addresses. Another class or set of classes could be defined for customer accounts, such as balances, payments, credit information, and all financial activity. If the customer information and the account information are somewhat limited, then they could all be combined into a single class with medium cohesiveness. Either medium or highly cohesive classes can be acceptable in system design.

separation of responsibility

design principle in which analysts divide a class into several highly cohesive classes

The common solution to classes with low cohesion is to divide a class into several classes, each of which is highly cohesive. This concept, called separation of responsibilities, is another principle of good object-oriented design. *Separation of responsibilities* states that in most cases, it is a better design to separate disparate tasks into distinct classes. Think of separation of responsibilities as another way to develop highly cohesive classes.

Developing the First-Cut Design Class Diagram

To start the design process, we develop a first-cut design class diagram based only on the domain model. Figure 11-9 repeats the domain model class diagram for RMO developed in Chapter 7. As you learned earlier, the focus during analysis was to identify the classes, their attributes, and the relationships between the classes.

FIGURE *11-9*

RMO domain model class diagram.

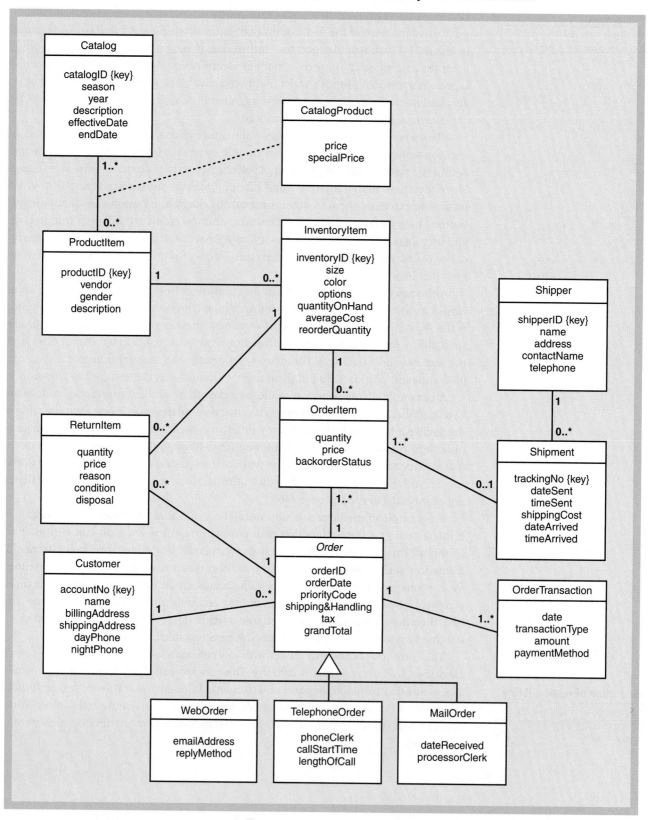

FIGURE *11-10*

First-cut RMO design class diagram.

The first-cut design class diagram is developed by extending the domain model class diagram. It requires two steps, (1) elaborating the attributes with type and initial value information and (2) adding navigation visibility arrows. Figure 11-10 is a design class diagram for RMO showing the results of these two steps.

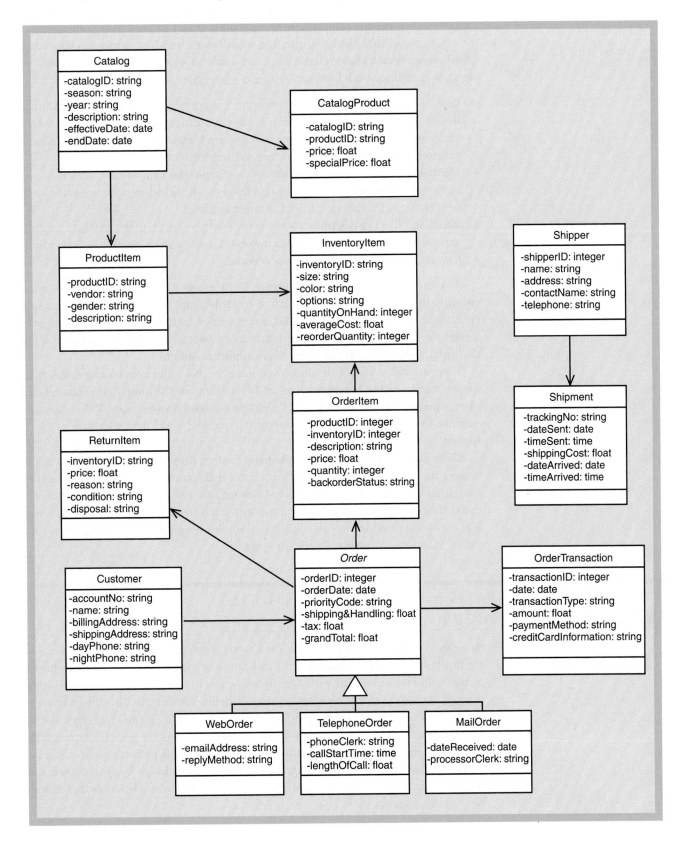

The elaboration of the attributes is fairly straightforward. The type information is determined by the designer based on his or her expertise. Finally, in most instances, all attributes are kept invisible or private, as indicated by the minus signs preceding them in the diagram.

Navigation visibility is a little more difficult to design. Remember that we are designing just the first-cut class diagram, so we may need to add or delete navigation arrows as the design progresses. The basic question we ask when building navigation visibility is, which classes need to have references to, or be able to access, which other classes? Here are a few guidelines—not hard-and-fast rules—but general guidelines.

- One-to-many relationships that indicate a superior/subordinate relationship are usually navigated from the superior to the subordinate, for example, from Order to OrderItem. Sometimes these relationships form hierarchies of navigation chains, for example, from Catalog to ProductItem to InventoryItem.
- Mandatory relationships, in which objects in one class cannot exist without objects of another class, are usually navigated from the more independent class to the dependent class, for example, from Customer to Order.
- When an object needs information from another object, then a navigation arrow may be required, pointing either to the object itself or to its parent in a hierarchy.
- Navigation arrows may also be bidirectional.

As indicated in the guidelines, Figure 11-10 shows that Customer has navigation visibility to Order. Order has navigation to OrderItem, to OrderTransaction, and to ReturnItem. Catalog, ProductItem, and InventoryItem form a hierarchy, with navigation going from top to bottom. CatalogProduct, as an association class, should also be visible from Catalog. Shipper has navigation visibility to Shipment.

There are still a couple of unresolved issues at this point concerning visibility between OrderItem and InventoryItem, and OrderItem and Shipment. It is not yet clear what is the best way to implement navigation between those classes. Those questions may need to wait until additional design is done and other design principles can be applied.

Three points are important to note here. First, as detailed design proceeds use case by use case, we need to ensure that the interaction diagrams support and implement the navigation that has been initially defined. Second, the navigation arrows will need to be updated as design progresses, to be consistent with the design details. Finally, method signatures will be added to each class based on the design decisions made when creating the interaction diagrams for the use cases.

INTERACTION DIAGRAMS—REALIZING USE CASES AND DEFINING METHODS

As we mentioned at the beginning of the chapter, developing interaction diagrams is at the heart of object-oriented design. The realization of a use case—determining what objects collaborate by sending messages to each other to carry out the use case—is done through the development of an interaction diagram. The objective of this section is to explain how to do object-oriented design using interaction diagrams. Two types of interaction diagrams are developed during design: sequence diagrams and collaboration diagrams. Design can be done using either type. Some designers prefer sequence diagrams, and others prefer collaboration diagrams. First we discuss design by utilizing sequence diagrams. Explanations of collaboration diagrams will follow to show how they can also be used to do design.

It is important to understand the difference between designing a system and documenting the results of the design process and decisions. Designers develop diagrams such as design class diagrams and interaction diagrams while doing software design. The diagrams communicate structural and behavioral details to programmers and other developers. But the diagrams are not an end in themselves. Instead, they represent the results of design decisions based on well-established design principles such as coupling, cohesion, and separation of responsibilities. Typically, designers develop rough drafts of diagrams and then evaluate their quality by evaluating how well they reflect principles of good design. The diagrams may be modified many times as designers refine them to improve their quality and correct errors. The diagrams are both a scratchpad for the designers' thinking and a means to communicate the final result of that thinking to developers.

Since you will be learning how to do design as you learn about the detailed interaction diagrams, we will intersperse the discussion with pointers on the principles of good design. We start with two more fundamental principles of good design—object responsibility and use case controllers. Figure 11-11, which is a partial design class diagram that contains all of the classes that will be needed for the use case *Look up item availability*, will be used to illustrate these two design principles.

FIGURE *11-11*

Partial design class diagram for the *Look up item availability* use case.

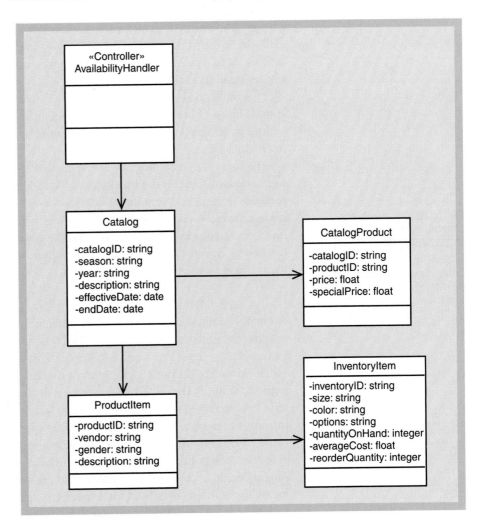

object responsibility

design principle in which objects are responsible for carrying out system processing

Object Responsibility

One of the fundamental principles of object-oriented development is the idea of *object responsibility*—objects are responsible for carrying out the system processing. One of the major activities of design is to define the responsibilities of classes of objects. These responsibilities are categorized in two major areas—knowing and doing—in other words, what is an object expected to know? and what is an object expected to do or to initiate?

Knowing includes such responsibilities as knowing about its own data and knowing about other classes with which it must collaborate to carry out use cases. Obviously, a class should know about its own data, what attributes exist and how to maintain the information in those attributes. It should also know where to go get information when required. For example, during the initiation of an object, data for some attributes may be passed in as part of the constructor being invoked. However, other data that is not passed in may be required. An object should know about, that is, have navigation visibility, to other objects that can provide the required information. For example, in Figure 11-7, the constructor for the Student class does not receive a studentID value as a parameter. The Student class takes responsibility for creating a new studentID value based on some rules it knows.

Doing includes all the activities an object does to assist in the execution of a use case. Some of those activities will be to receive and process messages. Another doing responsibility will be to instantiate, or create, new objects that may be required for completion of a use case. Classes must collaborate to carry out a use case. Some classes are responsible for coordinating the collaboration. For example, for the use case *Look up item availability* in Figure 11-11, the Catalog class has primary responsibility to see that the use case is carried out. Another class, such as InventoryItem, is only responsible for providing information about itself.

One of the most important activities of a systems designer is to determine object responsibility and build the system based on those decisions. In the following sections, which discuss design, we provide more examples of object responsibility and of assigning responsibility.

The concept of assigning object responsibilities has been part of object-oriented design for a long time. In the history of object-oriented design, one of the most popular methods of assigning responsibilities was developed using $3'' \times 5''$ index cards. These cards, called *CRC cards*—for class-responsibility-collaboration cards—were used to document the classes in the system, the ways the classes collaborated, and the responsibilities of each class for each use case collaboration. The CRC card technique is still used in many places to assist in the design process.

Use Case Controller

Normally, each use case can have many different input messages coming from the external actors. A system sequence diagram (SSD) developed as part of object-oriented analysis shows these input messages, but it only indicates that all of these messages go to the system. However, during design, we must decide which objects receive all of these messages. To simplify the collection and processing of all the messages for a use case, systems designers frequently make up a new class that can serve as a collection point for incoming messages. We call these classes *use case controllers*. For example, for the use case *Look up item availability*, there might be a controller class named AvailabilityHandler. A use case controller is a completely artificial class, created by the person doing the system design. Sometimes these classes that are just made up are called *artifacts*, which means something just created for a specific purpose because it is needed. We add this term because you will find it used frequently in industry. Essentially, an artifact is a person-made article. In one sense of the word, the design models we are creating in this chapter can be considered artifacts.

The use case controller acts as an intermediary between the outside world and the internal system. In Figure 11-1, we saw an Input window object, which sits on the system boundary, and a problem domain object called the Student object. What if that

---

CRC (class-responsibility-collaboration) cards

index cards that are used to document the classes in a system, the ways the classes collaborate, and the responsibilities of each class for each use case collaboration

use case controllers

a class systems designers create to serve as a collection point for incoming messages

artifact

a class invented by a system designer to handle a needed system function

Input window object needed to send messages to several objects? It would need references to all of the problem domain objects. The coupling between the Input window object and the system would be very high—there would be many connections. So, coupling between the user-interface objects and the problem domain objects could be reduced by making a single use case controller object to handle all of the input messages.

There are several ways to create use case controllers. A single use case controller could be defined for all use cases. The set of responsibilities assigned to such a use case would be very broad, and the cohesiveness of such a class would most likely be very low. So, a better design might result from defining several use case controllers, each with a specific set of responsibilities. It would also be possible to create a single use case controller for all the use cases in a single subsystem, such as the order-entry subsystem. Several use case controllers would raise the coupling between the user-interface classes and the internal classes, but it would result in highly cohesive classes with a defined set of responsibilities for each. A solution with one use case controller per use case is a viable option as well. Figure 11-11 illustrates a single use case controller for a single use case. The use case controller and the problem domain classes for the use case are included in one design class diagram. Note that in this design class diagram only the navigation visibility arrows are shown.

The thought process illustrated in the preceding paragraphs is precisely the process of systems design—to balance the design principles of coupling, cohesion, and class responsibility. We will elaborate on this process in the next sections.

DESIGNING WITH SEQUENCE DIAGRAMS

Interaction diagrams form the heart of the OO design process, and sequence diagrams are used to explain object interactions and document design decisions. First let's review the mechanics and syntax of sequence diagrams and then follow with some principles and procedures to design the system. You first learned about sequence diagrams in Chapter 7 when you learned how to develop a system sequence diagram (SSD). By now you should feel comfortable not only reading and interpreting a system sequence diagram but also developing one. Remember that an SSD is used to document the inputs to and outputs from the system for a single use case or scenario. An SSD captures the interactions between the system and the external world as represented by the actors. The system itself is treated as a single object named :System. The inputs to the system are messages from the actor to the system, and the outputs are usually return messages showing the data being returned. Figure 11-12 is an elaborated version of Figure 7-11, which shows the basic components of a sequence diagram and gives explanations. Let us start with a simple RMO sample use case, *Look up item availability*. As seen in the figure, each message has a source and a destination. In an SSD, since there are only two objects with lifelines, the source and destination are constrained. When we get to detailed sequence diagrams, some of the most critical decisions to be made are the source and destination objects for the messages.

Remember that the syntax of an input message discussed in Chapter 7 is

    * [true/false condition] return-value := message-name (parameter-list)

For the output message, we normally only provide the parameter list without parentheses.

A detailed sequence diagram uses all the same elements as an SSD. The difference is that the :System object is replaced by all of the internal objects and messages within the system. In other words, for an SSD, the system was treated as a black box, and we could not see the internal processing. The objective of design is to open up the black box and

FIGURE *11-12*

SSD for the *Look up item availability* use case.

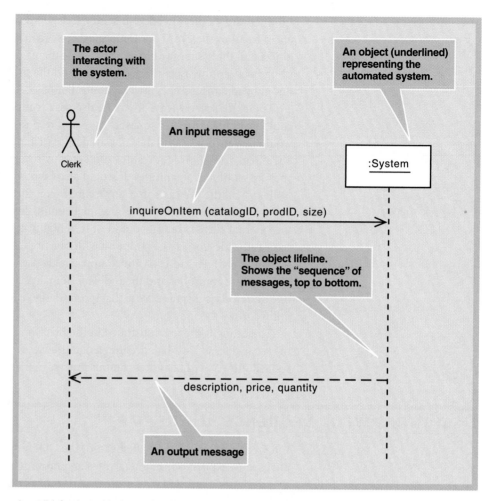

determine the internal processing that must occur within the automated system. As explained earlier and shown in Figure 11-1, we need to identify the internal objects and messages that are involved in the realization of the use case or use case scenario. Notice that just as earlier analysis activities were centered around use cases, so is detailed design.

First-Cut Sequence Diagram

Let's proceed with the detailed design for the *Look up item availability* use case based on the SSD in Figure 11-12. The first step in expanding an SSD is to determine which other objects may need to be involved to carry out the use case. The SSD indicates that information that is to be returned about an item includes the description, the price, and the quantity. Looking at the first-cut design class diagram, we see that description comes from the ProductItem, price from the CatalogProduct, and quantity from the InventoryItem. So, these three objects will be included in the first-cut sequence diagram. As explained earlier, we will also add a use case controller object. For this use case, the controller object will be called AvailabilityHandler. Remember that a controller object serves as a central collection point for all input messages for the use case, acting as a central switchboard for the messages for a use case. It accepts the input messages and distributes them to the correct internal objects. Part of the design is determining how these messages are distributed based on the navigation visibility you have defined in the design class diagram.

We start constructing the first-cut diagram with the elements from the SSD in Figure 11-12. Replace the :System object with the use case controller, :AvailabilityHandler. Then, add the other objects that need to be included in the use case. From the design

FIGURE *11-13*

Objects included in *Look up item availability*.

class diagram, we see that there is a hierarchy from Catalog to ProductItem and on to InventoryItem. Catalog also has navigation visibility to CatalogProduct. The first step, then, is to select an input message from the use case—in this example there is only one message. Add to the sequence diagram all of those objects that must collaborate. Figure 11-13 illustrates this first step.

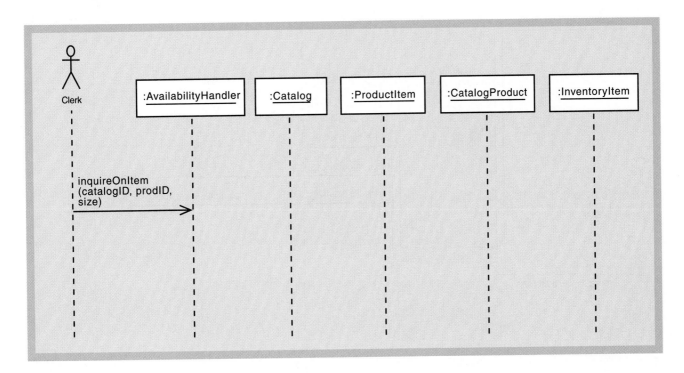

The next step is to determine which other messages must be sent, including which object should be the source and destination of each message to collect all the necessary information. Decisions about what messages are required and which objects are involved is based on the design principles described earlier—coupling, cohesion, responsibility, and controller.

As noted earlier, the Catalog object is the top of a navigation hierarchy to other objects that have the required information. So, the :AvailabilityHandler forwards the input message to the :Catalog object. The :Catalog object sends messages to :ProductItem and :CatalogProduct to get the description and price. However, :Catalog does not have direct navigation visibility to :InventoryItem, so it sends another message to :ProductItem to ask for help getting the quantity. The :ProductItem object knows that :InventoryItem has quantity information, so it forwards the request and collects quantity information. :Catalog collects all the information and returns it to the :AvailabilityHandler, which sends it back to the Clerk. Figure 11-14 shows the completed first-cut sequence diagram for the *Look up item availability* use case.

Before moving on, let's analyze this solution based on some of the design principles of good design that were previously discussed—coupling, cohesion, object responsibility, and use case controller.

The use case controller provides the link between the internal objects and the external environment. This limits the coupling to the external environment to that single object. The responsibilities assigned to :AvailabilityHandler are to catch incoming messages and distribute them to the correct internal domain objects and to return the required information to the external environment. By using a use case controller as the switchboard, overall coupling between the domain objects and the environment is limited. The :Availability-Handler class also is highly cohesive, with only two primary responsibilities.

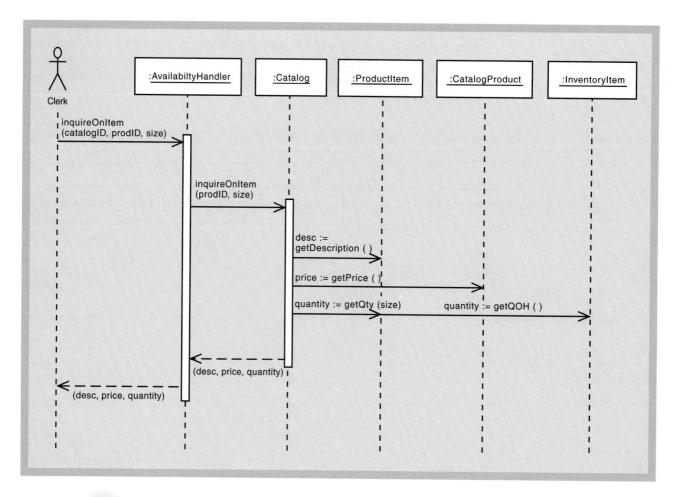

FIGURE *11-14*

First-cut sequence diagram for the *Look up item availability* use case.

activation lifelines

vertical rectangles in a sequence diagram that indicate when an object is executing a method

The responsibility assigned to :Catalog is to be in charge of collecting all product and inventory information in the product hierarchy. Since the catalog is at the top of the hierarchy, this appears to be a reasonable responsibility assignment. The :Catalog class is highly cohesive, at least for the messages seen in this use case. Coupling is straightforward, being basically vertical on the hierarchy. Thus, the assignment of responsibilities and corresponding messages seems to conform to good design principles.

One question we might ask is whether the request for the quantity should go from :ProductItem to :InventoryItem, or could it come directly from :Catalog? As illustrated in Figure 11-14, it does follow the hierarchy, and since this hierarchy will probably be used in other use cases, the design as shown is a good choice. The more direct solution would result in fewer messages, but it would increase overall coupling. Both :Catalog and :ProductItem would require navigation visibility and hence increase coupling. So, the design as shown is the better solution.

A new symbol is included within the diagram shown in Figure 11-14. The tall, narrow vertical rectangles on the :AvailabilityHandler and the :Catalog lifelines are called the *activation lifelines*. Remember from Chapter 7 that the lifeline of an object is represented by a vertical dashed line. An object can either be in an active or inactive state. An object is active while it is executing a method and inactive once that method terminates. An inactive object resides in memory awaiting further messages and remembers the values of its internal data attributes. The activation lifeline represents the period when the object is active and executing. In Figure 11-14, when :AvailabilityHandler receives the inquireOnItem message, it becomes active—begins executing a method—and it remains active until it sends a response back to the Clerk.

Obviously, with this preliminary design, we are focusing on problem domain classes and basic interactions. Many other details can be added to make the solution more complete. Such details as user-interface classes, indexes to find customers based on name, and maybe even a database to retrieve a customer could be added to the solution. That is the reason this is a preliminary solution.

Guidelines for Preliminary Sequence Diagram Development

Even though the example in Figure 11-14 was very simple, we can distill several tasks to help you learn to do design for a use case or scenario using sequence diagrams. Note that these tasks are not done sequentially, but they are done when necessary to build the sequence diagram. We only identify them as separate tasks to ensure that all three are completed.

- Take each input message and determine all of the internal messages that result from that input. For that message, determine its objective. Determine what information is needed, what class needs it—the destination—and what class provides it—the source. Determine whether any objects are created as a result of the input. This will help you to define internal messages, their origin objects, and their destination objects. In other words, you are trying to define which classes and which internal messages are needed to support the input message.
- As you work with each input message, be sure to identify the complete set of classes that will be affected by that message. In other words, select all the objects from the domain class diagram that need to be involved. In Chapter 7 you learned about use case preconditions and postconditions. Any classes that are listed in either of those preconditions or postconditions are classes that should be included in the design. Other classes to be included, even though they might not be listed in the preconditions or postconditions, are classes that are created, that are the creators of objects for the use case, that are updated during the use case, or that provide information that is utilized in the use case.
- Additionally, flesh out the components for each message. Add iteration, true/false conditions, return values, and passed parameters. The passed parameters should be based on the attributes that are found in the domain class diagram. Return values and passed parameters can be attributes, but they may also be objects from classes.

This list of steps will produce the preliminary design. Refinements and modifications may be necessary, and again, we have only focused on the problem domain classes involved in the use case.

Developing a Multilayer Design for
Look up item availability

The development of the first-cut sequence diagram focuses only on the classes in the domain layer. However, as explained previously, in systems design we must also design the user-interface classes, and, when appropriate, the data access classes. In this section, let's expand the design presented in Figure 11-14 and make it a multilayer design, including both a view layer and a data access layer.

In the early days of interactive systems and graphical user interfaces, tool developers invented languages and tools that made it easy to develop systems with graphical user interfaces, such as windows and buttons. Languages such as early versions of Visual Basic, Delphi, and PowerBuilder were designed to make it easy to build interactive, event-driven, graphical systems. However, in these languages, the program logic was attached to the windows and other graphical components. So, to move these systems to other environments, such as browser-based systems, designers had to completely rewrite the system. In fact, systems developed this way became good illustrations of the problems that follow when design principles such as highly cohesive classes and separating responsibilities are violated. When a class has both user-interface functions and business logic mixed together, upgrading and maintaining the system become more difficult.

As object-oriented programming became more prevalent and tools integrated both object-oriented programming and graphical interfaces, it became easier to build systems that could be partitioned and in which responsibilities could be separated. User-interface classes did not need to have business logic—other than edits on the input data. Designers could build multilayer systems that were more robust and easier to maintain, and they could apply the principles of good design. Tools such as Java and Visual Studio .NET provide the capability to easily build graphical user interfaces as well as sophisticated problem domain classes.

Designing the View Layer

The view layer involves human-computer interaction and requires designing the user interface for each use case. User-interface design is one of the key activities of the SDLC design phase and is discussed in detail in Chapter 14. The interface designer takes the steps in a use case description and begins to develop a dialog design for the use case, usually defining one or more Windows forms or Web forms the user will use to interact with the system. Sometimes the forms are sketched out and shown as storyboards describing the dialog. Sometimes prototypes of the forms are created so that users can try out the interface design. Recall that the design activities are being completed in parallel, so some members of the project team are working on user-interface design while others are developing sequence diagrams. These two design activities must be coordinated so that the view layer classes defined in the sequence diagrams are consistent with the forms developed by the user-interface design team. Chapters 14 and 15 explain this process of designing the forms, reports, and even the dialogs used in the interface between the users and the system. Once the electronic forms are designed, then an object-oriented window class can be defined for each form. Normally, the UML class definition for a form only specifies the data entry attributes. Other class attributes, such as buttons and menu items, are specified in the form description and not in the class definition.

Based on the progress completed on the dialog design for a use case, forms are added as windows classes to the sequence diagram—as the view layer. Typically, there is one input form for the messages entering the system for the use case. If the messages are unique, then each may require its own input form. In most instances, however, one form may be sufficient for all the related messages within a use case. Obviously, each message from an external actor must be entered into the system in some way, and output messages must be displayed. One logical way for this to happen is through a window form class that can accept input data and possibly display output data. Figure 11-15 illustrates the result of defining a window class for the input message named :ProductQuery from the Clerk.

Adding the user-interface classes to the sequence diagram is usually straightforward. As shown in Figure 11-14, we assumed that a message came directly from the external actor to the :AvailabilityHandler object. In reality, the data is entered into the electronic form via the keyboard, and the user-interface window object catches that information, formats a message, and transmits the message to the :AvailabilityHandler object. Thus, adding the user-interface objects only requires us to place those objects between the actor, the Clerk, and the domain objects and to include the appropriate message. Figure 11-15 shows this process of adding a user-interface object to the sequence diagram.

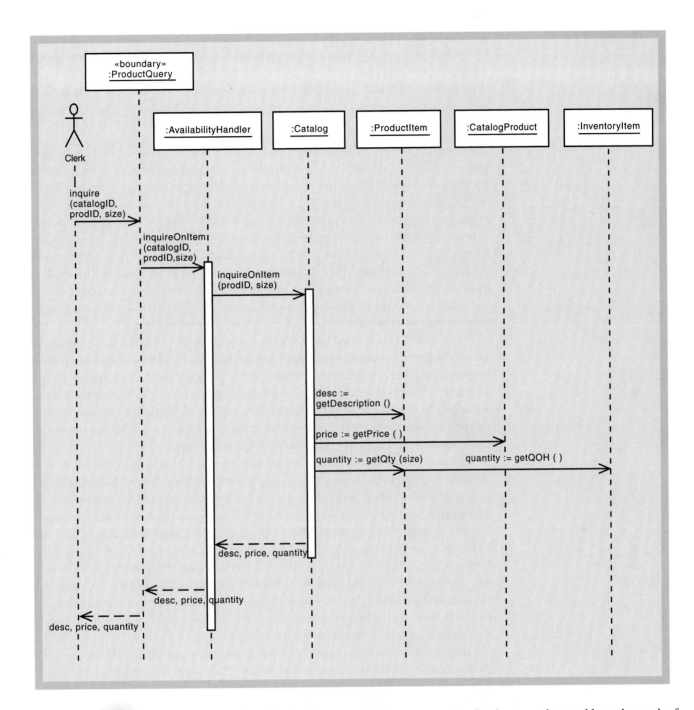

FIGURE *11-15*

Look up item availability use case with view layer and user-interface object.

It is critically important that you, as a system developer, understand how the work of the user-interface design team and their prototypes and dialog storyboards should affect the development of detailed sequence diagrams. In both Chapter 7 and this chapter, we have emphasized the benefits of capturing system requirements by building logical models. You should, by now, see the strengths of models and model building. However, building models based only on discussions with users has proven to be less than successful in constructing systems. The problem lies in the fact that users, as all human beings, may not understand exactly what they need unless they can see and experience the end product. As a result, the most effective approach to building systems—especially the parts involving the user interface—is to actively involve the users through the use of prototypes, mockups, and storyboards. A combination of prototyping and model building is the most effective approach. Sometimes developers are tempted to build a system only based on prototyping and skip the model building. This may work for very small systems, but as expressed at the beginning of the chapter, models are necessary to produce a truly robust and correct system. You are encouraged to use all the skills and techniques you have to ensure that the systems you build are high-quality, strong systems.

Designing the Data Access Layer

The principle of separation of responsibilities also applies to the data access layer. On smaller systems, two-layer designs exist, where the SQL statements to access a database are embedded within the business logic layer. In OO two-layer designs, this implies that SQL statements are included in methods of the problem domain classes. However, on larger, more complex systems, it makes sense to create classes whose sole responsibility is to execute database SQL statements, get the results of the query, and provide that information to the domain layer. As hardware and networks became more sophisticated, multilayer design became more important to support multitier networks in which the database server was on one machine, the business logic was on another server, and the user interface was on several desktop client machines. This new way of designing systems creates not only more robust systems but more flexible systems.

In Chapter 5, you learned how to build a domain model to describe the "things," or entities about which information is to be maintained. The domain model serves two purposes. First, of course, it is used to develop the database for the new system. Chapter 13 explains how to use the domain model to design the database. The second purpose, as we have seen, is to identify the internal classes that make up the new system. It should be apparent that there will be a very close correlation between the database tables and the design classes. Both come from the same domain model.

In your database course, you learn how to access the tables in the relational database by using Structured Query Language (SQL) statements. Executing SQL statements on a database enables a program to access a record or a set of records from the database. One of the problems with object-oriented programs that use databases is a slight mismatch between programming languages and database SQL statements. For example, in a database, tables are linked through the use of foreign keys, such as an Order having a CustomerID as a column so that the order can be joined with the customer in a relational join. However, in OO programming languages, the navigation is in the opposite direction, and the Customer class may have an array of references that point to the Order objects that are in the computer memory and are being processed by the system. In other words, design classes do not have foreign keys.

These differences between programming languages and database languages have partially driven the trend to a multilayer design. The design, programming, and maintenance of a system is easier if separate classes are defined to access the database and get the data in a form that is conducive to processing in the computer. Rather than mix the business logic with the data access logic, it is better to define separate classes and let each focus on its primary responsibility. This idea is an application of good design principles of highly cohesive classes with appropriate responsibilities.

In this chapter, we take a somewhat simplified approach to design in order to teach the basic ideas without getting embroiled in the complexities of database access. Let us assume that every domain object will have a table in a relational database. Let's also assume that each domain object is responsible for initializing itself by going to the database when necessary to read the data. Given these assumptions, the modifications to the *Look up item availability* use case that are required to add the data access layer can be precisely defined. Before the use case can be executed, the domain objects have to be initialized with the necessary data from the database. Figure 11-16 illustrates this process.

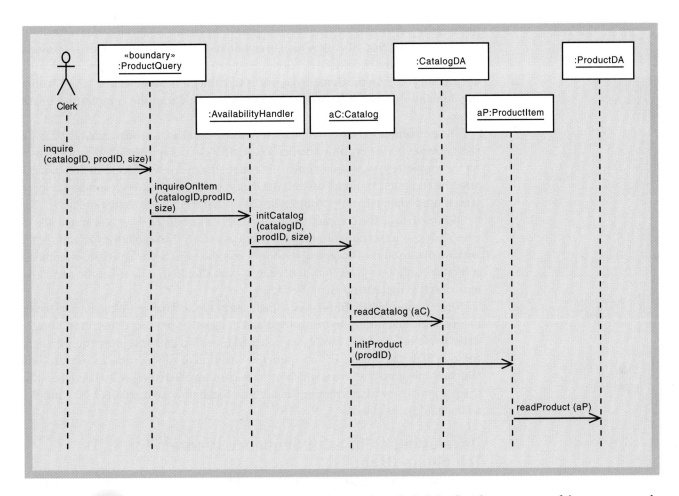

FIGURE *11-16*

Partial three-layer design for *Look up item availability.*

In Figure 11-16, we have only included the first few messages of the use case so that we can focus on the data access classes. Carefully note the changes between Figure 11-15 and 11-16. When the use case controller, :AvailabilityHandler receives the initial request for an item inquiry, it first begins a process to initialize the necessary objects in memory. The set of "init" messages can either create new objects, if necessary, or simply verify that existing objects are in memory. The two classes CatalogDA and ProductDA are the data access classes. The incoming message to the catalog object simply identifies which catalog is needed. The catalog object reads the database, that is, accesses the CatalogDA object, to get all of the necessary data. The process to complete the product data fields is the same, as shown in the figure.

Remember that within sequence diagrams the boxes refer to objects and not classes. The name within the box is underlined to indicate an object. In most cases, it is not important to specifically identify an object, so the notation would simply be <u>Person</u> or <u>:Person</u>. Sometimes, it is important to denote a specific object. In this case, the notation might be <u>Mary:Person</u>. When denoting a specific object, the colon serves as the divider between the object name and the specific object identifier. Specific identifiers are used when a reference to the object is required in another part of the diagram.

Note the description of each of the domain objects, for example <u>aC:Catalog</u>. The "aC" is the identifier for a particular catalog object. When a new object is instantiated, such as a new catalog object, it recognizes that it needs to go to the database to retrieve its data. It sends a message to the <u>:CatalogDA</u> object, with a reference to itself, that is, aC, to read and retrieve the data from the database. The data access class reads the database with a SQL statement and places the appropriate attribute information in the original object by using the passed reference parameter. In Figure 11-16, all of the domain objects are initiated first. After they have been initiated, the use case proceeds as previously defined. One of the benefits of developing sequence diagrams in two steps, with the first cut focusing only on the domain objects, is that the problem of identifying collaborating domain objects can be solved without worrying about the complexities of data access.

The domain objects that have data stored in a database are often referred to as persistent classes. As we explained earlier, a persistent class is an entity that must persist after the computer system is shut down. Of course, the memory object itself does not exist after the computer is turned off, but the data must persist between executions. So, the term is used when referring to objects or classes that require permanent storage.

Figure 11-17 is the complete sequence diagram for the *Look up item availability* use case. At first, it looks rather intimidating, but as you study it carefully, you will see that we have discussed the conceptual basis for every message in the figure. So, even though it is rather complex—it has lots of messages—you should be able to see the overall pattern as well as understand all of the details.

This example began simply—as a single query to the database. We identified the domain objects that needed to be involved and designed a solution. Other, more complex database solutions could have been designed based on database joins and with more complex SQL statements. Let's now turn our attention to a more complex use case—one with multiple input messages, which requires creating objects. We will again take a multistep approach by first designing the use case based only on domain objects and then adding multilayer objects.

Developing a First-Cut Sequence Diagram for RMO Telephone Order

Figure 7-17 from Chapter 7 presented an SSD for a telephone order from RMO. Figure 11-18 is another version of the telephone order. This SSD is for the telephone scenario of the *Create new order* use case. As before, we will do a design for each input message in the SSD. The design components for all the messages are combined to provide a comprehensive sequence diagram for the entire use case. As we did in Figure 11-4, information from the SSD and the first-cut design class diagram will again be used to develop the sequence diagram, with one exception. We define a controller object for this use case with the name of :OrderHandler. We anticipate that this controller may serve both for creating new orders and for maintaining existing orders. Whether this is the best design will be decided after other use cases have been designed and the design has been reviewed for good design principles.

The first input message is startOrder (accountNo), which comes from the Order Clerk to the :System. What does the system need to do to start an order? The system needs to create a new Order object and connect that Order object to a Customer object. Thus, a message to create a new order is needed. The destination of the message will be the Order object itself. In fact, if you remember your programming class, the create message invokes a constructor method on the object, which will create a new object. In UML, when a create message is sent to an object, it is often drawn directly to the object's box and not to the lifeline. That is an optional diagramming technique used for convenience.

What object should be the source object for the create message? Should it be the :OrderHandler itself, or should it be some other object? Information included in the domain model indicates that the Order object has a relationship or link with the Customer object. There are several ways this link could be built. One way would be to have the :OrderHandler object send the create message to the Order object, then send another

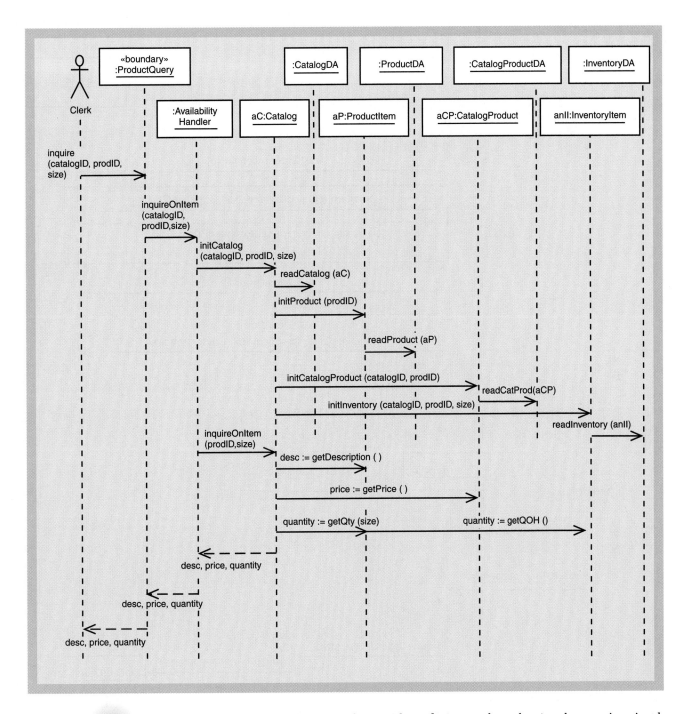

FIGURE *11-17*

Completed three-layer design for *Look up item availability.*

message to the Customer object with a reference to the order. Another way is to just let the Customer object create the Order object. Since Order objects are not allowed unless a Customer object exists, this is one way to ensure that the customer existence precondition is met. Figure 11-19 shows the results using the second approach. Note that a specific identifier is given to the newly created Order object—anOrd. That reference is passed back to the Customer object, which passes it back to the :OrderHandler. We will see the need for this in later steps.

The next input message is addItem (catalogID, prodID, size, quantity) and is a repeating message to add line items to the order. By referencing the OrderItem class in the domain model, we find the necessary attributes for an OrderItem are quantity, price, description, and back-order status. The price can be obtained from the CatalogProduct class. Description comes from the Product class. The quantity is input by the clerk,

SSD for the telephone order scenario
of the *Create new order* use case.

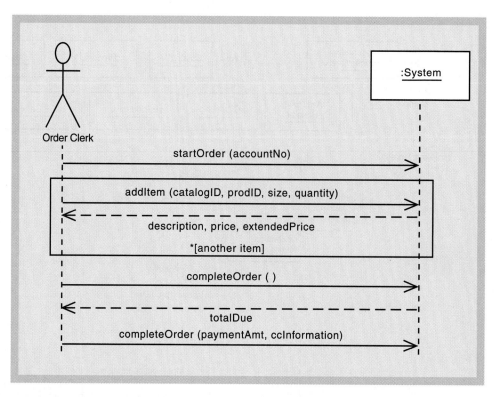

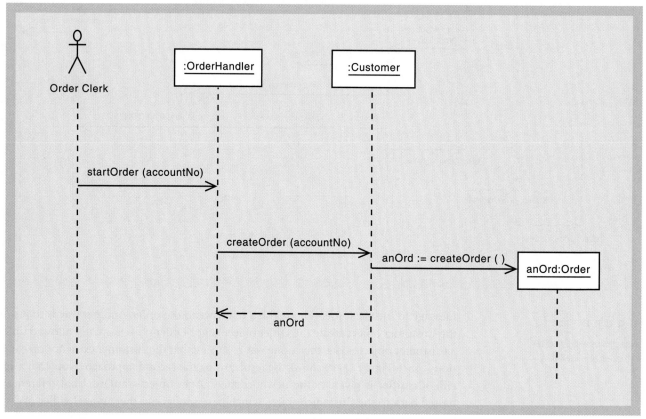

Partial sequence diagram for the
telephone order scenario.

although to see whether items are in stock, the system must check the InventoryItem class. (The detailed description of the use case also indicates that inventory should be checked by the system.) So, the sequence diagram will also need objects for :OrderItem, :CatalogProduct, :Product, and :InventoryItem.

As we identify the specific messages, along with source and destination as well as the passed parameters, we need to consider some critical issues. As before, an important decision is, which object is the source or initiator of a message? If the message is a query message, the source is the object that needs information. If the message is an update or create message, then the source is the object that controls the other object or that has the information necessary for its creation.

Another important consideration is navigation visibility—to send a message to the correct destination object, the destination object must be visible to the source object. Remember that the purpose of doing design is to prepare for programming. As a designer, you will need to think about how the program will work and consider programming issues. Given these two considerations, and the source considerations discussed in the previous paragraph, we have determined that the following internal messages will be required. For each message, a source object and a destination object have been identified.

- **addItem ( ).** Original message, from Clerk to :OrderHandler
- **addItem ( ).** A forwarded version of the input message from :OrderHandler to :Order. Since OrderItem objects are dependent to Order, Order is the logical object to create OrderItems. System has visibility to the Order from the previous return message when anOrd was returned to the system.
- **createOrdItem ( ).** The internal message from :Order to :OrderItem. Since the OrderItem will be responsible for obtaining the data for its attributes, it needs visibility to :CatalogProduct, :Product, and :InventoryItem. As a result, those keys are sent as parameters. An alternate approach is to let :Order collect the required information, such as price and backOrderStatus and send that to the :OrderItem as parameters. However, since the domain model indicates that there is a link between an OrderItem and an InventoryItem, the first approach is better.
- **getPrice ( ).** The message to get the price from the :CatalogProduct object. The :OrderItem initiates the message. It has visibility since it has the key values.
- **getDescription ( ).** The message, initiated by :OrderItem, to get the description from :Product.
- **updateQty ( ).** The message that checks to see that there is sufficient quantity on hand. This message also initiates the updating of the quantity on hand. The :OrderItem initiates the message. It does not have key visibility, but it does have enough information to search on an index of catalogID, productID, and size.

Figure 11-20 shows the results of the design for the addItem message. In the figure, the input parameters and return values have also been added. You should review the design, including the parameters, to ensure that you understand all aspects of this design.

After each item is added to the order, control is returned to the Order Clerk. The clerk will add another item or, at the end of the order, will send a completeOrder( ) message. This message has no parameters. Its purpose is simply to tell the order to calculate the total amount due. If we assume, as designers, that the Order object keeps a running total of the individual line items that were added, then it simply calculates the appropriate tax and shipping and sends back a total amount. This is a valid and solid design. Another alternative is that the Order object does not keep a running total but must query each of the line items and accumulate a total. The second design requires additional detailed messages to be sent to the :OrderItem objects.

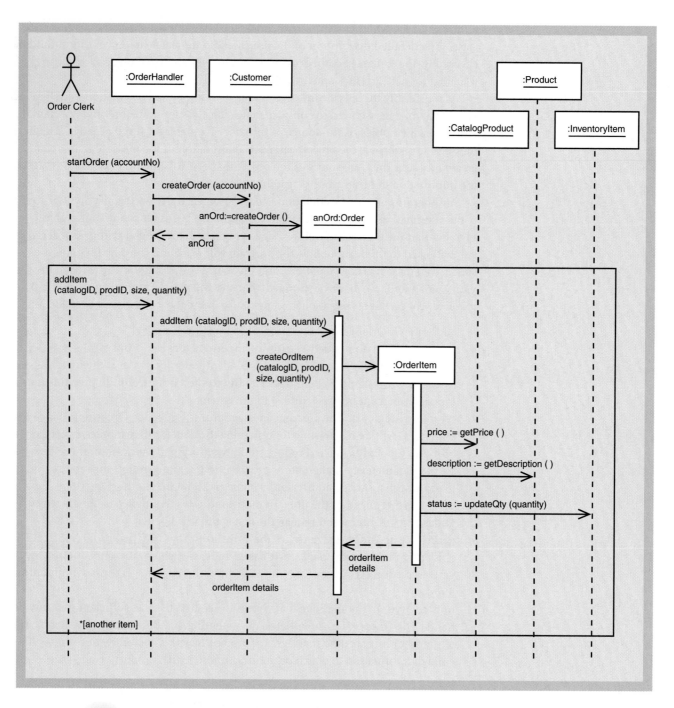

FIGURE *11-20*

Another partial sequence diagram
for the telephone order scenario.

The final message on the SSD is the makePayment (ccInformation) message. We have made the simplifying assumption that it is always a credit-card payment. As we review the domain model, we see that a new object must be created—the OrderTransaction object. Since transactions are connected to orders in the model, an Order object should create a transaction. Thus, the system forwards the completion message to the :Order, which in turn creates a payment for :OrderTransaction. These new messages are shown in Figure 11-21.

Figure 11-21 contains all of the internal messages for the final design. The figure contains the design of the domain model classes and the internal messages that are required to execute the use case. This section has focused only on the classes from the domain model, plus one additional object called :OrderHandler. By focusing only on the

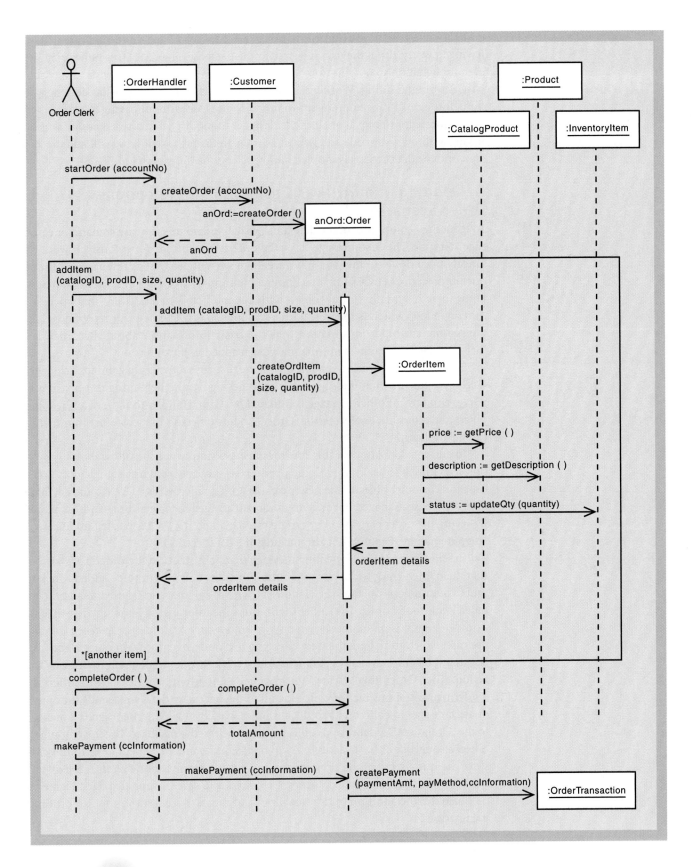

FIGURE *11-21*

Sequence diagram for the telephone scenario of the *Create new order* use case.

Chapter 11   The Object-Oriented Approach to Design: Use Case Realization   **415**

domain classes, we have been able to design the core processing for the use case without having to worry about the user interface or the database. Figure 11-21 is rather complex, even though it only contains domain objects. However, this design provides a very solid base for programming. Working with design models enables you—the designer—to think through all the requirements to process a use case without having to worry about code. More importantly, it enables the designer to modify and correct a design without having to throw away code and write new code. In the next section, we will add the view layer and the data access layer objects to the telephone order scenario.

Developing a Multilayer Design for the Telephone Order Scenario

To reduce the complexity of the diagrams, we will extend this use case scenario one message at a time. The first message starts the order. As mentioned previously, if work has already been done on the user interface, then we will use that as the basis for the sequence diagram. Let's assume that enough preliminary work has been done so that we know there is a :MainWindow object with a menu item—or a button—that opens another window, which is the new :OrderWindow object. The new :OrderWindow contains some customer information, has the basic information about the order, and includes places to list the items that have been added to the order.

Figure 11-22 illustrates the additions to the first-cut sequence diagram to include the view layer for this first message. We add a message at the beginning to open up the :OrderWindow. Another message is initiated by :OrderHandler to retrieve a :Customer object and return that object to the :OrderWindow object so that customer information can be displayed.

To add the data layer for this first message, we simply need to allow the :Customer object to initialize itself. As with the previous use case, we will use a single data access class for Customer objects. The newly created :Order object is also saved to the database. Many database management systems automatically generate keys for records in a table. By saving the order early, the key can be extracted and the data is backed up in case of computer failure. Figure 11-22 also includes the data access class.

The next message is a repeating message that adds items to the order. Let's assume that the :OrderWindow object has a button to allow a new item to be added. Clicking on the button pops up another window, :NewItemWin, that can be used to add the detailed items to the order. The Order Clerk enters the information about the new item to be ordered. At this point, the :NewItemWin window invokes the *Look up item availability* use case, which initializes the necessary :Product and :InventoryItem objects and verifies that the item is available. A note is added to the sequence diagram to indicate this action. The :Order then creates the required :OrderItem. As discussed earlier, the :OrderItem is responsible for collecting all of the necessary data, so it sends the appropriate messages to collect data for its attributes. Finally, the :OrderItem sends a message to the :OrderItemDA data access object to save itself to the database. The :InventoryItem object also invokes its data access class to update its quantity-on-hand field. In this case, since the initializing of the product and inventory objects is done by the *Look up item availability* use case, the only data access required is to save the updated data. Figure 11-23 illustrates the addition of the view layer and the data access layer to this portion of the use case.

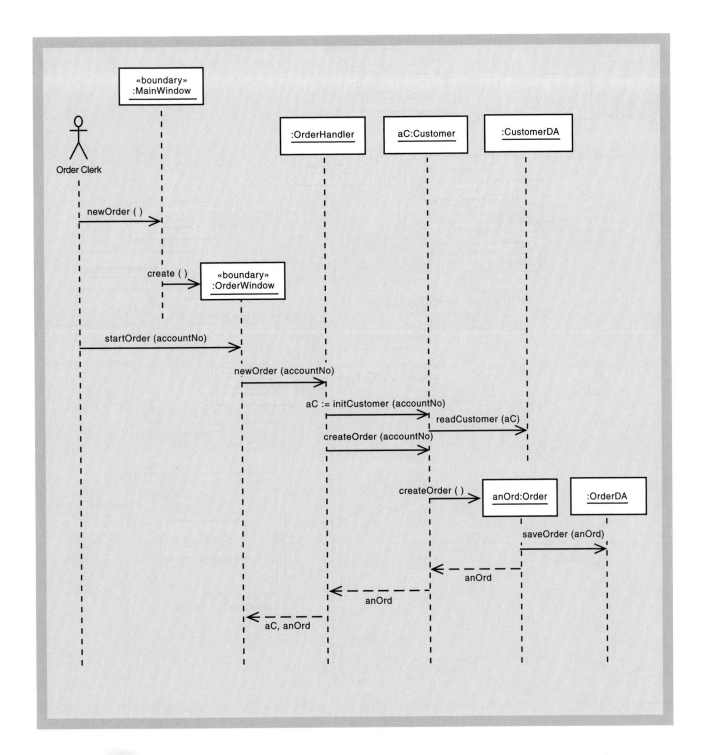

For the final two messages, completeOrder ( ) and makePayment ( ), no new window classes need to be defined. Let us assume that the :NewOrderWin window has appropriate fields to indicate the completion of the order and to enter payment amount. The order will need to be saved to the database using the :OrderDA data access class. One new data access class to save the :OrderTransaction object is needed. Figure 11-24 illustrates these final additions to the sequence diagram.

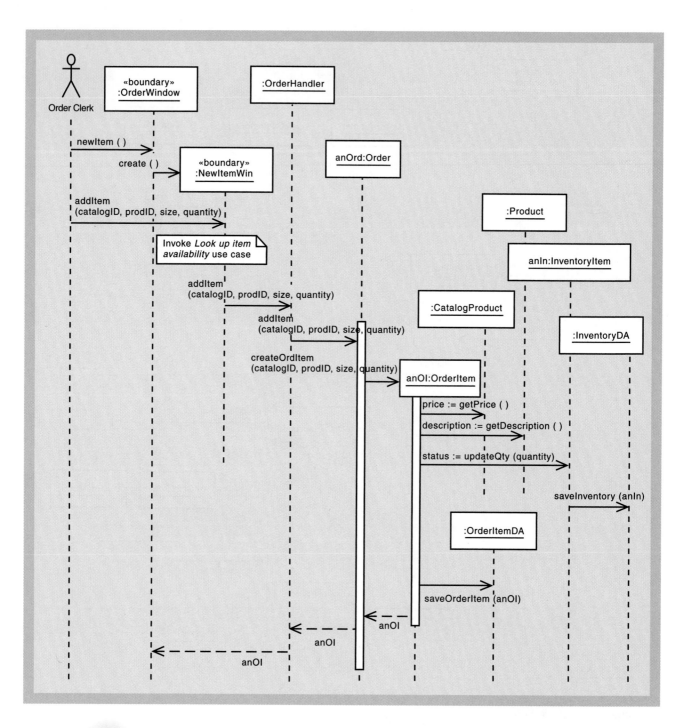

F I G U R E *11-23*

Telephone order sequence diagram
for the addItem message.

As you can see, these sequence diagrams do get somewhat busy and complicated. However, you should note that these diagrams provide an excellent foundation for programming the use case. By going through the process of detailed design, the designer can think through the complexities of each use case without programming complications. It should also be noted that no design elements have been added yet to cover error handling or failures in the use case. For example, what happens if a customer record

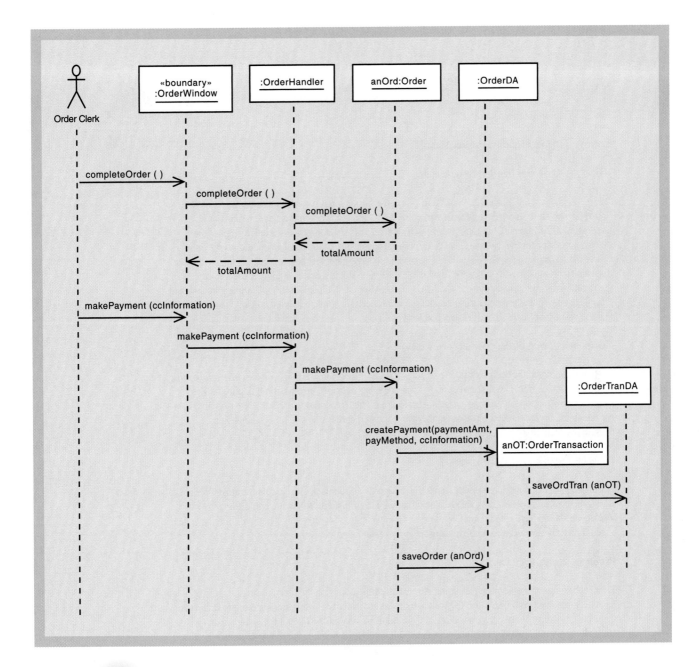

FIGURE *11-24*

Telephone order sequence diagram
for the final messages.

is not found? Or what does the system do if the input edits fail? The sequence diagram could be extended to cover those situations. Another approach is simply to add narrative description to indicate how the system should handle those exception conditions and leave the details to the programmer.

When developing a sequence diagram, it is often necessary to work on several tasks at the same time. For example, the database design and user-interface prototyping might be ongoing activities that are concurrent with detailed design. The RMO memo illustrates how this combination of design activities occurs.

May 15, 2005

To: John MacMurty

From: Barbara Halifax, Project Manager

RE: Customer Support System status

John, now that we are in design, we need to proceed along several fronts at the same time. As you know, the schedule shows several activities going on in parallel. The primary controlling activity is the design of the internal computer processes for each use case. Before we started designing the system processes for the use cases, we did a preliminary design of the database based on the domain class diagram that we developed earlier. Next, we took a use case and developed a detailed sequence diagram for that use case. Before we finalize the design of any use case, we build a few prototypes of the input forms and screens and review them with the users. It is a fairly time-consuming process, but I am pleased with the progress we are making. I am especially confident that the design is solid and that the workflows and screens are acceptable to the users.

Completed during the last period (two weeks)

We have completed the design of three use cases, *Update customer account*, *Look up item availability*, and *Create new order*. The first use case, *Update customer account*, includes both adding new customer information and updating existing customers. The *Create new order* use case only includes the telephone order scenario. (You remember we discussed earlier that we would focus on the telephone orders first.)

Plans for the next period (two weeks)

During the next two weeks, we will finish the design of all the use cases in this iteration. We will also start tomorrow on programming the use cases that we have finished. So, by the end of the next two weeks, we should have some complete use cases that we can show the users. They will only have been tested lightly, but we want to involve the users early in our testing.

Problems, issues, open items

There are no major problems at this point. Almost all of the issues on the Outstanding Items Log have been resolved. You might test the waters for us at the oversight committee meeting to make sure the users are happy. From our perspective, they seem to be very excited about the progress that we are making. I would just like to make sure that the feeling is the same at the senior executive level.

BH

cc. Steven Deerfield, Ming Lee, Jack Garcia

DESIGNING WITH COLLABORATION DIAGRAMS

Collaboration diagrams and sequence diagrams are both interaction diagrams, and they capture the same information. The process of designing is the same whether you are using collaboration diagrams or sequence diagrams. Which model is used for design is primarily a matter of a designer's personal preference. Many designers prefer to use sequence diagrams to develop the design because use case descriptions and dialog designs follow a sequence of steps. Collaboration diagrams are useful for showing a different view of the use case—one that emphasizes coupling.

A collaboration diagram uses the same symbols for actors, objects, and messages found in a sequence diagram. The lifeline and activation lifeline symbols are not used. However, a different symbol, the link symbol, is used. Figure 11-25 illustrates the four symbols used in most collaboration diagrams.

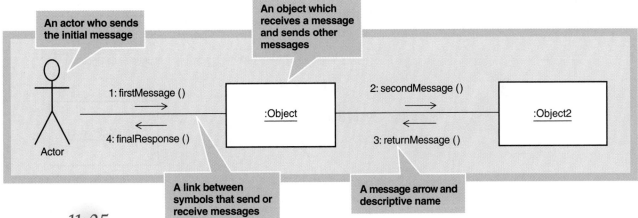

FIGURE *11-25*

The symbols of a collaboration diagram.

The format of the message descriptor for a collaboration diagram differs slightly from that for a sequence diagram. Since no lifeline shows the passage of time during a scenario, each message is numbered sequentially to indicate the order of the messages. The syntax of the message descriptor in a collaboration diagram is the following:

[true/false condition] sequence-number: return-value := message-name (parameter-list)

As you can see in Figure 11-25, a colon always directly follows the sequence number.

The connecting lines between the objects or between actors and objects represent *links*. In a collaboration diagram, a link shows that two items share a message—that one sends a message and the other receives it. The connecting lines are essentially used only to carry the messages, so you can think of them as the wires used to transmit the messages.

Figures 11-25 and 11-26 present collaboration diagrams for the same two RMO use cases shown earlier with sequence diagrams Figures 11-14 and 11-20, namely, *Look up item availability* and *Create new order*. These collaboration diagrams contain only domain model objects and not the view layer or data access layer. However, multilayer design can be done just as effectively with collaboration diagrams as with sequence diagrams.

The numbers on the messages indicate the sequence in which the messages are sent. Notice the messages numbered 5 and 5.1. The hierarchical dot numbering scheme is used when messages are dependent on other messages. In this instance, the primary message, 5: quantity := getQty (size), is sent to the :ProductItem, which then forwards a similar message, 5.1: quantity := getQOH ( ), to :InventoryItem. The second message is a direct result of the first, so it is numbered 5.1, as a subordinate to the primary message. Sometimes new designers struggle with knowing when to number messages as subordinate and when to number them at the same level. For example, you could argue that the entire sequence of messages is dependent on the very first one being sent and that the entire set should be subordinated to the initial message. One good way to think

links

notations in a collaboration diagram that carry messages between objects or between actors and objects

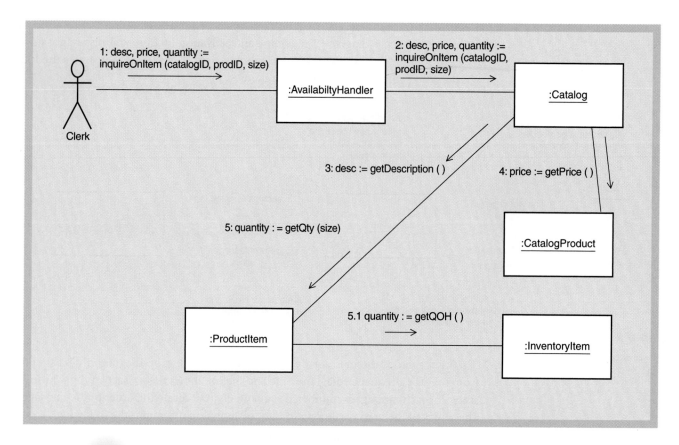

FIGURE *11-26*

A collaboration diagram for
Look up item availability.

about when and how to number with subordinated numbers is to construct the set of messages in an outline format, as you would before writing a paper. The messages that are at the same level for an outline should all be numbered as primary messages. Lower-level messages are like dependent paragraphs and headings in a paper outline. The hierarchy of messages can go as deep as required to indicate dependency. In Figure 11-27, the numbering sequence goes down several levels. Since multiple input messages initiate a series of other messages, this diagram is very explicit in using the hierarchical numbering scheme.

When comparing the collaboration diagrams with the sequence diagrams, it should be evident that the focus of a collaboration diagram is on the objects themselves. Drawing a collaboration diagram is an effective way to get a quick overview of the objects that work together. However, as you look at the diagrams, you should observe that it is more difficult to visualize the sequence of the messages. You have to hunt to find the numbers to see the sequence of the messages. On the other hand, to get a quick overview of the collaborating objects, a collaboration diagram is very effective.

Many designers use collaboration diagrams to sketch out a solution. If the use case is small and not too complex, a simple collaboration diagram may suffice. However, for more complex situations, a sequence diagram may be required to visualize the flow and sequence of the messages. It is not unusual to find a mix within the same set of specifications: some use cases described by collaboration diagrams and others shown with sequence diagrams. As a system developer, you should be comfortable using both types of diagrams.

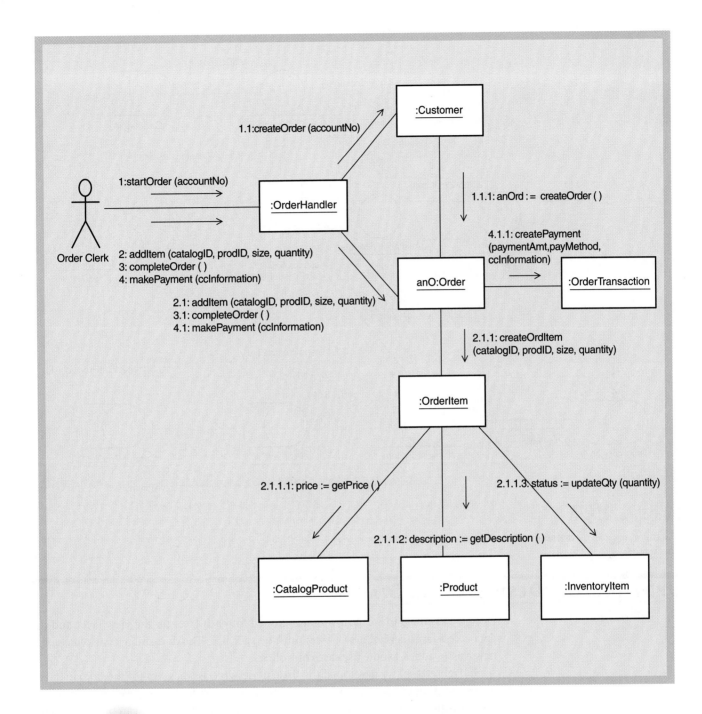

FIGURE *11-27*

A collaboration diagram for
Create new order.

An even more abbreviated form of collaboration diagrams can be used. Figure 11-5 showed two symbols that are used to specify objects. The iconic symbols show the object name, and the symbol itself indicates the stereotype that it is. Specific icons can be used to show the view layer, controller, domain layer, and data access layer. Figure 11-28 is an example of the multilayer design for the *Look up item availability* use case. This type of drawing can be used without messages to show collaborating objects or with messages to provide a shorthand notation for a complete collaboration diagram. The icons can also be used on sequence diagrams as shorthand notation for stereotypes.

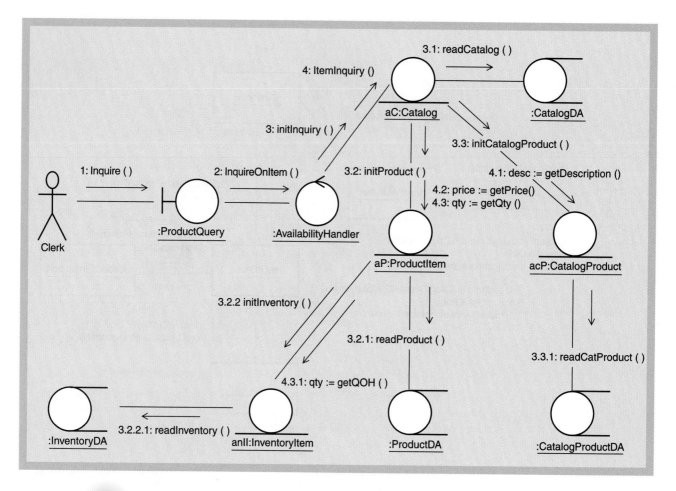

FIGURE *11-28*

Look up item availability use case using iconic symbols.

UPDATING THE DESIGN CLASS DIAGRAM

Design class diagrams can now be developed for each layer. In the view layer and the data access layer, several new classes must be specified. The domain layer also has some new classes added for the use case controllers.

In Figure 11-10 we developed the first-cut design class diagram for the domain layer. At that point in the development, no method signatures had been developed. Now that several sequence diagrams have been created, method information can be added to the classes. We also mentioned that the navigation arrows may also need updating from the decisions that were made during sequence diagram development.

First, we add method signatures. Three types of methods are found in most classes: (1) constructor methods, (2) data get and set methods, and (3) use case specific methods. Remember that constructor methods are the ones that create new instances of the objects. Get and set methods retrieve and update attribute values. Since every class must have a constructor, and most usually have get and set methods, it is optional to include those method signatures in the design class diagram. The third type of method must be included in the design class diagram. The following examples include all three types of method signatures as part of the process to find and document method signatures.

As in sequence diagrams, every message has a source object and a destination object. When a message is sent to an object, it must be prepared to accept that message and initiate some activity. This process is nothing more than invoking or calling a method on an object. In other words, every message that appears in a sequence diagram requires a method in the destination object. In fact, the syntax for a message looks very much like the syntax for a method. Thus, the process of adding method signatures to a design class is to go through every sequence diagram and find the messages sent to that class. Each message indicates a method.

Let's work through one example based on the Order class. The sequence diagrams that were completed during the examples in the chapter are shown in Figure 11-16 for the *Look up item availability* use case, and Figure 11-21 for the *Create new order* use case. Since we are defining method signatures for domain model classes, we use the sequence diagrams prior to adding the multilayer design.

In Figure 11-16, no messages are sent to Order. In Figure 11-21, there are four messages to Order. The first message is createOrder ( ), with no parameters. The next message, addItem (catalogID, prodID, size, quantity), contains several parameters and has a return message with orderItem details. Finally, there are two messages, completeOrder ( ) and makePayment (ccInformation), one of which, completeOrder, initiates a return value of the totalAmount of the order. Given these messages and the return values, the design class notation for Order is shown in Figure 11-29.

This process is continued for every class in the domain layer, including the added use case controller classes. Figure 11-30 contains the completed design class diagram for the domain layer classes. As you can see, this diagram provides excellent, thorough documentation of the design classes and serves as the blueprint for programming the system.

The two major additions to the domain layer classes are the two use case handlers. Additional navigation arrows have also been added to document which classes are visible from the controller classes. The other navigation arrows, which were defined during the first cut of the class diagram, have proved to be adequate for these two use cases. Additional use case development will enable us to add more navigation arrows, such as those to shipment and return items.

FIGURE *11-29*

Design class, with method signatures, for the Order class.

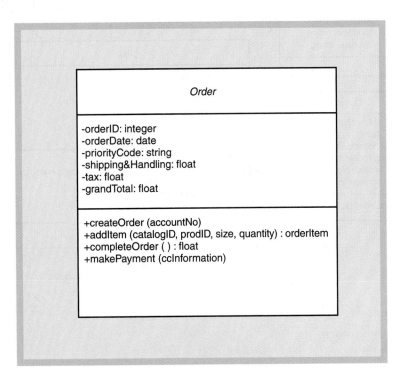

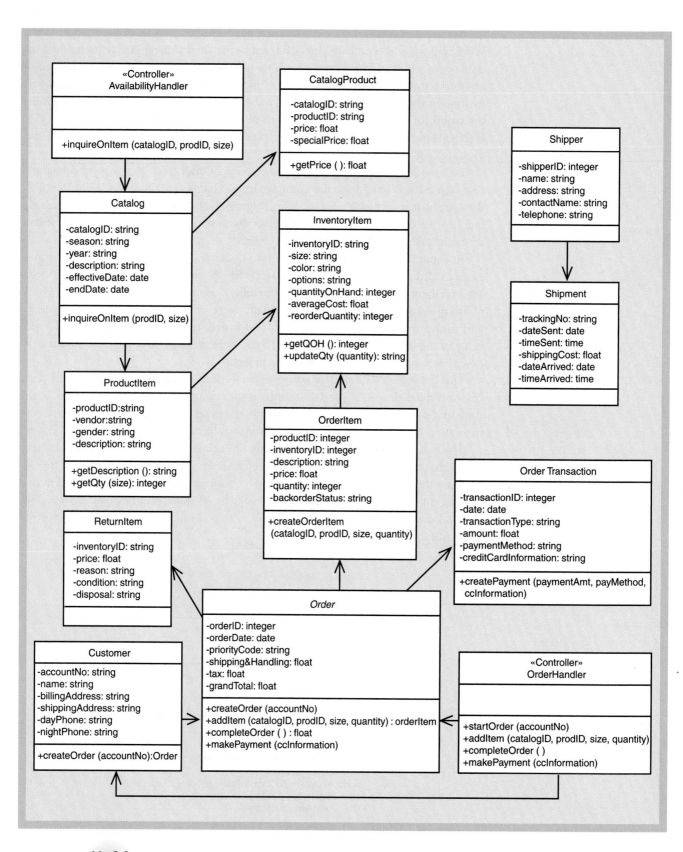

FIGURE *11-30*

Updated design class diagram
for the domain layer.

PACKAGE DIAGRAMS—STRUCTURING THE MAJOR COMPONENTS

A package diagram in UML is simply a high-level diagram that allows designers to associate classes of related groups. The preceding sections illustrated three-layer design, which includes the view layer, the domain layer, and the data access layer. In the interaction diagrams, the objects from each layer were shown together in the same diagram. However, at times designers need to document differences or similarities in the objects' relationships in these different layers—perhaps separating or grouping them based on a distributed processing environment. This information can be captured by showing each layer as a separate package. Figure 11-31 illustrates how these layers might be documented.

The package notation is a tabbed rectangle. The package name is usually shown on the tab, although for a very high level view, if no details are shown inside the package, the name can also be placed inside the package rectangle. In this instance, the classes that belong to that package are placed inside the package rectangle.

The classes are placed inside the appropriate package based on the layer to which they belong. Classes are associated with different layers as they are developed in the

FIGURE *11-31*

Partial design of a three-layer package diagram for RMO.

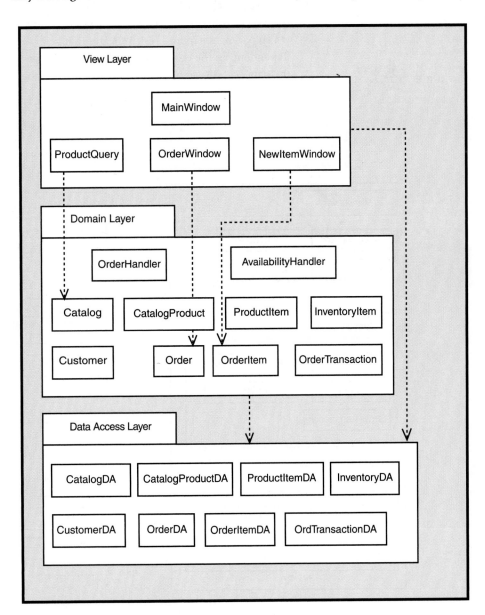

interaction diagrams. To develop this package diagram, we simply extracted the information from design class diagrams and interaction diagrams for each use case. Figure 11-31 is only a partial package diagram, because the packages contain only the classes from the use case interaction diagrams that were developed in this chapter.

The other symbol used on a package diagram is a dashed arrow, which represents a *dependency relationship*. The arrow's tail is connected to the package that is dependent, and the arrowhead is connected to the independent package. Dependency relationships are used on package diagrams, class diagrams, and even interaction diagrams. A good way to think about a dependency relationship is to note that if there is a change in one element (the independent element), the other element (the dependent element) may also have to be changed. Dependency relationships may be between packages or between classes within packages. Figure 11-31 indicates that two classes in the view layer are dependent on classes in the domain layer. So, for example, if a change is made in the Order class, then the OrderWindow class should be evaluated to capture that change. However, the reverse is not necessarily true. Changes to the view layer usually do not carry through to the domain layer.

Two examples of dependency relationships are given in Figure 11-31. The first, we have seen, is between classes. Another is less detailed and indicates a dependency between packages. Figure 11-31 indicates that both the view layer and the domain layer are dependent on the data access layer. Thus, changes to the data structures, as reflected in the data access layer, usually require changes at the domain layer and the view layer.

dependency relationship

relationship among elements in package diagrams, class diagrams, and interaction diagrams that indicates which elements affect other elements in a system so that designers can track the carry-through effects of changes

F I G U R E *11-32*

RMO subsystem packages.

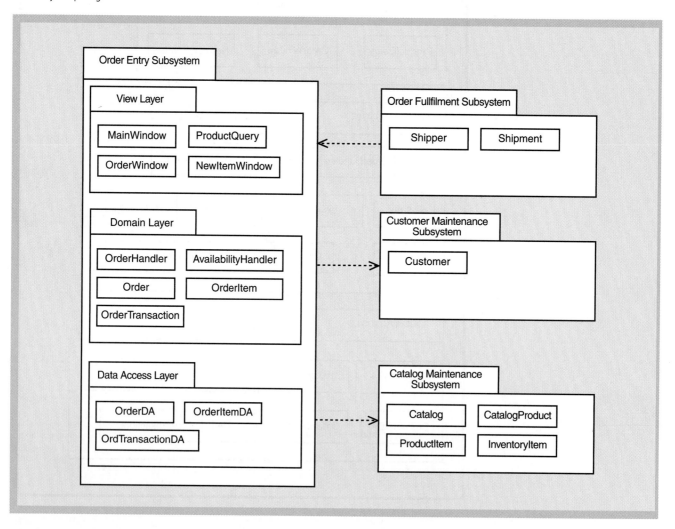

Package diagrams can also be nested to show different levels of packages. Figure 11-32 indicates that the packages, and some of the classes contained within them, are all part of the order-entry subsystem. As show in Figure 7-4, the RMO system can be divided into subsystems. One way to document these subsystems is with package diagrams. One major benefit of this documentation is that different packages can be assigned to different teams of programmers to program the classes. The dependency arrows will help them recognize where communication among teams must occur to ensure a totally integrated system.

As shown in Figure 11-32, dependency is indicated from order fulfillment to order entry. Also, order entry is dependent on the customer maintenance and catalog maintenance subsystems. The order fulfillment subsystem may use the order classes that are defined in the order-entry subsystem, or information may be sent from the order-entry subsystem. In any event, if anything changes in the order-entry subsystem, the order fulfillment subsystem may also require modification. As the classes are added to individual packages, the use of objects in a class will determine the dependencies. For example, dependency arrows can also be determined once the classes are assigned to packages. The order-entry subsystem obviously needs to access the Customer class; thus, it is dependent on the customer maintenance package.

In summary, package diagrams are used to show related components and dependencies. Generally, we use package diagrams to relate classes or other system components such as network nodes. The preceding figures show two uses of package diagrams—to divide a system into subsystems and to show nesting within packages.

IMPLEMENTATION ISSUES FOR THREE-LAYER DESIGN

Using design class diagrams, interaction diagrams, and package diagrams, programmers can begin to build the components of a system. So, implementation in this sense means constructing the system by programming with a language such as Java or VB. NET. Over the last few years, very powerful integrated development environment (IDE) tools have been developed to help programmers construct systems. Such tools as Jbuilder and Eclipse for Java, Visual Studio for Visual Basic and C#, and C++Builder for C++ provide a very high level of programming support, especially in the construction of the view layer classes—the windows and window components of a system. Unfortunately, these same tools have also propagated some bad programming habits in some developers.

The ease with which programmers can build the graphical user interface windows, and automatically insert programming code, has allowed some programmers to put all of the code in the windows. Each windows component has several associated events where code can be inserted. So, some programmers find it easy to build a window with an IDE tool, let the tool automatically generate the class definition, and merely insert business logic code. No new classes need to be defined, and very little other coding is required. Many of these tools also have database engines, so the entire system can be built only with windows classes. Taking such shortcuts exacts a price later on, however.

The problem with this approach is the difficulty of maintaining the system. Code snippets scattered throughout the graphical user interface classes are very hard to find and maintain. Plus, when the user-interface classes need to be upgraded, the programmer must also find and update the business logic. If a network-based system needs to be enhanced to include a Web front end, then a programmer must rebuild nearly the entire system. Or, if two user interfaces are desired, then all of the business logic is programmed twice. Finally, without the tool that generates the code, it is almost impossible to keep the system current. This problem is exacerbated by new releases of the IDE tools, which may not be compatible with earlier versions. Many programmers have had to completely rewrite the front end of a system because the new release of the IDE tool does not generate code the same way the previous release did. So, we advise would-be analysts and programmers to use good design principles in the development of new systems.

Based on the design principle of object responsibility, it is possible to define what program responsibilities belong to each layer. If you follow these guidelines in writing code, a new system will be much more easily maintained throughout its lifetime. Let's summarize the primary responsibilities of each layer.

View layer classes should have programming logic to:

- Display electronic forms and reports
- Capture input, such events as clicks, rollovers, and key entry
- Display data fields
- Accept input data
- Edit and validate input data
- Forward input data to the domain layer classes
- Start up and shut down the system

Domain layer classes should have the following responsibilities:

- Create problem domain (persistent) classes
- Process all business rules with appropriate logic
- Prepare persistent classes for storage to the database

Data access layer classes should include the following:

- Establish and maintain connections to the database
- Contain all Structured Query Language (SQL) statements
- Process result sets (the results of SQL executions) into appropriate domain objects
- Disconnect gracefully from the database

SUMMARY

Object-oriented design is the bridge between the user requirements, as expressed in the analysis models, and the final system, as constructed using a programming language. Systems design in the object-oriented approach is a highly technical activity that transforms the analysis models into a set of blueprints from which programmers can write code.

Design is driven by use cases. That is, design is done on a use case–by–use case basis. The two primary models that are developed during design are design class diagrams and sequence diagrams. Domain models—domain class diagrams—are transformed into design class diagrams by the addition of attribute type and visibility information and by the addition of method signatures. Sequence diagrams are extensions of system sequence diagrams and are developed by deriving the internal processing required to carry out a use case. A sequence diagram specifies the objects that collaborate and the way they collaborate, specifying the messages they send to each other to complete the processing for a use case.

To derive the correct set of messages and ensure a good design, certain object-oriented design principles must be applied. These principles include encapsulation, coupling, cohesion, navigation, and object responsibilities. Encapsulation is a standard OO principle that ensures that the data fields are placed in the correct classes and that there are sufficient methods to process that data. Coupling is a principle that applies to the entire set of classes and refers to the amount of connectivity between classes. Less connectivity is better when constructing a good design. Cohesion refers to the nature of an individual class. It is an expression of the focus of the class. A class that has methods that perform many different and disparate processes is said not to be cohesive—in other words, it is out of focus. Navigation refers to which classes have access or visibility to which other classes. A system with too many navigation links has too much coupling. Finally, object responsibility is a principle to help determine which classes should be receiving and sending which messages. Objects should be responsible for themselves, but they also might have other responsibilities. Those responsibilities should be carefully assigned so that the system maintains low coupling and so that the classes are cohesive.

An approach to designing systems that are maintainable is to design using three-layer design. Three-layer, or even multilayer, designs partition classes into groups based on their primary focus or responsibility. In the chapter you learned about three-layer design, which is made up of a view layer, consisting of those classes in the graphical user interface; the domain layer, consisting of business classes; and the data access layer, consisting of classes that access the database. A three-layer design is a very robust and flexible design for a system.

In the chapter, you also learned that design can be accomplished through the use of collaboration diagrams. Collaboration diagrams are an alternative to sequence diagrams. Whether you use sequence diagrams or collaboration diagrams is primarily a matter of personal preference.

Finally, a new type of notation was introduced for grouping components, particularly classes, together into related items. A package diagram, which is denoted by a tabbed rectangle with the grouped classes shown inside the rectangle, is analogous to taking items and putting them into a package or container. One way that a package diagram can be used is to group classes by subsystem. Another use is to show which classes belong to the view layer, which to the domain layer, and which to the data access layer.

KEY TERMS

activation lifelines, p. 405

artifact, p. 400

boundary class, p. 390

class-level method, p. 393

cohesion, p. 395

control class, p. 390

coupling, p. 394

CRC (class-responsibility-collaboration) cards, p. 400

data access class, p. 391

dependency relationship. p. 428

encapsulation, p. 393

entity class, p. 390

information hiding, p. 393

instantiation, p. 384

links, p. 421

method signature, p. 391

navigation visibility, p. 393

object responsibility, p. 399

object reuse, p. 393

overloaded method, p. 392

persistent class, p. 390

realization of use cases, p. 388

separation of responsibility, p. 395

stereotype, p. 389

use case controllers, p. 400

visibility, p. 391

REVIEW QUESTIONS

1. Which three models are most used to do object-oriented design?

2. Why do we say that design is "use case driven"?

3. There are four icons, or shortcuts, that can be used to depict different types of classes. List the four icons, tell what each means, and show the symbol for it.

4. List the elements included in a method signature. Give an example of a method signature with all elements listed correctly.

5. What is the notation used to indicate a stereotype? Show an example of a stereotyped class.

6. What is meant by navigation visibility? How is it shown in UML? How is it implemented in programming code?

7. What does coupling mean? Why is too much coupling considered bad?

8. What are some of the problems that occur from classes with low cohesion?

9. What is meant by *object responsibility*? Why is it such an important concept in design?

10. What is the objective of a use case controller class? What design principles does it typify?

11. What is three-layer design? What are the most common layers found in three-layer design?

12. Why is three-layer design a good principle?

13. What is the recommended way to carry out three-layer design? In other words, in what order are the layers designed?

14. To develop the first-cut sequence diagram, you should follow three steps. Briefly describe each of those three steps.

15. Briefly describe the major differences between a sequence diagram and a collaboration diagram.

16. Describe the message notation used on a collaboration diagram.

17. What is the purpose of a package diagram? What notation is used? Show an example.

18. How is dependency indicated on a package diagram? What does it mean?

19. List the primary responsibilities of classes in the view layer. Now do the same for the domain layer and the data access layer.

20. What is the difference between an Internet-based system and a network-based system?

THINKING CRITICALLY

Note: Exercises 1, 2, 3, and 4 are based on the solutions you developed in Chapter 7 for "Thinking Critically" exercises 1 and 2 based on the university library system. Alternatively, your teacher may provide you with a use case diagram and a class diagram.

1. Figure 11-33 is a system sequence diagram for the use case *Check out books* in the university library system. Do the following:

   a. Develop a first-cut sequence diagram, which only includes the actor and problem domain classes.

   b. Add the view layer classes and the data access classes to your diagram from part a.

   c. Develop a design class diagram based on the domain class diagram and the results of parts a and b.

   d. Develop a package diagram showing a three-layer solution with view layer, domain layer, and data access layer packages.

FIGURE *11-33*

System sequence diagram for
Check out books.

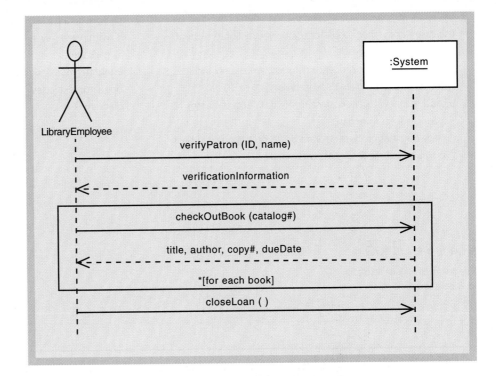

2. Figure 11-34 is an activity diagram for the use case *Return books* for the university library system. Do the following:

   a. Develop a first-cut sequence diagram, which only includes the actor and problem domain classes.

   b. Add the view layer classes and the data access classes to your diagram from part a.

   c. Develop a design class diagram based on the domain class diagram and the results of parts a and b.

   d. Develop a package diagram showing a three-layer solution with view layer, domain layer, and data access layer packages.

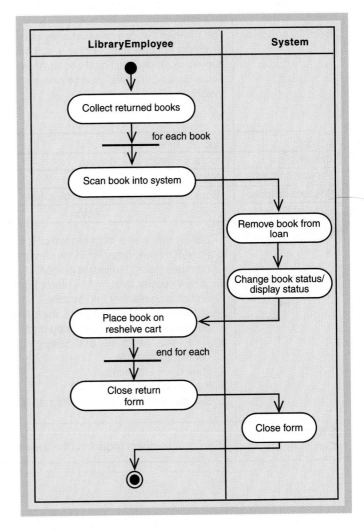

FIGURE *11-34*

Activity diagram for *Return books.*

3. Figure 11-35 is a fully developed use case description for the use case *Receive new book* for the university library system. Do the following:

a. Develop a first-cut sequence diagram, which only includes the actor and problem domain classes.

b. Add the view layer classes and the data access classes to your diagram from part a.

c. Develop a design class diagram based on the domain class diagram and the results of parts a and b.

d. Develop a package diagram showing a three-layer solution with view layer, domain layer, and data access layer packages.

4. Integrate your design class diagram solutions that you developed for exercises 1, 2, and 3 into a single design class diagram.

FIGURE *11-35*

Fully developed use case description for *Receive new book*.

Use Case Name:	Receive new book	
Scenario:	Receive new book	
Triggering Event:	Book arrives for newly purchased book	
Brief Description:	The librarian decides on purchases of new books and places order (prior to this use case). Shipments of new books arrive. Each new book is assigned a library catalog number. Some books are simply additional copies of existing titles. Some books are new editions of existing titles. Some books are new titles and new physical books. The new book information is added to the system.	
Actors:	Library Employee	
Stakeholders:	Library Employee, Librarian	
Preconditions:	None	
Postconditions:	Book Title exists, Physical Book exists	
Flow of Events:	Actor	System
	1. Collect new books from receipt of shipment. 2. For each book, research book category and catalog numbers. Assign tentative number. 3a. If new copy of existing title, enter book information and catalog number into system. 3b. If new edition of existing title, enter book information, edition information, and catalog number. 3c. If new title, assign general catalog number. Assign book copy number. 4. Mark book with number. 5. Place book on shelving cart. 6. Repeat for each book (back to step 2).	3a.1 Update catalog with new number. Verify that not duplicate. 3b.1 Update catalog with new number. Verify that not duplicate. 3c.1 Verify that catalog number not duplicate.
Exception Conditions:	Duplicate numbers require further research and reassignment of catalog numbers.	

Note: Exercises 5, 6, 7, and 8 are based on the solutions you developed for "Thinking Critically" exercises 3 and 4 in Chapter 7 on the the dental clinic system. Alternatively, your teacher may provide you with a use case diagram and class diagram.

5. Figure 11-36 is a system sequence diagram for the use case *Record dental procedure* in the dental clinic system. Do the following:

a. Develop a first-cut sequence diagram, which only includes the actor and problem domain classes.

b. Add the view layer classes and the data access classes to your diagram from part a.

c. Develop a design class diagram based on the domain class diagram and the results of parts a and b.

d. Develop a package diagram showing a three-layer solution with view layer, domain layer, and data access layer packages.

FIGURE *11-36*

System sequence diagram for *Record dental procedure.*

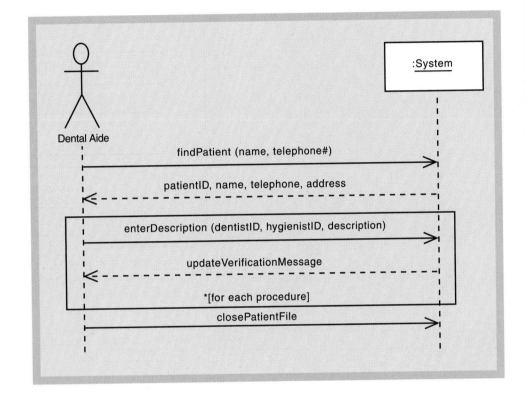

6. Figure 11-37 is an activity diagram for the use case *Enter new patient information* for the dental clinic system. Do the following:

   a. Develop a first-cut sequence diagram, which only includes the actor and problem domain classes.

   b. Add the view layer classes and the data access classes to your diagram from part a.

   c. Develop a design class diagram based on the domain class diagram and the results of parts a and b.

   d. Develop a package diagram showing a three-layer solution with view layer, domain layer, and data access layer packages.

FIGURE *11-37*

Activity diagram for *Enter new patient information*.

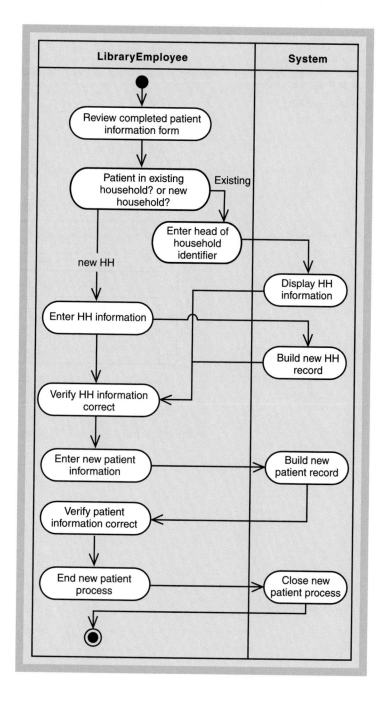

7. Figure 11-38 is a fully developed use case description for the use case *Print invoices* for the dental clinic system. Do the following:

a. Develop a first-cut sequence diagram, which only includes the actor and problem domain classes.

b. Add the view layer classes and the data access classes to your diagram from part a.

c. Develop a design class diagram based on the domain class diagram and the results of parts a and b.

d. Develop a package diagram showing a three-layer solution with view layer, domain layer, and data access layer packages.

8. Integrate your design class diagram solutions that you developed for exercises 5, 6, and 7 into a single design class diagram.

9. In Chapter 7 "Thinking Critically" exercise 6 you developed an activity diagram for each of two scenarios for purchases at Quality Building Supply. In exercise 9, you developed a system sequence diagram. Based on either your activity diagram or your system sequence diagram, and the following list of classes from the domain class diagram, develop a detailed collaboration diagram. Include only problem domain classes.

10. In Chapter 7 "Thinking Critically" exercises 7, 8, and 10, you developed a system sequence diagram for the *Add a new vehicle to an existing policy*. You were also provided a list of classes. Based on the SSD you created, develop a detailed collaboration diagram. Include only problem domain classes.

Use Case Name:	Print invoices	
Scenario:	Print invoices	
Triggering Event:	At the end of the month, invoices are printed	
Brief Description:	The billing clerk manually checks to see that all procedures have been collected. The clerk spot-checks, using the written records to make sure procedures have been entered by viewing them with the system. The clerk also makes sure all payments have been entered. Finally, he/she prints the invoice reports.	
Actors:	Billing clerk	
Stakeholders:	Billing clerk, Dentist	
Preconditions:	Patient records must exist, procedures must exist	
Postconditions:	Patient records are updated with last billing date	
Flow of Events:	Actor	System
	1. Collect all written notes about procedures completed this month. 2. View several patients to verify that procedure information has all been entered. 3. Review log of payments received and verify that payments have been entered. 4. Enter month-end date and request invoices. 5. Verify invoices are correct. 6. Close invoice print process.	2.1 Display patient information, including procedure records. 3.1 Display patient information including account balance and last payment transactions. 4.1 Review every patient record. Find unpaid procedures. List on report as aged or current. Calculate and break down by copay and insurance pay.
Exception Conditions:	None	

FIGURE *11-38*

Fully developed use case
description for *Print invoices.*

EXPERIENTIAL EXERCISES

1. Find a local company that is using UML and object-oriented development. Set up an interview with a member of the IS staff. Find out how they use UML. Ask about using domain models to do analysis. Find out if they use sequence diagrams and how they actually carry out the design of new systems. Also ask about the SDLC that they use. Is it an iterative approach? How closely does it follow the UP?

2. Find a system that was developed using Java. If possible, find one that has both an Internet user interface and a network-based user interface. Is it multilayer—three layer or two layer? Can you identify the view layer classes, the domain layer classes, and the data access layer classes?

3. Find a system that was developed using Visual Studio .NET (or Visual Basic). If possible, find one that has both an Internet user interface and a network-based user interface. Is it multilayer? Where is the business logic? Can you identify the view layer classes, the domain layer classes, and the data access layer classes?

4. Pick an object-oriented programming language that you are familiar with. Find a programming integrated development environment (IDE) tool that supports that language. Test out its reverse-engineering capabilities to generate UML class diagrams from existing code. Evaluate how well it does and how easy the models are to use. Does it have any capability to input UML diagrams and generate skeletal class definitions? Write a report on how it works and what UML models it can generate.

CASE STUDIES

The Real Estate Multiple Listing Service System

In Chapter 7, you developed a use case diagram, a class diagram, and a system sequence diagram for the real estate company's use cases. Based on those solutions, or others provided by your teacher, develop a first-cut sequence diagram for the problem domain classes. Next, add view layer and data access layer objects to the sequence diagram. Convert the domain class diagram to a design class diagram by typing the attributes and adding method signatures.

The State Patrol Ticket Processing System

In Chapter 7 you developed a use case diagram, a class diagram, and a system sequence diagram for the use cases *Recording a traffic ticket* and *Scheduling a court date*. Based on those solutions, or others provided by your teacher, develop a first-cut sequence diagram for the problem domain classes. Next, add view layer and data access layer objects to the sequence diagram. Convert the domain class diagram to a design class diagram by typing the attributes and adding method signatures.

The DownTown Videos Rental System

In Chapter 7, you developed a use case diagram, a class diagram, and system sequence diagrams for the use cases *Rent movies* and *Return movies*. Based on those solutions, or others provided by your teacher, develop a first-cut collaboration diagram for the problem domain classes. Next, add view layer and data access layer objects to the collaboration diagram. Convert the domain class diagram to a design class diagram by typing the attributes and adding method signatures.

TheEyesHaveIt.com Book Exchange

In Chapter 7, you developed a use case diagram, a class diagram, and a system sequence diagram for the use cases *Add a seller* and *Record a book order*. Based on those solutions, or others provided by your teacher, develop a first-cut collaboration diagram for the problem domain classes. Next, add view layer and data access layer objects to the collaboration diagram. Convert the domain class diagram to a design class diagram by typing the attributes and adding method signatures.

Rethinking Rocky Mountain Outfitters

This chapter presented the solutions for two use cases for RMO—*Look up item availability* and *Create new order* for the telephone order scenario. Design three-layer solutions for two more use cases, *Create order return* and *Record order fulfillment*. Update the design class diagram for the problem domain classes with method signatures from these use case designs. Often, the sequence diagram to produce a report can be quite interesting. Do a three-layer design for the use case *Produce order fulfillment report*. Since you do not have detailed user requirements for this use case, you will need to first lay out a sample fulfillment report.

Focusing on Reliable Pharmaceutical Service

In Chapter 7, you developed a use case diagram, a domain model class diagram, and detailed documentation for three use cases. In your detailed documentation, you generated a fully developed specification and a system sequence diagram. Based on that information and the guidelines in this chapter, design a three-layer architecture for each of those three use cases. Update the design class diagram with attribute type information and method signatures derived from the sequence diagrams.

FURTHER RESOURCES

Grady Booch, James Rumbaugh, and Ivar Jacobson. *The Unified Modeling Language User Guide*. Addison-Wesley, 1999.

Frank Buschmann, R. Meunier, H. Rohnert, P. Sommerlad, and M. Stal. *Pattern-Oriented Software Architecture: A System of Patterns*. John Wiley and Sons, 1996.

E. Reed Doke, J. W. Satzinger, and S. R. Williams. *Object-oriented Application Development Using Java*. Course Technology, 2002.

E. Reed Doke, J. W. Satzinger, and S. R. Williams. *Object-oriented Application Development Using Microsoft Visual Basic .NET*. Course Technology, 2003.

Ivar Jacobson, Grady Booch, and James Rumbaugh. *The Unified Software Development Process*. Addison-Wesley, 1999.

Philippe Kruchten. *The Rational Unified Process, An Introduction*. Addison-Wesley, 2000.

Craig Larman. *Applying UML and Patterns: An Introduction to Object-oriented Analysis and Design and the Unified Process*, 2nd edition. Prentice-Hall, 2002.

James Rumbaugh, Ivar Jacobsen, and Grady Booch. *The Unified Modeling Language Reference Manual*. Addison-Wesley, 1999.

CHAPTER 12

Advanced Topics in Object-Oriented Design

LEARNING OBJECTIVES

After reading this chapter you should be able to:

- Explain the importance of design patterns in object-oriented design

- Briefly describe and use the singleton, adapter, and observer design patterns

- Explain why enterprise-level systems require special design considerations

- Apply UML physical design notation to Web-based systems

- Explain how Web services work to obtain information on the Internet

- Explain how statecharts can be used to describe system behaviors

- Use statecharts to model object and system behaviors

CHAPTER OUTLINE

Design Principles and Design Patterns

Designing Enterprise-Level Systems

Modeling System and Object Behavior

The integrated customer account system project for New Capital Bank was now several months old. The project was proceeding steadily; it was well past the elaboration iterations and into construction iterations. Most of the team's work now consisted of developing detailed designs and implementing many of the more advanced features of the new system. Bill Santora, the project leader, was discussing some of the system's technical details with one of his team leaders, Charlie Hensen.

"Are you sure you have covered all the security issues with the Web interface for the system?" Bill asked. "Senior managers, including our boss, want to be sure we don't have any holes where hackers can break in. Building an enterprise-level system with both a network front end and a Web interface magnifies the security issues—they just seem to multiply like rabbits."

Charlie nodded. "I agree completely. The team and I have spent hours brainstorming every possible security risk we could think of. We recognize the danger to the back end of the system if we miss anything."

"Well, I did feel better after we had those long review sessions with the systems support staff. You did a good job of handling all the 'what if' questions they asked you. I especially liked your use of the statechart diagrams to show how the system would allow only certain transactions at certain times. Of course, the tricky part now is to write the code so that it conforms exactly to the diagrams, but the diagrams were a great piece of work."

"Thanks," said Charlie. "I have learned to appreciate the expressive power of statecharts. They really help you think about how the system should behave—what conditions should be allowed and what conditions shouldn't."

Bill continued, "Where did you learn about statecharts, and how did you learn to use them?"

"Remember that night course I took last fall on object-oriented development? We spent a couple of weeks learning how to do statecharts. At the time, I thought they were too complex for any of the stuff I was working on. But when the team got to the Web interface requirements for the system, they just fit the bill. It took a couple of days to get the rest of the team up to speed, but once they caught on, we were really able to look at a lot of different options quickly. The statecharts were invaluable to express the complexities we were facing."

"Well, I have to admit the systems programmers were impressed, not only with your solution but also with the elegance of how you had approached it. Congratulations on a job well done!"

OVERVIEW

In Chapter 11, you learned some fundamental principles of object-oriented design. However, designing, building, and deploying a complete system for an organization requires both broad and deep knowledge in many areas. In your other courses, such as programming, database management, and network administration, you extend your basic knowledge and develop many of those other necessary skills. But your learning does not end with formal schoolwork. You will need to pursue a path of lifelong learning to succeed as a systems developer.

This chapter provides you with a broader understanding of some important issues of systems design. It elaborates on some of the fundamental principles presented in Chapter 11, then it introduces the following new object-oriented design topics:

- Design principles and design patterns
- Designing enterprise-level systems, including Web-based systems
- Modeling system and object behavior

You learned some basic principles of good design in Chapter 11: object responsibility, coupling, and cohesion. We build on those principles in this chapter, presenting two additional principles to enhance the flexibility of and ease of maintaining a new system. Systems designers should consider protection from variations to ensure that systems are stable and can be updated easily. Another principle of good design, indirection, helps stabilize and protect systems.

Every engineering discipline has a set of standard design rules and templates. For example, civil engineers learn about the strength of triangular support structures. Such design rules are sometimes called *patterns* or *templates* for good design. Even though software development is a young discipline, computer scientists and engineers have now begun to identify and categorize good design patterns. These patterns are an attempt to establish best practices in the construction of software systems. We introduce some basic design patterns and present some examples of the most important patterns.

Enterprise-level systems must support large distributed and networked organizations. Many pieces of an enterprise system can be purchased and must be considered part of the environment of design. We introduce various deployment environments and note what issues must be considered in designing for these environments, whether they are client-server systems run on a network or Internet-based systems. To design for Web-based systems, designers must consider a few additional design principles. We extend the design principles and diagrams from Chapter 11, modifying them to enable you to design for Web-based systems.

Many objects in a system have complex behavior, such as user-interface objects or network objects. However, sometimes designers may need to track the status of even simple business objects in a system. For example, it might be important to know that an order is complete and ready to be shipped. The Unified Modeling Language (UML) includes a type of model, called a *statechart diagram*, which is very useful in capturing this information about objects—and about the system as a whole. We introduce you to statecharts and to their many uses in this chapter.

DESIGN PRINCIPLES AND DESIGN PATTERNS

In Chapter 11, as you learned the fundamental concepts of design, you were introduced to several principles of good design. One principle was the idea of object responsibility. As a designer, you must identify what the responsibilities of each class are. One basic object responsibility is that an object is responsible for maintaining its own attributes, including getting information from other objects, when needed, to fill in values. Another general object responsibility is to create objects that are dependents of it. Yet another general responsibility may be to be an expert on certain sets of information. An object can be an expert either by having the needed information itself or by having navigation visibility and access to the other objects that do have the information. The principle of object responsibility is one of the most important concepts that will help you to design logical, easily maintained systems.

Two other related principles that you learned were coupling and cohesion. A cohesive class is one whose set of responsibilities, and methods, are all closely related. The functions that are assigned to the class, and that are executed through the methods, are focused. Coupling, on the other hand, relates to the idea of links between classes. Coupling must always be considered from the perspective of an entire system. The objective of this principle is to couple the classes that need to work together, but not to have excessive links between classes. You will see the reason for this in the next section, when you learn about a principle called *protection from variations*.

Additional Design Principles

We have stated many times that developing software systems is difficult and complex. Historically, developers have not routinely been able to develop systems that work correctly and solve the right business problem. In addition, every system requires constant modification and upgrading. Perhaps functions that were not added in the first version need to be added later or errors need to be fixed. Or maybe the business requirements have changed, and the system needs to be upgraded. The cost of maintaining a system over its lifetime can be many times the cost of the original development. Not only are systems that are based on good design principles easier to develop and put into operation the first time, but they are also much easier to maintain. The following principles enhance the flexibility, ease of maintenance, and extensibility of a new system.

Protection from Variations

One of the underlying principles of good design is protection from variations. *Protection from variations* is a principle that says that parts of a system that are unlikely to change should be segregated (or protected from) those that will need to be changed. In other words, as you design systems, you should try to isolate the parts of the system that are likely to change from those that are more stable. You will notice in following sections that many design patterns are solutions to problems of changes in a system.

Protection from variations is one of the principles that drives the multilayer design pattern. Designers could mix all of the user-interface logic and business logic together in the same classes. In fact, in early user-oriented, event-driven systems, the business logic was put right in the view layer classes, such as in the windows input forms. For example, early versions of Visual Basic and Powerbuilder were very good at building graphical user interfaces, but they did place all of the business logic in the view layer classes themselves. The problem with this design is that when an interface needed to be updated, all of the business logic also had to be rewritten. A better approach is to decouple the user-interface logic from the business logic. Then, the user interface can be rewritten without affecting the business logic. In other words, the business logic, being more stable, is protected from variations in the user interface.

Also, what if the updates to the business required adding new classes and new methods? If the user-interface classes were tightly coupled to the business classes, then there could be a ripple effect of changes throughout the user-interface classes. However, since the user interface can simply send all of its input messages to the use case controller class, as discussed in Chapter 11, changes to the methods or classes in the business logic and domain layer are isolated to the controller class. As we describe design patterns in a later section, you will find that this principle affects almost every design decision. You should watch for and recognize the application of this principle in all design activities.

Indirection

Indirection is a frequently used object-oriented design principle that is used to protect stable components from variations and to reduce coupling. *Indirection* is a principle of decoupling two classes, or other system components, by placing an intermediate class between them to serve as a link. In other words, instructions don't go directly from A to B; they are sent through C first. Or in message terminology, don't send a message from A to B. Let A send the message to C, and then let C forward it to B.

Although there are many ways to implement protection from variations, indirection is one method that is frequently used. By inserting an intermediate object, any variations in one system can be isolated in that intermediate object. Indirection is also a principle that applies to many corporate security systems. Many companies have firewalls and proxy servers that receive and send messages from an internal network to the Internet. A proxy server appears as a real server, ready to receive messages such as e-mail and HTML page requests. However, it is a fake server, which catches all of the messages and redistributes them to the recipients. This step of indirection allows security controls to be put in place to protect the system.

Importance of Design Patterns

Templates and patterns are used repeatedly in everyday life. A chef uses a recipe, which is just another word for a pattern, to combine ingredients into a flavorful dish. A tailor uses a pattern to cut fabric for a great-fitting suit. Engineers take standard components and combine them into established configurations, or set patterns, to build buildings, stereos, and thousands of other products. Patterns are created to solve problems. Over time, and with many attempts, those working on a particular problem develop a set solution to the problem. The solution is general enough so that it can be applied over and over again. As time passes, the solution is documented and published and eventually becomes accepted as the standard.

In Chapter 11 we briefly introduced you to two standard design templates—use case controller and three-layer design. The use of standard design templates has become very popular among software developers because they can speed OO design work. The formal name for these templates is *design patterns*. Design patterns became a widely accepted object-oriented design technique in 1996, with the publication of the book *Elements of Reusable Object-Oriented Software* by Eric Gamma, Richard Helm, Ralph Johnson, and John Vlissides. The four authors of this book are now referred to as the Gang of Four (GoF). As you become better informed about design patterns, you will often see references to a design pattern as a GoF pattern. In the original book, the authors identified 23 basic design patterns. Today, scores of patterns have been defined—from low-level programming patterns, to mid-level architectural patterns, to high-level enterprise patterns. The two primary enterprise platforms, Java and .NET both have sets of enterprise patterns, which are described in various books and publications.

design patterns
templates used to speed OO design

Patterns exist at various levels of abstraction. At a concrete level, a pattern may be a class definition that is written in code to be used by any developer. At the most abstract level, a pattern might only be an approach to solving a problem. For example, the multilayer design pattern tends to be more abstract, stating that it is better to separate system functions into three layers of classes: the graphical user interface logic is placed in a set of view-layer classes that are separate and distinct from the domain layer and data access layer. So, multilayer design is an approach to building a system rather than a specific solution.

The use case controller pattern is more concrete. That pattern defines a specific class or classes that act as the switchboard for all incoming messages from the environment. As with all patterns, there are multiple ways to implement the controller pattern. A single controller class can be defined to handle all messages from the view layer to the domain layer. Alternatively, a class can be defined for each use case. Or some combination of the two can be used. Regardless of the specific approach, the controller pattern does require a separate, specified class.

The specification for a pattern should contain at least five main elements:

1. The pattern name
2. The problem that requires a solution
3. The solution or explanation of the pattern
4. Examples of the pattern
5. The benefits and consequences of the pattern

We show these five elements for the controller pattern in Figure 12-1.

Basic Design Patterns

The study of design patterns is an ongoing effort. Numerous books list hundreds of design patterns. The objective of this section is to introduce you to a few of the more basic patterns so that you will have a foundation from which to become more knowledgeable about design patterns as you mature in your career.

Name:	Controller
Problem:	Domain classes have the responsibility of processing use cases. However, since there can be many domain classes, which one(s) should be responsible for receiving the input messages? User-interface classes become very complex if they have visibility to all of the domain classes. How can the coupling between the user-interface classes and the domain classes be reduced?
Solution:	Assign the responsibility for receiving input messages to a class that receives all input messages and acts as a switchboard to forward them to the correct domain class. There are several ways to implement this solution: (a) Have a single class that represents the entire system, or (b) Have a class for each use case or related group of use cases to act as a use case handler.
Example:	The RMO order-entry subsystem accepts inputs from an OrderWindow. These input messages are passed to an OrderHandler, which acts as the switchboard to forward the message to the correct problem domain class. 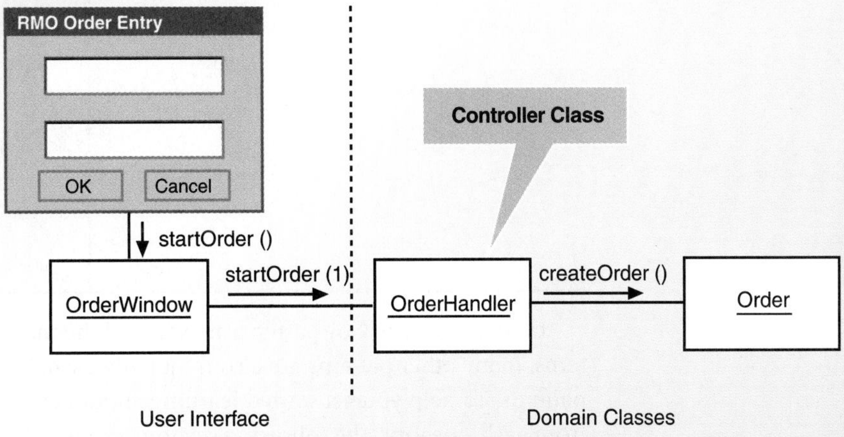 Other examples of the controller can be found for each RMO subsystem.
Benefits and Consequences:	Coupling between the view layer and the domain layer is reduced. The controller provides a layer of indirection. The controller is closely coupled to many domain classes. If care is not taken, controller classes can become incoherent, with too many unrelated functions. If care is not taken, business logic will be inserted into the controller class.

FIGURE *12-1*

Pattern description for the controller pattern.

In their pioneering work, the GoF developed a basic classification scheme for patterns. Figure 12-2 lists most of the original patterns, showing their classifications. The rows of the table identify the scope of the pattern—whether the design is a class-level or object-level pattern. Class-level patterns define solutions such as abstract classes that apply to static methods or that do not actually instantiate objects. Object-level patterns apply when the implementation of the pattern results in specific objects being instantiated from classes. The columns of the table classify the patterns as creational, structural, or behavioral. Creational patterns are patterns that help assign responsibilities to classes to instantiate new objects. Structural patterns provide solutions to meet the architectural needs of the system, that is, the set of classes and the ways they are related. Structural patterns help solve problems associated with indirection. Behavioral patterns provide solutions to problems related to the way internal system processes execute. For example, the iterator pattern provides a solution to the problem of how to process arrays and lists effectively.

F I G U R E *12-2*

Classification of design patterns.

SCOPE OF PATTERN	TYPE OF PATTERN		
	Creational	Structural	Behavioral
Class-level patterns	Factory Method	Adapter	Interpreter
			Template Method
Object-level patterns	Abstract Factory	Adapter	Chain of Responsibility
	Builder	Bridge	Command
	Prototype	Composite	Iterator
	Singleton	Decorator	Mediator
		Façade	Memento
		Proxy	Flyweight
			Observer
			State
			Strategy
			Visitor

Even though the GoF patterns are some of the most fundamental and important patterns, many other patterns are also frequently used. We will limit our discussion to GoF patterns. To help you get started learning about design patterns, we present one pattern from each category. The following sections explain the singleton, adaptor, and observer patterns.

Singleton

Some classes must have exactly one instance—for example, a class that starts up the system or the main window class. These classes have only one instance, but since they are instantiated from only one place, it is a simple matter to limit the logic to create only one object.

Other classes must have exactly one instance but cannot be easily controlled by having only one place to invoke the constructor. Usually, these classes are service classes that manage a system resource, such as a database connection. They are usually invoked by many other classes and from many locations throughout the system. This common problem has a standard solution: the singleton pattern.

Figure 12-3 presents the template of the pattern description for the singleton pattern. You should carefully read the contents of Figure 12-3, especially the example section to ensure that you understand how it works. The singleton pattern provides a solution in which the class itself controls the creation of only one instance.

The approach of the singleton solution is that the class has a static variable that refers to the object that is created. A method, such as getInstance, is defined that is used to get the reference to the object. The first time the getInstance method is called, it instantiates an object and returns a reference to it. On later calls to the method, it simply

Name:	Singleton
Problem:	Only one instantiation of a class is allowed. The instantiation (new) can be called from several places in the system. The first reference should make a new instance, and later attempts should return a reference to the already instantiated object. How do you define a class so that only one instance is ever created?
Solution:	A singleton class has a static variable that refers to the one instance of itself. All constructors to the class are private and are accessed through a method or methods, such as getInstance(). The getInstance() method checks the variable; if it is null, the constructor is called. If it is not null, then only the reference to the object is returned.

Example:	In RMO's system, the connection to the database is made through a class called Connection. However, for efficiency, we want each desktop system to open and connect to the database only once, and to do so as late as possible. Only one instance of Connection, that is, only one connection to the database, is desired. The Connection class is coded as a singleton. The following coding example is similar to C# and Java.
	
	``
	Class Connection
	{
	private static Connection conn = null;
	public synchronized static getConnection ( )
	      {
	        if (conn = = null) {
	            conn = new Connection ( ) ;}
	        return conn;
	      }
	}
	``
	
	Another example of a singleton pattern is a utilities class that provides services for the system, such as a factory pattern. Since the services are for the entire system, it causes confusion if multiple classes provide the same services.
	
	An additional example might be a class that plays audio clips. Since only one audio clip should be played at one time, the audio clip manager will control that. However, to do so, there must be only one instance of the audio clip manager.

Benefits and Consequences:	There are other times when only one instance of an object is needed, but if it is instantiated from only one place, then a singleton may not be required. The singleton object controls itself and ensures that only one instance is created—no matter how many times it is called and wherever the call occurs in the system. The code to implement the singleton is very simple, which is one of the desirable characteristics of a good design pattern.

FIGURE *12-3*

Singleton pattern template.

returns a reference to the already instantiated object. As shown in the figure, the code is simple and elegant. The example does not show the constructor; however, to ensure that only one instance is created, all constructors are specified as private—not accessible—so that no other class can accidentally invoke one.

In the singleton template, the pattern is represented by code. To specify this in your design, you should stereotype the class as a «singleton». Good programmers will recognize the stereotype and know exactly how to code that class.

Adapter

The basic idea behind the adapter pattern is the same as an electrical adapter that you use if you travel internationally. Say that you are traveling to the United Kingdom, and you want to take your hair dryer with you. Your hair dryer has a switch for either 110 volts or 220 volts, so you think you can run it on either voltage. However, the plug on the end of the power cord has two flat prongs. Unfortunately, in the United Kingdom, the wall plugs have three large prongs set at right angles. What you need is an adapter. You need something that can adapt the power cord's two prongs to the wall's three angled prongs. Figure 12-4 shows a typical electrical adapter you might use.

FIGURE *12-4*

Electrical adapter.

The adapter design pattern works just like the electrical adapter; it plugs an external class into an existing system. The method signatures on the external class are different from the method names that are being called from within the system, so the adapter class is inserted to convert the method calls from within the system to the method names in the external class.

This pattern is a standard solution for protection from variations. The external class could be a variable class—that is, it could be replaced at any time by an upgrade or an entirely different class. This situation often occurs when commercial software libraries are purchased to provide special services. For example, in an organization's internal payroll system, designers might purchase a set of classes to calculate all of the complex income tax deductions. Since tax deductions are a very specialized area, and since most companies do not have tax specialists on their IT staff, an easy solution is simply to purchase those classes. However, knowing that those classes could be replaced at a later time if the tax law changes, a designer would be wise to create an adapter class between the system and the tax calculation classes.

Figure 12-5 describes the details of the adapter design pattern. Be sure to read it carefully. In the sample diagram, there are four UML classes. The one labeled System represents the entire system. The classes within the system use method names such as getSTax ( ) and getUTax ( ) to access the tax routines. The tax calculator class has method names of findTax1 ( ) and findTax2 ( ). The two UML classes in the middle represent the adapter. The top class symbol is an interface class. An interface is useful to specify the names of the methods. Although not absolutely necessary, it is a simple way to specify and enforce the use of the correct method names. The adapter class then inherits those method names and provides the method logic for those methods. The body of each method simply extends a call to the final method name of findTax1 ( ) or findTax2 ( ). In other words, it "adapts," or translates, the method names from one to the other.

As you become familiar with this design pattern, you will find a multitude of uses for it. It is a very powerful and elegant solution to make a system more maintainable. Experienced developers use this pattern frequently, not only for foreign classes but also for classes that are written internally that may need to be upgraded often. It is an excellent way to insulate the system from frequently changing classes.

Observer

The observer pattern is a powerful approach to solving an interesting system problem. In fact, this solution is more than just a pattern. It is an entire approach to a particular problem and has been around for a long time. The observer pattern has been used with all types of system development, even before object-oriented techniques were used. It has several names—observer, listener, publish/subscribe. Sometimes you may even hear developers refer to it as the callback technique.

Name:	Adapter
Problem:	A class must be replaced, or is subject to being replaced, by another standard or purchased class. The replacing class already has a predefined set of method signatures that are different from the method signatures of the original class. How do you link in the new class with a minimum of impact so that you don't have to change the names throughout the system to the method names in the new class?
Solution:	Write a new class, the adapter class, which serves as a link between the original system and the class to be replaced. This class has method signatures that are the same as those of the original class (and the same as those expected by the system). Each method then calls the correct desired method in the replacement class with the method signature. In essence, it "adapts" the replacement class so that it looks like the original class.
Example:	There are several places in the RMO system where class libraries were purchased to provide special processing. These purchased libraries provide specialized services such as tax calculations and shipping and postage rates. From time to time, these service libraries are updated with new versions. Sometimes a service library is even replaced with one from an entirely different vendor. The RMO systems staff apply protection from variations and indirection design principles by placing an adapter in front of each replaceable class. 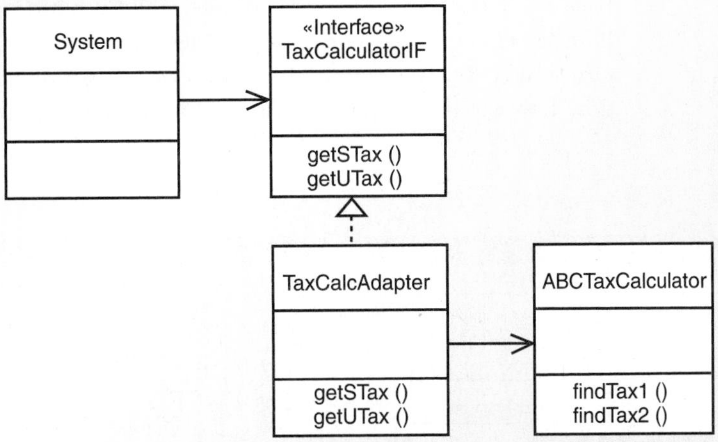
Benefits and Consequences:	The adaptee class can be replaced as desired, Changes are confined to the adapter class and do not ripple through the system. Two classes are defined, an interface class and the adapter class. Passed parameters may add more complexity, and it is difficult to limit changes to the adapter class.

FIGURE *12-5*

Adapter pattern template.

Let's first describe a scenario to illustrate the problem. As discussed in Chapter 11, generally the view layer classes have navigation visibility to the problem domain layer classes. In other words, the windows classes know about the domain layer classes, such as Customer and Order, and can send messages to those internal classes. However, it is better when developing the design if the domain layer classes are not coupled to the view layer classes. So, even though a Customer window can send a message to a Customer object, the Customer object should not have navigation visibility to the Customer window and should not be able to send a message to it.

In this hypothetical example, let's use three classes, a Customer window, an Order window, and an Order class. The first two are windows classes and part of the view layer. The Order class is a domain layer class. The use case is *Create new order*, which you learned about in Chapter 11. When the order was first created, navigation visibility was provided from both the Customer window and the Order window to the Order class. Figure 12-6 illustrates the three classes, with two windows and the Order class. The navigation arrows show how messages can be sent.

Notice on the Customer window that there are fields called Past Purchases, Current Order Amount, and Year-to-date Purchases. Since Rocky Mountain Outfitters (RMO) provides special discounts to customers when their year-to-date purchases exceed a certain amount, the telephone clerk watches these amounts as the order is being taken to see whether he or she can suggest additional purchases to take advantage of the discounts. As items are added to the order, the Current Order Amount and Year-to-date Purchases fields need to be updated. The data to update these fields is contained in the Order class. But since Order does not have navigation visibility to the Customer window, how can it send information to that window? A novice designer would probably

FIGURE *12-6*

Three classes in the *Create new order* use case.

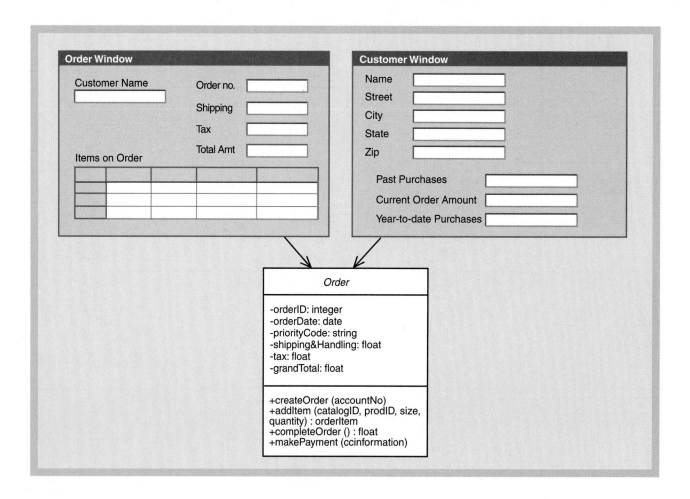

say, "Well let's just put a reference to the Customer window in the Order object." But that violates the coupling principle. Instead, the observer pattern has been developed to handle this problem. Let's see how that solution works.

The concept of the observer, or listener, pattern is to have the Customer window "listen" for any changes to the Order object. When it "hears" of a change in an order, it updates the appropriate fields. For the listening to work, the Order class must contain mechanisms, first to allow other classes to "subscribe" as listeners and second to "publish" the changes to the listeners when they occur.

The Order class must have the following components. First, the Order class must have an array of object references that holds the list of all objects that have subscribed as listeners. The type of the array is an object array. Second, the Order class must have a method, usually something like addOrderListener, which can be called by potential listeners to subscribe. When a class wants to subscribe as a listener, it just calls the addOrderListener method and sends a reference to itself as a parameter. The logic of addOrderListener is simply to add the passed reference parameter to the array. Third, the Order class has a method named *notify*, which iterates through the array and sends a message to each object referenced in the array. Normally, along with the Order class, a designer develops an interface class, which we call OrderEvents, to provide the method signature that will be called by the Order object—often this interface class is called the *publisher*. The listener class, that is, the CustomerWindow, can then inherit from the OrderEvents interface to define the method signatures.

Figure 12-7 illustrates the changes required to implement the observer pattern. The changes have been added in bold. The Order class has a new array variable called orderListeners to hold the subscriber objects. Order has three new methods—one each to add and remove listeners and one to notify OrderListener of any changes. The third method is private, or invoked internally. The interface class OrderListener is used to

FIGURE *12-7*

Implementation for the observer pattern.

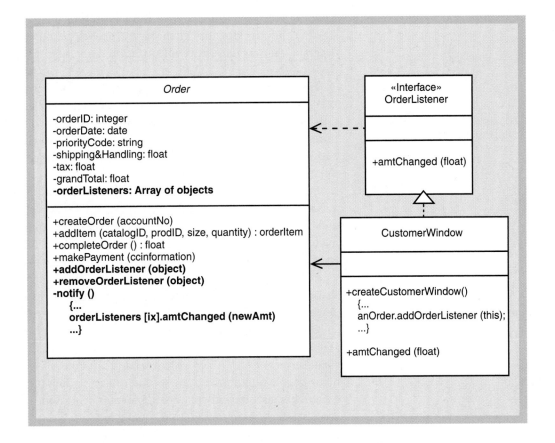

specify the name of the method that will be called when the listener is called back. (This callback approach is why the pattern is sometimes called the *callback pattern*.)

The CustomerWindow class has navigation visibility to the Order object—it is needed to be able to subscribe to it. The CustomerWindow also inherits from the OrderListener interface so that it knows which method to implement. The OrderListener interface is dependent on the Order class. So, if the Order class changes, the OrderListener might need to change, too. The figure also includes some snippets of code to illustrate how the methods are invoked between the classes. You should carefully read the code to ensure that you understand how this works.

Since the link from the Order to the CustomerWindow is dynamic and temporary, it has no negative side effects on the code. In other words, if the CustomerWindow was no longer there, it would not affect the Order class. Or, if multiple windows wanted to listen to the order amount changing, no changes would be required in the Order class. This type of dynamic linking is a powerful and effective technique to avoid permanent coupling in places that could cause problems.

Figure 12-8 is the pattern template for the observer pattern. The listener design pattern is used extensively as the technique to handle event processing by window objects. The class windows libraries for Java and .NET are all implemented using the listener pattern as the fundamental event-handling technique. The explanation in Figure 12-8 illustrates the names used in the window's events. For example, the method on a windows button to add a listener is addListener ( ). The method that is invoked on the subscriber is actionPerformed ( ). By using these standard names, the window's graphical user interface classes become standard classes that can be reused in any application.

DESIGNING ENTERPRISE-LEVEL SYSTEMS

In Chapter 11, and so far in Chapter 12, you have learned the principles, techniques, and patterns of object-oriented design. The general principles apply to all types of systems—whether they are desktop, client-server, or Web-based systems. But systems that are developed to serve an entire organization deserve special consideration, so we cover those topics here.

First, what is an enterprise-level system? The term can mean many different things. We define an *enterprise-level system* as a system that has shared components among multiple people or groups in an organization. In Chapter 11 you learned about two- and three-layer design as a good approach for designing maintainable systems. A multilayer design, however, does not necessarily imply an enterprise-level system. It is just as applicable to a single desktop system.

Enterprise-level systems almost always use multiple tiers of computers. You learned about *n*-layer or *n*-tiered architectures in Chapter 9. A typical example of this architecture is a client-server environment, where the client computers contain the view and domain layer programs, and the data access layer is on a central server. This configuration is a three-layer, two-tiered system. Because the central database is shared across the enterprise, it is placed on a central server that everyone using the application program can share. Since local client computers are often very powerful Macintosh or personal computers, both the view layer and domain logic can be executed locally.

Our definition of an enterprise-level system uses a broad definition for the term. Two major categories of systems fit this definition in relation to systems design: (1) client-server network-based systems and (2) Internet-based systems. You may find that many people only think of the second category when they talk about enterprise-level systems because so much new development is being done for the Web. Remember, however, that the broader definition is equally valid.

enterprise-level system

a system that has shared components among multiple people or groups in an organization

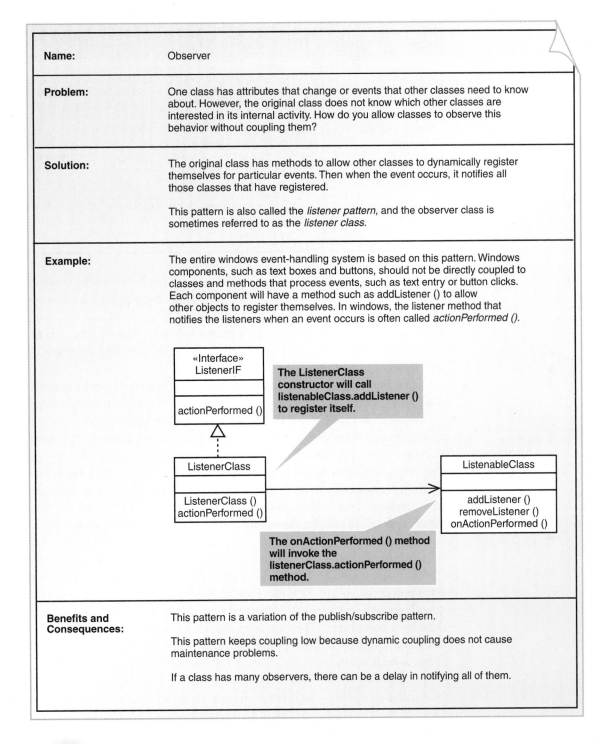

Name:	Observer
Problem:	One class has attributes that change or events that other classes need to know about. However, the original class does not know which other classes are interested in its internal activity. How do you allow classes to observe this behavior without coupling them?
Solution:	The original class has methods to allow other classes to dynamically register themselves for particular events. Then when the event occurs, it notifies all those classes that have registered. This pattern is also called the *listener pattern,* and the observer class is sometimes referred to as the *listener class.*
Example:	The entire windows event-handling system is based on this pattern. Windows components, such as text boxes and buttons, should not be directly coupled to classes and methods that process events, such as text entry or button clicks. Each component will have a method such as addListener () to allow other objects to register themselves. In windows, the listener method that notifies the listeners when an event occurs is often called *actionPerformed ().*

«Interface»
ListenerIF

actionPerformed ()

The ListenerClass constructor will call listenableClass.addListener () to register itself.

ListenerClass

ListenerClass ()
actionPerformed ()

ListenableClass

addListener ()
removeListener ()
onActionPerformed ()

The onActionPerformed () method will invoke the listenerClass.actionPerformed () method.

| **Benefits and Consequences:** | This pattern is a variation of the publish/subscribe pattern.

This pattern keeps coupling low because dynamic coupling does not cause maintenance problems.

If a class has many observers, there can be a delay in notifying all of them. |

FIGURE *12-8*

Observer pattern template.

These two methods of implementing enterprise-level systems have many similar properties. Both require a network. Both have central servers. Both have the view layer on the client machines. However, there are also some fundamental differences in the design and implementation of these two approaches. The primary difference is in how the view layer interacts with the domain and data access layers. It is important for us, as developers, to be able to distinguish between these two types of systems because we must consider important design issues. Figure 12-9 identifies three fundamental differences that affect the architectural design of the system.

FIGURE *12-9*

Differences between client-server
and Internet systems.

Design Issue	Client-Server Network System	Internet System
State	"Stateful" or state-based system, e.g., client-server connection is long term.	Stateless system, e.g., client-server connection is not long term and has no inherent memory.
Client Configuration	Screens and forms that are programmed are displayed directly. Domain layer is often on the client or split between client and server machines.	Screens and forms are displayed only through a browser. They must conform to browser technology.
Server Configuration	Application or data server directly connects to client tier.	Client tier connects indirectly to the application server through a Web server.

The concept of state relates to the permanence of the connection between the client view layer and the server domain layer. If the connection is permanent, as in a client-server system, values in variables can be passed back and forth and are remembered by each component in the system. The view layer has direct access to the data fields in the domain layer. For example, data in an order, such as all of the line items and their prices, are displayed in the forms.

In a stateless system, such as the Internet, the client view layer does not have a permanent connection to the server domain layer. The Internet was designed so that when a client requests a screen via a URL address typed in the browser, the server sends the appropriate document, and then the two disconnect. In other words, the client does not know the state of the server, and the server does not remember the state of the client. This transient connection makes it difficult to implement such things as an order in a shopping cart. To add more permanence to the stateless environment, Web designers have developed other techniques, such as cookies and session variables. As a system designer, you must consider these additional components when designing an Internet enterprise-level system.

Concerning client configuration, the client side of a network-based system contains the view layer classes and often the domain layer classes. Formatting, displaying, and event processing within the screens are all directly controlled by the view layer and domain layer program logic. There is great flexibility in the design and programming of these electronic screens. The view layer classes and domain layer classes can communicate directly with each other. Even if the domain layer is split across tiers, a permanent communication link can be established—all under the program's control.

In an Internet-based system, all electronic screens are displayed by a browser. The formatting, displaying, and event processing all must conform to the capabilities of the browser being used. Special techniques and tools, such as scripting languages and applets, have been developed to simulate the network-based capability. However, as a designer, you will need to design for the environment.

The server configuration in a network-based system consists of data access layer classes and sometimes domain layer classes. These classes collaborate through direct communication and access to each other's public methods. In an Internet-based system, all communications from the client tier must go through the HTTP server. Communication is not direct, and methods and program logic are invoked indirectly through passed parameters. This indirect technique of accessing domain layer logic is more complex and requires additional care in designing the system.

As you can see, at the enterprise level of design, we have started to become concerned about the physical components of the system—how the system is partitioned into executable components. Earlier design discussions focused on logical design. But as we begin identifying physical components, we move toward a system's physical design. Diagrams that depict physical components are called *implementation diagrams*. There are several types of implementation diagrams; however, we will use a specific implementation diagram that shows how the components are deployed across the various tiers. This type of diagram is called a *deployment diagram*. So, a deployment diagram is a type of implementation diagram whose purpose is to show the deployment of various physical components across different locations.

UML has specific notations for implementation diagrams. In the next section, we explain the UML notation for these diagrams. UML does not currently have specific notations for Internet elements, such as Web pages. It does, however, allow for extensions to the existing notation. Therefore, in the last two sections, we introduce an extension to UML that has been developed for depicting physical elements of Internet-based systems.

UML Notation for Deployment Diagrams

The two symbols we use for physical design as shown in deployment diagrams are the component symbol and the node symbol. A component is an executable module or program, and it consists of all the classes that are compiled into a single entity. It has well-defined interfaces, or public methods, that can be accessed by other programs or external devices. The set of all of these public methods that are available to the outside world is called the *application program interface*, or *API*. Figure 12-10 illustrates the UML notation for a component and its interfaces. It consists of a rectangle with two smaller rectangles protruding from one side. It is not necessary to list all of the interfaces on a single component. Only those that are pertinent to the context of the diagram are listed. The name of the component is written inside. Sometimes designers want to emphasize a single instance of a component, rather than a classification. In that situation, the name of the component is underlined.

A node can be thought of as a computer, or a bank of computers, representing a single computing resource. A node is a physical entity at a specific location. Figure 12-11 illustrates the symbol that is used for a node—a shaded rectangle. The shading is added to represent a real object that can cast a shadow. The name of the node is listed inside, either as a type or classification of node, or as a single instance with the name underlined.

FIGURE *12-10*

UML component notation.

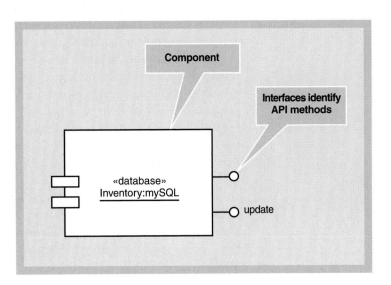

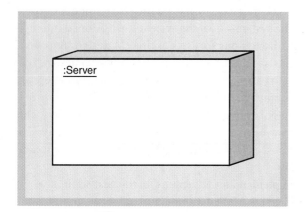

FIGURE *12-11* UML node notation.

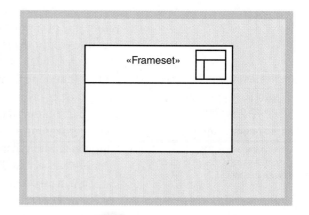

FIGURE *12-12* UML extension for frameset.

The other notation needed for physical design is a symbol to represent a Web page or frameset. A *frameset* is a high-level object that can hold items to be displayed by a browser. We will use a frameset notation and stereotype to indicate a Web page. You can think of a frameset as the window in a browser that can display frames or a set of frames. Figure 12-12 shows the notation for a frameset.

We now apply these physical design notations within deployment diagrams to show various configurations that are applicable for Internet-based systems.

Internet-Based Systems

Many colleges have courses in Web development. Most of those courses fall into two categories: Web site design or Web programming languages. Both classes are important and beneficial for your education as a systems developer.

The Web site design courses usually focus on usability. You learn how to design a set of attractive and effective Web pages to support a business application. Some of the critical issues addressed are graphical design, Web page layout, and navigation among pages. You might also learn how to program in HTML and how to use special development tools such as FrontPage or Dreamweaver. Knowing how to use these tools is vital if you plan to do Web site design.

Web programming courses teach you various programming languages and the ways to insert program logic into Web pages. You learn JavaScript, VBScript, PHP, and ASP. You may learn how to use advanced database tools, such as Cold Fusion, to access databases from your pages. You also learn how the browser and server work together to serve up pages that have sufficient programming logic to support the business application. Advanced versions of this course even teach you the Java or .NET environments so that you can configure an entire application.

We do not intend this short section to replace those courses. Instead, we introduce the architecture of these Web-based systems and provide a few principles of good design that you can apply as you develop skills in those other courses. In Chapter 11, we explained three-layer design as one effective approach to develop robust, easily maintainable systems. But how can designers implement a three-layer design in a Web-based architecture? This question is particularly important if an organization wants to use the same domain logic for both types of enterprise systems, a client-server system and an Internet-based system. We start the discussion with an example of Internet-based system architecture.

Simple Internet Architecture

Figure 12-13 illustrates a very simple Internet architecture. The two nodes represent two separate computers. On the client computer, the browser component is executing. On the server computer, the Internet server component is executing. The browser requests a page. The server sends the page. The pages, indicated by the framesets, reside permanently on the server but are transported for display to the client. Within the page, program logic can be inserted through the use of JavaScript, VBScript, Java Applets, or ActiveX controls.

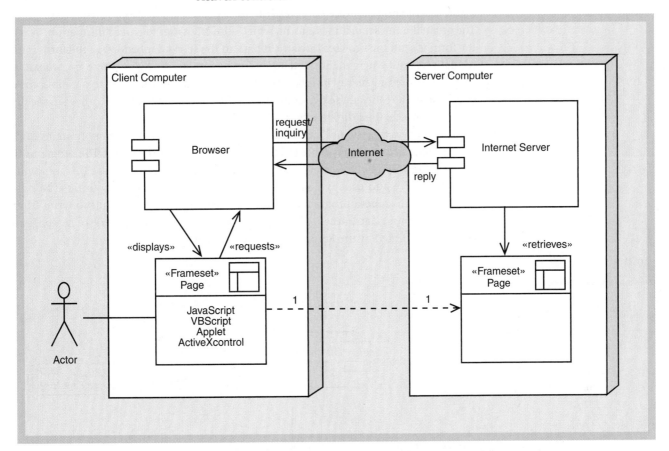

FIGURE *12-13*

Simple Internet architecture.

This simple architecture allows users to view static information on the Web or other Internet locations. The scripting languages, JavaScript and VBScript, and other program elements are useful for making the page more attractive with animation, rollovers, and so forth. However, users only interact with the system by clicking on links to other URLs and viewing other Web pages.

This set of components is used only for viewing information and is insufficient for creating a business application. Note, however, that this framework is the architecture within which Internet business systems must operate. The browser and server are key elements that transport all information between the client computer and the server computer.

Two-Layer Architecture for User Interaction

To support a business application, the system must be able to respond to other user requests, however—to accept inputs and events from the user, process those inputs, and respond with a meaningful output. The interaction diagrams developed in Chapter 11 also were designed to handle all such use cases—to accept inputs and requests from users; process those requests, including updating the database; and display the results. So, the structure for Internet-based systems should be similar.

As indicated in Chapter 11, many simple business systems can be designed as two-layer systems. These systems primarily capture information from the user and update a database. No complex domain layer logic is required. In those instances, the domain layer and data access layer are usually combined. The business logic in the domain layer frequently relates only to the formatting of the data and to deciding which database table to update. Many business applications fall in this category. A system such as a customer sales system could be easily designed as a two-layer system.

Let's expand the simple Internet architecture diagram to illustrate a two-layer design. Figure 12-14 shows a more complex Internet architecture to support user interaction. The original frameset includes a form, which allows the user to enter information. When the form is sent, the browser forwards it back to the server. Embedded in the form is information about where the input data should be sent. In the figure, there are two paths, one to a common gateway interface (CGI) component and the other to an application component. In reality, these two components are very similar. We divide them in the figure to emphasize their differences.

The CGI was the original way to process input data. The CGI directory contains compiled programs that are available to receive input data from the server. The programs in the CGI directory can be written in any compiled language, such as C++. This technique is very effective and usually has very quick response and processing times. The only downside is that these programs could be quite complex and difficult to write. They process the input data, access any required database, and format a response page in HTML, as indicated by the Response Page in the diagram.

FIGURE *12-14*

Two-layer Internet architecture.

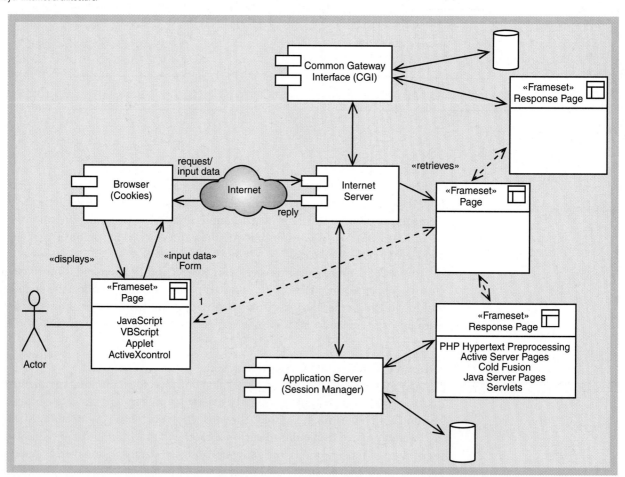

The other potential direction for input data is directly to a URL for a Web page with embedded program code. The extension shown on the Response Page indicates what type of program code is embedded in the page—ASP for Active Server Pages, PHP for PHP Hypertext Preprocessor, JSP for Java Server Pages, CFM for Cold Fusion Pages, and so forth. Depending on the type of extension, an application server—which is the language processor—is invoked to process the embedded code. The embedded code, via the application server, can process the code, including reading and writing to a database. The application server, working with cookies on the browser, can manage sessions with the user. Session variables are set up to maintain information about the user across multiple page requests and forms. The application server also formats the response page, based on the HTML statements and the code and forwards it back to the Internet server.

This architecture works well for two-layer applications that are not too complex. For example, when the response pages already have most of the HTML written. The embedded code performs functions such as validating the data and storing it in the database. Note that the business logic is minimal, so mixing it with the data access logic still provides a maintainable solution.

However, there are also some inherent complexities with this Internet system. The processing and data access code is embedded within the HTML pages, which are also user-interface pages. Since these response pages may also contain additional forms, they may also have other client-side code such as JavaScript and VBScript. So, a single entity—the HTML page—could potentially contain user-interface controls, user-interface logic, problem domain logic, and data access logic. All three layers are mixed together. Many Internet systems have been built with this architecture. As you might guess, if any of the three pieces of logic gets complex, testing and maintaining the system becomes very difficult.

Three-Layer Architecture for User Interaction

For systems that require more complex business logic, it is better to add a separate domain layer. Three-layer architecture is also more conducive to systems that need to support multiple user interfaces, both Internet based and network based. Figure 12-15 expands the diagram in Figure 12-14 to show how a three-layer approach can be implemented.

On the CGI leg, the three-layer approach is implemented by defining separate domain layer and data access layer classes. These classes are designed the same way that was shown in Chapter 11. A use case controller is identified for each input form. The use case controller then distributes messages to the individual objects of the system. The design follows the three-layer structure that was developed in Chapter 11.

On the application server side, the programmer has less direct control. However, tools make development faster and easier. Let's address two approaches—the Java approach and the .NET approach.

For Java server pages, which have the .jsp extension, the application server invokes a Java servlet when the input form is received. A Java servlet is a Java program that executes as any other program. The Java servlet identified for the input form can be the use case controller, which can then distribute the input message to other domain classes to process the request. After the request is processed, including any database access, the servlet takes control and formats the output response page. The sequence of messages flows exactly as indicated by the design sequence diagrams done in Chapter 11. The difference, of course, is that the output is in HTML statements and must flow through the server and browser before it is displayed back to the user.

For the .NET environment, the process is very similar. The input data form is sent to the ASP.NET application server. From there, the appropriate program module is called. In the new .NET environment, the program modules are compiled into a common

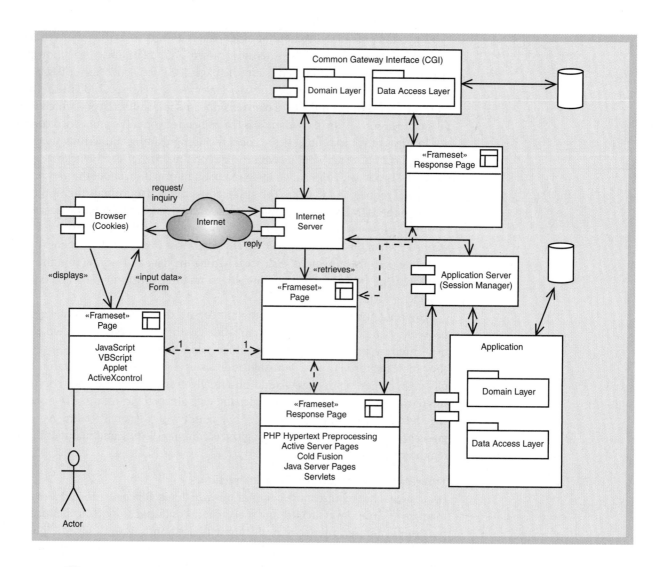

FIGURE *12-15*

Three-layer Internet architecture.

Common Language Runtime

a mid-level language that allows programs to be partially compiled and managed for faster execution

language called *Common Language Runtime*. Common Language Runtime is a mid-level language that allows programs to be partially compiled and managed for faster, more efficient execution. As with Java, the program structure can use a three-layer design based on the principles from Chapter 11.

Web Services

One of the new techniques being used in the development of Internet-based systems is the use of Web services. We expect to hear and see much more about Web services in the near future. So what are Web services? Figure 12-16 shows how a Web service might be used.

Suppose in processing an input form the application requires some data from an external database. Perhaps the application is a financial services system, and it needs the latest financial information about a particular company. Instead of trying to maintain this data itself and keep it current, the system developers decide to access it from some other service on the Internet. So, as the program is executing, it determines that it needs current financial data. It does not care where it gets this data, only that it is current. First, the program sends out a request for information. This request will go to a services directory called *Universal Discovery, Description, and Integration (UDDI)*, which is an indexing service to help locate Web services. The request will be based on certain keywords that describe exactly what is desired. The UDDI will provide an Internet address of a program that will provide that service. The application then requests the desired information over the Internet.

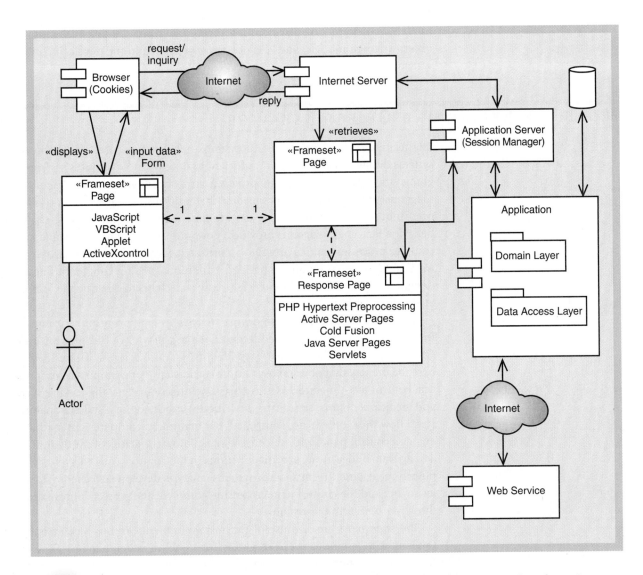

FIGURE *12-16*

Invoking a Web service.

This process sounds very easy, but it has never been possible before, for various reasons. One of the major obstacles was that the requestor and the provider both had to use exactly the same format of data exchange. Defining the same format is easy if only two programs are involved. But to have a general-purpose format that any program could use has been a major problem. In Web services, however, all communication is based on XML. XML (eXtensible Markup Language) is a text-based language much like HTML. The difference is that HTML has standard tags, whereas XML can include self-defining tags to describe any data desired. In other words, the sequence and format of the data are defined within the transmitted file itself. Thus, the recipient of the data can process the data no matter what the sequence. Examples of XML are shown in Chapter 15 in the discussion of interface design.

You should expect to see Web service capabilities expand very rapidly within the next few years. There are a lot of buzzwords associated with Web services, and many different ways to implement them. Just remember that in principle, Web services are a simple concept. In this chapter, we focused only on those Web services issues that apply to systems design. A more detailed explanation of how Web services work is given in Chapter 17 in the discussion of component-based development.

We conclude this chapter by introducing another object-oriented model that has a long history of use in computer science theory. The UML version of this model is called a statechart diagram, and it is based on the computer science state machine model. We deferred the discussion of this model since it is one of the more advanced object-oriented models, and it is not always required in the development of business systems.

MODELING SYSTEM AND OBJECT BEHAVIOR

The techniques and processes for systems analysis presented in Chapter 7 and for systems design in Chapter 11 are sufficient to develop most business systems. Class diagrams are used to understand program structure and database schemas. Use case diagrams and interaction diagrams describe basic business and system processes. However, another UML model can be used to describe more complex processes and object behavior. Statechart diagrams provide a powerful technique to understand and describe complex processes, particularly for sophisticated event-driven, real-time systems such as medical devices that monitor a patient's condition. Even for simpler business processes, designers find it sometimes beneficial to use statecharts to describe the activities associated with an object.

Statecharts can be used both during analysis and discovery and during design. During analysis, statecharts are used to describe the behavior of complex business objects. The primary focus of a statechart is to identify the various states of a business object and describe the way that object moves from state to state. For example, in computer systems that control real-time processes, such as satellite controllers, flight-control computers, and medical monitors, statecharts are critically important for a systems analyst to understand how these objects are controlled. For example, a computer system that links a jet engine's control panel with the engine must know the status of the engine at all times and activate or deactivate specific functions in the control panel to avoid dangerous inputs to the engine. To create such systems, analysts develop statechart diagrams for the jet engine and the control panel to ensure complete and correct coordination between those two objects in the system.

But statecharts are also useful for business systems. In business systems, analysts think more in terms of a status condition for an object rather than its state. However, the two terms are synonymous. An example of a business object for Rocky Mountain Outfitters is a customer order. A customer order has various status conditions, such as open for shopping, ready for shipment, shipped, and closed. Other potential status conditions might be partially shipped—if some items are on back order—and pending payment—if a payment is coming via check. Obviously, RMO would not want to start shipping an order that is still open for shopping and not yet complete. A statechart provides the modeling tools necessary to capture the business rules that apply to these business objects.

During design, statecharts also play an important role. When you are using your computer and you click your mouse on a menu item, some drop-down menu items on the screen are enabled and others may be disabled—appearing faded on screen. How does the designer of the menu user interface know how and when to enable and disable menu items? For most computer systems, designing the behavior of the user interface is quite complex. The tool used to design this system behavior is a statechart. In user interfaces, for example, the statecharts can be very complex, with nested states within other states and multiple parallel concurrent states.

Statecharts are also useful to help design an object's method logic. In other words, they provide components that allow the designer to describe the internal activity, or logic, of a system object. Specifying this logic provides additional documentation to help the programmer to write code later. In this chapter, we will not overwhelm you by trying to make you an expert with statecharts. However, we do provide the basics of this powerful tool so that you will be able to use it when needed.

This section begins with a discussion of statecharts as they are used to capture the behavior of business objects—for problem domain classes. First, the basic symbols are explained, and then rules and guidelines for developing statecharts are explained.

Statecharts for Problem Domain Classes

A state for a problem domain object is like a status condition. For example, for a customer order, a customer might ask, "What is the status of my order?" The customer wants to know whether the order is complete and whether it has been shipped. A manager might ask a similar question about an order and may want to know not only whether it was shipped but also whether it has been paid for. All of those status conditions can be described in a statechart as different states of an Order object. These status conditions, and hence the statechart, span many business events, such as creating a new order and shipping an order. Statecharts are not always needed, however. If an object in the problem domain class does not have status conditions that need to control the processing that is allowed for that object, a statechart is probably not necessary. For example, in the RMO class diagram, a class such as Order may need a statechart. However, another class such as OrderTransaction probably does not. An order transaction is created when the payment is made and then just sits there; it does not need to track other conditions. The class Customer may or may not need a statechart. If all orders are paid for at the time of the order and the system does not extend a customer credit, then a customer is simply a customer, and the system does not worry about customer statuses such as overdue or ineligible to purchase.

A statechart diagram is very similar to an activity diagram. It is composed of ovals representing statuses of an object and arrows representing its transitions. Figure 12-17 illustrates a simple statechart for a printer. Since it is a little easier to learn about statecharts by using tangible items, we start with a few examples of computer hardware. Once the basics are explained, we will illustrate modeling of software objects in the problem domain. The starting point of a statechart is a black dot, which is called a *pseudostate*. The first oval after the black dot is the first state of the printer. In this case, the printer begins in the *Off* state. A state is represented by a rectangle with rounded corners (almost like an oval, but more squared) with the name of the state placed inside.

pseudostate

the starting point of a statechart, indicated by a black dot

FIGURE *12-17*

Simple statechart for a printer.

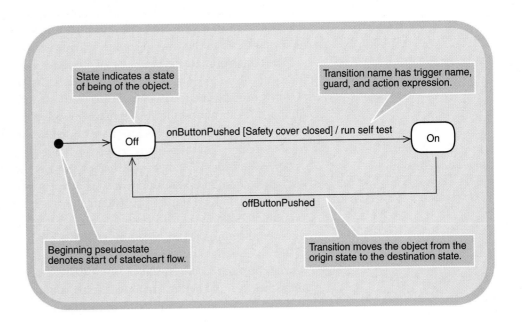

Defined precisely, a *state* of an object is a condition that occurs during its life when it satisfies some criteria, performs some action, or waits for an event. Each unique state has a unique name. A state is a semipermanent condition of an object, such as *On*, *Working*, or *Loading equipment*. States are described as semipermanent conditions because external events can interrupt them. An object remains in a state until some event causes it to move to another state.

The naming convention for status conditions helps identify valid states. A state may have a name of a simple condition such as *On* or *In repair*. Other states are more active, with names consisting of gerunds or verb phrases such as *Being shipped* or *Working*. For example, a specific Order object comes into existence when a customer orders something. Right after it is created, the object is in a state such as *Adding new order items*, then a state of *Waiting for items to be shipped*, and finally a state of *Order completed* when all items have been shipped. If you find yourself trying to use a noun to name a state, you probably have an incorrect idea about states or object classes. Rethink your analysis and describe only states of being of the object itself. Another way to help identify states is to think about status conditions that may need to be reported to management or to customers. A status condition such as *Shipped* for an order is something a customer may want to know.

The arrow leaving the *Off* state is called a *transition*. The firing of the transition causes the object to leave the *Off* state and make a transition to the *On* state. A *transition* is the movement of an object from one state to another state. It is the mechanism that causes an object to leave a state and change to a new state. Remember that states are semipermanent conditions. They are semipermanent because transitions interrupt them and cause them to end. Generally, transitions are considered to be short in duration compared with states and cannot be interrupted. In other words, once a transition begins, it runs to completion by taking the object to the new state, called the *destination state*. A transition is represented by an arrow from an *origin state*—the state prior to the transition—to a destination state and is labeled with a string to describe the components of the transition.

The transition label consists of three components:

transition-name (parameters, …) [guard-condition] / action-expression

In Figure 12-17, the transition-name is onButtonPushed. The transition is like a trigger that fires or an event that occurs. The name should reflect the action of a triggering event. In Figure 12-17, no parameters are being sent to the printer. The guard-condition is Safety cover closed. For the transition to fire, the guard must be true. The forward slash divides the firing mechanism from the actions or processes. Action-expressions indicate some process that must occur before the transition is completed and the object arrives in the destination state. In this case, the printer will run a self-test before it goes into the *On* state.

The transition-name is the name of a *message event* that triggers the transition and causes the object to leave the origin state. Notice that the format is very similar to a message in a system sequence diagram. In fact, you will find that the message names and transition-names use almost the same syntax. One other relationship exists between the messages and the transitions; transitions are caused by messages coming to the object. The parameter portion of the message name comes directly from the message parameters.

The *guard-condition* is a qualifier or test on the transition, and it is simply a true/false condition that must be satisfied before the transition can fire. For a transition to fire, first the trigger must occur, and then the guard must evaluate to true. Sometimes a transition has only a guard-condition and no triggering event. In that case, the trigger is constantly firing, and whenever the guard becomes true, the transition is taken.

Recall from the discussion of sequence diagrams that messages have a similar test, which is called a *true/false condition*. This true/false condition is a test on the sending side of the message, and before a message can be sent, the true/false condition must be true. In contrast, the guard-condition is on the receiving side of the message. The message may be received, but the transition fires only if the guard-condition is true. This combination of tests, messages, and transitions provides tremendous flexibility in defining complex behavior.

The *action-expression* is a procedural expression that executes when the transition fires. In other words, it describes the action to be performed. Earlier, we indicated that statecharts can be used to specify method logic. It is in the action-expressions that method logic is specified.

Any of the three components—transition-name, guard-condition, or action-expression—may be empty. If either the transition-name or the guard-condition is empty, then it automatically evaluate to true. Either of them may also be complex, with AND and OR connectives.

Nested States and Concurrency

Before moving into how to develop statecharts for business software objects, we present two related advanced concepts used in statechart modeling—nested states and concurrent states. Objects in the world, both system objects and physical objects, behave in more complex ways than simply being in one state at a time. The condition of being in more than one state at a time is called *concurrency*, or *concurrent states*. One way to show this is with a synchronization bar and concurrent paths, just as was done in activity diagrams (see Figure 4-13). A *path* is a sequential set of connected states and transitions. Another way to show concurrent states is to have states nested inside of other, higher-level, states. These higher-level states are called *composite states*.

A *composite state* represents a higher level of abstraction and can contain nested states and transition paths. For example, with the previous printer example, we may want to identify an *On* state. But while the printer is on, it may also be idle or working. To show these two states, we draw lower-level statecharts within the *On* state. The rounded rectangle for the *On* state is divided into two compartments. The top compartment contains the name, and the lower compartment contains the nested states and transition paths. Figure 12-18 illustrates the notation for composite states. When the printer enters the *On* state, it automatically begins at the nested black dot and moves to the *Idle* state. So, the printer is in both the *On* state and the *Idle* state. When the print message is received, the printer makes the transition to the *Working* state but also remains in the *On* state. Some new notation is also introduced for the *Working* state. In this instance, the lower compartment contains the action-expressions, that is, the activities that occur while the printer is in the *Working* state.

action-expression

a description of the activities to be performed

concurrency, or concurrent states

the condition of being in more than one state at a time

path

a sequential set of connected states and transitions

composite state

a state containing multiple levels and transitions

FIGURE *12-18*

Sample composite states for the printer object.

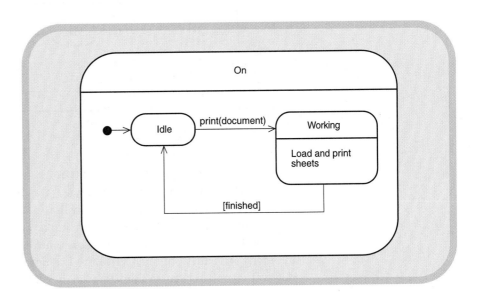

Object behavior is frequently even more complex than that expressed with simple nested states. Perhaps an object has entire sets of states and transitions, that is, multiple paths, that are active concurrently. To document concurrent behavior of a single object, we draw a composite state with the lower portion divided into multiple compartments, one for each concurrent path of behavior. For example, imagine a printer that has an input bin to hold the paper. This printer also cycles between two states in its work cycle of *Idle* and *Working*. We may want to describe two separate paths, one representing the states of the input paper tray and the other the states of the printing portion. The first path will have states of *Empty, Full,* and *Low*. The second path will contain the two states *Idle* and *Working*. These two paths are independent—the movement between states in one compartment is completely independent of movement between states in the other compartment.

The notation for composite states with nested paths can be expanded to represent multiple concurrent paths. Figure 12-19 extends the printer example from Figure 12-18. In this example, there are two concurrent paths within the composite state. The upper concurrent path represents the paper tray part of the printer. The two paths are completely independent, and the printer moves through the states and transitions in each path independently. When the Off button is pushed, the printer leaves the *On* state. Obviously, when the printer leaves the *On* state, it also leaves all of the paths in the nested states. It does not matter whether the printer is in a state or in the middle of a transition. When the Off button is pushed, all activity is stopped and the printer exists the *On* state. Now that you know the basic notation of statecharts, we turn next to how to develop a statechart.

FIGURE *12-19*

Concurrent paths for a printer in the *On* state.

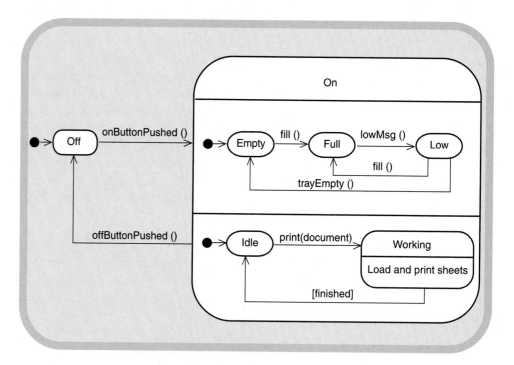

Rules for Developing Statecharts

Statechart development follows a set of rules. The rules help you to develop statecharts for classes in the problem domain. Usually the primary problem in building a statechart is to identify the right states for the object. It may be helpful to pretend that you are the object itself. It is easy to pretend to be a customer but a little more difficult to say, "I am an order," or, "I am a shipment. How do I come into existence? What states am I in?" However, if you can begin to think this way, it will help you develop statechart diagrams.

The other major area of difficulty for new analysts is to identify and handle composite states with nested threads. Usually, the primary cause of this difficulty is a lack of experience in thinking about concurrent behavior. The best solution is to remember that developing statecharts is an iterative behavior, more so than developing any other type of diagram. Analysts seldom get a statechart right the first time. They always draw it, then refine it again and again. Also, remember that when you are defining requirements during analysis, you are only getting a general idea of the behavior of an object. During design, as you build detailed sequence diagrams, you will have an opportunity to refine and correct important statecharts.

Finally, don't forget to ask about an exception condition—especially when you see the words *verify* or *check*. Normally, there will be two transitions out of states that verify something—one for acceptance and one for rejection.

Here is a list of steps that will help you get started in developing statecharts:

1. **Review the class diagram and select the classes that will require statecharts.** Then begin with the classes that appear to have the simplest statecharts, such as the OrderItem class.

2. **For each selected class in the group, make a list of all the status conditions you can identify.** At this point, simply brainstorm. If you are working in a team, have a brainstorming session with the whole team. Remember that you are defining states of being of the software classes. However, these states must also reflect the states for the real-world objects that are represented in software. Sometimes it is helpful to think of the physical object, identify states of the physical object, then translate those that are appropriate into corresponding system states or status conditions. It is also helpful to think of the life of the object. How does it come into existence in the system? When and how is it deleted from the system? Does it have active states? Does it have inactive states? Does it have states where it is waiting? Think of activities done to the object or by the object. Often, the object will be in a particular state as these actions are occurring.

3. **Begin building statechart fragments by identifying the transitions that cause an object to leave the identified state.** For example, if an Order is in a state of *Ready to be shipped*, then a transition such as beginShipping will cause the Order to leave that state.

4. **Sequence these state-transition combinations in the correct order.** Then aggregate these combinations into larger fragments. As the fragments are being aggregated into larger paths, it is natural to begin to look for a natural life cycle for the object. Continue to build longer paths in the statechart by combining the fragments.

5. **Review the paths and look for independent, concurrent paths.** When an item can be in two states concurrently, there are two possibilities. The two states may be on independent paths, such as the printer example of *Working* and *Full*. This occurs when the states and paths are independent, and one can change without affecting the other. Alternatively, one state may be a composite state and the two states should be nested, one inside the other. One way to identify a candidate for a composite state is if it is concurrent with several other states, and these other states depend on the original state. For example, the *On* state has several other states and paths that can occur while the printer is in the *On* state, and those states depend on the printer being in the *On* state.

6. **Look for additional transitions.** Often, during a first iteration, several of the possible combinations of state-transition-state are missed. One method is to take every pairwise combination of states and ask whether there is a valid transition between the states. Test for transitions in both directions.

7. **Expand each transition with the appropriate message event, guard-condition, and action-expression.** Include with each state appropriate action-expressions. Much of this may have been done as the statechart fragments were being built.
8. **Review and test each statechart.** We test statecharts by "desk-checking" them. Review each of your statecharts by doing the following:
   a. Make sure your states are really states of the object in the class. Ensure that the names of states truly describe states of being of the object.
   b. Follow the life cycle of an object from its coming into existence to its being deleted from the system. Be sure that all possible combinations are covered and that the paths on the statechart are accurate.
   c. Be sure your diagram covers all exception conditions as well as the normal expected flow of behavior.
   d. Look again for concurrent behavior (multiple paths) and the possibility of nested paths (complex states).

Developing RMO Statecharts

Let's practice these steps by developing two statecharts for RMO. Step 1 is to review the class diagram and select the classes that may have status conditions that need to be tracked. In this case, we select the Order and OrderItem classes. We assume that customers will want to know the status of their orders and the status of individual items on the order. Other classes that are candidates for statecharts are InventoryItem, to track in-stock or out-of-stock items; Shipment, to track arrivals; and possibly Customer, to track active and inactive customers. For our purposes here, we focus on the Order and OrderItem classes. We use the OrderItem class because it is simpler, and it is always best to start with the simplest class. Also, it is a dependent class—it depends on Order. Finally, it is best to use a bottom-up approach, starting with the lower items on a hierarchy, which usually have less ripple effect.

Developing the OrderItem Statechart

Start by identifying the possible status conditions that may be of interest. Some necessary status conditions are *Ready to be shipped, On back order,* and *Shipped.* An interesting question comes to mind at this point. Can an order item be partially shipped? In other words, if the customer ordered ten of a single item, but there are only five in inventory, should RMO ship those five and put the other five on back order? You should see the ramifications of this decision. The system and the database would need to be designed to track and monitor detailed information to support this capability. The problem domain class diagram for RMO (see Figure 7-20) indicates that an order item can be associated with either zero (not yet shipped) or one (totally shipped) shipment. Based on the current specification, the definition does not allow partial shipments of order items.

This is just another example of the benefit of building models. Had we not been developing the statechart model, this question may never have been asked. The development of detailed models and diagrams is one of the most important activities that a system developer can do. It forces analysts to ask fundamental questions. Sometimes new system developers think that model development is a waste of time, especially for small systems. However, truly understanding the user's needs before writing the program always saves time in the long run.

The next step is to identify exit transitions for each of the status conditions. Figure 12-20 is a table showing the states that have been defined and the exit transitions for each of those states. One additional state has been added to the above list, *Newly added,* which covers the condition when an item has been added to the order, but the order is not complete or paid for, so the item is not ready for shipping.

FIGURE *12-20*

States and exit transitions for OrderItem.

State	Transition causing exit from state
Newly added	finishedAdding
Ready to ship	shipItem
On back order	itemArrived
Shipped	No exit transition defined

The fourth step is to combine the state-transition pairs into fragments and to build a statechart with the states in the correct sequence. Figure 12-21 illustrates the partially completed statechart. The flow from beginning to end for an OrderItem is quite obvious. However, there does seem to be at least one transition missing. There needs to be some path to allow entry into the *On back order* state, so we recognize that this first-cut statechart needs some refinement. We will fix that in a moment.

FIGURE *12-21*

Partial statechart for OrderItem.

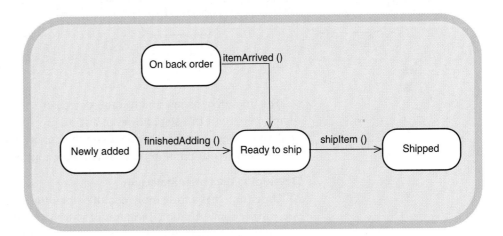

The fifth step is to look for concurrent paths. In this case, it does not appear that an order item can be in any two of the identified states at the same time. Of course, since we chose to begin with a simple statechart, that is what was expected.

Step 6 is to look for additional transitions. This step is where we will flesh out other necessary transitions. The first addition is to have a transition from *Newly added* to *On back order*. To continue, take every pair of states to see whether there are other possible combinations. In particular, look for backward transitions. For example, can an order item go from *Ready to ship* to *On back order*? This would happen if the shipping clerk found that there were not enough items in the warehouse even though the system indicated there should have been. Other backward loops, such as from *Shipped* to *Ready to ship*, or *On back order* to *Newly added*, do not make sense and are not included.

Step 7 is to complete all the transitions with correct names, guard-conditions, and action-expressions. Two new transition names are added. The first is the transition from the beginning black dot to the *Newly added* state. That transition is the transition that causes the creation, or in system terms the instantiation, of a new OrderItem object. It is named with the same name as the message into the system that adds it—addItem ( ). The final transition is the one that causes the order item to be removed from the system. This transition goes from the *Shipped* state to a final circled black dot, which is a final pseudostate. On the assumption that it is archived to a backup tape when it is deleted from the active system, that transition is named archive ( ).

FIGURE *12-22*

Final statechart for OrderItem.

Action-expressions are added to the transitions to indicate any special action that is initiated by the object or on the object. In this case, there is only one action that needs to be done. When an item that was *Ready to ship* moves to *On back order*, the system should initiate a new purchase order to the supplier to buy more items. So, on the mark-BackOrdered ( ) transition, an action-expression is noted to place a purchase order. Figure 12-22 illustrates the final statechart for OrderItem.

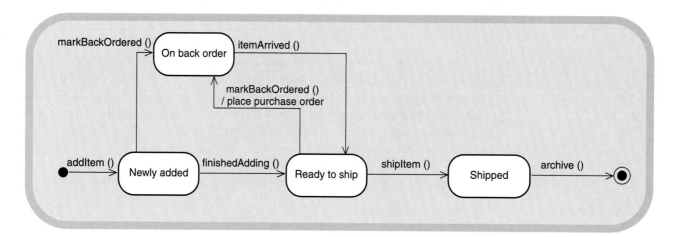

The final step, review and test the statechart, is the quality-review step. It is always tempting to omit this step; however, a good project manager ensures that the systems analysts have time in the schedule to do the quality check of their models. A structured walkthrough at this point in the project is very appropriate.

Developing the Order Statechart

An Order object is a little more complex than the OrderItem objects. In this example, you will see some additional features of statecharts that support more complex objects.

Figure 12-23 is a table of the defined states and exit transitions that, on first iteration, appear to be required. Reading from top to bottom, the states mirror the life cycle of an order. First, an order comes into existence and is ready to have items added to it—*Open for item adds*. The users in RMO indicated that they wanted an order to remain in this state for 24 hours in case the customer wanted to add more items. After all the items are added, the order is *Ready for shipping*. Next, it goes to shipping and is in the *In shipping* state. At this point, it is not quite clear how *In shipping* and *Waiting for back orders* relate to each other. That relationship will have to be sorted out as the statechart is being built. Finally, the order is *Shipped*, and after the payment clears, it is *Closed*.

FIGURE *12-23*

States and exit transitions for Order.

State	Exit transition
Open for item adds	completeOrder
Ready for shipping	beginShipping
In shipping	shippingComplete
Waiting for back orders	backOrdersArrive
Shipped	paymentCleared
Closed	archive

In step 4, fragments are built and combined to yield the first-cut statechart. Figure 12-24 illustrates the first-cut statechart. The statechart built from the fragments appears to be correct for the most part. However, we note that there are some problems with the *Waiting for back orders* state.

FIGURE *12-24*

First-cut statechart for Order.

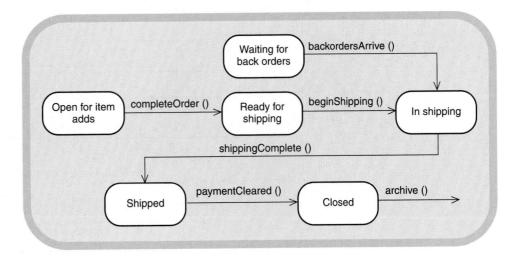

After some analysis, we decide that being *In shipping* and *Waiting for back orders* are concurrent states. And another state is needed for when the shipping clerk is actively shipping items, called *Being shipped*. One way to show the life of an order is to put it in the *In shipping* state when shipping begins. It also enters the *Being shipped* state at that point. The order can cycle between *Being shipped* and *Waiting for back orders*. Note that the exit out of the composite state only occurs from the *Being shipped* state, which is inside the *In shipping* state. Obviously, upon leaving the inside state, the order also leaves the composite *In shipping* state.

As we go through steps 5, 6, and 7, we note that new transitions must be added. The creation transition from the initial pseudostate is required. Also, transitions must be included to show when items are being added and when they are being shipped. Usually, we put these looping type of activities on transitions that leave a state and return to the same state. In this case, the transition is called addItem ( ). Note how it leaves the *Open for item adds* state and returns to the same state. Figure 12-25 takes the statechart to this level of completion.

The benefit of developing a statechart for an object, even a business object, is that it helps to capture and clarify business rules. From the statechart, we can see that shipping cannot commence while the order is in the *Open for item adds* state. New items cannot be added to the order after the order has been placed in the *Ready for shipping* state. The order is not considered shipped until all items are shipped. If the order has the status of *In shipping*, we know that it is either actively being worked on or waiting for back orders.

As always, the benefits of careful model building help us gain a true understanding of the system requirements. Let's now look at the big picture and pull the different models into a whole to see how they fit together.

Statecharts as Design Models

Once you learn how to use statecharts, you will find that they are a very effective tool for describing all kinds of system behaviors and constraints. As you learned in the preceding paragraphs, a statechart represents the life of the objects in a single class. However, since the system itself can be considered as an object (everything is an object in OO), then a statechart can also be developed for the system itself. As indicated earlier, this is a

FIGURE *12-25*

Second-cut statechart for Order.

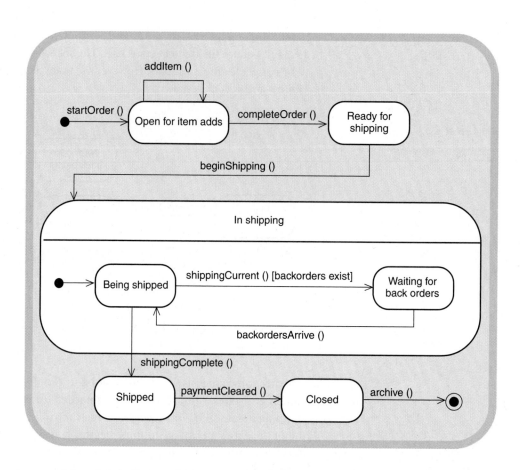

powerful tool that allows a designer to designate the states of a system, including which menu items are enabled and disabled, what windows are modal—must be responded to and closed before the system will respond—and which tool bars and icons are active.

Figure 12-26 illustrates an example of how a statechart is used to describe the states and rules of a system that requires a user to log on before it allows processing. The statechart in the figure is not meant to be a complete statechart of the system. It is only a fragment used to illustrate a single control process in the system. In fact, as a designer, you will find yourself sketching out many statechart fragments to describe and document processing controls. You can probably imagine that if we tried to document an entire system in a single statechart it would be massive. One of the strengths of models is that you can sketch out partial models containing only the immediately relevant points.

Let's review the statechart. The system opens up with the transition startProgram ( ) and immediately enters the *Locked* state. An action-expression is attached to the entry into the state, which indicates that the system is to display the logon window. The system sits in that state until the logonSent ( ) transition occurs. Notice that this statechart indicates that nothing else can happen until the logon is sent. That is what the designer intends to happen. Once the logonSent ( ) transition occurs, then the system enters the *Verifying user* state. Again, an entry action-expression indicates that the system should access the database to verify the user. Exit transitions out of this state either return to the *Locked* state or move forward to the *Unlocked* state. Notice the associated action-expressions on each transition, indicating what the system must do before completing each transition. If the logon is rejected, the system displays the logon window again with appropriate error messages. Otherwise, the menu (and possibly other items) are activated for the user to continue using the system.

FIGURE *12-26*

Logon statechart for a computer system.

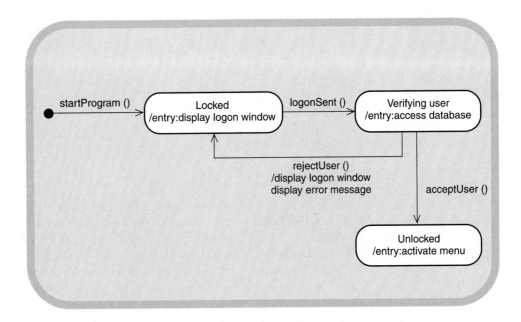

One other addition we would make to the design class that is controlling the logon is to define a new attribute to record a value of the current state. In programming jargon this is often called a *state variable*. In this instance, a variable called systemState can be created. It holds values such as *Locked, Verifying user*, and *Unlocked*. Based on the values in the state variable, the system can tell which menu items to enable and disable and which messages to process or to ignore.

During the final design of the menu items and Web user interface in RMO, the design team depended heavily on statecharts to document the possible options that were to be presented to the users (see Barbara Halifax's status memo). When designing network-based systems, it is normal to enable and disable menu items to control what the user is allowed to do. The same controls can be done with a Web interface, but instead of enabling and disabling menu items, links are either included or excluded from the Web page. Of course, removing links from a Web page must be done very carefully or the user can become frustrated or even lost.

Messages, Transitions, and Methods

As described earlier, the name of each transition represents a trigger that fires and causes the transition to occur. The trigger for a transition fires when it receives a message from some other object. This chain of events is reminiscent of sequence diagrams. Remember that a message on a sequence diagram goes to a destination object, and the destination object has a method to process that incoming message. In fact, the method signature was derived based on the name of the incoming message.

If a method signature is derived from an incoming message, and a transition is triggered by an incoming message, then there must be some sort of connection between methods and transitions. In essence, a transition, at least one with a message trigger, is the statechart version of what later will become a method. So, the information on a transition can help in programming a method. The action-expressions on a transition indicate what logic will be required for that method. Often, designers use the same name for messages, method signatures, and transition-names. Each item is a different "thing," but each has the same name to show that it is related to other items.

July 15, 2005

To: John MacMurty

From: Barbara Halifax, Project Manager

RE: Customer Support System status

John, during the last several weeks we have been working on the design of the Web interface for the system. This task has been very challenging. I knew it would be complex, but I had no idea of the level of detail that we needed to address. This has been a joint activity, involving my technical staff, the systems analysts, and the user-interface people. We have been identifying which functions are allowed from which Web pages and at the same time trying to keep the system very user-friendly and intuitive. As I mentioned to you, we have developed a statechart for every Web page to ensure that we know exactly what services we want to provide for that page. It has been a lot of work, but the team is very confident that the solution is solid.

Completed during the last period (two weeks)

We completed the design of the Web interface for the order shopping cart. This took us longer than expected, since Bill McDougal and JoAnn White were both very concerned that we make sure that the Web pages and underlying system did not have any security holes. Consequently, we are about a week behind schedule on this part of the design, but we are confident that the solution is correct and has good security.

Plans for the next period (two weeks)

During the next period, we are going to begin implementing the Web interface. With the iterative approach we are using with the UP, we will be programming the Web interface and inserting it into the portions that are already coded and running.

Problems, issues, open items

Nothing major at this point. There is just a lot of work to do. However, we are moving ahead rapidly and making good progress.

BH

cc: Steven Deerfield, Ming Lee, Jack Garcia

Let's return to the Order statechart and see how design class diagrams, sequence diagrams, and statecharts work together. Figures 12-27 and 12-28 are duplicates of Figures 11-21—a sequence diagram for processing an order—and 11-29—the design class for Order. In these two figures, we see messages, and corresponding methods for completeOrder ( ) and makePayment ( ). The statechart in Figure 12-25 contains a corresponding transition for completeOrder ( ), as expected. However, the statechart does not have a transition for makePayment. So, we need to enhance the statechart so that it is consistent with the sequence diagram. We add an appropriate transition for makePayment ( ) and a corresponding state, *Ready for payment*.

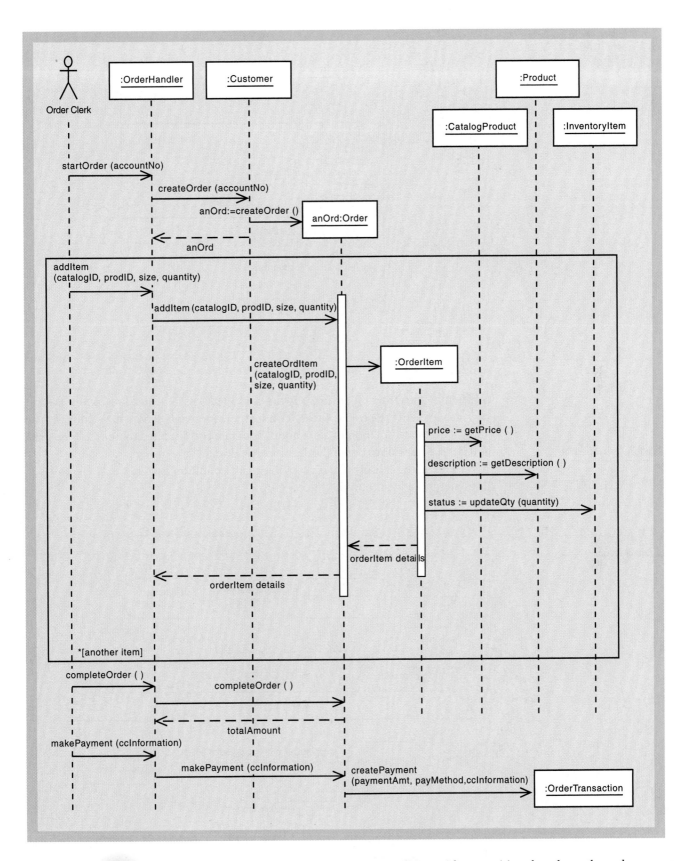

Let's also add an exception condition with a transition that closes the order process if the customer's credit card fails the verification process. There are several ways this could be done. In this instance, we add an action-expression to the makePayment ( ) transition to verify credit. Again, the action-expression means that an action will be

FIGURE *12-28*

Design class for Order.

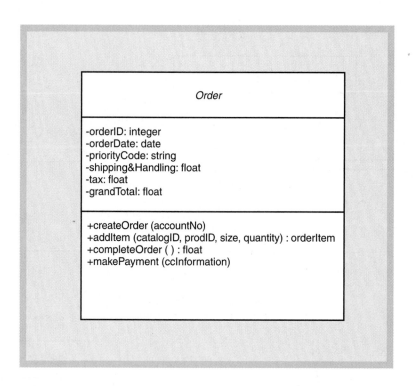

initiated by the Order object to verify the customer's credit. Instead of adding a whole new state to verify the credit, we do it on the transition and denote the results with an abbreviated state notation of a diamond. The state is considered to be a *decision pseudostate*. The transitions out of the decision pseudostate have only guard-conditions, that is, they fire automatically if and when the guard-condition becomes true. Figure 12-29 illustrates this added concept.

decision pseudostate

a diamond notation on a statechart indicating a decision point on a path

FIGURE *12-29*

Statechart for Order.

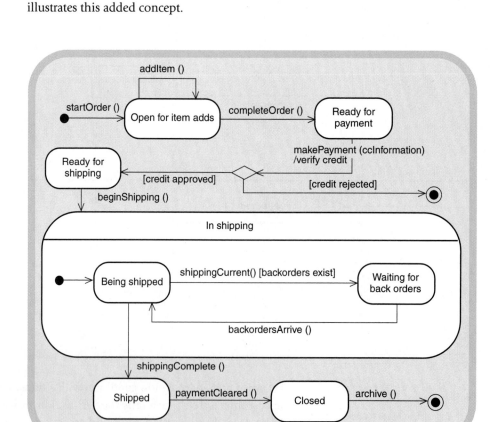

Since the makePayment ( ) transition is the statechart version of the method to be programmed, the programmer can look at the transition action-expressions to help write the method logic. In this instance, the programmer can tell that the makePayment ( ) method will receive input parameters for the credit-card information, and it must include logic to verify the customer's credit.

In this section, we have illustrated how statecharts are used during analysis to discover and document the various status conditions that must be tracked on business objects from the domain model. During design, statecharts are also used to specify system behavior that is identified as the detailed design diagrams are built. As you become proficient with the use of statecharts, you will discover that they are a powerful modeling tool. As mentioned earlier, there is an entire body of knowledge associated with the theory of statecharts and their use in describing system behavior.

SUMMARY

Systems that are based on good design principles are not only easier to develop but are also more maintainable. As object-oriented development becomes more mature, the principles that drive good design are becoming better understood. Good design is based on two factors: good principles and standard design patterns.

Two new design principles that were introduced in the chapter are protection from variations and indirection. Protection from variations is applied to a system by separating components or design elements that do not change from those that are more apt to change. The purpose is to make the system more maintainable by isolating the changes to well-identified and easily replaceable components. The other principle, indirection, is one way to implement protection from variations. Indirection is a technique to insert an intermediate object between two objects that need to be linked but that do not have a natural interface to each other. The intermediate object provides an interface between the other two elements. The advantage of this technique is that (1) neither of the two elements must be changed to interface with each other and (2) if one needs to be replaced, then the impact on the other is minimized.

In addition to design principles, the use of design patterns is becoming more popular to ensure that the design of a new system follows accepted approaches. A design pattern is like a template or a standard solution to a common problem. Patterns are used frequently in all engineering fields because many engineering problems are very similar. In software development, a set of commonly accepted design patterns is now providing standard solutions to speed design.

Three design patterns were presented as an introduction to this approach to design: singleton, adapter, and observer. The singleton pattern is a standard solution to the problem of ensuring that only one object of a given class is created. In some software situations, an object may have to maintain control over various system processes. To ensure that control is maintained, all processing is forced to go through one control point. To make sure that another object is not inadvertently subverting this control, only one control object is allowed. The singleton pattern is one way to implement this requirement.

The adapter pattern is an example of the indirection design principle. As with an electrical adapter, which adapts one set of plugs to another set, the adapter pattern converts one application program interface (API) into another API. Thus, an object that normally would not be compatible with a system can be "plugged into" the system through the use of an adapter object.

The observer pattern is a fairly complex pattern, but it is widely used and critically important. All of the windows GUI event-processing capability is built around the observer pattern. The observer pattern allows two objects to be linked dynamically. This pattern is important to keep coupling low in a system, yet still provide communication capability between objects that should not be permanently linked.

The final major topic of the chapter is a description and use of statecharts. A statechart is an effective technique to describe the behavior, or life cycle, of an object or of the system. A statechart comprises two major components: states and transitions. A state is a condition of the object. For business objects, a state is often synonymous with a status condition of an object. For example the state, or status condition, of a customer order might be "shipped." Transitions are the connectors between the states and permit an object to move from state to state. So, a statechart is a diagram that identifies all of the states of an object and those actions that cause it to change from state to state. Statecharts are useful during design to understand the behavior of real-world objects, which must be captured within the system. Statecharts are also useful during design to specify precise behavior and constraints on the various portions of the final system.

KEY TERMS

REVIEW QUESTIONS

1. What is meant by *protection from variations*? Why is it an important design principle?

2. What is meant by *indirection*? Why is it important in systems design?

3. What is a design pattern? How are design patterns used during systems design?

4. What problem does the singleton design pattern address?

5. Briefly explain how the singleton pattern works.

6. What problem does the adapter pattern solve? Briefly explain how it is implemented.

7. What are the basic elements of the observer pattern?

8. Give an example or two of when the observer pattern might be used.

9. What is meant by an enterprise-level system? Why is that an important consideration in systems design?

10. What is a deployment diagram used for?

11. Describe the two primary symbols used for deployment diagrams. What do they mean?

12. What are some of the differences between a network-based system and a Web-based system?

13. What are Web services? How are they implemented?

14. What is the purpose of a statechart?

15. Show the two symbols used in a statechart and describe their use.

16. List the primary steps for developing a statechart.

17. What is meant by the origin state? By a destination state?

18. What is a decision pseudostate?

19. List the component elements of a transition description. Which are optional?

20. What is a nested state? What is a composite state? What are they used for?

THINKING CRITICALLY

1. In Figure 12-30, the package on the left contains the classes in a payroll system. The package on the right is a payroll tax subsystem. What technique would you use to integrate the tax subsystem into the payroll system? Show how you would solve the problem. Show what modifications you would make to the existing classes (in either figure) or what new classes you would add. Use UML notation.

FIGURE *12-30* Payroll system classes.

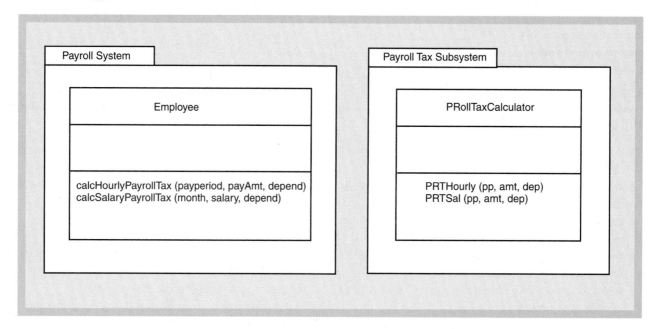

2. On the right side of Figure 12-31, a system is running that is simulating the manufacture of computer chips. The equations in the simulation system are based on statistical probabilities of failures in the manufacturing process. The package on the left illustrates a window and its associated class definition, which is to display these results dynamically as they occur. The values in the top five fields are obtained when the window is opened. However, the bottom three fields should be updated after every iteration, which takes about one second. It is not desirable to have the simulation system (on the right) coupled to the user-interface system (on the left). Show how you would solve this problem. Show any class methods for existing classes, new classes, and new definitions that you would use. Use UML notation.

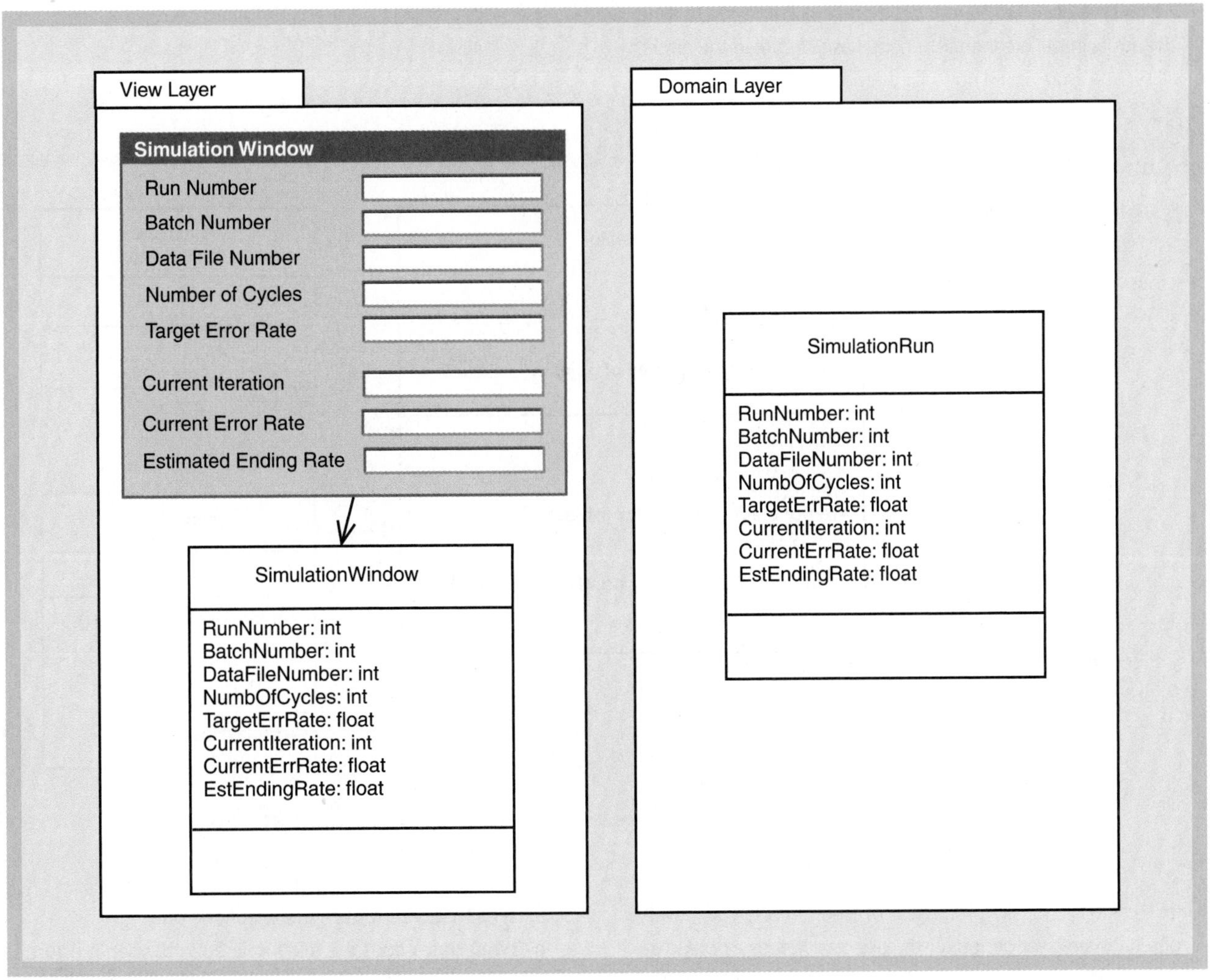

View Layer

Simulation Window

Run Number	
Batch Number	
Data File Number	
Number of Cycles	
Target Error Rate	
Current Iteration	
Current Error Rate	
Estimated Ending Rate	

SimulationWindow

RunNumber: int
BatchNumber: int
DataFileNumber: int
NumbOfCycles: int
TargetErrRate: float
CurrentIteration: int
CurrentErrRate: float
EstEndingRate: float

Domain Layer

SimulationRun

RunNumber: int
BatchNumber: int
DataFileNumber: int
NumbOfCycles: int
TargetErrRate: float
CurrentIteration: int
CurrentErrRate: float
EstEndingRate: float

FIGURE *12-31*

Simulation system classes.

3. Review the observer pattern description in Figure 12-32. Find the errors in the diagram. After you have identified the errors, state what you would do to fix them.

FIGURE *12-32*

Sample observer pattern with errors.

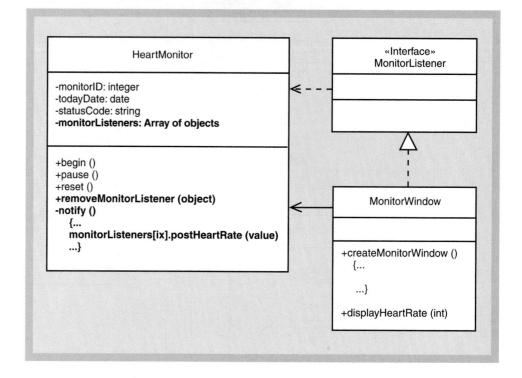

4. In Thinking Critically problem 2 of Chapter 11, you developed a three-layer design for the use case *Return a book*. Assume that the library has a main office and branch locations. The library staff would like to make this use case available with a Web interface so that various locations can access it. Take your three-layer design and superimpose it on the deployment diagram in Figure 12-15 for an enterprise-level solution. Show where you would put the domain classes (into which package) and your data access classes. Replace the «Boundary» with the Frameset classes that are appropriate for a Web solution.

5. In Thinking Critically problem 3 of Chapter 11, you developed a three-layer design for the use case *Receive new book*. Assume that the library has a main office and branch locations. The library staff would like to make this use case available with a Web interface so that various locations can access it. Take your three-layer design and superimpose it on the deployment diagram of Figure 12-15 for an enterprise-level solution. Show where you would put the domain classes (into which package) and your data access classes. Replace the «Boundary» classes with the Frameset classes that are appropriate for a Web solution.

6. Review the cellular telephone statechart in Figure 12-33, then answer the following questions:

a. What happens to turn the telephone on?

b. What states does the telephone go into when it is turned on?

c. What are the three ways that the telephone can be turned off? Identify what must be true for each case. Can the telephone turn off in the middle of the *Active (Talking)* state?

d. How can the telephone get to the *Active (Talking)* state?

e. Can the telephone be plugged in while someone is talking?

f. Can the telephone change battery states while someone is talking? Explain which movement is allowed and which is not allowed.

g. What states are concurrent with what other states? (Make a two-column table showing states with concurrent states.

7. Given the following description of a Union Package Shipments (UPS) shipment, first identify all of the various states and exit transitions, then develop a statechart.

A shipment is first recognized after it has been picked up from a customer. Once it is in the system, it is considered to be active and in transit. Every time it goes through a checkpoint, such as arrival at an intermediate destination, it is scanned and a record is created indicating the time and

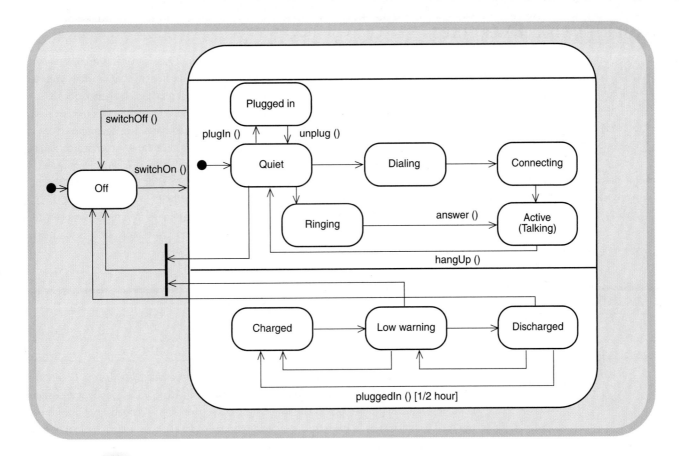

FIGURE *12-33*

Cellular telephone statechart.

place of the checkpoint scan. The status changes when it is placed on the delivery truck. It is still active, but now it is also considered to have a status of *delivery pending*. Of course, once it is delivered, the status changes again.

From time to time, the shipment has a destination that is outside the area serviced by UPS. In those cases, UPS has working relationships with other courier services. Once a package is handed off to another courier, it is noted as having a condition of *handed over*. In those instances, a tracking number for the new courier is recorded (if it is provided). UPS also asks the new courier to respond with a status change transaction once the package has been delivered.

Unfortunately, from time to time a package gets lost. In those cases, it remains in an active state for two weeks but also is categorized as *misplaced*. If after two weeks the package has not been found, it is considered *lost*. At that point the customer can initiate lost procedures to recover any damages.

8. The International Hospital Service (IHS) hospital is building a new scheduling system to help schedule beds and rooms in the hospital. To help understand how rooms are scheduled, the following narrative is provided. Based on the narrative,

develop a statechart that will help explain how a hospital room is scheduled.

A hospital room is either in-use or vacant. However, scheduling a room is a little more complicated. A room can be scheduled or unscheduled, independent of whether it is in-use or not. When a room is vacant, it becomes in-use when a patient is assigned to the room. If the patient is admitted as an emergency (unscheduled), then nothing else needs to happen. However, if it was a scheduled arrival, the room-scheduling system must note that fact and update the schedule to show it as being fulfilled.

Immediately after a room is vacated, it goes into a temporary state of dirty. That simply means that it must be cleaned before another patient can be moved in. So, even though it is in reality vacant, it is not considered vacant until it has been cleaned.

When a scheduled patient arrives, the room changes to an unscheduled state. Since it is impossible to ascertain exactly how long a patient will be in a room, the hospital does not allow a room to be scheduled by more than one patient at a time.

EXPERIENTIAL EXERCISES

1. Design patterns are a young, but rich field of research and study. Locate the original GoF book on design patterns and give a brief summary of two or three patterns discussed in that text. This exercise will begin your lifelong learning process of reading and understanding technical design material.

2. Find another book on design patterns (see the "Further Resources" list for suggestions) and report on two or three of the patterns listed in that book. Some patterns are for network designs. Books on enterprise-level designs are frequently oriented toward Internet design.

3. Do more research into the basics of Web services. Chapter 17 will also add to your understanding. Find some articles on the Internet that explain how Web services are implemented. Then find an example of a company that provides Web services and document what it provides and how it does it.

4. Statecharts are a variation of a basic computer model called a *finite state machine*, which has a rich theoretical body of knowledge. Using Google or another good search engine, find some articles about finite state machines. Identify and list several of the different names for this model, and describe some of the common uses for it.

5. Find some examples of system behavior that have been documented through the use of statecharts. Discuss the various ways that statecharts are used.

CASE STUDIES

The State Patrol Ticket Processing System

One of the most important classes in the State Patrol Ticket Processing System is the Ticket class. Based on the original case description provided in Chapter 5, identify the various states (status conditions) that might be appropriate for a traffic ticket. Consider all possible options from its being paid immediately to the recipient's waiting for results from a court ruling to other possible conditions. Develop a statechart for the traffic ticket class.

The DownTown Videos Rental System

Based on the case description provided in Chapter 7 and your understanding of how a video store might work, identify the possible states (status conditions) for a copy of a physical movie. In other words, even though the case does not include other states, such as damaged or in repair, you should try to expand the requirements to include every possible state. Based on your list of states, build a statechart for the movie copy class.

TheEyesHaveIt.com Book Exchange

In Chapter 11 you built a network-based solution for TheEyesHaveIt.com Book Exchange. Of course, the real intent of this system is to provide access via the Internet. Using Figure 12-15 and your solution in Chapter 11 as the starting point, show which classes and framesets you would add to Figure 12-15 to provide an enterprise-level solution. Place your classes in package diagrams, as appropriate, to group them according to layers.

Rethinking Rocky Mountain Outfitters

In Chapter 11 detailed three-layer sequence diagrams were provided for *Look up item availability* and the telephone order scenario of *Create new order*. Chapter 7 provided a system sequence diagram for the Web scenario of *Create new order*. Starting from Figure 12-15, add the appropriate classes and framesets to the diagram for each of the use cases. Create two separate drawings, one for *Look up item availability* and one for the Web scenario of *Create new order*. Add your classes to the appropriate packages depicted inside the components. Delete any components or packages that are not necessary. You will need to replace the windows boundary classes with Web framesets based on the design and the system sequence diagram.

Focusing on Reliable Pharmaceutical Service

In Chapter 11, you developed a three-layer design for three use cases that you had documented in Chapter 7. Based on the Web design principles given in this chapter, show how a Web solution would be structured. Begin with Figure 12-15 and place the classes you identified in your Chapter 11 solution in the appropriate packages. You will need to replace your windows classes with frameset classes as required for the Web interface.

FURTHER RESOURCES

Grady Booch, James Rumbaugh, and Ivar Jacobson. *The Unified Modeling Language User Guide*. Addison-Wesley, 1999.

Frank Buschmann, R. Meunier, H. Rohnert, P. Sommerlad, and M. Stal. *Pattern-Oriented Software Architecture: A System of Patterns*. John Wiley and Sons, 1996.

Erich Gamma, R. Helm, R. Johnson, and J. Vlissides. *Elements of Reusable Object-Oriented Software*. Addison-Wesley, 1995.

Craig Larman. *Applying UML and Patterns: An Introduction to Object-oriented Analysis and Design and the Unified Process*, 2nd Edition. Prentice-Hall, 2002.

James Rumbaugh, Ivar Jacobsen, and Grady Booch. *The Unified Modeling Language Reference Manual*. Addison-Wesley, 1999.

David S. Linthicum. *Next Generation Application Integration: From Simple Information to Web Services*. Addison-Wesley, 2004.

Mark Grand. *Patterns in Java*, Volumes I and II. John Wiley and Sons, 1999.

Alur Deepak, J. Crupi, and D. Malks. *Core J2EE Patterns: Best Practices and Design Strategies*. Sun Microsystems Press, 2001.

Designing Databases

LEARNING OBJECTIVES

After reading this chapter, you should be able to:

- Describe the differences and similarities between relational and object-oriented database management systems

- Design a relational database schema based on an entity-relationship diagram

- Design an object database schema based on a class diagram

- Design a relational schema to implement a hybrid object-relational database

- Describe the different architectural models for distributed databases

CHAPTER OUTLINE

Databases and Database Management Systems

Relational Databases

Object-Oriented Databases

Hybrid Object-Relational Database Design

Data Types

Distributed Databases

The project leaders for Nationwide Book's (NB) new Web-based ordering system were meeting with NB's database administrator to discuss how they were going to tackle the database design. Present at the meeting were Sharon Thomas (who had led the project since its inception), Vince Pirelli (a contractor who had completed most of the analysis tasks), and Bill Anderson (NB's database administrator, who hadn't directly participated in earlier phases of the project). Sharon started the meeting by saying, "When the project began, we planned to use the existing DB2 database. Maria Peña [the chief information officer] also wanted us to use this project to try out newer development methods and tools that we knew we'd need in the coming years. So we hired Vince to do the systems analysis using object-oriented [OO] methods. We also purchased a Java development tool and sent two of our programmers to a three-week training course."

Vince added, "I developed a traditional entity-relationship diagram as a basis for designing the interface between the new programs and the existing database. I checked the documentation for the current database schema. Most of the information needed by the new system is already there, although we need to add some new fields and we might have to change some table definitions. But now that we're looking at design and implementation, I'm concerned that we may be handicapping our new system with an outdated database management system."

Sharon added, "We decided to use this project as a pilot for OO development and implementation to speed up development. But I'm not sure that we'll actually achieve that if we use the existing DB2 database."

Bill said, "I understand your concerns about interfacing OO programs with a relational database. It sounds like trying to mix oil and water, but it's really not that difficult. We just hired Anna Jorgensen, a database developer who has experience writing Java programs that interface with relational databases. I had her look over the class diagrams and use cases for the new system, and she assures me that the interfaces to DB2 will be straightforward and simple. I can assign her to your project for a few months if you need the help."

Vince responded, "There's no question that an interface can be written, but it may not be as easy as Anna thinks. There are also problems with basic elements of an OO program such as inheritance and class methods. Those things simply can't be represented in a relational database. That'll force us to make some ugly compromises when designing and implementing our OO code. I'm afraid that it'll lengthen our design and implementation phases and make future system upgrades much more difficult."

"I've done some research, and there are quite a few commercial OO data packages available," said Sharon. "None of them have the track record of DB2, but then the technology is fairly new. An OO database management system would be the best fit with Java, and it would open the door to other new technologies."

Vince added, "It would also shorten the time we need to complete design, since an OO database is directly based on the class diagram."

Sharon continued, "I think that this project presents a good opportunity for us to make the leap to the next generation of database software."

Bill was taken aback by the suggestion but quickly replied, "Are you asking me to support two redundant databases based on two completely different database technologies? Management is already breathing down my neck about my budget. I've managed to save some money by switching to cheaper server hardware, but it's only a marginal improvement. Supporting another database management system will be a major effort. And we'd need new hardware to isolate the new DBMS from our existing database. I can't risk having a buggy new piece of software crash our operational databases. I'd also need to train my people to bring them up to speed on the new software. And how will we get data back and forth between the two databases?"

Sharon let the air clear for a bit before replying. "There's no question that it'll be a major undertaking. I'm not trying to downplay that fact or stretch your people and equipment to the breaking point. But there comes a time when we need to move on to newer technologies, and I think that time has arrived. I've already discussed the idea with Maria, and she thinks it deserves serious consideration."

Vince added, "I can do the database design either way. But we won't be building a base for the future if we use a relational database. I could design it to interface with indexed files on an old IBM mainframe if I had to. But why would we want to go backward instead of forward?"

OVERVIEW

Databases and database management systems are important components of a modern information system. Databases provide a common repository for data so that it can be shared by the entire organization. Database management systems provide designers, programmers, and end users with sophisticated capabilities to store, retrieve, and manage data. Sharing and managing the vast amounts of data needed by a modern organization simply would not be possible without a database management system.

In Chapter 5, you learned to construct conceptual data models. You also learned to develop entity-relationship diagrams (ERDs) for traditional analysis and class diagrams for object-oriented (OO) analysis. To implement an information system, developers must transform a conceptual data model into a more detailed database model and implement that model within a database management system.

The process of developing a database model depends on the type of conceptual model and the type of data management software that will be used to implement the system. This chapter will describe the design of relational and OO data models and their implementation using database management systems. We will use examples from Rocky Mountain Outfitters to show how information collected during analysis is used in database design.

DATABASES AND DATABASE MANAGEMENT SYSTEMS

database (DB)

an integrated collection of stored data that is centrally managed and controlled

A *database (DB)* is an integrated collection of stored data that is centrally managed and controlled. A database typically stores information about dozens or hundreds of entity types or classes. The information stored includes entity or class attributes (for example, names, prices, and account balances) as well as relationships among the entities or classes (for example, which orders belong to which customers). A database also stores descriptive information about data such as field names, restrictions on allowed values, and access controls to sensitive data items.

database management system (DBMS)

system software that manages and controls access to a database

A database is managed and controlled by a *database management system (DBMS)*. A DBMS is a system software component that is generally purchased and installed separately from other system software components (for example, operating systems). Examples of modern database management systems include Microsoft Access, Oracle, DB2, ObjectStore, and Gemstone.

DBMS Components

physical data store

the storage area used by a database management system to store the raw bits and bytes of a database

schema

a description of the structure, content, and access controls of a physical data store or database

Figure 13-1 illustrates the components of a typical database and its interaction with a DBMS, application programs, users, and administrators. The database consists of two related information stores: the physical data store and the schema. The *physical data store* contains the raw bits and bytes of data that are created and used by the information system. The *schema* contains descriptive information about the data stored in the physical data store, including the following:

- Access and content controls, including allowable values for specific data elements, value dependencies among multiple data elements, and lists of users allowed to read or update data element contents
- Relationships among data elements and groups of data elements (for example, a pointer from data describing a customer to orders made by that customer)
- Details of physical data store organization, including type and length of data elements, the locations of data elements, indexing of key data elements, and sorting of related groups of data elements

A DBMS has four key components: an application program interface (API), a query interface, an administrative interface, and an underlying set of data access programs and subroutines. Application programs, users, and administrators never access the physical data store directly. Instead, they tell an appropriate DBMS interface what data they need to read or write, using names defined in the schema. The DBMS accesses the schema to verify that the requested data exist and that the requesting user has appropriate access

FIGURE *13-1*

The components of a database and database management system and their interaction with application programs, users, and database administrators.

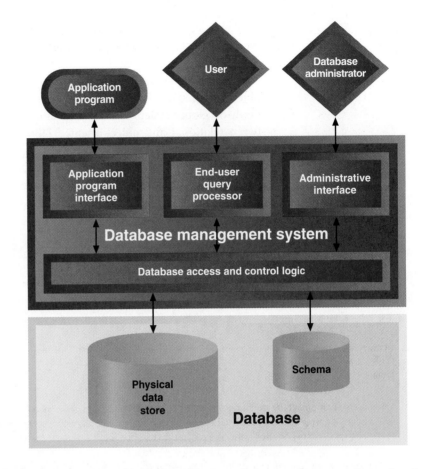

privileges. If the request is valid, then the DBMS extracts information about the physical organization of the requested data from the schema and uses that information to access the physical data store on behalf of the requesting program or user.

Databases and database management systems provide several important data access and management capabilities, including the following:

- Simultaneous access by many users and application programs
- Access to data without writing application programs (that is, via a query language)
- Application of uniform and consistent access and content controls

For these and other reasons, databases and DBMSs are widely used in modern information systems.

Database Models

DBMSs have evolved through a number of technology stages since their introduction in the 1960s. The most significant change has been the type of model used to represent and access the content of the physical data store. Four such model types have been widely used:

- Hierarchical
- Network
- Relational
- Object-oriented

The hierarchical model was developed in the 1960s. It represented data using sets of records organized into a hierarchy. The network model also grouped data elements into sets of records but allowed those records to be organized into more flexible network structures. Few new databases have been developed with the hierarchical and network models in the last two decades. However, a few older hierarchical and network databases are still in use today, especially in large-scale batch transaction processing applications.

The remainder of this chapter describes design issues for the relational and object-oriented database models—the most widely used models for both existing and newly developed systems. Rocky Mountain Outfitters faced the decision of choosing between relational and object-oriented DBMSs to support its new customer support system (see Barbara Halifax's memo). Design issues for the hierarchical and network models are not described, since few students of information systems are likely to encounter DBMSs based on these models by the time they enter the workforce.

April 2, 2005

To: John MacMurty

From: Barbara Halifax

RE: Customer support system database design and DBMS

John, we've completed the initial design based on the class diagram developed last month. Since we're still on the fence about whether to use a relational or an object-oriented DBMS, we've specified both a relational and object schema. I told you my opinion when the project first started, but I'll reiterate it here.

This project is our first big chance to break with our legacy technology. We've moved in that direction by doing OO analysis and using OO development tools and Web-based deployment for the CSS. Choosing an object DBMS will simply complete that transition, making future maintenance and enhancement much easier. There are several viable choices for object DBMS packages.

I'm aware of Mac's and Ann's concerns about supporting our existing relational DBMS along with a new object DBMS. I know Mac worries about vendor stability in the object DBMS market, but increasing standardization has made switching vendors much simpler. Nonetheless, I admit that there are significant costs and risks.

But the costs and risks of supporting our new OO software with a relational DBMS are even greater: performance bottlenecks, more expensive hardware, slow response during peak demand, and more difficult upgrades. We risk falling behind our competitors and having to rebuild the new system from scratch much sooner than should be necessary.

Well, enough of my salesmanship! We can meet before the next staff meeting to go over this in more detail (please pick a time and update my on-line schedule). The decision must be made within the next week or two, or we'll jeopardize our delivery deadline.

BH

cc: Mac Preston, Ann Hamilton

RELATIONAL DATABASES

relational database management system (RDBMS)

a database management system that stores data in tables

table

a two-dimensional data structure containing rows and columns; also called a *relation*

row

the portion of a table containing data that describes one entity, relationship, or object; also called *tuple* or *record*

field

a column of a relational database table; also called an *attribute*

field value

the data value stored in a single cell of a relational database table; also called an *attribute value* or *data element*

The relational database model was first developed in the early 1970s. Relational databases were slow to be adopted because of the difficulties inherent in converting systems implemented with hierarchical and network DBMSs and because of the amount of computing resources required to implement them successfully. As with other theoretical advances in computer science, it took many years for the cost-performance characteristics of data storage and processing hardware to catch up to the new theory. Relational DBMSs now account for the vast majority of DBMSs currently in use.

A *relational database management system (RDBMS)* is a DBMS that organizes stored data into structures called *tables*, or *relations*. Relational database tables are similar to conventional tables—that is, they are two-dimensional data structures of columns and rows. However, relational database terminology is somewhat different from conventional table and file terminology. A single row of a table is called a *row*, *tuple*, or *record*, and a column of a table is called a *field*, or *attribute*. A single cell in a table is called a *field value*, *attribute value*, or *data element*.

Figure 13-2 shows the content of a table as displayed by the Microsoft Access relational DBMS. Note that the first row of the table contains a list of field names (column headings) and that the remaining rows contain a collection of field values that each describe a specific product. Each row contains the same fields in the same order.

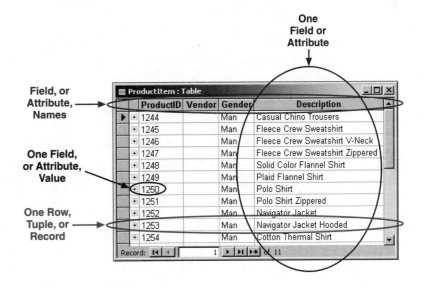

FIGURE *13-2*

A partial display of a relational database table.

key

a field that contains a value that is unique within each row of a relational database table

primary key

a key used to uniquely identify a row of a relational database table

Each table in a relational database must have a unique *key*. A key is a field or set of fields, the values of which occur only once in all the rows of the table. If only one field (or set of fields) is unique, then that key is also called the table's *primary key*. If there are multiple unique fields (or sets of fields), then the database designer must choose one of the possible keys as the primary key.

Key fields may be natural or invented. An example of a natural key field in chemistry is the atomic weight of an element in a table containing descriptive data about elements. Unfortunately, in business, few natural key fields are useful for information processing, so most key fields in a relational database are invented. Your wallet or purse probably contains many examples of invented keys, including your Social Security number, driver's license number, credit card numbers, and ATM card number. Some invented keys are externally assigned (for example, a Federal Express tracking number) and some are internally assigned (for example, ProductID in Figure 13-2). Invented keys are guaranteed to be unique because unique values are assigned by a user, application program, or the DBMS as new rows are added to the table.

Keys are a critical element of relational database design because they are the basis for representing relationships among tables. Keys are the "glue" that binds rows of one table to rows of another table—in other words, keys relate tables to each other. For example, consider the ERD fragment from the Rocky Mountain Outfitters example shown in Figure 13-3 and the tables shown in Figure 13-4. The ERD fragment shows an optional one-to-many relationship between the entities Product Item and Inventory Item. The upper table in Figure 13-4 contains data representing the entity type ProductItem. The lower table contains data representing the entity type InventoryItem.

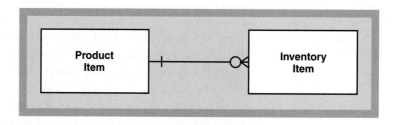

The relationship between the entity types Product Item and Inventory Item is represented by a common field value within their respective tables. The field ProductID (the primary key of the ProductItem table) is also stored within the InventoryItem table, where it is called a foreign key. A *foreign key* is a field that duplicates the primary key of a different (or foreign) table. In Figure 13-4, the existence of the value 1244 as a foreign key within the InventoryItem table indicates that the values of Vendor, Gender, and Description in the first row of the ProductItem table also describe inventory items 86779 through 86790.

foreign key

a field value stored in one relational database table that also exists as a primary key value in another relational database table

Designing Relational Databases

Relational database design begins with either an ERD or a class diagram. This section explains how to create a schema based on an ERD. Schema creation based on a class diagram is discussed later in this chapter.

To create a relational database schema from an ERD, follow these steps:

1. Create a table for each entity type.
2. Choose a primary key for each table (invent one, if necessary).
3. Add foreign keys to represent one-to-many relationships.
4. Create new tables to represent many-to-many relationships.
5. Define referential integrity constraints.
6. Evaluate schema quality and make necessary improvements.
7. Choose appropriate data types and value restrictions (if necessary) for each field.

The following subsections discuss each of these steps in detail.

Representing Entities

The first step to creating a relational DB schema is to create a table for each entity on the ERD. Figure 13-5 shows the ERD for the RMO customer support system. Eleven entities

FIGURE *13-5*

The RMO entity-relationship diagram.

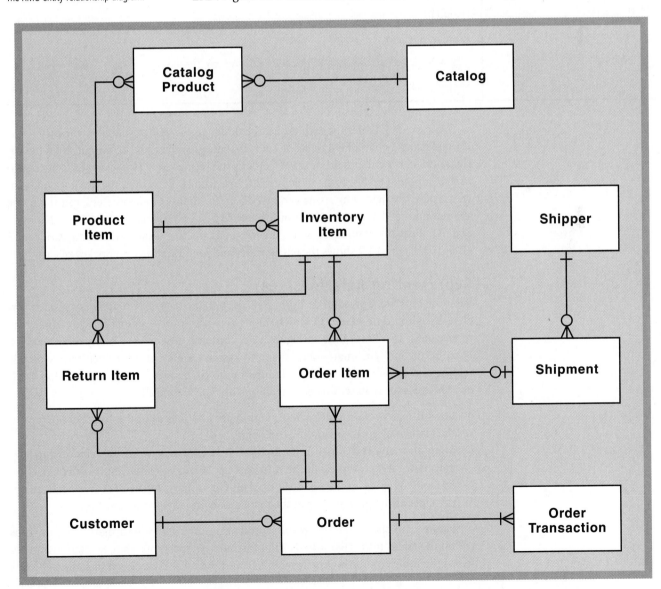

FIGURE *13-6*

An initial set of tables representing the entities on the ERD.

are represented, and a table is created for each entity. The data fields of each table will be the same as those defined for the corresponding entity on the ERD. To avoid confusion, table and field names should match the names used on the ERD and in the project data dictionary. Initial table definitions for the RMO case are shown in Figure 13-6.

Table	Attributes
Catalog	Season, Year, Description, EffectiveDate, EndDate
CatalogProduct	Price, SpecialPrice
Customer	AccountNo, Name, BillingAddress, ShippingAddress, DayTelephoneNumber, NightTelephoneNumber
InventoryItem	InventoryID, Size, Color, Options, QuantityOnHand, AverageCost, ReorderQuantity
Order	OrderID, OrderDate, PriorityCode, ShippingAndHandling, Tax, GrandTotal, EmailAddress, ReplyMethod, PhoneClerk, CallStartTime, LengthOfCall, DateReceived, ProcessorClerk
OrderItem	Quantity, Price, BackorderStatus
OrderTransaction	Date, TransactionType, Amount, PaymentMethod
ProductItem	ProductID, Vendor, Gender, Description
ReturnItem	Quantity, Price, Reason, Condition, Disposal
Shipment	TrackingNo, DateSent, TimeSent, ShippingCost, DateArrived, TimeArrived
Shipper	ShipperID, Name, Address, ContactName, Telephone

After creating tables for each entity, the designer selects a primary key for each table. If a table already has a field or set of fields that are guaranteed to be unique, then the designer can choose that field or set of fields as the primary key (for example, TrackingNo in the Shipment table). If the table contains no primary keys, then the designer must invent a new key field. Any name can be chosen for an invented key field, but the name should indicate that the field is a unique key field. Typical names include Code, Number, and ID, possibly combined with the table name (for example, ProductCode and OrderID). Figure 13-7 shows the entity tables and identifies the primary key of each.

Representing Relationships

Relationships are represented within a relational database by foreign keys. Which foreign keys should be placed in which tables depends on the type of relationship being represented. The RMO ERD in Figure 13-5 contains nine one-to-many relationships. There is one many-to-many relationship between Catalog and Product Item, which is represented by two one-to-many relationships and the associative entity Catalog Product. The rules for representing each relationship type are as follows:

- **One-to-many relationship.** Add the primary key field(s) of the "one" entity type to the table that represents the "many" entity type.
- **Many-to-many relationship.** If no associative entity exists for the relationship, create a new table to represent the relationship. If an associative entity does exist, use its table to represent the relationship. Use the primary key field(s) of the related entity types as the primary key of the table that represents the relationship.

Figure 13-8 shows the results of representing the nine one-to-many relationships within the tables from Figure 13-7. Each foreign key represents a single relationship between the table containing the foreign key and the table that uses that same key as its

Table	Attributes
Catalog	**CatalogID**, Season, Year, Description, EffectiveDate, EndDate
CatalogProduct	**CatalogProductID**, Price, SpecialPrice
Customer	**AccountNo**, Name, BillingAddress, ShippingAddress, DayTelephoneNumber, NightTelephoneNumber
InventoryItem	**InventoryID**, Size, Color, Options, QuantityOnHand, AverageCost, ReorderQuantity
Order	**OrderID**, OrderDate, PriorityCode, ShippingAndHandling, Tax, GrandTotal, EmailAddress, ReplyMethod, PhoneClerk, CallStartTime, LengthOfCall, DateReceived, ProcessorClerk
OrderItem	**OrderItemID**, Quantity, Price, BackorderStatus
OrderTransaction	**OrderTransactionID**, Date, TransactionType, Amount, PaymentMethod
ProductItem	**ProductID**, Vendor, Gender, Description
ReturnItem	**ReturnItemID**, Quantity, Price, Reason, Condition, Disposal
Shipment	**TrackingNo**, DateSent, TimeSent, ShippingCost, DateArrived, TimeArrived
Shipper	**ShipperID**, Name, Address, ContactName, Telephone

FIGURE *13-7*

Entity tables with the primary keys identified in bold.

primary key. For example, the field AccountNo was added to the Order table as a foreign key representing the one-to-many relationship between the entities Customer and Order. The foreign key ShipperID was added to the Shipment table to represent the one-to-many relationship between Shipper and Shipment. When representing one-to-many relationships, foreign keys do not become part of the primary key of the table to which they are added.

Table	Attributes
Catalog	**CatalogID**, Season, Year, Description, EffectiveDate, EndDate
CatalogProduct	**CatalogProductID**, Price, SpecialPrice
Customer	**AccountNo**, Name, BillingAddress, ShippingAddress, DayTelephoneNumber, NightTelephoneNumber
InventoryItem	**InventoryID**, *ProductID*, Size, Color, Options, QuantityOnHand, AverageCost, ReorderQuantity
Order	**OrderID**, *AccountNo*, OrderDate, PriorityCode, ShippingAndHandling, Tax, GrandTotal, EmailAddress, ReplyMethod, PhoneClerk, CallStartTime, LengthOfCall, DateReceived, ProcessorClerk
OrderItem	**OrderItemID**, *OrderID*, *InventoryID*, *TrackingNo*, Quantity, Price, BackorderStatus
OrderTransaction	**OrderTransactionID**, *OrderID*, Date, TransactionType, Amount, PaymentMethod
ProductItem	**ProductID**, Vendor, Gender, Description
ReturnItem	**ReturnItemID**, *OrderID*, *InventoryID*, Quantity, Price, Reason, Condition, Disposal
Shipment	**TrackingNo**, *ShipperID*, DateSent, TimeSent, ShippingCost, DateArrived, TimeArrived
Shipper	**ShipperID**, Name, Address, ContactName, Telephone

FIGURE *13-8*

Represent one-to-many relationships by adding foreign key attributes (shown in italics).

Figure 13-9 expands the table definitions in Figure 13-8 by updating the Catalog-Product table to represent the many-to-many relationship between Catalog and Product-Item. The primary key of the CatalogProduct becomes the combination of CatalogID and ProductID. The old primary key CatalogProductID is discarded. The two fields that

Table	Attributes
Catalog	**CatalogID**, Season, Year, Description, EffectiveDate, EndDate
CatalogProduct	***CatalogID, ProductID***, Price, SpecialPrice
Customer	**AccountNo**, Name, BillingAddress, ShippingAddress, DayTelephoneNumber, NightTelephoneNumber
InventoryItem	**InventoryID**, *ProductID,* Size, Color, Options, QuantityOnHand, AverageCost, ReorderQuantity
Order	**OrderID**, *AccountNo,* OrderDate, PriorityCode, ShippingAndHandling, Tax, GrandTotal, EmailAddress, ReplyMethod, PhoneClerk, CallStartTime, LengthOfCall, DateReceived, ProcessorClerk
OrderItem	**OrderItemID**, *OrderID, InventoryID, TrackingNo,* Quantity, Price, BackorderStatus
OrderTransaction	**OrderTransactionID**, *OrderID,* Date, TransactionType, Amount, PaymentMethod
ProductItem	**ProductID**, Vendor, Gender, Description
ReturnItem	**ReturnItemID**, *OrderID, InventoryID,* Quantity, Price, Reason, Condition, Disposal
Shipment	**TrackingNo**, *ShipperID,* DateSent, TimeSent, ShippingCost, DateArrived, TimeArrived
Shipper	**ShipperID**, Name, Address, ContactName, Telephone

F I G U R E *13-9*

The table CatalogProduct is modified to represent the many-to-many relationship between Catalog and ProductItem.

---

referential integrity

a consistent relational database state in which every foreign key value also exists as a primary key value

---

make up the primary key are also foreign keys. CatalogID is a foreign key from the Catalog table, and ProductID is a foreign key from the ProductItem table.

Enforcing Referential Integrity

Now that we've described how foreign keys are used to represent relationships, we need to describe how to enforce restrictions on the values of those foreign key fields. The term *referential integrity* describes a consistent state among foreign key and primary key values. Each foreign key is a reference to the primary key of another table. In most cases, a database designer wants to ensure that these references are consistent. That is, foreign key values that appear in one table must also appear as the primary key value of the related table. A referential integrity constraint is a constraint on database content—for example, "an order must be from a customer" and "an order item must be something that we normally stock in inventory."

The DBMS usually enforces referential integrity automatically once the schema designer identifies primary and foreign keys. Automatic enforcement is implemented as follows:

- When a row containing a foreign key value is created, the DBMS ensures that the value also exists as a primary key value in the related table.
- When a row is deleted, the DBMS ensures that no foreign keys in related tables have the same value as the primary key of the deleted row.
- When a primary key value is changed, the DBMS ensures that no foreign key values in related tables contain the same value.

In the first case, the DBMS will simply reject any new row containing an unknown foreign key value. In the latter two cases, a database designer usually has some control over how referential integrity is enforced. When a row containing a primary key is deleted, the DBMS can be instructed to delete all rows in other tables with corresponding keys. Or, the designer can instruct the DBMS to set all corresponding foreign keys to NULL. A similar choice is available when a primary key value is changed. The DBMS can be instructed to change all corresponding foreign key values to the same value or to set foreign key values to NULL.

Evaluating Schema Quality

After creating a complete set of tables, the designer should check the entire schema for quality. Ironing out any schema problems at this point ensures that none of the later design effort will be wasted. A high-quality data model has the following features:

- Uniqueness of table rows and primary keys
- Lack of redundant data
- Ease of implementing future data model changes

Unfortunately, there are few objective or quantitative measures of database schema quality. Database design is the final step in a modeling process, and as such, it depends on the analyst's experience and judgment. Various formal and informal techniques for schema evaluation are described in the following sections. No one technique is sufficient by itself, but a combination of techniques can ensure a high-quality database design.

Row and Key Uniqueness

A fundamental requirement of all relational data models is that primary keys and table rows be unique. Since each table must have a primary key, uniqueness of rows within a table is guaranteed if the primary key is unique. Data access logic within programs usually assumes that keys are unique. For example, a programmer writing a program to view customer records will generally assume that a database query for a specific customer number will return one and only one row (or none if the customer doesn't exist in the database). The program will be designed around this assumption and will probably fail if the DBMS returns two records.

A designer evaluates primary key uniqueness by examining assumptions about key content, the set of possible key values, and the methods by which key values are assigned. Internally invented keys are relatively simple to evaluate in this regard because the system itself creates them. That is, an information system that uses invented keys can guarantee uniqueness by implementing appropriate procedures to assign key values to newly created rows.

It is common for several different programs in an information system to be capable of creating new database rows. Each program needs to be able to assign keys to newly created database rows. However, the importance of key uniqueness requires that key-creation procedures be consistently applied throughout the information system.

Because key creation and management are such pervasive problems in information systems, many relational DBMSs automate key creation. Such systems typically automate a special data type for invented keys (for example, the AutoNumber type in Microsoft Access). The DBMS automatically assigns a key value to newly created rows and communicates that value to the application program for use in subsequent database operations. Embedding this capability in the DBMS frees the IS developer from designing and implementing customized key-creation software modules.

Invented keys that aren't assigned by the information system must be given careful scrutiny to ascertain their uniqueness and usefulness over time. For example, employee databases in the United States commonly use Social Security numbers as keys. Since the U.S. government has a strong interest in guaranteeing the uniqueness of Social Security numbers, the assumption that they will always be unique seems safe. But other assumptions concerning their use deserve closer examination. For example, will all employees who are stored in the database have a Social Security number? What if the company opens a manufacturing facility in Europe or South America?

Invented keys assigned by nongovernmental agencies deserve even more careful scrutiny. For example, Federal Express, UPS, and most shipping companies assign a tracking number to each shipment they process. Tracking numbers are guaranteed to be unique at any given point in time, but are they guaranteed to be unique forever (that is, are they ever reused)? Could reuse of a tracking number cause a primary key duplication in the RMO database? And what would happen if two different shippers assigned the same tracking number to two different shipments?

Uncertainties such as these make internally invented keys the safest long-term strategy in most cases. Although internally invented keys may initially entail additional design and development, they prevent one possible source of upheaval once the database is installed. Few changes have the pervasive and disruptive impact of a database key change in a large information system with terabytes of stored data and thousands of application programs and stored queries.

Data Model Flexibility

Database flexibility and maintainability were primary goals in the original specification of the relational database model. A database model is considered flexible and maintainable if changes to the database schema can be made with minimal disruption to existing data content and structure. For example, adding a new entity to the schema should not require redefining existing tables. Adding a new one-to-many relationship should only require adding a new foreign key to an existing table. Adding a new many-to-many relationship should only require adding a single new table to the schema.

Data redundancy plays a key role in determining the flexibility and maintainability of any database or data model. A truism of database processing is that "redundant storage requires redundant maintenance." That is, if data are stored in multiple places, then each of those places must be found and manipulated when data are added, changed, or deleted. Of course, performing any of those actions on multiple data storage locations is more complex (and less efficient) than performing them on a single location. Failure to update, modify, or delete multiple copies of the same information creates a condition called *inconsistency*. By definition, inconsistency cannot occur if information is stored only once.

The relational data model deliberately stores key values multiple times (that is, redundantly) and non-key fields only once. Key values are stored once as a primary key and again each time they are used as a foreign key. The model requires such redundancy because correspondence between the primary and foreign key is the only way to represent relationships among tables, but using redundant key values adds complexity to processes that manipulate key fields.

Relational DBMSs ensure consistency among primary and foreign keys by enforcing referential integrity constraints, but there are no automatic mechanisms for ensuring consistency among other redundant data items. Thus, the best way to avoid inconsistency in a relational database is to avoid redundancy in non-key fields. Database designers can avoid such redundancy by never introducing it into a schema—but it is all too easy to let redundancy slip in. If data redundancy is somehow introduced into the schema, then it must be identified and removed. The most commonly used process to detect and eliminate redundancy is database normalization.

Database Normalization

Normalization is a formal technique used to evaluate the quality of a relational database schema. It determines whether a database schema contains any of the "wrong" kinds of redundancy and defines specific methods to eliminate them. Normalization is based on a concept called *functional dependency* and on a series of normal forms:

- **First normal form (1NF).** A table is in *first normal form* if it contains no repeating fields or groups of fields.
- **Functional dependency.** A *functional dependency* is a one-to-one relationship between the values of two fields. The relationship is formally stated as follows: *Field A is functionally dependent on field B if for each value of B there is only one corresponding value of A.*

normalization

a technique that ensures relational database schema quality by minimizing data redundancy

first normal form (1NF)

a relational database table structure that has no repeating fields or groups of fields

functional dependency

a one-to-one correspondence between two field values

■ **Second normal form (2NF).** A table is in *second normal form* if it is in first normal form and if each non-key element is functionally dependent on the entire primary key.

■ **Third normal form (3NF).** A table is in *third normal form* if it is in second normal form and if no non-key element is functionally dependent on any other non-key element.

Let's explain these concepts further.

First Normal Form   First normal form defines a structural constraint on table rows. Repeating fields such as Dependent in Figure 13-10 are not allowed within any table in a relational database. Repeating groups of fields are also prohibited. In practice, this constraint is not difficult to enforce since relational DBMSs do not allow a designer to define a table containing repeating fields.

SSN	Name	Department	Salary	Dependent1	Dependent2	Dependent3 ...	DependentN
111-22-3333	Mary Smith	Accounting	40,000	John	Alice	Dave	
222-33-4444	Jose Pena	Marketing	50,000				
333-44-5555	Frank Collins	Production	35,000	Jane	Julia		

F I G U R E *13-10*

An employee table with a repeating field.

Functional Dependency   Functional dependency is a difficult concept to describe and apply. The most precise way to determine whether functional dependency exists is to pick two fields in a table and insert their names in the italicized portion of the definition shown previously. For example, consider the fields ProductID and Description in the ProductItem table (see Figure 13-4). ProductID is an internally invented primary key that is guaranteed to be unique within the table. To determine whether Description is functionally dependent on ProductID, substitute Description for field A and ProductID for field B in the italicized portion of the functional dependency definition:

> Description is functionally dependent on ProductID if for each value of ProductID there is only one corresponding value of Description.

Now ask whether the statement is true for all rows that could possibly exist in the ProductItem table. If the statement is true, then Description is functionally dependent on ProductID. As long as the invented key ProductID is guaranteed to be unique within the ProductItem table, then the preceding statement is true. Therefore, Description is functionally dependent on ProductID.

A less formal way to analyze functional dependency of Description on ProductID is to remember that the ProductItem table represents a specific product sold by RMO. If that product can have only a single description in the database, then Description is functionally dependent on the key of the table that represents products (ProductID). If it is possible for any product to have multiple descriptions, then the field Description is not functionally dependent on ProductID.

Second Normal Form   To evaluate whether the ProductItem table is in second normal form, we must first determine whether it is in first normal form. Since it contains no repeating fields, it is in first normal form. Then we must determine whether every non-key field is functionally dependent on ProductID (that is, consider each field in turn by substituting it for A in the functional dependency definition). If all the non-key fields are functionally dependent on ProductID, the ProductItem table is in 2NF. If one or more non-key fields are not functionally dependent on ProductID, then the table is not in 2NF.

Verifying that a table is in 2NF is more complicated when the primary key consists of two or more fields. For example, consider the RMO table CatalogProduct shown in

Figure 13-11. Recall that this table represents a many-to-many relationship between Catalog and ProductItem. Thus, the table representing this relationship has a primary key consisting of the primary keys of Catalog (CatalogID) and ProductItem (ProductID). The table also contains two non-key fields called Price and SpecialPrice.

FIGURE *13-11*

A simplified RMO CatalogProduct table.

CatalogID	ProductID	Price	SpecialPrice
23	1244	$15.00	$12.00
23	1245	$15.00	$12.00
23	1246	$15.00	$13.00
23	1247	$15.00	$13.00
23	1248	$14.00	$11.20
23	1249	$14.00	$11.20
23	1252	$21.00	$16.80
23	1253	$21.00	$16.40
23	1254	$24.00	$19.20
23	1257	$19.00	$15.20

If this table is in 2NF, then each non-key field must be functionally dependent on the *combination* of CatalogID and ProductID. We can verify the first functional dependency by substituting terms in the functional dependency definition and determining the truth or falsity of the resulting statement:

Price is functionally dependent on the combination of CatalogID and ProductID if for each combination of values for CatalogID and ProductID there is only one corresponding value of Price.

Analyzing the truth of the preceding statement is tricky, since you must consider all the possible combinations of key values that might occur in the CatalogProduct file. A simpler way to approach the question is to think about the underlying entities represented in the table. A product can appear in many different catalogs. If Price can be different in different catalogs, then the preceding statement is true. If a product's Price is always the same, regardless of the catalog in which it appears, then the preceding statement is false and the table is not in 2NF. The correct answer doesn't depend on any universal sense of truth. Instead, it depends on RMO's normal conventions for setting product prices in different catalogs.

If a non-key field is functionally dependent on only part of the primary key, then you must remove the non-key field from its present table and place it in another table. For example, consider a modified version of the CatalogProduct table as shown in the upper half of Figure 13-12. The non-key field CatalogIssueDate is functionally dependent only on CatalogID, not on the combination of CatalogID and ProductID. Thus, the table is not in 2NF.

To correct the problem, you must remove CatalogIssueDate from the CatalogProduct table and place it in a table that uses CatalogID alone as the primary key. Since the Catalog table in Figure 13-9 uses CatalogID alone as its primary key, you should add CatalogIssueDate to that table. If a Catalog table did not already exist, then you would need to create a new table to hold CatalogIssueDate, as shown in Figure 13-12.

Third Normal Form   To verify that a table is in 3NF, we must check the functional dependency of each non-key element on every other non-key element. This can be cumbersome for a large table since the number of pairs that must be checked grows quickly as the number of non-key fields grows. The number of functional dependencies to be checked is $N \times (N - 1)$, where N is the number of non-key fields. Note that functional dependency must be checked in both directions (that is, A dependent on B, and B dependent on A).

FIGURE *13-12*

Decomposition of a first normal
form table into two second
normal form tables.

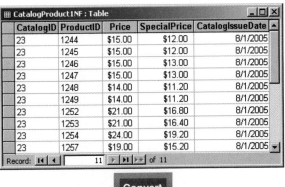

In practice, you can simplify finding 3NF violations by concentrating on two common types of problems:

- Tables that store attributes describing two or more entities
- Computable fields

Consider the simple table shown in the upper half of Figure 13-13. Assume that AccountNo is the primary key and that all customers live in the United States. Since there are three non-key fields, you must check six functional dependencies:

- Is State functionally dependent on StreetAddress?
- Is StreetAddress functionally dependent on State?
- Is ZipCode functionally dependent on StreetAddress?
- Is StreetAddress functionally dependent on ZipCode?
- Is ZipCode functionally dependent on State?
- Is State functionally dependent on ZipCode?

The answer to the first five statements is no, but the answer to the last is yes. In the United States, all zip codes are wholly contained within a single state. Thus, for each value of Zip Code there is only one corresponding value of State (for example, 87123 is always in New Mexico). Including both fields in this table is a form of redundancy. For example, if the addresses of 100 customers who live in the 87123 zip code are stored in the table, the fact that 87123 is located in New Mexico is (redundantly) stored 100 times.

In essence, the table combines information about two entities—Customer and Postal Delivery Area. Each entity has its own primary key (AccountNo for Customer and ZipCode for Postal Delivery Area) and its own non-key attributes (StreetAddress and Zip Code for Customer, and State for Postal Delivery Area).

FIGURE *13-13*

Converting a second normal form table into two third normal form tables.

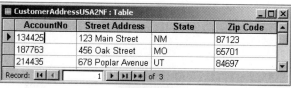

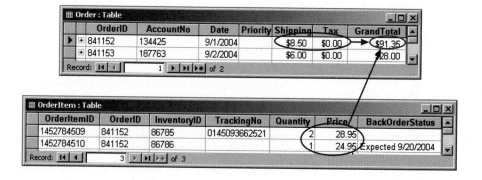

Because State is functionally dependent on Zip Code, the table is not in 3NF. To correct this problem, you must remove State from the table. The relationship between states and zip codes must be stored somewhere in the database, or the information system can't generate complete mailing labels. The solution is to create a new table containing only ZipCode and State (see Figure 13-13) and remove State from the Customer table. ZipCode is the primary key of the new table, and State is its only non-key field. Programs or methods that print or display a complete mailing address must now use the value of Zip Code in the CustomerAddress table to look up the corresponding value of State in the newly created table.

A computable field stores a value that can be computed by a formula or algorithm that uses other database fields as inputs. Common examples of computable fields include subtotals, totals, and taxes. For example, consider the field GrandTotal in the Order table in Figure 13-9 and the formula:

$$GrandTotal = (\Sigma\ Quantity \times Price) + Shipping + Tax$$

Note that all of the inputs to the formula are not stored in the same table (see Figure 13-14). Unlike 3NF violations involving multiple entities stored in the same table, problems with computable fields can involve multiple tables. Shipping and Tax are stored in the Order table, and Quantity and Price are stored in related rows of the OrderItem table. An algorithm that computes GrandTotal for a particular invoice needs to extract all matching rows in the OrderItem table using the OrderID foreign key.

FIGURE *13-14*

GrandTotal is computed from fields in two tables.

GrandTotal is functionally dependent on the combination of the other four fields. Computational dependencies are a form of redundancy because a change to the value of any input variable in the computation (for example, Shipping) also changes the result of the computation (in other words, GrandTotal).

The way to correct this type of 3NF violation is simple: Remove the computed field from the database. Eliminating the computed field from the database doesn't mean that its value is lost. For example, any program or method that needs GrandTotal can query the OrderItem table for matching values of Quantity and Price, sum the result of multiplying each Quantity and Price, and add Shipping and Tax.

Entity-Relationship Modeling and Normalization
Entity-relationship modeling and normalization are complementary techniques for relational database design. Note that the tables generated from the RMO ERD (see Figure 13-9) do not contain any violations of first, second, or third normal form. This is not a chance occurrence. Attributes of an entity are functionally dependent on any unique identifier (primary key) of that entity. Attributes of a many-to-many relationship are functionally dependent on unique identifiers of both participating entities. Thus, while creating an ERD, an analyst must directly or indirectly consider issues of functional dependency when deciding which attributes belong with which entities or relationships.

We now turn our attention to the second type of database commonly in use today—the object-oriented database.

OBJECT-ORIENTED DATABASES

**object database
management system
(ODBMS)**

a database management system
that stores data as objects or class
instances

Object database management systems (ODBMSs) are a direct extension of the OO design and programming paradigm. ODBMSs are designed specifically to store objects and to interface with object-oriented programming languages. It is possible to store objects in files or relational databases. But there are many advantages to using an ODBMS, including direct support for method storage, inheritance, nested objects, object linking, and programmer-defined data types.

ODBMSs first appeared as research prototypes in the 1980s and later as fledgling commercial products in the early 1990s. Current commercial ODBMSs include Gem-Stone, ObjectStore, and Objectivity. ODBMSs are the database technology of choice for newly designed systems implemented with OO tools, especially for scientific and engineering applications. ODBMSs are expected to supplant RDBMSs in more traditional business applications gradually over the next decade.

Because ODBMSs are relatively new, there are few widely accepted standards for specifying an object database schema. Some object database standards that are gaining wide acceptance are those proposed by the Object Database Management Group. One of these standards is the *Object Definition Language (ODL)*. ODL is a language for describing the structure and content of an object database. The schema examples in the sections that follow use ODL.

**Object Definition Language
(ODL)**

a standard object database description
language promulgated by the Object
Database Management Group

Designing Object Databases

To create an object database schema from a class diagram, follow these steps:

1. Determine which classes require persistent storage.
2. Define persistent classes.
3. Represent relationships among persistent classes.
4. Choose appropriate data types and value restrictions (if necessary) for each field.

Each of these steps is discussed in detail in the following sections.

Representing Classes

Objects can be classified into two broad types for purposes of data management. A *transient object* exists only during the lifetime of a program or process. In a design that follows three-layer architecture, view layer objects such as windows and forms are usually transient. Transient objects are created each time a program or process is executed and then destroyed when a program or process terminates.

A *persistent object* is not destroyed when the program or process that creates it ceases execution. Instead, the object continues to exist independently of any program or process. In a design that follows three-layer architecture, problem domain (or business) classes are usually persistent. Storing the object state to persistent memory (such as a magnetic or optical disk) ensures that the object exists between process executions. Objects can be persistently stored within a file or database management system.

An object database schema includes a definition for each class that requires persistent storage. ODL class definitions derive from the corresponding UML class diagram. Thus, classes already defined in UML are simply reused for the database schema definition.

For example, an ODL description of the RMO Customer class is:

```
class Customer {
  attribute string accountNo
  attribute string name
  attribute string billingAddress
  attribute string shippingAddress
  attribute string dayTelephoneNumber
  attribute string nightTelephoneNumber
}
```

This ODL class definition corresponds to the Customer table in Figure 13-9. A similar ODL class definition would be created for each class shown in Figure 13-15. After defining each class, the analyst must define relationships among classes.

Representing Relationships

Each object stored within an ODBMS is automatically assigned a unique object identifier. An *object identifier* may be a physical storage address or a reference that can be converted to a physical storage address at run time. In either case, each object has a unique identifier that can be stored within another object to represent a relationship.

An ODBMS represents relationships by storing the identifier of one object within related objects. Object identifiers provide navigation visibility among objects, as first described in Chapter 11. For example, consider a one-to-one relationship between the classes Employee and Workstation as shown in Figure 13-16. Each Employee object has an attribute called *computer* that contains the object identifier of the Workstation object assigned to that employee. Each Workstation object has a matching attribute called *user* that contains the object identifier of the Employee who uses that workstation.

The ODBMS uses attributes containing object identifiers to find objects that are related to other objects. The process of extracting an object identifier from one object and using it to access another object is sometimes called *navigation*. For example, consider the following query posed by a user:

List the manufacturer of the workstation assigned to employee Joe Smith.

An ODBMS query processor can find the requested employee object by searching all employee objects for the name attribute Joe Smith. The query processor can find Joe Smith's workstation object by using the object identifier stored in computer. The query processor can also answer the opposite query (list the employee name assigned to a

---

transient object

an object that doesn't need to store any attribute values between instantiations or method invocations

persistent object

an object that must store one or more attribute values between instantiations or method invocations

object identifier

a physical storage address or a reference that can be converted to a physical storage address at run time

navigation

the process of accessing an object by extracting its object identifier from another (related) object

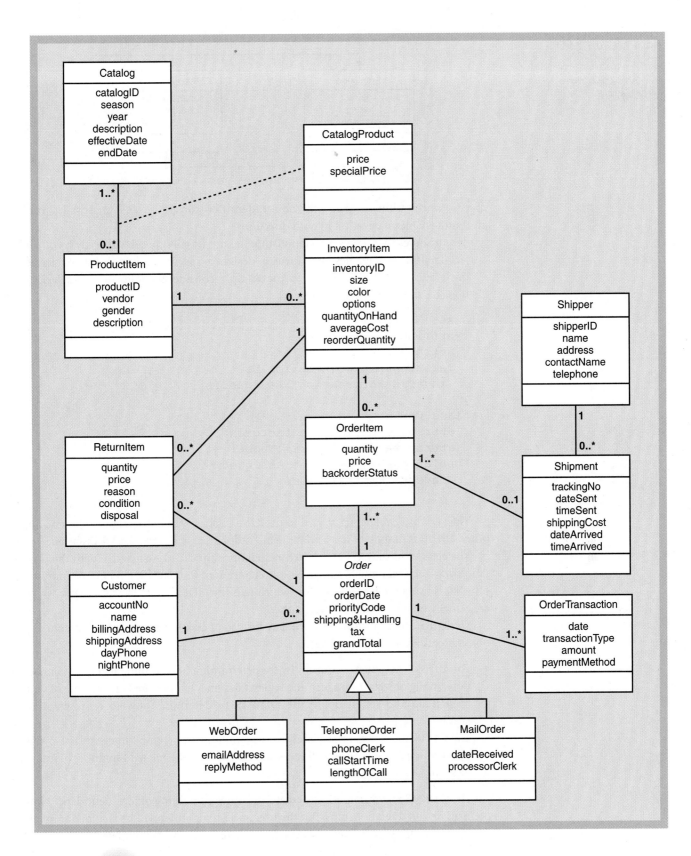

FIGURE *13-15*

The RMO class diagram.

Chapter 13   Designing Databases   **505**

FIGURE *13-16*

A one-to-one relationship represented with attributes (shown in color) containing object identifiers.

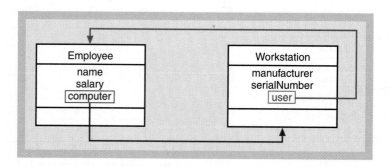

specific workstation) by using the object identifier stored in user. A matched pair of attributes enables navigation in both directions.

Attributes that represent relationships are not usually specified directly by an object database schema designer. Instead, designers specify them indirectly by declaring relationships between objects. For example, consider the following class declarations for the ODL schema language:

```
class Employee {
    attribute string name
    attribute integer salary
    relationship Workstation Uses
     inverse Workstation::AssignedTo
}
class Workstation {
    attribute string manufacturer
    attribute string serialNumber
    relationship Employee AssignedTo
     inverse Employee::Uses
}
```

The keyword *relationship* is used to declare a relationship between one class and another. The class Employee has a relationship called Uses with the class Workstation.

The class Workstation has a matching relationship called AssignedTo with the class Employee. Each relationship includes a declaration of the matching relationship in the other class using the keyword *inverse*, which tells the ODBMS that the two relationships are actually mirror images of one another.

Declaring a relationship as shown here instead of creating an attribute containing an object identifier has two advantages:

▪ The ODBMS assumes responsibility for determining how to implement the connection among objects. In essence, the schema designer has declared an attribute of type *relationship* and left it up to the ODBMS to determine how to represent that attribute.

▪ The ODBMS assumes responsibility for maintaining referential integrity. For example, deleting a workstation will cause the Uses link of the related Employee object to be set to NULL or undefined.

The ODBMS will automatically create attributes containing object identifiers to implement declared relationships. But the user and programmer will be shielded from all details of how those identifiers are actually implemented and manipulated.

One-to-Many Relationships Figure 13-17 shows the one-to-many relationship between the RMO classes Customer and Order. A Customer can make many different Orders, but a single Order can be made by only one Customer. A single object identifier represents the relationship of an Order to a Customer. Multiple object identifiers represent the relationship between one Customer and many different Orders, as shown in Figure 13-18.

FIGURE *13-17*

The one-to-many relationship between the Customer and Order classes.

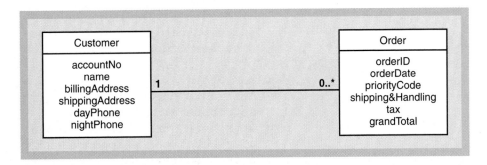

FIGURE *13-18*

A one-to-many relationship represented with attributes containing object identifiers.

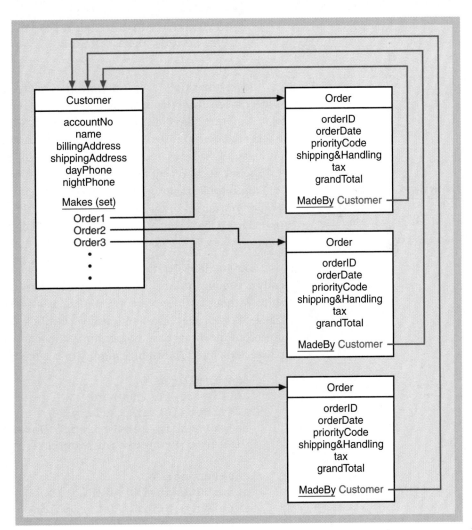

Partial ODL class declarations for the classes Customer and Order are as follows:

```
class Customer {
  attribute string accountNo
  attribute string name
  attribute string billingAddress
  attribute string shippingAddress
  attribute string dayPhone
  attribute string nightPhone
  relationship set<Order> Makes
    inverse Order::MadeBy
}
class Order {
  attribute string orderID
  attribute string orderDate
  attribute string priorityCode
  attribute real shipping&Handling
  attribute real tax
  attribute real grandTotal
  relationship Order MadeBy
    inverse Customer::Makes
}
```

The relationship Makes is declared between a single Customer object and a set of Order objects. By declaring the relationship as a set, you instruct the ODBMS to allocate as many Order object identifier attributes to each Customer object as are needed to represent relationship instances. The ODBMS dynamically adds or deletes object identifier attributes to the set as instances of the relationship are created or deleted.

The set of object identifier attributes can also be called a multivalued attribute. A *multivalued attribute*, also called a repeating group, is an attribute that contains zero or more instances of the same data type. Multivalued attributes are commonly supported in ODBMSs but are not supported in RDBMSs because they violate first normal form.

multivalued attribute

an attribute that contains zero or more instances of the same data type

Many-to-Many Relationships   A many-to-many relationship is represented differently depending on whether the relationship has any attributes. Many-to-many relationships without attributes are represented as a set of object attributes in both related classes. Both classes have a multivalued attribute containing object pointers to related objects of the other class. For example, the many-to-many relationship between Employee and Project shown in Figure 13-19 is represented as follows:

```
class Employee {
  attribute string name
  attribute string salary
  relationship set<Project> WorksOn
    inverse Project::Assigned
}
class Project {
  attribute string projectID
  attribute string description
  attribute string startDate
  attribute string endDate
  relationship set<Employee> Assigned
    inverse Employee::WorksOn
}
```

FIGURE *13-19*

A many-to-many relationship between
the Employee and Project classes.

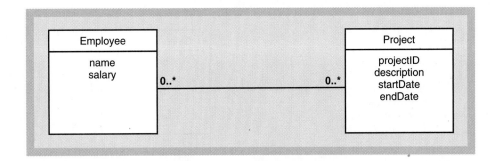

Representing a many-to-many relationship with attributes requires a more complex approach. The RMO class diagram has a many-to-many relationship between Catalog and ProductItem with an association class named CatalogProduct (see Figure 13-15). Recall from Chapter 5 that an *association class* is a class that stores the attributes of a many-to-many relationship.

To represent a many-to-many relationship with an association class, we must reorganize the relationship as shown in Figure 13-20. The many-to-many relationship between Catalog and ProductItem has been decomposed into a pair of one-to-many relationships between the original classes and the association class. The ODL schema descriptions are as follows:

```
class Catalog {
   attribute string season
   attribute integer year
   attribute string description
   attribute string effectiveDate
   attribute string endDate
   relationship set<CatalogProduct> Contains1
     inverse CatalogProduct::AppearsIn1
}
class ProductItem {
   attribute string productID
   attribute string vendor
   attribute string gender
   attribute string description
   relationship set<CatalogProduct> AppearsIn2
     inverse CatalogProduct::Contains2
}
class CatalogProduct {
   attribute real price
   attribute real specialPrice
   relationship Catalog AppearsIn1
     inverse Catalog::Contains1
   relationship ProductItem AppearsIn2
     inverse ProductItem::Contains2
}
```

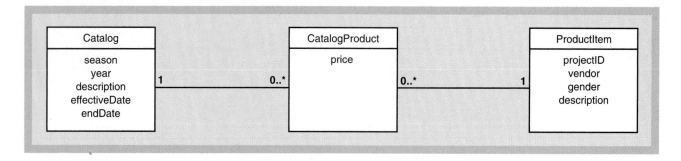

FIGURE *13-20*

A many-to-many relationship represented with two one-to-many relationships.

Generalization Relationships Figure 13-21 shows the order generalization hierarchy from the RMO class diagram. WebOrder, TelephoneOrder, and MailOrder are each more specific versions of the class Order. The ODL class definitions that represent these classes and their interrelationships are as follows:

```
class Order {
    attribute  string  orderID
    attribute  string  orderDate
    attribute  string  priorityCode
    attribute  real  shipping&Handling
    attribute  real  tax
    attribute  real  grandTotal
}
class WebOrder extends Order {
    attribute  string  emailAddress
    attribute  string  replyMethod
}
class TelephoneOrder extends Order {
    attribute  string  phoneClerk
    attribute  string  callStartTime
    attribute  integer  lengthOfCall
}
class MailOrder extends Order {
    attribute  string  dateReceived
    attribute  string  processorClerk
}
```

The keyword *extends* indicates that WebOrder, TelephoneOrder, and MailOrder derive from Order. When stored in an object database, objects of the three derived classes will inherit all of the attributes, methods, and relationships defined for the Order class.

FIGURE *13-21*

A generalization hierarchy within
the RMO class diagram.

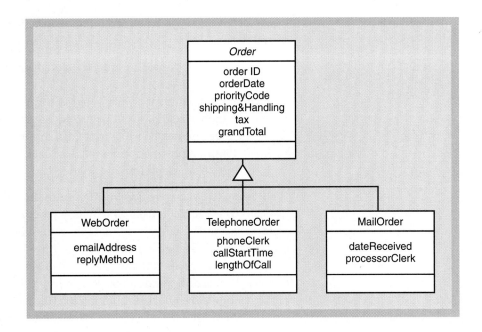

Key Attributes

Key attributes are not required in an object database since referential integrity is implemented with object identifiers. However, key attributes are useful in object databases for a number of purposes, including guaranteeing unique object content and providing a means of querying database contents. The ODBMS automatically enforces uniqueness of key attributes in an object database. Thus, declaring an attribute to be a key guarantees that no more than one object in a class can have the same key value.

In addition to relational and object-oriented databases, a third type exists that mixes elements of both the relational and OO approaches. We discuss this hybrid next.

HYBRID OBJECT-RELATIONAL DATABASE DESIGN

OO development tools were first widely employed in the mid- to late-1980s. During this same time period, RDBMSs were widely used and had reached a mature stage of development. Many OO tool developers exploited the existing base of RDBMS tools and knowledge by using an RDBMS to store persistent object states. This made sense both as an economy measure (there was one less OO tool to develop) and because many newer OO systems needed to manipulate data stored in existing relational databases. There is no widely accepted name to describe object storage using an RDBMS, so we will invent one to use for the remainder of the text: *hybrid object-relational DBMS* (or simply hybrid DBMS).

The hybrid DBMS approach is currently the most widely employed approach for persistent object storage. Designing a hybrid database is essentially two design problems in one. That is, the designer must develop a complete relational database schema and an equivalent set of classes to represent the relational database contents within the OO programs. This second task is complex because the designer must bridge the differences between the object-oriented and relational views of stored data.

hybrid object-relational DBMS

a relational database management system used to store object attributes and relationships; also called *hybrid DBMS*

Following are the most important mismatches between the relational and OO views of stored data:

- Class methods cannot be directly stored or automatically executed within an RDBMS.
- ODBMSs can represent a wider range of relationship types than RDBMSs, including classification hierarchies and whole-part aggregations. Relationships in an RDBMS can only be represented using referential integrity.
- ODBMSs can represent a wider range of data types than RDBMSs. New classes can be defined to store application-specific data.

Because RDBMSs were developed prior to the OO paradigm, they have no features that can represent methods or inheritance. Programs that access the database must implement methods internally. Inheritance cannot be directly represented in an RDBMS because a classification hierarchy cannot be directly represented.

Although the relational and OO views of stored data have significant differences, they also have significant overlaps. Recall from Chapter 5 that "things" within a system can be conceptually modeled using an ERD (the basis for a relational database schema), a class diagram (the basis for an OO database schema), or both. There is considerable overlap among the two representations, including the following:

- Grouping of data items into entities or classes
- Defining one-to-one, one-to-many, and many-to-many relationships among entities or classes

This overlap provides a basis for representing classes and objects within a relational database.

Classes and Attributes

Designers can store classes and object attributes in an RDBMS by defining appropriate tables in which to store them. For a completely new system, a relational schema can be designed based on a class diagram—essentially the same process as for an ERD. Figure 13-22 describes the correspondence among OO, ER, and relational database concepts. A table is created to represent each class, and the fields of each table are the same as the attributes of the corresponding class. Each row holds the attribute values of a single object.

FIGURE *13-22*

Correspondence among concepts in the object-oriented, entity-relationship, and relational database views of stored data.

Object-Oriented	Entity-Relationship	Relational Database
Class	Entity Type	Table
Object	Entity Instance	Row
Attribute	Attribute	Column

A key field (or group of fields) must be chosen for each table. As described earlier, a designer can choose a natural or invented key field from the existing attributes or add an invented key field. Primary key fields are needed to guarantee uniqueness within tables and to represent relationships using foreign keys.

Figure 13-23 shows a set of relational database tables that represent classes from the RMO class diagram in Figure 13-15. Note that the table definitions are identical to those in Figure 13-7, except for the addition of tables to represent the specialized classes MailOrder, TelephoneOrder, and WebOrder and corresponding changes to the Order table.

Table	Attributes
Catalog	**CatalogID**, Season, Year, Description, EffectiveDate, EndDate
CatalogProduct	**CatalogProductID**, Price, SpecialPrice
Customer	**AccountNo**, Name, BillingAddress, ShippingAddress, DayTelephoneNumber, NightTelephoneNumber
InventoryItem	**InventoryID**, Size, Color, Options, QuantityOnHand, AverageCost, ReorderQuantity
MailOrder	**MailOrderID**, DateReceived, ProcessorClerk
Order	**OrderID**, OrderDate, PriorityCode, ShippingAndHandling, Tax, GrandTotal,
OrderItem	**OrderItemID**, Quantity, Price, BackorderStatus
OrderTransaction	**OrderTransactionID**, Date, TransactionType, Amount, PaymentMethod
ProductItem	**ProductID**, Vendor, Gender, Description
ReturnItem	**ReturnItemID**, Quantity, Price, Reason, Condition, Disposal
Shipment	**TrackingNo**, DateSent, TimeSent, ShippingCost, DateArrived, TimeArrived
Shipper	**ShipperID**, Name, Address, ContactName, Telephone
TelephoneOrder	**TelephoneOrderID**, PhoneClerk, CallStartTime, LengthOfCall
WebOrder	**WebOrderID**, EmailAddress, ReplyMethod

FIGURE *13-23*

Class tables, with primary keys identified in bold.

Relationships

ODBMSs use object identifiers to represent relationships among objects. But RDBMSs do not create object identifiers, so relationships among objects stored in a relational database must be represented using foreign keys. Foreign key values serve the same purpose as object identifiers in an ODBMS. That is, they provide a means for one "object" to refer to another.

To represent one-to-many relationships, designers add the primary key field of the class on the "one" side of the relationship to the table representing the class on the "many" side of the relationship. To represent many-to-many relationships, designers create a new table that contains the primary key fields of the related class tables and any attributes of the relationship itself. Note that these methods of representing relationships among objects are the same as previously described for representing relationships among entities.

Figure 13-24 extends the table definitions in Figure 13-23 by adding foreign keys representing the relationships shown in Figure 13-15. For example, the one-to-many relationship between the Customer and Order classes is represented by the foreign key AccountNo stored in the Order table. The many-to-many relationship between the Catalog and ProductItem classes is represented by the table CatalogProduct that contains the foreign keys CatalogID and ProductID.

Table	Attributes
Catalog	**CatalogID**, Season, Year, Description, EffectiveDate, EndDate
CatalogProduct	***CatalogID, ProductID,*** Price, SpecialPrice
Customer	**AccountNo**, Name, BillingAddress, ShippingAddress, DayTelephoneNumber, NightTelephoneNumber
InventoryItem	**InventoryID**, *ProductID*, Size, Color, Options, QuantityOnHand, AverageCost, ReorderQuantity
MailOrder	***OrderID,*** DateReceived, ProcessorClerk
Order	**OrderID**, *AccountNo*, OrderDate, PriorityCode, ShippingAndHandling, Tax, GrandTotal,
OrderItem	**OrderItemID**, *OrderID*, *InventoryID*, *TrackingNo*, Quantity, Price, BackorderStatus
OrderTransaction	**OrderTransactionID**, *OrderID*, Date, TransactionType, Amount, PaymentMethod
ProductItem	**ProductID**, Vendor, Gender, Description
ReturnItem	**ReturnItemID**, *OrderID*, *InventoryID*, Quantity, Price, Reason, Condition, Disposal
Shipment	**TrackingNo**, *ShipperID*, DateSent, TimeSent, ShippingCost, DateArrived, TimeArrived
Shipper	**ShipperID**, Name, Address, ContactName, Telephone
TelephoneOrder	***OrderID,*** PhoneClerk, CallStartTime, LengthOfCall
WebOrder	***OrderID,*** EmailAddress, ReplyMethod

FIGURE *13-24*

Relationship information added to the class tables by adding foreign key attributes (shown in italics) to represent relationships.

Note that the tables in Figure 13-24 are identical to those in Figure 13-9 except for the content of the tables Order, MailOrder, TelephoneOrder, and WebOrder (these tables will be discussed shortly). The similarity is no accident; it follows from the similarity between the RMO entity-relationship and class diagrams. The diagrams are similar because they represent the same underlying reality. Thus, it should be no surprise that the relational database schemas derived from a class diagram and an ERD representing that underlying reality are similar. In fact, it would be surprising (and probably indicative of an error) if they weren't similar.

Classification relationships such as the relationship among Order, MailOrder, TelephoneOrder, and WebOrder are a special case in relational database design. Just as a child class inherits the data and methods of a parent class, a table representing a child class inherits some or all of its data from the table representing its parent class. This inheritance can be represented in two ways:

- Combine all the tables into a single table containing the superset of all class attributes but excluding any invented key fields of the child classes.
- Use separate tables to represent the child classes and substitute the primary key of the parent class table for the invented keys of the child class tables.

Either method is an acceptable approach to representing a classification relationship.

Figure 13-9 shows the definition of the Order table under the first method. All of the non-key fields from MailOrder, TelephoneOrder, and WebOrder have been added to the Order table. For any particular order, some of the field values in each row will be NULL. For example, a row representing a telephone order would have no values for the fields EmailAddress, ReplyMethod, DateReceived, and ProcessorClerk.

Figure 13-24 shows the table definitions for the RMO case using the second method for representing inheritance. The relationship among the three child order types and the parent Order table is represented by the foreign key OrderID in all three child class tables. The invented key of each table has been removed. Thus, in each case, the foreign key representing the inheritance relationship also serves as the primary key of the table representing the child class.

Data Access Classes

In Chapter 11 you learned how to develop an OO design based on three-layer architecture. Under that architecture, data access classes implement the bridge between data stored in program objects and in a relational database.

Figure 13-25 illustrates the interaction between the RMO problem domain class ProductItem, the data access class ProductItemDA, and the relational database. The data access class has methods that add, update, find, and delete fields and rows in the table or tables that represent the class. Data access class methods encapsulate the logic needed to copy values from the problem domain class to the database and vice versa. Typically, that logic is a combination of program code in a language such as C++ or Java and embedded relational database commands in Structured Query Language (SQL).

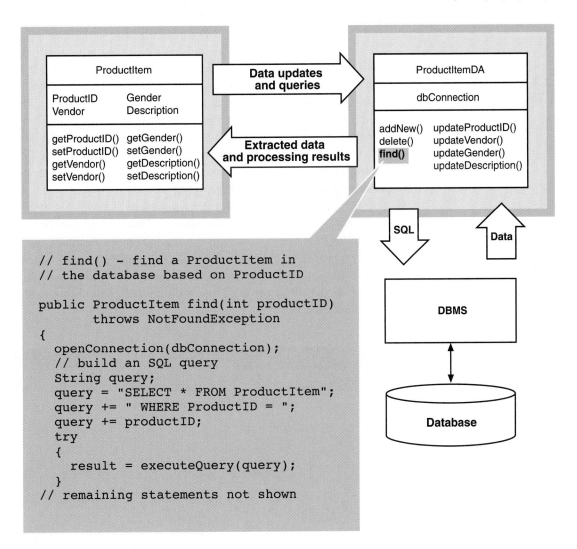

FIGURE *13-25*

Interaction among a problem domain class, a data access class, and the DBMS.

The lower-left part of Figure 13-25 shows a fragment of Java code with an embedded SQL statement that implements the find() method of ProductItemDA. Similar code is needed for all other methods in the data access class.

Now that we've covered the different approaches to database schema design, we consider the data types that are stored within that schema.

DATA TYPES

data type

the storage format and allowable content of a program variable or database field

primitive data type

a storage format directly implemented by computer hardware or a programming language

complex data type

a data type not directly supported by computer hardware or a programming language; also called *user-defined data type*

A *data type* defines the storage format and allowable content of a program variable, object state variable, or database field or attribute. *Primitive data types* are data types that are supported directly by computer hardware and programming languages. Examples include memory address (a pointer), Boolean, integer, unsigned integer, short integer (one byte), long integer (multiple bytes), single characters, real numbers (floating point numbers), and double precision (double-length) integers and real numbers. In some procedural programming languages (such as C) and most OO languages, programmers can define additional data types using the primitive data types as building blocks.

As information systems have become more complex, the number of data types used to implement them has increased. Examples of modern data types include dates, times, currency (money), audio streams, still images, motion video streams, and uniform resource locators (URL or Web links). Such data types are sometimes called *complex data types* because they are usually defined as complex combinations of primitive data types. They may also be called *user-defined data types* because they may be defined by users during analysis and design or by programmers during design and implementation.

Relational DBMS Data Types

The designer must choose an appropriate data type for each field in a relational database schema. For many fields, the choice of a data type is relatively straightforward. For example, designers can represent customer names and addresses using a set of fixed- or variable-length character arrays. Inventory quantities and item prices can be represented as integers and real numbers, respectively. A color can be represented by a character array containing the name of the color or by a set of three integers representing the intensity of the video-display colors red, blue, and green.

Modern RDBMSs have added an increasing number of new data types to represent the data required by modern information systems. Figure 13-26 contains a partial listing of some of the data types available in the Oracle RDBMS. Complex data types available in Oracle include DATE, LONG, and LONGRAW. LONG is typically used to store large quantities of formatted or unformatted text (such as a word-processing document). LONGRAW can be used to store large binary data values, including encoded pictures, sound, and motion video.

FIGURE *13-26*

A subset of the data types available in the Oracle relational DBMS.

Type	Description
CHAR	Fixed-length character array
VARCHAR	Variable-length character array
NUMBER	Real number
DATE	Date and time with appropriate checks of validity
LONG	Variable-length character data up to 2 gigabytes
LONGRAW	Binary large object (BLOB) with no assumption about format or content
ROWID	Unique six-byte physical storage address

Modern RDBMSs can also perform many validity and format checks on data as they are stored in the database. For example, a schema designer can specify that a quantity on hand cannot be negative, that a U.S. zip code must be five or nine digits long, and that a string containing a URL must begin with *http://*. All application programs that use the database then automatically share the validity and format constraints. Each program is simpler, and the possibility for errors from mismatches among data validation logic is eliminated. Application programs still have to provide program logic to recover from attempts to add "bad" data, but they are freed from actually performing validity checks.

Object DBMS Data Types

ODBMSs typically provide a set of primitive and complex data types comparable to those of an RDBMS. ODBMSs also allow a schema designer to define format and value constraints. But ODBMSs provide an even more powerful way to define useful data types and constraints. A schema designer can define a new data type and its associated constraints as a new class.

A class is a complex user-defined data type that combines the traditional concept of data with processes (methods) that manipulate that data. In most OO programming languages, programmers are free to design new data types (classes) that extend those already defined by the programming language. Incompatibility between system requirements and available data types is not an issue, since the designer can design classes specifically to meet the requirements. To the ODBMS, instances of the new data type are simply objects to be stored in the database.

Class methods can perform many of the type- and error-checking functions previously performed by application program code and/or by the DBMS itself. In essence, the programmer constructs a "custom-designed" data type and all of the programming logic required to use it correctly. The DBMS is freed from direct responsibility for managing complex data types and the values stored therein. It indirectly performs validity checking and format conversion by extracting and executing programmer-defined methods stored in the database.

The flexibility to define new data types is one reason that OO tools are so widely employed in non-business information systems. In fields such as engineering, biology, and physics, stored data is considerably more complex than simple strings, numbers, and dates. OO tools enable database designers and programmers to design custom data types that are specific to a problem domain.

Another issue that must be considered during database design is the locations where data are stored and accessed. In today's networked information systems, organizations often use distributed databases.

DISTRIBUTED DATABASES

Rarely does an organization store all of its data in a single database. Instead, organizations typically store data in many different databases, often under the control of many different DBMSs. Reasons for employing a variety of databases and DBMSs include the following:

- Information systems may have been developed at different times using different DBMSs.
- Parts of an organization's data may be owned and managed by different organizational units.
- System performance improves when data are physically close to the applications that use them.

Distributed Database Architectures

Chapter 9 described various approaches to organizing and computing information processing resources in a networked environment. Several architectures for distributing database services are possible, including the following:

- Single database server
- Replicated database servers
- Partitioned database servers
- Federated database servers

Combinations of these architectures are also possible.

Single Database Server

Figure 13-27 shows a typical single database server architecture. Clients on one or more LANs share a single database located on a single computer system. The database server may be connected to one of the LANs or directly to the WAN backbone (as shown in the figure). Connection directly to the WAN ensures that no one LAN is overloaded by all of the network traffic to and from the database server.

FIGURE *13-27*

A single database server architecture.

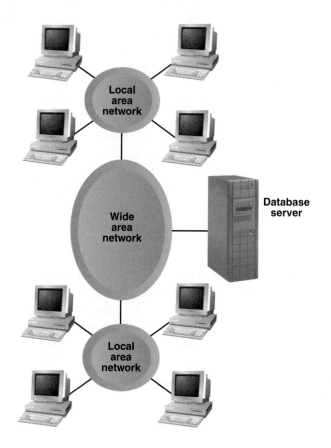

The primary advantage of single database server architecture is its simplicity. There is only one server to manage, and all clients are programmed to direct requests to that server. Disadvantages of the single database server architecture include susceptibility to server failure and possible overload of the network or server. A single server provides no backup capabilities in the event of server failure. All application programs that depend on the server are disabled whenever the server is unavailable (such as during a crash or during hardware maintenance). Thus, single database server architecture is poorly suited to applications that must be available on a seven-day, 24-hour basis.

Performance bottlenecks can occur within a single database server or in the network segment to which the server is attached. As transaction volumes grow, the capabilities of a single database server may become insufficient to respond quickly to all of the service requests it receives. In an attempt to improve performance, a designer may employ a more powerful computer system as the database server. But in an era of multiterabyte databases, it is not unusual for the size and transaction volume of larger databases to exceed the capabilities of any single computer system. Employing the largest mainframes may also be impractical because of cost, system management, or network performance considerations.

Requests to and responses from a database server may traverse large distances across local and wide area networks. Database transactions must also compete with other types of network traffic (such as voice, video, and Web site access) for available transmission capacity. Thus, delays in accessing a remote database server may result from network congestion or propagation delay from client to server.

One way to reduce network congestion is to increase capacity of the entire network. But this approach is expensive and often impractical. Another approach, specifically geared to improving database access speed, is to locate database servers physically close to their clients (for example, on the same LAN segment). This approach minimizes the distance-related delay for requests and responses and removes a large amount of traffic from the WAN.

Moving a database server closer to its clients is a relatively simple matter when all of the clients are located close to one another. But what happens when clients are widely dispersed, as in a multinational corporation? In this case, no single location for the database server can possibly improve database access performance for all clients at the same time. Thus, the "distant" clients must pay a greater performance penalty for database access.

Replicated Database Servers

Designers can eliminate delay in accessing distant database servers by using a replicated database server architecture (see Figure 13-28). Each server stores a separate copy of the needed data. Clients interact with the database server on their own LAN. Such an architecture eliminates database accesses from the WAN and minimizes propagation delay. Local network and database server capacity can be independently optimized to local needs.

Replicated database servers also make an information system more fault tolerant. Applications can direct access requests to any available server, with preference to the nearest server. When a server is unavailable, clients can redirect their requests across the WAN to another available server. Designers can also achieve load balancing by interposing a transaction server between clients and replicated database servers. The transaction server monitors loads on all database servers and automatically directs client requests to the server with the lowest load.

In spite of their advantages, replicated database servers do have some drawbacks. Data inconsistency is a problem whenever multiple database copies are in use. When data are updated on one database copy, clients accessing that same data from another database copy receive an outdated response. To counteract this problem, each database copy must periodically be updated with changes from other database servers. This process is called *database synchronization*.

Designers can implement synchronization by developing customized synchronization programs or by using synchronization utilities built into the DBMS. Custom application programs are seldom employed because they are difficult to develop and because they would need to be modified each time the database schema or number and location of database copies change. DBMS synchronization utilities are generally powerful and flexible but also expensive. Incompatibilities among synchronization methods make using DBMSs from different vendors impractical.

database synchronization

the process of ensuring consistency among two or more database copies

FIGURE *13-28*

A replicated database server architecture.

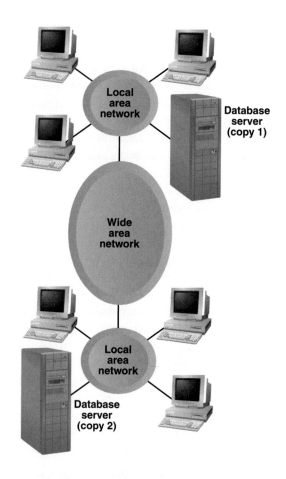

The time delay between an update to a database copy and the propagation of that update to other database copies is an important database design decision. During the time between the original update and the update of database copies, application programs that access outdated copies aren't receiving responses that reflect current reality. Designers can address this problem by reducing the synchronization delay. But shorter delays imply more frequent (or possibly continuous) database synchronization. Synchronization then consumes a substantial amount of database server capacity, and a large amount of network capacity among the related database servers must be provided. The proper synchronization strategy is a complex trade-off among cost, hardware and network capacity, and the need of application programs and users for current data.

Partitioned Database Servers

Designers can minimize the need for database synchronization by partitioning database contents among multiple database servers. Figure 13-29 shows the division of a hypothetical database schema into two partitions. A different group of clients accesses each partition. Figure 13-30 shows a partitioned database server architecture that maintains each partition on a separate database server. Traffic among clients and the database server in each group is restricted to a local area network.

Partitioned database server architecture is feasible only when a schema can be cleanly partitioned among client access groups. Client groups must require access to well-defined subsets of a database (for example, marketing data rather than production data). In addition, members of a client access group must be located in small geographic regions. When a single access group is spread among multiple geographic sites (for example, order processing at three regional centers), then a combination of replicated and partitioned database server architecture is usually required.

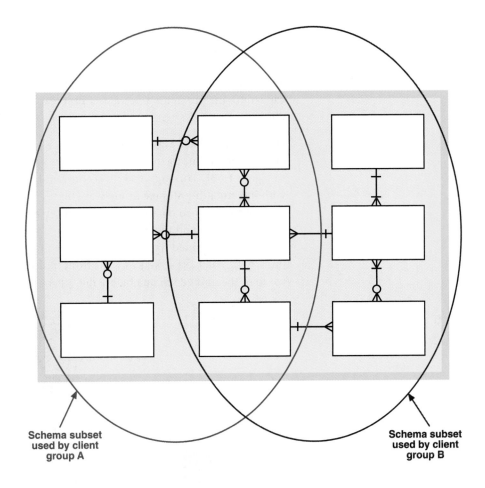

**Schema subset
used by client
group A**

**Schema subset
used by client
group B**

FIGURE *13-30*

A partitioned database server
architecture.

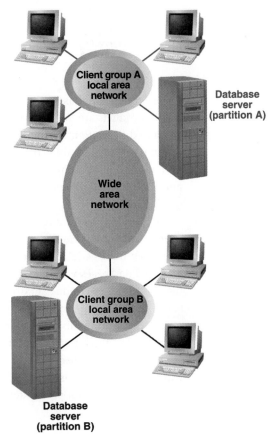

It is seldom possible to partition a database schema into mutually exclusive subsets. Some portions of a database are typically needed by most or all users, and those portions must exist in each partition. For example, data in the region of overlap in Figure 13-29 should be stored on each server with periodic synchronization. Thus, partitioning can reduce the problems associated with database synchronization, but it seldom eliminates them entirely.

Federated Database Servers

Some information systems are best served by a federated database architecture, as shown in Figure 13-31. This architecture is commonly used to access data stored in databases with incompatible storage models or DBMSs. A single unified database schema is created on a combined database server. That server acts as an intermediary between application programs and the databases residing on other servers. Database requests are first sent to the combined database server, which in turn makes appropriate requests of the underlying database servers. Results from multiple servers are combined and reformatted to fit the unified schema before the system returns a response to the client.

F I G U R E *13-31*

A federated database server architecture.

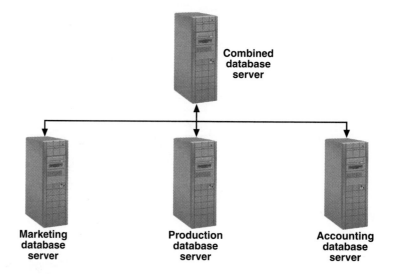

Federated database server architecture can be extremely complex. A number of DBMS products are available to implement such systems, but they are typically expensive and difficult to implement and maintain. Federated database architectures also tend to demand considerable computer hardware and network capacity, but their expense and management complexity are generally less than would be required to implement and maintain application programs that interact directly with all of the underlying databases.

A common use of a federated database server architecture is to implement a data warehouse. A *data warehouse* is a collection of data used to support structured and unstructured managerial decisions. Data warehouses typically draw their content from operational databases within an organization and multiple external databases (for example, economic and trade data from databases maintained by governments, trade industry associations, and private research organizations). Because data originate from a large number of incompatible databases, a federated architecture is typically the only feasible approach for implementing a data warehouse.

Now that we've discussed the basic issues underlying distributed database design, we will show how they come into play when making decisions for Rocky Mountain Outfitters' new customer support system.

data warehouse

a collection of data used to support structured and unstructured managerial decisions

RMO Distributed Database Architecture

The starting point for designing a distributed database architecture is information about the data needs of geographically dispersed users. Some of this information for RMO was gathered during the analysis phase (see Figures 6-37, 6-38, and 6-39) and is summarized here:

- Warehouse staff (Portland, Salt Lake City, and Albuquerque) need to check inventory levels, query orders, record back orders and order fulfillment, and record order returns.
- The phone-order staff (Salt Lake City) need to check inventory levels; create, query, update, and delete orders; query customer account information; and query catalogs.
- The mail-order staff (Provo) need to check inventory levels, query orders, query catalog information, and update customer accounts.
- Customers (location not yet determined) need the same access capabilities as phone-order staff.
- Headquarters staff (Park City) need to query and adjust orders, query and adjust customer accounts, and create and query catalogs and promotions.

RMO has already decided to manage its database using the existing mainframe computer in the Park City data center. Thus, a WAN will be required to connect the server to LANs in the warehouses, phone-order center, mail-order center, headquarters, and data center. A connection will eventually be required for the Web servers used for direct customer ordering, although they probably will be located at an existing site (such as the data center).

A single-server architecture for RMO is shown in Figure 13-32. This architecture requires sufficient WAN capacity to carry database (and other) traffic from all locations. The primary advantage of this architecture is its simplicity. There are no partitions or database copies to manage, and only a single server must be maintained. The primary disadvantages are relatively high WAN capacity requirements and the susceptibility of the entire system to failure of the single server.

A more complex alternative is shown in Figure 13-33. Each remote location employs a combination of database partitioning and replication. A server at each warehouse stores a local copy of the order and inventory portions of the database. Servers in the phone- and mail-order centers store local copies of a larger subset of the database. Corporate headquarters relies on the central database server in the data center.

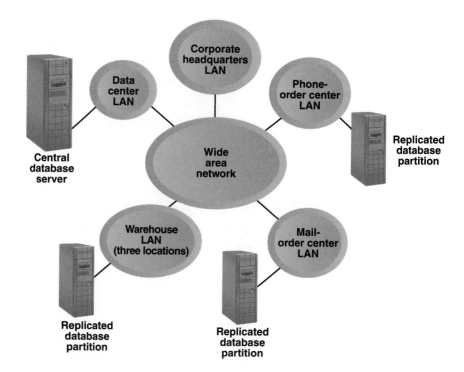

The primary advantages of this architecture are fault tolerance and reduced WAN capacity requirements. Each location could continue to operate if the central database server failed. However, as the remote locations continued to operate, their database contents would gradually drift out of synchronization. A synchronization strategy must be implemented to address both regular database updates and recovery from server failure. The strategy could vary by location.

The primary disadvantages to the distributed architecture are cost and complexity. The architecture saves WAN costs through reduced capacity requirements but adds costs for additional database servers. The cost of acquiring, operating, and maintaining the additional servers would probably be much higher than the cost of adding greater WAN capacity.

So which alternative makes the most sense for RMO? The answer depends on some data that hasn't yet been gathered and on answers to some questions about desired system performance. RMO management must also determine its goals for system performance and reliability. The distributed architecture would provide higher performance and reliability but does so at substantially increased cost. Management must determine whether the extra cost is worth the expected benefits.

Additional data about network traffic are needed to precisely determine LAN and WAN communication requirements between clients and database servers. Estimates of transaction and query volume, including normal and peak demand, are required for each location. Such estimates may be gathered during analysis or design. The estimates are required to determine an optimal configuration of LAN, WAN, and database server capacity. The analysis of the estimates and the actual design of the networks and database architecture are complex endeavors that require highly specialized knowledge and experience.

SUMMARY

Most modern information systems store data in a database and access and manage the data using a DBMS. Relational databases and DBMSs are most commonly used today, but object databases and DBMSs are increasing in popularity. One of the key activities of systems design is developing a relational or object database schema.

A relational database is a collection of data stored in tables. A relational database schema is normally developed from an entity-relationship diagram. Each entity is represented as a separate table. One-to-many relationships are represented by embedding foreign keys in entity tables. Many-to-many relationships are represented by creating additional tables containing foreign keys of the related entities.

An object database stores data as a collection of related objects. The design class diagram is the starting point for developing an object database schema. The database schema defines each class, and the ODBMS stores each object as an instance of a particular class. Each object is assigned a unique object identifier. Relationships among objects are represented by storing the object identifier of an object within related objects.

Objects can also be stored within a relational database. Object attributes and relationships among objects—including one-to-many, many-to-many, and generalization hierarchies—can be represented. However, an RDBMS cannot store methods and cannot directly represent inheritance.

Medium- and large-scale information systems typically use multiple databases or database servers in various geographic locations. Replicated database architecture employs multiple database copies on different servers, usually in different geographic locations. Partitioned database architecture employs partial database copies stored on different servers in proximity to a distinct user subset. Federated database architecture employs multiple databases (possibly of different types) and a special-purpose DBMS that provides a unified view of the databases and a single point of access.

KEY TERMS

REVIEW QUESTIONS

1. List the components of a DBMS and describe the function of each.

2. What is a database schema? What information does it contain?

3. Why have databases become the preferred method of storing data used by an information system?

4. List four different types of database models and DBMSs. Which are in common use today?

5. With respect to relational databases, briefly define the terms *row* and *field*.

6. What is a primary key? Are duplicate primary keys allowed? Why or why not?

7. What is the difference between a natural key and an invented key? Which type is most commonly used in business information processing?

8. What is a foreign key? Why are foreign keys used or required in a relational database? Are duplicate foreign key values allowed? Why or why not?

9. Describe the steps used to transform an ERD into a relational database schema.

10. How is an entity on an ERD represented in a relational database?

11. How is a one-to-many relationship on an ERD represented in a relational database?

12. How is a many-to-many relationship on an ERD represented in a relational database?

13. What is referential integrity? Describe how it is enforced when a new foreign key value is created, when a row containing a primary key is deleted, and when a primary key value is changed.

14. What types of data (or fields) should never be stored more than once in a relational database? What types of data (or fields) usually must be stored more than once in a relational database?

15. What is relational database normalization? Why is a database schema in third normal form considered to be of higher quality than an unnormalized database schema?

16. Describe the process of relational database normalization. Which normal forms rely on the definition of functional dependency?

17. Describe the steps used to transform a class diagram into an object database schema.

18. What is the difference between a persistent object and a transient object? Provide at least one example of each object type.

19. What is an object identifier? Why are object identifiers required in an object database?

20. How is a class on a class diagram represented in an object database?

21. How is a one-to-many relationship on a class diagram represented in an object database?

22. How is a many-to-many relationship without attributes represented in an object database?

23. What is an association class? How are association classes used to represent many-to-many relationships in an object database?

24. Describe the two ways in which a generalization relationship can be represented in an object database.

25. Does an object database require key fields or attributes? Why or why not?

26. Describe the similarities and differences between an ERD and a class diagram that models the same underlying reality.

27. How are classes and relationships on a class diagram represented in a relational database?

28. What is the difference between a primitive data type and a complex data type?

29. What are the advantages of having an RDBMS provide complex data types?

30. Does an ODBMS need to provide predefined complex data types? Why or why not?

31. Why might all or part of a database need to be replicated in multiple locations?

32. Briefly describe the following distributed database architectures: replicated database servers, partitioned database servers, and federated database servers. What are the comparative advantages of each?

33. What additional database management complexities are introduced when database contents are replicated in multiple locations?

THINKING CRITICALLY

1. The Universal Product Code (UPC) is a bar-coded number that uniquely identifies many products sold in the United States. For example, all copies of this textbook sold in the United States have the same UPC bar code on the back cover. Now consider how the design of the RMO database might change if all items sold by RMO were required by law to carry a permanently attached UPC (for example, on a label sewn into garments). How might the RMO relational database schema change under this requirement?

2. Assume that RMO plans to change its pricing policy. If two or more catalogs are in circulation at the same time, then all item prices in the catalogs must be the same. Prices can still rise or fall over time, and those changes will be recorded in the database and printed in newly issued catalogs. Any customer who makes an order will always be given the lower of the current price or the price in the current catalog. What changes to the tables shown in Figure 13-9 will be required to ensure that the RMO database is in 3NF after the pricing policy change?

3. Assume that RMO will begin asking a random sample of customers who order by telephone about purchases made from competitors. RMO will give customers a 15 percent discount on their current order in exchange for answering a few questions. To store and use this information, RMO will expand the ERD and class diagram with two new entities (classes) and three new relationships. The new entities (classes) are Competitor and ProductCategory. Competitor has a one-to-many relationship with ProductCategory, and the existing Customer entity (class) also has a one-to-many relationship with ProductCategory. Competitor has a single field (attribute) called Name. ProductCategory has four fields (attributes): Description, DollarAmountPurchased, MonthPurchased, and YearPurchased. Revise the relational database schema shown in Figure 13-9 to include the new entities and relationships. All tables must be in 3NF.

4. Assume that RMO is developing its database using object-oriented methods. Assume further that the database designers want to make some changes to the class diagram in Figure 13-15. Specifically, they want to make ProductItem an abstract parent class from which more specific product classes are specialized. Three specialized classes will be added: ClothingItem, EquipmentItem, and OtherItem. ClothingItem will add the attribute *color*, and that same attribute will be removed from the InventoryItem class. EquipmentItem will also add an attribute called *color* but will not have an attribute called *gender*. OtherItem will have both the *color* and *gender* attributes. Revise the relational database schema in Figure 13-24 to store the new ProductItem generalization hierarchy. Use a separate table for each of the specialized classes.

5. Assume that RMO will use a relational database as shown in Figure 13-9. Assume further that a new catalog group located in Milan, Italy, will now create and maintain the catalog. To minimize networking costs, the catalog group will have a dedicated database server attached to its LAN. Develop a plan to partition the RMO database. Which tables should be replicated on the catalog group's local database server? Update Figure 13-33 to show the new distributed database architecture.

6. Revisit the issues raised in the Nationwide Books (NB) case at the beginning of the chapter. Should NB adopt an ODBMS for the new Web-based ordering system? Why or why not?

EXPERIENTIAL EXERCISES

1. This chapter did not discuss network databases in detail, but some database textbooks discuss them. Investigate the network database model and its use of pointers to represent relationships among record types. In what ways is the use of pointers in a network database similar to the use of object identifiers in an object database? Does the similarity imply that object databases are little more than a renamed version of an older DBMS technology?

2. Access the Object Database Management Group Web site (www.odmg.org) and gather information on the current status of the ODMG standard.

3. Investigate the student records management system at your school to determine what database management system is used. What database model is used by the DBMS? If the DBMS isn't object oriented, find out what plans, if any, are in place to migrate to an ODBMS. Why is the migration being planned (or not being planned)?

4. Visit the Web site of an on-line catalog vendor similar to RMO (such as www.llbean.com) or an on-line vendor of computers and related merchandise (such as www.cdw.com). Browse the on-line catalog and note the various types of information contained therein. Construct a list of complex data types that would be needed to store all of the on-line catalog information.

Case Studies

Real Estate Multiple Listing Service System

Refer to the description of the Real Estate Multiple Listing Service system in the Chapter 5 case studies. Using the ERD and class diagram for that system as a starting point:

1. Develop a relational database schema in 3NF.
2. Develop an ODL database schema.

State Patrol Ticket Processing System

Refer to the description of the State Patrol ticket processing system in the Chapter 5 case studies. Using the ERD and class diagram for that system as a starting point:

1. Develop a relational database schema in 3NF.
2. Develop an ODL database schema.

Computer Publishing, Incorporated

In only a decade, Computer Publishing, Incorporated (CPI) had grown from a small textbook publishing house into a large international company with significant market share in traditional textbooks, electronic books, and distance education courseware. CPI's processes for developing books and courseware were similar to those used by most other publishers, but those processes had proven cumbersome and slow in an era of rapid product cycles and multiple product formats.

Text and art were developed in a wide variety of electronic formats, and conversions among those formats were difficult and error-prone. Many editing steps were performed with traditional paper-and-pencil methods. Consistency errors within books and among books and related products were common. Developing or revising a book and all its related products typically took a year or more.

CPI's president initiated a strategic project to reengineer the way that CPI developed books and related products. CPI formed a strategic partnership with Davis Systems (DS) to develop software that would support the reengineered processes. DS had significant experience developing software to support product development in the chemical and pharmaceutical industries using the latest development tools and techniques, including object-oriented software and databases. CPI expected the new processes and software to reduce development time and cost. Both companies expected to license the software to other publishers within a few years.

A joint analysis team specified the workflows and high-level requirements for the software. The team developed plans for a large database that would hold all book and courseware content through all stages of production. Authors, editors, and other production staff would interact with the database in a variety of ways, including traditional word-processing programs and Web-based interfaces. When required, format conversions would be handled seamlessly and without error. All content creation and modification would be electronic—no text or art would ever be created or edited on paper, except as a printed book ready for sale.

Software would track and manage content through every stage of production. Content common to multiple products would be stored in the database only once. Dependencies within and across products would be tracked in the database. Software would ensure that any content addition or change would be reflected in all dependent content and products, regardless of the final product form. For example, a sentence in Chapter 2 that refers to a figure in Chapter 1 would be updated automatically if the figure were renumbered. If a new figure were added to a book, it would be added automatically to the related courseware presentation slides. Related courseware and study material on the Web site would automatically reflect changes to the answer to an end-of-chapter question.

1. Consider the contents of this textbook as a template for CPI's database content. Draw an ERD and class diagram that represents the book and its key content elements. Which diagram is a more accurate representation of book content? Expand your diagrams to include related product content such as a set of PowerPoint slides, an electronic book formatted as a Web site, and a Web-based test bank.

2. Develop a list of data types required to store the content of the book, slides, and Web sites. Are the relational DBMS data types listed in Figure 13-26 sufficient?

3. What features of an ODBMS, beyond or different from RDBMS features, might be useful when implementing CPI's database? Give examples of how they might be used.

Rethinking Rocky Mountain Outfitters

The "Rethinking RMO" case in Chapter 5 asked you to consider additional things and relationships that would need to be modeled if RMO were to implement its own customer charge accounts. If you have not already done so, complete that exercise and update the ERD and class diagram accordingly, then complete the following tasks:

1. Update the RMO relational database design in Figure 13-9 based on the changes that you made to the ERD. Be sure that all your database tables are in 3NF.

2. Write ODL schema specifications for all new classes and relationships that you added to the class diagram.

3. Verify that the new classes and relationships are accurately represented in the updated relational database design that you developed for question 1.

Focusing on Reliable Pharmaceutical Service

Use the ERD that you developed in Chapter 5 and the design class diagram that you developed in Chapter 7 to complete the following tasks:

1. Develop a relational database schema in 3NF.

2. Develop an ODL database schema.

3. Discuss the pros and cons of distributed database architecture. Which architectural approach (or combination of approaches) should Reliable employ in their new system once it is fully implemented?

FURTHER RESOURCES

Peter Rob and Carlos Coronel, *Database Systems: Design Implementation and Management* (5th ed.) Course Technology, 2001.

The Object Database Management Group Web site, http://www.odmg.org.

Robert Orfali, Dan Harkey, and Jeri Edwards, *The Essential Client/Server Survival Guide* (3rd ed.) John Wiley & Sons, 1999.

Dirk Bartels and Greg Chase, "A Comparison Between Relational and Object-oriented Database Systems for Object-oriented Application Development," http://www.fastobjects.com/us/pdf/ODBMSvsRDBMS.pdf, 2001.

CHAPTER 14

Designing the User Interface

LEARNING OBJECTIVES

After reading this chapter, you should be able to:

- Understand the difference between user interfaces and system interfaces

- Explain why the user interface *is* the system to the users

- Discuss the importance of the three principles of user-centered design

- Describe the historical development of the field of human-computer interaction (HCI)

- Describe the three metaphors of human-computer interaction

- Discuss how visibility and affordance affect usability

- Apply the eight golden rules of dialog design when designing the user interface

- List the key principles used in Web design

- Define the overall system structure as a menu hierarchy

- Write user-computer interaction scenarios as dialogs

- Create storyboards to show the sequence of forms used in a dialog

- Use UML class diagrams and sequence diagrams to document dialog designs

- Design windows forms and browser forms that are used to implement a dialog

CHAPTER OUTLINE

Identifying and Classifying Inputs and Outputs

Understanding the User Interface

Guidelines for Designing User Interfaces

Documenting Dialog Designs

Guidelines for Designing Windows and Browser Forms

Guidelines for Designing Web Sites

Designing Dialogs for Rocky Mountain Outfitters

ob Crain was admiring the user interface for the manufacturing support system recently installed at Aviation Electronics (AE). Bob is the plant manager for AE's Midwest manufacturing facility, which is responsible for producing aviation devices used in commercial aircraft. These aviation devices provide guidance and control functions for flight crews, and they provide the latest safety and security features that pilots need when flying commercial aircraft.

The manufacturing support system is used for all facets of the manufacturing process, including product planning, purchasing, parts inventory, quality control, finished goods inventory, and distribution. Bob was involved extensively in the development of the system over a period of several years, including initial planning and development. The system reflected almost everything he knew about manufacturing. The information systems team that developed the system relied extensively on Bob's expertise. That was the easy part for Bob.

What particularly pleased Bob was the final user interface. Bob had insisted that the development team think "outside the box." He did not want just another cookie-cutter transaction processing system. He wanted a system that acted as a partner in the manufacturing process—with a look and feel that really fit the work the users were doing. After all, the facility produced devices whose major design goal was usability. Shouldn't the manufacturing support system be designed that way, too?

The first manager assigned to the project didn't want to discuss usability at all. "We'll add the user interface later, after we work out the accounting controls" was a typical comment. When Bob insisted that the project manager be replaced, the information systems department sent Sara Robinson to lead the project.

Sara had a completely different attitude; she started out asking about events affecting the manufacturing process and about cases where users need support from the system. Although she had a team of analysts working on the accounting transaction details right from the beginning, she always focused on how the user would interact with the system. Bob and Sara conducted meetings to involve users in discussions about how they might use the system, even asking users to act out the roles of the user and the system carrying on a conversation. That approach was outside the box.

At other meetings Sara presented sketches of screens and asked users to draw on them, placing right on the sketches information they wanted to see and options they wanted to be able to select. These sessions produced many ideas. For example, many users did not sit at their desks all day—they needed larger and more graphic displays they could see from across the room. Many users needed to refer to several displays, and they needed to be able to read them simultaneously. Several functions were best performed using graphical simulations of the manufacturing process. Users made sketches showing how the manufacturing process actually worked, and the team used these sketches later to define much of the interface. Sara and her team kept coming back every month or so with more examples to show, asking for more suggestions.

When the system was finally completed and installed, most users already knew how to use it since they had been so involved in its design. Bob knew everything the system could do, but he had his own uses for it. He sat at his desk and clicked the *review ongoing processes* button on the screen, and the manufacturing support system gave him his morning briefing.

OVERVIEW

Information systems capture inputs and produce outputs, and inputs and outputs occur where there are *interfaces* between the system and its environments. System interfaces handle inputs and outputs that require minimal human intervention. User interfaces handle inputs and outputs that involve a system user directly. This chapter differentiates between both types of interfaces and then focuses on the design of user interfaces. Then Chapter 15 focuses on system interfaces, system outputs, and systems controls.

One of the key systems design activities is to design the user interface for a system. Designing the user interface means designing the inputs and outputs involved when the user interacts with the computer to carry out a task. This chapter emphasizes the interaction between the user and the computer—called *human-computer interaction*, or HCI. For every input, a developer must consider the interaction between user and computer and design an interface to process the input. Similarly, for every output produced at the request of a user (an on-line report, for example), the developer must design the interaction. Because the interaction is much like a dialog between the user and the computer, user-interface design is often referred to as *dialog design*.

This chapter begins with discussion of the user interface by providing background on user-centered design, the development of the field of human-computer interaction, and several metaphors used to describe the user interface. Many guidelines are available to help ensure usability of the system, and some of the most important guidelines are discussed, including guidelines for Web-based development. Next, approaches to documenting dialog designs are presented, including the use of UML diagrams from the object-oriented approach to development. Examples are given throughout the chapter, including some dialog design examples for Rocky Mountain Outfitters that show Windows forms and Web pages.

IDENTIFYING AND CLASSIFYING INPUTS AND OUTPUTS

Inputs and outputs of the system are an early concern of any system development project. The project plan lists key inputs and outputs that the analyst identified when defining the scope of the system. During the analysis phase, analysts also discussed inputs and outputs early and often with system stakeholders to identify external agents and actors that affect the system and that depend on information it produces. Requirements models produced during analysis also emphasize inputs and outputs. For example, the event table includes a trigger for each external event, and the triggers represent inputs. Outputs are shown as responses to external, state, and temporal events.

Traditional and OO Approaches to Inputs and Outputs

In the traditional approach, inputs and outputs are shown as data flows on the context diagram, the data flow diagram (DFD) fragments, and the detailed DFDs. A data flow definition that lists all data elements describes each input and output in detail. During design, analysts add more detail about the data flows based on the choices they made when deciding on a design alternative. Whether an input is captured automatically or entered by a system user, for example, determines details about the design of the system. As discussed in Chapter 10, these details must be coordinated with the design of the application software.

In the object-oriented approach, inputs and outputs are defined by messages entering or leaving the system. Inputs and outputs are included in the event table as triggers and responses. Actors provide inputs for many use cases, and many use cases provide outputs to actors. The messages exchanged during a scenario define these inputs and outputs in more detail, and as the design of each scenario becomes more detailed, so does the specification of messages. They are reflected in interaction diagrams, in design class diagrams as methods, and in statecharts. The system sequence diagram introduced in Chapter 7 first showed these inputs and outputs.

User versus System Interfaces

In both the traditional and object-oriented approaches, a key step in systems design is to classify the inputs and outputs for each event as either a system interface or a user interface. *System interfaces* involve inputs and outputs that require minimal human intervention. They might be inputs captured automatically by special input devices such as scanners, electronic messages from another system, or batch processing transactions compiled by another system. Many outputs are considered system interfaces if they primarily send messages or information to other systems or if they produce reports, statements, or documents for external agents or actors without much human intervention.

User interfaces involve inputs and outputs that more directly involve a system user. A user interface enables a user to interact with the computer to record a transaction, such as when a customer service representative records a phone order for an RMO customer.

system interfaces

the parts of an information system involving inputs and outputs that require minimal human intervention

user interfaces

the parts of an information system requiring user interaction to create inputs and outputs

Sometimes outputs are produced after user interaction, such as the information displayed after a user query about the status of an order. In Web-based systems, a customer can interact directly with a system to request information, place an order, or look up the status of an order.

In most system development projects, analysts separate design of system interfaces from design of user interfaces because they require different expertise and technology. But as with the design of any system component, considerable coordination is required. This chapter discusses user interfaces. The next chapter deals with system interfaces and system controls. At Rocky Mountain Outfitters, Barbara Halifax's regular status memo updates John MacMurty on some of the activities of user-interface design for the customer support system.

Rocky Mountain Outfitters

MEMO

May 12, 2005

To: John MacMurty

From: Barbara Halifax

RE: User-interface design for the customer support system

John, we have been working on user-interface prototypes almost from the beginning of the project as we worked with various end-user groups. The Web-based components of the system have also been given a lot of attention (starting really when we were prototyping for feasibility). I have one team working on the user interface design and another team assigned to developing the detailed design of system outputs and controls. I'll have more on the outputs and controls later.

I just wanted to report briefly that we are on track to finalize the design of most interactive dialogs. I know you have seen the storyboards of key dialogs and have tried out many of the prototypes. We have gone a step further in documenting the dialogs using activity diagrams and sequence diagrams. These diagrams are helping us work through the technical issues for implementation, and they are providing templates to help make sure we provide a consistent look and feel from one dialog to another.

We have conducted usability tests on the designs and have held focus group meetings with users, particularly the phone-order representatives and customer groups recruited to help with the design of the Web components. The gift certificates we offered the customer focus group are really paying off.

Thanks for your input on the prototypes. I'll check in later before our next status meeting.

BH

cc: Steven Deerfield, Ming Lee

UNDERSTANDING THE USER INTERFACE

Many people think the user interface is developed and added to the system near the end of the development process. But the user interface of an interactive system is much more than that. The user interface is everything the end user comes in contact with while using the system—physically, perceptually, and conceptually (see Figure 14-1). To the end user of a system, the user interface *is* the system itself.

FIGURE *14-1*

Physical, perceptual, and conceptual aspects of the user interface.

Desk, chair, light, keyboard, mouse, touch screen, keypad, manuals, printed documents, paper forms.

Windows, menus, dialog boxes, buttons, lines, shapes, textures, colors, fonts, sounds, speech.

Customers, products, orders, catalogs, adding, deleting, updating, printing, select-click-drag-drop, double-click-escape-click-click.

Many system developers, particularly those who work on highly interactive systems, echo this point of view in claiming that to design the user interface is to design the system. Therefore, consideration of the user interface should come very early in the development process. The term *human-computer interaction (HCI)* is generally used to refer to the study of end users and their interaction with computers.

human-computer interaction (HCI)

the study of end users and their interactions with computers

Physical Aspects of the User Interface

Physical aspects of the user interface include the devices the user actually touches, including the keyboard, mouse, touch screen, or keypad. But other physical parts of the interface include reference manuals, printed documents, data-entry forms, and so forth that the end user works with while completing tasks at the computer. For example, a mail-order data-entry clerk at Rocky Mountain Outfitters works at a computer terminal but uses printed catalogs and hand-written order forms when entering orders into the system. The desk space, the documents, the available light, and the computer terminal hardware all make up the physical interface for this end user.

Perceptual Aspects of the User Interface

Perceptual aspects of the user interface include everything the end user sees, hears, or touches (beyond the physical devices). What the user sees includes all data and instructions displayed on the screen, including shapes, lines, numbers, and words. The user might rely on the sounds made by the system, even a simple beep or click that tells the user that the system recognizes a keystroke or selection. More recently, computer-generated speech makes it seem that the system is actually talking to the user, and with speech recognition software, the user can talk to the computer. The user "touches" objects such as menus, dialog boxes, and buttons on the screen using a mouse, but the user also touches objects such as documents, drawings, or records of transactions with a mouse when completing tasks.

Conceptual Aspects of the User Interface

Conceptual aspects of the user interface include everything the user knows about using the system, including all of the problem domain "things" in the system the user is manipulating, the operations that can be performed, and the procedures followed to carry out the operations. To use the system, the end user must know all about these details—not how the system is implemented internally, but what the system does and how to use it to complete tasks. This knowledge is referred to as the *user's model* of the system. Much of the user's model is a logical model of the system, as you learned in Chapters 5, 6, and 7. A logical model of the system requirements can be quite detailed, so the user must know quite a few details to operate the system. Recall also that a systems analyst relies on the end users to help define the requirements that the analyst captures in various models. The user's knowledge of the requirements for the system becomes the fundamental determinant of what the system is, and if the user's knowledge of the system is part of the interface, then the user interface must be much more than a component added near the end of the project.

User-Centered Design

Many researchers focus their attention on creating analysis and design techniques that place the user interface at the center of the development process because they recognize the importance of the user interface to system developers and system users. These techniques are often referred to collectively as *user-centered design*. User-centered design techniques emphasize three important principles:

- Focus early on the users and their work
- Evaluate designs to ensure usability
- Use iterative development

The early focus on users and their work is consistent with the approach to systems analysis in this text: Analysts must understand and identify the system users and their requirements for the system. The traditional approach to development focuses more on the requirements from the business point of view—what needs to be accomplished and what are the sources and destinations for data? The object-oriented approach, probably because most object-oriented systems are interactive, focuses more on users and their work by identifying actors, use cases, and scenarios followed when using the system. As discussed in Chapter 7, the automation boundary between user and the computer is defined very early during requirements modeling.

User-centered design goes much further in attempting to understand the users, however. What do they know? How do they learn? How do they prefer to work? What motivates them? The amount of focus on users and their work varies with the type of system being developed. If the system is a shrink-wrapped desktop application marketed directly to end users, the focus on users and their preferences is intense.

The second principle of user-centered design is to evaluate designs to ensure usability. *Usability* refers to the degree to which a system is easy to learn and use. Ensuring usability is not easy; there are many different types of users with differing preferences and skills to accommodate. Features that are easy to use for one person might be difficult for another. If the system has a variety of end users, how can the designer be sure that the interface will work well for all of them? If it is too flexible, for example, some end users may feel lost. On the other hand, if the interface is too rigid, some users will be frustrated.

user's model

what the user knows about using the system, including the problem domain "things" the user is manipulating, the operations that can be performed, and the procedures followed when carrying out tasks

user-centered design

a collection of techniques that place the user at the center of the development process

usability

the degree to which a system is easy to learn and use

But there is more to consider for ease of learning and ease of use. These concepts often conflict, because an interface that is easy to learn is not always easy to use. For example, menu-based applications with multiple forms, many dialog boxes, and extensive prompts and instructions are easy to learn—indeed, they are self-explanatory. Easy-to-learn interfaces are appropriate for systems that end users use infrequently. But if office workers use the system all day, it is important to make the interface fast and flexible, with shortcuts, hot keys, and information-intensive screens. This second interface might be harder to learn, but it will be easier to use once it is learned. Office workers (with the support of their management) are willing to invest more time learning the system to become efficient users.

Developers employ many techniques to evaluate interface designs to ensure usability. User-centered design requires testing all aspects of the user interface. Some usability testing techniques collect objective data that can be statistically analyzed to compare designs. Some techniques collect subjective data about user perceptions and attitudes. To assess user attitudes, developers conduct formal surveys, focus group meetings, design walkthroughs, paper and pencil evaluations, expert evaluations, formal laboratory experiments, and informal observation.

The third principle of user-centered design is to use iterative development—doing some analysis, then some design, then some implementation, and then repeating the processes. After each iteration, the project team evaluates the work on the system to date. Iterative development keeps the focus on the user by continually returning to the user requirements during each iteration and by evaluating the system after each iteration. Iterative development is discussed throughout this text as applicable to both traditional and object-oriented approaches to development.

Human-Computer Interaction as a Field of Study

User-interface design techniques and HCI as a field of study evolved from studies of human interaction with machines in general, referred to as *human factors engineering* or *ergonomics*. The formal study of human factors began during World War II, when aerospace engineers studied the effects on airplane pilots of rearranging controls in the cockpit. Pilots are responsible for controlling many devices as they fly, and the effectiveness of the interaction between the pilot and the devices is critical. If the pilot makes a mistake (that is, if he or she can't correctly use a device), the plane may crash. What the pilot does is the "human factor" that engineers realized was often beyond their control.

One story about the importance of the human factor involved a minor change to the design of the cockpit of a plane. The designers switched the locations of the throttle and the release handle for the ejection seat. The result was a dramatic increase in the number of unexplained pilot ejections. When under pressure, the pilots grabbed what they thought was the throttle and ejected themselves from the plane. Initially, designers dismissed the problem as the need for better training. But even with training, pilots under pressure continued to grab the wrong handle. It became apparent that the key to the "human factor" was to change the machine to accommodate the human rather than trying to change the human to accommodate the machine.

The field of human factors was first associated with engineering, since engineers designed machines. But engineers, who are generally used to precise specifications and predictable behavior, often find the human factor frustrating. Gradually, specialists emerged who began to draw on many disciplines to understand people and their behavior. These disciplines include cognitive psychology, social psychology, linguistics, sociology, anthropology, and others, as shown in Figure 14-2. Information systems specialists with an interest in human-computer interaction study computers plus all of these disciplines.

human factors engineering (ergonomics)

the study of human interaction with machines in general

FIGURE *14-2*

The fields contributing to the
study of HCI.

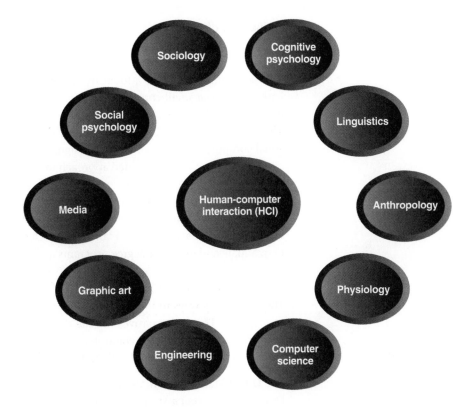

An important contribution to the development of the field of human-computer interaction began with the Xerox Corporation in the 1970s. Xerox produced high-speed photocopying machines that provided an increasing number of special options and capabilities that the human operator could specify. The designers of the photocopying machines recognized the importance of making the complex machines easy for the operators to learn and use. Xerox customers wanted minimal training time for their operators, and operator errors could be costly. For example, if a clerk began a large photocopying job but made a mistake in specifying details, it would be wasteful and delay the distribution of important documents. Therefore, Xerox emphasized the usability of its machines.

Xerox established a research and development laboratory, called the Xerox Palo Alto Research Center (Xerox PARC), to study issues that affect how humans operate machines. As a result of this investment, Xerox eventually offered photocopying machines with touch screen, menu-driven interfaces that displayed icons representing objects such as documents, stacks of paper, staples, and sorting bins.

Research and development at Xerox PARC also involved work on computers and object-oriented programming. The first pure object-oriented programming language, called Smalltalk, was created at Xerox PARC by Alan Kay and associates to facilitate the development of interactive user interfaces. In the early 1970s, Kay envisioned an advanced, portable personal computing platform (similar to today's ultralight notebook computers) called the Dynabook. Many researchers thought such a machine could not be built for three or four decades because the hardware required for the Dynabook was not available. Kay decided to work on the software that would run the machine in anticipation of the hardware, which led to Smalltalk.

Smalltalk includes classes that make up the key parts of windowing interfaces today—windows, menus, buttons, labels, text fields, and so forth. The design and programming philosophy used to describe and build these interfaces was developed along with the language—all 100 percent object oriented.

Because of the work at PARC, Xerox eventually developed and marketed one of the first general-purpose personal computers with a graphical user interface—the Xerox Star—in the late 1970s. Although it was ahead of its time and far too expensive, it is considered a landmark development in computing. Its key features were exploited in the early 1980s by a small company near Xerox PARC named Apple Computers. Apple first exploited the Xerox Star's features as the Apple Lisa and then as the Apple Macintosh. The work at Xerox PARC had a substantial impact on object-oriented programming, personal computers, and user-interface design.

Now that the object-oriented approach to system development is becoming more influential, user-interface design concepts and development techniques pioneered at labs such as Xerox PARC are becoming better integrated into system development methodologies used for business systems. The field of HCI has grown and now sponsors many academic journals, conferences, and book series devoted to research and practice. Undergraduate and graduate degree programs are also available to train HCI specialists.

Metaphors for Human-Computer Interaction

There are many ways to think about human-computer interaction, including *metaphors* or *analogies*. Three alternatives are the direct manipulation metaphor, the document metaphor, and the dialog metaphor. Since each metaphor provides an analogy to a different concept, each has implications for the design of the user interface.

The Direct Manipulation Metaphor

Direct manipulation assumes that the user interacts with objects on the screen instead of typing commands on a command line. Objects that the user can interact with are made visible on the screen so the user can point at them and manipulate them with the mouse or arrow keys. The earliest direct manipulation interfaces were word processors that allowed users to type in words directly where desired in a document. By the early 1980s, electronic spreadsheet applications (first VisiCalc, then Lotus 1-2-3) became available for IBM DOS PCs. These applications used a direct manipulation approach—the user typed numbers, formulas, or text directly into cells on a spreadsheet. The spreadsheet on the screen was conceptually similar to a paper spreadsheet that was familiar to people working in accounting and finance. The familiarity and direct manipulation features made these applications easy to understand and natural to use, and end users could speed their work by including formulas to do the calculations on the spreadsheets automatically. These early direct manipulation DOS applications were an important reason for the success of the personal computer. Even though they did not have graphical user interfaces, they were very popular because they made interacting with a computer straightforward, natural, and useful.

The Smalltalk language developed at Xerox PARC extended direct manipulation to all objects on the screen. Some of these objects are interface objects such as buttons, check boxes, scroll bars, and slider controls, but other problem domain objects such as documents, schedules, file folders, and business records were also displayed as objects that the user could directly manipulate. For example, an interface might include a trash can object; to delete a document file, the user clicks on the document with the mouse and drags the document to the trash can. By directly manipulating the objects in this way, the user tells the computer to delete the document file.

Direct manipulation coupled with object-oriented programming eventually evolved into the *desktop metaphor*, in which the display screen includes an arrangement of common desktop objects—a notepad, a calendar, a calculator, and folders containing documents. Many desktops now also include a telephone, an answering machine, a CD player, and even a video monitor. Interacting with any of these objects is similar to in-

direct manipulation

a metaphor of HCI in which the user interacts directly with objects on the display screen

desktop metaphor

a direct manipulation approach in which the display screen includes an arrangement of common objects found on a desk

FIGURE *14-3*

The desktop metaphor based on direct manipulation, shown on a display screen.

teracting with the real-world objects they represent (see Figure 14-3). End users now expect all applications, including business information systems, to be as natural to work with as objects on the desktop.

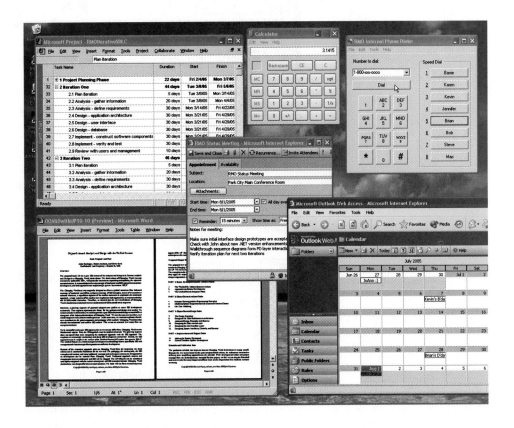

The Document Metaphor

document metaphor

a metaphor of HCI in which interaction with the computer involves browsing and entering data on electronic documents

hypertext

documents that allow the user to click on a link and jump to a different part of the document or to another document

hypermedia

technology that extends the hypertext concepts to include multimedia content such as graphics, video, and audio

Another view of the interface is the *document metaphor*, in which interaction with the computer involves browsing and entering data on electronic documents. These documents are much like printed documents, but because the documents are electronic, they are more interactive. Electronic versions of documents can be organized differently from paper versions because the reader can jump around from place to place. *Hypertext* documents allow the user to click on a link and jump to a different part of the document or to another document entirely.

Most common desktop applications create and edit electronic documents, which are not limited to text and usually include word processing, spreadsheets, presentations, and graphics. All of these applications produce documents, but any one document can contain words, numbers, and graphics produced by any of these applications, making documents collections of all sorts of interrelated media. *Hypermedia* extends the hypertext concept to include multimedia content such as graphics, video, and audio that can be linked for navigation by the user in a document.

The World Wide Web is based on the document metaphor, because everything at a Web site is organized as pages that are linked as hypermedia (note that HTML means *Hypertext* Markup Language). A Web site processes transactions by selecting information on a Web page document. The document metaphor and the browser interface function as useful ways of describing and designing interactive systems, and they will continue to affect user interface designs (see Figure 14-4).

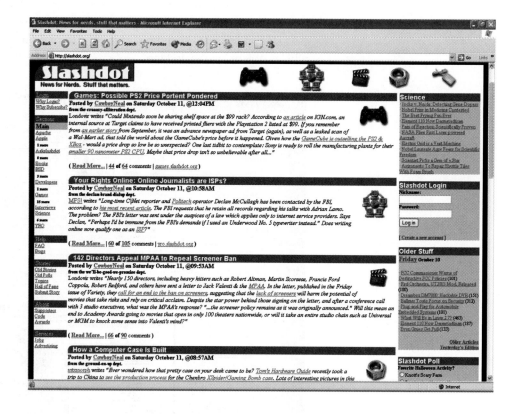

dialog metaphor

a metaphor of HCI in which interacting
with the computer is much like carrying on a conversation or dialog

The Dialog Metaphor

The direct manipulation and document metaphors emphasize objects in the computer
with which the user interacts. Another view of the interface is the *dialog metaphor*, in
which interacting with the computer is much like carrying on a conversation or dialog.
In fact, user-interface design is often referred to as *dialog design*. Carrying on a dialog, or
conversation, with someone requires each person to listen to and respond to questions
and comments from the other person, exchanging information in a sequence. The dialog metaphor is another way of thinking about human-computer interaction because
the computer "listens to" and responds to user questions or comments, and the user
"listens to" and responds to the computer's questions and comments. Like the direct
manipulation metaphor, the dialog metaphor is based on an object-oriented view of the
system because communication involves messages from one "object" to another. Figure
14-5 shows how the user and the computer communicate by sending messages to each
other.

FIGURE *14-5*

The dialog metaphor expresses the
concept that the user and computer
interact by sending messages.

Consider the following dialog between a manager and an assistant:

Manager: Did I get any messages while I was out?

Assistant: Yes, you have three messages—from Bob, Mary, and Lim.

Manager: What did Lim have to say?

Assistant: Lim left a message at 8:15 P.M. last night regarding the meeting next Monday about the inventory management system. The message is, "Can we change the time for the meeting to 10:30? I'll be delayed by the testing session."

Manager: I better respond. Say that the change is not a problem.

Assistant: Okay, I'll leave him that message. Do you want the next message?

Manager: What did Mary have to say?

Assistant: She left a message at 8:15 this morning regarding lunch. She said, "Lunch is still on, but Joe will not be able to join us."

Manager: Okay, no response. And that's all for now. Thanks.

Assistant: Okay, you still have one message from Bob. I'll remind you later.

This dialog involves the manager and the assistant carrying on a conversion about messages. The questions asked by the manager and the responses and follow-up questions asked by the assistant seem clear and natural. Would the basic dialog be any different if this were an automated phone-answering service that responded to voice commands and replied in a computer-generated voice? Probably not. Would it be any different if this were a computer application simulating an intelligent "assistant"? Probably not. The basic dialog followed would be the same: a question, a response, another question, a response that might include a request for clarification, a response to the request for clarification, and a final response.

The basic dialog is also the same for a typical e-mail application, even though the user and computer send messages in different ways. The user selects a menu item for *read new mail*. The computer lists the new mail messages for the user to choose, the user chooses one message, then the computer displays the message. It might seem odd to think of interaction with an e-mail application as being similar to the dialog just presented, but the basic information exchanged and the sequence of actions are the same.

The user and the computer both send messages, but each is forced to use a different language because of limitations of both the user and the computer. The user cannot understand cryptic binary codes or plug in directly to the computer to interpret the electrical impulses the computer uses to represent the binary codes. The natural language of the computer just won't work for people. The computer has to adapt to the user and provide its messages in a form that is natural for the user—text and graphics that the user can see and read.

Similarly, the computer cannot understand complex voice messages, facial expressions, and body language that are the natural communication cues of the user, so the user has to adapt to the computer and provide messages by clicking the mouse, dragging objects, and typing words on the keyboard. Advances in computer technology are making it possible for the user to communicate in more natural ways, but the typical user interfaces today still rely on the mouse and keyboard. One reason is the need for silence and also privacy in the office, so it is not clear whether voice commands will become common in computer applications.

The challenge of user-interface design is to construct a natural dialog sequence that allows the user and computer to exchange the messages required to carry out a task. Then the designer needs to develop the details of the language required for the user to send the messages to the computer (the user's language), plus the language needed for the computer to send messages to the user (the computer's language).

FIGURE *14-6*

The user's language and the computer's language used to implement an e-mail application based on the natural dialog between manager and assistant.

Figure 14-6 shows the earlier dialog between manager and assistant translated into the languages used by the user and the computer. Interface designers use a variety of informal diagrams and written narratives to model human-computer interaction. This is just one way the dialog design details can be modeled; you'll learn about additional techniques later in this chapter.

	Message	User's language	Computer's language
Manager	Did I get any messages while I was out?	Click the *read messages* menu item on the main menu.	
Assistant	Yes, you have three messages—from Bob, Mary, and Lim.		Look up new messages for the user and display a new message form with message headers listed in a list box.
Manager	What did Lim have to say?	Double-click the message from Lim in the list box.	
Assistant	Lim left a message at 8:15 P.M. last night regarding the meeting next Monday about the inventory management system. The message is, "Can we change the time for the meeting to 10:30? I'll be delayed by the testing session."		Look up the message body for the selected message and display it in message detail form.
Manager	I better respond. Say that the change is not a problem.	Click the Reply button on the message detail form. Type in the message, "Okay, that is not a problem." Click the Send Button.	Display the new message form addressed to the sender.
Assistant	Okay, I'll leave him that message. Do you want the next message?		Display the Message Sent dialog box and redisplay the new messages form with message headers listed in the list box.
Manager	What did Mary have to say?	Double-click the message from Mary in the list box.	
Assistant	She left a message at 8:15 this morning regarding lunch. She said, "Lunch is still on, but Joe will not be able to join us."		Look up the message body for the selected message and display it in message detail form.
Manager	Okay, no response. And that's all for now. Thanks.	Click the *close message* button. Click the *close new message form* button.	Redisplay a new message form with message headers listed in the list box.
Assistant	Okay, you still have one message from Bob. I'll remind you later.		Display the Closing Read New Mail dialog box, showing one unread message remaining.

GUIDELINES FOR DESIGNING USER INTERFACES

There are many published interface design guidelines to guide system developers. User-interface design guidelines range from general principles to very specific rules. This section describes some well-known guidelines for designing the user interface. Later, this chapter presents some of the guidelines and rules for designing windows forms and browser forms used with Web-based development. Some system development organizations adopt *interface design standards*—general principles and rules that an organization must follow when developing any system. Design standards help ensure that all user interfaces function well and that all systems developed by the organization have a similar look and feel.

interface design standards

general principles and rules that must be followed for the interface of any system developed by the organization

Visibility and Affordance

Donald Norman is a leading researcher in HCI who proposes two key principles to ensure good interaction between a person and a machine: visibility and affordance. These two principles apply to human-computer interaction just as they do for any other device.

Visibility means that a control should be visible so users know it is available, and that the control should provide immediate feedback to indicate it is responding. For example, a steering wheel is visible to a driver, and when the driver turns it to the left, it is obvious that the wheel is responding to the driver's action. Similarly, a button that can be clicked by a user is visible, and when it is clicked, it changes to look as though it has been pressed, to indicate it is responding. Some buttons make a clicking sound to provide feedback.

visibility

a key principle of HCI that states all controls should be visible and provide feedback to indicate that the control is responding to the user's action

Affordance means that the appearance of any control should suggest its functionality—that is, the purpose for which the control is used. For example, a control that looks like a steering wheel suggests that the control is used for turning. On the computer, a button affords clicking, a scroll bar affords scrolling, and an item in a list affords selecting. Norman's principles apply to any objects on the desktop, such as those shown previously in the examples in Figures 14-3 and 14-4.

affordance

a key principle of HCI that states the appearance of any control should suggest its functionality

If user-interface designers make sure that all controls are visible and clear in what they do, the interface will be usable. Most users are familiar with the Windows interface and the common Windows controls. However, designers should be careful to apply these principles of visibility and affordance when designing Web pages. Many new types of controls are now possible at Web sites, but these controls are not always as visible and their effects are not always as obvious as they are in a standard Windows interface. More objects are clickable, but it is not always clear what is clickable, when a control has recognized the click, and what the click will accomplish. For example, sometimes you click on an image and a new page opens in the browser. Other times you click on an image and nothing happens.

Eight Golden Rules

Ben Shneiderman, another leading researcher in HCI, proposes eight underlying principles that are applicable in most interactive systems (see "Further Resources" at the end of the chapter for Shneiderman's text). Although they are general guidelines rather than specific rules, he names them "golden rules" to indicate that they are the key to usability (see Figure 14-7).

1 Strive for Consistency
2 Enable Frequent Users to Use Shortcuts
3 Offer Informative Feedback
4 Design Dialogs to Yield Closure
5 Offer Simple Error Handling
6 Permit Easy Reversal of Actions
7 Support Internal Locus of Control
8 Reduce Short-Term Memory Load

Strive for Consistency

Designing a consistent-appearing and -functioning interface is one of the most important design goals. The way that information is arranged on forms, the names and arrangement of menu items, the size and shape of icons, and the sequence followed to carry out tasks should be consistent throughout the system. Why? People are creatures of habit. Once we learn one way of doing things, it is difficult to change. When we operate a computer application, many of our actions become automatic—we do not think about what we are doing. People who can touch-type do not have to think about each key press—their fingers just respond automatically. Consider what would happen to touch-typists if rows two and three on the keyboard were reversed. They would not be able to use the keyboard (and certainly wouldn't like it). If a new application comes along that has a different way of functioning, productivity suffers and users will not be happy.

The Apple Macintosh first emphasized the benefits of consistency in the 1980s. Apple provided applications for the Macintosh that set the standard for developers to follow when creating new applications. If new applications were consistent with these applications, Apple claimed that learning them would be easy. Apple also published a standards document to explain how to be consistent with the Macintosh interface. Similar examples and standards documents followed for the Microsoft Windows interface.

Business information systems are different from desktop applications originally produced for the Macintosh. Sometimes an application needs to be inconsistent with the original guidelines. For example, the original standards specified that every application include menus on the menu bar for File, Edit, and Format. All document-oriented applications—such as word processors, spreadsheets, and graphics—need those menus. But many business systems do not have file, edit, and format functions. Most other guidelines and standards do apply, however.

Research has also shown that inconsistent interfaces sometimes are beneficial. If the user is interacting with multiple applications in separate windows, a different visual appearance may help the user differentiate them. Additionally, when the user is learning several applications in one session, some differences in the interfaces may help the user remember which application is which. Inconsistencies introduced for these reasons should be subtle and superficial. The basic operation of the applications should be the same.

Enable Frequent Users to Use Shortcuts

Users who work with one application all day long are willing to invest the time to learn shortcuts. They rapidly lose patience with long menu sequences and multiple dialog boxes when they know exactly what they want to do. Therefore, shortcut keys reduce the number of interactions for a given task. Also, designers should provide macro facilities for users to create their own shortcuts.

Sometimes the entire interface should be designed for frequent users who do not need much flexibility. Consider the mail-order data-entry clerks for Rocky Mountain Outfitters. They enter orders into the system all day long from paper forms mailed by customers. These users need an interface that is simple, fast, and accurate. Long dialogs, multiple menus, and multiple forms would slow these users down.

Offer Informative Feedback

Every action a user takes should result in some type of feedback from the computer so the user knows that the action was recognized. Even keyboard clicks help the user, so an electronic "click" is included deliberately by the operating system. If the user clicks a button, the button should visually change and perhaps make a sound.

Feedback of information to the user is also important. If the Rocky Mountain Outfitters mail-order clerk enters a customer ID number in an order screen, the system should look up the customer to validate the ID number, but it should also display the name and address to the clerk so the clerk is confident the number is correct. Similarly, when the clerk enters a product ID for the order, the system should display a description of the product. As the clerk's attention shifts back and forth from the mail-order form to the computer screen, he or she compares the name and product description from the system with the information on the form to confirm that everything is correct. This sense of confirmation and the resulting confidence in the system are very important to users, particularly when they work with a system all day. But the system should not slow the user down by displaying too many dialog boxes to which the user must respond.

Sometimes feedback is provided to help the user in other ways. The phone-order representative at Rocky Mountain Outfitters needs information from the system just as the mail-order clerk does, but he or she also needs additional information. Phone customers may ask questions, so the information provided as feedback for the phone-order representative is more detailed and flexible. We discuss some designs for the phone-order representative at RMO later in this chapter.

Design Dialogs to Yield Closure

Each dialog with the system should be organized with a clear sequence—a beginning, middle, and end. Any well-defined task has a beginning, middle, and end, so users' tasks on the computer should also feel this way. If the user is thinking, "I want to check my messages," as in the earlier manager and assistant dialog example, the dialog begins with a request, exchanges information, and then ends. The user can get lost if it is not clear when a task starts and ends. Additionally, the user often focuses intently on a task, so when it is confirmed that the task is complete, the user can clear his or her mind and get ready to focus on the next task.

If the system requirements are defined initially as events to which the system responds, each event leads to processing of one specific, well-defined activity. In the traditional, structured approach, each activity is defined by data flow diagrams and structured English. With the object-oriented approach, each use case might be further defined as multiple scenarios, each with a flow of steps. Each scenario is a well-defined interaction; therefore, event decomposition sets the stage for dialogs with closure in both the traditional approach and the object-oriented approach.

Offer Simple Error Handling

User errors are costly, both in the time needed to correct them and in the resulting mistakes. If the wrong items are sent to a customer at Rocky Mountain Outfitters, it is a costly error. So, the systems designer must prevent the user from making errors whenever possible. A chief way to do this is to limit available options and allow the user to choose from valid options at any point in the dialog. Adequate feedback, as discussed previously, also helps reduce errors.

If an error does occur, the system needs mechanisms for handling it. The validation techniques discussed in Chapter 15 are useful for catching errors, but the system must also help the user correct the error. When the system does find an error, the error message should state specifically what is wrong and explain how to correct it. Error messages should not be judgmental. It is not appropriate to blame the user or make the user feel inadequate.

The system also should make it easy to correct the error. For example, if the user typed in an invalid customer ID, the system should tell the user that and then place the cursor in the customer ID text box with the previously typed number displayed and ready to edit. This way, the user can see the mistake and edit it rather than having to re-type the entire ID. Consider the following error message that occurs after a user has typed in a full screen of information about a new customer:

```
The customer information entered is not valid. Try again.
```

This message does not explain what is wrong or what to do next. Further, after this message appears, what if the system cleared the data-entry form and redisplayed it? The user would have to reenter everything previously typed, yet still have no idea what is wrong. The error message did not explain it, and now that the typed data have been cleared, the user cannot tell what might have been wrong. A better error message would say:

```
The date of birth entered is not valid. Check to be sure
only numeric characters in appropriate ranges are entered
in the date of birth field...
```

The input form should be redisplayed with all fields still filled in, and the cursor should be placed at the field with invalid data, ready for the user to edit the field.

Permit Easy Reversal of Actions

Users need to feel that they can explore options and take actions that can be canceled or reversed without difficulty. This is one way that users learn about the system—by experimenting. It is also a way to prevent errors; as users recognize they have made a mistake, they cancel the action. In the game of checkers, a move is not final until the player takes his or her fingers off the game piece; it should be the same when a user drags an object on the screen. Additionally, designers should be sure to include cancel buttons on all dialog boxes and allow users to go back one step at any time. Finally, when the user deletes something substantial—a file, a record, or a transaction—the system should ask the user to confirm the action.

Support Internal Locus of Control

Experienced users want to feel that they are in charge of the system and that the system responds to their commands. They should not be forced to do anything or made to feel as if the system is controlling them. Systems should make users feel that they are deciding what to do. Designers can provide much of this comfort and control through the wording of prompts and messages. Writing out a dialog like the manager and assistant message dialog given previously will lead to a design that conveys the feeling of control.

Reduce Short-Term Memory Load

People have many limitations, and short-term memory is one of the biggest. As discussed earlier in this book, people can remember only about seven chunks of information at a time. The interface designer cannot assume that the user will remember anything from form to form, or dialog box to dialog box, during an interaction with the system. If the user has to stop and ask, "Now what was the filename? the customer ID? the product description?" then the design places too much of a burden on the user's memory.

With these eight golden rules in mind, an interface designer can help ensure that user interactions are efficient and effective. We now turn to some basic techniques for documenting the design of the dialog.

DOCUMENTING DIALOG DESIGNS

Many techniques are available to help the designer think through and document dialog designs. The dialogs that must be designed are based on the inputs and outputs requiring user interaction, as discussed earlier. They are used to define a menu hierarchy that allows the user to navigate to each dialog. Storyboards, prototypes, and UML diagrams can be used to complete the designs.

Events, Subsystems, and the Menu Hierarchy

Inputs and outputs are obtained from data flow diagrams (in the traditional approach) or use cases and scenarios (in the object-oriented approach). Generally, each input *obtained interactively from a user* requires a dialog design. Additionally, each output produced *at the request of a user* requires a dialog design. So, each dialog is based on an event documented early during the analysis process that is classified as requiring a user interface rather than a system interface.

Dialog design must be done simultaneously with other design activities. As shown in Chapter 10, the structure charts for subsystems (transaction analysis) include details about menu structure of the interactive parts of the system. Additionally, the structure chart for each event (transform analysis of each DFD fragment) also includes details about the dialog with the user. The object-oriented approach also integrates dialog design very early, even during analysis tasks. Use case descriptions, activity diagrams, and system sequence diagrams (SSDs) include details about the dialog. Remember that menu design and dialog design are not done in isolation.

The available menus reflect the overall system structure from the standpoint of the user. Each menu contains a hierarchy of options, and they are often arranged by subsystem or by actions on objects. Rocky Mountain Outfitters' customer support system includes the order-entry subsystem, order fulfillment subsystem, customer maintenance subsystem, and catalog maintenance subsystem, as well as a reporting subsystem added during design. Menus might also be arranged based on objects—customers, orders, inventory, and shipments. Each menu might include duplicate functions, such as *Look up past orders*, under customers and under inventory.

Sometimes several versions of the menus are needed based on the type of user. For example, mail-order clerks at RMO do not need many of the options available—they process new orders only. The phone-order sales representatives need many more options, but they still do not need all system functions. And some options should be available only to managers, such as management reports and price adjustments.

Menus should also include options that are *not* activities or use cases from the event list—most important are options related to the system controls, which are discussed in Chapter 15. These include backup and recovery of databases in some cases, plus user account maintenance. Additionally, user preferences are usually provided to allow the user to tailor the interface. Finally, menus should always include help facilities.

All events that lead to dialogs in the RMO customer support system are listed and grouped by subsystem in Figure 14-8. These groupings form one set of menu hierarchies. In addition, there are menu hierarchies for utilities, preferences, and help. The list in the figure is only one of many possible menu hierarchy designs—a starting place.

FIGURE *14-8*

One overall menu hierarchy design for the RMO customer support system (not all users will have all of these options available).

Five menu hierarchies grouped by subsystem and based on events/use cases

Order Entry
    Check Availability
    Create New Order
    Update Order

Order Fulfillment
    Check Order Status
    Record Fulfillment
    Record Back Order
    Create Return

Customer Account Maintenance
    Provide Catalog
    Update Customer Account
    Adjust Customer Charges
    Distribute Promotional Package

Catalog Maintenance
    Update Catalog
    Create Special Promotion
    Create New Catalog

Reporting and Queries
    Order Summary Report
    Transaction Summary Reports
    Fulfillment Summary Reports
    Prospective Customer Lists
    Customer Adjustment Reports
    Catalog Activity Reports
    Ad Hoc Query Facility

Three menu hierachies added during design for controls, preferences, and help

System Utilities
    Printers and Devices
    Backup and Recovery
    User Accounts
        Maintain Accounts
        Change Current User Password

User Preferences
    Dialog Style
    Color and Font
    Shortcuts
    Macro Facility

Help
    Contents and Index
    Search for Help
    Task List
    About the RMO System

A dialog design is created for each of these menu options. After completing the dialog designs for all options, the designer can then redefine the structure of the menus for different types of users. Menu hierarchies can be rearranged easily as the design evolves.

Dialogs and Storyboards

After identifying all required dialogs, the designers must document the dialogs. Many options exist; there are no *de facto* standards. One approach is to list the key steps followed for the dialog with a written description of what the user and computer do at each step. The format for writing these steps can follow the activity diagram described in Chapter 4 or the use case description format in Chapter 7, or they can be more free-form.

Designers can also document dialog designs by writing out a dialog like the one presented earlier in this chapter (for the manager and assistant), then documenting how the user and computer express themselves, as shown previously in Figure 14-6. Sometimes the designer can go right from the dialog to sketching a series of screens that make the design more visible. Consider the DownTown Videos rental system described in a case study at the end of Chapter 7. One event is *Customer rents videos*, and the dialog for renting videos to a customer that goes on between the clerk and the system would follow this pattern:

System:	What DownTown Videos option do you want?
User:	I need to rent some videos.
System:	Okay, what is the customer's ID number for this rental?
User:	It is 201564.
System:	Fine, your customer is Nancy Wells at 1122 Silicon Avenue. Note that she has two kids, Bob and Sally, who can only rent PG-13 videos. Correct?
User:	Yes.
System:	What's the ID number of the first video she wants to rent?
User:	It is 5312.
System:	Fine, that is *Titanic* (the 1953 version with Barbara Stanwyck) with a rental charge of $1.00 for three days, rated PG. What's the ID number of the next video, or is that all?
User:	It is 8645.
System:	Fine, that is Titanic (the 2005 reissue version) with a rental charge of $3.00 for three days, rated PG-13. What's the ID number of the next video, or is that all?
User:	That's all.
System:	Okay, collect $4.00 plus $.24 tax, for a total of $4.24, and let me know when you are done.
User:	Yes, I'm done.
System:	Remind Nancy Wells that these videos are due back Wednesday, the ninth, before 11 P.M. closing.
User:	Okay. Thanks.

The designer thinks through the sequence of screens the user would need to carry on this dialog with the computer. The user needs to enter a customer ID and several video IDs. The computer would have to look up and display information on the customer and each video rented. Then the computer would calculate the total due and remind the user of the due date so the user can pass that information on to the customer.

storyboarding
a technique used to document dialog designs by showing a sequence of sketches of the display screen

One technique used to show the screens is called *storyboarding*—showing a sequence of sketches of the display screen during a dialog. The sketches do not have to be very detailed to show the basic design concept. Designers can implement a storyboard with a visual programming tool such as Visual Basic, but using simple sketches drawn with a graphics package can help keep the focus on the fundamental design ideas.

Figure 14-9 shows the storyboard for the *rent videos* dialog. The system has a menu hierarchy based on the event list plus needed controls, preferences, and help. The dialog uses one form and a few dialog boxes and adds more information to the form as the dialog progresses. Note that the prompt area at the bottom of the form displays the questions the computer asks, matching almost identically the phrases used in the written dialog. The user has a choice of either scanning or typing the few IDs that must be entered. Information provided to the user is shown in labels on the form. The information provided allows the user to confirm the identity of the customer, to see any restrictions that might apply, and to pass on to the customer any information about cost and return dates. In other words, the system helps the user do a better job of interacting with the customer by confirming information, providing feedback, and providing closure.

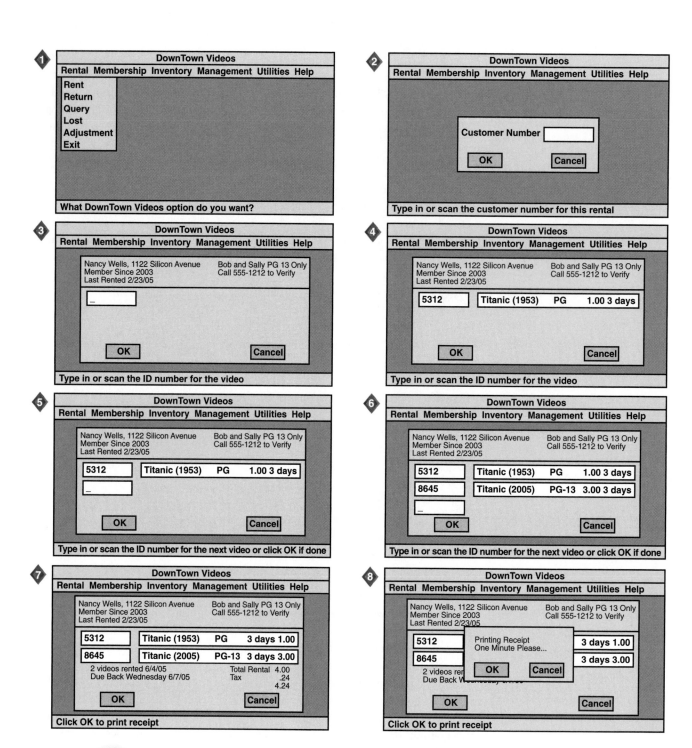

FIGURE *14-9*

Storyboard for the DownTown Videos
rent videos dialog.

These approaches to dialog design provide only a framework to work from, and the resulting design remains fairly general. As working prototypes are produced, many details still have to be worked out. As the design progresses, reviewing the golden rules and other guidelines will help you keep the focus on usability.

Dialog Documentation with UML Diagrams

The object-oriented approach provides UML diagrams that are useful for modeling user-computer dialogs. Use case descriptions, shown in Chapter 7, include a list of steps followed as the user and system interact. Activity diagrams, shown in Chapters 4 and 7, also document the dialog between user and computer for a use case. Both can be used to provide models of the user-computer interaction required in each dialog. In the object-oriented approach, objects send messages back and forth, listening to and responding to each other in sequence. People also send messages to objects and receive messages back from them. The system sequence diagram (SSD) described in Chapter 7 includes an actor (a user) sending messages to the system and the system returning information in the form of messages, shown in sequence. It basically shows a dialog between the user and the system. The SSD is based on the sequence of steps included in the use case description, so the dialog design for the use case begins very early and is refined continually.

The object-oriented approach involves adding more types of objects to class diagrams and interaction diagrams as the project moves from analysis to design, as discussed in Chapters 11 and 12. The additional classes of objects are packaged into three layers that contain user-interface classes, problem domain classes, and data access classes. Designers add user-interface classes and objects to these diagrams to show more detail about the design of the dialog between the user and computer. This design process was demonstrated in Chapter 11. The first step is to determine what window or Web forms are required for the dialog based on informal dialog design techniques described previously. Next, the sequence diagram for the scenario is expanded to show the user (an actor) interacting with the forms. The designer can then use a class diagram to model the user-interface classes that make up the forms. Finally, the sequence diagram can be further expanded to show the user interacting with specific objects that make up the form.

Recall the Rocky Mountain Outfitters use case *Look up item availability*. A system sequence diagram was shown for this use case in Chapter 11 (Figure 11-12) and it was expanded in a series of sequence diagrams to show the complete use case realization. Figure 14-10 shows a version of the sequence diagram that includes a form (or window) named ProductQueryForm. The form represents the user-interface layer of the three-layer design. The form is inserted between the actor and the use case controller. The actor interacts with the interface objects on the form to talk to the problem domain objects, which reply through the interface objects on the form. Recall that the requirements model of the interaction just shows the messages between the actor and the system. The design model inserts the user-interface layer to create a physical model of how the interaction is implemented.

FIGURE *14-10*

A sequence diagram for the RMO *Look up item availability* dialog with the ProductQueryForm added.

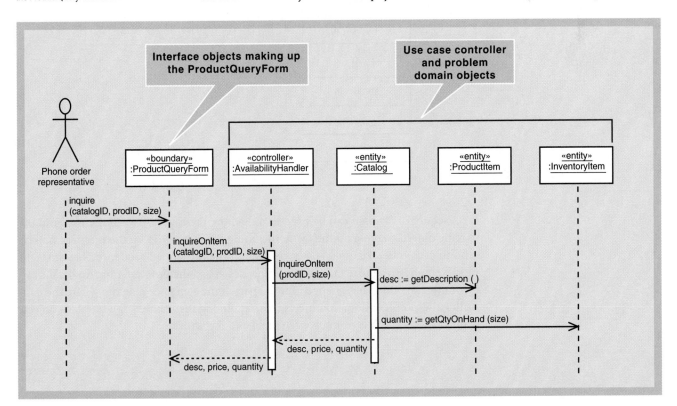

The form shown on the sequence diagram includes specific interface objects. Figure 14-11 shows a class diagram with the interface classes used to make up the form. The Frame class in Figure 14-11 represents the basic structure that contains other interface objects. A MenuBar is attached to the frame, a menu bar contains Menus, and a menu contains MenuItems. The associations between these classes are aggregations shown with the diamond symbol on the class diagram (aggregation relationships are discussed in Chapter 5). Other classes shown are List, Button, and Label, which also are parts of the frame. This example, based loosely on Java, enables the frame to "listen" for events that occur to interface objects, such as clicking a menu item or a button. The frame's actionListener() method is invoked when the frame "hears" that an event has occurred.

FIGURE *14-11*

A class diagram showing interface classes making up the ProductQueryForm.

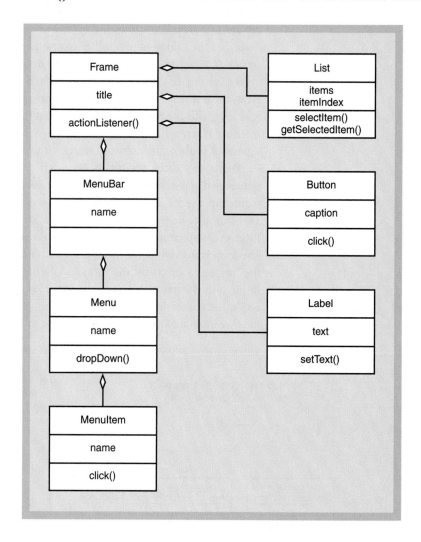

A sequence diagram can be used to model the messages between the user and the specific objects making up the form, including the messages that the interface objects send to each other. Figure 14-12 shows the further expanded sequence diagram. This model emphasizes the details of the form's design, so the problem domain details can be omitted. Nothing has changed for the part of the sequence diagram where the controller and problem domain objects interact among themselves. The interface objects simply have been plugged in between them and the actor.

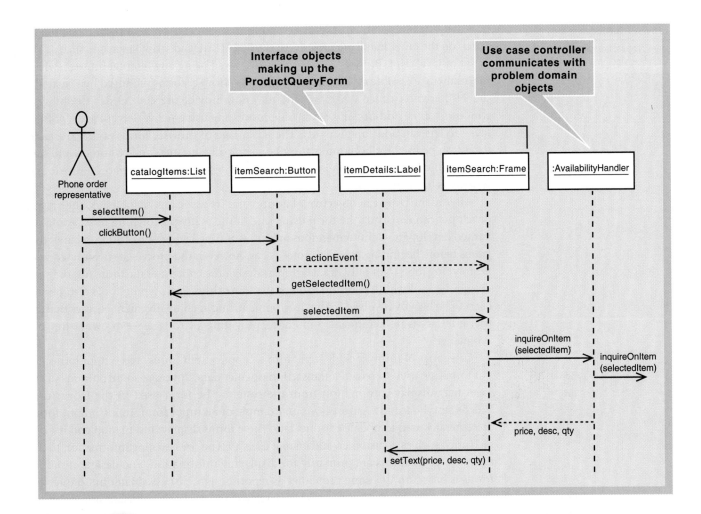

FIGURE *14-12*

A sequence diagram showing specific interface objects making up the Product-QueryForm for the *Look up item availability* dialog (not all problem domain objects are shown).

The user interacting with the *Look up item availability* form selects an item to look up in the list and then clicks the search button. The frame "hears" the click, asks the list for the selected item, and uses the selected item in the message to query the use case controller object, which interacts with other problem domain objects. When the use case controller object returns the requested information, the frame tells the label to display the information to the user. The designer would probably not specify the design of each form to this level of detail; the example is intended to illustrate how the interaction in the user-interface layer occurs.

GUIDELINES FOR DESIGNING WINDOWS AND BROWSER FORMS

As with the previous activities of user-interface design, analysts must take care in designing the forms that users see on the screen. Each dialog might require several windows forms, and each form must be designed for usability. Almost all of the new business systems today are developed for an interactive Microsoft Windows, X-Windows (UNIX), or Macintosh environment. The underlying principles are the same for forms in any of these environments. In this section, when we refer to forms or windows, we mean any of these three environments. Within each windows environment, however, we need to consider two types of forms: windows and browsers.

Windows forms are programmed in a full-featured programming language, such as Visual Basic, C++, or Java. Because of this, windows forms have the advantage of being extremely flexible and capable of accessing data directly on a workstation. *Browser forms*, on the other hand, are programmed using HTML and script languages such as VB-Script or JavaScript. Browser forms can be displayed using any Internet browser, which makes them accessible on a variety of platforms. Browser forms produced by Visual Studio .NET are now called *Web forms*, and they now provide the same design flexibility as windows forms. Additionally, server-side processing using Active Server Pages (ASP) or Java servlets can add functionality. The advantage of browser forms is that the same forms can be used for both internal staff on company intranets and customers and suppliers on the Internet. As a result, many firms are designing user interfaces for their new systems as browser forms.

After identifying the objective of a form and its associated data fields, the system developer can construct the form using one of the many prototyping tools available. Earlier, developers spent tremendous amounts of time laying out a form on paper diagrams before beginning programming. Today, however, the process is streamlined with prototyping tools. Not only can developers design the content of the form, but they can design the look and feel of the form at the same time. Prototyping tools also permit users to be heavily involved in the development, and such involvement ensures that the user interface is in fact the *user's* interface, providing a strong sense of ownership and acceptance.

Categories of forms include input forms, input/output forms, and output forms. Input forms are used to record a transaction or enter data, although some portions of the form may display information from the system. The form used in the DownTown Videos storyboard in Figure 14-9 is an example of an input form. Input/output forms are generally used to update existing data. These forms display information about a single entity, such as a customer, and enable users to type over existing information to update it. Output forms are primarily for displaying information. The design of output forms is based on the same principles as report design, discussed later in Chapter 15. Input and input/output forms are closely related and are designed using similar principles. Before developing a form, the designer should carefully analyze the integrity controls required for data input, which are discussed in Chapter 15.

There are four major issues to consider in the design of these forms:

- Form layout and formatting
- Data keying and entry
- Navigation and support controls
- Help support

Form Layout and Formatting

Form layout and formatting are concerned with the general look and feel of the form. You may have encountered systems with hard-to-use input forms—the font was too small, the labels were hard to understand, the colors were abrasive, the navigation buttons were not obvious, and so forth. These deficiencies can occur on both windows forms and browser forms. In contrast, forms that are easy to use are well laid out, with the fields easily identified and understood. One of the best methods to ensure that forms are well laid out is to prototype various alternatives and let users test them. Users will let you know which characteristics are helpful and which are distracting. As you design your input forms, you should think about the following:

- Consistency
- Headings, labels, and logos
- Distribution and order of data-entry fields and buttons
- Font sizes, highlighting, and colors

Consistency belongs at the top of the list because of its importance for ease of learning and use, as discussed previously. Some large systems require multiple input forms and several teams of programmers and analysts to develop them. Sometimes those teams don't coordinate their efforts, resulting in inconsistencies, and a system that is not consistent across all forms can be error-prone and difficult to use. To avoid these problems, all of the forms within a system need to have the same look and feel. A consistent use of function keys, shortcuts, control buttons, and even color and layout makes a system much more useful and professional looking. Cascading style sheets help designers control the consistency of Web forms. Design templates help designers control the consistency of windows forms. For example, with Microsoft Visual Studio .NET, a template form can be designed that is used as the superclass of all forms in the project.

The headings, labels, and logos on the form help to convey the purpose and use of the form. A clear, descriptive title at the top of the form helps to minimize confusion about a form's use. Labels should also be easy to identify and read.

The designer also should carefully place the data-entry fields around the form. Related fields are usually placed next to each other and can even be isolated with a fine-lined box. The designer also should carefully consider the tab order. If input is coming from a paper form, then the tab order should follow the order of the paper form—left to right, top to bottom. Blank space should be used throughout the form so that the fields do not appear crammed together and are easy to distinguish and read. Normally navigation buttons are at the bottom of the window. *De facto* traditions for the placement of buttons are developing based on standards of the large development firms such as Apple, Microsoft, Sun, Oracle, and others. It is a good idea to be sensitive to these traditions as they change with technology upgrades.

The purpose of font size, highlighting, and color is to make the form easy to read. A careful mix of large and small fonts, bold and normal type, and different color fonts or background can help a user find important or critical information on the form. Too much variation makes the form cluttered and difficult to use. However, judicious use of these techniques will make the form more easily understood. Column headings and totals can be made slightly larger or boldfaced. The form can highlight negative or credit balances by changing font color. However, font color and background color should be used in concert to ensure that the field is readable. For example, neither red type on green or black backgrounds nor a dark color on a dark background is a good choice—some people with colorblindness cannot distinguish red from green or black.

Figure 14-13 is an example of a windows form designed for the Rocky Mountain Outfitters customer support system. This form is used to look up information about a product and to add it to an order. Notice how the title and labels make the form easy to read. The natural flow of the form is top to bottom, with related fields placed together. Navigation and close buttons are easily found but are not in the way of data-entry activity.

Data Keying and Data Entry

The heart of any input form is the entry of the new data. Even here, however, a primary objective is to require as little data entry as possible. Any information already in the computer, or that can be generated by the computer, should not be reentered. A generous use of selection lists, check boxes, automatic retrieval of descriptive fields, and so forth will speed up data entry and reduce errors. The Product Detail form for RMO (Figure 14-13) shows many examples that reduce the need to enter data.

FIGURE *14-13*

The RMO Product Detail form used to look up information about a product, select size and color, and then add the product to an order.

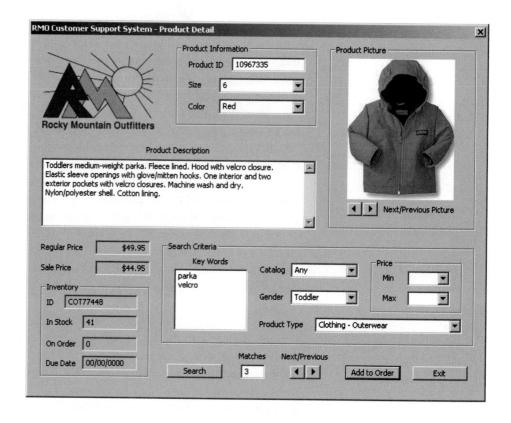

text box

an input control that accepts keyboard data entry

list box

an input control that contains a list of acceptable entries the user can select

spin box

a variation of the list box that presents multiple entries in a text box from which the user can select

combo box

another variation of the list box that permits the user to enter a new value or select from the entries

radio buttons (option buttons)

input controls that enable the user to select one option from a group

check boxes

input controls that enable the user to select more than one option from a group

Several types of data-entry controls are widely used in windows systems today. A *text box* is the most common element used for data entry. A text box consists of a rectangular box that accepts keyboard data. In most cases, it is a good idea to add a descriptive label to identify what should be typed in the text box. Text boxes can be designed to limit the entry to a specified length on a single line or to permit scrolling with multiple lines of data.

Variations of a text box consist of a list box, a spin box, and a combo box. A *list box* contains a list of the acceptable entries for the box. The list usually consists of a predefined list of data values, and the user selects one from the list. The list can be presented either within a rectangular box or as a drop-down list. A variation of a list box is a spin box. A *spin box* presents the possible values within the text box itself. Two spinner arrows let the user scroll through all the values. A *combo box* also contains a predefined list of acceptable entries but also permits the user to enter a new value when the list does not contain the desired value. Both a list box and combo box facilitate data entry by minimizing keystrokes and the corresponding possibility of errors.

Two types of input controls are used in groups: radio buttons (sometimes called *option buttons*) and check boxes. *Radio buttons* are associated as a group, and the user selects one and only one of the group. The system automatically turns off all other buttons in the group when one is selected. Since all of the possible values appear on the form, this control is used only when the list of alternatives is small and the values never change. *Check boxes* also work together as a group. However, check boxes permit the user to select as many values as desired within the group. Figure 14-14 shows a form with examples of these data-entry controls.

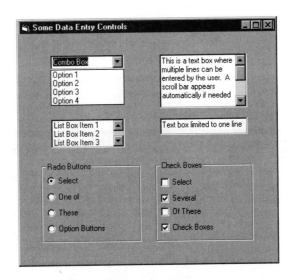

Browser forms contain similar controls. A major difference between a standard windows input form and a browser input form is that the windows form can easily perform edits field by field as the data are entered. In a basic browser input form, the edits are not performed until the entire form is transmitted to the server computer. However, as browser programs have become increasingly more sophisticated, more and more capabilities are being provided for data entry. Windows input forms and browser input forms now have very similar capabilities.

Navigation and Support Controls

Standard window interfaces provide several controls for navigation and window manipulation. For Microsoft applications, these controls consist of Minimize, Maximize, and Close buttons in the top-right corner, horizontal and vertical scroll bars, the record selection bar on the left panel, record navigation arrows at the bottom of the window, and so forth. To maintain consistency across systems, it is generally a good idea to utilize these navigation controls when possible. A well-designed user interface, however, should also include other controls or buttons. You can place buttons on the form to enable users to move to other relevant screens, to search and find data, and to close the open window. Browser forms also provide navigation and support controls. The browser contains navigation buttons and controls that the application should support using browser forms. Each page might include its own navigation buttons also.

Help Support

A primary objective in the design of each input form is for it to be intuitive so that users will not need help. However, even well-designed forms will be misunderstood, and access to on-line help is always recommended. Three types of help are common in today's systems: a tutorial that walks you through the use of the form, an indexed list of help topics, and context-sensitive help.

Most systems provide tutorial help to assist in training new users. Tutorials can be organized by task, in which case the tutorial generally includes one dialog with a set of related forms. Every new system also should have an indexed list of help topics. This list can be invoked either through a keyword search or, as with many Microsoft systems, with a help wizard. The help wizard is simply a program that does an automatic keyword search based on words found in a question or sentence. The wizard will return several alternative help topics based on the results of the keyword searches.

Context-sensitive help can be based on the indexed list of help, but it is invoked differently. Context-sensitive help automatically displays the appropriate help topic based on the location of the cursor. In other words, if the cursor is within a certain text field on a form, and the user invokes context-sensitive help, then the system displays the help for that text field.

GUIDELINES FOR DESIGNING WEB SITES

Web site design draws from the guidelines and rules for designing the windows forms and browser forms just presented. Many business systems today, including the RMO customer support system, make use of both technologies. Yet a business system such as the order-processing function for RMO is just part of the RMO Web site. Web sites also are used for corporate communication, customer information and service, on-line sales and distribution, and marketing. Because they are available 24 hours a day, 7 days a week, they need to interact seamlessly with customers. This section introduces some guidelines and lessons for Web design. A complete discussion of Web-site design principles is beyond the scope of this book. Many excellent books are available, and we list some of them in the "Further Resources" at the end of this chapter.

Ten Good Deeds in Web Design

Jacob Nielsen is an HCI researcher who now focuses specifically on Web design. Like many useful guidelines, Nielsen's guidelines focus on general issues, including these "Ten Good Deeds in Web Design."

1. Place the organization's *name and logo* on every page and make the logo a link to the home page.
2. Provide a *search* function if the site has more than 100 pages.
3. Write straightforward and simple *headlines and page titles* that clearly explain what the page is about and that will make sense when read out of context in a search engine listing.
4. Structure the page to *facilitate reader scanning* and help users ignore large chunks of the page in a single glance. For example, use grouping and subheadings to break a long list into several smaller units.
5. Instead of cramming everything about a product or topic into a single, huge page, use *hypertext to structure the content space* into a starting page that provides an overview and several secondary pages that each focus on a specific topic.
6. Use *product photos*, but avoid cluttered and bloated product family pages with lots of photos. The primary product page must load quickly and function fast, so it should be limited to a thumbnail product shot.
7. Use *relevance-enhanced image reduction* when preparing small photos and images. Instead of simply reducing the original image to a tiny and unreadable thumbnail, zoom in on its most relevant detail by cropping and resizing the image.
8. Use *link titles* to provide users with a preview of where each link will take them, *before* they have clicked on it.
9. Ensure that all important pages are *accessible for users with disabilities*, especially visually impaired users.
10. *Do the same as everybody else*, because if most big Web sites do something in a certain way, users will expect other sites to work similarly.

Web Site Design Principles

Because Web sites include so many facets, many designers take a broader view of Web site design principles. A Web design book by Joel Sklar suggests that the designer focus on three broad aspects of Web design: (1) designing for the computer medium, (2) designing the whole site, and (3) designing for the user.

Designing for the Computer Medium

It is important to remember that the Web site will be displayed on a computer screen and not on paper. Designers can select from a wide array of video display fonts, colors, and layouts, but the look of the site should flow from its function and the organization's goals. Hypermedia allows the user to navigate through the site in nonlinear ways, so the designer should take advantage of new ways to organize information. Five guidelines to consider include:

- Craft the look and feel of the pages to take advantage of the medium.
- Make the design portable since it will be accessed with a wide range of technology.
- Design for low bandwidth since users will not want to wait for a page to load.
- Plan for clear presentation and easy access to information to ease users' navigation through the site.
- Reformat information for on-line presentation when it comes from other sources.

Designing the Whole Site

The entire site must have unifying themes and a structure, and the theme should reflect the impression the organization wants to convey. For example, a site for adult, business-oriented users should use subdued colors, familiar business-oriented fonts, and structured linear columns. A site for children should combine bright colors, an open and friendly dynamic structure, and simple appealing graphics. Four guidelines to consider include:

- Craft the look and feel of the pages to match the impression desired by the organization.
- Create smooth transitions between Web pages so users are clear about where they have been and where they are going.
- Lay out each page using a grid pattern to provide visual structure for related groups of information.
- Leave a reasonable amount of white space on each page between groups of information.

Designing for the User

We discussed user-centered design previously in this chapter, and it is important to focus Web design efforts on the users and their needs. If a feature will annoy or distract users, do not include it. It is sometimes difficult to know who the Web users will be, but if the purpose and objectives of the whole site are defined carefully, the designer can make better judgments. Some guidelines to consider include:

- Design for interaction because Web users expect sites to be interactive and dynamic.
- Guide the user's eye to information on the page that is the most important.
- Keep a flat hierarchy so that the user does not have to drill down too deeply to find detailed information.
- Use the power of hypertext linking to help users move around and through the site.
- Decide how much content per page is enough based on the characteristics of the typical user; don't clutter the pages.
- Design for accessibility for a diverse group of users, including those with disabilities.

Now that dialog design concepts and techniques have been discussed, we can demonstrate the process of designing one specific dialog for Rocky Mountain Outfitters, as well as part of the RMO Web site.

Dialog Design for the RMO Phone-Order Representatives

The Rocky Mountain Outfitters customer support system includes support for the phone-order sales representatives who process orders for customers. The dialog corresponds to the event *Customer places an order* and more specifically to the scenario *Phone-order representative creates new order*. The target environment for this part of the system is the phone-order representative's desktop PC on a Windows platform.

The designer starts by referring to the models produced during analysis: either the data flow diagram fragment and corresponding detailed DFD for the activity or the UML sequence diagram for this scenario, depending on the approach used for analysis and design. The four basic steps followed in these models are as follows:

1. Record customer information.
2. Create a new order.
3. Record transaction details.
4. Produce order confirmation.

With the traditional approach to development, the designer would produce a structure chart, as shown in Chapter 10, to correspond to these steps. With the object-oriented approach, the system sequence diagram for this scenario would be expanded to include forms the user will need for interaction, as shown in Chapter 11. With either approach, the dialog design activity coordinates the user-interface design with the processing design for the activity or use case.

Based on the sequence of processing required, a basic dialog can be written to convey in more detail how the dialog will flow from the user's perspective. The details of the interface objects needed on each form should wait until this dialog is refined. The process of a new customer placing an order with a phone sales representative (the user) might resemble the following dialog:

Computer:	What customer support system option do you want?
User:	I need to create a new order.
Computer:	Okay, is it a new or existing customer?
User:	It's a new customer.
Computer:	Fine, give me the customer's name, address, phone, and so on.
User:	The customer is Ginny Decker, 11980 Visual Blvd. . . .
Computer:	Okay, what is the first item she wants to order?
User:	Boots, Women's, Hiking Supreme Line in the Spring Fling catalog.
Computer:	They come in these sizes and colors, and we have all in stock except size 9.
User:	Okay, one pair of size 8 in tan.
Computer:	That comes to $65.50 plus tax, shipping, and handling. Anything else?
User:	Yes, Raincoat, Woman's, On the Run Line in the Spring Fling catalog.
Computer:	They come in these sizes and colors, and we have all in stock except medium and large blue.
User:	Okay, one coat size small in brown.
Computer:	That comes to $87.95 plus tax, shipping, and handling. Anything else?
User:	That's it.
Computer:	How does the customer want the order shipped?

User:	UPS two-day air.
Computer:	Does the customer want to use a credit card, or is there another payment approach for this order?
User:	Use the Mastercard number xx674-22-xxxx expiring January 05.
Computer:	Okay, the order is recorded. To summarize, for customer Ginny Decker, ship by UPS two-day mail Boots, Women's, Hiking Supreme Line, size 8, tan, at $65.50, and Raincoat, Woman's, On the Run Line, size small, brown, at $87.95. Total cost, $153.45 plus $9.20 tax and $13.40 shipping and handling—$176.05 charged to Mastercard xx674-22-xxxx.
User:	Thanks.

While working on this dialog, the designer can begin to refine the forms that will be required for the user and the computer. A list of required forms might include the following:

- Main menu form
- Customer form
- Item search form
- Product detail form
- Order summary form
- Shipping and payment options form
- Order confirmation form

The designer can use the list of forms to define a design concept for the flow of interaction from form to form. One approach is to show potential forms in sequence, as shown in Figure 14-15. After the main menu form, the customer form appears first; the user fills it in or updates it, and then the item search form displays to let the user search for an item. Product details are shown for the item, and then the item search form is

FIGURE *14-15*

A design concept for the sequential approach to the *Create new order* dialog.

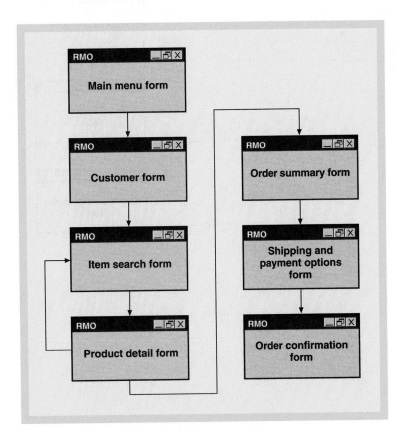

shown again. When all items are selected, the order summary form is shown, and so on. The designer should concentrate on highlighting parts of the dialog that occur through each form rather than worrying about the physical design of each form. After considering what information is needed on each form, the designer can create a more detailed storyboard or implement prototype forms using a tool such as Visual Basic.

This initial design is very sequential but reasonable for a first iteration. Phone sales representatives at RMO evaluated the storyboard and the prototype, and they suggested that the sequence was too rigid; it assumes the dialog always follows the same sequence. However, because phone-order representatives are on the phone with customers, they have to follow the customers' lead. Sometimes customers do not want to give out information about themselves until the order is processed and confirmed, for example. Sometimes customers want to know the totals for the order or to review details about something already included in the order. But the sequential design assumes that the customer information always comes first. Although the sequential approach might work well for mail-order clerks, more flexibility is required for the phone-order representatives.

To address the flexibility and information needs of the users, the project team developed a second design concept. This design concept makes the order form the center of the dialog, with options to switch to other forms at will. After each action, the order form is redisplayed to the phone-order representative with the current order details. The order-centered design concept is shown in Figure 14-16. It allows the same sequence to be followed as the basic dialog, but it also allows flexibility when needed. It also shows the user more information about the order throughout the dialog in case the user needs the information.

FIGURE *14-16*

A design concept for an order-centered approach to the *Create new order* dialog.

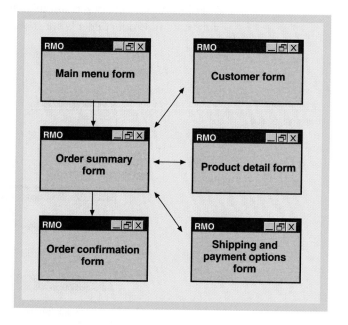

After adopting the order-centered concept, the project team designed the detailed forms. Some of the forms are shown in Figure 14-17. After the user selects *Place an order* from the main menu, the system displays the Order Summary form with a new order number assigned. The user can add the customer information immediately (either by searching for a previous customer based on customer number or name or by adding new customer information). The user can then search for a requested item and look up more details about the item on the Product Detail form. If the customer wants to order the item, the user adds it to the order, and the Order Summary form redisplays. The user can add another item to the order, make changes to the first item ordered, or select shipping and payment options. This flexibility and information display are what the phone-order representatives wanted. It took quite a few iterations and user evaluations to begin to achieve the best design.

FIGURE *14-17*

Prototype forms for an order-centered approach to the dialog.
(a) The Main Menu form for an order-centered approach to the dialog.

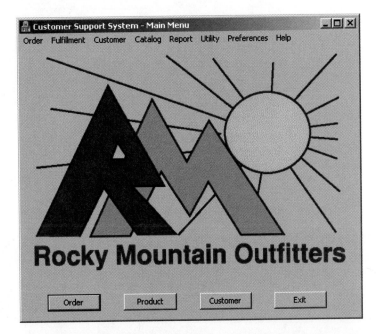

(b) The Order Summary form for an order-centered approach to the dialog beginning a new order.

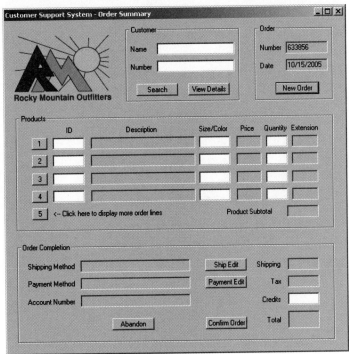

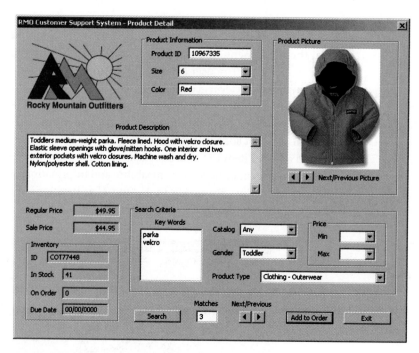

(c) The Product Detail form after the user
has searched for a product.

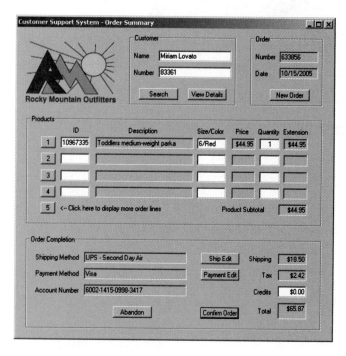

(d) The Order Summary form after the
user adds the product.

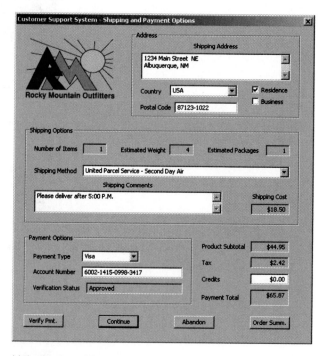

(e) The Shipping and Payment Option
form for the completed order.

Dialog Design for the RMO Web Site

The RMO customer support system event *Customer places order* requires several different dialog designs (one for each scenario), including a scenario for the phone-order representatives just discussed and a scenario for the mail-order clerks. But another key system objective is to allow customers to interact with the system directly to place orders via the World Wide Web. The design of a complete company Web site is beyond the scope of this text, but some general rules and guidelines apply to direct customer orders.

The basic dialog between user and computer will be about the same as for the phone-order representative, but the Web site will have to provide even more information for the user, be even more flexible, and be even easier to use. First, customers may want to browse through all possible information about a product. More pictures will be needed, including pictures showing different colors and patterns of items. Although the phone-order representative needs detailed information on the screen to be able to answer questions, the customer will probably want even more. The information will need to be displayed differently, too. For example, the phone-order clerk is accustomed to a dense display of information and knows where to look for the details, but customers will need organized information to make it easy to locate details the first time they use the system.

The system will need to be very flexible because customers will have different preferences for interacting with the system. As discussed for the phone-order scenario, the sequence should be flexible—some customers will want to browse first and even select items to order before entering any information about themselves. Such options as reviewing past orders and reviewing shipping and payment approaches will also be required. If a customer wants to do something that the system does not allow, the customer will become frustrated. Unlike phone-order and mail-order employees, who will work through their frustration, the customer can simply log off and shop elsewhere. Finally, the system must be so easy to learn and use that the customer does not even have to think about it. The initial dialog options need to be very clear, and once a sequence starts, all options should be self-explanatory. Customers cannot be expected to sit through training just to use a Web site, nor should they have to look up instructions or help (even though both should be available).

As mentioned previously regarding the guidelines for visibility and affordance, it is important that controls used on a Web page be clear in what they do and how they are used. Most users are more concerned about speed than fancy graphics and animations, but novice designers often go overboard on graphics and animation at the expense of speed. Additionally, since the Web site reflects the company image, it is important to involve graphic designers and marketing professionals in the design. A well-thought-out visual theme is important. Focus groups and other feedback techniques should be used in generating the design.

FIGURE *14-18*

Rocky Mountain Outfitters' home page.

The Rocky Mountain Outfitters' home page is shown in Figure 14-18. The main emphasis is direct customer interaction. The user can choose to learn more about RMO, contact RMO with an e-mail message, or request a catalog. There is no main menu option to place an order. Instead, the Web site offers the opportunity to browse through pages of RMO products. The customer can search for products based on keywords or product ID numbers, or the customer can select a category of a product from a list. Weekly specials are also offered. When customers find something they want, they can create an order. They can also change their minds at any time.

The RMO Web site uses the shopping cart analogy for orders. Once customers find a product they want, they select the quantity, size, and other options, and then they add the item to their shopping cart. They can view the shopping cart at any time and then continue browsing and adding items. When they are done, they check out, and the system confirms the order. Figure 14-19 shows a Product Detail page reached after a user navigates through the women's clothing option. Figure 14-20 shows the shopping cart with a summary of an order.

FIGURE 14-19

The Product Detail page from the Rocky Mountain Outfitters' Web site.

FIGURE 14-20

The shopping cart page from Rocky Mountain Outfitters' Web site.

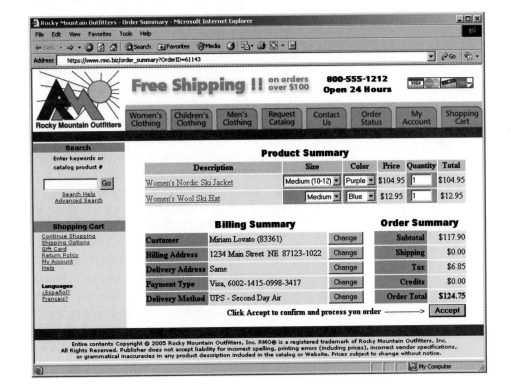

SUMMARY

Inputs and outputs can be classified as system interfaces or user interfaces. This chapter focuses on user interfaces and describes concepts and techniques for designing the interaction between the user and the computer—human-computer interaction (HCI). Chapter 15 describes system interfaces, including system outputs and system controls.

The user interface is everything the user comes into contact with while using the system— physically, perceptually, and conceptually. To the user, the user interface *is* the system. The knowledge the user must have to use the system (the user's model) includes information about objects and functions available in the system—the kind of information that defines the requirements model the analyst works hard to uncover during systems analysis.

User-centered design refers collectively to techniques that focus early on users and their work, evaluate designs to ensure usability, and apply iterative development. Usability refers to the degree to which a system is easy to learn and use. Ensuring usability is a complex task because design choices that promote ease of learning versus those that promote ease of use often conflict. Additionally, there are many different types of users to consider for any system. Human-computer interaction as a field of research grew out of human factors engineering (ergonomics) research that studies human interaction with machines in general.

There are many different ways to describe the user interface, including the desktop metaphor, the document metaphor, and the dialog metaphor. The dialog metaphor emphasizes the interaction between user and computer, and interface design is often called *dialog design* for that reason. Interface design guidelines and interface design standards are available from many sources. Norman's visibility and affordance guidelines state that controls should be visible, provide feedback indicating that they are working, and be obvious in their function. Shneiderman's eight golden rules are to strive for consistency, provide shortcuts, offer informative feedback, design dialogs to yield closure, offer simple error handling, permit easy reversal of actions, support internal locus of control, and reduce short-term memory load.

Dialog design starts with identifying dialogs based on events. Additional dialogs are needed for integrity controls added during design, for user preferences, and for help. Menu hierarchies can be designed for different types of users, and once the dialogs are designed, menu hierarchies can be rearranged easily. Writing a dialog sequence much like a script can help a developer work out the key information that needs to be exchanged during the dialog. A storyboard showing sketches of screens in sequence can be drawn to convey the design for review with users, or prototypes can be created using a tool such as Visual Basic. The object-oriented approach provides UML models that can document dialog designs, including sequence diagrams, collaboration diagrams, and class diagrams.

Each form used in a dialog needs to be designed, and there are guidelines for the layout, selection of input controls, navigation, and help. These guidelines apply to windows forms and to browser forms used in Web-based systems. Designing a dialog for a Web site is similar to creating any other dialog, except users need more information and more flexibility. Additional Web design guidelines apply to designing for the computer medium, designing the whole site, and designing for the user. Additionally, because a Web site reflects the company's image to customers, graphic designers and marketing professionals should be involved.

KEY TERMS

REVIEW QUESTIONS

1. Why is interface design often referred to as *dialog design*?

2. What are the three aspects of the system that make up the user interface for a user?

3. What term is generally used to describe the study of end users and their interaction with computers?

4. What are some examples of physical aspects of the user interface?

5. What are some examples of perceptual aspects of the user interface?

6. What are some examples of conceptual aspects of the user interface?

7. What collection of techniques places the user interface at the center of the development process?

8. What are the three important principles emphasized by user-centered design?

9. What term refers to the degree to which a system is easy to learn and use?

10. What is it about the "human factor" that engineers find difficult? What is the solution to human factor problems?

11. What are some of the fields that contribute to the field of human-computer interaction?

12. What research center significantly influenced the nature of the computers we use today?

13. What are the three metaphors used to describe human-computer interaction?

14. A desktop on the screen is an example of which of the three metaphors used to describe human-computer interaction?

15. What type of document allows the user to click on a link and jump to another part of the document?

16. What type of document allows the user to click on links to text, graphics, video, and audio in a document?

17. What is the name for general principles and specific rules that developers must always follow when designing the interface of a system?

18. What two key principles are proposed by Norman to ensure good interaction between a person and a computer?

19. List the eight golden rules proposed by Shneiderman.

20. What is the technique that shows a sequence of sketches of the display screen during a dialog?

21. What UML diagram can be used to show how the interface objects are plugged in between the actor and the problem domain classes during a dialog?

22. What UML diagram can be used to show the interface objects that are contained on a window or form?

23. What are the three basic types of windows and browser forms used in business systems?

24. What is the input control (interface object) used for typing in text?

25. What are some of the input controls that can be used to select an item from a list?

26. What two types of input controls are included in groups?

27. What are three requirements for usability of a direct customer access Web site beyond those of a windows interface used by employees?

28. What popular analogy is used for direct customer access with a Web site when customers shop on-line?

29. What are three principles of Web design that guide designers?

30. What are four of the 10 good deeds of Web design?

THINKING CRITICALLY

1. Think of all of the software you have used. What are some examples of ease of learning conflicting with ease of use?

2. Visit some Web sites and identify all of the controls used for navigation and input. Are they all obvious? Discuss some differences in visibility and affordance of the controls.

3. Consider the human factor solution that states it is better to change the machine than to try to change the human to accommodate the machine. Are there machines (or systems) that you use in your daily life that still have room for improvement? Are the current generations of Windows PC and Apple Mac as usable as they might be? If not, what improvements can you suggest? Is the World Wide Web as usable as it might be? If not, what improvements would help? Are we just beginning to see some breakthroughs in usability, or have most of the big improvements already been made?

4. Review the dialog between user and computer shown for DownTown Videos. Create a table like Figure 14-6 that shows how the dialog can be converted to the user's language and the computer's language. Discuss how moving from the dialog to the table starts with a logical model and then creates a physical model.

5. Refer again to the table shown in Figure 14-6. Create a storyboard for the e-mail system based on the information in the table.

6. Read through the following dialog, which shows a user trying to place an order with the system. Critique the dialog in terms of ease of learning and ease of use.

User:	I want to order a product.
System:	Okay. Enter your name and address.
User:	My name is Timothy Mudd, 5139 North Center Street, Los Angeles, CA 98210.
System:	Oh, we have all of that information on file, but thank you for entering it again.
User:	I want to order the Acme Drill Press with adjustable belt drive.
System:	Sure, continue with your request.
User:	I want the blue color and rubber feet but not the foot brake.
System:	Sure, anything else?
User:	I want it shipped priority with the special discount you offer.
System:	We hear you.
User:	Okay, that's all.
System:	We are sorry, but your transaction violated transaction code 312 and we must terminate the session.

7. Refer to the form shown in Figure 14-14, showing data-entry controls. Draw a UML class diagram that represents the objects that make up the form.

8. Review all of the controls that come with Visual Basic that are visible when added to a form. Discuss how well each satisfies the visibility and affordance requirements.

EXPERIENTIAL EXERCISES

1. Review the sequential design concept and the order-centered design concept for the Rocky Mountain Outfitters *Create new order* dialog. Consider what information and navigation options are not needed by the mail-order clerks. Then design the forms that would be appropriate for clerks to use when entering mail orders for RMO. Either sketch the forms or implement the forms using a tool such as Visual Basic. Use input controls that minimize data entry.

2. Evaluate the course registration system at your university. List the basic steps followed through a dialog with the system. What are some of the problems with the system from the perspective of ease of learning and ease of use? In what ways is the system inflexible? In what ways is needed information not available? Is too much information provided that distracts from the task at hand?

3. Evaluate the on-line catalog system at your university library (see some of the questions and issues in exercise 2). Write a dialog that shows the interaction between the user and the system. Rewrite the dialog to improve it. Create a storyboard to show how your design would look and feel.

4. Locate a system that users work with all day. Ask some of the users of the system what they like and dislike about it. Ask how they learned the system. Ask them what typical errors they make when using the system and how these errors are prevented or corrected. Observe the users working with the system. Does it appear that any of the interaction is "automatic"?

5. Find a Web site with direct customer ordering. Browse through some product descriptions and note the design of the dialog and the Web pages. What do you like and dislike about the design? Evaluate the Web site based on visibility and affordance and on the eight golden rules. Evaluate the design based on the 10 good deeds and three principles of Web design.

The 21st Century University System

Consider the following dialog for a user interacting with the course registration system at 21st Century University. The user is a student registering for classes at the university.

System: What course registration function do you want to perform?
User: *Preregistration.*
System: Okay, what is your account ID and password?
User: *My account ID is SATZ791, and my password is xxxxxxxx.*
System: Okay, welcome back, Kevin. You last adjusted your schedule on 7/7.
What preregistration function do you want to perform?
User: *To add courses.*
System: These are the courses you are preregistered for now:
(The system displays a list of course information for three courses.)

Since you are a CIS major, these are all of the CIS courses you might add that have no time conflicts with your present schedule (note, though, that prerequisites have not been checked for these courses).

(The system displays a list of CIS courses available.)
Your options are:
- Switch to a list of courses in another department
- List these courses again but ignore time conflicts
- List these courses again after checking prerequisites
- Provide details of the indicated course
- Add the indicated course to your schedule
- Cancel

User: *I want to switch to courses in another department.*
System: Fine. To what department do you want to switch?
User: *Art.*
System: These are the courses for which you are registered now:
(The system displays a list of course information for three courses.)

As requested, these are the art courses you might add that have no time conflicts with your present schedule (note, though, that prerequisites have not been checked for these courses).

(The system displays a list of art courses available.)
Your options are:
- Switch to a list of courses in another department
- List these courses again, but ignore time conflicts
- List these courses again after checking prerequisites
- Provide details of the indicated course
- Add the indicated course to your schedule
- Cancel

User: *Add the second course in the list.*
System: (The system continues the dialog.)

1. Is this dialog a logical or a physical model? Could it be a phone registration system? Could a clerk at a desk handle the dialog manually? Discuss.
2. Discuss how well this dialog follows the eight golden rules.

3. Complete the dialog by adding likely exchanges between the student and the system as the student adds one more course. The student should request more details about a course, and then decide to add the course. Include a request from the student to see final schedule details.

4. Design and implement either a storyboard or prototype for this dialog, using a tool such as Visual Basic and being as faithful to the dialog as possible. Make up some sample data to show in your design as needed.

The DownTown Videos Rental System

This chapter includes an example of a storyboard for DownTown Videos, a case study first introduced in Chapter 7. The storyboard showed the *Rent out videos* dialog. Revisit the DownTown Videos case and complete the following:

1. Implement the storyboard in this chapter as a prototype using a tool such as Visual Basic.
2. Write a dialog, and then create a storyboard for the event *Customer returns videos*. Consider that one or more videos might be returned and that one or more of them might be late, requiring a late charge.
3. Using a tool such as Visual Basic, implement the storyboard as a prototype, and then ask several people to evaluate it. Discuss the suggestions made.

The Waiters on Call System

Review the Chapter 5 opening case study that describes Waiters on Call, the restaurant meal-delivery service. The analyst found at least 14 events for the system, which are listed in the case.

1. Create a set of menu hierarchies for the system based on the events listed. Then add more menu hierarchies for system utilities based on controls, user preferences, and help.
2. The most important event listed in the case is *Customer calls in to place an order.* Write out a dialog between the user and the computer with a natural sequence and appropriate exchange of information.
3. Sketch a storyboard of the forms needed to implement the dialog.
4. Ask several people to evaluate the design and discuss any suggested changes.
5. Implement a prototype of the final dialog design using a tool such as Visual Basic.

The State Patrol Ticket Processing System

Review the State Patrol ticket processing system introduced as a case study at the end of Chapter 5.

1. Create a set of menu hierarchies for the system based on the events in the case. Then add more menu hierarchies for system utilities based on controls, user preferences, and help.
2. For the event *Officer sends in new ticket*, write out a dialog between the user and the computer for the dialog *Record new ticket* with a natural sequence and appropriate exchange of information.
3. Sketch a storyboard of the forms needed to implement the dialog. Be sure to use input controls that minimize data entry, such as list boxes, radio buttons, and check boxes.
4. Ask several people to evaluate the design and discuss any suggested changes.
5. Implement a prototype of the final dialog design using a tool such as Visual Basic.

Rethinking Rocky Mountain Outfitters

A few of the phone-order representatives at Rocky Mountain Outfitters were not completely satisfied with the order-centered dialog design presented in this chapter. They suggested that the design could be streamlined if only one form were displayed throughout the dialog—the order summary form. They thought the order summary form could expand to show additional information instead of switching to separate forms for customer information, product information, or shipping information. When the additional information is not needed, the form could contract. They thought this approach would be easier on the eyes, requiring less effort to focus and refocus on multiple forms that pop up. They requested that a prototype of the one-form design concept be created for their review. The interface might include both options, allowing users to choose the one they prefer.

1. Draw a storyboard to show how the one-form design concept might look as customer information, product information, and shipping information are added to the expanded form.
2. Implement a prototype of the storyboard using a tool such as Visual Basic.
3. Ask several people to evaluate the design, specifically comparing it with the order-centered design in the text, and discuss the results.
4. Can you describe or implement yet another alternative?

Focusing on Reliable Pharmaceutical Service

The Reliable Pharmaceutical Service system has users who process orders in the Reliable offices and users who place orders and monitor order information in the nursing homes. Consider the events and use cases that apply to Reliable employees versus nursing home employees.

1. Design two separate menu hierarchies, one for Reliable employees and one for nursing home employees.
2. Write out the steps of the dialog between the user and the system for the use case *Place new order* for nursing home employees.
3. Create a storyboard for the dialog for *Place new order* by making a sketch of the sequence of interaction with Web pages. Implement a prototype for the Web pages using a Web development tool such as Dreamweaver or Frontpage.

FURTHER RESOURCES

Merlyn Holmes, *Web Usability and Navigation*, McGraw-Hill Osborn, 2002.

Patrick J. Lynch and Sarah Horton, *Web Style Guide: Basic Design Principles for Creating Web Sites*. Yale University Press, 1999.

Deborah J. Mayhew, *Principles and Guidelines in Software User Interface Design*. Prentice Hall, 1992.

Jakob Nielsen, *Designing Web Usability: The Practice of Simplicity*. New Riders Publishing, 2000.

Donald Norman, *The Design of Everyday Things*. Doubleday, 1990.

Jenny Preece, Yvonne Rogers, David Benyon, Simon Holland, and Tom Carey, *Human Computer Interaction*. Addison Wesley, 1994.

Ben Shneiderman, *Designing the User Interface*, 3rd ed. Addison Wesley, 1998.

Joel Sklar, *Principles of Web Design*, 2nd ed. Course Technology, 2003.

CHAPTER 15

Designing System Interfaces, Controls, and Security

LEARNING OBJECTIVES

After reading this chapter, you should be able to:

- Discuss examples of system interfaces found in information systems

- Define system inputs and outputs based on the requirements of the application program

- Design printed and on-screen reports appropriate for recipients

- Explain the importance of integrity controls

- Identify required integrity controls for inputs, outputs, data, and processing

- Discuss issues related to security that affect the design and operation of information systems

CHAPTER OUTLINE

Identifying System Interfaces

Designing System Inputs

Designing System Outputs

Designing Integrity Controls

Designing Security Controls

Downslope Ski Company is a medium-sized manufacturer of skis and snowboards. The company started with the manufacture of downhill snow skis, hence its name. But a few years ago, it expanded into the production of snowboards and then into water skis. In the company's early years, ski manufacturing was fairly straightforward. However, with the introduction of advanced materials such as carbon-laced resins and other sophisticated compounds, manufacturing has become quite complicated, requiring exacting controls for the precise mixture of ingredients and temperature tolerances in baking furnaces. To maintain consistency in snowboard production, Downslope has been very demanding in the quality of the raw materials that it buys. Several times over the last few years, it has had to change suppliers to ensure an adequate supply of high-quality raw materials.

In addition to controlling the quality of materials for manufacturing, Downslope instituted a modified just-in-time (JIT) manufacturing process, which means that it does not stockpile a large quantity of raw materials. Generally, it keeps about a five-day supply on hand and depends on its suppliers to restock materials at least weekly. To facilitate quick ordering and delivery of its many raw materials, senior management at Downslope decided to permit its suppliers to access its inventory database. Depending on which types of boards were being produced, the various raw materials would be depleted at different rates. So, Downslope began developing and installing a complex inventory management system that was integral to the manufacturing process.

Nathan Lopez, Downslope's system development project manager, was fast recognizing that providing a system interface for suppliers to access the database was more complex than he originally thought. He reported to Downslope management the results of a two-month study to determine the feasibility of various alternatives for this electronic approach to supply chain management.

"I have met with each of our suppliers and determined what information they need and the formats they would like it in. As expected, there was little consistency in the desired formats. I have been able to consolidate some of their needs to narrow them down to three basic formats. Originally, we thought we could convince our suppliers to accept our output design—in other words, to make them conform to our output. However, that does not seem to be such a good idea anymore. If we do not build flexibility into our system, it will be more difficult for us to add or change suppliers. Instead, if we build the interface correctly, with several versions, it should be much easier for suppliers to gain access to the data that we allow them to see.

"Another critical issue that has surfaced is the integrity and security of our data and our systems. For example, our production process is unique and one of our competitive advantages. If the wrong company got access to our data, it could potentially analyze our usage patterns and not only discover what materials we used but perhaps even reverse-engineer our processes. To protect our data, we need to ensure that our computers allow only secure access. We also need to ensure that data are secure while being transmitted to our suppliers and are protected after arrival there. These security issues relate to both our systems and our suppliers."

The meeting lasted a long time, with considerable discussion about the opportunities and dangers of opening company systems to an outside system interface. The oversight committee finally decided that Nathan should study the situation for a couple more weeks and develop a list of every possible breach of security with potential solutions for each one. Only after these issues were addressed would the project move forward. Nathan knew that the development of this new type of system interface would be a very sticky problem. He hoped he would be able to find solutions for all of the issues that had been raised.

OVERVIEW

Most modern information systems involve extensive input and output (I/O), and many people and organizations require access to the data stored by a system. In Chapter 14, you read about human-computer interaction (HCI) and user interfaces, where I/O is the result of user interaction with the computer. But many system inputs and outputs do not involve much human interaction. Many of these system interfaces are not as obvious to end users. But systems analysts need a deep understanding of existing systems, databases, and network technologies where I/O occurs to design an information system that incorporates all I/O needs. We therefore discuss system interfaces separately from user interfaces in this chapter.

Many system interfaces are electronic transmissions or paper outputs to external agents, including reports, statements, and bills. They need to be identified and designed to suit their intended purpose. Frequently, the quality of the system outputs is a mark of the quality of the entire system and of the company that uses it. This chapter discusses the design of these outputs.

Because of the many and varied inputs and outputs, system developers need to design and implement integrity controls and security controls to protect the system and its data. Today more than ever, designing system controls is crucial because computer systems increasingly exist in an open environment. They are part of networks that provide broad access to many different people both within and outside an organization. So, one of the major considerations in systems design is how to ensure that errors or fraudulent transactions are not entered into a system. Integrity controls validate data when they are input and processed. Internal checks and cross-checks help ensure that data integrity is maintained. This chapter discusses techniques to provide the integrity controls to reduce errors, fraud, and misuse of system components.

Finally, outside threats to systems, business organizations, and individuals continue to be a major concern of companies that are connected to the Internet. Many companies use the Internet as a marketing and sales channel, so they need to let customers and prospective customers into their systems, yet keep intruders and malicious hackers out. In addition, e-commerce also entails transmission of private information, such as credit-card numbers and financial transactions. The last section of the chapter discusses security controls and explains the basic concepts of data protection, digital certificates, and secure transactions.

IDENTIFYING SYSTEM INTERFACES

The user interface, as described in the previous chapter, includes inputs and outputs that directly involve system users. But there are many other system interfaces. We define *system interfaces* broadly as any inputs or outputs with minimal or no human intervention. Included in this definition are standard outputs such as billing notices, reports, printed forms, and electronic outputs to other automated systems. Inputs that are automated or come from nonuser interface devices are also included. For example, inputs from automated scanners, bar-code readers, optical character recognition devices, and other computer systems are included as part of a system interface.

It often seems that user interfaces are the most common—and thus most important— interfaces to consider when analyzing and designing an information system. In fact, considerable progress has been made in understanding human-computer interaction and applying user-centered design principles to user-interface design. However, today's highly integrated and interconnected information systems increasingly go beyond user needs, requiring system interfaces to handle inputs and outputs faster, more efficiently, more accurately, and at any hour of the day or night.

System interfaces can process inputs, interact with other systems in real time, and distribute outputs with minimal human intervention. When researching and modeling the requirements for a system, analysts must look carefully for system interfaces that might not appear obvious at first. When designing the system, they should consider alternatives to HCI to automate the capture of inputs and the distribution of outputs. The full range of inputs and outputs in an information system is shown in Figure 15-1.

The following list provides some categories of system interfaces to aid in identifying I/O requirements and design possibilities:

- Inputs from other systems
- Highly automated inputs
- Inputs that are from data in external databases
- Outputs that are to external databases

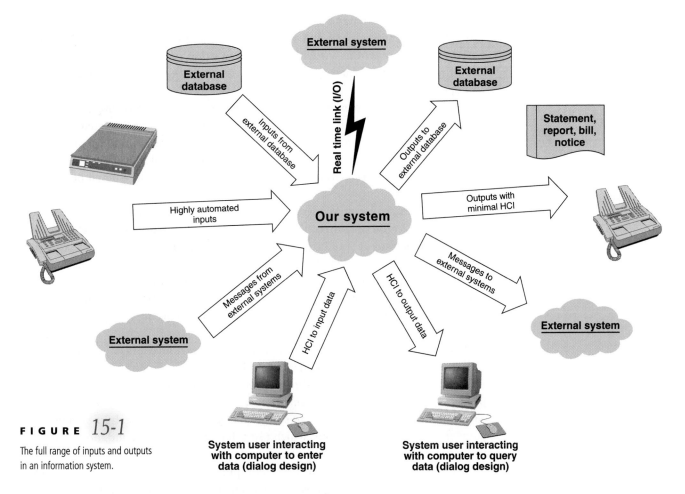

FIGURE *15-1*

The full range of inputs and outputs in an information system.

Inside the figure:

External system

External database

External database

External database

Statement, report, bill, notice

Inputs from external database

Real time link (I/O)

Outputs to external database

Highly automated inputs

Our system

Outputs with minimal HCI

Messages from external systems

HCI to input data

HCI to output data

Messages to external systems

External system

External system

System user interacting with computer to enter data (dialog design)

System user interacting with computer to query data (dialog design)

- Outputs with minimal HCI
- Outputs to other systems
- Real-time connections (both input and output)

Inputs can arrive directly from other information systems as network messages. Electronic data interchange (EDI) and many Web-based systems are integrated with other systems through direct messaging. The message received triggers system processing much the way that user interaction does. For example, in RMO's integrated supply chain management and customer support systems, the arrival of inventory items from a supplier might trigger the shipment of a back-ordered item to a customer. No human intervention is required, and as a result, the transaction can be processed immediately and with little chance of error. In Web-based systems, a separate shopping cart order-entry application might send a message to the order fulfillment system to process a new order. Analysts decide what is an input and what is an internal message by determining the scope of the system.

Highly automated input devices such as scanners can capture many system inputs. An item in a warehouse that is picked for shipment, for example, might have a machine-readable label that an attendant can scan. This highly automated process represents an input to a system. In some cases, a scanner might record the input as an item moves by on a conveyor belt—with no human interaction at all. We discuss specialized input devices later.

Many inputs may come from external databases. For example, one system might record transactions in a database, perhaps as a batch of transactions. Another system might periodically search those transactions and process one or more of them. For example, consider a charge-account system in which charges are stored in a database throughout the billing period. A separate billing and collection system might later process those transactions. As with inputs from other systems, whether these inputs come from an external database or are processed within one billing system is a question of scope.

Some inputs from an external database might occur during processing of another input, such as verifying credit history or verifying employment prior to extending credit. When a credit application is received, the user processing the application may have to rely on information from an external database before the application can be approved. Within the activity *Process credit application* is a requirement for a system input from an external database.

The output side of system interfaces mirrors the input side. Outputs to external databases might be required when the system produces large amounts of detailed data. Many system outputs are produced with minimal human intervention. Reports are produced and e-mailed to recipients or printed and distributed, but the user is not interacting with the system directly to obtain the output. Bills, notices, statements, form letters, and so on are similarly produced with minimal human intervention. They might be sent electronically or be printed. Messages sent to external systems, triggering processing, are also system outputs.

Sometimes system inputs and outputs must be real-time connections. RMO's real-time credit-card authorization is an example. Rather than accessing an external database, RMO's system establishes a real-time connection with another system that accepts inputs and provides outputs. A real-time connection is therefore both a system output and a system input much like a system-to-system dialog. In this aspect, real-time connections parallel user-interface functions, which use a dialog to enter data and look up data in the system (as shown in Figure 15-1).

Another mechanism for correct system input is to have a direct interface with another system. Electronic data interchange (EDI) reduces the need for user input. Purchase orders, invoices, inventory updates, and payment all are generated by one system sending transactions to another. With EDI, these transactions normally occur between systems in separate organizations. However, the same principle can be applied to systems within an organization.

One of the main challenges of EDI is in defining the format of the transactions. It is easy enough to design the format for a single type of transaction, but it becomes more complex when many different types of transactions are going to many different systems. The complexity and difficulty multiplies when many different companies are trying to work together. For example, General Motors, which was one of the early users of EDI, has literally thousands of suppliers with thousands of different transactions, each in a different format. To complicate the situation further, each of these suppliers may also be linked via EDI with tens or hundreds of customers, many of whom may also use EDI. So, a single type of transaction may have a dozen or more defined formats. It is easy to see why it is so costly to set up and maintain EDI systems. Even so, EDI is much more efficient and effective than paper transactions, which must be printed and reentered.

In recent years there has been a move to develop a standard communication method between systems based on Hypertext Markup Language (HTML) concepts. As you know, HTML embeds beginning and ending markup codes within a text-based document to define the characteristics, such as formatting, of text or of a figure. HTML embeds formatting information within the document itself. Thus, a program that can read HTML reads a text document and then, using the embedded markup codes, can format the document correctly. Although this system is not extremely efficient from a computational point of view, it has many advantages due to its simplicity and readability by human beings.

This new system-to-system interface that is gaining popularity is called *eXtensible Markup Language (XML)*. XML is an extension of HTML that embeds self-defining data structures within textual messages. So, a transaction that contains data fields can be sent with XML codes to define the meaning of the data fields. Many newer systems are using XML to provide a common system-to-system interface. Figure 15-2 illustrates a simple XML transaction that can be used to transfer customer information between systems.

```
<customer record>
        <accountNumber>RMO10989</accountNumber>
        <name>William Jones</name>
        <billingAddress>
                <street>120 Roundabout Road</street>
                <city>Los Angeles</city>
                <state>CA</state>
                <zip>98115</zip></billingAddress>
        <shippingAddress>
                <street>120 Roundabout Road</street>
                <city>Los Angeles</city>
                <state>CA</state>
                <zip>98115</zip></shippingAddress>
        <dayPhone>215.767.2334</dayPhone>
        <nightPhone>215.899.8763</nightPhone>
</customer record>
```

Just like HTML, XML is simple and readable by human beings. For it to work, both systems must recognize the markup codes, but once a complete set of codes is established, transactions can include many different formats and still be recognized and processed. The receiving system merely has to parse the incoming data stream and extract the values from the markup codes. XML is extremely scalable—it does not matter how many different companies or different transaction types there are. Every transaction can have its own format as long as it uses the standard markup codes.

Markup codes for XML are defined in a separate file called a *document type definition (DTD)* file or *XML schema*. Many industries and specialty groups have now formed standards committees who are defining sets of markup codes. Standard codes already exist for general business, retailing, railroad, news media, and medical transactions. The list of groups forming and defining standard codes is extensive.

Chapter 12 introduced the idea of Web services. Web services are based on XML, so these business transactions can be sent over the Internet. In fact, XML was designed to take advantage of the Internet. More details about how Web services work are provided in Chapter 17.

DESIGNING SYSTEM INPUTS

When designing inputs for a system, the system developer must focus on three areas:

- Identifying the devices and mechanisms that will be used to enter input
- Identifying all system inputs and developing a list with the data content of each
- Determining what kinds of controls are necessary for each system input

The first task, identifying devices and mechanisms, is a high-level review of the most up-to-date method of entering information into a system. In nearly all business systems, end users perform some input through electronic forms. However, in today's high-technology world, there are numerous ways to enter information into a system—among them are scanning, reading, and transmitting devices that are faster, more efficient, and less error-prone than user input.

The second task, developing the list of required inputs, provides the link between the design of the application software and the design of the user and system interfaces. As described earlier, the design of the system inputs and interfaces must be integrated with the design of the application. This activity accomplishes that purpose.

The third task is to identify the control points and the level of security required for the system being developed. The project team should develop a statement of policy and control requirements before beginning the detailed design of the electronic forms making up the system interface. These concepts are discussed in the last two sections of the chapter.

Input Devices and Mechanisms

Often when analysts begin developing a system, they assume that all input will be entered via electronic, graphical forms because they are now so common on personal computers and workstations. However, as the design of the user inputs commences, one of the first tasks is to evaluate and assess the various alternatives for entering information into the system.

The primary objective of any form of data input is to enter or update error-free data into the system. The key word here is *error-free*. Several good practices can help reduce input errors:

- Capture the data as close to the originating source as possible.
- Use electronic devices and automatic entry whenever possible.
- Avoid human involvement as much as possible.
- If the information is available in electronic form anywhere, use it instead of reentering the information.
- Validate and correct information at the time and location it is entered.

Today, many systems enable the data to be captured electronically at the point that they are generated. For example, the old way of selling a life insurance policy is to have the applicant or the insurance agent fill out a paper policy application. Then the agent sends the application to a central office to be entered into the system. With this method, numerous errors can occur from indecipherable writing, key-entry errors, missing fields, and so forth. Currently, agents often carry laptop computers with easy-to-use electronic forms, so applicants can fill in the data themselves. Or the agent can enter the data while the applicant looks over the agent's shoulder and verifies that the information is accurate and complete. A portable printer can even be attached to the laptop to print out the completed form for the applicant to review immediately. This new approach dramatically reduces the error rate and speeds the business process of new policy data entry.

The second and third practices, automating data entry and avoiding human involvement, are very closely related and often are essentially different sides of the same coin, although using electronic devices does not automatically avoid human involvement. When system developers think carefully about minimizing human input and using electronic input media, they can design a system with fewer electronic input forms and avoid some common data-entry problems. One of the most pervasive sources of erroneous data is users' typing mistakes in fields and numbers. A few of the more prevalent devices used to avoid human keystroking are:

- Magnetic card strip readers
- Bar-code readers

- Optical character recognition readers and scanners
- Touch screens and devices
- Electronic pens and writing surfaces
- Digitizers, such as digital cameras and digital audio devices

We have all seen new electronic input devices. At the grocery store, electronic scanners identify and price each item from the printed UPC codes. Machines weigh and price the produce automatically at the checkout. Cash registers now read your check, including the amount and customer and bank information. New self-service checkout stations depend almost entirely on automated data-entry devices.

Historically, paper contracts and ink signatures were necessary for legally binding contracts. Today, new laws and regulations allow paper documents and signatures to be digitized. Credit-card purchases now also record digitized signatures to eliminate the need for paper vouchers. This technique conforms to the good practices stated previously—the information is captured at its source in electronic form, which eliminates many of the sources of errors.

The next principle of error reduction is to reuse the information already in the computer whenever possible. Some outdated systems require reentry of the same information multiple times. This practice not only generates errors but also creates multiple copies of the same information, which require more checks and balances—as well as computer resources—to synchronize the various copies. And when an error is discovered, it is very difficult to know which copy is correct. Also, when a change is required, it must be made to all copies of the data. One current high-tech example of using existing information is found in car rental systems. When a customer rents a car, the rental agency's system captures the customer's name and credit-card, car mileage, and fuel information. Then when the customer returns the car, an agent in the parking lot simply scans the contract ID and enters the return mileage and fuel-gauge level. The system calculates the charges and prints a credit-card receipt for the customer right in the parking lot. This solution was designed primarily to provide a higher level of customer service, but it also eliminates many problems and errors with data entry.

Eliminating input errors through various techniques, one of which is using electronic devices for data input, is critical, but another potential source of problems is the input of fraudulent information. Two problems need to be addressed with fraudulent data: access control and input control. First, access to the system must be controlled so that only authorized persons or systems can gain access. Today, devices such as fingerprint readers, body temperature sensors, and iris scanners are used more and more often to provide additional security to standard password access. Second, input controls must be built into the system so that fraudulent data cannot easily be entered. Although it is impossible to completely eliminate the potential for fraud, the careful design of input controls will help minimize the risk. More discussion on security devices and input controls is provided later in the chapter.

Defining the Details of System Inputs

The objective of this task is to ensure that the designer has identified all of the required inputs to the system and specified them correctly. In the previous chapter, you learned various methods of defining user inputs based on analysis of user workflows. Those techniques defined the user interface through user-centered design. In this chapter, we focus on defining the system inputs, including user inputs, by analyzing models that were built during the analysis phase. As with other aspects of analysis and design, this task also provides a mechanism to cross-check the quality of both the user-centered design and the detailed information developed in the analysis models.

The fundamental approach that analysts use to identify user and system inputs is to identify all information flows that cross the system boundary. The idea is the same for both traditional structured models and object-oriented models; however, the detailed techniques are unique to each model. We begin with the structured approach.

Using Traditional Structured Models

During systems design with structured techniques, one of the first tasks is to define the automation boundary. Figure 15-3, duplicated from Figure 10-2, is an example of an automation boundary on a data flow diagram. Several of the inputs to the system based on this data flow diagram are:

- Time card information
- Updates to tax rate tables
- Updates to employee files

The updates to employee files are probably done with a graphical user interface screen. However, both the time card information and the updates to the tax tables are more likely done through a system interface that is not graphical. Time card information frequently comes through an automated employee check-in system. Tax updates may come directly from a government service bureau and may be provided electronically.

FIGURE *15-3*

An automation boundary on a system-level DFD.

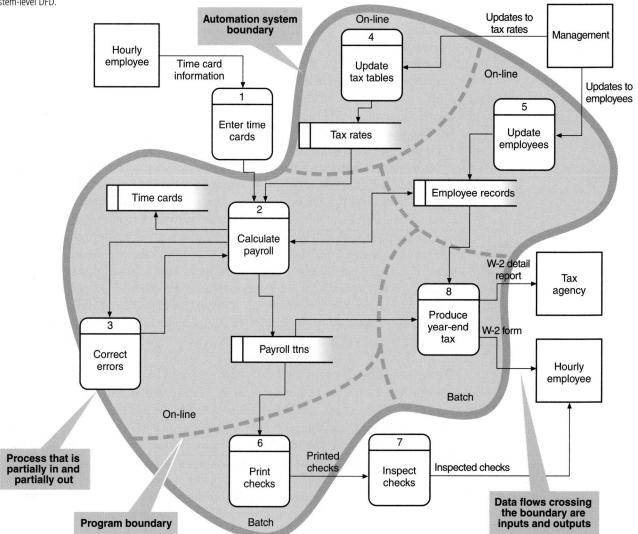

Even though it is possible to build an automation boundary on a high-level DFD and identify the inputs on this diagram, it usually is better to work from the DFD fragments or even more detailed DFDs. The high-level diagram frequently does not provide enough detail to discern many data flows and, hence, inputs to the system. For example, one of the processes on the diagram, *Correct errors*, is intersected by the system boundary. Thus, a view from a lower-level DFD, which provides more detailed process bubbles for the *Correct errors* process, would be necessary to discern what processes to include within the automated system and what data flows cross the boundary.

Chapter 6 explained how to build DFD fragments based on the events in the event table. For more complex models, you can define system inputs by looking at each DFD fragment and creating the system boundary on each fragment. The high-level DFD with an automation boundary gives a good overview, but the DFD fragments, or even the detailed DFD for each fragment, are easier to work with.

Figure 15-4 shows the *Create new order* detailed data flow diagram from RMO with the automation boundary superimposed. The input data flows crossing the boundary are clearly defined, so required inputs will be the new order information data flow and the real-time link from the credit bureau. The input for the user interface will be the new order information, and the real-time link to the credit bureau will be an electronic system interface.

FIGURE *15-4*

The *Create new order* DFD with an automation boundary.

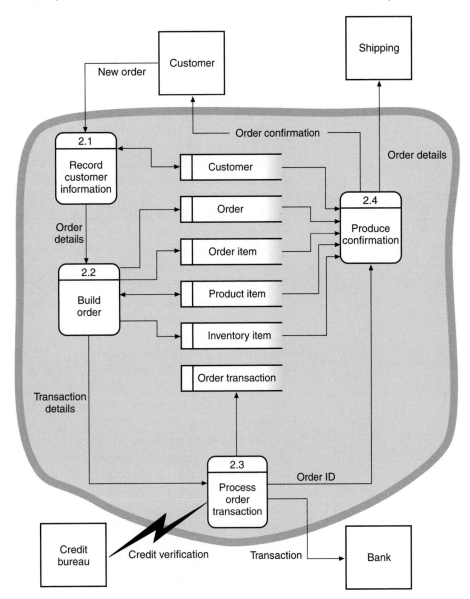

The designer analyzes each DFD fragment to determine the required inputs. The data flows that cross the boundary on the DFDs as inputs correspond to triggers for external events in the event table. The result of this task is a list of high-level inputs for the new system. Figure 15-5 is a list of inputs for RMO's customer support system as developed from the DFD fragments. To develop this list, the designer analyzes every DFD fragment in Figure 6-12, as well as all other DFD fragments for RMO. The purpose of this preliminary list is to provide a master control list of all system inputs and user inputs that need to be designed. It does not, however, provide quite enough detailed information to design the inputs themselves. The additional information that is needed is obtained from the data flow definitions, structure charts, and the user-centered design activities (for the user interface) as explained in Chapter 14.

FIGURE *15-5*

List of inputs (system and user) for the customer support system.

Item inquiry
New order information
Change order information
Order status inquiry
Order fulifillment notice
Back-order notice
Order return notice
Catalog request
Customer account update
Promotion package information
Customer change adjustment
Catalog update information
Special promotion information
New catalog information
Credit card authorization

While developing the structure charts, the designer defines individual program modules and their associated data couples. Chapter 10 discussed the process of defining the detailed data content of each data couple. Each input data flow on a data flow diagram may translate into one or more physical inputs on the structure chart. In Figure 15-6, which derives from Figure 10-15, input modules have been defined for getting customer information and for getting order information, including the details for several order line items. In this figure, the New order data flow on the DFD is expanded into four separate data couples on the structure chart. The structure chart identifies three modules that get data from outside the system. These three modules and their associated data couples are named *Get customer information, Get order information,* and *Get credit card information.* In other words, it requires three modules to provide all of the information from outside the system on the Customer information and New order information data flows.

The next step is to analyze each module and data couple and list the individual data fields for each data couple. This analysis consists of reviewing the elements in the data stores to ensure that all elements on the data stores can be built based on the input data couples. Figure 15-7 expands Figure 15-5 to include the data couples associated with each data flow as well as the data fields to be associated with each data couple. Since the identification of the detailed data elements is based on the analysis models, including the entity-relationship data model, it provides a cross-check for the approach explained in Chapter 14. Ensuring that the two approaches yield the same data elements is a powerful technique to verify the quality of the design.

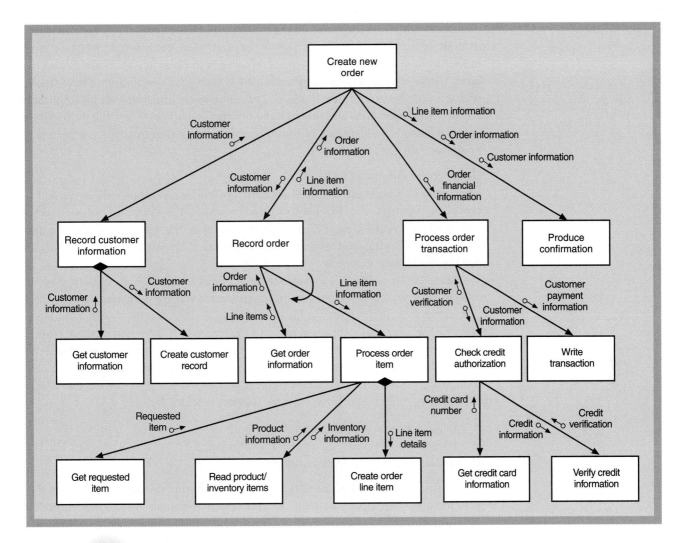

FIGURE *15-6*

Structure chart for *Create new order*.

Data Flow	Data couples	Data elements
New order	Customer information	Account number, Name, Billing address, Shipping address, Day phone, Night phone
	Order information	Order date, Priority code
	Line item information	Product ID, Color, Size, Quantity
	Credit card number	Credit card name, Credit card number, Expiration date
Credit authorization (real time)	Credit information (output)	Credit card number, Expiration date, Customer name, Amount
	Credit verification (input)	Accept/Reject code, Date/time, Authorization number, Amount

FIGURE *15-7*

Data flows, the data couples, and the
data elements making up inputs.

Each of the items identified as inputs made by end users listed in the data couple column of Figure 15-7 becomes part of an electronic input data form or an input/output form. An input/output form permits users to enter key blank fields and then query the database to display information such as product descriptions. These tasks associated with design of input forms are best done in conjunction with the application architecture design tasks and then refined through careful consideration of interaction between humans and computers. Let's now shift to the object-oriented approach to see how to identify the inputs.

Using Object-Oriented Models

Identifying user and system inputs with the object-oriented approach consists of the same tasks as with the traditional structured approach. The difference is that system sequence diagrams and design class diagrams are used. The system sequence diagrams identify the incoming messages, and the design class diagrams are used to identify and describe the input parameters.

Figure 15-8 is a simplified system sequence diagram for an object-oriented version of a payroll system (such as the traditional structured version shown in Figure 15-3). In this system sequence diagram, snippets from various use cases have been combined to illustrate the same major inputs as indicated in Figure 15-3. The messages that cross the system boundary identify inputs, both system inputs and graphical user interface (GUI) inputs. The identification of these inputs on the sequence diagram provides a cross-check with the GUI forms defined with the user-centered design as described in Chapter 14. The three system inputs that cross the system boundary are:

- updateEmployee (empID, empInformation)
- updateTaxRate (taxTableID, rateID, rateInformation)
- inputTimeCard (empID, date, hours)

The first input is part of the GUI and is detailed during the design of the user interface. The other two inputs, however, are from external systems and do not require user involvement. The information from the tax bureau may be sent as a set of real-time messages or in the form of an input file on a CD or some other electronic device. The time card information could come into the system in various formats. Perhaps physical time cards are entered via an electronic card reader. Or an input from a subsystem, such as an electronic employee ID card reader, may send time card information at the end of every workday. These last two input messages need to be precisely defined, including transmission method, content, and format.

Figure 15-9 is a variation of the system sequence diagram for the telephone order scenario of the *Create new order* use case, as originally given in Figure 7-17. In the figure, the Order Clerk actor and the BankSystem package are external to the system. The messages that go from the Order Clerk to the system are part of the user interface. The messages that go between the external BankSystem package and the system are system inputs. In the object-oriented models, the boundary between actors and external packages with the internal objects is more explicit than in structured models. In Figure 15-9, four messages go from the Order Clerk actor to the system, and one input message enters the system boundary from the BankSystem package. The input messages, along with the actual message signatures as shown in the figure, are:

- startOrder(accountNo)
- addItem(catalogID, prodID, size, quantity)
- completeOrder ( )
- makePayment(paymentAmt, ccInformation)
- returnVerification (creditCard#, verificationCode, amount)

FIGURE *15-8*

Partial system sequence diagram
for the payroll system use cases.

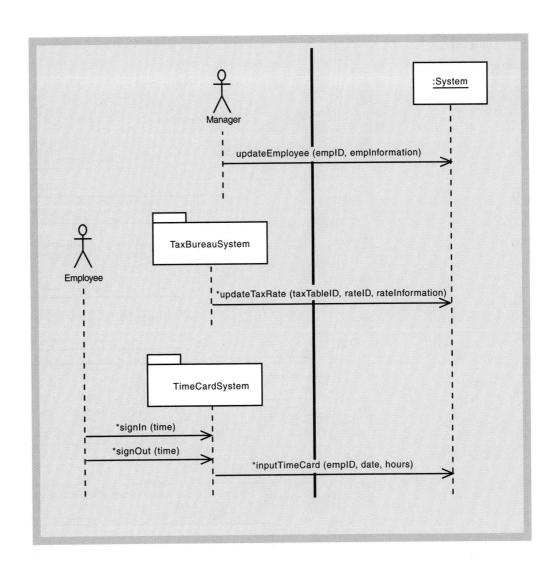

The system sequence diagram developed during the analysis phase identifies the series of steps that occur in the overall process to create a new order from a telephone call. Along with a sequence diagram, a detailed dialog is developed to highlight the communications between the user and computer, as explained in Chapter 14. The point to note here is that a sequence diagram provides a detailed perspective of the user and system inputs to support the use case and the corresponding business event.

Additional analysis of the messages themselves also supplies information about the data fields on the message. To obtain a more thorough analysis of the messages, the developer may need to consult the design class diagram. The actual parameters that are passed in on the messages need to be consistent with the attributes that are found in the design classes. Since the design class attributes are typed, the input parameters can also be typed to be consistent with the design class attributes.

Figure 15-10 lists each input message and the data fields that must be passed with the message. In the example, we show the data associated with every input message. However, the messages associated with the user interface may have already been precisely specified during the design of the user interface. In that case, only the system inputs need to be placed in the table. Not only is this analysis necessary to develop the details of system inputs, but it also provides a good check on the analysis. Notice that the table is more detailed to define the input parameters more precisely. Often this detail will be transferred back to the sequence diagram to make it more complete also.

FIGURE *15-9*

System sequence diagram for
Create new order.

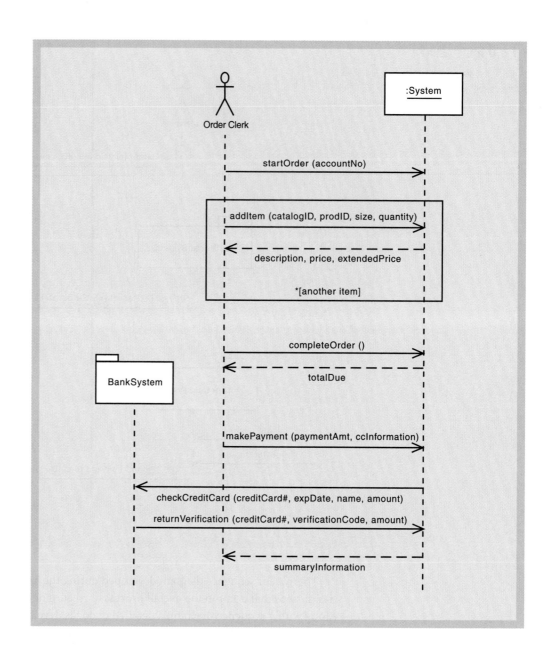

FIGURE *15-10*

Input messages and data parameters
from an RMO system sequence diagram.

Message	Data parameters
startOrder	accountNo
addItem	catalogID, prodID, size, quantity
completeOrder	—
makePayment	paymentAmt, creditCard#, expDate, name
returnVerification	creditCard#, verificationCode, amount

DESIGNING SYSTEM OUTPUTS

The primary objective of system outputs is to present information in the right place at the right time to the right people. Historically, the most common form of output information has been printed textual reports. Although straight text and tables are still used extensively, new formats, such as charts and diagrams, provide many more options to present, emphasize, and summarize information.

As with input design, the tasks in this activity accomplish four objectives:

- Determine the type of each system output.
- Make a list of specific system outputs required based on application design.
- Specify any necessary controls to protect the information provided in the output.
- Design and prototype the output layout.

The purpose of the first two tasks is to evaluate the various alternatives and to design the most appropriate approach for each needed output. The list of required output reports is normally specified during the analysis phase as part of modeling system requirements. During design, the task is to coordinate the production of those outputs with the modules (structured techniques) and methods (object-oriented techniques) that are identified during the application architecture design.

The third task ensures that the designer evaluates the value of the information to the organization and protects it. Frequently, organizations implement controls on the inputs and system access but forget that output reports often have sensitive information. The upcoming section "Designing Integrity Controls" identifies several important controls that all outputs should include.

Today users can develop their own reports using tools and preformatted templates. These reports, called *ad hoc reports* have not been designed by programmer/analysts. Instead, an ad hoc report is the result of a new user query to a database in response to a specific question or need. In Chapter 10, you learned about relational databases and Structured Query Language (SQL). Many systems provide a simplified graphical tool to permit users to formulate queries in SQL to produce ad hoc reports. Obviously, analysts do not design those reports during system development. However, the tools and capability to support user requests do need to be built into the system. The report-design activity is a good time to ask whether the system requires an ad hoc reporting capability and to add such capability if necessary.

ad hoc reports

reports that are not predefined by a programmer but designed as needed by a user

Defining the Details of System Outputs

The objective of this task is to ensure that the designer has identified and specified all of the outputs for the new system. The technique is the same as that used for the definition of the system inputs. This model-based approach utilizes the information in the event table and other models to identify and define the detailed specifications of the outputs. Although many system inputs may be defined as part of the user-interface design, many outputs do not require dynamic human-computer interaction—for example, printed reports, turnaround documents, or simple screen displays. So, analysts must look to the analysis models they developed earlier for many more system outputs. For analysts using traditional structured techniques, the data flows from an internal process to an external agent or external process identify the outputs. For object-oriented techniques, messages that originate on internal classes and whose destination is an actor or another external system are the outputs.

Using Traditional Structured Models

To identify the outputs in the traditional approach, analysts look at data flows coming out of the system across the system boundary in the data flow diagrams and fragments.

The payroll example shown in the DFD of Figure 15-3 contains several system outputs that are not normally part of the graphical user interface. Processes 6, *Print checks*, and 8, *Produce year-end tax*, have three outputs that are generated in a batch execution. The program to print checks prints all payroll checks at once for a specific payroll period. Similarly, the year-end tax program will print W-2 forms and a detail report. Batch-oriented reports are always classified as system outputs.

Figure 15-4, the *Create new order* DFD, shows three outputs: a confirmation to the customer, a notice to shipping, and a payment transaction that goes to the bank. As with system inputs, analysts should build a table of the DFD outputs, defining exactly what reports are needed and determining the data fields. Data flow definitions for each of these outputs should have been created and documented on a data flow diagram as part of the analysis of requirements. One additional system output shown in Figure 15-4, which is not a report, is the electronic credit verification going to the credit bureau.

As analysts build the table of DFD outputs, report definitions, and data fields, they add two more columns to the three identified in the input table. These two additional columns list (1) the files or data stores that are required to produce the report and (2) the number of records from which the report is generated—a single record or a set of records. Figure 15-11 is an example of the table of system outputs. The single/multiple record column generally indicates whether the report is printed immediately after each process or in batch at some other time. For example, from Figure 15-3, the W-2 reports will process multiple records to print the report.

DFD data flow	Structure chart data couple	Data content	Database files required	Single record/mutliple records
Order confirmation	Produce confirmation	Account-Number, Name, Billing-Address, Shipping-Address, Order-Number, Date, Priority-Code, Shipping-And-Handling, Tax, Total *(Product-Item-Number, Description, Size, Color, Options, Quantity, Price)	Customer, Order, Order Item	Single
Shipping order details	Produce confirmation	Account-Number, Name, Shipping Address, Order-Number, Date, Priority-Code	Customer, Order	Single
Transaction details	Write transaction	Account-Number, Shipping-And Handling, Tax, Total	Customer, Transaction	Single
Credit verification (real time)	Credit information	Credit-Card-Number, Expiration-Date, Customer-Name, Amount	Customer, Order	Single

FIGURE *15-11*

A table of system outputs based on the traditional structured approach.

To verify that the structure chart modules are consistent with the structure of the output report, analysts again look at the data couples and the report data requirements. An analysis of the data couple being sent to the module and the data fields on the output report will verify that the application has been designed correctly to generate the report.

Using Object-Oriented Models

In the object-oriented approach, outputs are indicated by messages in sequence diagrams that originate from an internal system object and are sent to an external actor. In Figure 15-9, the output message with the parameters of description, price, extendedPrice is an example of an output message. This message is generated as a result of the input message addItem(catalogID, prodID, size, quantity) to the internal order object. A review of all the output messages generated across all sequence diagrams provides the consistency check against the required outputs identified during the activities of the analysis phase.

Output messages that are based on an individual object (or record) are usually part of the methods of that object class. To report on all objects within a class, a class-level method is used. A class-level method is a method that works on the entire class of objects, not a single object. For example, a customer confirmation of an order is an output message that contains information for a single order object. However, to produce a summary report of all orders for the week, a class-level method looks at all the orders in the Order class and sends output information for each one with an order date within the week's time period.

The system sequence diagram in Figure 15-9 shows four output messages—three from the system to the Order Clerk actor, and one to the BankSystem package. Each of these messages must have a list of parameters that are transmitted with it, even though the displaySummary () message in the diagram does not show them. Output design is a good time for analysts to elaborate the messages to include all of the required parameters. Similar to the traditional approach, Figure 15-12 shows a table listing the output messages, the database files or tables required, and the number of objects (a single object or a set of objects) to be included in the report. Comparing this figure with Figure 15-11, you may notice that some outputs are not identified. Since Figure 15-12 derives from a single use case, the *Create order* use case, it does not include other messages such as shipping messages that are on other use cases.

FIGURE *15-12*

A table of system outputs based on object-oriented messages.

Output message	Data parameters	Classes or database tables	Single record/ multiple records
Response to addItem ()	description, price, extendedPrice	CatalogProduct, ProductItem, InventoryItem, OrderItem	Single
Response to completeOrder ( )	totalDue	Order, OrderItem	Single
checkCreditCard ( )	creditCard#, expDate, name, amount	Customer, OrderTransaction	Single
summary Information— response to makePayment ( )	customerName, billingAddress, shippingAddress, orderNumber, date, s&h, tax, totalAmount	Order, OrderItem, Customer	Single

Designing Reports, Statements, and Turnaround Documents

With the advent of office automation and other business systems, businesspeople originally thought that paper reports would no longer be needed. In fact, just the opposite has happened. Business systems have made information much more widely available, with a proliferation of all types of reports, both paper and electronic. In fact, one of the major challenges to organizations and the designers of their information systems today is to organize the overwhelming amount of information so that it is meaningful. One of the most difficult aspects of output design is to decide what information to provide and how to present it to avoid a confusing mass of complex data.

Type of Output

Before looking at the different formats that analysts use in designing reports, let's discuss four types of output reports that users require: detailed reports, summary reports, exception reports, and executive reports.

Detailed reports are used to carry out the day-to-day processing of the business. They contain specific information on business transactions. Sometimes a report may be for a single transaction, such as an order confirmation sheet with details of a particular customer order. Other detailed reports may list a set of transactions—for example, a list of all overdue accounts, with each line of the report presenting information about a particular account. A clerk could use this report to research overdue accounts and determine actions to collect past-due amounts. The purpose of detailed reports is to provide working documents for people in the company.

Summary reports are often used to recap periodic activity. An example of this report is a daily or weekly summary of all sales transactions, with a total dollar amount of sales. Middle managers often use this type of report to track departmental or division performance. Exception reports are also used to monitor performance. An *exception report* is produced only when a normal range of values is exceeded. When business is progressing normally, then no report is needed. But when something exceeds an expected range, a report is produced to alert staff. An example is a report from a production line that lists rejected parts. If the reject rate is above a set threshold, then a report is generated. Sometimes exception reports are produced regularly. The rejected parts report might be produced every day if the production line usually has some rejected parts. An aged accounts receivable report might be produced each month showing the accounts that are past due. Unfortunately, the organization may always have some accounts to list in such reports, so they are produced regularly.

Top management uses *executive reports* to assess overall organizational health and performance. These reports thus contain summary information from activities within the company. They may also contain comparative performance with industrywide averages. Using these reports, executives can assess competitive strengths or weaknesses of their company.

Chapter 1 discussed the various types of information systems that systems analysts develop. You will note that some types of information systems focus on producing a particular type of report. Although there is no strict requirement for a system to produce only one type of report, we often categorize a system based on the type of report it produces. The next section looks at some examples of printed reports.

Internal versus External Outputs

Printed outputs are classified as either internal outputs or external outputs. *Internal outputs* are produced for use within the organization. The types of reports just discussed fall under this category. *External outputs* include statements, notices, and other documents that are produced for people outside the organization. Because they are official business documents for an outside audience, they need to be produced with the highest-quality graphics and color. Some examples include monthly bank statements, late notices, order confirmation and packing slips (such as those provided to Rocky Mountain Outfitters' customers), and legal documents such as insurance policies. Some external outputs are referred to as *turnaround documents* because they are sent to a customer but include a tear-off portion that is returned for input later, such as a bill that contains a payment stub to be returned with a check. All of these printed outputs must be designed with care, but organizations have many more options for printed output. Today's high-speed color laser printers enable all types of reports and other outputs to be produced.

An example of a detailed report for an external output is shown in Figure 15-13. This report is produced from a Web order similar to that shown in Figure 14-20. Good user-interface design specifies that when a customer places an order over the Web, the system will be able to print the order information as a confirmation. Of course, a user can always print the Web screen display using the browser's print capability, but doing so is time-consuming since it includes all of the graphics and index links on the page. It is much more user-friendly to provide shoppers a formatted order confirmation only. Figure 15-13 illustrates such a report. This type of report is based on the information of a single order. The data required to print this order is a customer record, an order record, and all of the line-item records for ordered items. Notice that it is nicely formatted for easy reading. Different pieces of information are grouped together and placed within boundaries. The report is comprehensive; it contains complete and current information about the report, including today's date, items on the order, payment details, and shipping details.

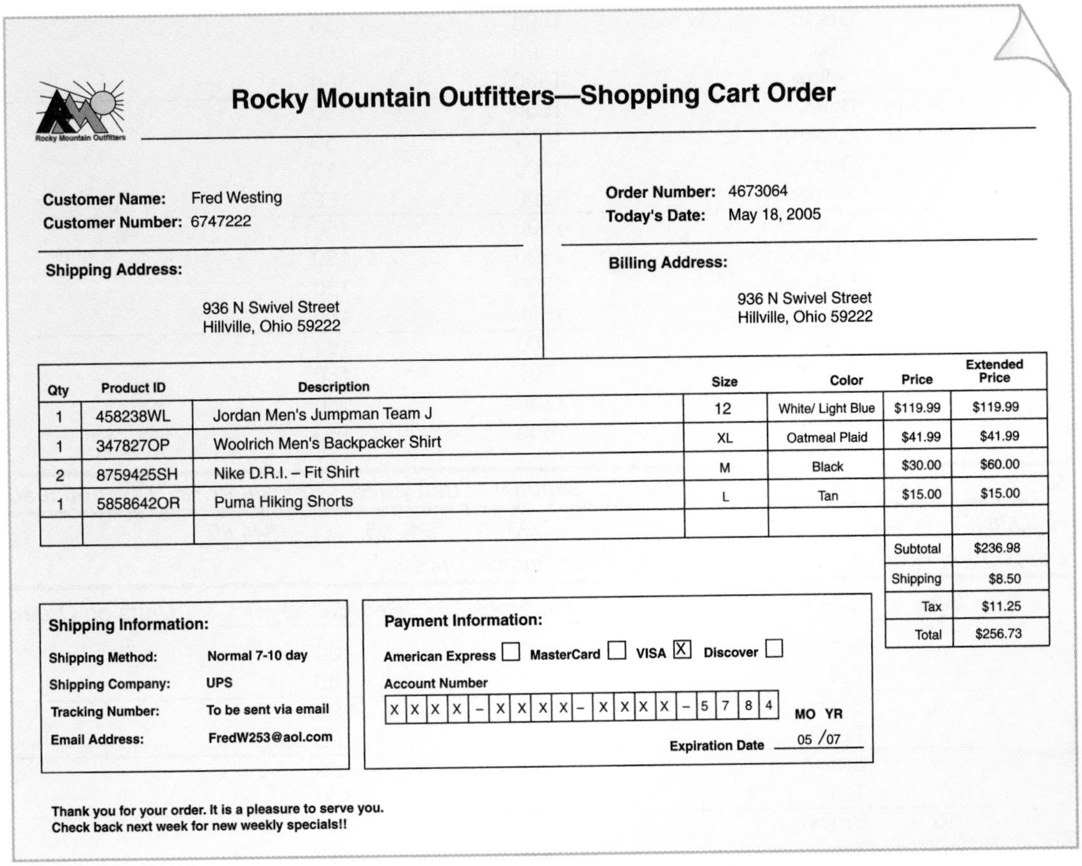

Rocky Mountain Outfitters—Shopping Cart Order

| Customer Name: | Fred Westing | | Order Number: | 4673064 |
| Customer Number: | 6747222 | | Today's Date: | May 18, 2005 |

Shipping Address:

936 N Swivel Street
Hillville, Ohio 59222

Billing Address:

936 N Swivel Street
Hillville, Ohio 59222

Qty	Product ID	Description	Size	Color	Price	Extended Price
1	458238WL	Jordan Men's Jumpman Team J	12	White/ Light Blue	$119.99	$119.99
1	347827OP	Woolrich Men's Backpacker Shirt	XL	Oatmeal Plaid	$41.99	$41.99
2	8759425SH	Nike D.R.I. – Fit Shirt	M	Black	$30.00	$60.00
1	5858642OR	Puma Hiking Shorts	L	Tan	$15.00	$15.00

Subtotal	$236.98
Shipping	$8.50
Tax	$11.25
Total	$256.73

Shipping Information:

Shipping Method:	Normal 7-10 day
Shipping Company:	UPS
Tracking Number:	To be sent via email
Email Address:	FredW253@aol.com

Payment Information:

American Express ☐   MasterCard ☐   VISA ☒   Discover ☐

Account Number

| X | X | X | X | – | X | X | X | X | – | X | X | X | X | – | 5 | 7 | 8 | 4 |

MO YR

Expiration Date ___ 05 /07

Thank you for your order. It is a pleasure to serve you.
Check back next week for new weekly specials!!

FIGURE *15-13*

RMO shopping cart order report.

control break report

a report that includes detailed and summary information

In contrast to the Web shopping cart order report, Figure 15-14 is an example of an internal output. This report is different in several ways. First, it is based on an entire set of records from the inventory database, whereas the shopping cart order is for a single order. The report includes both a detailed and summary section, sometimes called a *control break report*. A control break is the data item that divides the report into groups. In this example, the control break is on the product item number—called *ID* on the report. Whenever a new value of the ID is encountered on the input records, the report begins a new control break section. The detailed section lists the transactions of records from the database, and the summary section provides totals and recaps of the information. The report is sorted and presented by product. However, within each product is a list of each inventory item showing the quantity currently on hand.

External outputs can consist of complex, multiple-page documents. A well-known example is the set of reports and statements that you receive with your car insurance statement. This statement is usually a multipage document consisting of detailed automobile insurance information and rates, summary pages, turnaround premium payment cards, and insurance cards for each automobile. Another example is a report of

ID	Season	Category	Supplier	Unit Price	Special Price	Discontinued
RMO12587	Spr/Fall	Mens C	8201	$39.00	$34.95	No

Description Outdoor Nylon Jacket with Lining

Size	Color	Style	Units in Stock	Reorder Level	Units on Order
Small	Blue		1500	150	
	Green		1500	150	
	Red		1500	150	
	Yellow		1500	150	
Medium	Blue		1500	150	
	Green		1500	150	
	Red		1500	150	
	Yellow		1500	150	
Large	Blue		1500	150	
	Green		1500	150	
	Red		1500	150	
	Yellow		1500	150	
Xlarge	Blue		1500	150	
	Green		1500	150	
	Red		1500	150	
	Yellow		1500	150	

ID	Season	Category	Supplier	Unit Price	Special Price	Discontinued
RMO28497	All	Footwe	7993	$49.95	$44.89	No

Description Hiking Walkers with Patterned Tread Durable Uppers

Size	Color	Style	Units in Stock	Reorder Level	Units on Order
7	Brown		1000	100	
	Tan		1000	100	
8	Brown		1000	100	
	Tan		1000	100	
9	Brown		1000	100	
	Tan		1000	100	
10	Brown		1000	100	
	Tan		1000	100	
11	Brown		1000	100	
	Tan		1000	100	
12	Brown		1000	100	
	Tan		1000	100	
13	Brown		1000	100	
	Tan		1000	100	

FIGURE *15-14*

RMO inventory report.

employment benefits with multiple pages of information customized to the individual employee. Sometimes the documents are printed in color with special highlighting or logos. Figure 15-15 is one page of an example report for survivor protection from an employee benefit booklet. The text is standard wording, and the numbers are customized to the individual employee.

FIGURE *15-15*

A sample employee benefit report.

Survivor Protection

In the event of your death while working for a participating employer, your designated beneficiaries could receive:

Lump Sum Benefits

$50,000	Basic Life Insurance
$230,000	Supplemental Life Insurance
$148,677	Thrift Plan
$31,686	Tax Sheltered Annuity (TSA) Plan
$255	Social Security for your eligible dependents
$460,618	Total*

You have not elected Universal Life Insurance. If you would like more information on this plan, please call 1-800-555-7772.

*Refer to page 7 for additional information on the amount of coverage needed to provide ongoing replacement income.

Accidental Death Benefits

If your death is due to an accident, your designated beneficiaries will receive the above benefits plus:

$100,000	24-Hour Accidental Death and Dismemberment Insurance
$100,000	Occupational Accidental Death and Dismemberment Insurance, if the accident is work related

Monthly Death Benefits

If you die before receiving the Master Retirement Plan benefits and you are vested and have a surviving spouse, your spouse may be eligible for a Qualified Pre-Retirement Survivor Annuity.

In addition, your family may be eligible for the following estimated monthly benefits from Social Security, not to exceed a maximum of $2,591 based on:

$1,110	for each child under age 18
$1,110	for a spouse with children under age 16; or
$1,058	for a spouse age 60 or older

Electronic Reports

Organizations use various types of electronic reports, each serving a different purpose, and each with its respective strengths and weaknesses. Electronic reports provide great flexibility in the organization and presentation of information. In some instances, screen output is formatted like a printed report but displayed electronically. However, electronic reports can also present information in many other formats: Some have detailed and summary sections, some show data and graphics together, others contain boldface type and highlighting, others can dynamically change their organization and summaries, and still others contain hotlinks to related information. An important benefit of electronic reporting is that it is dynamic—it can change to meet the specific needs of a user in a particular situation. In fact, many systems provide powerful ad hoc reporting capabilities so that users can design their own reports on the fly. For example, an electronic report can provide links to further information. One technique, called *drill down*, allows the user to activate a "hot spot hyperlink" on the report, which tells the system to display a lower-level report, providing more detailed information. For example,

drill down

to link a summary field to its supporting detail and enable users to view the detail dynamically

Figure 15-16 contains a summary valuation report of inventory on hand. The report provides a summary valuation for each product item. However, if the user clicks on the hotlink for any product, a detailed report pops up with the list of inventory items, the quantities on hand, and the valuation for each inventory item.

Monthly Sales Summary

Year **2005** *Month* **January**

Category	Season Code	Web Sales	Telephone Sales	Mail Sales	Total Sales
Footwear	All	$ 289,323	$ 1,347,878	$ 540,883	$ 2,178,084
Men's Clothing	Spring	$ 1,768,454	$ 2,879,243	$ 437,874	$ 4,691,484
	Summer	213,938	387,121	123,590	724,649
	Fall	142,823	129,873	112,234	384,930
	Winter	2,980,489	6,453,896	675,290	10,109,675
	All	1,839,729	4,897,235	349,234	7,086,198
Totals			747,368	$ 1,698,222	$ 23,391,023
Women's Clothing	Spring				965,610
	Summer				
	Fall				
	Winter				
	All				
Totals					

Monthly Sales Detail

Year **2005** *Month* **January** *Category* **Men's Clothing** *Season* **Winter**

Product ID	Product Description	Web Sales	Telephone Sales	Mail Sales	Total Sales
RMO12987	Winter Parka	$ 1,490,245	$ 3,226,948	$ 337,640	$ 5,054,833
RMO13788	Fur-Lined Gloves	149,022	322,695	33,765	505,482
RMO23788	Wool Sweater	596,097	1,290,775	135,058	2,021,930
RMO12980	Long Underwear	298,050	645,339	68,556	1,003,005
RMO32998	Fleece-Lined Jacket	447,075	1,258,079	100,271	1,805,425
Total		$ 2,980,489	$ 6,743,836	$ 675,290	$ 10,394,615

FIGURE *15-16*

An RMO summary report with drill down to the detailed report.

Another variation of this hotlink capability lets the user correlate information from one report to related information in another report. Most people are familiar with hotlinks from using their Internet browsers. In an electronic report, hotlinks can refer to other information that correlates or extends the primary information. This same capability can be very useful in a business report, which, for example, links the annual statements of key companies in a certain industry.

Another dynamic aspect of electronic reports is the capability to view the data from different perspectives. For example, it might be beneficial to view sales commission data by region, by sales manager, by product line, or by time period or to compare the data with last season's data. Instead of printing all these reports, you can use an electronic format to generate the different views only as needed. Sometimes long or complex reports include a table of contents with hotlinks to the various sections of the report. Some report-generating programs provide electronic reporting capability that includes all of the functionality that is found on pages on the Internet, including frames, hotlinks, graphics, and even animation.

Graphical and Multimedia Presentation
The graphical presentation of data is one of the greatest benefits of the information age. Tools that permit data to be presented in charts and graphs have made information reporting much more user-friendly for printed and electronic formats. Information is being used more and more for strategic decision making as businesspeople examine

their data for trends and changes. In addition, today's systems frequently maintain massive amounts of data—much more than people can review. The only effective way to use much of these data is by summarizing them and presenting them in graphical form. Figure 15-17 illustrates a bar chart and a pie chart, which are two common ways to present summary data.

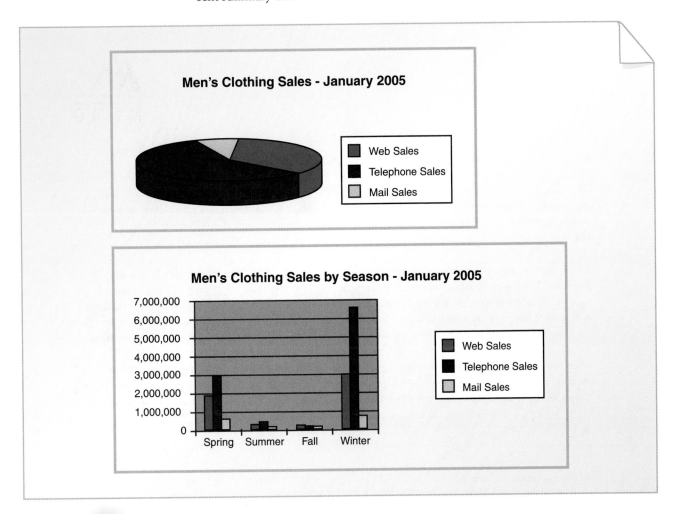

FIGURE *15-17*

Sample bar chart and pie chart reports.

Multimedia outputs have become available recently as multimedia tool capabilities have increased. Today it is possible to see a graphical, and possibly animated, presentation of the information on a screen and have an audio description of the salient points. Combining visual and audio output is a powerful way to present information. (Of course, video games are pushing the frontier of virtual reality to include visual, audio, tactile, and olfactory outputs.)

As the design of the system outputs progresses, it is beneficial to evaluate the various presentation alternatives. Reporting packages can be designed into the system to provide a full range of reporting alternatives. Developers should carefully analyze each output report to determine the objective of the output and to select the form of the output that is most appropriate for the information and its use.

The previous examples illustrated some reports for RMO's customer support system. RMO has many options to present data to satisfy the needs of the different users. In Chapter 4, the RMO organization chart showed that many upper and middle managers had an interest in the new system. Each of the departments represented needs information from the system, and in most cases, each will want the information to be presented in a format unique to its needs. Barbara Halifax emphasized to her team the importance of this flexibility at various times during the development of the outputs (see the accompanying memo).

To:	John MacMurty
From:	Barbara Halifax
Date:	May 19, 2005
RE:	Customer Support System Project Status

John, as you requested, I am writing this memo to update you on the information reports that the various departments throughout the company wanted. One of your concerns was that the definition of the reports could potentially uncover information requirements that we had not anticipated. In other words, there is a risk that the definition of the reports will cause substantial scope creep.

At this point, we have defined essentially all of the reports. I am happy to report that none of the report definitions has caused changes to the database structure or to the system architecture. We have had several requests for new reports, as well as more dynamic updating of the reports. For example, marketing wants not only printed reports but also on-line reports that are linked to provide drill-down capability. However, the basic information is all in the database, and we have been able to define the reports without major problems.

Would you please indicate in the next oversight committee meeting that we are still on schedule? Please thank all the members of the committee for the instructions they gave to their departments of the need to keep the project on schedule and to work within the boundaries of the previously defined functions and data structures. I think every department's information requirements have been satisfied.

BH

cc: Steven Deerfield, Ming Lee

Formatting Reports

With all of the choices available today for output format, system designers have more flexibility in what they can offer users. But sifting through those options to provide workable reports can be challenging. Analysts must keep three principles in mind during the design of output reports:

- What is the objective of the report?
- Who is the intended audience?
- What is the media for presentation?

The importance of these principles cannot be overemphasized when designing reports. In some instances users only need the reports to monitor progress. In others, however, the report may be a critical element in a strategic decision. As a system designer, you should be sure you understand who is going to use the report and how he or she is going to use it. Both the content and the format of the report should be decided based on the audience and the use of the report. Without considering these factors, you could easily omit critical information or present it in an unwieldy format.

Often designers must decide on the level of detail for the format of the report. It can be tempting to produce reports that mirror the structure and format of the data in the database. Newer systems, however, maintain a tremendous amount of detail in the database. Without careful consideration, designers can easily produce reports that suffer from information overload. Information overload occurs when so much data are provided that it becomes difficult for the user to find and focus on the information that is important. Many people have this problem when searching the Internet—the search engine often returns an overwhelming number of results for the search. Careful design and presentation are required to prevent the same problem with output reports.

The format of the report is also important. Every report should have a meaningful title to indicate the data content. The heading should also list both the date the report was produced and a separate date indicating the effective date of the underlying information; sometimes the two dates may be different. Reports should also be paginated. In earlier systems, when reports were printed on continuous forms, page numbers were not as critical. With today's sheet-fed printers, however, it is easy for pages to be misplaced, and results can then be misinterpreted.

Labels and headings should be used to ensure the correct interpretation of the report data. Charts should be clearly labeled with the identification of the axes and units of measure, and a legend should be provided. In Figure 15-14, notice the headings and labels on the report, which help to ensure that the reader does not misinterpret the data. Control breaks are used to divide the data into meaningful pieces that can be easily referenced. Use of lines, boldfacing, and different size fonts makes the report easy to read. Generally, report design is not difficult if you remember that the objective of any report is to provide meaningful information, not just data, and to provide it in a format that is easy to read.

Designers often assume that reports will be printed on standard stock paper. However, that assumption may not be correct. As we just saw, electronic reports are also a very powerful method of producing output information, and the forms of electronic presentation are becoming more and more diverse, ranging from standard computer screens to wireless portable devices. Designers need to carefully consider whether output information will be accessed from nonstandard devices and transmitted via limited bandwidth channels.

DESIGNING INTEGRITY CONTROLS

Information system controls are mechanisms and procedures that are built into a system to safeguard both the system and the information within it. Let's describe a few scenarios to illustrate the need for controls.

- A furniture store sells merchandise on credit with internal financing. An error was made to a customer balance. How do we ensure that only a manager, someone with authority to make adjustments to credit balances, can make the correction?
- A person in accounts payable uses the system to write checks to suppliers. How does the system ensure that the check is correct and that it is made out to a valid supplier? How does the system ensure that no one can commit fraud by writing checks to a bogus supplier? How does the system know that a given payment has been authorized?
- Many companies now have internal LAN networks or intranets. How does a company protect its sensitive data from being accessed by outsiders or even from disgruntled employees?
- Electronic commerce is expanding exponentially, and many companies are now providing e-commerce sites. How does a company ensure that the financial transactions

of its customers are protected and secure? How does a company make sure that its systems and databases are protected from hackers who use the Internet access paths to break in?

■ Many companies are now connected to the Internet to provide on-line access to external employees, such as salespeople, or to customers and suppliers. How does a company safeguard its systems from viruses, worms, and other malicious attacks?

All of these situations involve common business and system activities. Since a company's information is one of its most valuable assets, developers of a new system must consider how to protect and maintain information integrity. As illustrated in Figure 15-18, various locations must be protected with security measures and controls. Some of the controls must be integrated into the application programs that are being developed and the database that supports them. Other controls are part of the operating system and the network. Generally, controls that are integrated into the application and database are called *integrity controls*. The controls in the operating system and network are often referred to as *security controls*. This section explains integrity controls. Later sections discuss security controls.

integrity control

mechanisms and procedures that are built into an application system to safeguard information contained within it

FIGURE *15-18*

Points of security and integrity controls.

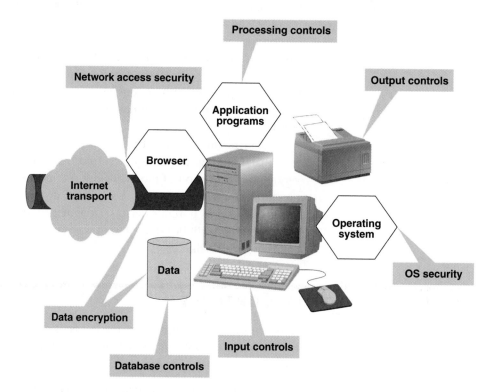

Usually when considering integrity controls, system developers focus on avoiding problems with the application systems and the employees who rightly have access to those systems. Thus, the primary focus is internal—inside the organization. The primary objectives of integrity controls are to:

■ Ensure that only appropriate and correct business transactions occur
■ Ensure that the transactions are recorded and processed correctly
■ Protect and safeguard the assets of the organization (including hardware, software, and information)

The first objective, to ensure that only appropriate and correct business transactions occur, focuses on the identification and capture of input transactions. Integrity controls must make sure that all important business transactions are included—that is, that none are lost or missing and that no fraudulent or erroneous transactions are entered.

The second objective, to ensure that the transactions are recorded and processed correctly, also relates to errors and fraudulent activities. Controls need to detect and alert users to data-entry errors and system bugs that cause problems in processing and recording data. An example of a fraudulent activity is a user who changes the dollar amount on an otherwise valid transaction.

The third objective, to protect and safeguard the assets of the organization, addresses loss of information from computer crashes or catastrophes. It also includes protection of important information on computer files that could be destroyed by a disgruntled employee or possibly even a hacker.

Frequently, system developers are so focused on designing the system software itself that they forget to develop the necessary controls. Because computer systems are so pervasive and companies depend on information systems so heavily, a development project that does not specifically include integrity controls is inviting disaster. The system will be subject to errors, fraud, and deceptive practices, making it unusable. One of the primary control points for ensuring correct data is at the points of data input.

Input Integrity Controls

Input integrity controls are used with all input mechanisms, from electronic devices to standard keyboard inputs. Input controls are an additional level of verification that helps reduce errors on input data. For example, a system may need a certain amount of information for a valid entry, but an input device cannot ensure that all the necessary fields have been entered. An additional level of verification, a *control*, is necessary to check for completeness.

The old computer systems adage of "garbage in, garbage out" relates to input controls, where the objective is to reduce bad data within the system by limiting erroneous input. Historically, the most common control method to ensure correct input was to enter data twice. This technique, called *keypunch and verify*, was first developed for batch entry of large amounts of data. One person would enter the data, and a second person would reenter them on equipment that would then verify that the two inputs were the same. Today, that technique is not used as much because many high-volume transactions are scanned for data. On-line systems also validate input as it is being entered. Here are the more common control techniques in use today:

field combination control

an integrity control that verifies the data in one field based on data in another field or fields

value limit control

an integrity control that identifies when a value in a field is too large or too small

completeness control

an integrity control to ensure that all necessary fields on an input form have been entered

data validation control

an integrity control to validate the input data for correctness and appropriateness

- *Field combination controls* review various combinations of fields to ensure that the correct data are entered. For example, on an insurance policy, the application date must be prior to the date the policy is placed in force.
- *Value limit controls* check numeric fields to make sure that the amount entered is reasonable. For example, the amount of a sale or the amount of a commission usually falls within a certain range of values.
- *Completeness controls* ensure that all the necessary fields are completed. This check can be executed as input occurs so that, depending on which fields are entered, additional required fields must also be entered. For example, if a dependent is entered on an insurance form, then that person's birthday must also be entered.
- *Data validation controls* ensure that numeric fields that contain codes are correct. For example, bank account numbers might be created with a seven-digit field and a trailing check digit to make an eight-digit account number. The check digit is calculated based on the previous seven digits, and the system recalculates it as the data-entry person enters the account number with the check digit. If results do not match, then an input error has occurred. Other data validation can be done on-line against internal tables or files. For example, a customer number can be validated against the

customer file at the time a new order is entered. The systems designer can reduce the need for this type of control by designing a system to obtain the data for a particular field from other information already in the system.

Database Integrity Controls

Most database management systems include integrity controls and security features providing an additional layer of control. Five major areas of security and control can be implemented at the database level:

- Access control
- Data encryption
- Transaction control
- Update control
- Backup and recovery protection

Access Controls

Access controls refer to the ability of a user to get access to the data. An operating system typically applies security and access controls on a file-by-file basis. A DBMS can apply these controls at a much finer level of detail. Controls can be defined on schema subsets such as groups of related tables or objects, single tables or objects, or single fields or attributes. For example, different controls might be applied to the name, Social Security number, and salary fields of an employee table. Also, controls on a single field might differ for read and write access.

A DBMS stores security access information within the schema and applies controls each time data are read or written. When the DBMS enforces security controls, it automatically enforces them for application programs that access the database. Some DBMSs rely on the operating system to identify the user who is attempting to access data, which relieves the user from having to identify himself or herself multiple times. Other DBMSs implement security controls independently of the operating system.

Encryption

Encryption is used both for data within a database and the transmission of data, especially over public carriers. Data within a database are normally encrypted with a single-key encryption method. More details on the various types of encryption are explained in the section on security controls.

Transaction Controls

Transaction logging is a technique by which any update to the database is logged with audit information such as user ID, date, time, input data, and type of update. The fundamental idea is to create an audit trail of all updates to the database that can trace any errors or problems that occur. The more advanced database systems—such as those that run on servers, workstations, and mainframes—include transaction logging as part of the DBMS software. However, several smaller DBMSs, particularly those that run on personal computers, do not include this capability, so design teams must add it directly to those applications.

Transaction logging achieves two objectives. First, it helps discourage fraudulent transactions. If a person knows that every transaction is logged, then that person is less apt to attempt a fraudulent transaction. For example, if a person knows that his or her ID will be associated with every check request, then that person is not likely to request a bogus payment.

The second objective of a logging system is to provide a recovery mechanism for erroneous transactions. A mid-level logging system maintains the set of all updates. The system can then recover from errors by "unapplying" the erroneous transactions. More sophisticated logging systems can provide a "before" and "after" image of the fields that

transaction logging

a technique whereby all updates to a database are recorded with the information of who, when, and how the update was performed

are changed by the transaction, as well as the audit trail of all transactions. These sophisticated systems are typically used only for highly sensitive or critical data files, but they do represent an important control mechanism that is available when necessary.

Update Controls

Database management systems are designed to support many application programs simultaneously. Thus, several programs may want to access and update a record or field at the same time. Update controls within a DBMS provide record locking to protect against multiple updates that might conflict with or overwrite each other.

In addition, some transactions that are applied to the database have multiple parts, such as a financial transaction that must credit one account and debit a different account. Delaying commitment of the update until all updates have been verified is a technique used to protect the data from partial updates of these complex transactions.

Backup and Recovery

Backup and recovery procedures are designed to protect the database from all other types of catastrophes. Many database management systems provide various levels of backup and recovery. Partial or incremental backups are used to capture changes to the database during the time periods between total backups. A total backup is used only periodically to archive a complete copy of all the data. Frequently, this archive is placed in a secure off-site location to protect it against catastrophic threats such as fire, earthquake, or terrorist attacks.

Another popular security measure used for systems that rely on up-to-the-minute data is a mirror database or mirror site. This technique completely duplicates the database and all transactions as they occur. Obviously, this approach can be expensive, but it is becoming more important as information becomes more and more critical to the daily operations of organizations.

Output Integrity Controls

As already discussed, output from a system comes in various forms, such as output that is used by other systems, printed reports, and data output on computer screens. The purpose of output controls is to ensure that output arrives at the proper destination and is accurate, current, and complete. It is especially important that reports with sensitive information arrive at the proper destination and that they not be accessed by unauthorized persons.

Destination Controls

In the past, when most output was in printed form, a distribution control desk collected all the printed reports from the nightly processing and distributed them to the correct departments and people. This control desk was important because some of the reports had sensitive, confidential information, and it was important to keep those reports secure. Systems with good controls printed destination and routing information on a report cover page along with the report. Today, businesses accomplish the same function of a control desk by placing printers in each of the locations that need printed reports. It is still a good idea to print a cover sheet with destination and report heading information. Destination codes and routing capabilities are included during the design process to handle the distribution of reports to separate printing facilities. Controlling access to these reports then becomes an issue of physical access. These types of controls are called *destination controls*.

destination controls

integrity controls to ensure that output information is channeled to the correct persons

Electronic output to other systems is usually provided in two forms: either an on-line transaction-by-transaction output or a single data file with a batch of output transactions. Each form has its own type of controls. If the system produces on-line transactions, then it must ensure that each transaction includes the routing codes identifying the correct destination. Both systems need to work together to ensure that each transaction is sent and received correctly. The output transaction will have verification codes and bits to permit the receiving system to verify the accuracy of the transaction. The receiving system also responds with an acknowledgment of a successful receipt of the transaction. Many of these controls are now built into the network transmission protocols. However, during the design activities, the systems designers need to be aware of the network and operating system capability and supplement it where necessary to ensure that the data are received successfully.

Controls for output data files carefully identify the contents, version, date, and time of the file. Normally, a system produces a data file, either on magnetic tape or disk, and another system must find that data file and use it. The major control issue is how to ensure that the second system uses the correct data file. For example, to avoid serious problems, we want to make sure that Friday's transactions aren't run twice. Or, if, by some processing quirk, two data files are produced for the same day—one for the first half of the day and another for the second—the system must use both data files. Or, if the second system had processing errors and needs to be rerun, it must be able to find the correct file to use on its rerun. Controls for this situation generally have special beginning and ending records that contain date, time, version, record counts, dollar control totals, processing period, and so forth. During systems design, provision must be made to accumulate the appropriate totals and to produce the necessary control records.

Destination controls for computer screen output are not as widely used as those for printed reports. Normally, the previously discussed user access controls manage the availability of information on computer screens. In some instances, however, destination controls limit what information can be displayed on which terminal. This extra safeguard is used primarily for military or other systems that house computer terminals in secure areas and provide access to the system's information to anyone who has access to the area. The design of these systems requires close coordination between the application program and the network security control system.

Completeness, Accuracy, and Correctness Controls

The completeness, accuracy, and correctness of output information are a function primarily of the internal processing of the system rather than any set of controls. System developers ensure completeness and accuracy by printing control fields on the output report. For example, every report should have a date and time stamp, both for the time the report was printed and for the date of the underlying data. Frequently, they are the same, but not always, especially when a report is reprinted because of a previous error. The following items are controls that should be printed on reports:

- Date and time of report printing
- Date and time of data on the report
- Time period covered by the report
- Beginning header with report identification and description
- Destination or routing information
- Pagination in the form "page __ of __"
- Control totals and cross footings
- An "End of Report" trailer
- The report version number and version date (such as those for special printed forms)

Integrity Controls to Prevent Fraud

The preceding sections have identified several types of integrity controls that support the three control objectives. Many of those techniques are focused on preventing errors and protecting the system from foreign intrusion. However, an equally serious problem is the use of the system by authorized people to commit fraud against an organization.

Fraud is a problem that is reaching epidemic proportions in the United States and around the world. Almost every week we see newspaper articles describing fraud and other white-collar crime. The economic losses caused by fraudulent activity around the world are staggering. These losses reach into the billions of dollars and far exceed those from violent and personal crimes. In the last few years, several major corporations have been forced into bankruptcy or closure due to the fraudulent behavior of key executives. Obviously, software and system controls will not completely eliminate fraud. However, systems developers should be aware of the fundamental elements that make fraud possible and incorporate system controls to combat it. The controls that we discussed previously—input controls, database controls, and output controls—are critical components in the battle against fraud, but several additional techniques should be considered in system design to further increase protection.

Research into the perpetration of fraud indicates that three conditions are present in almost all fraud cases:

- Personal pressure, such as the desire to maintain an extravagant lifestyle
- Rationalization, such as a person's thoughts that "I will repay this money"
- Opportunity, such as unverified cash receipts

The objective of integrity controls is to reduce or eliminate the opportunity for fraud by having adequate manual controls and automated records of money and assets. Control of fraud requires both manual procedures and computer integrity controls. Neither component is sufficient by itself to reduce the opportunities for fraud. System developers need to work closely with business users who are knowledgeable about accounting principles to prevent fraud.

Sometimes system developers may think that integrity controls are not necessary, since the system in development is not a financial or accounting system. However, an opportunity for fraud exists in almost every business system. Since most business systems track an organization's assets, someone could manipulate those assets, writing checks for incorrect amounts or to fictitious parties. Hence, almost every system requires some type of integrity controls.

Figure 15-19 contains several of the more important factors that increase the risk of fraud. This list is not comprehensive, but it does provide a foundation from which

FIGURE *15-19*

Fraud risks and prevention techniques. *Source:* Information in the table was provided by Dr. Marshall Romney of the School of Accountancy and Information Systems at Brigham Young University.

Factors affecting fraud risk	Techniques to reduce risk
Separation of duties	Design separate electronic forms, with separate access controls, for request, approval, and generation of expenditures.
Inadequate audit trails	Include transaction logging. Avoid, or very tightly control, manual override capability that circumvents logs.
Inadequate records	Implement a comprehensive database with sufficient detail and logs.
Inadequate monitoring	Include manual procedures and automated routines to monitor patterns and out-of-bound conditions. Include exception reports. Implement third-person audit capability.
Easily removable assets	Include an easy-to-use capability to cross-check physical counts with automated records.
Inadequate security system	Supplement operating system security features with additional program and data level security. Include automatic shutdown and lockup features. Include routines to analyze access patterns.

developers can design a computer system that reduces the opportunity for fraud. As a system developer, you should include discussions both with your users and within the project teams to ensure that adequate controls have been included to reduce fraud.

Now that we have an overview of input and output integrity controls, we turn our attention to security controls.

DESIGNING SECURITY CONTROLS

security control

mechanisms usually provided by the operating system or environment to protect the data and processing systems from malicious attack

Although the objective of *security controls* is to protect the assets of an organization from all threats, as indicated earlier, the primary focus is generally on external threats. In addition to the objectives enumerated earlier for integrity controls, security controls also have the following two objectives:

- Maintain a stable, functioning operating environment for users and application systems (usually 24 hours a day, seven days a week)
- Protect information and transactions during transmission outside the organization (public carriers)

The first objective, to maintain a stable operating environment, focuses on security measures to protect the organization's systems from external attacks such as from hackers, viruses, worms, and message overloads. Most organizations today have gateways between their internal systems and the Internet. Every time someone in an organization sends a communication to or receives one from the Internet, there is the potential for a security violation and for undesirable access that could disrupt the internal systems. So, eliminating and controlling any undesirable access help avoid disruption of the system.

The second objective, to protect transactions during transmission, focuses on the information that is sent or received via the Internet. More and more organizations utilize the Internet as a portal to their customers and to their suppliers. Once a transaction is sent outside the organization, it could be intercepted, destroyed, or modified. So, security controls use techniques to protect data while they are in transit from the source to the destination.

Security controls can be implemented within different types of software, including the network and computer operating system, the database management system, or the application programs. The most common security control points are network and computer operating systems because they exercise direct control over assets such as files, application programs, and disk drives. All modern operating systems contain extensive security features that can identify users, restrict access to files and programs, and secure data transmission among distributed software components. Operating system security is the foundation of security for most information systems.

On some occasions, developers may implement security controls directly within application software. Developers may define their own security controls over individual data items or records when data are stored in files instead of a database. Developers may also implement security controls to prevent unauthorized users from performing certain functions such as deleting existing data or creating backup copies on removable storage media.

Most developers avoid implementing security controls within application software because of the complexity and importance of security functions. Most operating system and DBMS developers have a large programming staff dedicated exclusively to developing and maintaining security software. It is difficult for application developers to dedicate sufficient resources to implement system security controls correctly and fully. Thus, security-related implementation tasks in a typical information system development project are usually limited to configuring security software in the underlying operating system or DBMS.

Security for Access to Systems

Modern operating systems, networking software, and Internet access all need to implement control mechanisms. These mechanisms can be used to control access to any resource managed by the operating system or network—including hardware, application programs, and data files.

System access controls are mechanisms that are established to restrict what portions of the computer system a person can use. This category includes controls to limit access to certain applications or functions within an application, restrict access to the computer system itself, and limit access to certain pieces of data.

With proper design and implementation, an information system can use access control functions embedded in system software. The advantage to this approach is that a consistent set of access controls is then applied to every resource on a hardware platform or network. Thus, the systems designer can implement a single access control scheme and apply it to every resource or information system.

The systems designer can also add controls over and above those already provided by system software. However, designing and implementing effective application-based access controls require technical expertise. Operating system and network software developers expend considerable energy and resources to develop reliable and efficient access controls, and it is difficult and expensive for a typical organization to duplicate these efforts. For these reasons, most information systems build on the access control already within system software.

Types of Users

System developers must consider different types of users when designing access controls. Figure 15-20 illustrates various types of users and the access that is appropriate for each. The following paragraphs explain the types of system access available to users.

To begin development of access controls, designers first must identify and consider all three of these user categories: unauthorized users, registered users, and privileged users. *Unauthorized users* are people who are not allowed access to any part or functions of the system. Such users include employees who are prohibited from accessing the system, former employees who no longer are permitted to access the system, and outsiders such as hackers and intruders. Controls must be able to identify and exclude access from these people.

Registered users are those who are authorized to access the system. Normally, various levels of registered users are set up depending on what they are authorized to view and update. The different levels of access are defined during the design of the new system. For example, some users may be allowed to view data but not update them, and other users can update only certain data fields. Some screens and functions of the new system may be hidden from other levels of registered users. The important point for systems designers to recognize is that there may be multiple levels of registered users. *Authorization* is the process of determining whether a user is permitted to access a specific resource for a particular purpose. In other words, it is the process of deciding whether a user should be a registered user. The security system stores an access control list for each protected resource. An *access control list* is a list of users or user groups that can access a resource and the permitted access type(s).

Privileged users include people who have access to the source code, executable program, and database structure of the system. These people include system programmers,

unauthorized user

a person who does not have authorized access to a system

registered user

a user who is registered or known to the system and is authorized to access some part of it

authorization

the process of determining whether a user is permitted to have access to the system and data

access control list

the list of users who have rights to access the system and data

privileged user

a user who has special security access privileges to a system

FIGURE *15-20*

Users and access roles to
computer systems.

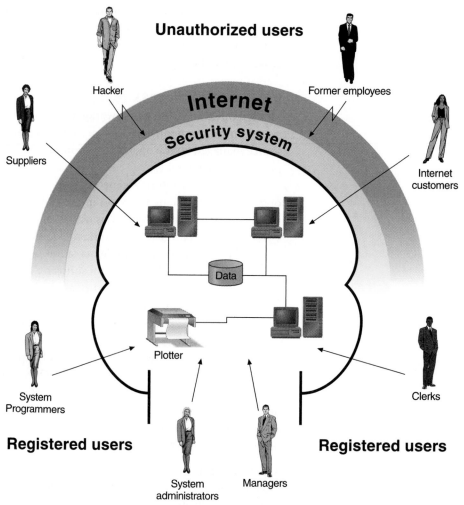

application programmers, operators, and system administrators, and they may also have differing levels of security access. Usually system programmers have full access to all components of the systems and data. Application programmers have access to the applications themselves but often not to the secure libraries and data files used for the systems in production. System administrators have access to all functions of the system and can control and establish the various levels of registration and register users. A system administrator also usually has software programs to help control access and to monitor access attempts.

Passwords and Smart Cards

Authentication is the process of identifying users (that is, the authorized or registered users) who request access to sensitive resources. Authentication is the basis of all security because security controls are useless unless the user is correctly identified. In many operating systems, authentication requires the user to enter a user name and password. The user is authenticated if the password he or she enters matches the password stored in the security database.

Two techniques are used to define passwords. The computer can randomly generate and assign passwords, or each user can define his or her own password. There are advantages to both techniques. The first creates passwords that are usually longer and more random, but they tend to be hard for users to remember. Most users would have a hard time trying to remember a password such as a3x7869bts21. User-developed passwords

authentication

the process of identifying a user to verify that he or she can have access to the system

are easier to remember, but they are usually not as complex and therefore not quite as secure. Some restrictions can be placed on the syntax of the password to ensure at least a minimum level of security.

Of course, one of the problems with passwords is remembering what they are. It is not uncommon for heavy computer users to have 5 or even 10 different passwords for different systems that they access. One alternative is to use the same password for all systems, but if someone determines the password, then all the systems are compromised. Most often, the security system should be organized so that all resources can be accessed with the same unique identifier and password combination. In other words, only one user ID/password combination should be required for access to the different systems throughout the organization. When users have to remember different IDs and passwords to access different systems, they often write them down and post them near the computer. Obviously, this practice defeats the purpose of user verification security.

A *smart card* is a computer-readable plastic card with a small amount of stored security data that can be read by a card scanner in much the same way a credit or debit card can be read at a supermarket checkout counter. The smart card stores an encrypted version of the user's password, fingerprint, retinal scan, or voice characteristics. To authenticate himself or herself, the user scans the card and then enters the password or submits to a fingerprint, retinal, or voice scan. Such a system enhances security because the user must possess both the card and the appropriate identifying information to be authenticated. Only the security subsystem knows the key, which prevents potential intruders from using cards with altered data.

A final security step is to make sure the system keeps a record of attempted logons, especially unsuccessful ones. An unsuccessful logon may simply indicate that the user mistyped or forgot a password, but it may also indicate an attempted breach of security, which should be investigated.

Biometric Devices

Authentication can also be based on other forms of personal identification, including keystroke patterns, fingerprints, retinal scans, and voice characteristics. When a user enters a password or other keystroke sequence, the timing and force of each keystroke are unique. Some security systems use both the password and the keystroke pattern to authenticate the user, which prevents someone with a stolen password from accessing system resources.

Many companies are now experimenting with a new form of security based on biometric devices. The principle behind use of a biometric device is that the person himself becomes the password or gateway into a secure system. These more sophisticated security systems can scan fingerprints, retinal blood vessels, or voices, which are unique for every person. With the advent of very small computer chips with very high memory densities and logic circuitry, biometric devices can be built into almost any of the normal hardware components of a computer. In addition, the complex logic necessary to do sophisticated pattern matching of fingerprints, hand vein patterns, retinas, iris patterns, or complete facial patterns can be located right in the micro-sized biometric device itself.

Biometric fingerprint devices are now being embedded in such components as a computer mouse, computer keyboard, and small touch pads. Other biometric scanners, such as very small cameras, can be embedded in the computer monitor. Such a device might do an iris or facial scan of the person looking at the monitor. Figure 15-21 illustrates a computer mouse with an embedded touch pad to test fingerprints. Other types of mouse devices have the sensor on the side so that the thumb must be placed on it and authorization can be performed before every mouse action.

FIGURE *15-21*

Biometric mouse.
Source: Used with permission of onClick Corporation, Houston, Texas.

Security based on biometric devices can also be multilevel. Security verification can be done when the user first tries to log on. Higher levels of security can later be activated within a given program to obtain additional authorization to access specific forms or database records. Obviously, each individual must be authorized and appropriate information stored for the level of security allowed.

Data Security

In addition to the need for controlling access to an organization's systems and internal network, it is frequently important to make the data themselves secure. For example, user IDs and passwords are important information that must be secret. Frequently, the password information is even kept secret from the system administrators. They can assign a new password to a user, but they cannot read or access the current password. So, if a user forgets his or her password, the administrator assigns a new one.

Many other types of files are also kept confidential. Some examples include files that contain the following:

- Financial information
- Credit-card numbers, bank account numbers, payroll information, and other personal data
- Strategies and plans for products and other mission-critical data
- Government and sensitive military information

Some operating systems, especially UNIX and its derivations, have built-in security for each file in the system. Each UNIX file has security corresponding to three types of users: the owner of the file, other members of the owner's workgroup, and all other users. The security for each user is also further divided into three levels: read access, update (create, update, and delete) access, and execute access. Execute access determines whether the file is executable (such as an .exe file in Windows), and the security level determines who is allowed to execute the file.

Data that reside on an internal system need to be protected, but data that are being transmitted outside the organization are especially subject to snooping and even modification. With the increasing acceptance of electronic communications, more and more organizations are transmitting and receiving transactions via the Internet. On the sales and distribution side of business, customers are viewing catalogs, ordering products, making payments, and tracking shipments all via the Internet. On the supply side, organizations are ordering inventory, monitoring receivables, sending purchase orders, and making financial transactions through the Internet. Since this information is being transmitted via the public Internet, the raw data are available to anybody who has tools to listen and intercept information packets.

The primary method of maintaining the security of data, both on internal systems and transmitted data, is by encrypting the data. *Encryption* is the process of altering data so that unauthorized users cannot view them. *Decryption* is the process of converting encrypted data back to their original state. Data stored in files or a database on hard drives or other storage devices can be encrypted to protect them against theft. Data sent across a network can be encrypted to prevent eavesdropping or theft during transmission. A thief or eavesdropper who steals or intercepts encrypted data receives a meaningless group of bits that are difficult or impossible to convert back into the original data.

An *encryption algorithm* is a complex mathematical transformation that encrypts or decrypts binary data. An *encryption key* is a binary input to the encryption algorithm—typically a long string of bits. The encryption algorithm varies the data transformation based on the encryption key so that data can be decrypted only with the same key or a compatible decryption key. Many encryption algorithms are available, and a few, includ-

encryption

the process of altering data so that they are unreadable by unauthorized users

decryption

the process of converting encrypted data back into a readable format

encryption algorithm

a complex mathematical formula and process that encrypts or decrypts data

encryption key

a binary field that the encryption algorithm uses to transform the data

ing data encryption standard (DES) and several algorithms developed by RSA Security, are widely deployed governmental or Internet standards. An encryption algorithm must generate encrypted data that are difficult or impossible to decrypt without the encryption key. Decryption without the key becomes more difficult as key length is increased. Both sender and receiver must use the same or compatible algorithms.

Figure 15-22 is an example of *symmetric key encryption*, where the same key encrypts and decrypts the data. A significant problem with symmetric key encryption is that both sender and receiver use the same key, which must be created and shared in a secure manner. Security is compromised if the key is transmitted over the same channel as messages encrypted with the key. Also, sharing a key among many users increases the possibility of key theft.

symmetric key encryption

an encryption process that uses the same key to encrypt and to decrypt the data

FIGURE *15-22*

Symmetric key encryption.

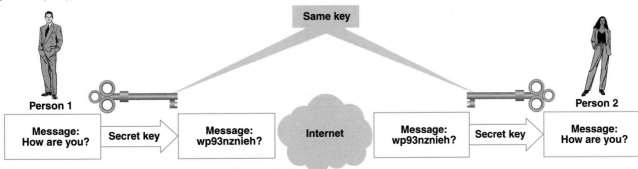

asymmetric key encryption

an encryption process that uses one key to encrypt and a different key to decrypt the data

public key encryption

an asymmetric key method where one key is publicized and the other key is kept private

Asymmetric key encryption uses different but compatible keys to encrypt and decrypt data. *Public key encryption* is a form of asymmetric key encryption that uses a public key for encryption and a private key for decryption. The two keys are like a matched pair. Once information is encrypted with the public key, it can be decrypted only with the private key. It cannot be decrypted with the same public key that encrypted it. Organizations that use this technique broadcast their public key so that it is freely available to anybody who wants it. Then when some entity—for example, someone who wants to order something from the vendor—wishes to transmit a secure message to a vendor, that customer would read the vendor's public key from a public source such as a Web site. The customer would encrypt the message with the public key and send the message to the vendor. The vendor would decrypt the message with the private key. Since no one else has the private key, no one else can decrypt the message.

Some asymmetric encryption methods can encrypt and decrypt messages in both directions. That is, in addition to encrypting a message with the public key that can be decrypted with the private key, an organization can also encrypt a message with the private key and decrypt it with the public key. Notice that both keys must still work as a pair, but the message can go forward or backward through the encryption/decryption pair. This second technique is the basis for digital signatures and certificates, which are explained in the next section. Figure 15-23 illustrates an asymmetric key encryption transmittal.

You may ask, "How can an encryption algorithm go one direction (with one key) and not be able to come back the same way (be decrypted with the same key)?" The mathematics of this type of algorithm is beyond the scope of this text. However, you should be able to understand a simple example: multiplication and factoring. If someone gives you two or three numbers, even big numbers, and asks you to multiply them, you can do that fairly easily. However, if someone gives you one very big number and asks you to factor it (that is, find the numbers that were originally multiplied to get that number), you would not be able to do that easily. It would take you a long time. Algorithms based on this one-directional mathematical characteristic form the basis of many asymmetric key encryption routines.

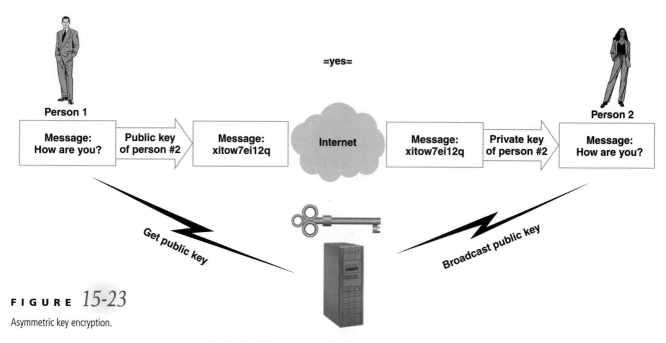

FIGURE *15-23*

Asymmetric key encryption.

Digital Signatures and Certificates

The encryption of messages is an effective technique to enable a secure exchange of information between two entities who have appropriate keys. However, how do you know that the entity on the other end of the communication is really who you think it is? A *digital signature* is a technique in which a document is encrypted using a private key to verify who wrote the document. If you have the public key of an entity, and that entity sends you a message with its private key, you can decode it with the public key. You know that the party is the one you want to communicate with because that entity is the only one who can encode a message with that private key. The encoding of a message with a private key is called *digital signing*.

Taking the example one step further, you can ask the question, "How do I know that the public key I have is the correct public key and not some counterfeit key?" In other words, maybe someone is impersonating another entity and is passing out false public keys to be able to intercept encoded messages (such as financial transactions) and steal information. In essence, the problem is ensuring that the key that is purported to be the public key of some institution is in fact that institution's public key. The solution to that problem is a certificate.

A *certificate*, or *digital certificate*, is an institution's name and public key (plus other information such as address, Web site URL, and validity date of the certificate) that is encrypted and certified by a third party. Many third parties are very well known and widely accepted *certifying authorities*, such Verisign or Equifax. In fact, they are so well known that their public keys are built right into Netscape and Internet Explorer. As shown in Figure 15-24, you can know that the entities with whom you are communicating are in fact who they say they are and that you do have their correct public key.

An entity who wants a certificate with its name and public key goes to a certifying authority and buys a certificate. The certifying authority encrypts the data with its own private key (signs the data) and gives the data back to the original entity. Now when someone, such as a customer, asks the entity for its public key, it sends the certificate. The customer receives the certificate and opens it with the certifying authority's public key. Again, the certifying authority is so well known that its public key is built into everyone's browser and is essentially impossible to counterfeit. Now the customer can be sure that he or she is communicating with the original entity and can do so with encrypted messages using the entity's public key.

digital signature

a technique in which a document is encrypted using a private key to verify who wrote the document

certificate, or digital certificate

a text message that is encrypted by a verifying authority and used to broadcast an organization's name and public key

certifying authority

a well-known third party that sells digital certificates to organizations

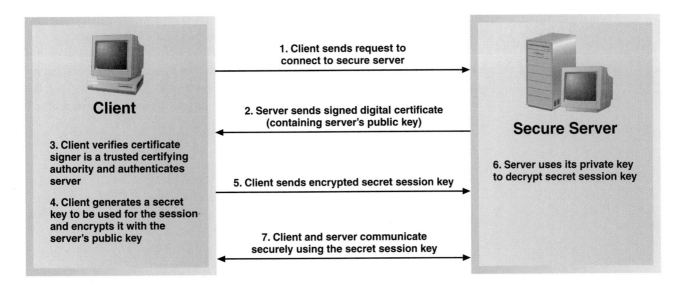

1. Client sends request to connect to secure server

2. Server sends signed digital certificate (containing server's public key)

Client

3. Client verifies certificate signer is a trusted certifying authority and authenticates server

4. Client generates a secret key to be used for the session and encrypts it with the server's public key

5. Client sends encrypted secret session key

Secure Server

6. Server uses its private key to decrypt secret session key

7. Client and server communicate securely using the secret session key

FIGURE *15-24*

Using a digital certificate.

Secure Sockets Layer (SSL)

a standard protocol to connect and transmit encrypted data

Transport Layer Security (TLS)

an updated version of SSL

Secure Hypertext Transport Protocol (HTTPS or HTTP-S)

an Internet standard for transmitting Web pages securely

A variation of this scenario occurs when the buyer and seller transmit their certificates to one another. Each participant can decrypt the certificate using the certifying authority's public key to extract information such as name and address. However, to ensure that the public key contained within the certificate is valid, the certificates are transmitted to the certifying authority for verification. The authority stores certificate data, including public keys, within its database and verifies transmitted certificates by matching their content against the database.

Secure Transactions

Secure electronic transactions require a standard set of methods and protocols that address authentication, authorization, privacy, and integrity. Netscape originally developed the *Secure Sockets Layer (SSL)* to support secure transactions. SSL was later adopted as an Internet standard and renamed *Transport Layer Security (TLS)*, though the original name, SSL, is still widely used.

TLS is a protocol for a secure channel to send messages over the Internet. Sender and receiver first establish a connection using ordinary Internet protocols and then ask each other to create a TLS connection. Sender and receiver then verify each other's identity by exchanging and verifying identity certificates as explained previously. At this point, either or both have exchanged public keys, so they can send secure messages. Since asymmetric encryption is quite slow and difficult, the two entities agree on a protocol and encryption method, usually a single-key encryption method. Of course, all of the messages to establish a secure connection are sent using the public key/private key combination. Once the encryption technique has been decided and the secret, single key has been transmitted, then all subsequent transmission is done using the secret, single key.

IP Security (IPSec) is a newer Internet standard for secure message transmission. IPSec is implemented at a lower layer of the network protocol stack, which enables it to operate with greater speed. IPSec can replace or complement SSL. Both protocols can be used at the same time to provide an extra measure of security. IPSec supports more secure encryption methods than SSL, but these methods are not yet fully deployed on the Internet.

Secure Hypertext Transport Protocol (HTTPS or HTTP-S) is an Internet standard for securely transmitting Web pages. HTTPS supports several types of encryption, digital signing, and certificate exchange and verification. All modern Web browsers and servers support HTTPS. It is a complete approach to Web-based security, though security is enhanced when HTTPS documents are sent over secure TLS or IPSec channels.

Security is an important consideration in the development and deployment of information systems in today's networked environment. Fortunately, many tools and programs are available and can be integrated into new systems as part of the total solution. System developers need to be aware of the need to include security measures and to be familiar with the latest tools and techniques.

Summary

The chapter began with a discussion of identifying and then designing system interfaces. System interfaces include all inputs and outputs except those that are part of the graphical user interface (GUI).

Designing the inputs to the system is a three-step process:

- Identify the devices and mechanisms that will be used to enter input.
- Identify all system inputs and develop a list with data content of each.
- Determine what kinds of controls are necessary for each system input.

To develop a list of the inputs to the system, designers use diagrams that were developed during the analysis and application design activity. For the traditional structured approach, DFDs, data flow definitions, and structure charts are used. For the object-oriented approach, sequence diagrams are the primary source of information, but the design class diagram is used to ensure that the correct data fields and the correct methods that produce the outputs are provided.

The process to design the outputs from the system consists of the same steps as for input design. For output design, the DFDs and sequence diagrams are used to identify data flows and messages that exit the system. New technology provides numerous ways to present output with charts, graphs, and multimedia. Before deciding on an output media, the designer should carefully consider the intended audience and the purpose of the output.

This chapter next discussed the concepts of integrity controls in systems. The objectives of integrity controls are to:

- Ensure that only appropriate and correct business transactions occur
- Ensure that the transactions are recorded and processed correctly
- Protect and safeguard the assets (including information) of the organization

Integrity controls are concerned with defining who has access to the various components of the system and the database. Access controls identify various classifications of users—such as unauthorized users, registered users, and privileged users—to ensure that systems are safeguarded. Additional integrity controls are concerned with reducing errors, preventing fraud, and maintaining the correctness of the data in the system.

The last section of the chapter introduced the basic concepts of security for systems that have access to public networks (primarily the Internet). Security is becoming more and more important, and various techniques should be considered when developing new information systems. The underlying technology in many of the security approaches is based on public key systems that have public and private key components. Encryption and public key systems are the basis for digital signatures, digital certificates, secure connection, and secure transaction implementations.

Key Terms

access control, p. 602

access control list, p. 607

ad hoc reports, p. 589

asymmetric key encryption, p. 611

authentication, p. 608

authorization, p. 607

certificate, or digital certificate, p. 612

certifying authority, p. 612

completeness control, p. 601

control break report, p. 593

data validation control, p. 601

decryption, p. 610

destination controls, p. 603

detailed report, p. 592

digital signature, p. 612

drill down, p. 595

encryption, p. 610

encryption algorithm, p. 610

encryption key, p. 610

exception report, p. 592

REVIEW QUESTIONS

1. What does XML stand for? Explain how XML is similar to HTML. Also discuss the differences between XML and HTML.

2. Compare the strengths and weaknesses of using a DFD to define inputs with using a system sequence diagram to define inputs. Which do you like the best? Why?

3. Explain the system boundary. Why was one used on a DFD but not used on a system sequence diagram?

4. What additional information does the structure chart provide that is not obtained from a DFD in the development of input forms?

5. How are the data fields identified using the traditional structured approach?

6. How are the data fields identified using UML and the object-oriented approach?

7. Explain four types of integrity controls for input forms. Which have you seen most frequently? Why are they important?

8. What protection does transaction logging provide? Should it be included in every system?

9. What are the different considerations for output screen design and output report design?

10. What is meant by *drill down*? Give an example of how you might use it in a report design.

11. What is the danger from information overload? What solutions can you think of to avoid it?

12. Describe what kinds of integrity controls you would recommend to place on all output reports. Why?

13. What are the objectives of integrity controls in information systems? In your own words, explain what each of the three objectives means. Give an example of each.

14. What are the four types of input controls used to reduce input errors? Describe how each works.

15. Explain what is meant by update controls for a database management system.

16. What is the basic purpose of transaction logging? Microsoft Access does not have automatic transaction logging. Is this a deficiency, or is it not really an important consideration in database integrity?

17. On a printed output report, what is the difference between the date the report was printed and the date of the data?

18. What are the two primary objectives of security controls?

19. Explain the three categories of user access privileges. Is three the right number, or should there be more or fewer than three? Why or why not?

20. How does single-key (symmetric) encryption work? What are its strengths? What are its weaknesses?

21. How does public key (asymmetric) encryption work? What are its strengths? What are its weaknesses?

22. What is a digital certificate? What role do certifying authorities play in security systems?

23. What is a digital signature? What does it tell a user?

THINKING CRITICALLY

1. The chapter described various situations that emphasized the need for controls. In the first scenario presented, a furniture store sells merchandise on credit. Based on the descriptions of controls given in the chapter, identify the various controls that should be implemented in the system to ensure that corrections to customer balances are made only by someone with the correct authorization.

2. In the second scenario illustrating the need for controls, an accounts payable clerk uses the system to write checks to suppliers. Based on the information in the chapter, what kinds of controls would you implement to ensure that checks are written only to valid suppliers, that checks are written for the correct amount, and that all payouts have the required authorization? How would you design the controls if different payment amounts required different levels of authorization?

3. The executives of a company have asked for a special decision support system report on corporate financials. They want this report to be based on actual financial data for the past several years. The report is to have several input parameters so that the executives can do "what-if" analysis of future sales based on past performance. They would like the report to be viewable on-line and in printed form. What kinds of controls would you implement to ensure that (1) only authorized executives can request the report, (2) the executives understand the basis (past and projected data) for a given report, and (3) executives are aware of the sensitive nature of the information and treat it as confidential?

4. A payroll system has a data-entry subsystem that is used to enter time card information for hourly employees. What kinds of controls would you implement to ensure that the data are correct and error-free? What other controls would you include to ensure that a data-entry clerk (who may be a friend of an employee) not inflate the hours on the time card (after it was approved by a supervisor)?

5. Based on the DFD (Figure 10-26) given in Chapter 10, "Thinking Critically" problem 3, *Add class to schedule*, and the structure chart you developed there, identify the set of input and output screens for the system. Include the data fields that will be required.

6. Based on the DFD (Figure 10-27) given in Chapter 10, "Thinking Critically" problem 5, *Special-order purchasing*, and the structure chart you developed there, identify the set of inputs and outputs required. Develop the list of data fields for each screen and report.

7. A university library system is depicted in Figure 15-25, with partial system sequence diagrams for two use cases, *Check out a book* and *Return a book*. Based on the figure, construct four tables showing inputs and outputs as shown in Figures 15-10 and 15-12: (1) Inputs for the Library System, (2) Outputs for the Library System, (3) Inputs for the Student Record System, and (4) Outputs for the Student Record System.

8. You work for a grocery chain that always has many customers in the stores. To facilitate and speed checkout, the company would like to develop self-service checkout stands. Customers can check their own groceries and pay by credit card or cash. How would you design the checkout register and equipment? What kinds of equipment would you use to make it easy and intuitive for the customers, make sure that prices are entered correctly, and ensure that cash or credit-card payments are done correctly? In other words, what equipment would you have at the checkout station? In your solution, you may use existing state-of-the-art solutions or invent new devices.

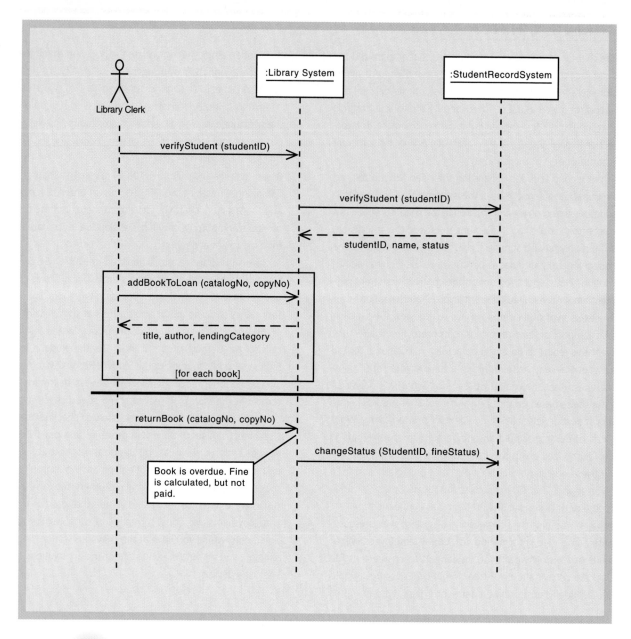

FIGURE *15-25*

Partial system sequence diagram
for the university library system.

EXPERIENTIAL EXERCISES

1. Look on the Web for an e-commerce site (for example, Amazon.com or eBay). Evaluate the effectiveness of the screens. What kind of security and controls are integrated into the system? Do you see potential problems with the integrity controls? Evaluate the design of the individual screens. How easy are they to read and use? What suggestions would make them easier to use? How effective are they in minimizing data-entry errors?

2. Examine the information system of a local business (a fast-food restaurant, doctor's office, video store, grocery store, and so forth). Evaluate the screens (and reports, if possible) for ease of use and effectiveness. What kind of integrity controls are in place? How easy are the screens to use? What kinds of improvements would you make?

3. Find and research a system that is being constructed or has recently been constructed. You may work for a company that has a development project in progress or have a friend who works for such a company. Another source of development projects is your university or college itself. Interview one of the developers. Ask about integrity controls, methodology for screen design, and guidelines to ensure consistency across the user interface. Ask about the number and scope of the input and output design tasks (for example, how many screens or hours required) and the method used to lay out the screens and reports (such as prototyping, CASE tools, and so forth).

4. If your university uses Java, find out about the JSwing class library. Write a one-page description of the JSwing library, its purpose, and ways to use it. Your objective is to demonstrate that you understand the concept of JSwing and the way it is used to build windows and input screens in a Windows environment.

5. If your university uses Studio .NET from Microsoft, find out about using the .NET class library to build user interfaces. Write a one-page description of the .NET library, its purpose, and ways to use it. Your objective is to demonstrate that you understand the concept of .NET forms design and the way it is used to build windows and input screens in a Windows environment.

6. Go to the Internet and find out what you can about Pretty Good Privacy. What is it? How does it work? Research what you can about a passphrase. What does it mean? Here are two sites that you can use to start your research: http://www.pgpi.org/ and http://web.mit.edu/network/pgp.html.

CASE STUDIES

All-Shop Superstores

All-Shop Superstores is a regional chain of superstores in the Boston, New York, and Washington, DC, corridor. These stores compete with other giants, such as Wal-Mart, Kmart, Target, and other budget retailers. The stores contain large grocery stores as well as domestics, clothing, automobile, and home improvement products. Overall, the margins in this portion of the retail industry are very small. Grocery profits have always been small, in the range of 5 to 10 percent. The margin for domestics, clothing, and other goods is a little higher, but to compete with Wal-Mart, All-Shop must keep all margins low.

To reduce operating costs as much as possible, All-Shop has decided to move very heavily into electronic data interchange (EDI) with its suppliers. All-Shop is aware that several of its more advanced competitors allow their suppliers to manage inventory levels in the stores themselves. For example, paper hygiene products such as disposable diapers and toilet paper are high-volume products that require very close monitoring of inventory levels. All-Shop has already installed sophisticated sales and inventory systems that track activity of each individual item (using the UPC code) daily. These systems not only capture daily activity but also maintain histories in a data warehouse to support on-line data analysis.

The first step for All-Shop was to enable its major suppliers to have access to its daily sales and inventory database. That way, the suppliers could monitor sales activities and check inventory to ensure that deliveries are made on time to maintain optimal inventory levels. The system should also permit each supplier to access and check the status of its individual accounts and a history of past payment activity. Obviously, All-Shop must control all of this information so that suppliers cannot observe each other's information.

1. Based on what you have learned in this and previous chapters, develop a use case diagram identifying the use cases that apply to the supplier as an actor. Even though this is really a system-to-system interface, the supplier system can be considered an actor. Identify two lists of controls that you consider necessary for this interface. On the first list, identify overall controls for the entire EDI interface. Then, for the second list, for each identified use case, develop a specific set of controls that will be necessary. Base your analysis on the types of controls discussed in the chapter as well as the three primary objectives of integrity controls. In other words, your assignment is to develop a statement of required controls that the system developers can use to ensure that the system adequately protects the assets and information of All-Shop.

2. All-Shop is considering a plan to provide supplier access to its data warehouse to enable executives to analyze past trends and help design promotions to increase overall sales and those of individual products. In other words, All-Shop is building partnerships with its suppliers to maximize its presence in the retail marketplace. One major concern of All-Shop executives is how to ensure that the suppliers treat this information with maximum security and not damage All-Shop. How can they ensure that the suppliers do not use this information to benefit All-Shop's competitors inadvertently, since suppliers also work with these competitors?

3. Do you think this second step is a wise move for All-Shop? If not, why not? If so, what kinds of controls and contractual arrangements should be made to protect All-Shop? You may see how a narrow focus on integrity controls may be inadequate to protect proprietary information. A broader view and understanding of controls and their objectives are required in this instance.

Real Estate Multiple Listing Service System

Based on the DFD fragments you developed in Chapter 6 and the structure charts from Chapter 10, develop a table of inputs along with the associated data couples and data fields for each input. Also, develop a table of outputs with the required data fields.

TheEyesHaveIt.com Book Exchange System

Based on the system sequence diagrams you developed in Chapter 7, develop a list of inputs and outputs required for this system. Also, identify any specific controls that may be necessary to ensure that information is entered accurately.

DownTown Videos Rental System

Using the system sequence diagrams you developed in Chapter 7, develop a list of inputs and outputs, along with the necessary data fields, for the system.

Rethinking Rocky Mountain Outfitters

The RMO event table lists six system reports that are part of the new system:

- Order summary
- Transaction summary
- Fulfillment summary
- Prospective customer activity
- Customer adjustments
- Catalog activity

For each of these six reports, answer the following questions:

1. Identify the data fields that each report should include.
2. What questions will users want each report to answer?
3. What type of report is it: detailed, summary, or exception?

4. How might graphics be used? What about drill-down capabilities?
5. How would you prepare a mock-up of each report, assuming a printed output and also an on-line output?
6. What output controls should be associated with each report?

Focusing on Reliable Pharmaceutical Service

One of the challenges of a pharmaceutical company is keeping current with new drugs and changes to existing drugs. New drugs are continually being developed and approved. In addition, generic drugs are often available to compete with brand-name drugs. One of the services that Reliable provides is to try to find the least expensive alternative to fulfill a prescription. This cost-saving service is one of the marketing advantages that the nursing homes can use to promote their services. Obviously, this service builds tremendous loyalty between Reliable and its customers.

To keep current with these changes, Reliable subscribes to an on-line drug-update service. The service provides updates in several formats, one of which is an XML file.

1. Based on the content of your design class diagrams that you developed in Chapter 11, illustrate a sample XML input file that could be used to update drug information in the Reliable database.
2. In earlier chapters, the case description indicated that a case manifest was produced for each patient whenever prescriptions needed to be filled and delivered. Based on the data found in your class diagrams, design a case manifest. Consider that a patient may have multiple prescriptions that are being filled on the same delivery.
3. Each month, Reliable produces a statement for each nursing home. The statement lists each patient who received prescriptions during the month. All the filled prescriptions are listed. For each prescription, the following information is listed: the price, the amount billed to the patient's insurance provider, the amount paid by the insurance provider, and the co-pay amounts due from the patients. Design this monthly statement. Also, identify and highlight output controls that you believe are appropriate for this type of report.
4. In the preceding chapter, you defined an input form to be used to collect orders from the nursing homes. Go back and analyze that input form and identify all of the input controls that you think are necessary to ensure that the prescriptions are correct. What other procedures or controls would you recommend to make sure that there are no mistakes on the prescriptions?

FURTHER RESOURCES

David Benyon, Diana Bental, and Thomas Green, *Conceptual Modeling for User Interface Development*. Springer-Verlag, 1999.

Elfriede Dustin, Jeff Rashka, Douglas McDiarmid, and Jakob Nielson, *Quality Web Systems: Performance, Security, and Usability*. Addison-Wesley, 2001.

Simson Garfinkel, Gene Spafford, and Debby Russell, *Web Security, Privacy, & Commerce*. O'Reilly Publishing, 2001.

Anup K. Ghosh, *E-Commerce Security: Weak Links, Best Defenses*. John Wiley & Sons, 1997.

IS Audit and Control Association, *IS Audit and Control Journal*, Volume I. 1995.

Brenda Laurel, *The Art of Human-Computer Interface Design*. Addison-Wesley, 1990.

Ben Shneiderman, *Designing the User Interface: Strategies for Effective Human-Computer Interaction*. Addison-Wesley-Longman, 1998.

Donald Warren Jr., and J. Donald Warren, *The Handbook of IT Auditing*. Warren Gorham & Lamont, 1998.

Donald A. Wayne and Peter B. B. Turney, *Auditing EDP Systems*. Prentice Hall, 1990.

Implementation and Support

CHAPTER 16

Making the System Operational

LEARNING OBJECTIVES

After reading this chapter, you should be able to:

- Describe implementation and support activities

- Choose an appropriate approach to program development

- Describe various types of software tests and explain how and why each is used

- List various approaches to data conversion and system installation and describe the advantages and disadvantages of each

- Describe different types of documentation and the processes by which they are developed and maintained

- Describe training and user support requirements for new and operational systems

CHAPTER OUTLINE

Program Development

Quality Assurance

Data Conversion

Installation

Documentation

Training and User Support

Maintenance and System Enhancement

It was 8:30 a.m. on Monday morning, and Maria Grasso, Kim Song, Dave Williams, and Rajiv Gupta were just about to begin the weekly project status review meeting. Tri-State Heating Oil had started developing a new customer order and service call scheduling system five months earlier. The target completion date was 10 weeks away, but the project was behind schedule. Analysis and design had taken eight weeks longer than anticipated because key users had disagreed on what new system requirements to include and the system scope was larger than expected.

Maria began the meeting by saying, "We've gained a day or two since our last meeting due to better-than-expected unit testing results. All of the methods developed last week sailed through unit testing, so we won't need any time this week to fix errors in that code."

Kim spoke, "I wouldn't get too cocky just yet. All of the nasty surprises in my last object-oriented project came during integration testing. We're completing the user-interface classes this week, so we should be able to start integration testing with the business classes sometime next week."

Dave nodded enthusiastically and said, "That's good! We have to finish testing those user-interface classes as quickly as possible because we're scheduled to start user training in three weeks. I need that time to develop the training materials and work out the final training schedule with the users."

Rajiv replied, "I'm not sure that we should be trying to meet our original training schedule with so much of the system still under development. What if integration testing shows major bugs that require more time to fix? And what about the unfinished business and database classes? Can we realistically start training with a system that's little more than a user interface with half a system behind it?"

Dave replied, "But we have to start training in three weeks. We contracted for a dozen temporary workers so that we could train our staff on the new system. Half of them are scheduled to start in two weeks and the rest two weeks after that. It's too late to renegotiate their start dates. We can extend the time they'll be here, but delaying their starting date means we'll be paying for people we aren't using."

Maria spoke up and said, "I think that Rajiv's concerns are valid. It's not realistic to start training in three weeks with so little of the system completed and tested. We're at least five weeks behind schedule, and there's no way we'll recapture more than four or five days of that during the next few weeks. I've already looked into re-arranging some of the remaining coding to give priority to the work most critical to user training. There are a few batch processes that can be back-burnered for awhile. Kim, can you rearrange your test plans to handle all of the interactive applications first?"

Kim replied, "I'll have to go back to my office and take another look at the dependencies among those programs. Offhand, I'd say yes, but I need a few hours to make sure."

Maria replied, "Okay, let's proceed under the assumption that we can rearrange coding and testing to complete a usable system for training in five weeks. I'll confirm that by e-mail later today as soon as Kim gets back to me. I'll also schedule a meeting with the CIO to deliver the bad news about temporary staffing costs."

After a few moments of silence, Rajiv asked, "So what else do we need to be thinking about?"

Maria replied, "Well, let's see. . . . There's user documentation, hardware delivery and setup, operating system and DBMS installation, importing data from the old database, the network upgrade, and stress testing for the distributed database accesses."

Rajiv smiled and said to Maria, "You must have been a juggler in your youth, and it was good practice for keeping all of these project pieces up in the air. Does management pay you by the ball?"

Maria chuckled and replied, "I do think of myself as a juggler sometimes. And if management paid me by the ball, I could retire as soon as this project is finished!"

OVERVIEW

This chapter focuses on the activities of the implementation and support phases of the systems development life cycle (SDLC). Activities that occur before the system is turned over to its users are collectively called *implementation*. Activities that occur after the system becomes operational are collectively called *support*. Figure 16-1 lists the activities of each phase.

Implementation and support activities are often considered straightforward and dull—they don't attract the same attention or enthusiasm as analysis and design activities. The situation is analogous to the difference between architecture and construction. An architect gets most of the credit for creating a new building, even though his or her job essentially ends with the blueprints. Yet, the vast majority of the effort that goes into

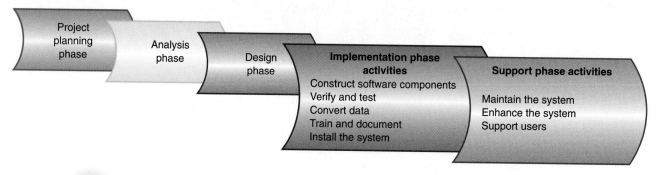

FIGURE *16-1*

Activities of the implementation and support phases.

making the building a reality occurs after the blueprints are finished. The same is true of information system development.

Implementation consumes substantially more time and resources than earlier phases of the SDLC. A large number of people are required to perform implementation activities—particularly software construction and testing. In addition, implementation activities are highly interdependent. Project management complexity is at its greatest during the implementation phase because so many people and activities must be coordinated. The RMO project progress memo illustrates the complexity of project management and the many tasks that must be completed.

July 10, 2005

To: John Blankens

From: Barbara Halifax

MEMO

RE: Customer support system implementation milestones

The implementation team met earlier this week to develop a master schedule. Completing the system by November will be difficult, but I think we have a workable plan. Application software programming and testing will be ongoing from now through early October. Key milestones and target dates for other tasks include:

September 1	Complete hardware and operating system installation
September 15	Complete DBMS installation
October 1	Complete database initialization
October 20	Complete performance testing and tuning
October 25	Begin user training
November 1	Begin parallel operation with existing systems
December 1	Terminate existing systems

The last date assumes that parallel operation uncovers no serious problems and that senior management signs off on final acceptance tests sometime during November. If necessary, we'll continue parallel operation through the holiday shopping season, though that would require substantial overtime or hiring more temporary staff.

The schedule is ambitious and leaves little room for mistakes. The development team will meet every Monday and Thursday morning at 8:00 to review recent test results and upcoming work. I'd like you to attend all of the Monday meetings, if possible, to monitor our progress and to underscore the strategic importance of the new system to the staff.

BH

Information systems are the lifeblood of a modern organization. Thus, supporting those systems is one of the most important jobs in an organization. Support activities ensure that the system and its users function efficiently and effectively for years after installation. Most organizations spend much more money maintaining and supporting existing systems than they do building new ones.

PROGRAM DEVELOPMENT

Developing a complex system is an inherently difficult process. Consider the complexity of manufacturing automobiles. Tens of thousands of parts must be fabricated or purchased. Parts are assembled into small subcomponents (such as dashboard instruments, wiring harnesses, and brake assemblies), which are in turn assembled into larger subcomponents (such as instrument clusters, engines, and transmissions) and eventually into a complete automobile. Parts and subcomponents must be constructed, tested, and passed on to subsequent assembly steps. There are tens or hundreds of thousands of individual production steps. The effort, timeliness, cost, and output quality of each step depend on all of the preceding steps.

Program development is similar in many ways to automobile manufacturing. Requirements and design specifications have already been determined. What remains is a complex production and assembly process that must ensure efficient resource use, minimal construction time, and maximum product quality. But unlike automobile manufacturing, the process is not designed once and then used to build thousands of similar units. Instead, a software manufacturing process must be redeveloped for each new project to match that project's unique characteristics.

When most people think of system development, they primarily think of programming. Programming isn't the only development phase activity, but it is clearly one of the most important. Its importance arises from several factors, including:

- Required resources
- Managerial complexity
- System quality

Program development consumes more resources than any other system development activity. Program development (including unit testing) typically accounts for at least one-third of all development labor. Program development also accounts for between one-third and one-half of the project development schedule. The magnitude of resources and time consumed during program development clearly warrants careful planning and management attention.

Order of Implementation

One of the most basic decisions to be made about program development is the order in which program components will be developed. Several orders are possible, including:

- Input, process, output
- Top-down
- Bottom-up

Each project must adapt one or a combination of these approaches to specific project requirements and constraints.

Input, Process, Output Development Order

The *input, process, output (IPO) development* order is based on data flow through a system or program. Programs or modules that obtain external input are developed first.

Programs or modules that process the input (that is, transform it into output) are developed next. Programs or modules that produce output are developed last.

For structured designs and programs, an analyst can determine IPO ordering by examining the system flowchart and structure charts. For example, consider the payroll system flowchart in Figure 16-2. The programs Maintain tax tables and Maintain employee database obtain and modify data inputs for other programs, so they would be the first programs implemented. The Payroll program combines input and processing, so it would be implemented next. The Check printing and Year-end tax programs produce system outputs, so they would be implemented last.

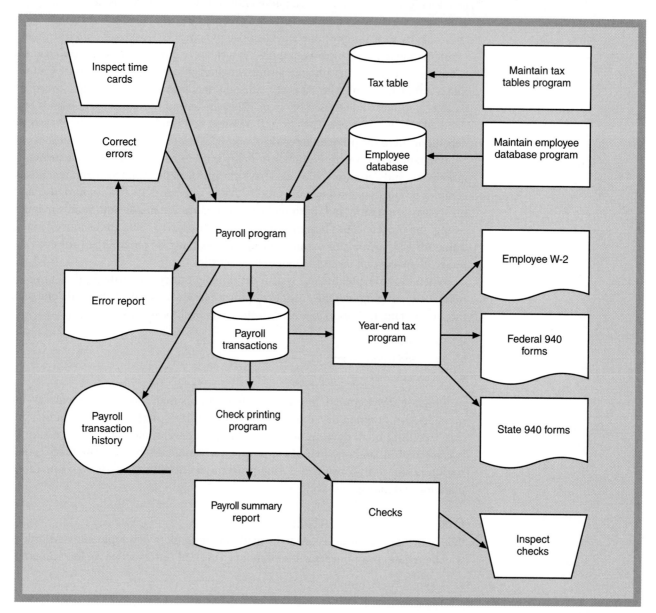

FIGURE *16-2*

A system flowchart for a payroll system.

Figure 16-3 shows a structure chart for the Payroll program. An analyst can apply IPO implementation order to modules within a program by classifying them as input, process, and output modules. If the analyst developed the structure chart using transform analysis, then modules will be clearly organized into afferent (input), central transform (process), and efferent (output) "legs." (See the section "Developing a Structure Chart" in Chapter 10 for a review of structure chart organization.) For this program, the modules in the afferent leg of the structure chart (those below and including Enter time

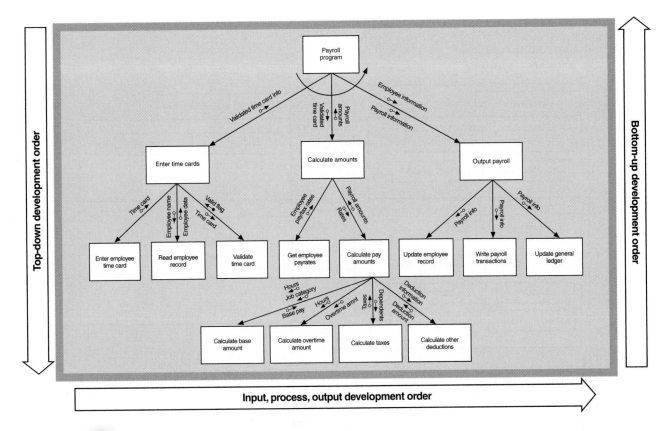

FIGURE *16-3*

A structure chart for the Payroll program in Figure 16-2.

cards) would be implemented first, followed by the modules in the transform leg (those below and including Calculate amounts), and finally the modules in the efferent leg (those below and including Output payroll).

IPO development order can also be applied to object-oriented (OO) designs and programs. The key issue to analyze is dependency—that is, which classes and methods capture or generate data that are needed by other classes or methods. Dependency information is documented in package diagrams and may also be documented in a class diagram. Thus, either or both diagram types can guide implementation order decisions.

For example, the package diagram in Figure 16-4 shows that the customer and catalog maintenance subsystems are not dependent on each other or on either of the other two subsystems. The order-entry subsystem is dependent on both the customer and catalog maintenance subsystems, and the order fulfillment subsystem is dependent on the order-entry subsystem.

Data dependency among the packages (subsystems) implies data dependency among their embedded classes. Thus, the classes Customer, Catalog, and Package have no data dependency on the remaining RMO classes. Under IPO development order, those three classes are implemented first.

The chief advantage of the IPO development order is that it simplifies testing. Because input programs and modules are developed first, they can be used to enter test data for process and output programs and modules. The need to write special-purpose programs to generate or create test data is reduced, thus speeding the development process.

IPO development order is also advantageous because important user interfaces (for example, data-entry routines) are developed early. User interfaces are more likely to require change during development than other portions of the system, so early development allows for early testing and user evaluation. If changes are needed, there is still plenty of time to make them. Early development of user interfaces also provides a head start for related activities such as training users and writing documentation.

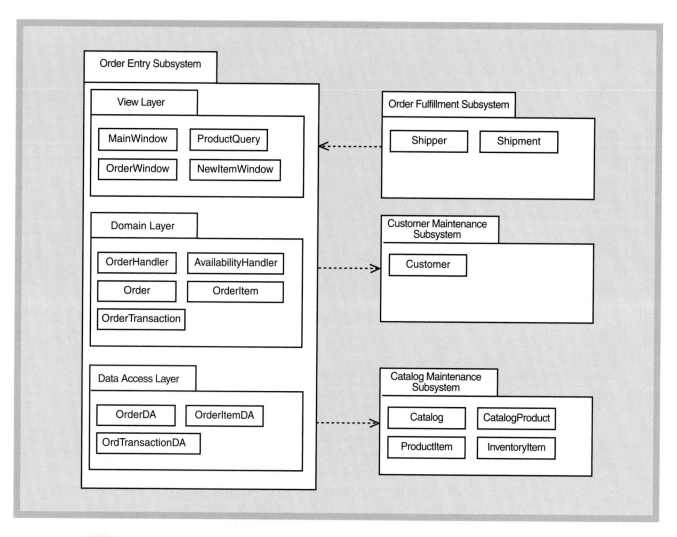

A disadvantage of IPO development order is the late implementation of outputs. Output programs are useful for testing process-oriented modules and programs; analysts can find errors in processing by manually examining printed reports or displayed outputs. IPO development defers such testing until late in the development phase. However, analysts can usually generate alternate test outputs by using the query processing or report writing capabilities of a database management system (DBMS). If such outputs can be quickly and easily defined, then the disadvantage of late implementation of output routines is substantially mitigated.

Top-down and Bottom-up Development Order

The terms *top-down* and *bottom-up* have their roots in traditional structured design and structured programming. Both terms describe the order of implementation with respect to a module's location within a structure chart. For example, consider the structure chart in Figure 16-3. *Top-down development* begins with the module at the top of the structure chart (Payroll program). *Bottom-up development* begins with the set of modules at the lowest level of the structure chart.

Top-down and bottom-up program development can also be applied to OO designs and programs, although a visual analogy is not as obvious with OO diagrams as with structure charts. The key issue is method dependency—that is, which methods call which other methods. Within an OO subsystem or class, method dependency can be examined in terms of navigation visibility, as discussed in Chapter 11.

top-down development

a development order that implements
modules at the top of a structure
chart first

bottom-up development

a development order that implements
modules at the bottom of a structure
chart first

For example, consider the three-layer design of part of the RMO order-entry subsystem shown in Figure 16-5. The arrows among packages and classes show navigation visibility requirements. Methods in the view layer call methods in the domain layer, which, in turn, call methods in the data access layer. Top-down implementation would implement the view layer classes and methods first, the domain layer classes and methods next, and the data access layer classes and methods last. Bottom-up implementation would reverse the top-down implementation order.

FIGURE *16-5*

A package diagram for a three-layer OO design.

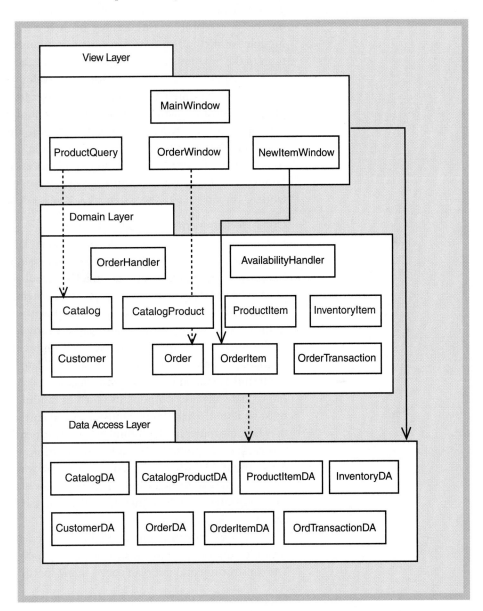

The primary advantage of top-down development is that there is always a working version of a program. For example, top-down development of the program in Figure 16-3 would begin with a complete version of the topmost module and dummy (or stub) versions of its three subordinate modules (stub modules are discussed later in the "Testing" section). This set of modules is a complete program that can be compiled, linked, and executed, although at this point it wouldn't do very much when executed. Top-down development of the three-layer design in Figure 16-5 would begin with a complete version of the view layer classes and dummy (or stub) versions of the domain layer classes.

Once the topmost module or layer is completed, development proceeds downward to the next level of the structure chart or package diagram. As each module or class is implemented, stubs for the modules or classes on the next lower level are added. At every stage of development, the program should be complete (that is, it should be able to be compiled, linked, and executed). Its behavior becomes more complex and realistic as development proceeds.

The primary disadvantage of top-down development order is that it doesn't use programming personnel very efficiently at the beginning of software development. Development has to proceed through two or three iterations before a significant number of modules or methods can be developed simultaneously. However, if the first few iterations of the program can be completed quickly, then the disadvantage is minimal.

The primary advantage of bottom-up development is that many programmers can be put to work immediately. In addition, lower-level modules are often the most complex and difficult to write, so early development of those modules allows more time for development and testing. Unfortunately, bottom-up development also requires writing a large number of driver programs to test bottom-level modules, which adds additional complexity to the development and testing process (this issue is discussed further in the "Testing" section). Also, the entire system isn't assembled until the topmost modules are written. Thus, testing of the system as a whole is delayed.

Other Development Order Considerations

IPO, top-down, and bottom-up development are only a starting point for creating a program development plan. Other factors that must be considered include user feedback, training, documentation, and testing. User feedback, training, and documentation all depend heavily on the user interfaces of the system. Early implementation of user interfaces enables user training and the development of user documentation to begin early in the development process. It also gathers early feedback on the quality and usability of the interface. Note that interface development is substantially advanced if user-interface prototypes were constructed and validated during analysis or design.

Testing is also an important consideration when determining development order. As individual software components (such as modules or methods) are constructed, they must be tested. Programmers must find and correct errors as soon as possible because they become much harder to find and more expensive to fix as the construction process proceeds. It's important both to identify portions of the software that are susceptible to errors and to identify portions of the software where errors can pose serious problems that affect the system as a whole. These portions of the software must be built and tested early regardless of where they fit within the basic approaches of IPO, top-down, or bottom-up development.

Testing and construction are highly interdependent. For this reason, a formal plan covering both testing and construction is normally created before either activity begins. The construction and test plan covers many specifics, including:

- Development order
- Testing order
- Data used to test modules, module groups, methods, classes, programs, and subsystems
- Acceptance criteria
- Personnel assignments (construction and testing)

Testing is discussed in detail later in this chapter. But for now, keep in mind that construction and testing go hand in hand. Their interdependence and complexity necessitate formal planning and regular comparisons between the plan and actual performance.

Framework Development

When implementing a large OO system, it is not unusual to build an object framework (or set of foundation classes) that covers most or all of the domain and data access layer classes. For example, when implementing an OO account maintenance system for a bank, developers might build a set of classes to represent and store customers and various types of bank accounts (for example, savings accounts, checking accounts, and certificates of deposit).

Foundation classes are typically reused in many parts of the system and across many different applications. Because of this reuse, they are a critical system component. Errors in a foundation class can affect every program in the system. In addition, later changes to foundation classes may require significant changes throughout the system.

Foundation classes are typically implemented first to minimize the impact of errors and changes. They are typically assigned to the best programmers and are tested more thoroughly than other classes. Early and thorough testing guarantees that bugs or other problems will be discovered before other code that depends on the foundation classes has been written.

Team-Based Program Development

A team of programmers normally works on program development. Using multiple programmers compresses the development schedule by allowing many portions of the system to be developed simultaneously. However, team-based program development introduces its own set of management issues, including:

- Organization of programming teams
- Task assignment to specific teams or members
- Member and team communication and coordination

There are many different ways of organizing an implementation team. Some commonly used organizational models include the following:

- Cooperating peer
- Chief developer
- Collaborative specialist

Figure 16-6 summarizes the characteristics of each team type, and the types of projects and tasks best suited to each.

FIGURE *16-6*

A comparison and summary of development team types.

Team type	Team characteristics	Task and project types
Cooperating peers	Equal skill levels Overlapping specialities Consensus-based decision making	Experimentation Creative problem solving
Chief developer	Organized as a military platoon or squad One leader makes all important decisions	Well-defined objectives Well-defined path to completion
Collaborative specialists	Wide variation in skill and experience Minimal overlap in technical specialities Leader is primarily an administrator Consensus-based decision making	Diagnosis or experimentation Creative and integrative problem solving Wide range of technology

A *cooperating peer team* includes members of roughly equal skill and experience with overlapping areas of specialization. Members are considered equals, although they may be assigned tasks of varying importance or complexity. Decisions are primarily made by consensus, and the team frequently meets to exchange information and build consensus.

A *chief developer team* is similar to a small military unit. An assigned leader performs a number of functions, including technical consulting, team coordination, and task assignment. In this type of team, there is much less communication than with a cooperating peer team. The chief developer makes most of the important decisions, although he or she may seek input from members individually or collectively.

A *collaborative specialist team* is similar to a cooperating peer team, but its members have wide variation in and minimal overlap of skills and experience. Such teams are often composed of members from different organizational subunits. The team may have an appointed leader, but his or her leadership covers only administrative functions such as scheduling, coordinating, and interacting with external constituencies. Technical decisions are generally made by consensus, although member opinions usually carry extra weight within the member's own area of expertise. In large projects, a collaborative specialist team may be formed to "float" among other teams to deal with complex problems as they arise.

Some common principles of team organization underlie all development projects and organizational structures. One is that team size should be kept relatively small (no more than 10 members). Larger teams tend to be inefficient because of the inherent complexity of communication and coordination in large groups. When more than 10 developers are assigned to a project, it is best to break them up into small teams (approximately five members each). Each team should be assigned a relatively independent portion of the project. One member of each team should be designated to handle coordination and communication with other teams. Having a single point of contact simplifies communication and provides for some specialization of functions within each team.

Another common principle of team organization is that team structure should be matched to the task and project characteristics. Teams with a well-defined implementation task that does not push the limits of member knowledge or technical feasibility are usually best organized as chief programmer teams. A chief programmer team operates very efficiently in such an environment.

Teams assigned to tasks that require experimentation or a high level of creativity are better served by a cooperating peer or collaborative specialist model. Because of their more open communication, cooperating peer teams are especially well suited to tackling tasks that require generating and evaluating a large number of ideas. Overlap in skill, specialty, and experience allows thorough evaluation of each idea.

Collaborative specialist teams are well suited to projects that span a wide range of cutting-edge technology. They are also well suited to tackling projects that require integrated problem solving (for example, diagnosing and fixing bugs in an existing complex system). However, success depends on a true collaborative process, which is sometimes difficult to achieve among members with wide variation in skill and experience.

Member skills must be appropriately matched to the tasks at hand. Skill matching is a fairly obvious requirement with respect to technical skills such as database management, user interfaces, and numeric algorithms. Skill matching is less obvious but no less important for nontechnical skills. Teams need a mix of nontechnical skills and traits, including the ability to generate new ideas, build consensus, manage details, and communicate with external constituencies. The project manager should perform a skills inventory early in the project so that gaps can be filled to avoid project delays and inappropriate personnel assignments.

Source Code Control

source code control
system (SCCS)

an automated tool for tracking source
code files and controlling changes to
those files

Development teams need tools to help coordinate their programming tasks. A *source code control system (SCCS)* is an automated tool for tracking source code files and controlling changes to those files. An SCCS stores project source code files in a repository. The SCCS acts the way a librarian would—it implements check-in and checkout procedures, tracks which programmer has which files, and ensures that only authorized users have access to the repository.

Programmers can manipulate files in the repository as follows:

- Check out a file in read-only mode
- Check out a file in read/write mode
- Check in a modified file

A programmer checks out a file in read-only mode when he or she wants to examine the code without making changes (for example, to examine a module's interfaces to other modules). When a programmer needs to make changes to a file, he or she checks out the file in read/write mode. The SCCS allows only one programmer to check out a file in read/write mode. The file must be checked back in before another programmer can check it out in read/write mode.

Figure 16-7 shows the main display of Microsoft Visual SourceSafe. Various source code files from the RMO customer support system are shown in the display. Some files are currently checked out by programmers. For each file checked out in read/write mode, the program lists the programmer who checked it out, the date and time of checkout, and the current location of the file. The icon for each checked-out file is displayed with a red border and check mark.

FIGURE *16-7*

Project files managed by a source code control system.

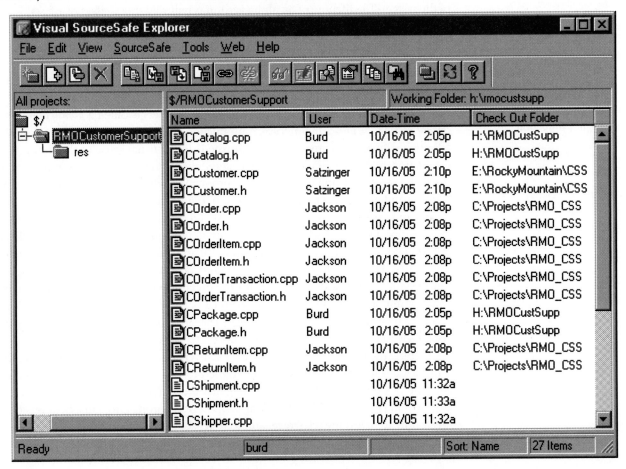

An SCCS prevents multiple programmers from updating the same file at the same time, thus preventing inconsistent changes to the source code. Source code control is an absolute necessity when programs are developed by multiple programmers. It prevents inconsistent changes and automates coordination among programmers and teams. The repository also serves as a common facility for backup and recovery operations.

Versioning

Medium- and large-scale systems are complex and constantly changing. Changes occur rapidly during implementation and more slowly afterward. System complexity and rapid change create a host of management problems—particularly for testing and support. Testing is problematic in such an environment because the system is a moving target. By the time an error is discovered, the code that caused the error may already have been moved, altered, or deleted. Support is complex for similar reasons. Support personnel need to know the state of the system as it is installed on a user's computer system to respond properly to a bug report or request for help.

Complex systems are developed, installed, and maintained in a series of versions to simplify testing and support. It is not unusual to have multiple versions of a system deployed to end users and more versions in different stages of development. A system version created during development is called a *test version*. A test version contains a well-defined set of features and represents a concrete step toward final completion of the system. Test versions provide a static system snapshot and a checkpoint to evaluate the project's progress.

An ***alpha version*** is a test version that is incomplete but ready for some level of rigorous testing. Multiple alpha versions may be built depending on the size and complexity of the system. The lifetime of an alpha version is typically short—days or weeks.

A ***beta version*** is a test version that is stable enough to be tested by end users. A beta version is produced after one or more alpha versions have been tested and known problems have been corrected. End users test beta versions by using them to do real work. Thus, beta versions must be more complete and less prone to disastrous failures than alpha versions. Beta versions are typically tested over a period of weeks or months.

A system version created for long-term release to users is called a ***production version, release version,*** or ***production release.*** A production version is considered a final product, although software systems are rarely "finished" in the usual sense of that term. Minor production releases (sometimes called ***maintenance releases***) provide bug fixes and small changes to existing features. Major production releases add significant new functionality and may be the result of rewriting an older release from the ground up.

Figure 16-8 shows a series of possible test and production versions for the RMO customer support system. Each version is described in Figure 16-9. The system is delivered in two major production releases—versions 1.0 and 2.0. Each initial production release is preceded by one or more alpha and beta test versions. Each version adds or updates functionality and includes bug fixes for the previous version. Version 1.1 is a maintenance or minor production release of version 1.0. Note that the time line for developing version 2.0 overlaps maintenance changes to version 1.0. Overlapping older production versions with test versions of future production releases is typical.

Keeping track of versions is complex. Each version needs to be uniquely identified for users and testers. In applications designed to run under Windows, users typically view the version information by choosing the About item of the standard Help menu, as shown in Figure 16-10. Users seeking support or reporting errors in a beta version use this feature to report the system version to testers or support personnel.

---

alpha version

a system that is incomplete but ready for some level of rigorous testing

beta version

a system that is stable enough to be tested by end users

production version, release version, or production release

a system that is formally distributed to users or made operational

maintenance release

a system update that provides bug fixes and small changes to existing features

FIGURE *16-8*

A time line of test and production versions for the RMO customer support system.

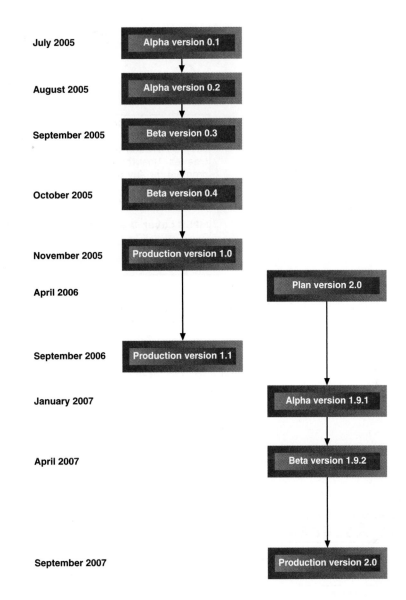

July 2005	Alpha version 0.1
August 2005	Alpha version 0.2
September 2005	Beta version 0.3
October 2005	Beta version 0.4
November 2005	Production version 1.0
April 2006	Plan version 2.0
September 2006	Production version 1.1
January 2007	Alpha version 1.9.1
April 2007	Beta version 1.9.2
September 2007	Production version 2.0

FIGURE *16-9*

Description of versions in Figure 16-8 for the RMO customer support system.

Alpha 0.1—Basic database functionality with simple CRUD capabilities. Handles regular transactions only, no reports or printing capability.

Alpha 0.2—Full CRUD for all transaction types with screens in near final form, no reports or printing capability. Includes bug fixes for version 0.1.

Beta 0.3—Screens in final form with simple on-line help, simple printing of screen contents. Includes bug fixes for version 0.2.

Beta 0.4—Adds reports and formatted printing. Includes bug fixes for version 0.3.

Production 1.0—Includes all bug fixes for version 0.4.

Production 1.1—Adds keystroke shortcuts for experienced users. Includes all bug fixes for version 1.0.

Alpha 1.9.1—Adds simple database extraction into a DSS tool for version 1.0.

Beta 1.9.2—Adds user-friendly database navigation and downloads into a DSS tool to version 1.1. Includes all bug fixes for version 1.9.1.

Production 2.0—Includes all bug fixes for version 1.9.2.

FIGURE *16-10*

The About box of a typical Windows application.

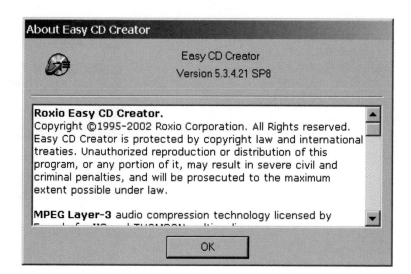

Controlling multiple versions of the same system requires sophisticated version control software. Version control capabilities are normally built into an SCCS. Programmers and support personnel can extract the current version or any previous version for execution, testing, or modification. Modifications are saved under a new version number to protect the accuracy of the historical snapshot.

Beta and production versions must be stored as long as they are installed on any user machines. Stored versions are used to evaluate future bug reports. For example, when a user reports a bug in version 1.0, support personnel will extract that release from the archive, install it, and attempt to replicate the user's error. Feedback provided to the user will be specific to version 1.0 even if the most recent production release is a higher-numbered version.

QUALITY ASSURANCE

quality assurance (QA)

the process of ensuring that an information system meets minimal quality standards

As with any business procedure or system, quality is a major concern with information systems. *Quality assurance (QA)* is the process of ensuring that an information system meets minimal quality standards as determined by users, implementation staff, and management. QA is sometimes equated with finding bugs in program code, but this view is narrow and incomplete. QA is a set of activities that are performed throughout the SDLC to build systems correctly from the start or to detect and fix errors as soon as possible. Integrating quality assurance into early project phases allows many programming errors to be completely avoided. It also ensures that the system that is actually developed meets the needs of the users and the organization.

QA activities during analysis concentrate on identifying gaps or inconsistencies in system requirements. QA activities during design concentrate on satisfying stated requirements and on making design decisions that will lead to easily implemented, bug-free programs. QA activities during implementation consist primarily of testing. However, design and implementation overlap in many projects. Thus, quality assurance activities for design are typically integrated with testing activities.

QA activities are often shortchanged during design and especially during implementation. This lapse occurs for several reasons, including the following:

- Schedule pressures can build with each successive phase. QA and testing activities may be bypassed in an ill-fated attempt to speed up the project.
- QA activities require development personnel to open their work to thorough examination and criticism by others. Many people are reluctant to do this.

■ Many people view testing and test personnel as the bearers of bad news. They mistakenly believe that no news is good news and that bypassing testing is a way of avoiding bad news.

There is a simple way to prevent human nature and schedule pressure from derailing QA activities: Formally integrate QA into the project and schedule from the beginning and never abandon it. Quality standards should be clearly stated, be measurable, and require a product that doesn't meet those standards to be fixed no matter what the effect on the schedule or budget. This approach to QA requires an organizational commitment from top management to the lowest levels. Unfortunately, top management is often the source of pressure to short-change QA activities to speed up a project.

Another key factor in firmly establishing QA in the development process is to build an environment of openness, collegiality, and mutual respect among project participants. Personnel must be receptive to suggestions and constructive criticism and be willing to provide suggestions and criticisms to others. QA cannot be effective if it is allowed to devolve into an exercise in destructive criticism, finger-pointing, and blame assignment.

The cost of fixing an error rises during each development phase. Errors are best detected during analysis or design and, thus, never committed to program code. Errors in programs are much easier to fix in the early stages of implementation than during later acceptance testing or, in the worst case, after the system is operational. This economic reality makes QA efforts throughout the development life cycle well worth their cost.

Technical Reviews

Most programmers have had many experiences in which they were unable to correct an error because they "couldn't see it." But when the source code containing the error is shown to another programmer, the other programmer spots it immediately. Common examples include misspelled keywords, malformed if statement conditions, and illegal or spurious characters in source code. Such errors happen to programmers at all skill and experience levels.

technical review
a formal or informal review of design or construction details by a group of developers

A *technical review* opens the design and implementation process to input from other people. Technical reviews provide an opportunity for other people to find problems and offer constructive criticism. Technical reviews vary widely from one organization to another and sometimes among projects within an organization. Some organizations use informal processes, while others adopt formal procedures.

A *walkthrough* is a review by two or more people of the accuracy and completeness of a model or program. Walkthroughs are most often used during analysis and design, although they can be used during implementation. During design and implementation, a walkthrough is a technical review in which two or more developers review a design or implementation to assess and improve its quality. Typically, one of the participants has already created a model or module before the walkthrough. The developer describes its underlying assumptions and operation, and the other participants provide comments and suggestions.

inspection
a formal review of design or construction details by a group of developers, where each person plays a specific role

An *inspection* is a more formal version of a walkthrough. Participants review and analyze materials before they meet as a group. Review materials include the model or code to be inspected, related models (for example, a structure chart or class diagram), and notes on specific types of errors that could occur. Group meetings usually follow a standard format.

When the group meets, participants play specific roles, including presenter, critic, and secretary. The presenter (usually the developer of the model or code) summarizes the material being inspected. The critics describe errors or concerns they found before the meeting, and the errors are discussed by all members of the group. Additional errors or problems may be uncovered during the discussion. The participants discuss possible solution strategies and agree on a specific approach. The secretary records all of the errors and the agreed-upon solution strategies.

Walkthroughs and inspections are important QA processes because they can detect errors *before* code has been written. Studies have shown that technical reviews accomplish the following:

- Reduce the number of errors that reach testing by a factor of 5 to 10
- Reduce testing costs by approximately 50 percent

Technical reviews reduce development costs and shorten the development schedule because a large number of errors never are passed along to be coded, tested, diagnosed, or fixed.

Testing and technical reviews each find between 50 and 75 percent of errors. But some errors are more easily detected by one method or the other. Some errors are rarely found by one technique but are easily found by the other. Thus, the two techniques are more effective jointly than individually.

Testing

Testing is the process of examining a product to determine what defects it contains. To conduct a test, programmers must have already constructed the software and have in hand well-defined standards for what constitutes a defect. The developers can test products by reviewing their construction and composition or by exercising their function and examining the results. This section concentrates on the latter type of testing. This process is shown in Figure 16-11.

FIGURE *16-11*

A generic model of software testing.

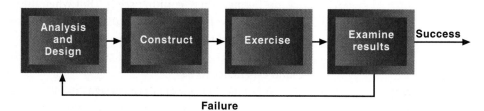

Software Testing

An information system is an integrated collection of software components. Components can be tested individually or in groups, or the entire system can be tested as a whole. Testing components individually is called *unit testing*. Testing components in groups is called *integration testing*. Testing entire systems is called *system testing*. Each type of testing is described in detail later in this section.

The three testing types are each correlated to a specific phase of the SDLC, as shown in Figure 16-12. A system test examines the behavior of an entire system with respect to technical and user requirements. These requirements are determined during the analysis phase of the SDLC. During high-level design, the division of the system into high-level components and the structural design of those components are determined. Integration testing tests the behavior of related groups of software components. Low-level design is concerned with the internal construction of individual components. Unit testing tests each individual software component in isolation.

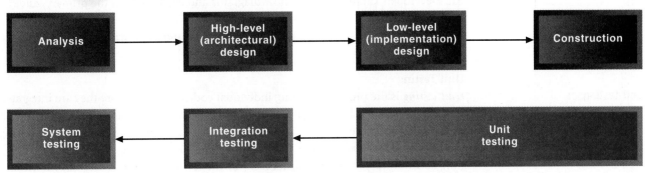

Because each testing level is related to a specific phase of the SDLC, testing activities can be spread throughout the life cycle, as illustrated in Figure 16-13. Planning for each type of testing can occur during its related SDLC phase, and development of specific tests can occur once the planning is complete. Tests cannot be conducted until relevant portions of the system have been constructed, however.

FIGURE *16-13* SDLC phases and the testing activities that can be performed within each phase.

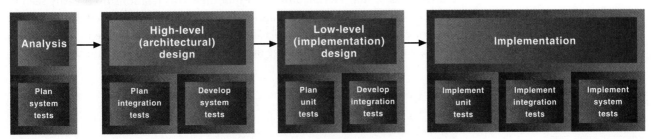

test case

a formal description of a starting state, one or more events to which the software must respond, and the expected response or ending state

test data

a set of starting states and events used to test a module, group of modules, or entire system

An important part of developing tests is specifying test cases and data. A **test case** is a formal description of the following:

- A starting state
- One or more events to which the software must respond
- The expected response or ending state

Both starting state and events are represented by a set of **test data.**

For example, the starting state of a system may represent a particular set of database contents (such as the existence of a particular customer and order for that customer). The event may be represented by a set of input data items (such as a customer account number and order number used to query order status). The expected response may be a described behavior (such as the display of certain information) or a specific state of stored data (such as a canceled order).

Preparing test cases and data is a tedious and time-consuming process. At the program or module level, every instruction must be executed at least once. Ensuring that all instructions are executed during testing is a complex problem. Fortunately, automated tools based on proven mathematical techniques are available to generate a complete set of test cases. See the Watson and McCabe article in the "Further Resources" section for a thorough discussion of this topic.

Analysis phase documentation is useful when preparing test cases. If the system was analyzed and designed with OO techniques, developers prepare test cases for each use case and scenario. Many test cases representing both normal and exceptional processing situations should be prepared for each scenario.

The correspondence between traditional analysis models and test cases is less clearcut. Developers can use the data flow diagrams and event table as the primary guide to prepare test cases. Developers should prepare multiple test cases for each event, and every process on every detailed data flow diagram should be exercised by at least one test case.

Unit Testing

Unit testing is the process of testing individual code modules before they are integrated with other modules. Unit testing is sometimes called *module testing*, although that term implies that software units are structured programming modules. In fact, unit testing can be applied to structured or OO software, and the unit being tested may be a function, subroutine, procedure, or method (for the remainder of this section, we will use the term *module* to refer to any of these programming constructs). Units may also be relatively small groups of interrelated modules that are always executed as a group. The goal of unit testing is to identify and fix as many errors as possible before modules are combined into larger software units (such as programs, classes, and subsystems). Errors become much more difficult and expensive to locate and fix when many modules are combined.

Few modules are designed to operate in isolation. Instead, groups of modules are designed to execute as an integrated whole. Modules may call other software units to perform tasks or may be called by other modules. This relationship is easily seen in a structure chart (see Figure 16-14), although it also exists among methods in OO software. For example, the module *Calculate pay amount* is called by the module *Calculate amounts*, which in turn calls the three modules below it in the structure chart.

**unit testing, or
module testing**

testing of individual code modules or
methods before they are integrated
with other modules

FIGURE *16-14*

A portion of a structure chart for a
program to calculate payroll.

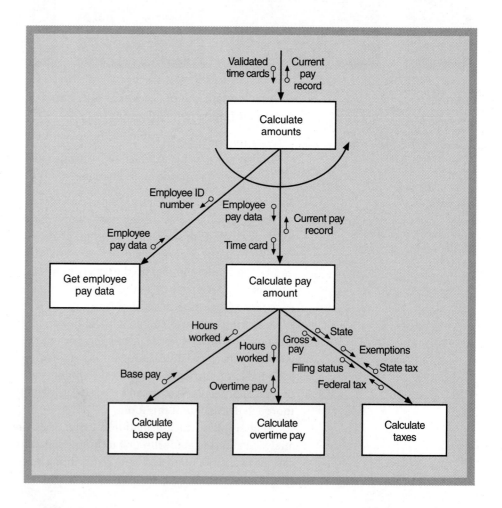

driver

a module, developed for unit testing, that simulates the calling behavior of a module that hasn't yet been developed

If *Calculate pay amount* is being tested in isolation, then two types of testing modules are required. The first module type is called a *driver*. A ***driver*** simulates the calling behavior of a module. A driver module implements the following functions:

- Sets the value of input parameters
- Calls the tested module, passing it the input parameters
- Accepts return parameters from the tested module and prints or displays them

Figure 16-15 shows a simple driver module for testing *Calculate pay amount*. A more complex driver module might use test data consisting of hundreds or thousands of module inputs and correct outputs stored in a file or database. The driver would loop through the test data and repeatedly call *Calculate pay amount*, check the return parameter against the expected value, and print or display warnings of any discrepancy.

FIGURE 16-15

A driver module for testing *Calculate pay amount.*

```
module main()
// Driver Module to Test CalculatePayAmount()
{
   //  Declare Module Parameters

   record EmployeePayData    {
      integer EmployeeIDNumber;
      boolean SalariedEmployee;
      real PayRate;
      char[2] State;
      integer FilingStatus;
      integer Exemptions;
   }
   record TimeCard    {
      integer EmployeeIDNumber;
      date StartDate;
      array[7] of real HoursWorked;
   }
   record CurrentPayRecord    {
      real BasePay;
      real OvertimePay;
      real FederalTax;
      real StateTax;
   }

   // Set Input Parameter Values

   EmployeeData.EmployeeNumber=123456789;
   EmployeeData.SalariedEmployee=false;
   EmployeeData.PayRate=32.50;
   EmployeeData.State="AZ";
   EmployeeData.FillingStatus=1;
   EmployeeData.Exemptions=5;
   TimeCard.EmployeeIDNumber=123456789;
   TimeCard.StartDate=05/21/2005;
   TimeCard.HoursWorked[0]=0.0;
   TimeCard.HoursWorked[1]=0.0;
   TimeCard.HoursWorked[2]=8.5;
   TimeCard.HoursWorked[3]=7.5;
   TimeCard.HoursWorked[4]=8.0;
   TimeCard.HoursWorked[5]=8.0;
   TimeCard.HoursWorked[6]=9.0;

   // Call Tested Module

   call CalculatePayAmount (EmployeeData, TimeCard, CurrentPayRecord);

   // Print Results

   print(EmployeeData, TimeCard, CurrentPayRecord);
}
```

Using a driver allows a subordinate module to be tested before modules that call it have been written. Drivers are used extensively in bottom-up development since child modules (or methods) are developed and unit-tested before their parents are developed.

The second type of testing module used to perform unit tests is called a *stub*. A *stub* simulates the behavior of a called module that hasn't yet been written. A unit test of *Calculate pay amount* would require three stub modules, one for each of the modules that appear below it in Figure 16-14. Stubs are relatively simple modules that usually have only one or two lines of executable code. Each of the stubs used to test *Calculate pay amount* can be implemented as a statement that simply returns a constant regardless of the parameters passed as input. Figure 16-16 shows sample code for each of the three stub modules.

stub

a module, developed for testing, that simulates the execution or behavior of a module that hasn't yet been developed

```
Module CalculateBasePay(HoursWorked,BasePay)

// Stub Module

array[7] of real HoursWorked;
real BasePay;
{
    BasePay=1000.00;
    return;
}

Module CalculateOvertimePay(HoursWorked,OvertimePay)

// Stub Module

array[7] of real HoursWorked;
real OvertimePay;
{
    OvertimePay=125.00;
    return;
}

Module CalculateTaxes(GrossPay,FilingStatus,State,Exemptions,FederalTax,StateTax)

// Stub Module

real GrossPay;
integer FilingStatus;
char[2] State;
integer Exemptions;
real FederalTax;
real StateTax;
{
    FederalTax=275.00;
    StateTax=75.00;
    return;
}
```

FIGURE *16-16*

Stub modules used for testing *Calculate pay amount.*

integration test

a test of the behavior of a group of modules or methods

Stubs are needed for top-down development. In fact, top-down development often begins by writing a stub for every module or method in a program or class. Individual stub modules and methods are then replaced with fully implemented code as it is developed.

Integration Testing

An *integration test* tests the behavior of a group of modules or methods. The purpose of an integration test is to identify errors that were not or could not be detected by unit-testing individual modules or methods. Such errors may result from a number of problems, including:

■ **Interface incompatibility.** For example, a caller module passes a variable of the wrong data type to a subordinate module.

- **Parameter values.** A module is passed or returns a value that was unexpected (such as a negative number for a price).
- **Run-time exceptions.** A module generates an error such as "out of memory" or "file already in use" due to conflicting resource needs.
- **Unexpected state interactions.** The states of two or more modules interact to cause complex failures (such as an order class method that operates correctly for all possible customer object states except one).

These are some of the most common integration testing errors, but there are many other possible errors and causes.

Once an integration error has been detected, the responsibility for incorrect behavior must be traced to a specific module or modules. The person responsible for performing the integration test is generally also responsible for identifying the cause of the error. Once the error has been traced to a particular module, the programmer who wrote the module is asked to rewrite it to correct the error.

Integration testing of structured software is straightforward but not necessarily easy to implement. Most structured modules are called by only a single parent module. In addition, most structured modules do not store a permanent state within themselves. Internal variables are always reinitialized to the same values each time the module is called. The combination of these characteristics allows test personnel (often with the assistance of automated testing tools) to generate test cases and data that exercise all possible control paths through the software being tested. Confidence that testing has revealed important errors increases with the number of control paths that are tested.

In contrast, integration testing of OO software is much more complex and not as well understood. There is no clear hierarchical structure to an OO program. An OO program consists of a set of interacting objects that can be created or destroyed during execution. Object interactions and control flow are dynamic and complex.

Additional factors that complicate OO integration testing include the following:

- Methods can (and usually are) called by many other methods, and the calling methods may be distributed across many classes.
- Classes may inherit methods and state variables from other classes.
- The specific method to be called is dynamically determined at run time based on the number and type of message parameters.
- Objects can retain internal variable values (that is, the object state) between calls. The response to two identical calls may be different due to state changes that result from the first call or occur between calls.

The combination of these factors makes it difficult to determine an optimal testing order. The factors also make it difficult to predict the behavior of a group of interacting methods and objects. Thus, developing and executing an integration test plan for OO software are much more complex than for structured software. Specific methods and techniques for dealing with that complexity are well beyond the scope of this textbook. See the "Further Resources" for OO software testing references.

System Testing

system test

a test of the behavior of an entire system or independent subsystem

A *system test* tests the behavior of an entire system or independent subsystem. System testing is normally first performed by developers or test personnel to ensure that the system does not malfunction in obvious ways and that the system fulfills the developers' understanding of user requirements. Later testing by users confirms whether the system does indeed fulfill their requirements. If a system is developed in many iterations, system testing is usually performed at the end of each iteration to identify significant issues such as performance problems that will need to be dealt with in the next iteration.

build and smoke test

a system test that is performed daily

A *build and smoke test* is a system test that is typically performed daily. The system is completely compiled and linked (built), and a battery of tests is performed to see whether anything malfunctions in an obvious way ("smokes"). Build and smoke tests are commonly associated with iterative or rapid development. However, build and smoke tests can also be used in more traditional projects if top-down development is employed.

Build and smoke tests are valuable because they provide rapid feedback regarding significant problems. Any problem that occurs during a build and smoke test must be the result of code modified or added since the previous test. Daily testing ensures that errors are found quickly and that they can be easily tracked to their source. Less frequent testing provides rapidly diminishing benefits because more code has changed and errors are more difficult to track to their source.

performance test

a system test that determines whether a system can meet time-based performance criteria

response time

the desired or maximum allowable time limit for software response to a query or update

throughput

the desired or minimum number of queries and transactions that must be processed per minute or hour

A *performance test* is a system test that determines whether a system can meet time-based performance criteria such as response time or throughput. *Response time* requirements specify desired or maximum allowable time limits for software responses to queries and updates. *Throughput* requirements specify the desired or minimum number of queries and transactions that must be processed per minute or hour.

Performance testing may be performed as part of unit or integration testing, but it is more commonly performed during system testing. Performance tests are complex because they can involve multiple programs, subsystems, computer systems, and network infrastructure. They require a large suite of test data to simulate system operation under normal or maximum load. Diagnosing and correcting performance test failures are also complex. Bottlenecks and underperforming components must first be identified. Corrective actions may include application software tuning or reimplementation, hardware or system software reconfiguration, and upgrade or replacement of underperforming components.

acceptance test

a system test performed to determine whether the system fulfills user requirements

An *acceptance test* is a system test performed to determine whether the system fulfills user requirements. Acceptance testing is typically the last round of testing before a system is handed over to its users. Acceptance testing is a very formal activity in most development projects. Details of acceptance tests are sometimes included in the request for proposal (RFP) and procurement contract when a new system is built by or purchased from an external party.

Who Tests Software?

There are many participants in the testing process. Their exact number and role depend on the size of the project and other project characteristics. Specific participants include these people:

- Programmers
- Users
- Quality assurance personnel

Programmers are generally responsible for unit-testing their own code prior to integrating it with modules written by other programmers. In some organizations, programmers are assigned a *testing buddy* to help them test their own code. The name derives from programmers who are assigned the specific responsibility for testing their buddy's code prior to integration testing. Having a different programmer test the code usually results in more errors being found.

testing buddy

a programmer assigned to test code written by another programmer

Users are primarily responsible for beta testing and acceptance testing. When beta versions are developed, they are distributed to a group of users for testing over a period of days, weeks, or months. Volunteers are frequently used, although they are not always desirable since they tend to be more computer literate and have a higher tolerance for malfunctions than ordinary users. These characteristics may result in higher-quality feedback for some problems but a lack of feedback for other problems.

Acceptance testing is normally conducted by users with assistance from IS development or operations personnel. The rigor and importance of acceptance tests require participation by a large number of users across a wide range of user levels (for example, data-entry clerks and the managers who will "own" the system). Although IS personnel may perform system setup and troubleshooting functions, it is ultimately up to the users to accept or reject the system.

In a large system development project, a separate quality assurance group or organization is usually formed. The QA group is responsible for all aspects of testing except unit testing and acceptance testing. The QA group's responsibilities and activities typically include the following:

- Developing a testing plan
- Performing integration and system testing
- Gathering and organizing user feedback on alpha and beta software versions and identifying needed changes to the system design or implementation

To maintain objectivity and independence, the QA group normally reports directly to the project manager or to a permanent IS manager.

DATA CONVERSION

An operational system requires a fully populated database to support ongoing processing. For example, the RMO order-entry subsystem relies on stored information about catalogs, products, customers, and previous orders. Implementation staff must ensure that such information is present in the database at the moment the subsystem becomes operational.

Data needed at system startup can be obtained from the following sources:

- Files or databases of a system being replaced
- Manual records
- Files or databases of other systems in the organization
- User feedback during normal system operation

Reusing Existing Databases

Most new information systems replace or augment an existing manual or automated system. In the simplest form of data conversion, the old system database is used directly by the new system with little or no change to the database structure. Reusing an existing database is fairly common because of the difficulty and expense of creating new databases from scratch, especially when a single database often supports multiple information systems as in today's enterprise application systems.

Although old databases are commonly reused in new or upgraded systems, some changes to database content are usually required. Typical changes include adding new classes or entities, adding new attributes or relationships, and modifying existing attributes or relationships. Modern database management systems (DBMSs) usually allow database administrators to modify the structure of a fully populated database. Simple changes such as adding new attributes or changing attribute types can be performed entirely by the DBMS.

Reloading Database Contents

More complex changes to database structure may require reloading data after the change. In that case, implementation staff must develop programs to alter data after the database has been modified. Figure 16-17 shows two possible approaches to reloading data. The first approach initializes a new database and copies the contents of the old database to it. The conversion program translates data stored within the former database structure into the newly modified database structure.

FIGURE *16-17*

Two approaches to reloading database content after a structural modification.

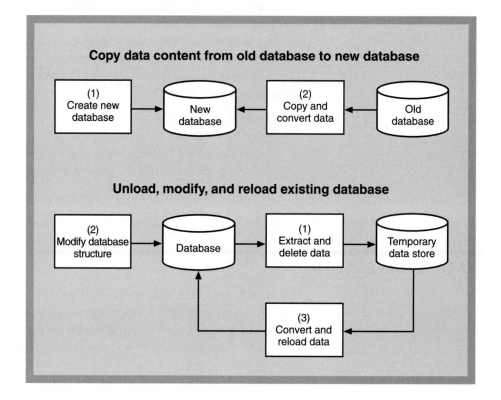

The second approach uses a program or DBMS utility to extract and delete data from an existing database and store it in a temporary data store. The database structure is then modified, and a second DBMS utility or program is used to reload the modified database. The first approach is simpler than the second, but it requires sufficient data storage to hold both databases temporarily. The second approach is required if there is insufficient data storage to hold two complete sets of data.

Many DBMSs provide a rich set of import utilities to extract and load data from existing databases, files, or scanned documents. DBMS developers provide such utilities because system developers are more likely to adopt a DBMS that eases the process of importing data from other sources. If DBMS import and export utilities are inadequate for data conversion, then developers must construct conversion programs that will be used only once. Although conversion programs are not part of the operational system, they must be constructed and tested in the same manner as operational software.

Creating New Databases

If the system being developed is entirely new or if it replaces a manual system, then initial data must be obtained from manual records or from other automated systems in the organization. Data from manual records can be entered using the same programs being developed for the operational system. In that case, data-entry programs are usually developed and tested as early as possible. Initial data-entry can be structured as a user training exercise. In addition, data from manual records can also be scanned into an optical character recognition program and then entered into the database using custom-developed conversion programs or a DBMS import utility.

Some data may already be stored in other automated systems within the organization. For example, when implementing a new order-entry system, some product data may already be present in a manufacturing planning and control system, and some customer data may already be present in an existing billing system. Copying such data to a new database is similar to reloading a modified database from an old database or backup data store.

Figure 16-18 shows a complex data-conversion process that draws input from a variety of sources. Data are input using a mix of manual data entry, optical character recognition, conversion programs, and DBMS import and export utilities. Data-conversion processes of this complexity are common in large system development projects.

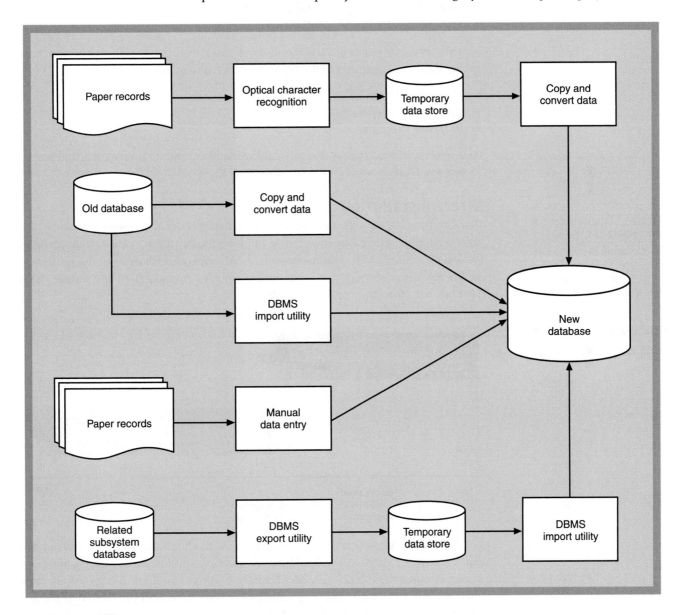

FIGURE *16-18*

A complex data-conversion example.

In some cases it may be possible to begin system operation with a partially or completely empty database. For example, a customer order-entry system need not have existing customer information loaded into the database. Customer information could be added the first time a customer places an order, based on a dialog between a telephone order-entry clerk and the customer. Adding data as they are encountered reduces the complexity of data conversion but at the expense of slower processing of initial transactions.

INSTALLATION

Once a new system has been developed and tested, it must be installed and placed into operation. Installing a system and making it operational are complex because there are many conflicting constraints, including cost, customer relations, employee relations, logistical complexity, and overall exposure to risk. Some of the more important issues to consider when planning installation include the following:

- Incurring costs of operating both systems in parallel
- Detecting and correcting errors in the new system
- Potentially disrupting the company and its IS operations
- Training personnel and familiarizing customers with new procedures

Different approaches to installation represent different trade-offs among cost, complexity, and risk. The most commonly used installation approaches are:

- Direct installation
- Parallel installation
- Phased installation

Each approach has different strengths and weaknesses, and no one approach is best for all systems. Each approach is discussed in detail here.

Direct Installation

In a *direct installation*, the new system is installed and quickly made operational, and any overlapping systems are then turned off. Direct installation is also sometimes called *immediate cutover*. Both systems are concurrently operated for only a brief time (typically a few days or weeks) while the new system is being installed and tested. Figure 16-19 shows a time line for direct installation.

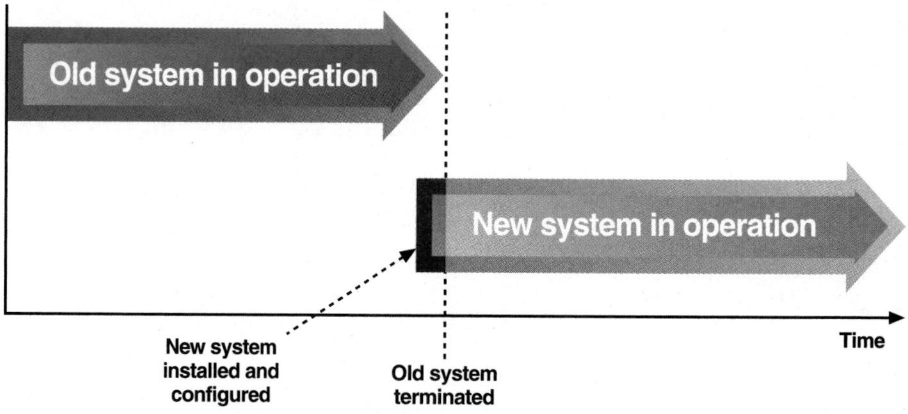

direct installation, or immediate cutover

an installation method that installs a new system, quickly makes it operational, and immediately turns off any overlapping systems

FIGURE *16-19*

Direct installation and cutover.

The primary advantage of direct installation is its simplicity. Since the old and new systems aren't operated in parallel, there are fewer logistical issues to manage and fewer resources required. The primary disadvantage of direct installation is its risk. Because older systems are not operated in parallel, there is no backup in the event that the new system fails. The magnitude of the risk depends on the nature of the system, the cost of workarounds in the event of a system failure, and the cost of system unavailability or less-than-optimal system function.

Direct installation is typically used under one or both of the following conditions:

- The new system is not replacing an older system (automated or manual).
- Downtime of days or weeks can be tolerated.

If neither condition applies, then parallel or phased installation is usually preferable to minimize the risk of system unavailability.

Parallel Installation

In a *parallel installation*, the old and new systems are both operated for an extended period of time (typically weeks or months). Figure 16-20 illustrates the time line for parallel installation. Ideally, the old system continues to operate until the new system has been thoroughly tested and determined to be error-free and ready to operate independently. As a practical matter, the time allocated for parallel operation is often determined in advance and limited to minimize the cost of dual operation.

FIGURE *16-20*

Parallel installation and operation.

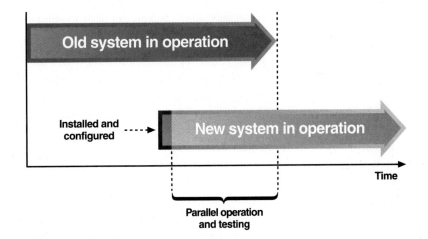

The primary advantage of parallel installation is a relatively low risk of system failure and the negative consequences that might result from that failure. If both systems are operated completely (that is, using all data and exercising all functions), then the old system functions as a backup for the new system. Any failure in the new system can be mitigated by relying on the old system.

The primary disadvantage of parallel installation is cost. During the period of parallel operation, the organization pays to operate both systems. Extra costs associated with operating two systems in parallel include:

- Hiring temporary personnel or temporarily reassigning existing personnel
- Acquiring extra space for computer equipment and personnel
- Increasing managerial and logistical complexity

Unless the operational costs of the new system are substantially less than that of the old system, the combined operating cost is typically 2.5 to 3 times the cost of operating the old system alone.

Parallel operation is generally best when the consequences of a system failure are severe. Parallel operation substantially reduces the risk of a system failure through redundant operation. The risk reduction is especially important for "mission-critical" applications such as customer service, production control, basic accounting functions, and most forms of on-line transaction processing. Few organizations can afford any significant downtime in such important systems.

Full parallel operation may be impractical for any number of reasons, including the following:

- Inputs to one system may be unusable by the other, and it may not be possible to use both types of inputs.

- The new system may use the same equipment as the old system (for example, computers, I/O devices, and networks), and there may not be sufficient capacity to operate both systems.
- Staffing levels may be insufficient to operate or manage both systems at the same time.

When full parallel operation is not possible or feasible, a partial parallel operation may be employed instead. Possible modes of partial parallel operation include the following:

- Processing only a subset of input data in one of the two systems. The subset could be determined by transaction type, geography, or sampling (for example, every 10th input transaction).
- Performing only a subset of processing functions (such as updating account history but not printing monthly bills).
- Performing a combination of data and processing function subsets.

Partial parallel operation always entails the risk that significant errors or problems will go undetected. For example, parallel operation with partial input increases the risk that errors associated with untested inputs will not be discovered.

Phased Installation

phased installation

an installation method that installs a new system and makes it operational in a series of steps or phases

In a *phased installation*, the system is installed and brought into operation in a series of steps or phases. Each phase adds components or functions to the operational system. During each phase, the system is tested to ensure that it is ready for the next phase. Phased installation can be combined with parallel installation, particularly when the new system will take over the operation of multiple existing systems.

Figure 16-21 shows a phased installation with both direct and parallel installation of individual phases. The new system replaces two existing systems. The installation is divided into three phases. The first phase is a direct replacement of one of the existing systems. The second and third phases are different parts of a parallel installation that replace the other existing system.

There is no single method for performing phased installation. Installation details such as the composition of specific phases and their order of installation vary widely from one system to another. These specifics determine the number of installation phases, the order of installation, and the parts of the new system that are operated in parallel with existing systems.

The primary advantage of phased installation is reduced risk. Risk is reduced because failure of a single phase is less problematic than failure of an entire system. The primary disadvantage of phased installation is increased complexity. Dividing the installation into phases creates more activities and milestones, thus making the entire process more complex. However, each individual phase contains a smaller and more manageable set of activities. If the entire system is simply too big or complex to install at one time, then the reduced risks of phased installation outweigh the increased complexity inherent in managing and coordinating multiple phases.

Phased installation is most useful when a system is large, complex, and composed of relatively independent subsystems. If the subsystems are not substantially independent, then it is difficult or impossible to define separate installation phases. System size and complexity may also be too great for an "all at once" installation to be feasible. In that case, there is really no choice but to use phased installation.

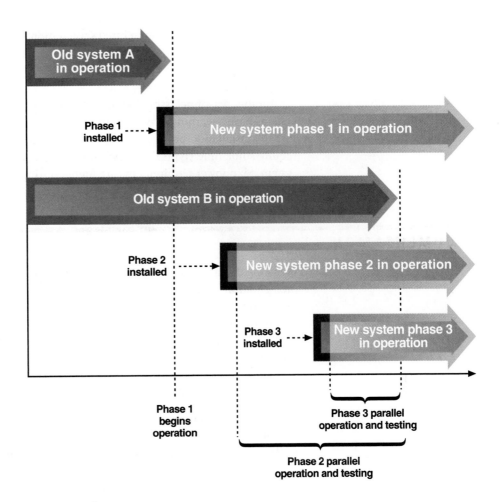

Personnel Issues

Installing a new system places significant demands on personnel throughout an organization. Installation typically involves demanding schedules, rapid learning and adaptation, and high stress. Planning should anticipate these problems and take appropriate measures to mitigate their effects.

New system installation usually stretches IS personnel to their limits. Many tasks must be performed in little time. The problem is most acute in parallel installation, where personnel must operate both the old and new systems. Often, development and customer support personnel must be temporarily reassigned to provide sufficient manpower to operate both systems. Reassignment may reduce progress on other ongoing projects and also reduce support and maintenance activities for other systems.

Temporary and contract personnel may be hired to increase available manpower during an installation. Two types are particularly useful:

- Personnel with experience in hardware and software installation and configuration
- Personnel with experience (or who can be trained) to operate the old system

Installation may require technical skills that are in short supply within the organization. In that case, hiring contractors to assist in hardware and software installation is a necessity. Hardware and software specialists can be contracted directly from vendors or from IS consulting firms.

Hiring temporary employees to operate the old system during a parallel installation has several benefits. First, it provides the extra manpower needed to operate both systems. Second, it frees permanent employees for training and new system operations.

Temporary personnel are often hired several months in advance, trained to operate the old system, and employed for the duration of parallel operation. If problems occur with the new system, then employment contracts can be extended until the old system can be safely phased out.

Another personnel issue that must be considered is employee productivity. All new systems have a learning curve for users and system operators. Both require training before the installation begins, but no matter how good the training, users and operators require some time (typically a few months) to reach their peak efficiency with a new system. Manpower requirements are higher during that time, as is the general level of employee stress.

DOCUMENTATION

Preparing documentation is an important but frequently overlooked activity during implementation. Documentation provides information to users on how to operate and maintain a system. Documentation also provides information needed for future modifications or reimplementation.

Rapid technology changes over the last two decades have altered the nature of documentation. Prior to the 1980s, most documentation was printed on paper and organized into bound or loose-leaf books. Automated documentation is now the norm, and formats include the following:

- Electronic documents, such as documents stored in Microsoft Word or Adobe Acrobat files
- Hyperlinked documents, such as documents formatted for Web browser viewing with embedded links among documentation components
- On-line documentation stored on a vendor Web site that can be viewed with a Web browser or downloaded and installed on a local computer system
- Embedded documentation, such as manuals, tutorials, and multimedia presentations included on a CD and installed as an integral part of an application
- Electronic system models, such as text and graphics formatted and stored in graphics file formats such as GIF, JPEG, and Visio
- Tool-specific system models, such as those developed with integrated programming environments, DBMSs, and CASE tools

Electronic documents can be distributed in a number of standard formats, including Adobe Acrobat, Windows help files, and standard Web pages. Choosing a standard format ensures that users have or can easily obtain the software to view the documentation. Hyperlinked documents enable users to navigate rapidly among related topics. Storing documentation on Web sites allows vendors to make updates easily and allows users to share a single copy. Embedded documentation enables users to access information through the application and provides features such as context-sensitive help.

Electronic and tool-specific system models are primarily intended for software developers' use. Generic model formats (such as ordinary text and GIF images) can be formatted as any type of electronic format. Tool-specific models must generally be accessed via specific software tools (for example, viewing a model generated by a CASE tool usually requires a viewer supplied with the CASE tool). However, most development tools allow models to be exported to other formats (such as Acrobat or Microsoft Word).

Documentation can be loosely classified into two types:

- *System documentation*—descriptions of system functions, architecture, and construction details
- *User documentation*—descriptions of how to interact with and maintain the system

system documentation

descriptions of system functions, architecture, and construction details, as used by maintenance personnel and future developers

user documentation

descriptions of how to interact with and maintain the system, as used by end users and system operators

System documentation is generated throughout the SDLC as outputs of each life cycle phase and activity. User documentation is created during the implementation phase of the SDLC. The development team cannot create user documentation earlier because many details of the user interface and system operation either haven't yet been determined or may change during development.

System Documentation

System documentation serves one primary purpose: providing information to designers and developers who will maintain or reimplement the system. Most or all of the documentation needed for this purpose is generated as a by-product of analysis, design, and implementation activities. Figure 16-22 shows three phases of the SDLC and the system documentation produced or modified in each phase. The documentation produced in each phase is useful for future maintenance or upgrades, although documentation produced in later phases is used more frequently than documentation produced in earlier phases.

Life cycle phase	System documentation	
	Traditional approach	**Object-oriented approach**
Analysis	Entity-relationship diagram Data flow diagram Process description Data flow and element definition	Class diagram Use case Activity diagram System sequence diagram
	Event list	
Design	System flowchart Structure chart	Design class diagram Interaction diagram Collaboration diagram Package diagram Statechart
	Module or method pseudocode Database schema diagram	
Implementation	Program source code Database schema source code Test data	

FIGURE *16-22*

Life cycle phases and system documentation generated in each phase.

Source code is the most frequently used documentation since it is the most direct link to the system's executable software. Direct changes to binary code are complex and expensive, so changing and recompiling source code is the only realistic method of altering a system's behavior. After a system is changed, test data are used to check the system. Rerunning old tests with old test data helps determine whether a change in one part of the system has accidentally "broken" some other part of the system.

Source code can be difficult (and thus inefficient) for human beings to use as documentation because it is entirely textual and often poorly commented. Important types of system information—such as how programs interact and what user needs a program satisfies—are usually not documented within source code. Yet such information is needed when evaluating significant changes in systems design or function and when tracing errors that flow from one program to another via shared data. Information needed to perform these tasks is readily available in analysis and design models.

Design models tend to be used more frequently than analysis models because design parameters change more often than system requirements. Examples of maintenance changes that require design models but not analysis models include redeploying existing programs or databases to new hardware, fixing bugs in individual programs, and optimizing the performance of an existing distributed system. Such changes alter the corresponding design models (for example, the system flowchart or package diagram) but do not change analysis phase models.

Analysis models do change when user requirements are altered. For example, adding a new transaction type or an entirely new processing subsystem changes the data flow diagram or class diagram. It also changes other analysis phase models such as the entity-relationship diagram, event list, and use cases.

System documentation must be actively managed to remain effective. It must be stored in an accessible location and form, retrieved when necessary for maintenance changes, and updated once changes have been implemented. In large organizations with many information systems, managing the documentation is a very formal process. Large organizations typically have one person responsible for archiving and retrieving documentation and for enforcing documentation standards.

Failure to adequately maintain system documentation compromises the value of a system. Systems with inadequate documentation are difficult or impossible to maintain, thus increasing the likelihood that a system will be prematurely scrapped or reimplemented. Maintaining documentation extends the useful life of a productive asset.

System documentation mirrors the system itself. That is, any information contained within system documentation can also be obtained by directly examining the system. For example, programmers can determine the entities and relationships of a relational database by examining the SQL statements that describe the database schema. Programmers can also determine the modular structure of a traditional program or the classes within an OO program by directly examining the program source code. If the source code is unavailable, then program structure can be also be determined from executable code, although the process is much more difficult.

As changes are made to the system, its documentation must also be updated, however. If documentation is not updated, then it is inconsistent with the system and useless to future designers and maintenance programmers. Making documentation an integral part of the installed system minimizes or eliminates inconsistency because updates to the system automatically update the documentation. Some tools—in particular, CASE tools and reverse-engineering tools—can simplify documentation and help ensure its accuracy.

With a CASE tool, the system is built automatically from design models and stored by the CASE tool. Design models, in turn, are built automatically (or nearly so) based on analysis models. To implement a system change, a programmer modifies an analysis or design model and then regenerates the installed system. The CASE tool automatically maintains consistency among the installed system and the models. As long as only the models are changed (instead of the source or executable code), the models and system will always be consistent.

A reverse-engineering tool can generate system models by examining source code. For example, such a tool can generate a class diagram by examining OO programs, and it can generate a structure chart by examining a program written in a procedural programming language. If a reverse-engineering tool is powerful and reliable enough to generate all types of system documentation, then there is no need to maintain a separate store of documentation. The source code itself is the documentation, and the reverse-engineering tool generates other forms of documentation on demand.

Both CASE and reverse-engineering tools are highly specialized to specific operating environments (such as programming languages, database management systems, and operating systems). They also tend to be expensive and to have steep learning curves. As a result, they aren't used as often as you might think. Thus, for many systems, system documentation must still be maintained separately and manually.

USER DOCUMENTATION

User documentation provides ongoing support for end users of the system. It primarily describes routine operation of the system, including functions such as data entry, output generation, and periodic maintenance. Topics typically covered include the following:

- Software startup and shutdown
- Keystroke, mouse, or command sequences required to perform specific functions
- Program functions required to implement specific business procedures (for example, the steps followed to enter a new customer order)
- Common errors and ways to correct them

For ease of use, user documentation includes a table of contents, a general description of the purpose and function of the program or system, a glossary, and an index.

User documentation for modern systems is almost always electronic and is usually an integral part of the application. Most modern operating systems provide standard facilities to support embedded documentation. Figure 16-23 shows electronic user documentation of a typical Windows application. The left pane displays the table of contents, and the user can access an index or search engine by clicking the appropriate tab at the top. The right pane displays individual pages of user documentation. The sample page shows an embedded glossary definition (in green) and two hyperlinks (in blue).

FIGURE *16-23*

Sample Windows Help display.

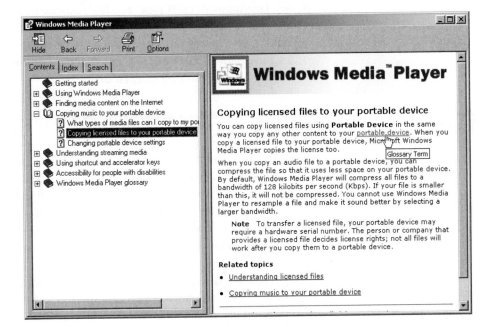

User documentation is an important organizational asset. Unfortunately, many organizations fail to prepare comprehensive high-quality user documentation for internally developed systems. Some of the reasons for this problem include the following:

- The assumption that trained programmers can examine source code, figure out how the system works, and train users as needed
- The assumption that the users trained during system implementation will informally pass on their knowledge to future users
- The lack of resources and special skills required to develop documentation and keep it up to date

As discussed in the previous section, source code is a poor form of system documentation. But it is an even worse form of indirect user documentation. Although source code provides a detailed instruction-by-instruction view of how pieces of a system work, it provides little or no information about how those pieces interact and how the entire system functions within a specific context. Supplementing source code with other forms of system documentation does provide other critical information. But even then, figuring out how a system works based only on system documentation is slow and error-prone.

Knowledge of how to use a system is as important an asset as the system itself. Once initial training is completed, that practical knowledge is stored in the minds of end users. But experience such as that is difficult to maintain or effectively transfer to other users. Employee turnover, reassignment, and other factors make direct person-to-person transfer of operational knowledge difficult and uncertain. In contrast, written or electronic documentation is easier to access and far more permanent.

Developing good user documentation requires special skills and considerable time and resources. Writing clearly and concisely, developing effective presentation graphics, organizing information for easy learning and access, and communicating effectively with a nontechnical audience are skills for which there is high demand and limited supply. Development takes time, and high-quality results are achieved only with thorough review and testing. Unfortunately, preparing user documentation is often left to technicians lacking in one or more necessary skills. Also, preparation time, review, and testing are often shortchanged because of schedule overruns and the last-minute rush to tie up all the loose ends of implementation.

TRAINING AND USER SUPPORT

Good documentation can reduce training needs as well as the frequency of support requests. But some training before and support after installation is almost always required. Remember, users are part of the system, too! Without training, users would slowly work their own way up the learning curve, error rates would be high, and the system would operate well below peak efficiency. Training allows users to be productive as soon as the system becomes operational. Support activities ensure continuing user productivity long after installation.

There are two classes of users—end users and system operators—who must be considered for documentation, training, and support. End users are people who use the system from day to day to achieve the system's business purpose. System operators are people who perform administrative functions and routine maintenance to keep the system operating. Figure 16-24 shows representative activities for each role. In smaller systems, a single person may fill both roles.

FIGURE *16-24*

Typical activities of end users and system operators.

End user activities	System operator activities
Creating records or transactions	Starting or stopping the system
Modifying database contents	Querying system status
Generating reports	Backing up data to archive
Querying database	Recovering data from archive
Importing or exporting data	Installing or upgrading software

Training and support activities vary with the target audience. Audience characteristics that affect training include the following:

- Frequency and duration of system use
- Need to understand the system's business context
- Existing computer skills and general proficiency
- Number of users

In general, end users use the system frequently and for extended periods of time, and system operators interact with the system infrequently and usually for short periods. End users solve a particular business problem with the system or implement specific business procedures. System operators are usually computer professionals with limited knowledge of the business processes that the system supports. End-user computer skill levels vary widely, while system operators typically have higher and more uniform skill levels. Also, the number of end users is generally much larger than the number of system operators.

Training for end users must emphasize hands-on use and application of the system for a specific business process or function, such as order entry, inventory control, or accounting. If the users are not already familiar with those procedures, then training must also include them. Widely varying skill and experience levels call for at least some hands-on training, including practice exercises, questions and answers, and one-on-one tutorials. Self-paced training materials can fill some of this need, but complex systems usually require some face-to-face training also. The relatively large number of end users makes group training sessions feasible, and a subset of well-qualified end users can be trained and then pass on their knowledge to other users.

System operator training can be much less formal when the operators are not end users. Experienced computer operators and administrators can learn most or all they need to know by self-study. Thus, formal training sessions may not be required. Also, the relatively small number of system operators makes one-on-one training feasible, if it is necessary.

Determining the best time to begin formal training can be difficult. On one hand, starting relatively early in the implementation phase provides plenty of time for learning and can ensure that users "hit the ground running." On the other hand, starting early can be frustrating to both users and trainers because the system may not be stable or complete. End users can quickly become upset when trying to learn on a buggy, crash-prone system with features and interfaces that are constantly changing.

In an ideal world, training doesn't begin until the interfaces are finalized and a test version has been installed and fully debugged. But the typical end-of-project crunch makes that approach a luxury that is often sacrificed. Instead, training materials are normally developed as soon as the interfaces are reasonably stable, and end-user training begins as soon as possible thereafter. It is much easier to provide training if system interfaces are developed early and if the top-down modular development approach is employed.

Ongoing Training and User Support

The term *user support* covers training and user assistance that occur after the system is up and running. Some of the activities are the same as preinstallation training activities. For example, new users must be trained periodically due to employee turnover. Other activities such as refresher training and help desk operation are unique to support.

User support can be provided by a number of methods, including the following:

- On-line documentation and troubleshooting
- Resident experts
- A help desk
- Technical support

On-line documentation and troubleshooting have surged as a support method in recent years. Much of this support is built into the application, although Web sites are also commonly employed. The goal of on-line support is to minimize the need for human support by putting useful information into the hands of end users when they need it. Achieving that goal, however, requires well-designed support materials that are comprehensive and easy to use.

Resident experts are the most common form of user support, and their help is usually provided informally. A resident expert can be an on-site IS staff member or (more frequently) a business area staff member or user who assists other users. The position of resident expert is often informal. A person frequently grows into that position simply by displaying exceptional computer literacy or knowledge of software. Over time, all other users begin to approach that person first with questions or problems.

A help desk is a permanent IS department that provides end-user support for a wide range of systems and software. Help desks are staffed by personnel trained to install, operate, and troubleshoot application software, including off-the-shelf products (such as word processors). A help desk serves as a central contact point for users. Help desk staff are trained to handle the majority of user problems and questions. Those who require further assistance are forwarded to technical support.

Technical support is typically a specific function or department within IS maintenance because of the close relationship between user support, change requests, and system error reporting. If help desk personnel can't solve a user's problem, then there's a good chance that an error has been discovered or that there is a gap between system capability and user needs. If the problem is a system error, then maintenance needs to be notified quickly to investigate the cause and correct it if it is critical. Noncritical errors and unmet user needs must also be brought to the attention of maintenance, but timeliness is less critical. In either case, technical support is the bridge between users and maintenance activities.

MAINTENANCE AND SYSTEM ENHANCEMENT

software maintenance

modification of a software product after delivery to correct faults, improve performance or other attributes, or adapt the product to a changed environment

The Institute of Electrical and Electronics Engineers (IEEE) and the American National Standards Institute (ANSI) have defined *software maintenance* as modification of a software product after delivery to accomplish at least one of the following objectives:

- Correct faults
- Improve performance or other attributes
- Adapt the product to a changed environment

The term *maintenance* covers virtually everything that happens to a system after delivery except total replacement or abandonment.

In most organizations, the cost of maintaining existing systems is at least as great as the cost of developing new ones. Existing systems are an organizational asset and must be actively managed to preserve their value and utility. In that sense, maintaining software is similar to maintaining other types of capital assets such as buildings and equipment.

Maintenance involves change—to adapt to a new environment, to adapt to changing user requirements, and to fix problems as they occur or are discovered. But change is risky. Making changes to an operational system is much more difficult than making changes to a system under development. When a change causes a developmental system to crash, there are no frantic calls to the support desk and no immediate financial impact. But changes to operational systems have an immediate impact on users, customers, and the organization as a whole.

Failure of an operational system can be disastrous. Thus, software maintenance differs greatly from new system development. New system development generally occurs in a relatively open environment where change is expected, new ideas are tried out, and risk taking is tolerated, if not encouraged. In contrast, maintenance is very conservative—change is tolerated as a necessary evil, and risk taking is strongly discouraged.

Maintenance activities include the following:

- Tracking modification requests and error reports
- Implementing changes
- Monitoring system performance and improving performance or increasing capacity
- Upgrading hardware and system software
- Updating documentation to reflect maintenance changes

Maintenance and new system development do have many activities in common, including analysis, design, construction, testing, and documentation. However, implementation of those activities differs in many ways, including scope and detail, triggering events, and implementation constraints. Each maintenance activity is described in detail in the following sections.

Submitting Change Requests and Error Reports

To manage the risks associated with change, most organizations adopt formal control procedures for all operational systems. Formal controls are designed to ensure that potential changes are adequately described, considered, and planned before being implemented. Typical change control procedures include the following:

- Standard change request forms
- Review of requests by a change control committee
- Extensive planning for design and implementation

Figure 16-25 shows a sample change request form that has been completed by a user or system owner and submitted to the change control committee for consideration. The change control committee reviews the change request to assess the impact on existing computer hardware and software, system performance and availability, security, and operating budget. The recommendation of the change control committee is formally recorded in a format such as the sample shown in Figure 16-26. Approved changes are added to the list of pending changes for budgeting, scheduling, planning, and implementation.

Change Request				
Request Date	2/1/2004	**Change Type**	☐	**Error Correction**
Requested By	Wen-Hsu Chang, Comptroller		☒	**Modification**
Target System	Customer Accounts - Refunds		☐	**New Function**

Change (or Error) Description

U.S. check formats will soon change due to a recently enacted federal law. The new format reserves an area to the right of the current routing number to be used for a security bar code checksum.

The law requires the new checksum to be printed on checks dated on or after January 1, 2005.

We currently use a portion of the area in question to print a multicolored security symbol. The security symbol will need to be moved or eliminated, and the security bar code checksum will have to be added.

Change Request ID	2002-11		
Date Received	2/2/2004	**Review Date**	2/7/2004, 0930-1100
Review Participants	W. Chang (Comptroller), R. Brooks (IS Operations), J. Hernandez (IS Security), G. Weeks (IS Change Coordinator)		

FIGURE *16-25*

A sample change request form.

Bugs can be reported using a standard change request form, but many computing organizations use a different form and procedure because they need to fix such bugs immediately. Bug reports can come from many sources, including end users, computer operators, or IS support staff. Bug reports are typically routed to a single person or organization for logging and follow-up.

Implementing a Change

Change implementation follows a miniature version of the system development cycle. Most of the same activities are performed, although they may be reduced in scope or sometimes completely eliminated. In essence, a maintenance change is an incremental development project in which the user and technical requirements are fully known in advance. Thus, analysis phase activities are typically skimmed or skipped.

Planning for a change includes the following activities:

- Identify what parts of the system must be changed
- Secure resources (such as personnel) to implement the change
- Schedule design and implementation activities
- Develop test criteria and a testing plan for the changed system

Change Review				
Change Request ID	2002-11	Date Reviewed	2/7/2004	
Priority	☐ **Critical**	☒ **Necessary**		☐ **Optional**

Hardware Implications
need to verify ability of current printers to write a security bar code in mandated area

Software Implications
database will need to be modified to store the security bar code with other check information
check writing program must be modified to generate and print the security bar code

Performance Implications
none

Operating Budget Implications
none

Other Implications
none

Disposition	☒ **Approved**	☐ **Rejected**	☐ **Suspended**
Reason			

Latest Implementation Date	12/31/2004	
Reevaluation Date	n/a	**Signature**

FIGURE *16-26*

A sample change review form.

production system

the version of the system used from day to day

test system

a copy of the production system that is modified to test changes

System documentation is reviewed by design, development, and operations staff to determine the scope of the change. Test criteria and plans for the existing system are the starting point for testing the new system. The testing plan is simply updated to account for changed or added functions, then the modified plan and test data are archived for use in future change projects.

Design may be combined with planning if the change is relatively simple. For more complex changes, a separate design phase is used. The existing system design is evaluated and modified as necessary to implement the proposed changes. As with test plans and data, the revised design is archived for use in future change projects.

Implementation activities are normally performed on a copy of the operational system. The *production system* is the version of the system used day to day. The *test system* is a copy of the production system that is modified to test changes. The test system may be developed and tested on separate hardware or on a redundant system. The test system becomes the operational system only after complete and successful testing.

Upgrading Computing Infrastructure

Computer hardware, system software, and networks must be periodically upgraded for many reasons, including:

- Software maintenance releases
- Software version upgrades
- Declining system performance

Like application software, system software such as operating and database management systems must periodically be changed to correct errors and add new functions.

System software developers typically distribute maintenance releases several times per year. The frequency of maintenance update distribution has increased in recent years in part because of the convenience of Internet-based software distribution. In some cases (such as virus checkers and operating system security subsystems), updates may be released weekly or even more frequently.

As with internally generated changes, system software updates are risky. Application software that worked well with an older software version may fail when that software is updated. For this reason, system software updates are extensively tested before they are applied to operational systems. In many cases, maintenance and version updates are simply ignored to reduce risk. Unless errors related to system software have already been encountered, there is little immediate benefit to an upgrade. Operational system maintenance usually follows the old engineering maxim, "If it isn't broken, don't fix it!"

Increases in transaction volume or support of new systems on existing hardware and networks sometimes reduce performance to unacceptable levels. So, an infrastructure upgrade may be required to add capacity or address a performance-related problem. Infrastructure upgrades are implemented like any other change. The primary difference is how a performance upgrade is initiated.

Input from user or IS staff may indicate the need for a performance upgrade. But a final determination of whether an upgrade is needed and what exactly should be upgraded requires thorough investigation and research. Computer and network performance is complex and highly technical, so what appears to be a performance problem may have little to do with hardware or network capacity. If the problem is ultimately traced to hardware or networks, then the specific cause must be identified and a suitable upgrade chosen.

Performance problems require careful diagnosis to determine the best approach to address the problem. Staff with solid technical backgrounds who can understand all of the relevant trade-offs should diagnose the problem. Larger IS organizations may have permanent staff with such skills, but many organizations must rely on contract personnel or consultants to diagnose performance problems, recommend corrective measures, and install or configure those measures. The skills come at a high price, but they can prevent an organization from wasting larger amounts of money buying hardware or network capacity that isn't really needed.

Summary

System development activities that occur after design and before system delivery are collectively called *implementation*. Implementation is complex because it consists of many interdependent activities, including programming, quality assurance, hardware and software installation, documentation, and training. Implementation is difficult to manage because activities must be properly sequenced and progress must be continually monitored. Implementation is risky because it requires significant time and resources and because it often affects systems vital to the daily operation of an organization.

Programming and testing are two of the most interdependent implementation activities. Software components must be constructed in an order that minimizes the use of development resources and maximizes the ability to test the system and correct errors. Unfortunately, those two goals often conflict. Thus, a program development plan is a trade-off among available resources, available time, and the desire to detect and correct errors prior to system installation.

Data conversion, installation, documentation, and training are activities that normally follow program development. They are highly interdependent since an installed and documented system is a prerequisite for complete training, and a fully populated database is needed to begin operation. Manpower utilization and the number of directly affected personnel generally peak during these activities.

Support activities occur after a system is made operational. User support activities ensure that an organization realizes the full benefit of the system. Maintenance and system enhancement activities ensure that the system functions at peak efficiency and that needed changes are implemented with minimal disruption to the organization. For most systems, the resources required for support are greater than the resources required to develop the system. Because of high resource requirements and greater operational risk, support activities are normally implemented in a formal and carefully managed fashion.

Key Terms

acceptance test, p. 646

alpha version, p. 636

beta version, p. 636

bottom-up development, p. 630

build and smoke test, p. 646

chief developer team, p. 634

collaborative specialist team, p. 634

cooperating peer team, p. 634

direct installation, or immediate cutover, p. 650

driver, p. 643

input, process, output (IPO) development, p. 627

inspection, p. 639

integration test, p. 644

maintenance release, p. 636

parallel installation, p. 651

performance test, p. 646

phased installation, p. 652

production system, p. 663

production version, release version, or production release, p. 636

quality assurance (QA), p. 638

response time, p. 646

software maintenance, p. 660

source code control system (SCCS), p. 635

stub, p. 644

system documentation, p. 654

system test, p. 645

technical review, p. 639

test case, p. 641

test data, p. 641

test system, p. 663

testing buddy, p. 646

throughput, p. 646

top-down development, p. 630

unit testing, or module testing, p. 642

user documentation, p. 654

REVIEW QUESTIONS

1. List and briefly describe the three basic approaches to program development order. What are the advantages and disadvantages of each?

2. How can the concepts of top-down and bottom-up development order be applied to object-oriented software?

3. Describe three approaches to organizing programming teams. For what types of projects or development activities is each approach best suited?

4. What is a source code control system? Why is such a system necessary when multiple programmers build a program or system?

5. Define the terms *alpha version*, *beta version*, and *production version*. Are there well-defined criteria for deciding when an alpha version becomes a beta version or a beta version becomes a production version?

6. List and briefly describe implementation phase QA activities other than software testing. What is the effect on software testing of not performing nontesting QA activities?

7. What are the characteristics of good test cases?

8. Define the terms *acceptance test*, *integration test*, *system test*, and *unit test*. In what order are these tests normally performed? Who performs (or evaluates the results of) each type of test?

9. What is a driver? What is a stub? With what type of test is each most closely associated? With what development order is each most likely to be used?

10. What factors make testing object-oriented programs more complex than testing structured programs?

11. List possible sources of data used to initialize a new system database. Briefly describe the tools and methods used to load initial data into the database.

12. Briefly describe direct, parallel, and phased installation. What are the advantages and disadvantages of each installation approach?

13. Why are additional personnel generally required during the later stages of system implementation?

14. What are the differences between documentation for end users and system operators?

15. How or why is system documentation redundant with the system itself? What are the practical implications of this redundancy?

16. List the types of documentation needed to support maintenance activities. Which documentation types are needed most frequently? Which are needed least frequently?

17. How do training activities differ between end users and system operators?

18. How does implementing a maintenance change differ from developing a new system? How are they similar?

19. Why might system software upgrades not be installed? What are the costs of not installing them?

THINKING CRITICALLY

1. Examine the system flowchart for Rocky Mountain Outfitters in Figure 10-5. Develop a preliminary development plan based on IPO development order. Which programs are difficult to classify as input, process, or output? Is a straightforward application of IPO development order appropriate for this system? If not, what changes should be made to the preliminary development plan?

2. This chapter discussed top-down and bottom-up development order for transform-oriented structure charts. Can these development orders also be applied to transaction-oriented structure charts such as the one shown in Figure 10-10? If so, how?

3. Describe the process of testing software developed using both top-down and bottom-up development order. Which method results in the fewest resources required for testing? What types of errors are likely to be discovered earliest under each development order? Which development order is best as measured by the combination of required testing resources and ability to capture important errors early in the testing process?

4. Assume that the Rocky Mountain Outfitters' customer support system will be developed as described in your answer to question 1. Assume that 14 people are available for programming and testing. What sizes and types of teams are best suited to the project?

5. Consider the issue of documenting a system using only electronic models developed with a full life cycle CASE tool. The advantages are obvious (for example, the analyst modifies the models to reflect new requirements and automatically generates an updated system). Are there any disadvantages? Hint: The system may be maintained for a decade or more.

6. Some types of system documentation (such as models developed during the analysis phase) are seldom looked at once the system is made operational. What are the advantages and disadvantages of not keeping such documentation types?

EXPERIENTIAL EXERCISES

1. Assume that you and five of your classmates are charged with developing the Customer order program shown in Figure 10-16. Create a development and testing plan to write and test the required modules. Assume that you have three weeks to complete all tasks.

2. Implement a formal QA process in one of your programming or system development classes (be sure to obtain permission from your instructor first). Form a group of students to implement an inspection process. Have one or all members prepare presentation materials for code that they've written and distribute them prior to a group meeting. If possible, create a buddy system for testing each other's code. Evaluate the results in terms of time required for code development and quality of the final product.

3. Examine the end-user documentation supplied with a typical personal or office productivity package such as Microsoft Office. Compare the documentation to the categories described in this chapter. Which categories of documentation are supplied? How might documentation for a business application differ in content or format?

4. Talk with a computer center or IS manager about the testing process used with a recently installed system or subsystem. What types of tests were performed? How were test cases and test data generated? What types of teams developed and implemented the tests?

5. Talk with an end user at your school or work about the documentation and training provided with a recently installed or distributed business application. What types of training and documentation were provided? Did the user consider the training to be sufficient? Does the user consider the documentation to be useful and complete?

CASE STUDIES

HudsonBanc Billing System Upgrade

Two regional banks with similar geographic territories merged to form HudsonBanc. Both banks had credit-card operations and operated billing systems that had been internally developed and upgraded over three decades. The systems performed similar functions, and both operated primarily in batch mode on IBM mainframes. Merging the two billing systems was identified as a high-priority cost-saving measure.

HudsonBanc initiated a project to investigate how to merge the two billing systems. Upgrading either system was quickly ruled out because the existing technology was considered old, and the costs of upgrading the system were estimated to be too high. HudsonBanc decided that a new system should be built or purchased. Management preferred the purchase option since it was assumed that a purchased system could be brought on-line more quickly and cheaply. An RFP was prepared, many responses were received, and after months of analysis and investigation, a vendor was chosen.

Hardware for the new system was installed in early January. Software was installed the following week, and a random sample of 10 percent of the customer accounts was copied to the new system. The new system was operated in parallel with the old system for two months. To save costs involved with complete duplication, the new system computed but did not actually print billing statements. Payments were entered into both systems and used to update parallel customer account databases. Duplicate account records were checked manually to ensure that they were the same.

After the second test billing cycle, the new system was declared ready for operation. All customer accounts were migrated to the new system in mid-April. The old system was turned off on May 1, and the new system took over operation. Problems occurred almost immediately. The system was unable to handle the greatly increased volume of transactions. Data-entry slowed to a crawl, and payments were soon backed up by several weeks. The system was not handling certain types of transactions correctly (for example, charge corrections and credits for overpayment). Manual inspection of the recently migrated account records showed errors in approximately 50,000 accounts.

It took almost six weeks to adjust the incorrect accounts manually and to update functions to handle all transaction types correctly. On June 20, the company attempted to print billing statements for the 50,000 corrected customer accounts. The system refused to print any information for transactions more than 30 days old. A panicked consultation with the vendor concluded that fixing the 30-day restriction would require more than a month of work and testing. It was also concluded that manual entry of account adjustments followed by billing within 30 days was the fastest and least risky way to solve the immediate problem.

Clearing the backlog took two months. During that time, many incorrect bills were mailed. Customer support telephone lines were continually overloaded. Twenty-five people were reassigned from other operational areas, and additional phone lines were added to provide sufficient customer support capacity. System development personnel were reassigned to IS operations for up to three months to assist in clearing the billing backlog. Federal and state regulatory authorities stepped in to investigate the problems. HudsonBanc agreed to allow customers to spread payments for late bills over three months without interest charges. Setting up the payment arrangements further aggravated the backlog and staffing problems.

1. What type of installation did HudsonBanc use for its new system? Was it an appropriate choice?
2. How could the operational problems have been avoided?

The DownTown Videos Rental System

Using the design class diagram you developed in Chapter 11 for the DownTown Videos rental system, develop an implementation and testing plan. Specify the order in which classes and their methods will be implemented and the groups of methods and classes that will be tested during integration testing.

Rethinking Rocky Mountain Outfitters

Assume that it is currently late April 2005 and that the analysis phase of the customer support system (CSS) is nearly completed. Design is scheduled to finish by June 15 and implementation is scheduled to finish by November 1. RMO wants to use the new CSS during the holiday sales peak—roughly between Thanksgiving and Christmas—during which 40 percent of annual sales normally occur. RMO would like to announce the availability of Web ordering in an upcoming catalog. New catalog mailings are scheduled for June 15, September 1, October 31, and December 10.

1. Describe the risks associated with planning the new CSS implementation and announcing the availability of Web ordering to customers. Remember that the new CSS will replace the current telephone and mail-order systems in addition to handling Web orders. How conservative should RMO be with respect to testing, installation, and customer announcements? What is the cost of being too conservative?
2. What fallback strategies should be developed, if any? What should the "drop dead" date be for deciding whether to use the new CSS to process holiday orders?
3. Develop an installation plan and schedule. Justify your approach(es) and your timetable based on your previous risk analysis.
4. Analyze the training requirements and develop a training plan and schedule. How can training, data conversion, and testing activities be overlapped or combined? What about training and support for customers using the Web ordering system?

Focusing on Reliable Pharmaceutical Service

Using the structure chart you developed in Chapter 10 for Reliable Pharmaceutical Service, develop an implementation and testing plan. Specify the order in which modules will be implemented and the groups of modules that will be tested during integration testing.

FURTHER RESOURCES

Robert V. Binder, *Testing Object-Oriented Systems: Models, Patterns, and Tools*. Addison-Wesley, 2000.

Barry Boehm, *Software Engineering Economics*. Prentice Hall, 1981.

Robert G. Ebenau and Susan H. Strauss, *Software Inspection Process*. McGraw-Hill, 1994.

William Horton, *Designing and Writing Online Documentation: Hypermedia for Self-Supporting Products* (2nd ed.). John Wiley & Sons, 1994.

William Horton, *Designing Web-Based Training: How to Teach Anyone Anything Anywhere Anytime*. John Wiley & Sons, 2000.

William Horton, *Illustrating Computer Documentation: The Art of Presenting Information Graphically on Paper and Online*. John Wiley & Sons, 1991.

International Association of Information Technology Trainers (ITrain) Web site—http://itrain.org/.

Edward Kit, *Software Testing in the Real World: Improving the Process*. ACM Press, 1996.

David Kung, Jerry Gao, Pei Hsia, Yasufumi Toyoshima, Chris Chen, Young-Si Kim, and Young-Kee Song, "Developing an Object-Oriented Software Testing and Maintenance Environment," *Communications of the ACM*, volume 38:10, October, 1995, pp. 75–87.

Steve McConnell, *Code Complete*. Microsoft Press, 1995.

Arthur H. Watson and Thomas J. McCabe, "Structured Testing: A Testing Methodology Using the Cyclomatic Complexity Metric," NIST Special Publication 500-235, National Institute of Standards and Technology, September, 1996, http://hissa.ncsl.nist.gov/HHRFdata/Artifacts/ITLdoc/235/mccabe.html.

CHAPTER 17

Current Trends in System Development

LEARNING OBJECTIVES

After reading this chapter, you should be able to:

- Describe rapid application development and the prototyping, spiral, eXtreme Programming, and Unified Process development approaches

- Compare the prototyping, spiral, eXtreme Programming, and Unified Process development approaches with more traditional development approaches

- Choose an appropriate development approach to match project characteristics

- Implement a risk management process

- Describe rapid development techniques, including joint application design, tool-based development, and code reuse

- Describe components, the process by which they are developed and deployed, and their impact on the systems development life cycle

CHAPTER OUTLINE

Rapid Application Development

The Prototyping Approach to Development

The Spiral Approach to Development

eXtreme Programming

The Unified Process

Rapid Development Techniques

Components

EUROBANC: FASTER, BETTER, AND CHEAPER SYSTEM DEVELOPMENT?

Erik Maastricht waited nervously, lost in thought, outside the office of Andrea Ramos, CEO of EuroBanc. He was thinking, "I wonder if today's the day I'll be cleaning out my desk?" Just then, a voice startled him back to reality. "Ms. Ramos will see you now."

As he entered the office, Andrea rose and walked over to shake Erik's hand and said, "Come in. Make yourself comfortable." The serious look in Andrea's eyes only heightened Erik's apprehension, so he sat quickly and quietly.

"I'll get right to the point," Andrea said. "You were at the last board meeting, and you know the importance of our next few IT projects. The board wants them all deployed within 12 months. The pace of our business gets faster every month, and those projects are critical for catching up to our competitors and . . . ," Andrea paused and looked Erik squarely in the eyes, "completing our rather late start into 21st century e-business."

Andrea continued, "What you're probably not aware of is how close you came to losing your job during the closed part of the meeting. You had exactly one supporter in the room—me! And it was difficult for me to defend your recent performance in delivering new systems and upgrades on time and within budget."

Erik waited a moment, then said, "I know that you've been unhappy with the delays in some of our recent projects, and I suspected that a few board members had me lined up in their sights."

Andrea replied, "The board members quiz me about IT project progress at nearly every meeting. Two of the members presented an analysis of our last six big projects that showed average budget overruns of 25 percent and late deliveries ranging from 4 to 12 months. How can I defend a record like that?"

Erik swallowed hard before speaking. "I'm a firm believer in the old engineer's saying—'faster, better, cheaper, pick two.' You and the board have always been generous with resources, so money isn't the problem. But with the last few projects, I've found myself squarely faced with the trade-off between system quality and development speed. I think you'll agree that the systems I've delivered have consistently been of good quality."

"I know that the pyramids weren't built in a day, but I also know that construction methods have advanced considerably since then," replied Andrea. "I know very little about building systems, but I do know that other companies, including our biggest competitors, regularly introduce new systems faster than we do. They've been gaining market share and improving their stock price at our expense. I'm worried that a few more late deliveries will put us so far behind them we'll never catch up."

Erik replied, "I've been experimenting with rapid development and deployment techniques on some of our smaller projects. I guess it's time to stop experimenting and use those approaches in all of our projects."

With an exasperated wave of her hand, Andrea said, "I don't care if you change your development tools, hire Martian consultants, or figure out a way to alter the space-time continuum. Just speed up those projects, or you'll have one less friend at the next board meeting."

OVERVIEW

The pace of change in business and other organizations accelerated throughout the 20th century, and no one expects that pace to slow during the 21st century. Rapid change in both business practices and information technology has created a significant management problem: how to quickly develop and deploy information systems that implement the latest business practices and employ cutting-edge technologies. Yet, changing technology is a two-edged sword. One edge provides system developers with new tools and techniques to improve their own productivity. The other increases users' expectations for system functionality. Organizations that can quickly develop systems to exploit new technologies thrive and prosper. Those that don't stagnate and flounder.

The tools and techniques described in earlier chapters provide a foundation for rapid system development, but additional tools and techniques can be applied to speed up the process. This chapter presents a small but important subset of those tools and techniques. The first half of the chapter concentrates on rapid development techniques. The chapter begins with a discussion of what rapid application development is and isn't. We then describe alternative approaches to system development. Then we discuss a set of techniques that can be used with the spiral and other life cycles to increase development speed or reduce the risk of schedule overruns.

The latter half of the chapter focuses on two technologies—object frameworks and components. Both technologies can significantly increase development speed through code reuse and flexible software construction. We will describe the technologies and ways they can be successfully integrated into the systems development life cycle (SDLC) to speed software development.

Think of the concepts covered in this chapter as a set of ingredients for fine cuisine at a five-star restaurant. Every menu item is a unique combination of ingredients, skillfully combined by the chef. As the accompanying RMO progress memo illustrates, alternative development approaches, tools, and techniques are a set of ingredients that may or may not be applicable to a particular project. A project leader must choose those that best match the project's characteristics and skillfully apply them to rapidly produce a high-quality system.

August 4, 2005

To: John Blankens

From: Barbara Halifax

RE: Update on programming and testing

MEMO

I wanted to give you an update on our programming and testing efforts and the effectiveness of the rapid application development techniques we're using. We're currently a bit ahead of schedule on our programming and testing. Approximately 40% of the code has been developed and unit tested. Integration testing on the first major system is in progress, and the problems we've discovered thus far are all minor.

As we discussed before the project began in earnest, a traditional development was a better fit to this project than a spiral approach. But we have structured our development activities as a series of iterations—borrowing techniques from both the spiral development and extreme programming as needed. Pair programming, in particular, has been a boon to productivity. Everyone on the project is amazed at how quickly code is being developed and how few errors are being discovered during integration testing.

Purchasing components for credit authorization and shipment preparation has also saved us considerable time. Searching for those components, designing around their infrastructure needs, and testing them did add a few weeks to analysis and design. But that investment has been repaid severalfold by not having to write code to perform those functions. We shortened the entire project by a month or more and I expect similar benefits during future upgrades of the system.

I know it's a bit early to talk about life after this project, but I think that we should consider conducting a postmortem analysis as soon as the project is completed. Perhaps we could start right after the New Year holiday. We could learn a lot by examining the successes and failures from this project and deciding what tools and techniques should be incorporated into future projects.

BH

RAPID APPLICATION DEVELOPMENT

Rapid application development (RAD) is an overused and poorly understood term. RAD is something that most software developers claim they do but cannot precisely define. Often, RAD has been equated with tools and techniques such as prototyping, fourth-generation programming languages, CASE tools, and object-oriented (OO) analysis, design, and development. Methodologies have also been developed that are either called

RAD or claim to result in RAD. Tool vendors frequently include the term *RAD* in their product descriptions. Given the blizzard of competing and confusing claims, it's little wonder that few people can precisely define what RAD is.

For the moment, we'll avoid defining the term. Instead, we'll first describe some of the factors that influence development speed. After we have provided that background, we'll return to a discussion of RAD.

Reasons for Slow Development

The reasons for slow development are far too numerous to list or describe exhaustively in this text. Steve McConnell, in his book *Rapid Development: Taming Wild Software Schedules*, provides many pages of reasons. This chapter focuses on three broad categories that have delayed many a project: rework, shifting requirements, and inadequate or inappropriate tools and techniques. The following sections briefly describe the problem areas. Methods of addressing the problems are discussed later in the chapter.

Rework

Think of the process of building a house. A foundation is laid, the house is framed, the roof and exterior walls are covered, windows and doors are installed, interior mechanical systems are installed, interior fixtures are installed, and finally finishing touches such as paint and flooring are added. Now imagine that after each stage is initially completed, the builder decides that the construction is poor quality. So the work is ripped out and done over. The result is that each construction stage is performed twice and the project schedule doubles.

This scenario may sound contrived, but similar scenarios are relatively common in system development. It is not uncommon for software construction tasks to be performed multiple times to correct earlier problems. Clearly, one way to avoid lengthening the development schedule is to ensure that each construction activity is performed only once.

You can avoid rework by ensuring that:

- The right software (and only the right software) is constructed or procured.
- The development process always produces software that meets minimum quality standards.

The only way to ensure that the right software is built is to be certain that requirements and overall design constraints are fully known before design and construction begins. There are many ways to ensure that construction always produces high-quality results, such as fully specifying important design parameters before beginning construction, assigning well-trained and motivated people to construction tasks, and providing system builders with the right tools for the job.

Shifting Requirements

One reason that many projects fall behind schedule is that requirements change during the project. Changes to requirements necessitate corresponding changes to design and construction. The later in the project a change occurs, the more costly it is to incorporate it into the system. For example, changes to a house design (such as enlarging a room and putting in additional windows) before details are added require relatively little effort. Changes after detailed design require drawing up new blueprints and modifying the materials list and construction schedule. Changes during construction usually require undoing some already completed construction (that is, rework).

Changes to software requirements are also subject to increasing costs as development progresses. A commonly used estimate is that a change that could be implemented for $1 during analysis costs $10 during architectural design, $100 during detailed design, $1,000 during coding and testing, and $10,000 after the system is operational (see Figure 17-1).

FIGURE *17-1*

The cost of implementing a requirements change increases in each life cycle phase.

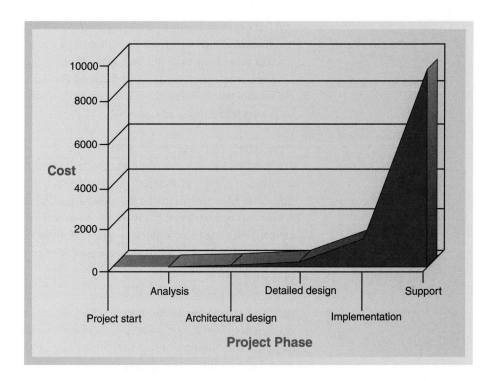

Some shift in requirements is to be expected in most software development projects. Users aren't always sure what they want at the beginning of the project. Also, a rapidly changing environment may necessitate changes before the system is fully developed. Ignoring important requirements changes only guarantees that the system will not meet the users' needs when it's delivered. To avoid schedule extensions while satisfying user requirements, a developer should anticipate and incorporate changes into the development process.

Tools and Techniques

No one set of tools and techniques is best suited to all system development projects. Different system requirements, development methodologies, and target operating environments may require different tools for analysis, design, and construction. For example, programming languages such as assembly and C are typically used for systems that require maximum execution speed. Object-oriented development and programming techniques are generally best for systems that will evolve over relatively long periods of time (such as many business applications).

Using the wrong tool or technique for a given project can reduce quality, increase development time, or both. Building a house with power tools is fast, and the result is generally high quality. Building a house with nothing more than hammers and hand saws will take much longer and probably result in a lower-quality product. Finishing a 100-story building on schedule requires not only the right tools and skills but also efficient construction and project management techniques. The parallels to software development are obvious. A development team can avoid slow development and poor quality by matching tools and techniques to the project at hand.

What Is RAD?

The time required to develop a system depends on a number of factors, including user requirements, budget, development approach, development tools and techniques, the developer's experience with similar projects, and management of the development process. Assume that both user requirements and budget are fixed (they usually are). Assume further that the budget is sufficient to buy the right tools and hire a well-trained and experienced staff. Given these assumptions, the only variables that affect development speed are the development approach, techniques, and management. Choosing the right approach and techniques and managing the development process efficiently and effectively will result in the shortest possible schedule.

Rapid application development (RAD) is a collection of development approaches, techniques, tools, and technologies, each of which has been proven to shorten development schedules under some conditions. Rapid application development is not a silver bullet that shortens development schedules for every project. In addition, no universal RAD approach shortens every project schedule, nor does any one technique, tool, or technology fit every project. No single approach or technique by itself is sufficient, and none will succeed for every project.

The key to shortening development schedules is to identify the overall development approach and the set of techniques, tools, and technologies most suitable to that approach and the specific project. For some projects, RAD may require an unconventional development approach and a host of newer techniques. For other projects, a conventional development approach supplemented by a few specific techniques and tools will yield the shortest development schedule.

The remainder of this chapter explores a diverse set of RAD concepts. As you read the material, keep in mind that the approaches, tools, techniques, and technologies need to be mixed and matched to specific projects. They are all related by their ability to speed development under some circumstances, but they are a set of alternatives that must be individually evaluated for their effectiveness with a given development project.

rapid application development (RAD)

a collection of development approaches, techniques, tools, and technologies, each of which has been proven to shorten development schedules under some conditions

RAD in Perspective

Previous chapters presented a relatively conventional approach to system development. We divided the SDLC into phases, and each phase into a set of activities. We described a linear flow of activities and a series of models, moving from requirements, to architectural design, and then to detailed design. Presenting development activities in sequential order is a logical way to *teach* systems analysis and design. But it isn't necessarily the best way to *do* analysis and design for a particular project.

The conventional approach to system development tends to be sequential. In particular, it stresses completely defining requirements before design and making major design decisions before implementation begins. The conventional approach does provide some opportunities for parallel activities, particularly within major life cycle phases, but significant use of parallelism and iteration is not the norm in conventional development.

The oldest approaches to software development were purely sequential. They had their roots in an era when:

- Systems were relatively simple and independent of one another.
- Computer hardware resources were very expensive.
- Software development tools were relatively primitive.

A sequential approach to system development is well suited to such an environment.

Underlying older development approaches are these assumptions: Simple systems are simple to analyze and model, and it is reasonable to assume that requirements can be captured fully and accurately before any design or construction activity begins. Complete analysis prior to design and construction is not only reasonable but an economic necessity. The high cost of computer resources and the labor-intensive nature of programming make building the "wrong" system a very expensive and time-consuming mistake. Complete design prior to construction is also necessary. Issues of technical feasibility and operational efficiency are paramount when computing resources are expensive, and those issues are best addressed by designing the entire system all at once.

Now consider the current validity of the assumptions just listed. Information systems today often consist of millions of lines of code and are interconnected with many other systems. Computer power has doubled every two to three years at no increase in cost. Labor costs have risen, and there is a shortage of skilled personnel in most areas of software development. Software development tools leverage greater hardware power to allow programs to be developed quickly and cheaply. Clearly, none of the original assumptions that motivated sequential system development apply today. But does that mean that sequential development should never be used? The answer is no, but the reasons are complex and interdependent.

As described earlier, shifting requirements are common because of changes in the external environment and uncertainties about requirements at the beginning of a project. For example, consider the following scenario. A competitor has successfully implemented a new customer support system based on new technology. Another firm responds by initiating a project to develop a similar system, but it has no experience with that type of system or with the underlying technology.

In this scenario, it isn't reasonable to assume that all system requirements can be specified completely and accurately before design and construction begin. Also, technical, operational, and economic feasibility cannot be completely determined at the beginning of the project. Thus, a sequential development approach is poorly suited to the project. An evolutionary approach—with experimentation and learning—is more appropriate. Such an approach allows requirements to be discovered and gradually refined. It also allows both users and developers to work their way up the learning curve of the new technology gradually.

Now let's revisit an earlier premise—the shortest possible schedule cannot be achieved if software is constructed multiple times. To ensure that you don't build software more than once, you must ensure that you build the right software and that your efforts produce quality products. The first of these requirements is essentially a restatement of the need for complete and accurate analysis and design prior to construction. But how is it possible if requirements are complex and shifting or feasibility is uncertain?

Obviously, you can't avoid all rework. Thus, you must accept some as a consequence of uncertainty about requirements or feasibility, and you must restructure your approach to system development to accommodate these uncertainties. The exact way in which you restructure your approach will depend on:

- Project size
- The degree of (un)certainty of requirements or feasibility at the start of the project
- The expected rate of change in user requirements during the life of the project
- The experience and confidence that developers have in the proposed implementation technology

A large project size, uncertain or shifting requirements, and new technologies are all indicators of a need to depart from sequential development. Projects combining several of these characteristics require the most significant SDLC modifications, as summarized in Figure 17-2.

This is not to suggest that you completely abandon sequential or conventional development. Sequential or conventional development is still the fastest way to develop software *when uncertainty about requirements and feasibility is low*. We're also not suggesting that you abandon the techniques described in earlier chapters. All of the techniques described in earlier chapters are useful and efficient for many types of projects (otherwise, we wouldn't have spent hundreds of pages describing them!). But when technology is new or requirements can't be completely specified, then techniques must be reorganized within an iterative development approach.

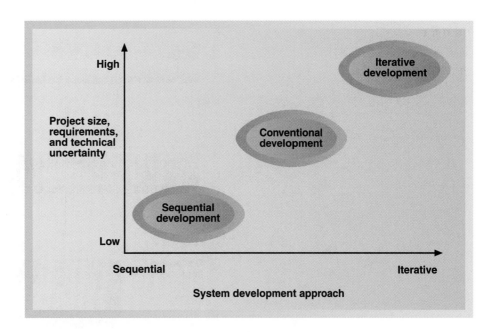

THE PROTOTYPING APPROACH TO DEVELOPMENT

Prototyping was described earlier in Chapter 4 in the section "Build Prototypes." Here we briefly review that material and then describe a complete life cycle based on prototyping. Prototyping is the process of building a partially or fully functioning system model that looks and acts as much as possible like a real system. A prototype is a system in its own right, but it may be incomplete or may not exactly match the final set of user requirements.

discovery prototype

a prototype system used to discover or refine system requirements or design parameters

developmental prototype

a prototype system that is iteratively developed until it becomes the final system

prototyping development approach

a development approach based on iterative refinement of a developmental prototype

Two types of prototypes are commonly used in software development. A *discovery prototype* is often used during analysis and occasionally during design. A *developmental prototype* is a prototype system that is not intended to be thrown away. Developmental prototypes become all or part of the final system. Developmental prototypes are primarily used in iterative software development. Typically, a simple prototype is developed, and more features are added in subsequent development phases. Or, a series of independent prototypes may be developed and later combined to form a complete system.

Steps in the Prototyping Development Approach

Figure 17-3 shows a *prototyping development approach*. The planning phase is as described in Chapter 2. The analysis phase can be implemented with traditional or OO techniques, provided that the techniques are compatible with those used for design and implementation. Analysis models may be less detailed than with conventional development approaches, under the assumption that some requirements will be discovered or fully specified by prototyping.

Design and implementation start out similarly to conventional design phases but differ substantially after architectural design is completed. Architectural design establishes the overall top-level design parameters and implementation environment. The analyst plans the remaining design and implementation activities by defining a series of prototypes or modifications to them. Typically, the requirements are divided into subsets based on specific functions (for example, order entry, order fulfillment, and purchasing) and architectural system boundaries (such as client and server). These subsets may be further divided into a series of implementation steps, starting with a core set and adding functions in each step.

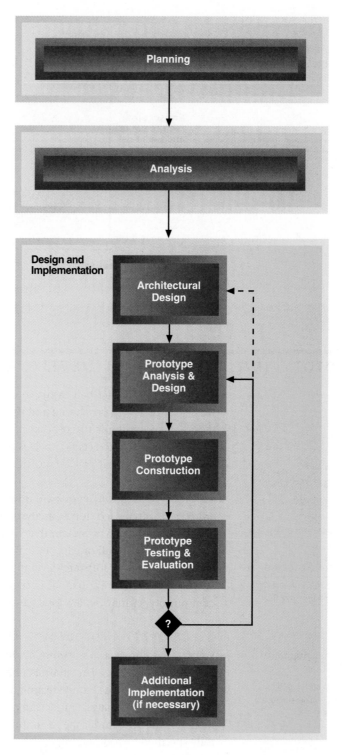

Once the prototype series has been defined, detailed design and implementation proceed in an iterative fashion. Each cycle begins by defining the requirements and design of a single prototype or prototype version, including aspects of analysis that were left unfinished earlier (such as detailed user-interface specifications). Once the developer determines the requirements and detailed design of a prototype, it is developed, tested, and evaluated. Testing and evaluation determine whether the prototype meets the objectives for the current cycle and whether any additional cycles are required. Information uncovered during testing and evaluation may require the developer to modify the development plan or, in unusual cases, the architectural design.

Some requirements of the system may be left out of the prototype development cycle and completed after the final cycle. Examples of such requirements include backup and recovery operations and systemwide security measures. Requirements with no interactive interfaces and those with a systemwide impact are generally good candidates for postprototype implementation.

When to Use a Prototyping Approach

The prototyping development approach usually results in faster system development than a conventional development approach under some or all of the following conditions:

- Some portion of the requirements cannot be fully specified independently of architectural or detailed design.
- Technical feasibility for some system functions is unknown or uncertain.
- Prototype development tools are powerful enough to create a fully functional system.

As discussed in Chapter 4, prototyping is a good means of solidifying uncertain requirements. Users can generally provide better feedback about requirements when examining a prototype than when examining graphical or textual models. A prototyping development approach allows full specification of some requirements that can be combined with design and implementation. Some of the requirements take longer to specify than others, and those can be prototyped to avoid delays in the entire project. However, a prototyping approach does work best when most of the requirements are understood in advance. If significant uncertainty exists, other techniques and development approaches may be better for the project.

A prototyping development approach also works well to ascertain the technical feasibility of a design. Development and testing of early prototypes help sort out problems in design or implementation. Unsuccessful attempts are simply discarded, and the cycle is retried with a different set of parameters. Once success is achieved, the developer can revisit the architectural design and revise the development plan of subsequent prototypes. As with requirements, successful prototyping requires that most aspects of architectural design and technical feasibility be specified in advance. Otherwise, the initial prototype is simply too big and cumbersome for rapid development and testing.

Some functions or requirements may not be suited to prototyping. Examples include systems or portions of a system that are:

- Noninteractive (such as a program that automatically generates orders to suppliers)
- Internally complex (such as a module that schedules deliveries for fastest delivery time or minimal cost using a complex algorithm)
- Subject to stringent performance or security requirements (such as a program that generates thousands of electronic payments per hour)

Noninteractive programs and systems exhibit little observable behavior to test directly or to validate. Thus, a developer can derive little value from prototyping their requirements. Internally complex programs or systems often have relatively well-defined processing requirements that can be formally stated (for example, mathematically). Such requirements are more easily implemented by using a conventional development approach based on structured or OO requirements and design models. Iterative development actually wastes development time for those systems because the repeated design, testing, and evaluation are unnecessary. Finally, software with stringent performance requirements must be constructed with tools that are optimized to produce efficient executable code. Unfortunately, such tools are often not optimal for development speed.

Prototyping Tool Requirements

Successful prototyping requires system developers to employ tools with power, flexibility, and developmental efficiency. Many of the modern "visual" programming tools (such as Microsoft Visual Studio .NET and PowerBuilder) satisfy these requirements. Many more specialized tools (such as Oracle Forms) satisfy the requirements for specialized applications and technical environments.

The key requirement for a successful prototyping tool is development speed. The ability to build and examine many prototypes is crucial to elicit complete and accurate user requirements and to determine technical feasibility through experimentation. The development tools must make it possible to construct, modify, or augment a prototype in a matter of minutes, hours, or at most a few days. Prototyping tool requirements are sometimes summarized by the phrase *FPF principle:* Make it *Functional*, make it *Pretty*, and make it *Fast*.

Flexibility and power are necessary for rapid development. Successful prototyping tools employ a highly interactive approach to application development. Other typical techniques and capabilities include:

- WYSIWYG (what you see is what you get) development of user-interface components
- Generation of complete programs, program skeletons, and database schema from graphical models or application templates
- Rapid customization of software libraries or components
- Sophisticated error-checking and debugging capabilities

Figure 17-4 shows an example of WYSIWYG development of a user-interface prototype for the RMO customer support system. The dialog box appears exactly as it would when displayed by an application program. To add user-interface components to the dialog box, you drag templates from the tool set on the left. You can customize features by clicking on a component and altering values in its properties list. To edit the underlying application program code, you can double-click the component and edit the statements in the pop-up code editor window that appears.

Different prototyping tools provide varying capabilities. No one tool is best in all categories, and many tools are specialized for specific technical environments (such as operating or database management systems) or application types (such as data entry and query to or from a DBMS and Web site development). Thus, it is important to choose the tool that best matches the characteristics of the project at hand.

Prototyping tools must also be chosen according to specific project conditions, including:

- Suitability to the technical environment in which the system will be deployed
- Ability to implement all system requirements
- Ability to interface with software developed with other tools

For example, a developer may use a simple PC software package such as Microsoft Office to develop the prototype, but if the deployment environment is an IBM mainframe supporting thousands of interactive users, then the prototype is unlikely to function reliably in that environment. When prototyping and deployment environments differ substantially, it is typically more feasible to redevelop the application from scratch, using the prototype only as a tangible representation of system requirements.

Some prototyping tools are designed to develop software components that can be easily incorporated into (or extended by) other software modules. Others are designed to produce stand-alone software, which is fine as long as the tool is capable of building software for all of the required system features. If there are any gaps in the tool's capabilities, then that tool is infeasible because the prototype will not be able to function with software developed by other tools.

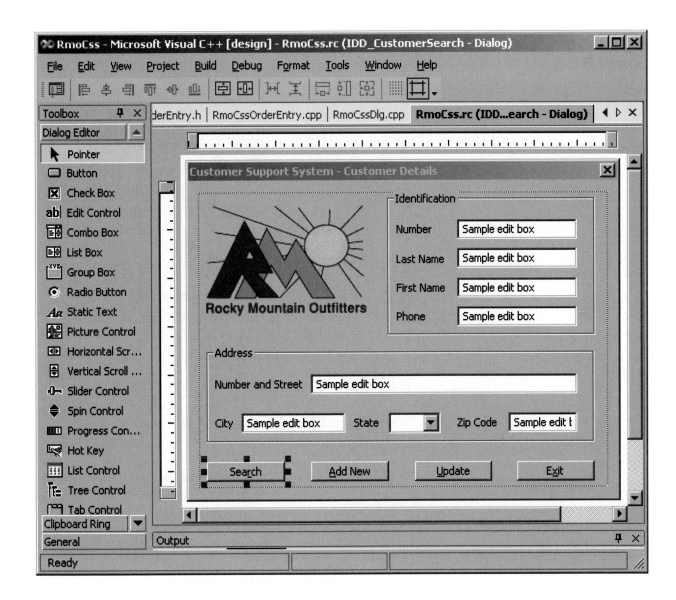

F I G U R E *17-4*

Development of a user-interface prototype using a WYSIWYG dialog box editor.

Fortunately, in recent years, incompatibility of tools has become less of a problem. In particular, object-oriented, component-based, and Web service technologies and standards have made software interoperability more achievable and someday perhaps will make it the norm. For the present, incompatibility is a potential problem that must be recognized and avoided.

THE SPIRAL APPROACH TO DEVELOPMENT

spiral development approach

an iterative development approach in which each iteration may include a combination of planning, analysis, design, or development steps

The *spiral development approach* is an iterative development approach in which each iteration may include a combination of planning, analysis, design, or development steps. First described by Barry Boehm (see "Further Resources"), the approach has since become widely used for software development. There are many different ways of implementing the spiral development approach. Figure 17-5 illustrates one version that uses developmental prototypes (for other versions, see the Boehm and McConnell references in the "Further Resources" section). It is a more radical departure from traditional development than the prototyping development approach described earlier.

FIGURE 17-5

The spiral life cycle.

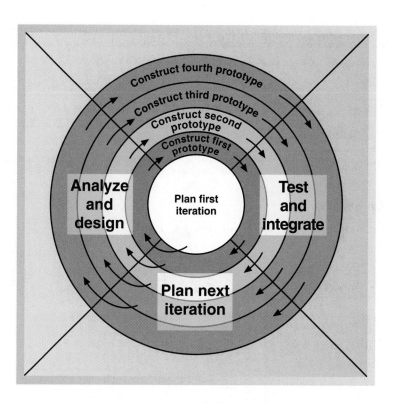

Steps in the Spiral Development Approach

The spiral development approach begins with an initial planning phase, as shown in the center of Figure 17-5. The purpose of the initial planning phase is to gather just enough information to begin developing an initial prototype. Planning phase activities include a feasibility study, a high-level user requirements survey, generation of implementation alternatives, and choice of an overall design and implementation strategy. A key difference between the spiral approach and the prototyping approach is the detail of user requirements and software design. The spiral approach uses less detail than the prototyping approach.

As in the prototyping approach, you can also develop preliminary plans for a series of prototypes. But these plans must be very flexible because analysis and design activities are very limited. Because this approach bypasses many analysis and design details, you often must alter plans later in the project. Thus, you should avoid detailed planning for future prototypes.

After the initial planning is completed, work begins in earnest on the first prototype (the blue ring in Figure 17-5). For each prototype, the development process follows a sequential path through analysis, design, construction, testing, integration with previous prototype components, and planning for the next prototype. When planning for the next prototype is completed, the cycle of activities begins again. Although the figure shows four prototypes, you can adapt the spiral development approach for any number of prototypes.

Which features to implement in each cycle's prototype is an important decision. Developers should choose features based on a number of criteria, including:

- User priorities
- Uncertain requirements
- Function reuse
- Implementation risk

How these criteria are evaluated varies widely from one project to another.

System requirements are typically prioritized into categories such as "must have," "should have," and "nice to have." One way to minimize schedule length is to include the "must have" and "should have" requirements in the earliest prototypes. Doing so greatly increases the probability that you can deliver a usable system to the customer while minimizing development time. If work takes longer than expected, lower-priority requirements can be delayed until a future system upgrade, added after installation, or ignored entirely.

As described earlier, prototyping is an excellent tool for firming up uncertain or poorly defined user requirements, and including them in early prototypes allows you to explore them and specify them fully as soon as possible. That knowledge can then guide subsequent prototype development. Delaying the development of poorly defined portions of the system may result in unanticipated rework if they are incompatible with portions that are already implemented.

Functions that will be used many times are excellent candidates for inclusion in early prototypes. Typical examples include data-entry screens, data lookup and retrieval functions, and problem domain classes. High-risk functions are also excellent candidates for inclusion in early prototypes. Risk analysis is discussed further in the next section.

Benefits and Risks of Spiral Development

The spiral development approach has many advantages over traditional and prototyping approaches, including:

- **High parallelism.** Many opportunities arise to overlap activities both within and among prototyping cycles. For example, planning of the next prototype can usually overlap testing and integration of the previous prototype.
- **High user involvement.** Users can be involved at each planning, analysis, and testing stage. Frequent and continual user participation produces a better system and higher user satisfaction.
- **Gradual resource commitment.** Resource consumption is much more evenly spread out over a spiral life cycle, which may lead to more efficient utilization of some resources (such as personnel). However, the total development cost is generally higher. Figure 17-6 compares typical expenditures for traditional and spiral development approaches.
- **Frequent product delivery.** Every prototype is a working system in its own right. With sufficient planning, you can put these prototypes to work immediately. Frequent product delivery also leads to more testing, thus improving the product by catching more bugs.

So why isn't every system developed with the spiral development approach? The primary drawbacks of the spiral approach are management and design complexity. Projects that use spiral development are more complex to manage than traditional projects because more activities occur in parallel and more people are working on the project at earlier stages. Also, because not all analysis and design occurs before construction, some rework is more likely to be necessary.

In a sequential development approach, most or all design activities are completed before construction begins. When a design is specified completely for an entire system, the result is generally of higher quality than if the same system were designed one piece at a time. In the spiral approach, fewer high-level design activities are performed before construction begins, but design decisions that may seem optimal when looking at a portion of the system may be less desirable when looking at the entire system. In mathematics and management science, this situation is sometimes described as "locally optimal but globally suboptimal."

FIGURE *17-6*

Cumulative cost plotted against time for spiral and sequential development.

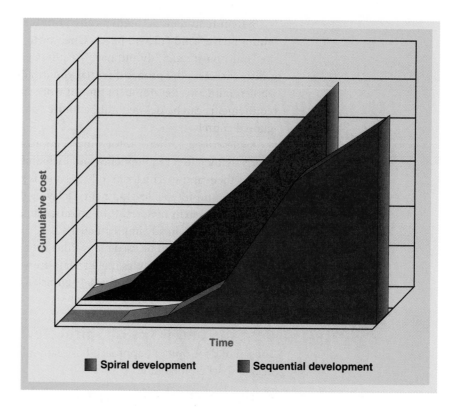

A designer using the spiral development approach runs a much greater risk of making decisions that are globally suboptimal (possibly resulting in lower performance, a greater number of bugs, or more difficult maintenance). Whether or not these problems are significant depends on several factors, including:

- Interdependence among system functions
- Expected life of the system
- Design team experience and skill
- Luck!

Systems with highly interdependent functions run a greater risk of design problems when the spiral approach is employed because design decisions for one function also affect many other parts of the system. Parts of the system that are constructed in later iterations may inherit a large number of design constraints that prove difficult or impossible to resolve.

Systems with a long expected lifetime (for example, more than five years) need more careful and all-encompassing design than those with short lives. Because almost all information systems are modified during their lifetimes, the longer the life, the greater the number of modifications. Modifications have a cumulative effect on software quality. As you add more features to the system, it becomes less efficient, less reliable, and more difficult to maintain. Eventually, these problems build to the point that you must scrap or reimplement the system.

Although no one can anticipate all future needs, a designer can allow for future modifications and enhancements. Systems can be designed for rapid development, maximum ease of modification, or both. But designing for future upgrades and flexibility is a task best performed on an entire system, not iteratively on its subsystems. Traditional development approaches provide greater flexibility in the design because more design activities are completed earlier in the project life cycle.

It is difficult to decide how much analysis and design to do in the initial planning phase and what can be left for later phases. The experience and skill of the analysis and design team are critical to this decision. Luck is also a factor. Even the most skilled and experienced software developers can misidentify a critical analysis or design area. The eventual discovery of its importance may require redoing or throwing away some already completed work.

EXTREME PROGRAMMING

eXtreme Programming (XP)

a rapid development approach focused on creating user stories, delivering releases of a system, and quickly testing

EXtreme Programming (XP) is a system development approach created by Kent Beck in the mid-1990s. XP borrows heavily from earlier development approaches and techniques such as prototyping, object-oriented development, and pair programming. However, XP combines elements and techniques borrowed from other approaches in a unique way.

XP Activities

Figure 17-7 shows an overview of the XP system development approach. The XP development approach is divided into three levels—system (the green ring), release (the red ring), and iteration (the orange ring). System-level activities occur once during each development project. A system is delivered to users in multiple stages called *releases*. Each release is a fully functional system that performs a subset of the full system requirements. A release is developed and tested within a period of no more than a few weeks or months. Releases are divided into multiple iterations. During each iteration, developers code and test a specific functional subset of a release. Iterations are coded and tested in a few days or weeks.

FIGURE *17-7*

The eXtreme Programming system development approach.

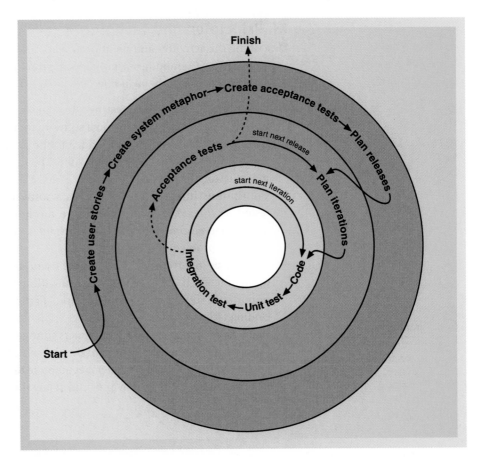

user story

a use case; used in eXtreme Programming

system metaphor

a class diagram that represents a system design in eXtreme Programming

The first XP development activity is creating *user stories*, which are similar to use cases in OO analysis. A team of developers and users quickly documents all of the user stories that will be supported by the system. Developers then create a class diagram to represent objects of interest within the user stories. In XP, the class diagram is called a *system metaphor*.

Developers and users then create a set of acceptance tests for each user story. Releases that pass the acceptance tests for the user stories that they support are considered finished. The final system-level activity is to create a development plan for a series of releases. The first release supports a subset of the user stories, and subsequent releases add support for additional stories. Each release is delivered to users and performs real work, thus providing an additional level of testing and feedback.

The first release-level activity is planning a series of iterations. Each iteration focuses on a small (possibly just one) system function or user story. The iterations' small size allows developers to code and test them within a few days. A typical release is developed using a few to a few dozen iterations.

Once the iteration plan is complete, work begins on the first iteration-level activity. Code units are divided among multiple programming teams, and each team develops and tests its own code. XP recommends a test-first approach to coding. Test code is written before system code. As code modules pass unit testing, they are combined into larger units for integration testing. (Testing was covered in more detail in Chapter 16.) When an iteration passes integration testing, work begins on the next iteration.

When all iterations of a release have been completed, the release undergoes acceptance testing. If a release fails acceptance testing, the team returns it to the iteration level for repair. Releases that pass acceptance testing are delivered to end users, and work begins on the next release. When acceptance testing of the final release is completed, the development project is finished.

XP Principles and Techniques

Describing the activities and iteration levels of XP doesn't paint a complete picture of the approach. XP embodies a number of principles and techniques that are woven throughout the approach, including:

- **Continuous automated testing.** Testing activities occur every day in the XP approach. When work begins on a software module or system function, tests related to that module or function are added to the permanent test suite, and all iterations and releases are tested against that suite. The intensive level of testing requires automated testing software.
- **Continuous integration.** As soon as a software module passes unit testing, it is tested in concert with other software modules. That way, errors are discovered quickly. Development of additional modules, iterations, and releases does not proceed until problems are resolved.
- **Heavy user involvement.** One or more users are permanently assigned to the development team. If the organization is unwilling to do so, then perhaps the project's importance should be questioned. Users participate in all important decisions, including technical ones.
- **Team programming.** Two programmers sitting in front of one workstation and sharing a single display and keyboard develop all software. This method is called *pair programming*. Also, programmers regularly review each other's work. Any programmer is allowed to change any program at any time.

pair programming

a programming method in which software is developed by two programmers sitting in front of one workstation sharing a single display and keyboard

- **Specific attention to human interactions and limitations.** The entire development team works within a common room—as large as needed. Programming workstations are placed in the center, and cubicle work spaces are placed around the perimeter. Every team member can see and communicate with every other member. Forty-hour work weeks are the norm. Overtime is prohibited to avoid burnout and excessive mistakes.

XP Compared with Other Development Approaches

XP techniques are inherently object-oriented, although the terminology for those activities differs from that used in the Unified Modeling Language and Unified Process. Other aspects of XP are borrowed from other development approaches and techniques, including the spiral development approach and joint application development.

Figure 17-8 compares the number of iterations and the relative effort expended on planning, analysis, design, and implementation for traditional development, spiral development, and XP. Moving from traditional to spiral to XP, the number of iterations increases and the total effort expended on planning, analysis, and design decreases. Within implementation, XP also differs from traditional and spiral development in that coding is more tightly interwoven with testing.

FIGURE *17-8*

Comparison of traditional development, spiral development, and eXtreme Programming.

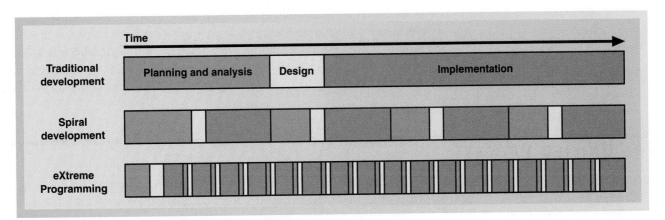

The merits and limitations of XP have been debated in many articles, books, and Web sites over the last several years. A clear consensus has not emerged yet, but XP does enjoy a substantial following in the software development community.

Arguments in favor of XP include:

- It borrows many proven techniques and principles from other development approaches.
- It cuts out the "fat" of long analysis and design phases. The potential negative implications of less stringent analysis and design are mitigated by continuous testing and occasional shuffling of existing code.
- It builds group ownership of and enthusiasm about the project.
- It has been proven successful in published case studies.

Arguments against XP include:

- Its team-based approach works well for smaller projects but scales up poorly to larger projects.
- It places too little emphasis on analysis and architectural design, creating the possibility that later iterations and releases will be suboptimal implementations.

When to Use XP

The published case studies and commentaries to date paint a glowing picture of XP's success under some very specific conditions:

- Small development teams of a dozen or fewer members
- Talented development personnel with a broad range of modeling, technical, and communication capabilities
- Scope limited to stand-alone systems, new systems, and systems with minimal interfaces to legacy hardware and software
- Extensive use of high-quality OO development and testing tools

For development projects with these resources and parameters, XP is a proven way to shorten the development life cycle.

XP is least successful in larger projects with many assigned staff. The close-knit teams and instant communication required by XP are simply not possible in large projects. As an analogy, compare a dozen people building a research prototype in a lab with the personnel and effort required to manufacture thousands of prototype copies per week. Large-scale projects, whether for manufacturing or software development, require specialization of personnel, hierarchical organization, formal methods of communication, and written documentation—all of which XP purposely avoids.

Systems that extend the capabilities of or interface with legacy systems may or may not be candidates for development with XP. Because of XP's reliance on OO methodologies and tools, it is not ideal for extending a legacy system built with other technology. Also, systems involving legacy hardware and software leave developers little freedom. In general, such systems require more thorough analysis and documentation than XP provides.

THE UNIFIED PROCESS

Unified Process

an OO development approach that emphasizes frequent iteration, risk management, testing, and user feedback

The *Unified Process (UP)* is a comprehensive development approach that combines practices from other development approaches and adapts them to OO models, tools, and techniques. It was originally developed by Jacobsen, Booch, and Rumbaugh in the late 1990s. The UP is currently the dominant approach for developing software with OO models and tools.

The UP Compared with Other Development Approaches

The UP adopts its most important principle—iteration—from the prototyping and spiral development approaches. Development proceeds through a series of iterations, each of which can incorporate planning, analysis, design, and construction activities. Each iteration produces a working but incomplete subset of the final system, with each iteration adding to or modifying the outputs of the previous iteration so that models and working software grow to become the finished product.

The primary differences between the UP and earlier iterative approaches are its exclusive reliance on OO models and tools and specific restrictions on iteration length and activities. Use cases are the primary means of documenting requirements, and other OO analysis and design models fill a necessary but supporting role. The UP is not dogmatic about the exact number and type of models that are developed. Rather, the mix of models is chosen based on specific project characteristics.

Compared with XP, the UP is a more formal process and generally includes more up-front planning, analysis, and design activities, which are always carried out within the context of well-defined iterations. As in XP, models are developed only as needed to build working software—the models themselves are not an end product. Even though the UP is similar to XP in its spare use of models, XP avoids modeling to a greater extent than does the UP. In essence, XP and the UP are moderately different descendants of the prototyping and spiral development approaches, one specialized for speed with tight-knit development groups and the other formalized for larger projects.

How the UP Organizes Software Development

To make a clean break with older, sequential development approaches, the UP adopts new terms to describe SDLC activities. The UP's SDLC includes four high-level activities:

- Inception
- Elaboration
- Construction
- Transition

Inception is a high-level activity similar to project planning as described in Chapter 3. During inception, key project parameters such as scope, participants, business purpose, and initial budget and schedule estimates are defined. Scope is defined primarily through use cases, though they are not necessarily complete or exhaustive.

Note that the definitions of inception and other high-level activities avoid using the word *phase*. The UP avoids that term because many people believe that it implies a strict sequence of activity types. In the UP, there are no phases per se, only iterations. But different iterations have different emphases, as shown in Figure 17-9, and incorporate different mixes of activities. The term *inception* describes the emphasis of one or more early project iterations.

inception

earliest activity of the UP's systems development life cycle, encompassing strategic planning

FIGURE *17-9*

A UP development project is a series of iterations in which emphasis shifts from inception, to elaboration, to construction, to transition.

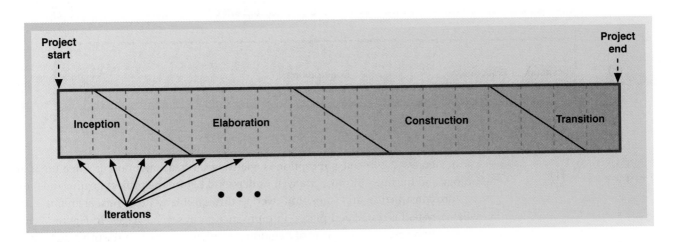

elaboration

second activity of the UP's systems development life cycle, encompassing planning, analysis, design, and construction of the highest-risk portions of a system

construction

third activity of the UP's systems development life cycle, encompassing programming, installation, and testing of lower-risk and simpler portions of a system

Elaboration is a high-level activity that embodies aspects of planning, analysis, design, and construction. The purpose of elaboration is to move beyond inception by defining the requirements and scope in more detail, estimating the budget and schedule with greater precision, and designing and constructing key architectural aspects of the final system. Elaboration concentrates on the highest-risk portions of the system. By dealing with these aspects in early iterations, the UP moves quickly to minimize overall project risk and reduce the uncertainty associated with later iterations.

The common interpretation of the term *construction* implies a large and pervasive set of activities such as programming, installation, and testing. Within the UP, *construction* iterations do include such activities, but only for lower-risk and simpler portions of the system that were not addressed in earlier iterations. In addition, construction iterations may incorporate analysis and design activities that were not performed earlier. All iteration types include planning activities for subsequent iterations. The key difference between elaboration and construction is the amount of requirements discovery and the degree of risk. Elaboration iterations focus on higher-risk aspects of the system and typically embody more discovery activities than construction iterations.

Transition is a high-level activity that moves a system from development into production. Transition iterations shift the focus from adding and testing incremental features to testing the system as whole and deploying it within its operational environment.

As shown in Figure 17-10, different projects may have very different arrangements of high-level activities. For example, the arrangement depicted in Figure 17-10a might be appropriate for a project that develops cutting-edge software to support an entirely new business paradigm. Inception and elaboration cover many iterations due to the significant number of risks and discovery activities. Construction is relatively short because most software is developed and tested during elaboration.

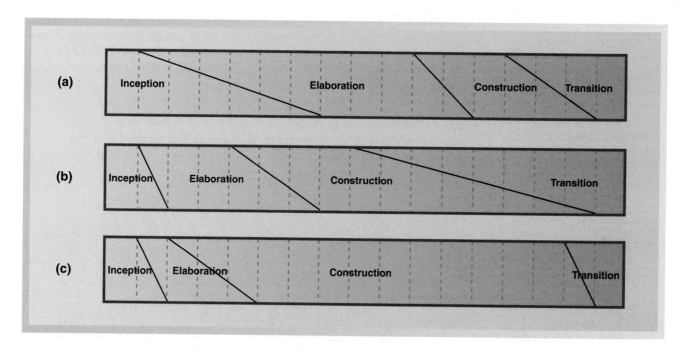

In Figures 17-10b and c, inception is relatively short, as might occur when the technology and business purpose are well understood before the project begins. In Figure 17-10b, construction and transition overlap through many iterations, as might occur when a system is tested and placed into production in several stages. In Figure 17-10c, construction covers the majority of iterations, as might occur when rewriting an existing application in a new programming language.

Iterations and Disciplines

Note that the dashed boxes representing iterations in Figures 17-9 And 17-10 are all the same size. This reflects a key UP concept called *timeboxing*, which simply means that all iterations are the same length. Development work is packaged to fit the time boxes, not vice versa. If it appears that an iteration will exceed its schedule, then the scope of the work is reduced to fit the remaining allotted time.

The benefits of iterative development are best realized when iterations are relatively short and when each iteration produces a concrete result. The UP recommends short iterations—from several weeks to a few months. Short iterations result in more frequent testing, more frequent and immediate feedback from users, and higher motivation of and a greater sense of accomplishment by all project participants.

discipline
a set of functionally related lower-level UP activities

So far, we have focused primarily on iterations and higher-level organization of the UP's SDLC. But what about the detailed activities that occur within iterations? To describe those activities, the UP defines more terms, the most important of which is discipline. A ***discipline*** is a set of functionally related activities that can occur in many different iterations. Jacobsen, Booch, and Rumbaugh define a core set of UP disciplines including:

- Business modeling
- Requirements
- Design
- Implementation
- Testing
- Deployment
- Configuration and change management
- Project management
- Environment

Like the high-level activities of inception, elaboration, construction, and transition, the disciplines are actually categories of lower-level activities, organized by functional specialty instead of by the time in the SDLC at which they are emphasized. Because the entire SDLC is based on iterations, activities from many different disciplines are performed within each iteration. Some disciplines, such as business modeling, tend to occur in early iterations, and others, such as deployment, tend to occur in later iterations. Other disciplines, such as project management and testing, occur in nearly every iteration although their relative emphasis and specific details may vary.

Figure 17-11 shows the distribution of effort for five disciplines across inception, elaboration, construction, and transition for a sample project. Color intensity increases as the relative effort for a discipline increases. In this project, activities related to the business modeling discipline occur primarily during inception and decrease after the project moves into elaboration. The exact mix of disciplines within each iteration varies from project to project and among iterations within a project.

FIGURE 17-11
Distribution of disciplines across high-level activities in a sample project.

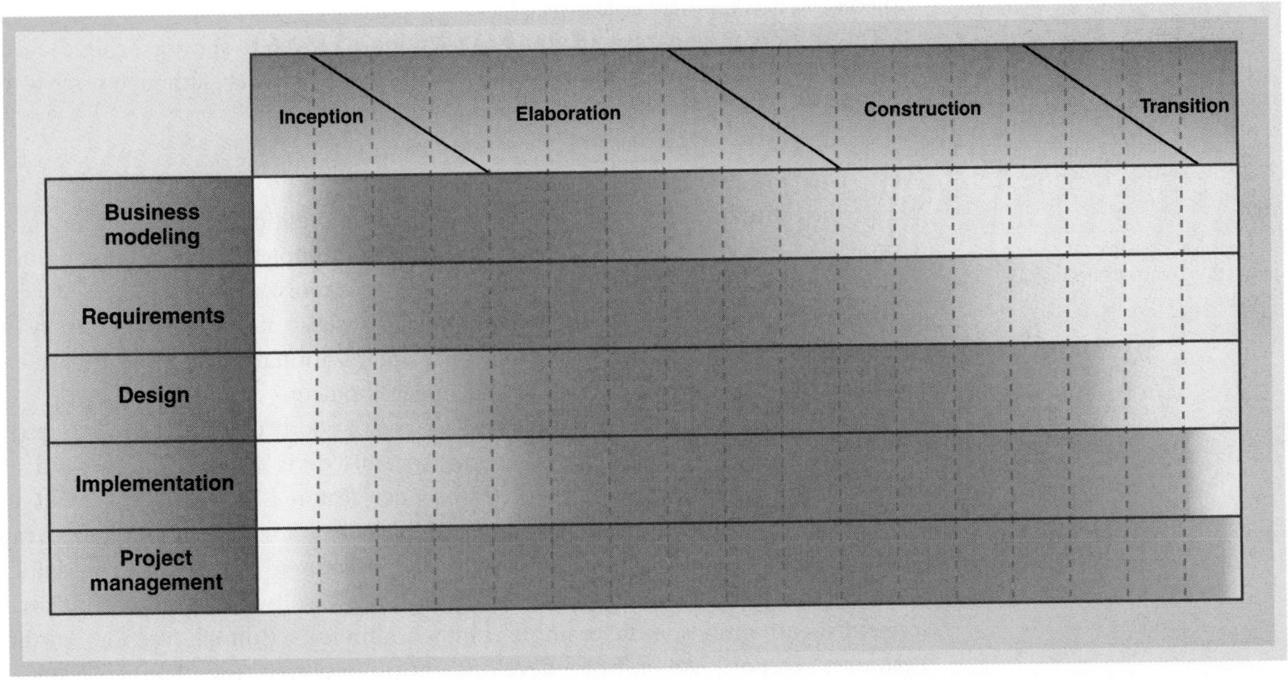

When to Use the Unified Process

The benefits and risks of the UP mirror those described earlier for spiral development. Since the UP was formalized in the late 1990s, there have been numerous studies of its effectiveness, and the consensus is that it is an effective process capitalizing on the theoretical benefits of iterative development. The major obstacles to its adoption include complex project management (compared with sequential development) and the need to adopt OO models, tools, and techniques throughout the project. Some early projects were hampered by development staff's lack of experience with OO analysis and design methods. But many developers are now trained in those methods, so that problem has diminished.

In deciding whether to use the UP or its primary competitor, XP, the critical trade-off is between speed and formality. XP avoids process formality and model development and directs the saved effort toward developing and testing software. As described earlier, this approach has demonstrated success with small and talented developer teams, systems with limited external interfaces, and projects that use high-quality development and deployment tools. Considerable debate still rages over whether and how XP practices can be adapted to larger projects.

The formality embedded within the UP is a reflection of its bias toward larger projects, particularly those with big, geographically dispersed development teams and projects that are completed under contract by external developers. In such projects, the UP's formal steps, well-defined roles, and significant attention to model building and validation address issues of project control and communication, which are handled much less formally under XP. Many developers and most managers of large-scale development projects believe that they need the formality of UP in their development environment.

RAPID DEVELOPMENT TECHNIQUES

Chapter 2 defined a *technique* as a collection of guidelines that help an analyst complete a system development activity or task. Many techniques have been developed over the years to speed development. Some live up to their claims, although none is suited to all projects and development scenarios.

This section presents a small group of techniques proven to shorten project schedules. None is unique to a particular system development approach, although some work better with (or are required by) a particular approach.

Risk Management

risk

a condition or event that can cause a project to exceed its shortest possible schedule

Within the context of software development, a *risk* is a condition or event that can cause a project to exceed its shortest possible schedule. Examples include changing user requirements; failure of hardware, support software, or tools; and loss of needed development resources such as funding or personnel. Risks also arise from dependency on others, including clients, suppliers, and other organizational units. Risks are present regardless of what approach is taken to system development.

Conditions are states that exist but are unknown or poorly understood at the present time. For example, when upgrading an existing system, a developer may become aware of bugs in critical software components only after a new feature is implemented. Events are things that may occur but haven't yet. Examples include user requirement changes, reassignment of key personnel, and failure of a hardware vendor to meet a delivery deadline.

Figure 17-12 describes several important categories of software development risks. A longer list with more specific examples can be found in McConnell (see the "Further Resources" section). All software development projects face risks that can result in schedule delays.

FIGURE *17-12*

Major categories of development
schedule risk.

System requirements
- Requirements are poorly understood when the schedule is created.
- Customer adds or changes features late in development.
- IS staff insists on unnecessary features (e.g., security or robustness).

Tools and technology
- Target deployment environment is poorly suited to the application.
- Development tools are poorly suited to the target deployment environment.
- New tools have unknown productivity and/or learning curve.
- New technology is unstable.
- Technology standards are not yet fully developed.

People
- Users are not adequately involved in all stages of development.
- Project management is poor.
- Developers are poorly trained or underperforming.
- Developer or manager skills are poorly matched to specific tasks.

Environment
- Upper management's commitment to the project is weak or variable.
- Competitive environment is rapidly changing.
- Technological environment is rapidly changing.
- Governmental action (or inaction) is affecting the project.

Steps in Risk Management

Risk management is a systematic process of identifying and mitigating software development risks. The underlying principles of risk management are based on the following ideas:

risk management

a systematic process of identifying and mitigating software development risks

- Most risks can be identified if specific attention is directed to them.
- Risks appear, disappear, increase, and decrease as the development process proceeds.
- Small risks should be monitored, whereas large risks should be actively mitigated.

Figure 17-13 contains a flowchart for risk management. The process begins at the start of the development project and continues until it is completed. The first step is to identify all risks that are likely to affect the project. This task is relatively difficult but critically important since risks that aren't identified can't be managed. A group of project participants (including users) usually identifies risks, since more heads are better than few to generate and evaluate a large number of ideas.

For each risk, the next step is to estimate its probability, determine possible outcomes, and estimate the probability and schedule delay associated with each outcome. For example, the risk of failing a performance test might be assigned a probability of 0.1 (10 percent) and determined to have two possible outcomes:

- Reprogram key software modules to improve performance.
- Purchase and install faster hardware.

Reprogramming might be assigned an estimated probability of 0.8 (80 percent) and schedule delay of four weeks. Purchasing and installing faster hardware might be assigned an estimated probability of 0.2 (20 percent) and a schedule delay of two weeks. Note that the probability assigned to an outcome is a conditional probability for a specific risk. For example, the 20 percent probability that faster hardware will need to be acquired only is relevant if the system actually fails a performance test.

FIGURE *17-13*

Steps in risk management.

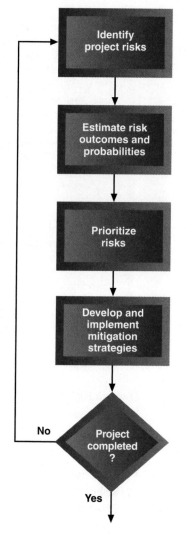

Developers can compute an expected schedule delay for each outcome by multiplying the risk probability, outcome probability, and outcome schedule delay. For example, the expected schedule delay for reprogramming to improve performance is $0.1 \times 0.8 \times 4$ weeks = 0.32 weeks. If all outcomes for a risk are mutually exclusive, developers can add the expected schedule delays for all risk outcomes to produce an expected schedule delay for the risk.

Once outcomes have been analyzed, the risks must be prioritized. Highest priority should go to risks with long expected schedule delays. Risks with long outcomes and low probabilities are also given high priority since probability estimates are inherently inaccurate. The goal of prioritizing risks is to generate a list of "high" risks and a list of "low" risks. Low-priority risks are simply monitored to ensure that they don't become higher-priority risks. High-priority risks call for more active management approaches.

Risk mitigation is any steps taken to minimize an expected schedule delay. Specific mitigation steps can vary widely. For example, the risk of a customer's adding significant requirements late in the project might be mitigated by developing an early throwaway prototype to validate requirements, actively involving the customer in all phases of project planning, or requiring the customer to sign a contract rigidly specifying the system features. Risk mitigation for a failed performance test might include detailed benchmarking of the tool and hardware, development of a test prototype, or parallel development with multiple tools.

risk mitigation

any step(s) taken to minimize the expected schedule delay of a risk

As a project proceeds, more information about risks becomes available. The additional information may arise as a natural by-product of development activities, or it may arise from specific risk mitigation efforts. In either case, new and revised information should be periodically evaluated. Thus, the entire risk management process is actually a loop whose frequency varies from project to project. In projects with few risks and moderately long schedules, developers may reevaluate risks infrequently (for example, after completion of each SDLC phase or after one or more iterations of the spiral, XP, or UP development approaches). Projects with many risks and longer development schedules may require formal reexamination of risks more frequently (for example, at the end of every week).

When to Implement Risk Management
Risk management should always be used—regardless of development approach. But inherently riskier projects call for more thorough and active risk management. Unknown requirements and feasibility are principal elements of overall project risk. Since such unknowns normally motivate developers to use iterative development, active and thorough risk management should be key elements of those approaches. When possible, developers should incorporate mitigation measures for the highest-priority risks into the earliest development activities.

Joint Application Design

Joint application design (JAD) was described in Chapter 4 in the section "Conduct Joint Application Design Sessions." JAD is an effective technique for quickly defining system requirements. It is primarily used as a systems analysis technique and occasionally to specify some higher-level design parameters. JAD shortens the time needed to specify system requirements by including all key decision makers in one or more intensive sessions. It shortens the project schedule by concentrating the efforts of many people in a short space of time. Participation by all key decision makers ensures that the project gets off to a speedy start with solid planning and clearly defined objectives.

Any development approach can incorporate JAD, but it is especially well suited to iterative development because both emphasize prototypes and development speed. JAD can be used to implement most or all of the initial planning phase of the spiral development approach. The user-interface prototypes developed during the JAD session can form the starting point for the first full-fledged developmental prototype in the spiral development approach. JAD is not a formal part of XP or UP, although many JAD principles underlie both approaches.

When used with a conventional development approach, a JAD session does not normally address all of the design issues that the initial planning phase of the spiral development approach does, such as choice of implementation tools, overall architectural design, and initial definition of the evolutionary prototypes. However, these decisions are next in line after those normally considered in a JAD session. The JAD session can be extended to include these decisions, or the results of the JAD session can be the input to a later phase.

Tool-Based Development

Chapter 2 defined the term *tool* as software that helps create models or other deliverables required in a project. The intervening chapters have discussed a large number of tools, including CASE tools, project management tools, database management systems, and integrated development environments. This chapter focuses on tools that are used directly to build software. Such tools include compilers, code generators, and object frameworks but exclude tools such as project management and model development software, unless they are included within a CASE tool or integrated development environment.

No one tool is best for producing all types of software. Tools have different strengths and weaknesses. Some parts of a system can be built very quickly because they match development tool capabilities very well. Other parts are much more difficult to build because of limitations or gaps in tool capabilities. Building these parts of the system requires much more time because human ingenuity and experimentation are required to get the tool to do something it does poorly or was never designed to do.

The premise of *tool-based development* is simple: Choose a tool or tools that best match the requirements and don't develop any requirements that aren't easily implemented with the tool(s). In essence, tool-based development applies the generic 80/20 or 90/10 rule: Resources are best used to construct a system that satisfies the 80 to 90 percent of the requirements most easily implemented, and the 10 to 20 percent of the requirements that are difficult to implement with the tool(s) are discarded.

Although tool-based development is easy to describe, it is very difficult to implement in an organization because it requires a developer to say no early and often. The developer and the customer must agree on what requirements the system will or won't include. The developer must be willing and able to say no to future requests to add difficult requirements. Saying no so often may not be possible because of user needs or project politics.

User satisfaction may be much lower than with other development methods because some desired functions are left out of the system. The developer can mitigate dissatisfaction by clearly describing the schedule and cost impacts of all requirements that don't directly match tool capabilities. On the plus side, design and development can proceed much more quickly, and reductions in coding and testing time are dramatic. Thus, tool-based implementation forces a clear and direct trade-off between development speed and system functionality.

Figure 17-14 shows a simple process for tool-based development in the context of a sequential development approach. New activities are added to the end of the analysis phase, including:

- Investigating development tools that support the system requirements
- Selecting tools and modifying system requirements to match the tool set

Note that requirement priorities are assumed to drive the tool selection, not the other way around. The underlying capability and methodology of the chosen tool set drive subsequent development phases. The developer can also incorporate tool-based development into the prototyping or spiral development approaches by embedding a tool selection activity within the analysis or initial planning phase of those approaches.

If a new tool is chosen, then project costs will increase because the new tool must be purchased and developers must be trained. Project managers must plan carefully to avoid having a new tool (or tools) lengthen the project schedule. Tool acquisition and installation must be completed before implementation activities begin. Training should be concurrent with systems analysis, and it should be completed prior to architectural or detailed design.

A developer must take great care when choosing multiple tools to build a new system. Tools that aren't designed to work together may have hidden incompatibilities that don't become apparent until late in the implementation phase. For example, development tools may use incompatible methods for representing data. Such incompatibilities may not become apparent until integration testing, when data stored in a program developed with one tool are passed via a function call or method invocation to software developed with another tool. Smoothing over these incompatibilities can add as much work and time to a project as adding system features that aren't well matched to a specific tool.

FIGURE *17-14*

Tool-based development applied within a traditional sequential life cycle.

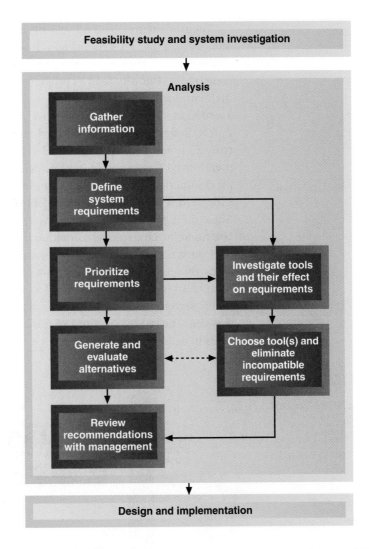

Another danger of tool-based development is the tendency to choose a different tool set for every project. This is particularly problematic when users hear success stories about a new tool set and they want a system similar to the one described in the success story. The problem is that each new tool adopted by a development team dramatically decreases productivity. Every tool has a learning curve, and productivity gains can't be realized until the development team has worked its way up the curve. The first project that uses the new tool typically does not achieve high productivity. Tools less suited to the project's requirements may actually deliver a better system on a shorter schedule if the development team has extensive experience with the tool.

Software Reuse

software reuse, or code reuse

any mechanism that allows software used for one purpose to be reused for another

Software reuse (or *code reuse*) is any mechanism that allows software used for one purpose to be reused for another purpose. Software reuse can significantly shorten a development schedule by reducing the effort, time, and money required to develop or modify information systems. Whether software reuse actually reduces development effort depends on many specifics, including:

- The effort required to identify potentially reusable software
- The extent to which existing software functions require modification to suit a new purpose
- The extent to which existing software must be repackaged into a form that can be plugged into a new system

There are many different ways of reusing software. At one extreme, an existing system (for example, an accounting software package) can be purchased, thus avoiding the vast majority of implementation phase activities. At the other extreme, small portions of code from previously written programs can be reused in another program. Of course, there are many kinds of software reuse between these extremes, and many different approaches, techniques, tools, and technologies can be used.

OO design and programming techniques have provided new tools to address software reuse. One of the reasons that OO development methods have become so popular is the premise that classes and objects are easier to reuse than programs built with traditional programming tools. Note the use of the word *premise* in the previous sentence. There is still too little hard evidence to say definitively that OO development is faster than traditional development. But most software developers do believe that OO development is faster, and ongoing research should soon close the gap between belief and knowledge.

Figure 17-15 shows two particular methods of OO software reuse: source code and executable code reuse. A project team can reuse OO source code by deriving new classes from existing classes or entire object frameworks. Executable objects are sometimes called *components*. A component can be a single object, an entire system, or anything in between. The last two sections of this chapter describe object frameworks and components in detail.

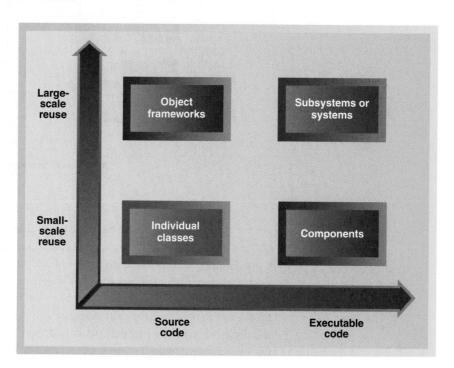

Software reuse is a technique applicable to any system development approach. Analysts and developers must actively search for reusable software. For large-scale reuse, the search must begin before architectural design; however, it can be time-consuming and expensive. The effort needed to search, evaluate, adapt, and integrate existing software for a new purpose may be greater than the effort to build a new system from scratch.

The search for smaller-scale reusable software can occur during systems design or development and is usually less time intensive. In many cases, developers need look no further than their programming toolkit. Most modern programming toolkits reuse software in one or more forms. Examples include object frameworks, program templates, code generators, and component libraries. Modern programming toolkits also provide tools for adapting and integrating reusable code.

Object Frameworks

Similar functions are embedded in many different types of systems. For example, the graphical user interface (GUI) is nearly ubiquitous in modern software. Many features of a GUI—such as drop-down menus, help screens, and drag-and-drop manipulation of on-screen objects—are used in many or most GUI applications. Other functions—such as searching, sorting, and simple text editing—are also common to many applications. Reusing software to implement such common functions is a decades-old development practice. But such reuse was awkward and cumbersome with older programming languages. Object-oriented programming languages provide a simpler method of software reuse.

An *object framework* is a set of classes that are specifically designed to be reused in a wide variety of programs. The object framework is supplied to a developer as a precompiled library or as program source code that can be included or modified in new programs. The classes within an object framework are sometimes called *foundation classes*. Foundation classes are organized into one or more inheritance hierarchies. Programmers develop application-specific classes by deriving them from existing foundation classes. Programmers then add or modify class attributes and methods to adapt a "generic" foundation class to the requirements of a specific application.

Object Framework Types

Object frameworks have been developed for a variety of programming needs. Examples include:

- **User-interface classes.** Classes for commonly used objects within a graphical user interface, such as windows, menus, toolbars, and file open and save dialog boxes.
- **Generic data structure classes.** Classes for commonly used data structures such as linked lists, indices, and binary trees and related processing operations such as searching, sorting, and inserting and deleting elements.
- **Relational database interface classes.** Classes that allow OO programs to create database tables, add data to a table, delete data from a table, or query the data content of one or more tables.
- **Classes specific to an application area.** Classes specifically designed for use in application areas such as banking, payroll, inventory control, and shipping.

General-purpose object frameworks typically contain classes from the first three categories. Classes in these categories can be reused in a wide variety of application areas. Application-specific object frameworks provide a set of classes for use in a specific industry or type of application. Third parties usually design application-specific frameworks as extensions to a general-purpose object framework. Many large organizations have moved aggressively to develop their own application frameworks. Smaller firms usually do not do so because the resource requirements are substantial. An application- or company-specific framework requires a significant development effort typically lasting several years. The effort is repaid over time through continuing reuse of the framework in newly developed systems. The effort is also repaid through simplified maintenance of existing systems. But the payoffs occur far in the future, often making them less valuable than the current funds required to build the frameworks.

The Impact of Object Frameworks on Design and Implementation Tasks

Developers need to consider several issues when determining whether to use object frameworks. Object frameworks affect the process of systems design and development in several different ways:

- Frameworks must be chosen before detailed design begins.
- Systems design must conform to specific assumptions about application program structure and operation that the framework imposes.

- Design and development personnel must be trained to use a framework effectively.
- Multiple frameworks may be required, necessitating early compatibility and integration testing.

The process of developing a system using one or more object frameworks is essentially one of adaptation. The frameworks supply a template for program construction and a set of classes that provide generic capabilities. Systems designers adapt the generic classes to the specific requirements of the new system. Frameworks must be chosen early so that designers know the application structure imposed by the frameworks, the extent to which needed classes can be adapted from generic foundation classes, and the classes that cannot be adapted from foundation classes and thus must be built from scratch.

Of the three object layers typically used in OO system development (view, business logic, and data), the view and data layers most commonly derive from foundation classes. User interfaces and database access tend to be the areas of greatest strength in object frameworks, and they are typically the most tedious classes to develop from scratch. It is not unusual for 80 percent of a system's code to be devoted to view and data classes. Thus, constructing view and data classes from foundation classes provides significant and easily obtainable code reuse benefits. Adapting view classes from foundation classes has the additional benefit of ensuring a similar look and feel of the user interface across systems and across application programs within systems.

Successful use of an object framework requires a great deal of up-front knowledge about its class hierarchies and program structure. That is, designers and programmers must be familiar with a framework before they can successfully use it. Thus, a framework should be selected as early as possible in the SDLC, and developers must be trained in use of the framework before they begin to implement the new system.

COMPONENTS

component

a standardized and interchangeable software module that is fully assembled and ready to use and that has well-defined interfaces to connect it to clients or other components

In addition to using object frameworks, developers often use components to speed system development. A *component* is a software module that is fully assembled, is ready to use, and has well-defined interfaces to connect it to clients or other components. Components may be single executable objects or groups of interacting objects. A component may also be a non-OO program or system "wrapped" in an OO interface. Components implemented with non-OO technologies must still implement objectlike behavior. In other words, they must implement a public interface, respond to messages, and hide their implementation details.

Components are standardized and interchangeable software parts. They differ from objects or classes because they are binary (executable) programs, not symbolic (source code) programs. This distinction is important because it makes components much easier to reuse and reimplement than source code programs.

For example, consider the grammar-checking function in most word processing programs. A grammar-checking function can be developed as an object or as a subroutine. Other parts of the word processing program can call the subroutine or object methods via appropriate source code constructs (for example, a C++ method invocation or a BASIC subroutine call). The grammar-checking function source code is integrated with the rest of word processor source code during program compilation and linking. The executable program is then delivered to users.

Now consider two possible changes to the original grammar-checking function:

- The developers of another word processing program want to incorporate the existing grammar-checking function into their product.

- The developers of the grammar-checking function discover new ways to implement the function that result in greater accuracy and faster execution.

To integrate the existing function into a new word processor, the source code of the grammar-checking function must be provided to the word processor developers. They then code appropriate calls to the grammar checker into their word processor source code. The combined program is then compiled, linked, and distributed to users.

When the developers of the grammar checker revise their source code to implement the faster and more accurate function, they deliver the source code to the developers of both word processors. Both development teams integrate the new grammar-checking source code into their word processors, recompile and relink the programs, and deliver a revised word processor to their users.

So what's wrong with this scenario? Nothing in theory, but a great deal in practice. The grammar-checker developers can provide their function to other developers only as source code, which opens up a host of potential problems concerning intellectual property rights and software piracy. Of greater importance, the word processor developers must recompile and relink their entire word processing programs to update the embedded grammar checker. The revised binary program must then be delivered to users and installed on their computers. This is an expensive and time-consuming process. Delivering the grammar-checking program in binary form would eliminate or minimize most of these problems.

A component-based approach to software design and construction solves both of these problems. Component developers, such as the developers of the grammar checker, can deliver their product as a ready-to-use binary component. Users, such as the developers of the word processing programs, can then simply plug in the component. Updating a single component doesn't require recompiling, relinking, and redistributing the entire application. Perhaps applications already installed on user machines could query an update site via the Internet each time they started and automatically download and install updated components.

At this point, you may be thinking that component-based development is just another form of code reuse. But structured design, object frameworks, and client-server architecture all address code reuse in different ways. What makes component-based design and construction different are the following:

- Components are reusable packages of executable code. Structured design and object frameworks are methods of reusing source code.
- Components are executable objects that advertise a public interface (that is, a set of methods and messages) and hide (encapsulate) the implementation of their methods from other components. Client-server architecture is not necessarily based on OO principles. Component-based design and construction are an evolution of client-server architecture into a purely OO form.

Components provide an inherently flexible approach to systems design and construction. Developers can design and construct many parts of a new system simply by acquiring and plugging in an appropriate set of components. They can also make newly developed functions, programs, and systems more flexible by designing and implementing them as collections of components. Component-based design and construction has been the norm in the manufacturing of physical goods (such as cars, televisions, and computer hardware) for decades. However, it has only recently become a viable approach to designing and implementing information systems.

Component Standards and Infrastructure

Interoperability of components requires standards to be developed and readily available. For example, consider the video display of a typical IBM-compatible personal computer. The plug on the end of the video signal cable follows an interface standard. The plug has a specific form, and each connector in the plug carries a well-defined electrical signal. Years ago, a group of computer and video display manufacturers defined a standard that describes the physical form of the plug and the type of signals carried through each connector. Adherence to this standard guarantees that any video display unit will work with any compatible personal computer and vice versa.

Components may also require standard support infrastructure. For example, video display units are not internally powered. Thus, they require not only a standard power plug but also an infrastructure to supply power to the plug. A component may also require specific services from an infrastructure. For example, a cellular telephone requires the cellular service provider to assign a transmission frequency with the nearest cellular radio tower, to transfer the connection from one tower to another as the user moves among telephone cells, to establish a connection to another person's telephone, and to relay all voice data to and from the other person's telephone via the public telephone grid. All cellular telephones require these services.

Software components have a similar need for standards. Components could be hard-wired together, but this reduces their flexibility. Flexibility is enhanced when components can rely on standard infrastructure services to find other components and establish connections with them.

In the simplest systems, all components execute on a single computer under the control of a single operating system. Connection is more complex when components are located on different machines running different operating systems and when components can be moved from one location to another. In this case, a network protocol independent of the hardware platform and operating system is required. In fact, a network protocol is desirable even when components all execute on the same machine because such a protocol guarantees that systems can be used in different environments—from a single machine to a network of computers.

Modern networking standards have largely addressed the issue of common hardware and communication software to connect distributed software components. Internet protocols are a nearly universal standard and thus provide a ready means of transmitting messages among components. Internet standards can also be used to exchange information among two processes executing on the same machine. However, Internet standards alone do not fully supply a component connection standard. The missing pieces are:

- Definition of the format and content of valid messages and responses
- A means of uniquely identifying each component on the Internet and routing messages to and from that component

To address these issues, some organizations have developed and continue to modify standards for component development and reuse.

CORBA

The *Common Object Request Broker Architecture (CORBA)* was developed by the Object Management Group (OMG), a consortium of computer software and hardware vendors. CORBA was designed as a platform- and language-independent standard. The standard is currently in its third revision and is widely used.

The core elements of the CORBA standard are the *Object Request Broker (ORB)* service and the *Internet Inter-ORB Protocol (IIOP)* for component communication. A component user contacts an ORB server to locate a component and determine its

---

Common Object Request Broker Architecture (CORBA)

a standard for software component connection and interaction developed by the Object Management Group (OMG)

Object Request Broker (ORB)

a CORBA service that provides component directory and communication services

Internet Inter-ORB Protocol (IIOP)

a CORBA protocol for communication among objects and object request brokers

capabilities and interface requirements. Messages sent between a component and its user are routed through the ORB, which performs any necessary translation services.

COM+

The *Component Object Model Plus (COM+)* is a Microsoft-developed standard for component interoperability. It is widely implemented in Windows-based application software, and it is often used in three-tier distributed applications based on Microsoft Internet Information Server and Transaction Server. Most Windows office suites, such as Microsoft Office, are constructed as a cooperating set of COM+ components.

COM+ components are registered by individual computer systems within the Windows registry, which limits COM+ components to computer systems running Windows operating systems. Once components locate one another through the registry, they communicate directly using a network protocol or Windows interprocess communication facilities.

Enterprise JavaBeans

Java is an OO programming language developed by Sun Microsystems. Most people have heard of Java in connection with applets that execute on Web pages. Java differs from other OO programming languages in several important ways, including the following:

- Java programs are compiled into object code files that can execute on many hardware platforms under many operating systems.
- The Java language standard includes an extensive object framework, called the Java Development Kit (JDK), which includes classes for GUIs, database manipulation, and internetworking.

The JDK defines a number of classes and naming conventions that support component development. One class enables a Java object to convert its internal state into a sequence of bytes that can be stored or transmitted across a network. Other classes allow components to enumerate a Java object's internal variables. Naming conventions allow components to deduce the names of methods that manipulate those variables. An object of a class that implements all of the required component methods and follows the required naming conventions is called a *JavaBean*.

An *enterprise JavaBean (EJB)* is a JavaBean that can execute on a server and communicate with clients and other components using CORBA. EJBs provide additional capabilities beyond JavaBeans including:

- Multicomponent transaction management
- Packaging of multiple components into larger run-time units
- Sophisticated object storage and retrieval in relational or object DBMSs
- Component and object access controls

The JavaBean and EJB standards have created new opportunities for software developers to use component-based technologies. Applications built on these standards can be easily deployed across a wide range of software and hardware platforms. The platform independence of JavaBean components makes them more portable and scalable.

SOAP and .NET

Both CORBA and COM+ have some significant disadvantages for building distributed component-based software. For CORBA, the primary problem is the complexity of the standard and the need for ORB servers. For COM+, the primary problem is dependence on proprietary technology and limited support outside Microsoft products.

Simple Object Access Protocol (SOAP) is a standard for distributed object interaction that attempts to address the shortcomings of both CORBA and COM+. Unlike CORBA, SOAP has few infrastructure requirements, and its programming interface is relatively

---

Component Object Model Plus (COM+)

a standard for software component connection and interaction developed by Microsoft

JavaBean

an object that implements the required component methods and follows the required naming conventions of the JavaBean standard

enterprise JavaBean (EJB)

a JavaBean that can execute on a server and communicate with clients and other components using CORBA

Simple Object Access Protocol (SOAP)

a standard for component communication over the Internet using HTTP and XML

simple. SOAP is an open standard developed by the World Wide Web Consortium (W3C). Perhaps the best evidence of SOAP's long-term potential for success is that Microsoft has adopted it as the basis of its .NET distributed software platform.

SOAP is based on existing Internet protocols, including Hypertext Transport Protocol (HTTP) and eXtensible Markup Language (XML). Messages among objects are encoded in XML and transmitted using HTTP, which enables the objects to be located anywhere on the Internet. Figure 17-16 shows two components communicating with SOAP messages. The same transmission method supports server-to-client and component-to-component communication. The SOAP encoder/decoder and HTTP connection manager are standard components of a SOAP programmers' toolkit. Applications can also be embedded scripts that use a Web server to provide SOAP message-passing services.

FIGURE *17-16*

Component communication using SOAP.

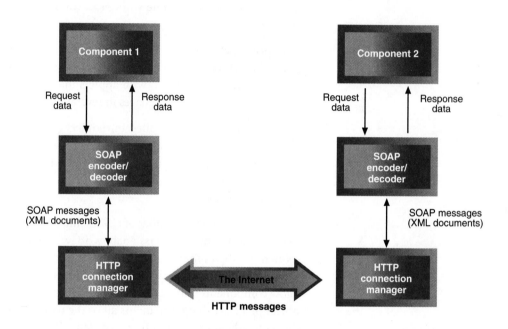

Although SOAP is a promising component-communication standard, it is still in its formative period. Current standard development activity is addressing many missing or underdeveloped parts of the initial standard, including security, message delivery guarantees, and specific conversion rules between programming language and CPU data types and XML. SOAP is often considered a "lighter-weight" version of CORBA because it is simpler and requires little supporting infrastructure to enable component communication. However, the standard may "gain weight" as it evolves to include capabilities needed by high-availability mission-critical applications.

SOAP and XML have enabled a new era of component-based applications, commonly described by the phrase *Web services*. Simply put, a Web service is a component or entire application that communicates using SOAP. Because SOAP components communicate using XML, they can be easily incorporated into applications that use a Web-browser interface. Complex applications can be constructed using multiple SOAP components that communicate via the Internet. We are only now beginning to see the potential of such applications.

Web services

a component or entire application that communicates using SOAP

Components and the Development Life Cycle

Component purchase and reuse is a viable approach to speed completion of a system. Two development scenarios involve components:

- Purchased components can form all or part of a newly developed or reimplemented system.
- Components can be designed in-house and deployed in a newly developed or reimplemented system.

Each scenario has different implications for the SDLC, as explored in the following sections.

Purchased Components

Figure 17-17 shows activities added to SDLC phases when purchased components form part of a new system. Components change the project planning phase because they may alter the analyst's estimate of the project schedule and his or her evaluation of the project's financial and technical feasibility. Purchasing and using components is generally cheaper and takes much less time than building equivalent software. Purchased components may also solve technical problems that developers could not easily or inexpensively solve themselves.

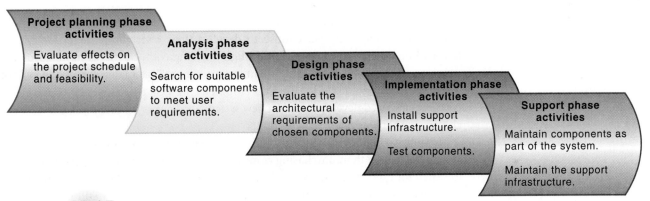

Project planning phase activities

Evaluate effects on the project schedule and feasibility.

Analysis phase activities

Search for suitable software components to meet user requirements.

Design phase activities

Evaluate the architectural requirements of chosen components.

Implementation phase activities

Install support infrastructure.

Test components.

Support phase activities

Maintain components as part of the system.

Maintain the support infrastructure.

FIGURE *17-17*

Activities added to SDLC phases when components are purchased.

The search for suitable components must begin during the analysis phase, but it cannot begin until user requirements are understood well enough to evaluate their match to component capabilities. When developers purchase entire software packages, the match between component capabilities and user requirements is seldom exact. Thus, developers may need to refine user requirements based on the capabilities of available components, particularly if the development project has a short schedule.

Components operate within an extensive infrastructure based on standards such as CORBA or SOAP. Many system software packages implement key parts of each standard. Thus, choosing a component isn't simply a matter of choosing an application software module. Developers must also choose compatible hardware and system software to support components.

The reliance of purchased components on a particular infrastructure has several implications for SDLC activities, including:

- The standards and support software required by purchased components must become part of the technical requirements defined during the analysis phase.
- A component's technical support requirements restrict the options considered during architectural design.
- Hardware and system software that provide component services must be acquired, installed, and configured before testing begins during the implementation phase.
- The components and their support infrastructure must be maintained after system delivery.

Many development projects, particularly large ones, may use components from many different vendors, which raises compatibility issues. The component search and selection process must carefully consider compatibility—often eliminating some choices and altering the desirability of others. Preliminary testing activities may have to be added to the end of the analysis phase to verify component performance and compatibility before the architectural design is structured around those components and their support infrastructure. Support and maintenance are also more complicated because significant portions of the system are not under the direct control of the system owner or the in-house IS staff.

Internally Developed Components

System developers can also choose to develop their own components for systems that will be developed internally. Although building components is more costly and time-consuming than purchasing them, savings are realized later during the support phase, system upgrades, and other development projects that reuse the components. Component-based development of a new system also makes it easier to incorporate purchased components later.

In-house component development has far fewer impacts on systems analysis and architectural design than purchasing components from outside. Developers do not need to search for external components during the analysis phase, and their infrastructure requirements are not carried into architectural design. However, they still must choose a suitable component infrastructure, and that choice may influence other aspects of the system design, including the choice of hardware, operating systems, and database management systems.

Components and Object-Oriented Techniques

It is possible to develop component-based applications without using OO analysis, design, or development techniques, but it isn't recommended. A component is a distributed object that passes messages to (or calls methods of) other components. Since objects are the basis of component construction and interaction, the proper analysis and design tools and techniques are object-oriented. Traditional structured analysis and design techniques are poorly matched to component-based systems.

From the users' viewpoint, the requirements for a component-based system are no different from the requirements of a system implemented with more traditional technology. That is, the behavioral aspects of the system—such as data inputs, the user interface, and basic processing functions—are the same. Thus, analyzing and documenting user requirements for a component-based system are the same as for any other OO system.

Design phase issues for a newly developed component-based system are similar to those of any system developed with object-oriented techniques. A suitable class hierarchy must be defined, and the messages and methods must be fully specified. The techniques and models used to do this are the same as for any other OO analysis and design effort.

Designing Components for Reuse

Most of the advantages of component-based systems arise from their reusability. Thus, object (component) reuse deserves extra attention in the analysis and design phases of the SDLC. Component reuse also has significant implications for maintaining information systems.

OO analysis and design concentrate on developing generally useful objects by modeling problem domain objects and their behavior. To the extent the problem domain is accurately modeled, the developed objects can be reused in other systems and across programs within a system. A component-based system exploits object reuse by sharing executable objects among many applications.

When a system reuses objects by sharing source code, a perfect fit between an old object and a new system is not required. The designers of the new system are free to modify existing objects as needed to suit new requirements. By using inheritance appropriately and overriding existing methods, designers can make such modifications relatively quickly, with maximum reuse of existing code.

In contrast, modifying an existing component to accommodate the requirements of a new system is much more complex. An existing component has already been installed and is in use as part of one or more information systems. Any change to the component for use in a new system may require:

- Modifying the existing component
- Developing a new component with many redundant functions
- Developing a new component that encapsulates the existing component and adds new functions

Modifying an existing component can be risky because it may break existing applications. The modified component may have subtle differences that cause errors in older applications. Thorough testing can determine whether such problems exist. But the effort required to test older systems can be substantial.

Both CORBA and COM+ support multiple component versions. Thus, two components can use the same name but a different version number. Older applications can use the older version, and new applications can use the newer version. This ensures that each application uses the component it was originally designed to use. It also eliminates the possibilities of introducing new bugs into old applications.

But component versioning has its own set of problems. Developers have more components to keep track of, and redundancy among their functions can create maintenance problems. If a redundant function does have a bug, then a developer must modify multiple components to fix the problem. One way around this problem is to design a new component version to call methods in the older version. The new component thus uses (wraps) the old one and adds new or modified functions. However, this approach creates a complex dependency chain among component versions, which can be difficult to manage if there are many versions. They can also be inefficient because of excessive message passing among versions.

The best long-term strategy is usually to modify or wrap existing components, thus avoiding the problem of redundant maintenance. Old systems can be tested with the new components and converted to use them if testing succeeds. Applications that do not function can be forced to use older components until they (or the new components) are repaired. Once all applications using the old component have been repaired, the old component can be deleted or merged with the newer version.

Designers are sometimes tempted to bypass future maintenance problems by designing components with forward-looking features. They try to guess what features not currently needed might be useful in future applications. They then add these features in the hope that future applications will be able to use existing components without modification.

Designing for the future may sound like a good strategy, but it introduces several problems, including:

- **Inaccuracy and waste.** It's difficult to guess what component behavior will be needed in future applications. If the designer guesses wrong, then the extra design and implementation effort is wasted.
- **Cost.** The extra effort adds cost to the current project and lengthens its schedule.

Component designers must walk a fine line between developing components that are general enough to be reused in the future but specific enough to be useful and efficient in a current application.

System Performance

Component-based software is usually deployed in a distributed environment. Components typically are scattered among client and server machines and among local area network (LAN) and wide area network (WAN) locations. Distributing components across machines and networks raises the issue of performance. System performance depends on the location of the components (that is, component topology), the hardware capacity of the computers on which they reside, and the communication capacity of the networks that connect computers. Performance also depends on demands on network and server capacity made by other applications and communication traffic, such as telephone, video, and interactions among traditional clients and servers.

The details of analyzing and fine-tuning computer and network performance are well beyond the scope of this text. But anyone planning to build and deploy a distributed component-based system should be aware of the performance issues. These issues must be carefully considered during systems design, implementation, and support.

Steps developers should take to ensure adequate performance include the following:

- Examine component-based designs to estimate network traffic patterns and demands on computer hardware.
- Examine existing server capacity and network infrastructure to determine their ability to accommodate communication among components.
- Upgrade network and server capacity prior to development and testing.
- Test system performance during development and make any necessary adjustments.
- Continuously monitor system performance after installation to detect emerging problems.
- Redeploy components, upgrade server capacity, and upgrade network capacity to reflect changing conditions.

Implementing these steps requires a thorough understanding of computer and network technology as well as detailed knowledge of existing applications, communications needs, and infrastructure capability and configuration. Applying this knowledge to real-world problems is a complex task typically performed by highly trained specialists.

SUMMARY

Rapid application development (RAD) is a broad term that covers a variety of tools, techniques, and development approaches with the goal of speeding application development. RAD techniques include risk management, joint application design, tool-based design, and software reuse. RAD approaches include prototyping, eXtreme Programming, spiral development, the Unified Process, and (under the right circumstances) traditional development life cycles. RAD tools include object frameworks and components and their supporting infrastructure.

No RAD tool, technique, or development approach speeds development for all projects. Developers must carefully examine project characteristics to determine which RAD concepts are most likely to speed development. Some parts of RAD (such as risk management and software reuse) will speed development for most projects. Other parts (such as tool-based design and spiral development) are applicable to far fewer projects.

Software reuse is a fundamental approach to rapid development. It has a long history, although it has been applied with greater success since the advent of object-oriented programming, object frameworks, and component-based design and development. Object frameworks provide a means of reusing existing software through inheritance. They provide a library of reusable source code, and inheritance provides a means of quickly adapting that code to new application requirements and operating environments.

Components are units of reusable executable code that behave as distributed objects. They are plugged into existing applications or combined to make new applications. Like the concept of software reuse, component-based design and implementation are not new, but the standards and infrastructure required to support component-based applications have only recently emerged. Thus, components are only now entering the mainstream of software development techniques.

KEY TERMS

Common Object Request Broker Architecture (CORBA), p. 702

component, p. 700

Component Object Model Plus (COM+), p. 703

construction, p. 689

developmental prototype, p. 677

discipline, p. 691

discovery prototype, p. 677

elaboration, p. 689

enterprise JavaBean (EJB), p. 703

eXtreme Programming (XP), p. 685

foundation classes, p. 699

inception, p. 689

Internet Inter-ORB Protocol (IIOP), p. 702

JavaBean, p. 703

object framework, p. 699

Object Request Broker (ORB), p. 702

pair programming, p. 686

prototyping development approach, p. 677

rapid application development (RAD), p. 675

risk, p. 692

risk management, p. 693

risk mitigation, p. 694

Simple Object Access Protocol (SOAP), p. 703

software reuse, or code reuse, p. 697

spiral development approach, p. 681

system metaphor, p. 686

timeboxing, p. 690

tool-based development, p. 696

transition, p. 690

Unified Process (UP), p. 688

user story, p. 686

Web services, p. 704

REVIEW QUESTIONS

1. During what life cycle phase is it least expensive to implement a requirements change? During which phase is it most expensive?

2. Is RAD a single approach or technique? Why or why not?

3. What factors determine the fastest development approach for a specific project?

4. Under what condition(s) is the sequential development approach likely to be faster than alternative development approaches?

5. Under what condition(s) is the spiral development approach likely to be faster than alternative development approaches?

6. Under what condition(s) is the UP likely to be faster than alternative development approaches?

7. How does the process (composition and order of phases and activities) of prototype-based system development differ from a sequential development approach? How does it differ from a spiral development approach? How does it differ from the UP?

8. How does the process (composition and order of phases and activities) of spiral system development differ from a sequential development approach?

9. What are the common characteristics of the prototyping, spiral, UP, and XP development approaches? What are their differences?

10. Which approach to system development (conventional, prototyping, spiral, UP, or XP) is likely to result in the shortest possible development time when user requirements or technical feasibility are poorly understood at the start of the project? Why?

11. What development tool characteristics are required for successful use of the prototyping, spiral, UP, and XP development approaches?

12. Define the following terms and phrases: *risk*, *risk management*, and *risk mitigation*.

13. With which development approaches should risk management be used?

14. How should JAD be incorporated into the prototyping or spiral approaches to software development?

15. Describe tool-based development. With which development approaches (conventional, prototyping, spiral, XP, or UP) can it be used?

16. What is an object framework? How is it different from a library of components?

17. What are the differences between general-purpose and application-specific foundation classes?

18. For which layers of an OO program are off-the-shelf components most likely to be available?

19. What is a software component?

20. Why have software components only recently come into widespread use?

21. In what ways do components make software construction and maintenance faster?

22. Describe four interoperability standards for software components. Compare and contrast the standards.

THINKING CRITICALLY

1. Consider the capabilities of the programming language and development tools used in your most recent programming or software development class. Are they powerful enough to implement developmental prototypes for single-user software on a personal computer? Are they sufficiently powerful to implement developmental prototypes in a multiuser, distributed, database-oriented, and high-security operating environment? If they were being used with a tool-based development approach, what types of user requirements might be sacrificed because they didn't fit language or tool capabilities?

2. Consider XP's team-based programming approach in general and its principle of allowing any programmer to modify any code at any time in particular. No other development approach or programming management technique follows this particular principle. Why not? In other words, what are the possible negative implications of this principle? How does XP minimize these negative implications?

3. The Object Data Management Group (ODMG) was briefly described in Chapter 10. Visit the Web sites of the OMG (www.omg.org) and ODMG (www.odmg.org) and gather information to answer the following questions. What is the goal or purpose of each standard? What overlap, if any, exists among the two standards? Are the standards complementary?

4. Read the article by Scott Lewandowski listed in the "Further Resources" section. Compare and contrast the CORBA and Microsoft COM+ approaches to component-based development. (Note: COM+ is called *DCOM* in the article.) Which approach appears better positioned to deliver on the promises of component-based design and development? Which approach is a true implementation of distributed objects? Which approach is likely to dominate the market in the near future?

5. Visit the Web site of the World Wide Web Consortium (www.w3.org) and review recent developments related to

the SOAP standard. What new capabilities have been added, and what is the effect of those capabilities on the standard's complexity and infrastructure requirements?

6. Compare and contrast object frameworks and components in terms of ease of modification before system installation, ease of modification after system installation, and overall cost savings from code reuse. Which approach is likely to yield greater benefits for a unique application system (such as a distribution management system that is highly specialized to a particular company)? Which approach is likely to yield greater benefits for general-purpose application software (such as a spreadsheet or virus protection program)?

7. Assume that a project development team has identified risks, outcomes, and probabilities for a new Web-based insurance-pricing system, as summarized in the table below. Compute the expected schedule delay for each risk and for the entire project. Which risks should be actively managed? What mitigation measures might be appropriate? Why do the outcome probabilities for some risks sum to more than 100 percent?

8. Consider the similarities and differences between component-based design and the construction of computer hardware (such as personal computers) and the design and construction of computer software. Can the "plug-compatible" nature of computer hardware ever be achieved with computer software? Does your answer depend on the type of software (for example, system or application software)? Do differences in the expected lifetime of computer hardware and software affect the applicability or desirability of component-based techniques?

Risk	Risk probability %	Possible outcomes	Outcome probability %	Outcome delay (weeks)
A change in state insurance laws or regulations requires changes to pricing algorithms.	100	Update pricing for 1–10 states Update pricing for 11–20 states Update pricing for 21–30 states Update pricing for 31–40 states Update pricing for 41–50 states	60 25 10 4 1	1 2 3 4 5
Internet security standards change significantly.	30	Update server software Minor content update Major content update	100 70 30	2 4 10
Management decides to negotiate a preferred service provider agreement with a major ISP (e.g., AOL).	25	No format/content changes required Cosmetic format/content changes required Significant format/content changes required Minor capacity upgrade Major capacity upgrade	70 20 10 50 50	0 2 6 5 12
A change in federal insurance law or regulations requires changes in pricing algorithms.	10	Minor recoding Major recoding	70 30	2 12

EXPERIENTIAL EXERCISES

1. Talk with someone at your school or place of employment about a recent development project that was canceled because of slow development. What development approach was employed for the project? Would a different development approach have resulted in faster development?

2. Talk with someone at your school or place of employment about a recent development project that failed to satisfy user or technical requirements. Review the reasons for failure and the risk management processes that were used. What changes (if any) in the risk management process might have prevented the failure or mitigated its effects?

3. Consider a project to replace the student advising system at your school with one that employs modern features (for example, Web-based interfaces, instant reports of degree program progress, and automatic course registration based on a long-term degree plan). Now consider how such a project would be implemented using tool-based development. Investigate alternative tools such as Visual Studio, PowerBuilder, and Oracle Forms, and determine (for each tool) what requirements would need to be compromised for the sake of development speed if the tool were chosen.

4. Examine the capabilities of a modern programming environment such as Microsoft Visual Studio .NET, IBM WebSphere Studio, or Borland Enterprise Studio. Is an object framework or component library provided? Does successful use of the programming environment require a specific development approach? Does successful use require a specific development methodology?

5. Examine the technical description of a complex end-user software package such as Microsoft Office. In what ways was component-based software development used to build the software?

CASE STUDIES

Midwestern Power Services

Midwestern Power Services (MPS) provides natural gas and electricity to customers in four Midwestern states. Like most power utilities, MPS is facing the prospect of significant federal and state deregulation over the next several years. Federal deregulation has opened the floodgates of change but provided little guidance or restriction on the future shape of the industry. State legislatures in two of the states MPS serves have already begun deliberations about deregulation, and the other two states are expected to follow shortly.

The primary features of the proposed state-level deregulation are:

- Separating the supply and distribution portions of the current regulated power utility business
- Allowing customers to choose alternate suppliers regardless of what company actually distributes electricity or natural gas

The deregulation proposals seek to increase competition in electricity and natural gas by regulating only distribution. Natural gas extraction and electricity generation would be unregulated, and consumers would have a choice of wholesale suppliers. The final form of deregulation is unknown, and its exact details will probably vary from state to state.

MPS wants to get a head start on preparing its systems for deregulation. Three systems are most directly affected—one for purchasing wholesale natural gas, one for purchasing wholesale electricity, and one for billing customers for combined gas and electric services. The billing system is not currently structured to separate supply and distribution charges, and it has no direct ties to the natural gas and electricity purchasing systems. MPS's general ledger accounting system is also affected because it is used to account for MPS's own electricity-generating operations.

MPS plans to restructure its accounting, purchasing, and billing systems to match the proposed deregulation framework:

- Customer billing statements will clearly distinguish between charges for the supply and distribution of both gas and electricity. The wholesale suppliers of each power commodity will determine prices for supply. Revenues will be allocated to appropriate companies (such as distribution charges to MPS and supply charges to wholesale providers).

- MPS will create a new payment system for wholesale suppliers to capture per-customer revenues and to generate payments from MPS to wholesale suppliers. Daily payments will be made electronically based on actual payments by customers.
- MPS will restructure its own electricity-generating operations into a separate profit center, similar to other wholesale power providers. Revenues from customers who choose MPS as their electricity supplier will be matched to generation costs.

MPS's current systems were all developed internally. The general ledger accounting and natural gas purchasing systems are mainframe based. They were developed in the mid-1980s, and incremental changes have been made ever since. All programs are written in COBOL, and DB2 (a relational DBMS) is used for data storage and management. There are approximately 50,000 lines of COBOL code.

The billing system was rewritten from the ground up in the mid-1990s and has been slightly modified since that time. The system runs on a cluster of servers using the UNIX operating system. The latest version of Oracle (a relational DBMS) is used for data storage and management. Most of the programs are written in C++, although some are written in C and others use Oracle Forms. There are approximately 80,000 lines of C and C++ code.

MPS has a network that is used primarily to support terminal-to-host communications, Internet access, and printer and file sharing for microcomputers. The billing system relies on the network for communication among servers in the cluster. The mainframe that supports the accounting and purchasing systems is connected to the network, although that connection is primarily used to back up data files and software to a remote location. The company has experimented with Web-browser interfaces for telephone customer support and on-line statements. However, no functioning Web-based systems have been completed or installed.

MPS is currently in the early stages of planning the system upgrades. It has not yet committed to specific technologies or development approaches. MPS has also not yet decided whether to upgrade individual systems or replace them entirely. The target date for completing all system modifications is three years from now, but the company is actively seeking ways to shorten that schedule.

1. Describe the pros and cons of the traditional, prototyping, spiral, UP, and XP development approaches to upgrading the existing systems or developing new ones. Do the pros and cons change if the systems are replaced instead of upgraded? Do the pros and cons vary by system? If so, should different development approaches be used for each system?
2. Is tool-based development a viable development approach for any of the systems? If so, identify the system(s) and suggest tools that might be appropriate. For each tool suggested, identify the types of requirements likely to be sacrificed because of a poor match to tool capabilities.
3. Assume that all systems will be replaced with custom-developed software. Will an object framework be valuable for implementing the replacements? Is an application-specific framework likely to be available from a third party? Why or why not?
4. Assume that all systems will be replaced with custom-developed software. Should MPS actively pursue component-based design and development? Why or why not? Does MPS have sufficient skills and infrastructure to implement a component-based system? If not, what skills and infrastructure are lacking?
5. List the risks and risk outcomes that may affect the planned upgrade or replacement. Which risks require active management? What mitigation measures should be pursued?

Rethinking Rocky Mountain Outfitters

Now that you've studied most of the material in this textbook, you'll be able to make more informed and in-depth choices regarding development approach and techniques for the RMO customer support system (CSS). Review the CSS development project charter in Figure 3-4, the detailed scope description in Figure 3-7, the RMO memo at the beginning of Chapter 2, and the "Rethinking RMO" case at the end of Chapter 2. You may also need to look at other RMO material from Chapters 2 and 3 to answer the following questions:

1. Consider the criteria discussed in this chapter for choosing among the traditional, prototyping, and spiral approaches to system development. Which CSS project characteristics favor traditional development? Which favor spiral development? Which approach is best suited to the CSS development project?
2. RMO had no experience developing Web-based systems prior to undertaking the CSS development project. What are the implications of RMO's inexperience for risk management within the CSS development project? What types of risks should system developers actively mitigate during design and implementation?
3. Should tool-based development or joint application design be used in the CSS development project? Why or why not?
4. Should RMO consider using purchased components within the new CSS? If so, when should it begin looking for components? How will a decision to use components affect the analysis, design, and implementation phases? If purchased components are used, should the portions of the system developed in-house also be structured as components? Will a decision to pursue component-based design and development make it necessary to adopt OO analysis and design methods?

Focusing on Reliable Pharmaceutical Service

Reread the Reliable Pharmaceutical Service cases in Chapters 8 and 9. Armed with the new knowledge that you've gained from reading this chapter, revisit the questions posed in the last paragraph of the Chapter 9 case:

1. Which of the development approaches described in this chapter seem best suited to the project? Why? Plan the first six weeks of the project under your chosen development approach?
2. What role will components play in the system being developed for Reliable? Does it matter on which component-related standards they're based? Why or why not?

FURTHER RESOURCES

Kent Beck, *Extreme Programming Explained: Embrace Change*, Addison-Wesley Publishing Company, 1999.

Barry W. Boehm, "A Spiral Model of Software Development and Enhancement," *Computer*, May 1988, pp. 61–72.

Frederick P. Brooks, Jr., "No Silver Bullet: Essence and Accidents of Software Engineering," *Computer*, April 1987, pp. 10–19.

Cetus Team, "Links on Objects & Components," http://www.cetus-links.org.

Craig Larman, *Applying UML and Patterns: An Introduction to Object-Oriented Analysis and Design and the Unified Process* (2nd ed.). Prentice-Hall, 2002.

Scott M. Lewandowski, "Frameworks for Component-Based Client/Server Computing," *ACM Computing Surveys*, Volume 30:1, March 1998, pp. 3–27.

Steve McConnell, *Rapid Development: Taming Wild Software Schedules*. Microsoft Press, 1996.

Object Data Management Group (developer of an object database management standard) home page, http://www.odmg.org.

Robert Orfali, Dan Harkey, and Jeri Edwards, *Client/Server Survival Guide* (3rd ed.). John Wiley & Sons, 1999.

Jawed Siddiqi, "An Exposition of XP But No Position on XP," IEEE Computer Society, http://computer.org/seweb/dynabook/Index.htm.

Software Productivity Consortium (www.software.org), *Component Evaluation Process*. SPC-98091-CMC, May 1999.

Steve Sparks, Kevin Benner, and Chris Faris, "Managing Object-Oriented Framework Reuse," *Computer*, Volume 29:9, September 1996, pp. 53–61.

Jane Wood, and Denise Silver, *Joint Application Development* (2nd ed.). John Wiley & Sons, 1995.XProgramming.com, http://www.xprogramming.com.

CHAPTER 18

Packaged Software and Enterprise Resource Planning

LEARNING OBJECTIVES

After reading this chapter, you should be able to:

- Discuss three major analysis and design issues associated more often with packages than with custom-developed software

- Name the three types of package customization and explain when each would be necessary

- Identify and explain the four major characteristics of ERP systems and discuss several reasons for adopting ERP

- Identify and explain several critical success factors for ERP implementation

- Distinguish the three major approaches to ERP system development

- Discuss how methods used for ERP systems deployment differ from those used for conventional system development

- Identify and explain the four major steps to ERP package selection

- Discuss the major areas of expansion for ERP

- Identify the four high-level work areas of SAP R/3

- Identify and discuss the five types of master data in SAP R/3

- Describe the function of R/3 Basis software

- Describe how transactions are entered in SAP R/3

CHAPTER OUTLINE

Packaged Software

Enterprise Resource Planning

A Closer Look at One ERP Package: SAP R/3

Enterprise resource planning (ERP) systems have become notorious for problems encountered during installation and final implementation. Perhaps the best example of the pitfalls is illustrated by the Premier Candy Corporation of Roanoke, Virginia. The company decided to implement an ERP system during early summer, right at the time retailers were placing large orders for back-to-school sales and Halloween. The new $120 million enterprise information system, which also included a customer relationship management (CRM) package and a production forecasting and scheduling application (both from other third-party vendors), was designed to replace scores of legacy systems that were running everything from inventory to order processing to human resources. The system was installed with high expectations for improved efficiency and effectiveness.

Instead, the end result was nearly disastrous. By late summer, the company was still trying to fix glitches in its order-processing and shipping functions. During the busiest season of the year, big retail chains turned from Premier to its major competitors to meet the extra demand for candy. Inventory levels in Premier warehouses mushroomed. The bottom line was that third-quarter sales dropped significantly compared with the previous year, and earnings plummeted.

What went wrong? The ERP vendor said that Premier's initial difficulty wasn't due to any problems with the ERP software. Like most companies with snarled ERP projects, Premier didn't offer many details. But outsiders point to two notable errors the company made. The first involves timing. The busiest season of the year is not the time to install a famously complex product like an ERP system and take it live. Snags always arise, and it's far easier to iron them out during less busy periods of the year.

Second, Premier attempted to do too much at once. Installing ERP software is complicated enough. Throw in a CRM program and a logistics package, and the project becomes dangerously complex. Premier probably felt rushed to get all three integrated systems up and running. The ERP vendor noted that Premier tried to implement "a complex system on a very aggressive project plan." Premier and four different teams of consultants squeezed into just 30 months the massive ERP system that was originally estimated to be a four-year project. Most companies install ERP systems in stages, especially when applications from multiple vendors are involved. An all-at-once rollout is tempting because it's faster and potentially cheaper, but it is also extremely risky.

Yet another potential problem at Premier involved massive changes to business processes. The new system required enormous changes in the way Premier's workers did their jobs. Premier drastically underestimated the effects of an ERP system on its people. For example, instead of completing all shipping documents manually, long-time Premier employees (who had never used a computer) had to learn to navigate through complex software to print them.

Is that the end of the story? No. A full year after the debacle, Premier reported that revenue and profits in the third quarter rose dramatically. The ERP systems as well as a revamped distribution facility in the eastern United States were credited with the turnaround. Premier officials declined to elaborate on the steps the company took to straighten out the system. While a complex ERP system requires extreme care in installation and sound implementation strategy and execution, the results can be very rewarding.

OVERVIEW

The information systems of many companies are often very complex. Even though Rocky Mountain Outfitters (RMO) is a relatively small company, it has many of the same needs as large corporations, needs that possibly can be addressed with stand-alone software packages, or even ERP software, instead of custom-developed software. This chapter begins by looking at the key issues of working with application packages in general, followed by a survey of ERP—what it is, why it is an important option for system improvement, and how it is implemented. Finally, one ERP package, SAP R/3, will be examined to give you a better understanding of how ERP works.

Figure 18-1 illustrates a computer-based information system consisting of many tightly linked components (hardware, software, data, networks, procedures, and people) working together to provide the information needs of one or more business functions. The process of systems analysis and design must address adequately all these components to satisfy the needs of users—the ultimate judges of system success. The focus of this text has been on the analysis and design of business application software— computer programs created to solve specific business problems. At issue in this chapter are two major questions related to business application software. First, what is the best way to procure this software? Second, after obtaining the software, how does a firm best utilize it to achieve the firm's objectives? The way in which individual companies, large or small, answer these questions has a tremendous impact on the future of IS organizations, the future of the enterprises they serve, and the future of the global economy.

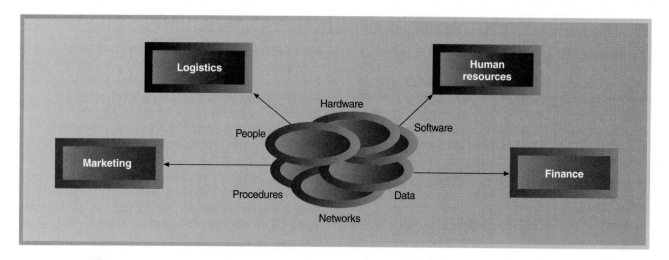

FIGURE *18-1*

IS components serving multiple business functions.

Let's consider the first question: How is software best procured? Throughout the early history of business computing, applications consisted almost entirely of custom-developed software—software created on demand according to the unique specifications of the intended users. The two parties primarily involved in system creation were in-house developers and in-house users. As the field of information systems matured, enterprising individuals realized that certain business applications, such as financial accounting functions, were fairly standardized. As third-party developers outside the business firm, they began to create packaged software to perform common or standardized functions. The intention was and is to sell these packages to a large number of customers with little or no need for modification. Today, the continuing demand for rapidly deployable, functional, and affordable software still drives third-party developers to explore new applications founded on various industries' "best practices," fairly standard procedures commonly accepted as the optimal way of performing a given function. Such packages enable firms to avoid reinventing the wheel every time they need a new application. Some systems analysts maintain that the best way to procure software is by purchasing a package, provided the package can meet most user requirements.

Now for the second question: How is application software best utilized in a business environment? Should software be used only to satisfy the needs of individual, isolated departments, or is the entire organization best served when the information produced in one department is readily available for use in another? *Software integration*, a process

software integration

the process of combining separately produced components or subsystems and addressing problems in their interactions

that combines separate components or subsystems and addresses problems in their interactions, has long been a goal of many organizations. A rudimentary form of integration is often achieved through batch processing, but that approach delays the availability of critical information. Recent advances in client-server technology and telecommunications have fueled the integration of business applications in real time across departments, functions, and divisions of corporations, and even between a firm and its external suppliers and customers. The combination of the availability of packaged software and the push toward real-time integration has spawned a new branch of management information systems called *enterprise resource planning (ERP)*. As discussed earlier, ERP uses software to manage companywide business processes in real time using a common database and shared management reporting tools. Several system development firms—such as SAP AG, Oracle, PeopleSoft, SAGE, and Microsoft—offer ERP systems. The key components in enterprise computing are shown in Figure 18-2.

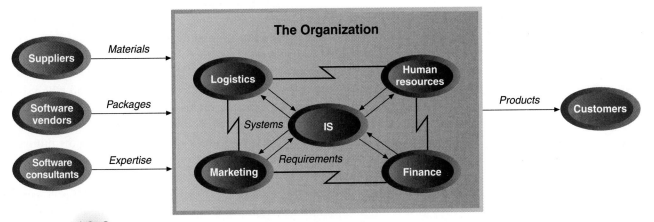

FIGURE *18-2*

The key components in enterprise computing.

As chief information officer at Rocky Mountain Outfitters, Mac Preston is keenly aware that an ERP approach to system development is a definite possibility for RMO. In fact, Mac had been reading some trade journals that discuss the feasibility of purchasing a customer relationship management (CRM) package, either used as a stand-alone application or possibly integrated with an ERP system. Mac decided to alert his boss, JoAnn White, of this alternative to the in-house development of the proposed customer support system (see the accompanying memo).

August 22, 2004

To: JoAnn White

From: Mac Preston

RE: Possible Alternative for Customer Support System

JoAnn, since we have just started organizing key personnel for the CSS project, I thought I would alert you to a possible alternative to developing the system in-house. As you may already know, many vendors offer customer relationship management (CRM) software packages that address many of our concerns. For example, Siebel Systems is currently the market leader in CRM software with offerings such as Siebel Call Center. SAP offers a competing product, mySAP CRM. These are just a few examples. Most of these products can run on traditional client-server platforms or over the Internet as part of an e-business solution.

A major advantage of going with a package is that it could free up much of our IT staff to work on other projects, and it could possibly be up and running much faster. One downside could be that a package may not do everything that we would like to do. Let me know if you want to pursue this idea further, or maybe call a meeting of the oversight committee. Thanks.

MP

cc: John MacMurty, Barbara Halifax

PACKAGED SOFTWARE

As we've already discussed, packaged software traditionally focused on performing isolated business functions. More recently, the emergence of packaged suites, such as ERP systems, has completely reengineered the information systems function in many large corporations. Stand-alone packages cannot share information unless forced to do so through the efforts of *system integrators*, individuals who are specially trained in integrating application software across various computing platforms. With an ERP system, however, the advanced application software for all functional departments in an organization can easily share information, improving effectiveness and efficiency tremendously. Organizations are now faced with critical decisions concerning the futures of their information systems. Should they continue to maintain their existing legacy systems or scrap them entirely? If they plan to acquire new application software, should it be through continued custom development or by purchasing packages? If packages are involved, should firms integrate separate vendors' products, or should they purchase a comprehensive ERP system? Or, should organizations pursue a combination of custom development, packages, and an ERP system? Recent changes in licensing practices among software package vendors (such as Microsoft) are making upgrades of packages much more expensive. In the case of RMO, it plans to develop custom software and purchase packaged software. The company also plans to integrate several of its key applications. However, RMO has not yet addressed a possible need for ERP. Obviously, the choices are not easy or simple.

The Trend Toward Packaged Software

As discussed in Chapter 8, custom development may be preferred when a package cannot provide the required functionality. However, current trends appear to favor the packaged software solution. A recent survey reported that during a one-year period, spending on new in-house application development had dropped 34 percent, while spending on software packages had doubled. If this trend continues, how will in-house system development be affected? Obviously, the decision to purchase a software package, in lieu of developing a custom program, can have far-reaching consequences for an organization's information technology (IT) function and for the organization as a whole.

What are the key considerations behind the acquisition of packaged software? First and foremost, an enterprise is concerned with the *effectiveness* of its information systems. A firm wants *quality* information to make better business decisions, perhaps to create or sustain a competitive advantage, or just to survive in a turbulent economic environment. Second, an enterprise is concerned with the *efficiency* associated with both the development and use of its information systems. The efficiency with which a firm creates, operates, and maintains its information systems contributes greatly to its cost of doing business, which ultimately affects its ability to compete or survive. The primary goal of packaged software is both to improve the quality and reduce the cost of information. Whether packaged software actually achieves this two-part goal is debatable.

Implementation and Support of Packaged Software

Chapter 8 introduced the issues of analysis and design when packaged software is part of the solution. The analysis phase defines user requirements for the system. If a package can be found that meets all the critical requirements and the vast majority of the users' preferences, it will be seriously considered. The design phase addresses many more detailed issues, such as what computing platform is required for the package, how users will be trained to use the new package, and how the changeover from the existing system will take place. Such issues actually differ little from those that arise from custom development.

---

system integrator

an individual who is specially trained in integrating application software across various computing platforms

---

FIGURE *18-4*

Obstacles to integration.

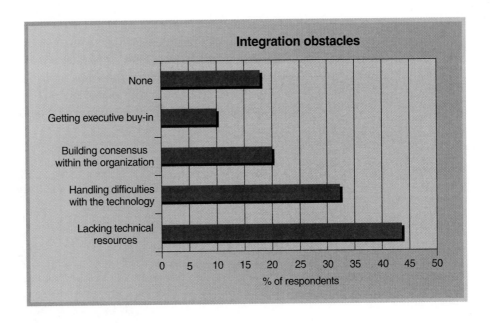

Upgrading Packages

Software vendors are notorious for releasing frequent upgrades of their packages. An upgrade usually adds functions or makes the software easier to use. Sometimes a new release, or patch, is needed to fix a bug in the previous version of the package. Upgrades are often fairly straightforward for in-house staff or vendors, because the upgrade is designed to be compatible with the previous version. However, when extensive customization and integration have occurred with previous versions at the user's site, upgrades can be particularly challenging. This is why customization and integration must be carefully planned and documented during system development.

ENTERPRISE RESOURCE PLANNING

The previous section discussed general issues of software packages. Once a package is acquired as part of an information system, it may be customized, integrated, or upgraded within a single function or department of an organization. This section explores information systems that are designed to reach beyond the needs of a single department. The goal of such systems is to improve the operation and competitiveness of an entire organization. Such software is at the hub of ERP. This section explores what ERP is, why firms choose ERP, how ERP is implemented, and how ERP functions.

RMO plans to integrate its customer support system, inventory management system, and retail store system, but it has not addressed whether to integrate these systems with its new accounting/finance and human resources packages. Successfully doing so would essentially constitute an ERP system.

ERP and the Business Environment

Many organizations today have turned to ERP to help them compete locally and globally in the electronic marketplace. Such well-known firms as Boeing and Federal Express have implemented ERP systems for both traditional and Internet applications. Organizations worldwide are continuing to spend tens of billions of dollars on ERP in an attempt to replace outdated legacy systems, integrate applications, save on IT costs, and achieve competitive advantage. Figure 18-5 illustrates the basic differences between nonintegrated systems and systems integrated using ERP: In the former, only one function segregates and accesses data; in the latter, all functions can quickly and easily share a common database. With today's global organizations and widespread Internet use, ERP offers extraordinary advantages due to the tight integration of systems and easy accessibility to employees, customers, and strategic partners. Spending on ERP systems is expected to grow substantially in the new millennium, fueled by expansion into areas such as customer relationship management (CRM) and supply chain management (SCM). However, ERP implementation is not without its challenges.

FIGURE *18-5*

(a) Nonintegrated IS.

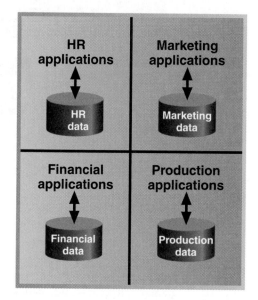

(b) Integrated IS in an ERP system.

Objectives of ERP

ERP is designed to provide best practice functionality within departments and a high level of integration across all enterprise functions, such as human resources, finance, and manufacturing. Not only is ERP designed to integrate functions, but it is also designed to integrate organizational units, such as plants and divisions, across the country or around the world. Going one step further, ERP is capable of integrating previously separate and independent organizations—that is, integrating a firm with its suppliers

value chain

the sequence of all processes within an
organization that add value to a product or service offered

and customers. Thus, ERP is intended to help integrate the entire *value chain*, the sequence of all processes within (and possibly between) organizations that adds value to the product or service.

Additionally, ERP can bridge the gaps in the value chain in real time, making critical data available to a host of different users (inside or outside the firm) exactly when needed. Thus, ERP is designed to break down significant barriers of function, space, and time. RMO could use ERP to link its financial, human resource, manufacturing, inventory, retail, and distribution systems. Managers could instantly see sales and inventory at the Park City and Denver retail outlets or examine the performance of mail-order, phone-order, or Internet sales. ERP could tie manufacturing plants in Salt Lake City and Portland directly to all these sales systems and to the inventory systems for the Salt Lake City, Portland, and Albuquerque distribution centers. To plan the training and movement of personnel within the various manufacturing, distribution, and sales functions, RMO could integrate these functions with the human resource system and instantly update the bottom line by integrating all these systems with the financial system. An ERP approach to doing business can be a powerful competitive weapon.

ERP and Business Information Systems

As discussed in Chapter 1, there are many types of business information systems within organizations, including transaction processing systems (TPS), management information systems (MIS), and decision support systems (DSS). ERP is often used first to integrate the core TPSs in human resources, finance, and manufacturing and later to integrate the MISs and DSSs of various departments. Not only does ERP integrate functional departments, but it does so on-line in real time, not a day or week later. ERP can span functional departments within a single facility as well as geographical boundaries of large, multinational corporations. It can also reach beyond the boundaries of a firm to include external suppliers and customers in the information system, further extending the firm's supply chain. These aspects of ERP are illustrated in Figure 18-6. The integrated, real-time, global, and extended characteristics of the information obtained through the TPS of an ERP system make the MIS and DSS of the organization more valuable. Through ERP, the firm can make faster and better decisions to become more effective, efficient, and, ultimately, competitive.

FIGURE *18-6*

Four major characteristics of an ERP system.

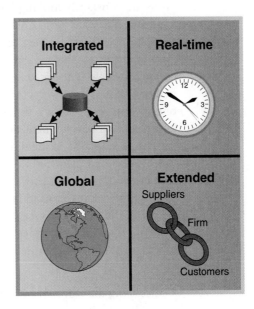

Major ERP Vendors

The major players in the packaged ERP market are (in descending market-share order) SAP AG, Oracle, PeopleSoft, SAGE, and Microsoft Business Solutions. But SAP AG represents more than three and one-half times the share of its next leading competitor. These companies have exhibited phenomenal growth in recent years, although that growth now appears to be slowing or even declining due to the state of the economy in general and published reports of failed implementations. These ERP vendors began offering integrated packages for managing the three core business functions: finance, manufacturing, and human resources. Recently, they have expanded by offering added functions in the following areas: customer relationship management (meeting the needs of all customers from initial inquiry to support after the sale), supply chain management (managing all resources and processes involved from raw-material acquisition to finished goods delivery), *sales-force automation* (helping sales personnel to service customer accounts), e-commerce (buying and selling goods and services on the Internet, especially the World Wide Web), and *business intelligence* (acquiring and using information designed to improve competitiveness).

Why Consider ERP?

The changing climate of business and advances in information technology are leading many firms away from the legacy systems of the past and toward the packaged systems of today. The recent increases in mergers, government deregulation, and globalization have pressured firms to search for new ways to compete. These competitive measures include *total quality management (TQM)*, the process of achieving the highest levels of quality throughout the entire organization, and business process reengineering (BPR), radical, large-scale improvements in business processes to improve efficiency and effectiveness significantly. Also, recent advancements in client-server technology, object-orientation, telecommunications, and the Internet have opened new opportunities for applying software solutions to business problems. All these changes have combined to open the door for advanced software packages to take over wherever homespun legacy systems may be falling short.

Reasons for Implementing ERP

Firms are constantly looking for ways to survive and thrive in a complex competitive environment. Many firms are losing their competitive edge by relying on obsolete systems. From a broad perspective, there are several reasons that firms are turning to ERP. First, many believe that information technology (IT) is an excellent way to enhance their competitiveness. They may, however, question their ability or desire to continue to maintain a staff that is expert in all areas of increasingly complex technologies. ERP is one means of updating a firm's information systems without investing in costly in-house software development. Second, some firms go beyond the goal of simply improving IT and subscribe to the philosophy of business process reengineering. These firms may choose to implement ERP and reengineer their processes to match the model of the ERP system. Third, reorganization through mergers or expansion into global markets has motivated some firms to integrate geographically dispersed systems within one ERP system. All these reasons essentially point to the need to replace outdated technologies.

From a business's perspective, there are several reasons to embrace ERP. Based on several case studies, the most common reasons given for ERP projects in major companies are:

- Reducing the workforce in core transaction processing systems (such as order processing and inventory management) by eliminating manual tasks and duplication
- Supporting global business operations (such as having one order-processing system for all customers worldwide)

---

sales-force automation

processes that make it easier for sales personnel to service customer accounts

business intelligence

the acquisition and use of information designed to improve competitive advantage

total quality management (TQM)

the process of achieving the highest levels of quality throughout the entire organization

- Achieving economies of scale by replacing division-level systems (such as replacing plantwide purchasing with corporatewide purchasing)
- Improving communication and information exchange among departments
- Reducing information system development staff by using packaged solutions
- Improving customer service through better logistics based on more complete and accessible information
- Improving data integrity through a common database
- Improving decision support via more timely reports and drill-down capability

Of course, some firms are moving toward ERP simply because they desire to keep pace with competitors that have already done so.

Costs and Benefits

Investing in an ERP package may cost upwards of $5 million for a midsize firm (one with annual sales of about $200 million to $1 billion) and upwards of $50 million for a large corporation (one with annual sales exceeding $1 billion). Midsize firms can expect implementation to take up to two years to complete; it takes several more years for a large firm. The cost of the software itself is usually very competitive among the major vendors, but the cost of consulting services can be quite high, primarily due to the shortage of qualified people. The payback is usually estimated to take between 6 and 30 months, depending on the size and complexity of the system. Vendors usually estimate a 30 percent reduction in administrative and information systems costs. Payback can also be calculated in terms of the ability to continue to compete in the long term against firms that are currently upgrading their information systems with ERP. An enterprise should be able to generate revenue more effectively by providing better data more quickly for decision support, by creating products at lower cost, and by offering better service to suppliers and customers.

A firm must ultimately gauge its need for ERP by comparing the current information system with the proposed ERP system. The annual cost of operation and maintenance of the existing system should be compared with the initial cost of installing the ERP system plus its annual cost of operation and maintenance. Although benefits are difficult to quantify, the ERP system should add to revenue by enabling better and faster decisions and providing better customer service. RMO may be unable to survive in the highly competitive outerwear industry without extremely fast, effective, and integrated information systems, such as those provided by ERP.

Implementing ERP

Since ERP systems are so complex and they affect the entire organization, it is critical to implement them systematically and carefully. So, what are the factors that can help ensure success?

Critical Success Factors of ERP

ERP projects can be extremely complex, time-consuming, and expensive, and there are always high levels of risk involved in overhauling an information system. Based on several case studies to date, some of the most critical factors for successful implementation are:

- A realistic ERP budget and proper allocation of resources (hardware, software, people, knowledge, and tools)
- Education of all key management personnel about ERP and consensus that ERP is really needed
- Strong top management support
- Centralized project management with a *business* leader (as opposed to an IT leader) in charge

- Strong IT management and staff support
- Selection of the right approach to ERP
- Heavy user involvement (from *all* affected areas) in project management and implementation
- Business process reengineering to standardize on the capabilities of the software, rather than customization of the software to existing business processes
- Retraining of existing software developers in ERP implementation and maintenance
- Extensive training of end users in the new system
- Use of consultants (knowledgeable in specific business functions and experienced in specific ERP modules) to lead in implementation and training
- A respected and effective champion for ERP within the organization
- Effective and continuous communication among all parties involved in the ERP project
- Top-notch systems analysts with excellent business knowledge and technical skills
- Retention of top-notch systems analysts throughout the project and beyond
- Sensitivity to user resistance to new systems

Approaches to ERP

In addition to recognizing the critical success factors for ERP implementation, an organization should understand that there are essentially three approaches to ERP (see Figure 18-7). One option is to purchase a comprehensive packaged system, such as those offered by the leading ERP vendors. Such systems attempt to cover the information needs of all critical business functions (finance, manufacturing, human resources, marketing, and distribution). This approach may be best for the enterprise as a whole, but some modules for certain departments may lack needed functions. The comprehensive package may also not be flexible enough to accommodate rapid changes that the firm needs to undergo to compete.

FIGURE *18-7*

Three approaches to ERP.

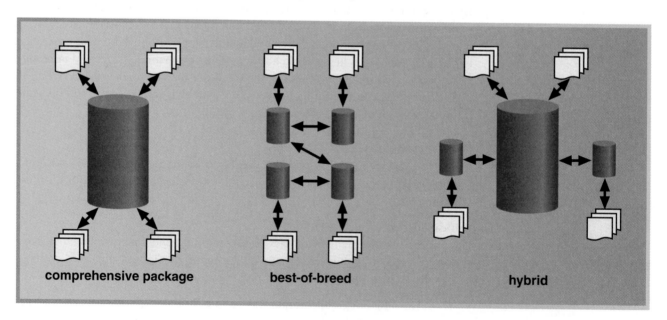

comprehensive package   best-of-breed   hybrid

best-of-breed

a software package developed for a particular application that is usually considered to incorporate the most current and best practices within the industry

A second approach is known as *best-of-breed*, in which an organization selects the presumed best available software (custom or packaged) for a particular application (such as production scheduling or e-commerce) and integrates it with other best-of-breed solutions in other functional areas. This approach provides individual functions or departments with the latest and greatest software solution. Although each business function may be extremely well served under the best-of-breed approach, the company as a whole may suffer due to a lack of integration, since disparate packages cannot often be integrated easily. However, the availability of middleware can make best-of-breed a viable option. With the use of middleware and a skilled systems integrator (either an in-house developer or a consultant), functional areas can use the best software for their application and still share important data with other functions in the firm. Currently, RMO is banking on this approach by purchasing packages for accounting/finance and human resources; developing its own systems for customer support, inventory management, and retail stores; and integrating the latter three systems.

A third approach is a hybrid of the comprehensive package and best-of-breed approaches. A particular comprehensive package may have many desirable features but may not provide all the functionality an organization needs. For example, a firm may choose a well-known ERP vendor to provide a backbone system for human resources, manufacturing, and finance but plug in various best-of-breed solutions for supply chain or sales forecasting applications. Instead of connecting many best-of-breed applications, the firm limits the number of applications it must link to the ERP backbone system, thus simplifying integration. The hybrid approach may be necessary if a company cannot afford to modify certain key business processes significantly to fit the process supported by a comprehensive package. The hybrid approach is also recommended if a firm must develop its own software in certain areas to provide unique effectiveness and to achieve a competitive edge. The firm can add this custom software to the package of core applications that provide basic efficiencies.

Organizational Structure for an ERP Project

Chapter 3 discussed the roles and responsibilities of the oversight committee and project teams for a system development project. The same principles of project management apply to an ERP project. Since ERP would usually be considered mission critical, an oversight committee, consisting of clients and key executives, should be formed. The members of the committee should represent all functions affected by ERP—finance, marketing, human resources, manufacturing, distribution, engineering, and information systems.

The project team usually consists of in-house IS personnel (managers, analysts, programmers, and technical specialists) and key in-house users. For an ERP project, the team might add vendor representatives and consultants because of the complexity of ERP software. In-house IS personnel must often be hired and trained to help ensure successful implementation. With the comprehensive package approach, the focus of IS personnel is on understanding, implementing, and supporting the system. Under the best-of-breed approach, IS personnel need a better understanding of how to create custom middleware or use packaged middleware to integrate the application packages. A hybrid approach requires both an understanding of the comprehensive package and of middleware. System users should also be educated and trained in all aspects of ERP that affect their jobs.

Since ERP projects are often large and extensive, an organization should provide comfortable meeting rooms to accommodate committee and team meetings; workspaces for project managers, team members, and consultants; computers, telephones, desks, chairs, bookshelves, and file cabinets; and administrative support to make arrangements for travel, lodging, meals, and meetings. A hypothetical organizational structure for an ERP project is provided in Figure 18-8.

RMO, with its $100 million annual sales, would be considered a small company. If RMO considers an ERP project, the oversight committee would consist of its president and three vice presidents. The project team could include the assistant vice president of production, director of operations, director of catalog sales, assistant vice president of

FIGURE *18-8*

Organizational structure for an ERP project.

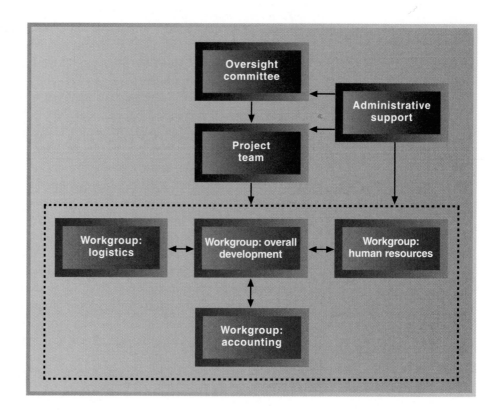

accounting and finance, and assistant vice president of systems. Various workgroups could include systems personnel and managers and operators from manufacturing, distribution, purchasing, warehousing, retail sales, telephone sales, human resources, and accounting. The project team and workgroups could also include vendor representatives and consultants.

An ERP System Development Methodology

There are many horror stories of failed attempts to implement ERP packages, regardless of vendor. Such failures are usually due to poor project management. As with any other system development project, a well-defined methodology is essential for successful implementation and continued, problem-free operation. The systems development life cycle, presented in Chapter 2, is just as applicable to a comprehensive or best-of-breed ERP system as to any custom-developed system. An ERP project differs from a more typical system development project in two important ways. First, an ERP project has a much greater *scope* than a typical project, involving virtually every function or department in the organization. Second, the ERP project often uses packaged software extensively, either a comprehensive package or several best-of-breed packages, instead of custom software. These important differences affect the systems development life cycle in many ways.

Phase 1: Planning
An oversight committee is formed to initiate the project. The primary task of the oversight committee is to identify the general purpose and overall scope of the ERP project and to appoint a project manager and project team members to develop the system.

The project team clearly defines the problem to be addressed by the ERP system and determines the detailed scope of the project. The team must then evaluate alternative approaches to ERP—custom solution, comprehensive package, integrated best-of-breed packages, or some combination—and make at least an initial selection. After selecting an approach, the project team develops a proposed project schedule and budget, addresses feasibility, and presents its findings to the oversight committee, both orally and in writing.

is needed when current versions of packages are upgraded or when comprehensive package modules, best-of-breed packages, or custom software programs are added or removed to meet changing requirements. Perfective maintenance may be needed to fine-tune the system for optimal performance. A system audit could be undertaken periodically to determine whether the objectives of the ERP system are being achieved.

Special Topics in ERP Development

Now that we have presented general ERP system development methodology, we turn to some details critical to ERP development: selecting a comprehensive package, customizing a package, linking applications within a system, and using packages for application development.

Selecting a Comprehensive ERP Package

Guidelines for selecting individual software packages were presented earlier in Chapter 8. But a firm has additional considerations when it is contemplating the purchase of a comprehensive ERP package. Both the costs and the risks are much higher, so the selection process should be thorough. Following are some detailed steps for evaluating comprehensive packages.

Step 1: Preliminary Evaluation   To begin, a firm evaluates ERP packages by looking at broad, general requirements. Four or five packages are typically considered at this preliminary stage. The decision makers could be members of the project team or the oversight committee, depending on a firm's management style. Package vendors and consultants may both be involved to provide needed information. Independent sources of information such as trade journals and personal contacts can provide insight also. Some of the key items to consider in a preliminary evaluation of an ERP package follow:

- Critical functionality, such as ability to deal with multiple currencies, to do simulations, and to support e-commerce.
- Up-front cost of the package. Initial quotes may vary widely among vendors, but they are often willing to negotiate.
- Cost of ongoing support for the package (upgrades, add-ons, and so on).
- The long-term viability of the vendor. How long will the company be around?
- The fit between the business model of the package and the actual business processes of the firm. Often it is easier for a firm to consider reengineering its processes before significantly altering the software.
- Satisfactory, *guaranteed* implementation time.
- A sizable installed base (at least 20 successful installations) so the firm is not one of the "guinea pigs."
- Local support (in multiple countries, if applicable) with a guaranteed, acceptable level of service.
- Proven success in the international environment (if applicable).
- Availability of customization tools to bolt on third-party packages or link to legacy systems.
- The best accommodation of the firm's management style (for example, whether the firm prefers to see the big picture or to drill down to detailed data).

Sometimes, a given package exhibits what may be called a *showstopper*—a missing feature or unsupported business process that transforms an otherwise good fit into a complete mismatch.

The result of this preliminary evaluation is a short list of perhaps two competing ERP packages.

Step 2: Detailed Evaluation   Once the field of packages is narrowed, lower-level managers, staff, or even operating personnel may assist in the selection process through meetings or workshops. Such workshops could be organized by major function (for example, finance or manufacturing) or by department (such as accounts payable within finance or production scheduling within manufacturing) to consider detailed evaluation criteria, based on information from vendors and advice from consultants. Just a few examples of the detailed criteria are the package's ability to accomplish the following:

- Set prices according to different criteria
- Provide financial information in multiple currencies
- Control commissions of sales representatives
- Support just-in-time (JIT) production

Such workshops provide two important benefits. First, staff can determine the candidate's ability to address both mandatory and optional functionality. Second, users of the new system are involved in the evaluation process, which increases their understanding and acceptance of the package and ERP in general. The results of the detailed evaluation process should also be documented.

Step 3: Vendor Presentations   After the detailed selection criteria have been applied to all candidate ERP packages, the project team can schedule vendor presentations for both the oversight committee and for the lower-level managers, staff, and operators. Both the form and substance of these presentations can assist staff in making the final recommendation.

Step 4: On-Site Visits   An additional evaluation method is on-site visits to different firms that use the packages under consideration. The names of such firms may be supplied by the vendor or, better yet, uncovered by the project team. During an on-site visit, information about the quality and quantity of vendor support, the amount of customization required for the package, the performance and reliability of the software, requirements for hardware and additional software, and implementation time can be obtained and added to the set of evaluation criteria.

Step 5: The Final Decision   After all previous steps have been completed, the project team usually performs a formal and structured evaluation based on (1) weights assigned to the criteria and (2) vendor scores on these criteria. The result is a total weighted score for each ERP candidate. Other factors may also influence the decision, especially where total scores are extremely close. One vendor may have a very high total score but may also have one glaring weakness that could make the package unsuitable. Intangible factors may also come into play. Whatever the considerations, the project team must select a vendor to continue the project. Consensus helps ensure the future success of the project. The project team's recommendation of a vendor is usually submitted to the oversight committee for final approval.

Linking Applications to an ERP Package

A popular approach to ERP is the hybrid approach: implementing a comprehensive ERP package from a major vendor and "bolting on" various best-of-breed software solutions to the main ERP backbone. One strategy for increasing success with a hybrid approach is to select a package that is already partnered with the ERP vendor. However, true partnering implies actual shared development, not just joint marketing, to ensure that the applications will work together smoothly.

Once a best-of-breed package is selected, some firms have the vendor integrate the package with the ERP system. Other firms have their own developers create the interfaces between the ERP system and the third-party application, often using tools pro-

8. The inventory withdrawals at the subassembly plants automatically trigger purchase orders directed to their suppliers of raw materials.

9. The purchase orders at the raw materials plants automatically trigger shipping orders, inventory withdrawals, production orders, purchase orders, and so on.

This entire hypothetical process could actually take only a few minutes (or seconds) in real time, linking the customer's order backward through the value chain to the most basic raw materials, using ERP and computer networking technologies. The rapid, real-time response of the ERP system helps to ensure a smooth, constant flow of goods and materials, minimizing the need for excessive inventories and providing excellent customer service as well. Firms that use ERP systems to their fullest capabilities could gain significant competitive advantage in the marketplace of the future.

A CLOSER LOOK AT ONE ERP PACKAGE: SAP R/3

The first part of this chapter discussed packaged software in general, and then discussed packaged software in the context of ERP. This section examines a particular ERP package, SAP R/3. The purpose of this section is not to empower you to become an SAP consultant but simply to provide a better understanding and appreciation of ERP systems through the exploration of SAP R/3.

What Is SAP?

SAP AG is a German software development firm (see Figure 18-9) that created and markets the SAP R/3 enterprise resource planning system. SAP (pronounced S-A-P, not "sap") began creating software in 1972. In 1983, it introduced the R/2 (meaning real-time system, version 2) system of integrated business application software for mainframe architectures. In 1989, SAP AG released SAP R/3, which utilized the increasingly popular client-server architecture and relational database technologies. SAP is currently the third-largest software vendor in the world and by far the market leader in ERP systems. There are now more than 60,000 installations of SAP in over 120 countries.

FIGURE *18-9*

SAP's home page.

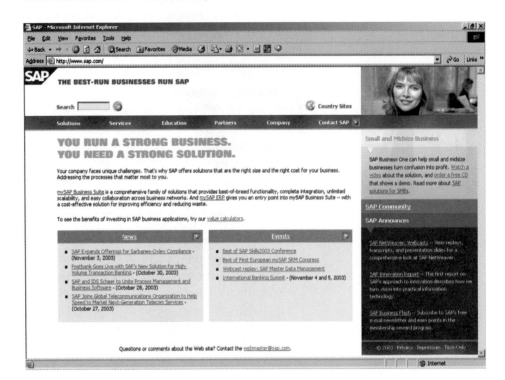

ERP systems are perhaps the most complicated information systems that a firm can implement, and SAP R/3 is considered to be the most complex ERP system available. SAP R/3 requires more than 15 gigabytes of disk space just to install and contains nearly 10 million lines of code. At www.sap.com, you can find an almost dizzying array of products, technologies, and services for small, medium, and large organizations, including the basic SAP R/3 solution, SAP R/3 Enterprise, and mySAP Business Suite. Although the complexity of SAP R/3 provides it with a high degree of flexibility and utility, it also makes for challenging implementation.

Organization of SAP R/3

Implementation of SAP R/3 within a company consists of a set of modules tailored to the needs of the enterprise. *Tailoring* means configuring several related application modules, not modifying or enhancing existing code. The modules within SAP R/3 are organized into four high-level work areas:

- Logistics
- Accounting
- Human Resources
- Business Tools

The first three areas represent the major functional groupings within most enterprises. Within Logistics are functions such as Materials Management, Production Scheduling, and Plant Maintenance. The Accounting area contains modules such as Financial Accounting, Controlling, and Project Management. Under the Human Resources area are modules such as Personnel Administration, Payroll, and Benefits. Not every firm, however, would choose to implement all available modules in these three areas. The Business Tools area provides the systems integrator the means to perform R/3 system installation and administration, including configuration and programming activities.

SAP R/3's integrated software modules are designed to meet information needs across the entire enterprise. The modules selected by a given firm become its SAP R/3 configuration. A firm just starting out with SAP R/3 may choose to implement only the core modules: Financial Accounting, Sales and Distribution, Materials Management, and Production Planning. After a firm successfully implements and uses these modules, it may choose to add others over time, such as Controlling, Warehouse Management, Plant Maintenance, Quality Management, Project System, Treasury Management, and Asset Management.

SAP Industry Solutions

SAP R/3 is a highly configurable system designed to meet the ERP needs of a wide variety of industrial segments. The SAP system includes the basic core functionality common to most organizations: finance, human resources, and logistics. To better serve the needs of different industries, SAP has created Industrial Business Units (IBUs) to provide system solutions for nearly two dozen distinct industries, based on what are generally considered "best practices." A brief overview of some of these industry-specific solutions can provide a better understanding of how SAP R/3 may be applied:

- **The automotive industry.** SAP's solution for the automotive industry includes a customer-pulled supply chain and a global infrastructure for the value chain, including component suppliers, after-market parts suppliers, and wholesalers. Additional functionality includes flow manufacturing, Kanban, shipping scheduling, and global available-to-promise capability.
- **The electronics industry.** Time to market is a key to success in this industry. New products with high quality and low prices are in demand. A real-time, integrated system can provide many benefits. SAP R/3 offers product data configuration man-

agement, configure-to-order quoting and estimating, product data management, bill of material, and routing development. The SAP High-Tech module provides a knowledge- and rule-based product configuration engine to forecast and monitor dynamic pricing and supply scenarios.

- **The public sector industry.** Public organizations should operate as efficiently and effectively as private enterprises. The SAP solution for this industry includes a Human Resources module that provides employee self-service (ESS), benefits administration, and event management capabilities, among others. The Purchasing and Materials Management module enables automatic day-to-day purchasing, supplier selection, and electronic data interchange (EDI).
- **The retail industry.** This industry is characterized by a wide variety of distribution possibilities (including walk-in traffic, catalog sales, and e-commerce) and small margins. Retailers must carefully manage inventory and monitor customer preferences. SAP provides Merchandise Logistics Tools to track customer buying behavior. The Retail module includes software for goods receipt, labeling, inventory management, and promotions management. RMO falls within this category and would be most interested in SAP's offerings within this IBU.

Other SAP IBUs include insurance, telecommunications, aerospace and defense, banking, chemicals, consumer products, health care, oil and gas, pharmaceuticals, and utilities. Obviously, SAP is endeavoring to tailor its ERP system to an extremely wide variety of business applications.

Master Data

The SAP R/3 system revolves around several standardized databases utilized by various application programs (see Figure 18-10). At the core of SAP R/3 are five master databases: the Material Master, the Vendor Master, the Customer Master, the Human Resources Master, and the General Ledger. The Material Master is used to store data relating to all materials tracked in the system. Logistics modules such as Materials Management, Sales and Distribution, Production Planning, Plant Maintenance, and Quality Management use these data. Materials within a firm are defined in SAP R/3 as one of several types, such as Finished Products, Semi-Finished Products, Raw Materials, Operating Supplies, and Packaging. The types of fields associated with an item of material may include material number, unit of measure, procurement type, temperature conditions, shelf life, and price.

FIGURE *18-10*

The five master databases for SAP R/3.

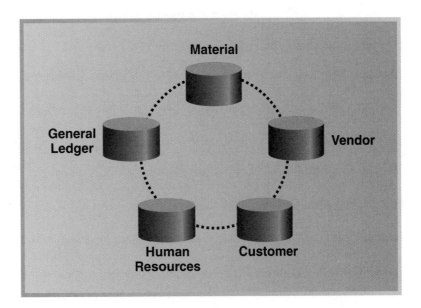

The Vendor Master contains information about external suppliers, such as address, telephone, tax codes, bank accounts, payment terms, and order currencies. The Customer Master stores data about a firm's business partners that buy products or services. Customers may be one of several types, including sold-to partners, ship-to partners, payers, or bill-to partners. For example, when you buy a gift for a friend, you may be the sold-to partner, your friend may be the ship-to partner, and your credit-card company may be the payer. All the information associated with these various parties—such as names, goods receiving hours, preferred shipping method, payment terms, and tax classifications—must be stored in the Vendor Master.

The Human Resources Master contains information about a firm's employees. This information is stored in various infotypes—logical groupings of similar data fields. For example, the Personal Data infotype contains name, birth date, marital status, and number of children. The Address infotype consists of various subtypes such as Permanent, Temporary, and Mailing. The Events infotype tracks important events for an employee, such as hire date, pay change, or change in organizational assignment.

The General Ledger accounts are the backbone of the entire SAP system and are used to produce reports to meet legal requirements. These accounts reflect all transactions that affect external reporting requirements. For example, billing a customer will increase the receivables account, and shipping an item will reduce the inventory account.

Implementing SAP R/3

As mentioned earlier, implementing SAP R/3 can be complex and time-consuming. This section highlights general issues of architecture, middleware, configuration, and accelerated installation.

Architecture of SAP R/3

The R/3 system is designed for a three-tier client-server architecture (see Figure 18-11). The three tiers, or levels, are presentation, application, and database. The user of R/3 works on a client machine at the presentation level with SAPGUI (SAP graphical user interface) software, enabling the user to work in a familiar Windows environment. Through SAPGUI, the user sends and receives data from what is called R/3 System software on various SAP application servers. It is at this application level that specific modules, such as Payroll or Materials Management, function. The R/3 System at the application level communicates with myriad SAP data tables within a relational database management system (RDBMS) on a database server. The language SAP applications use to interface with the RDBMS is standard SQL.

FIGURE *18-11*

The three-tier architecture of SAP R/3.

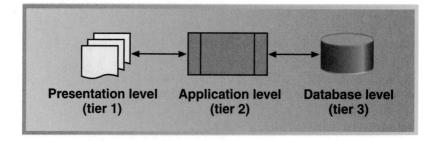

Presentation level   **Application level**   **Database level**
(tier 1)   (tier 2)   (tier 3)

R/3 Basis

R/3 Basis is the middleware that enables the integration of various SAP R/3 application programs on various computing platforms. Basis also provides R/3 administration tools (the data dictionary and the function library). Basis allows the R/3 system to run on a variety of platforms (such as AS/400, UNIX, or Windows NT) and to utilize different database systems (such as Oracle or DB2). Basis controls three different types of interfaces: user, programming, and communications.

The user interface is a graphical user interface (GUI) that makes work easier for users, developers, and administrators. For example, a developer may use the GUI of the data dictionary to create a table by pointing and clicking, rather than using SQL directly. The programming interface utilizes the ABAP/4 Development Workbench, which is required to create or change database tables and programs in the R/3 system. The communications interface provides the link among internal programs, as well as between R/3 and third-party programs. SAP uses standard relational database technology to store all three types of R/3 data: master, transactional, and configuration.

Accelerated SAP (ASAP)

Over the past decade, SAP and many other vendors have acquired a reputation for very expensive and slow ERP implementations. To address this issue, SAP has developed a methodology for a quicker, more efficient implementation called Accelerated SAP, or just ASAP. ASAP is designed for small to midsized companies and should require only about six months to implement, as opposed to one to two years for a normal SAP implementation in a large firm. ASAP follows the same basic phases just presented but places added emphasis on implementing the system with virtually no modification. Under this plan, the company may be forced to change its processes to fit the software. However, such BPR should be much more feasible for a small to midsized company than for a large one. RMO would be a very good candidate for ASAP.

Using SAP R/3

This section gives you a feel for how a user would navigate through SAP R/3. SAPGUI is loaded on a client computer that is networked to the server on which the SAP R/3 system has been installed. In a Microsoft Windows environment, the user starts the SAP R/3 program like any other program (by selecting *Start, Programs, SAP Frontend, SAP R3*). The logon screen then appears requiring the user to provide a user ID and password. Menus and buttons are available on this window for such functions as logging off or creating a new password.

Once logged on, the user will see the main SAP R/3 screen. The menu bar on this screen contains several options: Office, Logistics, Accounting, Human resources, Information systems, Tools, System, and Help. The Logistics, Accounting, and Human resources options represent the basic core of the SAP R/3 system. Selecting the Logistics option displays a pull-down menu with the following options: Materials management, Sales/distribution, Production, Plant maintenance, Service management, Quality management, and Project management. Selecting Accounting displays the following options: Financial accounting, Treasury, Controlling, Enterprise control, Capital investment management, and Project management. Selecting Human resources displays the following options: Personnel administration, Time management, Incentive wages, Payroll, Benefits, Planning, Recruitment, Travel expenses, and Information system. Ultimately, the user navigates through the various menus to accomplish the desired task, such as entering an order for a customer (see Figure 18-12) or viewing a materials requirements planning record (see Figure 18-13). Of course, the availability of all such menu options is contingent on having the associated modules configured and installed in the system.

FIGURE *18-12*

SAP R/3's Create Standard Order: Overview screen.

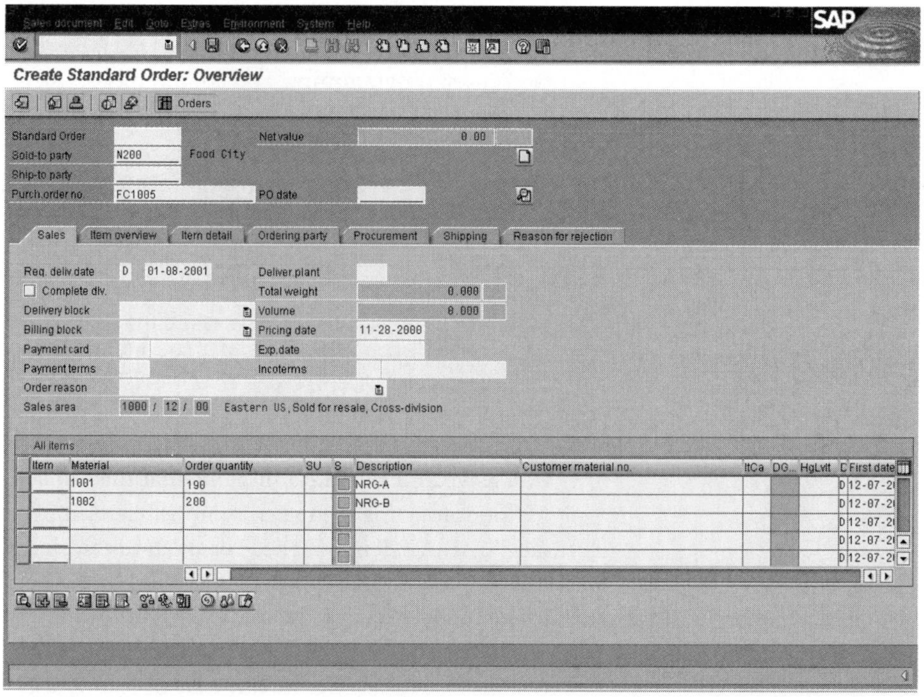

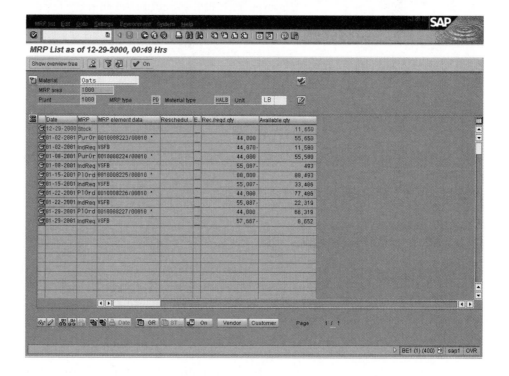

FIGURE *18-13*

SAP R/3's MRP List screen.

At the heart of the SAP R/3 system is the entry of business transaction data, followed by the use of those data to operate the firm efficiently and effectively. At the main SAP R/3 screen, the user can enter a four-digit transaction code in lieu of traversing a lengthy menu path to enter a valid SAP transaction. For example, to obtain a screen for confirming production work performed, the user can enter the code C011 instead of selecting *Logistics, Production, Production control, Confirmation, Enter, For operation,* and *Time ticket.* Obviously, a user well acquainted with transaction codes in a particular functional area would much rather enter the codes than use the menu. Figure 18-14 shows a screen called the Stock/Requirements List, which shows a real-time version of the MRP process. This output shows forecast demand and planned orders plus any dynamic changes such as completed production orders or material withdrawals.

FIGURE *18-14*

SAP R/3's Stock/Requirements List screen.

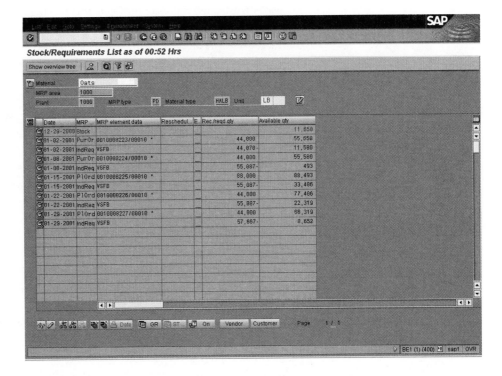

A button bar is also available in SAP R/3. Clicking the Back button will move you backward through the path just traversed. The Exit Session button closes down the current session. The Help button provides on-line help. The Dynamic Menu button displays a listing of menu bar options in a tree structure until it terminates in a particular task's transaction code. The user can click that option or enter the transaction code to access the desired task.

After a user has logged on to SAP R/3, he or she may create up to six different sessions in SAP R/3 at the same time. To create a session, the user selects *System, Create Session* from the main screen menu. Multiple sessions, available on the taskbar in Windows, allow the user to work on different tasks without spending extra time moving around through menus. A user trying to access the same record in a database in different sessions will be locked out of that record until the original update is completed.

To log off SAP R/3 when finished for the day, the user either selects *System, Log off* from the menu system or clicks the Close button in the SAP R/3 window.

SUMMARY

The issue of how best to procure business application software—custom-developed versus packaged—is extremely critical for the future of IS organizations and business enterprises. Equally important is the question of how best to deploy business application software—integrated versus stand-alone. The most recent trend in IS development is the deployment of integrated packages to increase both the efficiency and effectiveness of information technology as it serves to boost the competitiveness of the entire enterprise.

Software vendors design packaged software to be very easy to use and highly functional for its intended users. Application packages are now becoming more popular as both business processes and information systems requirements are becoming more complex. The debate rages whether a firm should custom-develop its own applications to match many of its unique system requirements or whether the firm should alter its processes, through business process reengineering, to match many of the "best practices" supported by recent packages. The three major issues that differentiate application packages from custom software are customization, integration, and upgrades. The systems analyst must address these three issues during the systems analysis and design phases.

A major class of packaged software is the comprehensive enterprise resource planning (ERP) system. ERP is the process of running virtually the entire organization with highly integrated application software. The approaches to deploying an ERP system include creating the system in-house, assembling the system from a diversified set of "best-of-breed" application packages, and purchasing the system as a comprehensive package. One of the most popular approaches to ERP is the comprehensive package. ERP is purported to increase efficiency throughout the organization, reduce IS staff, and improve customer service. However, implementing such a comprehensive, enterprisewide system can be extremely challenging, with many potential pitfalls. Properly organizing people (employees, vendors, and consultants) for an ERP project is a major key for success.

The current market leader in comprehensive ERP systems is the German company SAP, which offers its R/3 system worldwide. SAP R/3 software is organized around three major application areas (Logistics, Accounting, and Human Resources) plus Business Tools, which allow for the continuing administration of the system. SAP R/3 can be configured for a wide range of industries, such as automotive, electronics, public sector, and retail. SAP R/3 is usually installed on a client-server platform by system integrators with the assistance and guidance of in-house systems analysts. While very large corporations typically install more of the available modules with some customization (requiring more time and expense), SAP offers the scaled-down ASAP system to midsized and small firms for more rapid deployment.

KEY TERMS

best-of-breed, p. 728

business intelligence, p. 726

configuration, p. 721

enhancement, p. 722

enterprise application integration (EAI), p. 722

modification, p. 721

sales-force automation, p. 726

software integration, p. 718

system integrator, p. 720

total quality management (TQM), p. 726

value chain, p. 725

REVIEW QUESTIONS

1. What is meant by a "best-of-breed" software package?
2. What are the two most critical success factors for the development of software packages?
3. What are the three most common types of software packages?
4. What are the advantages and disadvantages of using an ASP as an alternative to purchasing a software package?
5. What are the three types of package customization?
6. What are the four major characteristics of ERP systems?
7. What are the three types of employee teams or committees that would usually be involved in an ERP project?
8. What are the major areas into which ERP vendors are expanding?
9. How can a firm use ERP as a strategic weapon to achieve competitive advantage?
10. What does the *R/3* in SAP R/3 stand for?
11. What are the four major high-level work areas of R/3?
12. What are the five master data sets found in R/3?
13. What are the three tiers found in the R/3 architecture?

THINKING CRITICALLY

1. Under what conditions would a firm prefer to develop custom software rather than purchase one or more packages?
2. How have the purposes of business information systems changed over the past 50 years?
3. Discuss when BPR would and would not be necessary when implementing an ERP system.
4. Discuss how the role of a traditional systems analyst will change when his or her firm adopts an ERP system.
5. Give several examples of how both efficiency and effectiveness are improved with the implementation of ERP.
6. Discuss how ERP could make substantial improvements in all four types of business information systems.
7. Which should come first: ERP or BPR? Defend your position.
8. Discuss the major differences between a system development methodology for a custom project and that for an ERP project. Cover all five phases of the SDLC.
9. What does ASAP stand for in the context of SAP software? Under what conditions would you recommend ASAP to a company?

EXPERIENTIAL EXERCISES

1. Do a search on the Web for *ERP* or *enterprise resource planning*. Find an article that discusses a recent development in ERP or provides more detail about a topic introduced in this chapter. Prepare a written or oral presentation of this article.
2. Form two teams of students to debate the issue of custom versus packaged software development. Provide arguments and evidence, pro and con, for the statement, "Custom development is essential for creating and sustaining competitive advantage."
3. Form two teams of students to debate the issue of best-of-breed versus comprehensive ERP systems. Provide arguments and evidence, pro and con, for the statement, "A comprehensive ERP package is superior to a best-of-breed approach."
4. Do some Web research on the term *middleware*. Prepare a written or oral report on some recent developments or other interesting aspects of this subject.
5. Perform the following role play for the class: You are the IS manager. It is your job to explain to four other students in the class (who are playing the roles of IS analysts and programmers) why the company is adopting ERP and how it will affect their jobs. The IS staff have heard rumors that at least 50 percent of them will probably be laid off.
6. Prepare a hypothetical cost/benefit analysis for an ERP system, showing the expected payback.
7. Select a particular type of enterprise (such as electronics manufacturer, governmental agency, or pharmaceutical company). Assume that your company is either small (less than $200 million sales), midsized ($200 million to $1 billion sales), or large (greater than $1 billion sales) and operates either domestically or globally. By simply reviewing the Web sites of several (at least five) ERP vendors, draw up a preliminary evaluation reducing the number of candidates to two. Make sure you provide sound reasons and evidence for your decision.

CASE STUDIES

Tools-R-Us, Inc., and Sales Quota Automation

Software packages are now being designed to perform very specialized tasks. One challenge for a marketing organization is to set sales quotas that are realistic enough to be attained and to motivate salespeople to excellence but not so easily attained as to create a financial burden for compensating salespeople for such excellence. It's a process that's too often unscientific and inexact.

Tools-R-Us, Inc., a maker of tooling machinery for General Motors, John Deere, United Technologies, and other manufacturers, starts from scratch each year by collecting spreadsheets and paper reports from field salespeople. It formerly set sales quotas by simply adding a percentage on top of last year's sales. Tools-R-Us now plans to install the quota module in Ockham Technology's SalesRazor 3.0 software, which automates the quota-setting process by analyzing new product launches and buying trends in different regions, industry sectors, and the broader U.S. economy. Tools-R-Us is considering plans to deploy modules for sales tracking in the fall.

The resulting quotas make sure salespeople in a historically strong region don't have an unfair advantage over others. They also give the $2 billion company better control over what percentage of salespeople meet or exceed quotas, thereby containing compensation costs. "Setting accurate quotas helps us not blow the sales budget," said Jack Hammer, Tools-R-Us's manager of marketing information systems. "If they're set consistently and everyone understands the methodology, you end up with a sales force that understands how to reach that goal."

Tools-R-Us is coupling SalesRazor with data warehouse software from SAS Institute. The data warehouse gathers information from the company's SAP software and an Oracle database that houses point-of-sale information from partners such as distributors and integrators. Tools-R-Us chose SalesRazor in part because of Ockham's willingness to host the application while Tools-R-Us's 100 IT staffers focus on upgrading the company's SAP enterprise software. Tools-R-Us plans to bring the application in-house in the near future. The licensed version of SalesRazor starts at $165,000 for a 250-person sales force.

1. What kinds of challenges does Tools-R-Us face as it attempts to integrate SalesRazor with software from SAS, SAP, and Oracle?
2. Do you believe this type of software should be purchased in a package or developed in house? Why?
3. Who should be responsible for getting SalesRazor up and running at Tools-R-Us: Ockham, SAP, or Tools-R-Us IT staffers?

Springfield General Life Insurance Co.

ERP installations don't always go smoothly. Springfield General Life Insurance Company contracted with Enterprise Solutions, Inc. (ESI) to install a finance system for $5 million. Springfield General later claimed that it had to develop its own homegrown software after ESI failed to deliver the new system as promised. In its lawsuit, the company also claimed that ESI had more than doubled its original price for the software and indicated the cost could go even higher. ESI countered that there was nothing wrong with the software it provided, but that Springfield General wasn't able to adopt the system due to its own internal problems.

In mid-2003, Springfield General paid the $5 million up front to ESI for software licenses, technical support and implementation, and customization services. Late in 2003, Springfield General personnel began to suspect that the ESI team was having difficulty with the customization and implementation services. Springfield General further claimed that ESI "failed to properly customize and implement" the promised software, leaving it "inoperable and entirely useless." ESI replaced the workers assigned to the project and sent in a crisis management team. But then early the next year, Springfield General charged that ESI said the original project bid was flawed and raised the price to $12 million, with the possibility of another 20 percent increase on top of that amount.

1. What are some of the possible causes in the breakdown of the relationship between Springfield General and ESI?
2. What are some of the possible "internal problems" that could cause problems adopting such software?
3. Do you agree with the steps apparently taken by ESI, such as sending in a crisis management team and subsequently adjusting the price to $12 million? Why or why not?

Integration Specialists, Inc. (ISI)

Integration Specialists, Inc. (ISI) has recently introduced its latest customer relationship management (CRM) package, the myISI CRM 3.0 suite. However, some analysts are warning customers to be wary of the applications unless they install or already use ISI X/4 enterprise resource planning software. Carol Oakley, senior vice president of Worldwide CRM Business Development for ISI, says customers can take advantage of ISI's latest CRM suite even if they haven't installed ISI on the back end. But Lance Speer, service director for IT research, says the only way to achieve the full functionality of the CRM software is to link to ISI's financial and order-management applications, because the CRM software is dependent on business processes and functionality built into ISI's ERP applications.

Even if users are willing to forgo some capabilities in the CRM suite, they will still have to tackle some tough integration issues. Speer says "bits and pieces" of the CRM applications are written in Java, but more than 50 percent of the CRM code is still written in ISI's proprietary language. That makes it difficult, and costly, to extend the suite to the Web or other vendors' applications.

One current ISI customer, Business International Corporation (BIC), is looking forward to installing the new release. BIC, a machine manufacturer in Stillwater, New Jersey, last week rolled out the ISI call center as part of a year-long deployment of its myISI 2.0 CRM release. Tony Shaeffer, director of MIS, plans to begin upgrading his company to CRM 3.0 early next year, primarily to leverage enhanced usability and functionality in the new call-center applications. ISI's Speer says that for existing ISI customers, the latest version of ISI's CRM software delivers better-than-average analytics, data-mining, order-management, and field-service functionality. However, Speer adds, the suite lags behind Clarity Systems in sales-force automation.

1. Should customers of myISI CRM 3.0 be very concerned that much of it is written in ISI's proprietary language? Why or why not?
2. What are some of the potential hazards of using myISI CRM 3.0 on the front end (the user interface) without having ISI X/4 ERP software on the back end (applications that perform detailed processing)?
3. What are the pros and cons if Business International decides to go with the current market leader, Clarity Systems, for its CRM applications?

Rayco Products Corporation

Dave Chen works as a production planner for Rayco Products Corporation (RPC), a very successful multinational, *Fortune* 500 company. It is Dave's responsibility to create monthly production plans for five plants in the United States and Canada. To be closer to the action, Dave works at one of the five plants in Texas instead of at corporate headquarters in Ohio. One of the primary inputs to Dave's production planning process is an 18-month, item-level sales forecast that groups literally thousands of distinct part numbers into dozens of product families. One of Dave's major problems is creating the production plans in a timely manner so that the plants can adjust their material, manpower, and equipment requirements to meet expected customer demand.

The process works something like this: Corporate IS in Ohio runs the sales forecast program based on historical sales figures and input received from marketing personnel. Several copies of the forecast data tape are produced and mailed overnight to the plant IS departments where the three production planners are located, including Dave in Texas. When the IS department at the Texas plant receives the tape, it runs a program on the mainframe to produce a very large print-out, and a data file on disk, of the forecast for all product families and their individual part numbers. When the file is created, Dave runs a program to download the file to his PC. He then imports the text file into an Excel spreadsheet. From there, Dave runs several macros to manipulate the data into an initial production plan.

The entire process of creating, mailing, and running the forecast tapes, then downloading the data to the PC and running the spreadsheet macros, takes at least three days (barring any glitches in computer operations or in the shipping of tapes). This puts Dave in an extreme bind to interpret the data and create the production plans in time for the scheduled monthly meeting at corporate headquarters. The plants are also in a hurry to receive the final production plans so they can begin ordering material, scheduling overtime, and shifting production to other plants, if necessary. The entire process—from the monthly close of business to the creation of material, manpower, and machine requirements at the plants—takes at least seven business days, or about one-third of a business month.

1. What major courses of action would you recommend for RPC in terms of ERP, and why?
2. What specific benefits would RPC derive for production planning via an ERP implementation?
3. What potential problems would RPC face during and after an ERP implementation?

Rocky Mountain Outfitters

Rethinking Rocky Mountain Outfitters

William McDougal, vice president of marketing and sales, just returned from a seminar on customer relationship management (CRM) sponsored by ISI, the ERP software vendor mentioned in the earlier case study. At the seminar, ISI promoted its most recent release, myISI CRM 3.0. Upon returning, William read some recent articles about packaged CRM software, particularly ISI's product. The articles pointed out the many possible advantages of using CRM software but also warned of integration problems.

As a result of this investigation, William was beginning to wonder whether RMO's customized approach to the CSS project was indeed the best option. He called in Mac Preston, chief information officer, to get an up-to-the-minute progress report on the CSS project, and to get his opinion on the possibility of switching to a packaged solution. The CSS project team was still in the process of identifying system requirements.

Mac reminded William that a packaged CRM product had been considered about six months ago but that a customized approach was judged best by the entire oversight committee, including William. William countered that he had not fully understood the capabilities of a CRM package at that time and that the latest versions of CRM software are much more powerful than previous versions. William now says he is leaning very strongly toward reopening the question of a CRM package with RMO president John Blankens.

1. What would be the major advantages and disadvantages of using a CRM package rather than a customized CSS?
2. What impact would using a CRM package, instead of a customized solution, have on various RMO personnel? Consider both user and developer groups.
3. Should RMO consider the myISI CRM 3.0 package without first having an ERP system, such as ISI R/3, already in place? Would your answer change if RMO decided to proceed with the customized CSS instead of a CRM package?

4. Based on what you know about RMO and ERP, what would you recommend regarding an ERP system for RMO? Completely discuss all important aspects of the reasoning behind your recommendation.

Focusing on Reliable Pharmaceutical Service

In Chapter 8, you considered whether Reliable should (a) contract for custom development, (b) purchase a "stock" prescription system and contract to modify the package, or (c) opt for some combination of these approaches. Suppose that Reliable is indeed leaning toward packaged software and contemplating one further step—purchasing an ERP system that would enable it to integrate nearly all functions of the company. Based on what you know about Reliable from previous chapters, perform the following:

1. Discuss Reliable's potential need for an ERP system specific to its operations.
2. List and discuss the most likely benefits that Reliable would achieve using an ERP system. Also list and discuss the most significant risks that Reliable would face with such a system.
3. Identify and discuss the most important critical success factors to ERP implementation in Reliable's case. Be specific.
4. Explore SAP's Web site at http://www.sap.com to learn what specific products and services SAP might recommend to Reliable. Report your findings.
5. Using the ERP system development methodology and special topics in ERP development as presented in this chapter, create and discuss an overall plan, start to finish, of how you see Reliable deploying a successful ERP system. Create a rough Gantt chart to illustrate your plan.

FURTHER RESOURCES

George W. Anderson, *SAP Planning: Best Practices in Implementation*, Pearson Education, 2003.

Joseph A. Brady, Ellen F. Monk, and Bret E. Wagner, *Concepts in Enterprise Resource Planning*. Course Technology, 2001.

Erin Callaway, *Enterprise Resource Planning: Integrating Applications and Business Processes across the Enterprise*. Computer Technology Research Corporation, 1999.

Steven Harwood, *ERP: The Implementation Cycle*, Butterworth-Heinemann, 2003.

Jose Hernandez, *Roadmap to mySAP.com*, Course Technology, 2001.

Bradley D. Hiquet, *SAP R/3 Implementation Guide: A Manager's Guide to Understanding SAP*. Macmillan Technical Publishing, 1998.

Daniel E. O'Leary, *Enterprise Resource Planning Systems: Systems, Life Cycle, Electronic Commerce, and Risk*, Cambridge University Press, 2000.

Murrell G. Shields, *E-Business and ERP: Rapid Implementation and Project Planning*. John Wiley & Sons, 2001.

Avraham Shtub, *Enterprise Resource Planning (ERP): The Dynamics of Operations Management*. Kluwer Academic Publications, 1999.

Norbert Welti, *Successful SAP R/3 Implementation: Practical Management of ERP Projects*. Addison Wesley Longman, Inc., 1999.

Liane Will, *SAP R/3 System Administration: The Official SAP Guide*. Sybex, 1999.

Principles of Project Management

Chapter 3 discussed the tasks associated with the management of a system development project. The chapter also detailed specific skills and techniques that are used by effective project managers. Finally, it explained the context of project management in an organization—how a project fits within an organization and particularly how the project manager must work with others in the organization to achieve success.

This appendix focuses on the fundamental principles of project management that underlie specific project management tasks. With in-depth understanding of the various project management areas, you can develop a broad foundation to prepare for your career as an IT professional, as a knowledgeable participant in a project, and someday as a project manager. However, participation in project management activities is not limited to the project manager. In successful teams, every member of the team helps with project management. This appendix introduces project management techniques that are presented throughout the textbook. By studying this appendix early in the course, you will be better prepared to learn the details of the various project management techniques.

PROJECT MANAGEMENT KNOWLEDGE AREAS

This appendix is based on the Project Management Body of Knowledge (PMBOK) that has been developed by the Project Management Institute. We discuss eight primary knowledge areas of project management:

- Project scope management
- Project time management
- Project cost management
- Project quality management
- Project human resource management
- Project communications management
- Project risk management
- Project procurement management

To learn more about project management and find additional resources, you should visit the Project Management Institute Web site at http://www.pmi.org/.

Project Scope Management

Need and Objectives
One common thread throughout almost every failed development project is vague or constantly changing system requirements. Probably the most pervasive problem of every development project is *scope creep*. Scope creep is the addition of new system requirements after the discovery activities and system specifications are completed. It can occur throughout a project, during analysis, during design, or even while the system is being

scope creep

the addition of new functions to the scope of a system that cause the project to increase in size

implemented, although it is more problematic during the later phases of a project. Unfortunately, neither the user nor the developer realizes the effect that one little addition can have on a project. Depending on when the additional request is added to the project, even a simple change can have far-reaching effects. For example, a simple field addition can affect the database, input forms, output reports, control algorithms, update algorithms, system testing plans and tests, data conversion, system interfaces, and system and user documentation. Suddenly, a simple change is not so simple.

Of course, requirements are defined during the analysis phase. But requests for changes can occur throughout a project—during design, programming, and even implementation. The danger of scope creep is in how fast it can cause a project to get off schedule and off budget.

Thus, scope management is a key project management task. The objectives of project scope management are the following:

- To precisely define the functions and capabilities to be included in the new system
- To verify that the identified capabilities are necessary and are priorities for the project at hand
- To control the set of functions so that it does not grow inappropriately

Scope management includes managing the original definition of the system requirements so that they can be accomplished within the allotted time and budget of the project. Obviously, the requirements must be comprehensive to include the needed functions; however, they must also be realistic in not including unnecessary functions. Project managers must verify that the selected functions are indeed essential for the intended use of the new system. Finally, they must control additional requests to avoid the problems associated with scope creep.

Techniques

There are three main categories of scope management techniques: (1) defining the scope, (2) verifying the scope, and (3) controlling the scope. Within those categories are supporting techniques.

Defining the Scope Scope definition should occur at two times in a project's life cycle. The scope of the new system is delineated when the project is first initiated. A high-level document will provide information about expected business benefits and system capabilities.

Details of the project scope are further defined as the system specifications are developed. During the analysis phase, team members gather information from the users and develop system requirements. As these processing requirements are developed and documented in a formal document, they become system specifications. As noted in Chapter 3, one of the major causes of project failure is lack of a precise definition of system requirements.

Verifying the Scope One technique for verifying a project's scope is to hold a review and approval meeting with users and project team members at the end of the scope definition stage. In preparation for this meeting, all parties review every defined system capability and assign each a priority such as "nice to have," "important," or "critical." This verification step usually does not require much time, but it is important to ensure that the scope is precisely defined and that all participants—users, clients, and developers— agree on what is to be included.

change control log

a log or list of requested changes to the existing set of functions

Controlling the Scope   As indicated earlier, controlling the project's scope is just as important as initially defining it. A formal mechanism should be established to review additional requests for new system functions. One effective way to track and control these changes is to set up a *change control log* and change control committee. They are typically instituted after the scope is verified and as the project moves from scope definition to design and implementation. Any proposed change is added to the log, along with a statement of its importance and priority. Team members estimate the impact of the change on the workload, schedule, resources, and budget. The change control committee then meets regularly and decides which, if any, of the additions should be made.

Formalizing the changes has several obvious benefits. First, only important and critical functions are added to the system. Second, everyone knows and agrees on the changes. Third, the client is not surprised by delays to the schedule. Fourth, groups of changes can be made at the same time, which frequently can reduce the overall impact.

Project Time Management

Need and Objectives
Possibly the most-asked question of any project team is, "When can I have it?" Another common variation on this question is the statement "We need the system to be operational by June first." The response to these requests and demands all too often is a schedule that is built to the demand, rather than to realistic workloads. Of course, when the project is not completed on time, everyone wonders what went wrong.

The objectives of project time management are multifaceted:

■ To ensure that the project schedule accurately accommodates the work to be done
■ To accurately measure completed work to correctly assess the percentage completed
■ To effectively use resources and techniques to accelerate the overall time to completion

Techniques
The core element in project time management is the schedule, and the development of the schedule is one of the most important tasks of the project manager. Time management techniques include building the project schedule, making changes to it, monitoring progress based on the scheduled dates, and shortening the project schedule using optimization techniques.

Building the Project Schedule   Building the project schedule is not an isolated activity. System functionality (scope), available resources, cost, and system quality all affect the details of the schedule. The project team uses the work breakdown structure to develop a comprehensive list of all the project activities, estimating the size of each activity and determining activity dependencies. The objective is an accurate estimate of the work to be done. The two most critical areas to ensure accuracy are identifying all activities and realistically estimating the size of each activity.

Some techniques are available to help team members validate the time estimates for the project. Such estimation techniques as function point analysis and *Constructive Cost Model* (COCOMO) can be used to estimate the total time to completion for specific development projects. These estimation techniques are based on analyzing the functions to be included in the new system in detail and estimating the total effort by summing the detailed estimates. Many companies also maintain a history of past system development projects to help estimate the size of new projects.

Modifying the Project Schedule   There are three main reasons why schedules need to be modified: (1) additional activities are added to the project, (2) the time estimates of certain activities are expanded because of complexity, and (3) resources are not available at the expected time.

Project schedules are often built in pieces. For example, when the analysis phase is winding down, the design phase planning tasks may be started. The detailed work breakdown structure for the design phase and implementation phases may be completed after the project has begun.

Occasionally, the time estimates for the activities need to be modified because of unanticipated complexity in the new system. The development team should analyze major changes to activity duration to identify all potential problems and unanticipated complexities that could have a negative impact on the project. Project managers and technical staff consider one of the toughest jobs of time management to be estimating the complexity of the problems that need to be solved.

Monitoring Progress   One of the early responsibilities of a project manager is to identify specific milestones or deliverables for the project. Decisions concerning the development method to use, the structure and form of a central repository of documentation and components, and the milestones to set for the project all affect the progress of the project. For example, if prototyping is used, then appropriate milestones can be associated with the completion of the prototypes for certain functions.

One cardinal rule for identifying and measuring milestones is to insist that milestones be measured either as "complete" or "not complete." Many new project managers get bogged down in the old *80/20 syndrome*, which states that 80 percent of the progress only requires 20 percent of the time. Therefore, a report of "we are 80 percent done" really can mean that "the last 20 percent will take 80 percent of the time." In fact, some activities reach 80 percent completion and never finish the last 20 percent.

80/20 syndrome
80 percent of the progress takes 20 percent of the effort, so the last 20 percent can take 80 percent more time

Optimizing or Improving the Project Schedule   One of the most valuable skills of an effective project manager is being able to move a project along. We call this *optimizing*, or shortening the schedule. There are numerous books on techniques such as joint application design (JAD), rapid application development (RAD), prototyping, rapid testing, and eXtreme Programming that can help a project manager expedite projects.

When a project gets in trouble, the first corrective step is to recalculate the project schedule—to determine the realities. Then the team can take corrective measures to optimize or improve the schedule. Knowing whether to add staff, reduce scope, juggle resources, or simply live with the delays shown in the recalculated schedule is based on the experience and judgment of the project manager.

Project Cost Management

Need and Objectives

Earlier we said that the most frequent question from clients is, "When can I have it?" Because schedule and cost are so closely linked, there is often another part to the question: "When can I have it *and how much will it cost*?" Many project costs involve human resource costs, so projects requiring much more time and effort naturally cost more.

Projects are initiated after estimated benefits are compared with anticipated costs. During the project, the effectiveness of cost control can mean the difference between profitability and nonprofitability. Cost control is closely associated with all the other areas of project management, and in many ways it is simply a reflection of the entire

management of the project. Projects that get off schedule cost more, and projects that suffer from scope creep cost more. Cost control is as critical for outsourced projects as it is for in-house development.

The specific objectives associated with project cost management are:

- To accurately estimate the anticipated project costs
- To accurately predict the cash flow and timing of expenditures
- To control actual project expenditures to those that are included within the plan
- To capture and record actual project expenditures correctly

Techniques

Project cost management techniques are based on fundamental accounting principles, which you learn in your cost accounting and managerial accounting classes. This section limits the discussion to identifying some of these basic techniques.

Estimating Costs   Every company has its own set of rules to determine which costs should be included in project costs. Direct costs are those that can be directly attributed to the project, such as salaries of project team members, costs of software licenses, and fees paid to contract personnel such as programmers. Other costs may be partly or fully allocated to the project based on company rules. For example, new computer equipment may be shared by programmers working on multiple projects (concurrently or over time). Project managers must ensure that all expenditures that will be directly assigned to the project are identified in the original estimate. A project's budget can be severely affected by unexpected costs, such as the cost of licensing software, if they are not anticipated at project initiation.

Once a company has developed the estimated project cost, it performs an economic feasibility analysis for the new system. Chapter 3 explains the details of calculating net present value and other measures to determine project feasibility. One of the major difficulties of project cost management is that sometimes costs must be estimated before the exact scope of the system is known. That is, an organization wants to know whether a project is economically feasible before it knows the detailed system requirements. To respond to this need, the project manager first develops a rough order of magnitude (ROM) estimate and indicates the possible range of the costs. Ranges can vary widely—it is not unusual to see a ROM estimate with a range of 100 percent. Although a doubled cost range may seem incredible, when you consider that some projects are over budget by factors of 3, 5, and even 10, an initial ROM with a factor of 2 may be acceptable.

Predicting Cash Flows   Part of the approval and initiation process of a project is to provide an estimate of the timing of expenditures so that the organization can budget for monthly and yearly expenses. Cash flows can be estimated based on the project schedule, including work to be done and purchases to be made.

At the conclusion of an initial funding period, a project manager needs to prepare and present a more precise estimate of the project costs for the remaining schedule. Two factors must be considered: additional tasks and their durations. As explained earlier, tasks may be added based on a more accurate understanding of the project's scope and complexity. One way to determine added costs is to develop an efficiency factor—the actual time to complete each task divided by the original estimate. An efficiency factor greater than one indicates that tasks are taking longer to complete than originally estimated, and it indicates that the project is over budget. Project managers then must do more detailed analysis to determine the cause for this lower efficiency and higher costs.

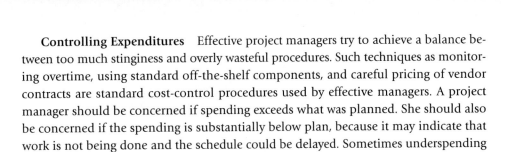

Controlling Expenditures   Effective project managers try to achieve a balance between too much stinginess and overly wasteful procedures. Such techniques as monitoring overtime, using standard off-the-shelf components, and careful pricing of vendor contracts are standard cost-control procedures used by effective managers. A project manager should be concerned if spending exceeds what was planned. She should also be concerned if the spending is substantially below plan, because it may indicate that work is not being done and the schedule could be delayed. Sometimes underspending is one of the early warning signs that the project is not progressing as planned.

Project Quality Management

Need and Objective

One of the most important responsibilities of a project manager is to ensure the quality of the final product. But software is especially subject to errors and failures. The failure of a single line of code can bring down an entire system. Many software systems are used in critical life situations—air traffic control, patient monitoring systems, and space exploration. Even noncritical systems demand extremely high levels of accuracy and performance—for example, banking and customer account systems, telephone and communication systems, and point-of-sale retail systems. Modern society functions more and more on technology, and these systems have become the fabric of our daily lives. Without technology, companies could not function. So, effective project management always includes substantial quality control.

The single objective of project quality management is the following:

To produce a system that is—

- Easy to use
- Fit for its intended purpose
- Robust
- Reliable

One problem with project quality management is that these characteristics are difficult to measure. How do you measure "ease of use"? Even though a precise metric may not exist, project managers must still attempt to provide high-quality software.

Ease of use is a measure of how easily and quickly the users understand and become proficient with the system. It includes the intuitive nature of the system, the efficiency of performing specific tasks, and the supporting tools and help systems.

Fitness for purpose gauges how well a system satisfies the business need for which it was developed. Fitness is also hard to measure because it depends not only on how well the system meets the user requirements but also on how complete and accurate the user requirements were. The key question to ask here is, "Does the system support all the activities and procedures of the defined business function?"

Robustness means how tough the system is in handling adverse situations such as bad data or even equipment. Is it free of errors that cause it to crash? Is it available 24/7 (or whatever the requirement)?

Reliability relates to the ability to always give correct results. With all different types of input data and combinations of scenarios, does it provide results that are correct? Can the users depend on the results to be absolutely correct all the time?

Techniques

All too frequently, project managers focus exclusively on program testing to ensure a high-quality system. But quality control should be planned from the beginning, and procedures should be integrated throughout all activities of the project. Every milestone and intermediate delivery should contain specific quality reviews and measurements to test for quality. As shown in Figure A-1, the automated system depends on the design; the design depends on the specifications; the specifications depend on the requirements. Errors and problems anywhere along the chain will reduce the quality of the final system. So, quality control needs to begin with the user requirements and continue with every milestone.

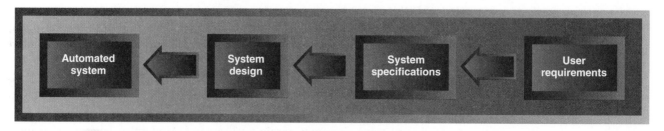

FIGURE *A-1*

Dependencies of intermediate project deliverables.

The specific techniques to ensure quality in each of these intermediate steps are well documented and widely described. The real problem for a project manager is to integrate quality reviews into the schedule and ensure that they are done. Quality reviews do not happen automatically. Specific checks and safeguards (with appropriate documentation) need to be included as part of the project plan.

Two types of quality checks need to be done for each deliverable. The first type, often called *validation*, checks to be sure that the deliverable conforms to external requirements. Take user requirements, for example. Do the user requirements accurately describe the needs of the users? Are they complete? Are they comprehensive? This type of quality check needs to involve people who provided the external requirement. In this case, the users must review the requirements to verify their accuracy and completeness.

The other type of quality check is for internal consistency. It is frequently referred to as *verification*. Again, considering the user requirements, project managers should review requirements to make sure that some parts do not contradict other parts. Another element of this review is to make sure that the system is complete—that there are no omissions in the description of the deliverable.

Specific techniques vary depending on the type of project and the deliverable that is being reviewed. Intermediate deliverables use techniques such as structured walkthroughs, desk checking of models and documents, and even prototyping. Final system components must be checked with various levels of program and system testing. These techniques are explained throughout the text. Again, the role of the project manager is to plan for quality reviews and ensure that they are carried out.

Project Human Resource Management

Need and Objectives

Many companies say, "Our employees are our most important asset." But many of those same companies do not treat their employees that way. On a development project, the project team *is* the most important part. Even though the project manager's time is important, the project team does most of the work to develop most of the new system.

Therefore, another important responsibility of a project manager is to establish an environment so that team members can work as rapidly and efficiently as possible. In other words, a project manager clears roadblocks so that the project team can develop the intermediate deliverables and the final system.

There are six primary objectives for project human resource management:

- **Staff acquisition.** To ensure that the project team is staffed at the right time with people who have adequate skills in the right mix
- **Personnel development.** To provide appropriate training for members of the team
- **Team organization.** To organize the project team and subteams for effective work
- **Team building.** To encourage work teams to become effective working units
- **Team member motivation.** To provide the leadership and vision necessary to encourage and motivate members of the team
- **Work environment optimization.** To ensure the working environment, including facilities, tools, and support, is conducive to getting work accomplished

Techniques

Human resource management is a broad and well-developed area. It is outside the scope of this short appendix to treat the multitude of theories and techniques for team management extensively. Numerous courses, books, and other resources that thoroughly treat this topic are available.

Given this limitation, this section simply provides a few tips and fundamental ideas that a project manager should keep in mind. Project managers should always remember that their primary responsibility is to enable the team members to do their best work.

Staff Acquisition   The various phases of system development require different levels of staffing and different sets of skills. One of the key staffing factors in a development project is the formation of teams that have diverse members who have complementary skills. A development team frequently starts with a few core members and grows throughout the project. Figure A-2 illustrates a typical staffing scenario for the phases of a project.

FIGURE *A-2*

Staffing levels of a typical project.

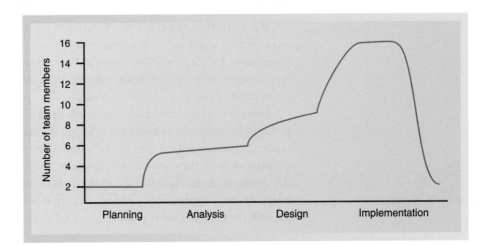

During the planning phase, the team consists of only a few members, typically a project manager and one or two experienced systems analysts. Experienced analysts and developers usually have the required management expertise to staff the project in this phase. The analysis phase requires team members with good analysis skills and strong problem domain knowledge, so the project team during this phase expands by adding more systems analysts, business analysts, and selected key users. Design is a more technical activity requiring the addition of members with technical expertise, such as networking and database professionals. If the systems analysts are also experienced in technical issues, they may carry out the design activities alone.

During implementation, more programmers and quality-control team members are usually added. Technical writers may also be added to complete the documentation. Also, the project team may be enhanced with additional users to assist them in training on the new system.

If the available team members do not have adequate skills or experience to carry out their functions, the project manager must ensure that they receive training. Experienced project managers do not risk project failure by asking team members to function in areas in which they have no experience or training. It is better to add more team members, hire consultants, provide training, or do all three to reduce the risk that the team will make poor technical decisions.

Team Organization   There are two main considerations with team organization: (1) the overall hierarchy of the project and (2) the structure of each work team. Larger projects may be divided into teams by subsystem or by functional department. Many structures and compositions are possible. The only caveat is to keep the overhead low; that is, do not insert intermediate team leaders. "Lean and mean" is usually the best project team organization philosophy.

Team Building   Any project, especially a large one, consists of groups of work teams. Frequently, people are assigned to the teams with little thought about team dynamics for effective work relationships. Teams that can form a strong work unit become extremely productive. Team building includes training, social interactions, personality profiling, and other activities to teach team members to work together smoothly.

Team Member Motivation   Novice project managers often assume that all members are motivated by money, particularly wage increases and bonuses. Yet, different people are motivated in different ways. Job satisfaction and motivation can come from work responsibilities, rank advancement, acquisition of new skills, performance recognition, and satisfaction with the job itself.

One of the most important contributions to team motivation is a clear vision of the project and its potential benefits to the organization. A key role of project managers is leadership, which entails sharing the vision of the project with all team members. Communication with the team must be frequent and explicit. As team members internalize the vision, their motivation and commitment to the project increase. On the other hand, a lack of communication is a strong demotivator. If people do not know what is going on with the project, they become less committed and less motivated to work hard.

Project Communications Management

Need and Objectives

Probably the most fundamental tool to effective teamwork is the ability to communicate. Communication is essential for every member of the project team but even more critical for the project manager. For simplicity, this section organizes the discussion of communications based on incoming, outgoing, and internal communication.

First is the need to receive communication or to gather information. One of the most critical elements of project success is careful gathering of information to understand the user requirements. Information is also collected from vendors and other project stakeholders. In addition, the project manager must gather and understand information from the team on status, milestones, problems, and technical issues. So, procedures need to be established for the entire team to ensure that the gathering of information is timely and accurate.

Second, the project team must disseminate information. Each member of the team should report his or her status, progress, and results. In addition, team members must explain technical issues to users and other stakeholders. The project manager has the overall responsibility to report the status of the project, explain problems, and discuss additional project needs with all stakeholders.

Finally, internal communication among the members of the team is extremely important. All team members must keep each other informed of progress, of technical issues and decisions, of specifications, and of a multitude of details. Specific procedures should be established to ensure that internal communication is timely and appropriate to communicate vision and provide motivation and encouragement to team members.

Project communication management can also be organized to accomplish the following objectives:

- To ensure that the necessary information is gathered in a timely manner and is complete and accurate
- To ensure that project information is disseminated frequently and is an accurate representation of the project
- To ensure that members of the project team have current information
- To capture and record important project information in a central information repository

Techniques

There are a multitude of techniques and approaches to achieving good communications. Often, project managers outline a plan for project communications early to ensure a smooth flow of information throughout the project. The four objectives just listed provide a starting point.

The plan for gathering information should address such questions as:

- What information is to be gathered?
- Who should gather the information?
- What methods should they use?
- How should information be recorded?
- How is it verified for accuracy?

The plan for internal communications should answer these questions:

- What internal information must be maintained and tracked?
- Who needs to be included in which types of internal information?
- What internal procedures are needed to ensure information is disseminated accurately and in a timely fashion?
- What methods should be used to capture and record internal information?
- What kinds of meetings are necessary (and unnecessary)?

Planning for dissemination of information about the project should consider these questions:

- How is status and progress information collected?
- How is it reported—how frequently and in what format?
- Who needs to receive progress information?
- What types of information should be disseminated?

The results of communication planning will identify the specific communication techniques, meetings, reports, tracking logs, forms, and templates that may be needed to maintain effective project communications. Electronic techniques, such as e-mail, bulletin boards, Web sites, and CASE tools as central repositories, have facilitated many types of communication. However, face-to-face meetings and reviews are still necessary and should not be eliminated. Communication is such an integral part of every project that careful consideration should be given to ensure a comprehensive approach to gathering, recording, and disseminating information.

Project Risk Management

Need and Objectives

Software development is a high-risk activity, and information systems history is strewn with failed projects. Experienced project managers in all industries have concluded that risk management is critical for success.

We define *risk* as the possibility of failure or loss. Since projects are organized to accomplish an end result, project risks are any hindrances or obstacles to achieving the result or a portion of it. Risk is such a pervasive problem in software development that entire development approaches have been created to reduce it. The spiral SDLC approach was invented to include risk assessment as a primary ingredient. The prototyping approach to software development is used to reduce the risk associated with incomplete or inaccurate understanding of user requirements. Phased delivery and conversion approaches have as their primary motivation the reduction of risk.

Risk management is unique, because it does vary tremendously with the type of project undertaken. We identify three primary objectives of risk management:

- To determine the potential areas of high risk for the project
- To develop strategies and plans of actions to reduce the identified risks
- To carry out the plans of action to monitor and control the project risks

Techniques

Risk management techniques vary from project to project. Probably the toughest part of risk management is to be realistic in assessing the risks and potential problems of a

given project. Human beings tend not to see problems in projects with which they are integrally involved, so one technique of risk assessment is to put an outsider on the risk assessment team.

Risk Identification and Tracking   A common technique to identify project risks is to have an open session, including senior project members, outsiders, and even team members, on risk identification and tracking. Many times team members will observe risks that senior management may not see.

Risk identification and tracking meetings should be held from time to time, perhaps monthly and certainly at the beginning of each new project phase. At each meeting, attendees should compile a list of the top project risks, along with a trigger that will help identify when the risk has become a problem. In addition to the identification of risks, they should develop a strategy and plan of action to reduce the risks or to solve the problems. As later meetings are held, the list is updated by removing risks that are resolved and adding new risks that appear. Figure A-3 illustrates an example of a top-10 risk list.

FIGURE *A-3*

Top-10 risk list example.

Rank	Risk description	Contact person	Reduction strategy	Reduction progress
1	Undefined objectives	R. Jones	Meet with VP of marketing to define	None yet
2	New networking software	B. Hardy	Send staff to training or hire expert	Training scheduled
3	Late delivery of contractor subsystem	T. Hansen	Closer monitoring of contractor milestones	Begun this week
4	Poor quality of contractor subsystem	T. Hansen	Begin testing prior to normal test plan	Scheduled to start next month
5	Project team understaffed	A. Wilson	Expedite hiring and training or find additional contract staff	HR director made it highest priority; no solution in sight
6	Response time of high-volume inquiry transactions	B. Hardy	Write database access routines in low-level code	Approach defined; prototype needs to be built to test it
. . .	. . .	. . .	. . .	. . .
Off list	No users assigned to data conversion	B. Marble		Fixed—one user full time, two part time

Project Procurement Management

Need and Objectives

As the complexity of information systems increases, more and more systems are built by integrating separate systems. A team consisting of inside developers, hardware vendors, software vendors, and consulting firms often builds these systems. In fact, in many companies today, this mix of in-house staff and outside consultants is now normal for developing new systems. One of the main benefits of seeking outside assistance is to tap into the knowledge and expertise of others.

For these types of projects, there are two issues for project managers: (1) handling all the normal concerns such as cost, scope, and quality and (2) finding and working with a provider or providers. We use the term *provider* to mean an outside firm that offers specific hardware and software or services such as programming. So, project procurement management includes both finding the needed goods or services and managing and controlling the performance of the provider.

Specific objectives of project procurement management are:

provider

an organization that sells systems or consulting services to other organizations

- To plan the procurement process
- To ensure that solicitation documents are complete and accurate
- To evaluate and select alternative providers
- To ensure that contracts are adequate, with sufficient performance controls and metrics
- To monitor and control deliverables

Techniques

Project procurement management is a complex and wide-ranging area, from the development of a request for proposal (RFP), to contract negotiations, to performance assessment, to daily work relationships and chains of authority, and even to arbitration and litigation. Unfortunately, little training is available to strengthen these skills. In addition, each company has its own procedures for working with providers, and this knowledge is often disseminated through many levels and departments within the organization. Project managers need to seek out this information wherever it is.

Procurement Planning   Planning for procuring system components frequently begins as soon as the project is initiated. As the project manager develops the schedule, he or she should assess whether it is feasible to purchase components or services for the project. Since it is less risky and often less expensive to buy a solution rather than build one, the project manager should survey the industry to see whether total or partial solutions are available for purchase. The results of this survey help determine the direction of the rest of the project.

Developing Solicitation Documents   The solicitation documents that are needed vary with the types of components that are purchased. A request for proposal (RFP) is a comprehensive document specifying the requirements that must be met for the new system. Developing an RFP is a major effort that should be included in the project schedule. Users should be involved to ensure that requirements are specified correctly, and with adequate detail, in the request. Companies frequently have their own internal procedures for RFP development. Project managers must ensure that the document is a complete and comprehensive description of the requirements.

Evaluating Providers and Alternatives   Providers and development alternatives are usually evaluated by measuring their relative merits. Chapter 8 explains how to evaluate alternatives based on a weighting scheme that measures performance in the areas of technical requirements, functional requirements, and general provider performance. By using a quantitative weighting scheme, project managers can ensure a more objective evaluation of the various alternatives.

Developing the Contract   Writing contracts is a fairly complex process that requires technical expertise, an understanding of working relationships, and knowledge of legal issues. The best technique to ensure a good contract is to involve experienced technical team members, managers, the project manager, and legal staff. Bad contracts result

from not consulting interested stakeholders. A good contract provides protection for both the provider and the purchaser. A win-win contract is best to ensure that cooperative relationships are established at the outset.

Monitoring and Controlling Delivery   The success of monitoring and controlling provider delivery is a direct consequence of a solid contract. Generally, timing and quality of intermediate deliverables, achievement of milestones, and review processes should be established early and occur throughout the project just as they are for in-house staff. It is always better to identify problems early, when they can be corrected easily and at less cost.

PROJECT MANAGEMENT TASKS BY PHASE

FIGURE *A-4*

Project management tasks by knowledge area and project phase.

The previous sections discussed, in general terms, the techniques for each of the eight knowledge areas of project management. Figure A-4 identifies many of the detailed tasks associated with these techniques. The figure provides a list of detailed tasks for each knowledge area that is associated with each SDLC project phase. This table should provide you with an effective checklist of steps to follow as you begin a development project.

Knowledge Area	Planing phase (including project initiation)	Analysis phase	Design phase	Implementation phase
Scope management	Identify objectives Identify business need Identify major functions	Defined detailed requirements Verify requirements Solidify scope Plan implementation approach	Control scope creep Monitor request log Evaluate change requests	Control scope creep Monitor request log Evaluate change requests
Time management	Build WBS Build schedule Identify milestones	Adjust/optimize schedule Monitor progress	Adjust/optimize schedule Monitor progress	Adjust/optimize schedule Monitor progress
Cost management	Do cost/benefit analysis Build budget (cash flow)	Monitor ongoing costs Update cost/benefit	Monitor ongoing costs Review budget/cash flow	Monitor ongoing costs Review budget/cash flow
Quality management	Identify quality metrics Identify project success metrics Establish quality control procedures	Control quality with procedures and reviews	Control quality with reviews	Control and monitor testing Monitor testing error log Ensure quality of final system
Human resource management	Identify project manager Develop staffing plan Recruit and staff	Organize teams Conduct team building Identify/add staff	Provide team training Identify/add resources Conduct performance reviews	Provide training Conduct performance reviews
Communication management	Identify stakeholders Develop communication plan Establish communication mechanisms	Conduct status reviews Report status Monitor internal communications	Conduct status reviews Report status Monitor communications	Conduct status reviews Report status Monitor communications
Risk management	Analyze feasibility and risks Define alternative plans	Reassess risks Monitor risk areas	Reassess risks Monitor risk areas	Reassess risks Monitor risk areas
Procurement management	Identify buy options Identify potential providers	Develop RFPs Elicit bids Select providers Develop contracts	Monitor deliverables	Monitor deliverables

KEY TERMS

change control log, p. 752

80/20 syndrome, p. 753

provider, p. 762

scope creep, p. 750

REVIEW QUESTIONS

1. What are the eight knowledge areas in the project management body of knowledge?
2. What are the specific objectives of project scope management?
3. What is meant by *scope creep*?
4. What techniques would you use to control the project scope?
5. What are the specific objectives of project time management?
6. What is the 80/20 rule? Why is it a problem in project management?
7. What techniques could you use to estimate a project's duration?
8. What are the specific objectives of project cost management?
9. What techniques would you use to estimate a project's cost?
10. What are the characteristics of quality in a new system?
11. Why is it important to monitor quality throughout a project?
12. What techniques, besides programming testing, can be used to monitor quality?
13. What are the objectives of project human resource management?
14. What are the important factors that affect team member motivation?
15. Name the three types of communication that project teams must manage.
16. What techniques would you use to ensure that information is gathered correctly?
17. What would you do to be sure information is disseminated thoroughly?
18. Discuss the importance of risk management in information systems development.
19. What techniques would you use to manage project risk?
20. What are the objectives of project procurement management?
21. What document do you use to elicit bids for a new system?

THINKING CRITICALLY

1. Are there other areas of project management that you would include besides the eight knowledge areas listed? What are they, and why would you include them?
2. Discuss ways you might measure the quality of a system. What specific metrics would you use?
3. Discuss the importance of team-building exercises to strengthen team performance. Find and summarize in one page an article on team building. Describe some examples of your own experiences working in teams.
4. This appendix focuses primarily on responsibilities of a project manager with regard to managing the team and the project. What other responsibilities might a project manager have, particularly concerning the relationship of the project with the sponsoring organization?
5. Find a project manager who has developed some successful projects. Discuss what techniques he considers to be most important in good project management. Write a report of your discussion.
6. Access the Project Management Institute's (PMI) Web site to learn more about the details of the Project Management Body of Knowledge (PMBOK). Write a one-page report about the PMI, including its background and its objectives. You should also access the *PMBOK Guidebook*, review some of the excerpts, and include a summary of what you learned. The Web site is http://www.pmi.org.
7. Learn about function point analysis. Write a one-page report on function point analysis. You can research it in textbooks on software engineering. Some good tutorials are also available on the Internet. There is an active function point users group at http://www.ifpug.org/.
8. Learn about COCOMO as an estimation technique. Write a one-page report on how COCOMO works. You can research it in textbooks on software engineering. Some good tutorials are also available on the Internet. Software tools can be used to calculate estimates automatically. One tool is located at http://www.softstarsystems.com/.

statecharts
  developing RMO's, 469–477
  rules for developing, 466–468
statements, designing, 591–598
states
  decision pseudostate, 476
  described, 464
steering committees, 77
stereotypes, 389–390
storyboards, 547–550
strategic planning
  and application development
    environment, 292
  determining system automa-
    tion levels, 287–288
  and system requirements, 122
  systems analysts and, 14–16
structure charts
  for *Create new order* (fig.), 585
  Create new order program (fig.),
    363
  described, 50–51, 354
  developing, 375–378
  high-level, for *Customer order*
    program (fig.), 360
  payroll system (fig.), 357
  purpose, using, 355–366
  symbols used in, 355–356
  three-layer architecture for
    activity (fig.), 371
structured analysis, 52
structured analysis and design
  technique (SADT), 49
structured programming, 49–54,
  357
Structured Query Language
  (SQL), 333
structured systems development,
  49
structured walkthrough, 143
stubs and modules, 644
subclasses and superclasses, 181,
  183
subsystems
  in CSS use case diagram (fig.),
    247
  described, 6
  design requirements, 325
  and information systems, 7
  and the menu hierarchy,
    547–548
  in system flowcharts, 351
summary reports, 592

Sun Microsystems, 703
superclasses and subclasses, 181,
  183
supersystems, 6–7
supply chain management
  (SCM), 19, 81–83, 724, 735
support
  described, 625
  ERP system development
    methodology, 732–733
  help desk, 42
  help for input forms, 557
  packaged software, 720–721
  user, 658–660
support phase of SDLC, 41–42,
  625
support systems decision, communication, office, 9
swimlane, activity diagram, 134
symbols
  See also notation
  collaboration diagrams, 421
  design class, 389–393
  guillemets (« »), 389
  package diagram notation,
    427–429
  structure charts, 355–356
  system flowchart (fig.), 352
symmetric key encryption, 611
synchronization bar, activity diagram, 134
synchronizing databases,
  519–520
system controls, 164–165
system documentation, 655–656
system flowcharts
  described, 348
  described, using, 351–354
  payroll system (fig.), 628
  symbols used in, 352–353
system implementation alternatives, 296–305
system interfaces
  described, 532
  designing system inputs,
    579–588
  designing system outputs,
    589–599
  identifying, 576–579
system maintenance, programmer analyst task, 41
system metaphor, 686
System Modeler tool, 65

system requirements
  analysis phase, conducting
    interviews, 129–132
  controlling system's scope and
    automation level, 284–287
  defining, prioritizing, 39,
    115–116
  described, 119–120
  evaluating alternatives for,
    281–282
  events and, 158–169
  information gathering, analysis
    phase, 125–142
  models and modeling,
    154–158
  object-oriented approach,
    overview, 241
  perfect technology assump-
    tion, 165
  stakeholders and, 120–125
  things and, 169–176
  validating, 142–145
system scope document, 84–85
system sequence diagrams
  (SSDs)
  dental procedure (fig.), 435
  developing, 261–266
  and dialog design, 547,
    551–553
  identifying inputs and outputs,
    258–266
  notation, 259–261
  using, 244
  for Web order scenario (fig.),
    266
systems
  described, 6
  enterprise-level. *See* enterprise-
    level systems
  Internet-based. *See* Internet-
    based systems
  range of inputs and outputs
    (fig.), 577
  statechart modeling capabili-
    ties, 471
  testing, 644
  user and access roles to (fig.),
    608
  user's model of, 535
systems analysis
  described, 3
  key skills for, 113–114
  tasks, 27